CAMBRIDGE LIBRARY COLLECTION

Books of enduring scholarly value

Life Sciences

Until the nineteenth century, the various subjects now known as the life sciences were regarded either as arcane studies which had little impact on ordinary daily life, or as a genteel hobby for the leisured classes. The increasing academic rigour and systematisation brought to the study of botany, zoology and other disciplines, and their adoption in university curricula, are reflected in the books reissued in this series.

Flora Australiensis

George Bentham (1800–84) was one of Britain's most influential botanists, whose own collection of plant specimens numbered more than 100,000. Although he donated his herbarium to the Royal Botanic Gardens, Kew in 1854, he continued to make significant contributions to the field, including this exhaustive, seven-volume work detailing the plant life of Australia, which was published from 1863 to 1878. It was part of a series of works commissioned by the British government to document the flora in its colonies. Using the extensive numbers of specimens at Kew – and with the help of Ferdinand Mueller (1825–96), a German botanist in Australia – Bentham was able to compile descriptions of more than 8,000 species of Australian plants, making these volumes the first completed compendium of the flora of any large continental area. Volume 2, published in 1864, gives descriptions of seven orders of the dicotyledon class of flowering plant.

Cambridge University Press has long been a pioneer in the reissuing of out-of-print titles from its own backlist, producing digital reprints of books that are still sought after by scholars and students but could not be reprinted economically using traditional technology. The Cambridge Library Collection extends this activity to a wider range of books which are still of importance to researchers and professionals, either for the source material they contain, or as landmarks in the history of their academic discipline.

Drawing from the world-renowned collections in the Cambridge University Library, and guided by the advice of experts in each subject area, Cambridge University Press is using state-of-the-art scanning machines in its own Printing House to capture the content of each book selected for inclusion. The files are processed to give a consistently clear, crisp image, and the books finished to the high quality standard for which the Press is recognised around the world. The latest print-on-demand technology ensures that the books will remain available indefinitely, and that orders for single or multiple copies can quickly be supplied.

The Cambridge Library Collection will bring back to life books of enduring scholarly value (including out-of-copyright works originally issued by other publishers) across a wide range of disciplines in the humanities and social sciences and in science and technology.

Flora Australiensis
A Description of the Plants of the Australian Territory

VOLUME 2: LEGUMINOSAE TO
COMBRETACEAE

GEORGE BENTHAM
FERDINAND VON MUELLER

CAMBRIDGE
UNIVERSITY PRESS

CAMBRIDGE UNIVERSITY PRESS

Cambridge, New York, Melbourne, Madrid, Cape Town,
Singapore, São Paolo, Delhi, Tokyo, Mexico City

Published in the United States of America by Cambridge University Press, New York

www.cambridge.org
Information on this title: www.cambridge.org/9781108037396

© in this compilation Cambridge University Press 2011

This edition first published 1864
This digitally printed version 2011

ISBN 978-1-108-03739-6 Paperback

FLORA AUSTRALIENSIS.

FLORA AUSTRALIENSIS:

A DESCRIPTION

OF THE

PLANTS OF THE AUSTRALIAN TERRITORY.

BY

GEORGE BENTHAM, F.R.S., P.L.S.,

ASSISTED BY

FERDINAND MUELLER, M.D., F.R.S. & L.S.,

GOVERNMENT BOTANIST, MELBOURNE, VICTORIA.

VOL. II.

LEGUMINOSÆ TO COMBRETACEÆ.

PUBLISHED UNDER THE AUTHORITY OF THE SEVERAL GOVERNMENTS
OF THE AUSTRALIAN COLONIES.

LONDON:
LOVELL REEVE & CO., 5, HENRIETTA STREET, COVENT GARDEN.
1864.

PRINTED BY JOHN EDWARD TAYLOR,
LITTLE QUEEN STREET, LINCOLN'S INN FIELDS.

CONTENTS.

—◆—

CONSPECTUS OF THE ORDERS CONTAINED IN THE SECOND VOLUME.

Class I. **DICOTYLEDONS.**

Subclass I. POLYPETALÆ.

(Continued from Vol. I. p. 4.)

Series III. Calyciflora.—Stamens and petals usually inserted on the margin of a thin disk lining the base or the whole of the calyx-tube, and free from the ovary unless the calyx-tube is also adnate to it. Stamens definite or indefinite. Ovary either free and superior, or enclosed in the calyx-tube, or inferior and adnate to the calyx-tube.

(In *Mimoseæ* and a few genera of *Papilionaceæ* and *Cæsalpinieæ*, as well as in isolated genera of *Saxifrageæ* and some other Orders, the stamen-bearing disk is reduced to a narrow ring or disappears altogether. The distinction between *Calyciflora* and *Thalamiflora* is therefore general only, not absolute.)

(The whole of the *Calyciflora* not having yet been gone through, either for the present work or for our 'Genera Plantarum,' the subdivision of the Series into Alliances is, for the present, deferred.)

XL. Leguminosæ. Trees, shrubs, or herbs. Leaves alternate or rarely opposite, often compound. Stipules rarely wanting. Gynœcium free, consisting of a single excentrical carpel with a terminal style, the ovules inserted along the upper or inner angle of the cavity. Albumen usually scanty or none.

Suborder 1. Papilionaceæ.—Flowers irregular. Petals usually 5, imbricate, the upper one or standard outside. Stamens 10, rarely fewer by abortion. Radicle curved and accumbent, rarely straight.

Suborder 2. Cæsalpinieæ. Flowers irregular or nearly regular. Petals 5 or fewer, imbricate, the upper one inside. Stamens (in Australian genera) 10 or fewer. Radicle straight.

Suborder 3. Mimoseæ. Flowers regular, small, in spikes or heads. Petals 5, 4, or rarely 3, valvate or rarely slightly imbricate. Stamens definite or indefinite. Radicle straight.

XLI. Rosaceæ. Shrubs or herbs. Leaves alternate, with stipules. Flowers regular. Stamens usually indefinite. Carpels of the gynœcium 1 or several, free and distinct, or, if adnate to the calyx-tube, either distinct or combined into a single ovary. Styles distinct. Albumen usually none.

XLII. Saxifrageæ. Shrubs or herbs. Leaves various, with or without stipules. Flowers regular or nearly so. Stamens definite or rarely indefinite. Carpels of the gynœcium usually united into a 1- or several-celled ovary, at least at the base, free or more or less adnate or inferior. Styles usually distinct or readily separable. Albumen usually copious.

XLIII. CRASSULACEÆ. Herbs with succulent leaves, without stipules. Flowers regular and perfectly isomerous. Stamens in 1 or 2 series. Gynœcium superior, with distinct carpels. Seeds albuminous.

XLIV. DROSERACEÆ. Herbs. Leaves fringed with glandular cilia. Stipules scarious or none. Flowers regular. Stamens definite. Ovary free, 1-celled, with parietal placentas. Styles distinct (except *Byblis*). Seeds albuminous.

XLV. HALORAGEÆ. Herbs, aquatic or terrestrial. Leaves opposite or alternate, without stipules. Flowers small, regular, often much reduced. Stamens definite. Ovary inferior, with as many cells and ovules as styles or rarely fewer, the ovule pendulous from the apex of the cell. Styles or sessile stigmas 1 to 4, distinct. Seeds albuminous.

XLVI. RHIZOPHOREÆ. Trees or shrubs, often maritime, with opposite leaves. Stipules deciduous. Flowers regular. Calyx-lobes valvate. Petals usually notched or jagged. Stamens twice as many as petals or more. Ovary usually inferior, several-celled, with 2 or more ovules pendulous from the apex of each cell. Style undivided. Seeds usually solitary, with or without albumen.

XLVII. COMBRETACEÆ. Trees, shrubs, or woody climbers. Leaves opposite or alternate, without stipules. Flowers regular or nearly so. Stamens definite or rarely indefinite. Ovary inferior, 1-celled, with 2 or more (1 in *Gyrocarpus*) ovules pendulous from the apex of the cavity. Style undivided. Seed solitary, without albumen. Cotyledons convolute.

FLORA AUSTRALIENSIS.

ORDER XL. LEGUMINOSÆ.

Calyx of 5 or rarely fewer, usually united sepals, campanulate or tubular, more or less divided into 5 or fewer teeth or lobes, or rarely the sepals entirely distinct. Corolla of 5 or rarely fewer petals, perigynous or rarely hypogynous, very irregular in the first suborder, less so in the second, small regular and the petals often united in the third. Stamens twice the number of petals, rarely fewer or sometimes indefinite, inserted with the petals. Ovary single (consisting of a single carpel), with 1, 2, or more ovules arranged along the inner or upper angle of the cavity; style simple. Fruit a pod, usually flattish and opening round the margin in 2 valves, but sometimes follicular or indehiscent, or variously shaped. Seeds with 2 large cotyledons, a short radicle, and, with few exceptions, little or no albumen.—Herbs shrubs trees or climbers. Leaves alternate or (chiefly in some Australian genera) opposite, usually furnished with stipules, compound or reduced to a single leaflet or to a dilated petiole, or in a few cases really simple, the leaflets or leaves entire or rarely toothed or lobed. Flowers in axillary or terminal racemes spikes or clusters, when terminal often becoming leaf-opposed by the growth of a lateral shoot, rarely solitary and axillary.

The largest Natural Order of Phænogamous plants next to *Compositæ*, and widely distributed over the whole surface of the globe. Out of 92 Australian genera, 33 are dispersed over the warmer, chiefly tropical regions of both the New and the Old World; of 20 other tropical genera, 13 are in Africa and Asia but not in America, 2 in America and Asia but not in Africa, 4 in Asia alone, 1 (*Erythrophlœum*) only in Africa; 4 more of the Australian genera belong to the temperate regions of the northern hemisphere, 1 (*Clianthus*) extends only to New Zealand, and 84 are endemic in Australia.

The genera marked with an asterisk in the following table are those which are mentioned only as introduced, not described as indigenous.

SUBORDER I. **Papilionaceæ.**—*Flowers 5-merous. Corolla very irregular, papilionaceous, or very rarely (in a few Sophoreæ) nearly regular, the petals imbricate, the upper one (or standard) always outside in the bud. Stamens 10, or very rarely 9 or 5.*

TRIBE I. **Podalyrieæ.**—*Shrubs, rarely herbs or small trees. Leaves simple or digitately compound (except in a few Gompholobiums and Burtonias), without stipellæ. Stamens all free or scarcely united at the base. Pod not articulate.*

Standard small or narrow. Ovules 4 or more.
 Flowers in heads of 4, surrounded by an involucre. Upper lobes
 of the calyx very small. Leaves simple, opposite . . . .   1. JANSONIA.
 Flowers not in heads. Upper lobes of the calyx as large as or
 larger than the others. Leaves simple, opposite alternate or
 none . . . .   2. BRACHYSEMA.
Standard orbicular or reniform, large. Ovules 4 or more (except
 iu a few *Mirbelias*).
 Calyx-lobes shorter or scarcely longer than the tube. Leaves
 simple or none
 Ovary and pod divided by a longitudinal partition . . . .   5. MIRBELIA.
 Ovary not divided longitudinally.
 Keel about as long as the wings. Leaves usually more or
 less opposite or verticillate . . . . ▸ . . . .   3. OXYLÒBIUM.
 Keel shorter than the wings or beaked. Leaves alternate .   4. CHORIZEMA.
 Calyx-lobes much longer than the tube.
 Calyx-lobes imbricate. Ovary sessile. Funicles short or
 slender. Pod oblong-linear. Leaves simple or unifoliolate   6. ISOTROPIS.
 Calyx-lobes valvate. Ovary stipitate. Funicles long and
 thick, all folded or curved downwards. Pod globular.
 Leaves pinnate, digitately 3- to 5-foliolate, or simple . .   7. GOMPHOLOBIUM.
Standard orbicular or reniform, large. Ovules 2 (4 to 6 in *Jack-
 sonia piptomeris*).
 Calyx-lobes much longer than the tube, valvate.
 Funicles long and thick, one folded or curved upwards, the
 other downwards. Pod globular or nearly so. Leaves
 pinnate, digitately 3- to 5-foliolate or simple . . . . .   8. BURTONIA.
 Funicles short and slender. Pod flattened or oblong. Leaves
 none or very rarely 1-foliolate . . . .   9. JACKSONIA.
 Calyx-lobes shorter than the tube, or, if longer, imbricate or
 open in the bud.
 Pod nearly globular, usually stipitate. Strophiole none.
 Calyx upper lip very large. Petals nearly sessile. Leaves
 simple and narrow or none . . . .   10. SPHÆROLOBIUM.
 Pod sessile, ovoid, small, and indehiscent. Calyx shortly 5-
 toothed. Leaves reduced to a long petiole, with or without
 1 or 3 digitate leaflets . . . .   11. VIMINARIA.
 Pod triangular, 2-valved. Seeds strophiolate. Calyx shortly
 5-toothed. Leaves simple, alternate or none . . . . .   12. DAVIESIA.
 Pod ovate or oblong, 2-valved. Calyx 5-lobed or toothed or
 2-lipped. Leaves simple, sessile or shortly petiolate.
 Leaves flat or folded lengthwise, or with revolute margins,
 or, if terete, channelled underneath.
 Bracteoles none or very deciduous.
 No strophiole. No stipules. Flowers 1 to 3 in each
 axil . . . .   13. AOTUS.
 Seeds strophiolate. Stipules often present. Flowers
 in racemes or dense axillary clusters . . . . .   15. GASTROLOBIUM.
 Bracteoles persistent close under the calyx, or adnate
 to it.
 No strophiole. No stipules. Filaments some or all
 united with the petals at the base . . . . . .   14. PHYLLOTA.
 Seeds strophiolate. Stipules usually (not always) pre-
 sent. Filaments free . . . .   16. PULTENÆA.
 Leaves concave or with incurved or involute margins, or, if
 terete, channelled above. Seeds strophiolate.
 Bracteoles persistent close under the calyx or adnate to
 it. Stipules usually present . . . .   16. PULTENÆA.

Bracteoles none or at a distance from the calyx, and
usually very small. Stipules none or minute.
Calyx-teeth or lobes equal. Pod flat, usually oblong . 17. LATROBEA.
Calyx more or less 2-lipped, or the upper lobes broad.
Pod ovate, flat or turgid.
Leaves all opposite . . . . 18. EUTAXIA.
Leaves alternate or crowded. Standard usually very
broad . . . . 19. DILLWYNIA.

(73, *Barklya*, has simple, or rather 1-foliolate, leaves, and the stamens free ; but it is a
large tree, with small nearly regular flowers.)

TRIBE II. **Genisteæ.**—*Shrubs or herbs, very rarely small trees. Leaves simple or
with 1 or 3 or more digitate leaflets (except* Goodia). *Stamens all united in a sheath open
on the upper side in all the Australian genera (except in one species of* Hovea).

Leaves all simple or none. Flowers axillary, solitary or clustered.
Seeds strophiolate.
Anthers uniform. Pod very flat.
Upper suture of the pod bordered by a narrow wing and not
splitting, the valves rolling back upon it elastically. Leaves
opposite . . . . 20. PLATYLOBIUM.
Pod not winged, opening at both sutures. Leaves opposite or
alternate . . . . 21. BOSSIÆA.
Anthers alternately longer and shorter. Leaves alternate or none.
Pod at least twice as long as broad, with coriaceous convex
valves. Flowers red, yellow, or reddish-purple . . . . 22. TEMPLETONIA.
Pod turgid, scarcely longer than broad. Flowers blue or
bluish-purple. . . . . 23. HOVEA.
Leaves pinnately 3-foliolate. Flowers in terminal or leaf-opposed
racemes. Seeds strophiolate . . . . 24. GOODIA.
Leaves digitate or simple. Flowers or racemes terminal or leaf-
opposed. Seeds not strophiolate.
Anthers alternately longer and shorter. Style (often very
minutely) bearded under the stigma.
Keel acute or beaked. Pod turgid . . . . 25. CROTALARIA.
Keel obtuse. Pod flat . . . . , ... 26. PENTADYNAMIS.
Anthers uniform. Style beardless. Pod linear, follicular.
Flowers minute, solitary or in short racemes . . . . . 27. ROTHIA.

(32, *Ptychosema*, has the flowers of *Genisteæ*, but the pinnate leaves of *Galegeæ*.)
(65, *Flemingia*, and a very few species of other genera of *Phaseoleæ*, have 3 digitate
leaflets, but may readily be distinguished from *Genisteæ*, either by their upper stamen free
or by the twining herbaceous stems. Some species of 30, *Psoralea*, and 31, *Indigofera*,
with digitate leaves, may be known, the former by their ovary and pod, the latter by the
stamens and anthers.)

TRIBE III. **Trifolieæ.**—*Herbs, very rarely shrubs. Leaflets usually 3, pinnate or
rarely digitate, the veinlets extending to the edge and often produced into minute teeth.
Peduncles racemes or flower-heads axillary (or apparently terminal by the reduction of
the upper floral leaves), never leaf-opposed. Upper stamen free (except* Ononis), *the
others united in a sheath. Pod not articulate.*

Keel beaked. Stamens all united . . . . *ONONIS.
Keel obtuse. Upper stamen free.
Petals free from the staminal tube.
Pod straight, or falcate, or undulate, linear, or flat, or beaked 28. TRIGONELLA.
Pod spiral (rarely small, curved and 1-seeded) . . . . . *MEDICAGO.
Pod small, thick, straight, and indehiscent . . . . . . *MELILOTUS.
All the petals, or the 4 lower ones, with their claws adnate to
the base of the staminal tube. Pod usually included in the
calyx . . . . *TRIFOLIUM.

TRIBE IV. **Euloteæ.**—*Herbs, rarely shrubs. Leaves pinnate, leaflets entire. Flowers capitate or umbellate on axillary peduncles. Upper stamen usually free, at least at the base, the others united in a sheath ; filaments either all or 5 only dilated towards the end. Pod not articulate.*

Leaflets 5, the 2 lowest taking the place of stipules.   Keel beaked   29. LOTUS.

TRIBE V. **Galegeæ.**—*Herbs not twining, shrubs, or rarely trees or tall woody climbers. Leaves pinnate, rarely reduced to 3 or 1 leaflets. Stipellæ none, or setaceous in a few pinnate genera. Upper stamen usually free, at least at the base, the others united in a sheath, very rarely all united ; filaments filiform. Ovules 2 or more (except in* Indigofera linifolia *and in* Psoralea). *Pod not articulate, 2-valved (except* Psoralea).

Ovule 1. Fruit small, the pericarp adhering to the seed. Herbs
  or shrubs with black glandular dots. Leaflets (in Australia) 1
  or 3, sometimes toothed . . . . 30. PSORALEA.
Ovules 2 or more (1 in *I. linifolia*). Anthers tipped with a small
  gland. Pod 2-valved. Herbs or shrubs, sometimes glandular.
  Leaflets entire. Hairs often appressed and attached by the
  centre. . . . . 31. INDIGOFERA.
Ovules 2 or more. Anthers without glands.
  Racemes or flowers terminal or leaf-opposed. Herbs or shrubs.
    Stamens all united in a sheath, open on the upper side.
      Leaflets 3. Flowers in racemes. Seeds strophiolate . . 24. GOODIA.
      Leaflets several, small. Flowers solitary, on long peduncles 32. PTYCHOSEMA.
      Leaflets few. Petals not exceeding the deeply lobed calyx.
        Seeds strophiolate, with a straight radicle . . . . . 33. LAMPROLOBUM.
    Upper stamen usually free, or all united in a closed tube . . 34. TEPHROSIA.
  Racemes in a terminal panicle. Tall woody climbers. Pod
    hard . . . . 35. MILLETTIA.
  Racemes or flowers axillary. Herbs or shrubs.
    Style not bearded (rarely a small tuft of hairs on the stigma
      in *Tephrosia*).
      Pod linear, rarely short and oblong ; valves thin or coria-
        ceous, flat or convex when ripe . . . . 34. TEPHROSIA.
      Pod long, narrow and thick, the endocarp continuous with
        the transverse partitions between the seeds. . . . . 36. SESBANIA.
      Pod 1-seeded (ovules 2), muricate. Plant glandular. An-
        thers with confluent cells opening in unequal valves . . 39. GLYCYRRHIZA.
    Style bearded under the stigma. Pod turgid, membranous or
      coriaceous.
      Petals acuminate . . . . 37. CLIANTHUS.
      Petals obtuse . . . . 38. SWAINSONA.

(66, *Abrus*, and a very few pinnate-leaved *Phaseoleæ*, may have the technical characters
of *Galegeæ*, but are distinguished by their herbaceous more or less twining stems.)

TRIBE VI. **Hedysareæ.**—*Pod separating into 1-seeded articles, or the whole pod 1-seeded and indehiscent (except* Pycnospora, *and rarely* Desmodium). *Foliage and inflorescence, in the Australian genera, either of* Galegeæ *or of* Phaseoleæ.

Leaves pinnate with several leaflets, as in *Galegeæ.* Stamens
  united in a sheath, or in 2 bundles of 5 each.
    Tall shrubs. Articles of the pod oblong, striate . . . . . 40. ORMOCARPUM.
    Herbs. Articles of the pod square or semi-orbicular, flat . . 41. ÆSCHYNOMENE.
    Herbs. Articles of the pod folded over each other within the
      calyx . . . . 42. SMITHIA.
Leaves with 2 leaflets. Bracteoles large, enclosing the flowers.
    Stamens all united. Anthers alternately long and short . . . 43. ZORNIA.

Leaves pinnately 3-foliolate or 1-foliolate, with stipellæ as in *Phaseoleæ*. Stipules usually dry.
Pod flat, not folded . . . . 44. DESMODIUM.
Pod turgid, not articulate, but with transverse lines . . . . 45. PYCNOSPORA.
Pod-articles folded over each other within the calyx.
Calyx-tube small, lobes subulate . . . . 46. URARIA.
Calyx campanulate, enlarged after flowering, with short broad
lobes . . . . 47. LOUREA.
Pod-articles globular, oblong-terete, or slightly flattened but
thick. Calyx narrow, dry, deeply lobed. Leaves 1-foliolate. 48. ALYSICARPUS.
Leaves pinnately 3-foliolate, rarely 1-foliolate, without stipellæ.
Ovule 1. Pod 1-seeded, flat, indehiscent . . . . 49. LESPEDEZA.

TRIBE *Vicieæ.—*Herbs. Leaves abruptly pinnate, the common petiole usually ending in a tendril or fine point. Flowers and fruit of Phaseolcæ. Peduncles or racemes axillary.*

Style with a tuft of hairs at the top on the outside or all round
(not bearded longitudinally inside) . . . . *VICIA.
Style not bearded. Upper stamen wanting . . . . 66. ABRUS.

TRIBE VII. **Phaseoleæ.**—*Herbs usually twining or prostrate, rarely erect or shrubby at the base, very rarely trees. Leaves pinnately 3-foliolate or 1-foliolate, rarely 5- or 7-foliolate, with stipellæ (digitate in Flemingia and a very few species of other genera, stipellæ minute or none in Rhynchosia and its allies). Upper stamen usually free, at least at the base or all but the base. Anthers uniform or nearly so (except in Mucuna). Pod not articulate, 2-valved.*

Flowers in axillary short clusters with persistent striate bracts and
bracteoles. Seeds not strophiolate.
Calyx tubular . . . . 50. CLITORIA.
Calyx campanulate . . . . 60. DOLICHOS.
Flowers pedunculate umbellate or racemose, the rhachis not nodose.
Bracts persistent or deciduous. Seeds strophiolate.
Flowers red, in 1 or 2 pairs or in umbels or in short racemes . 53. KENNEDYA.
Flowers small, blue or purple, in loose racemes. Keel usually
small . . . . 52. HARDENBERGIA.
Flowers small, single, scattered in a loose raceme, the rhachis not
nodose. Bracts small, deciduous. (Lower flowers often solitary
in the axils.) Seeds not strophiolate . . . . 51. GLYCINE.
Trees or tall erect herbs with conical prickles. Flowers large, red.
Wings very short . . . . 54. ERYTHRINA.
Anthers alternately long and short. Flowers large, purple yellow
or white. Standard short. Keel acuminate . . . . . . 55. MUCUNA.
Twining or erect at the base, not glandular. Flowers in pairs or
clusters along or at the top of a common peduncle, the rhachis
of the cluster gland-like or forming a protruding node. Bracts
deciduous or none. Anthers uniform.
Style beardless.
Calyx-lobes 4 (the upper one of 2 united), acuminate . . . 56. GALACTIA.
Calyx 2 upper lobes united in a large upper lip, the 3 lower
minute . . . . 57. CANAVALIA.
Style bearded under the stigma.
Stigma oblique or lateral.
Keel spirally twisted . . . . 58. PHASEOLUS.
Keel straight, or with a curved beak not forming a complete spire . . . . 59. VIGNA.
Stigma small, terminal . . . . 60. DOLICHOS.

Twining or erect. Flowers racemose umbellate or solitary, the rhachis not nodose. Bracts usually membranous and deciduous. Stipellæ usually minute or none. Style beardless. Upper stamen free.

Ovules 4 or more.

Pod very flat, obliquely acuminate . . . . 61. DUNBARIA.

Pod flattened, very obtuse, with transverse lines or depressions between the seeds . . . . 62. ATYLOSIA.

Ovules 2 or rarely 1.

Pod flattened. Hilum of the seed parallel to the suture with a central funicle . . . . 63. RHYNCHOSIA.

Pod flattened. Seed obliquely transverse, the funicle attached to one end of the hilum. Standard usually very silky . . 64. ERIOSEMA.

Pod turgid. Leaflets digitate . . . . 65. FLEMINGIA.

Erect. Flowers small, in terminal racemes or panicles. Stipules dry, and habit of *Desmodium* . . . . 45. PYCNOSPORA.

Twiners. Leaves abruptly pinnate with small leaflets. Upper stamen wanting, the other 9 united in a sheath open on the upper side . . . . 66. ABRUS.

TRIBE VIII. **Dalbergieæ.**—*Trees or woody climbers. Leaves pinnate with 5 or more leaflets or sometimes 1 leaflet, very rarely 3. Stipellæ none or small and subulate. Stamens all united in a sheath or tube or into two parcels of 5, very rarely the upper one free. Pod indehiscent.*

Anthers small, erect, didymous, opening at the top. Flowers small, in cymes or short panicles. Pod flat and thin . . . . 67. DALBERGIA.

Anthers opening longitudinally. Flowers racemose.

Pod flat and thin, not winged . . . . 68. LONCHOCARPUS.

Pod flat, thin or coriaceous, one or both sutures edged with a narrow wing . . . . 69. DERRIS.

Pod flattened but thick, with obtuse sutures . . . . . . 70. PONGAMIA.

(35, *Millettia*, is closely allied to *Lonchocarpus*, but has a dehiscent 2-valved pod.)

TRIBE IX. **Sophoreæ.**—*Trees, woody climbers, or rarely tall shrubs or almost herbaceous. Leaves pinnate, with several leaflets, without stipellæ, or reduced to a large leaflet. Stamens all free or scarcely united at the base.*

Leaves pinnate.

Corolla papilionaceous. Pod terete or 4-angled, moniliform . 71. SOPHORA.

Corolla papilionaceous or nearly so. Pod large, hard, almost woody, spongy inside . . . . 72. CASTANOSPERMUM.

Leaves 1-foliolate. Corolla small, nearly regular, the upper petal outside. Pod flat and thin . . . . 73. BARKLYA.

(A few species of 7, *Gompholobium*, and 8, *Burtonia*, have pinnate leaves, but with the habit and small leaflets of *Podalyrieæ*.)

SUBORDER II. **Cæsalpinieæ.**—*Flowers usually 5-merous. Corolla irregular or nearly regular, imbricate in the bud, the upper petal inside. Stamens (in the Australian genera) 10 or fewer and all free. Radicle usually straight.*

Leaves twice pinnate, often with hooked prickles. Stamens 10, all bearing anthers.

Ovules 2 or more. Stigma small.

Pod ovate, compressed, covered with prickles . . . . . 74. GUILANDINA.

Pod ovate-oblong or linear-falcate, compressed, without prickles or wings, 2-valved . . . . 75. CÆSALPINIA.

Pod flat, thin or coriaceous, the upper suture winged . . . 76. MEZONEURUM.

Ovule 1. Pod samara-like, with a terminal wing . . . . . 77. PTEROLOBIUM.
Ovules 2 or more. Stigma large, peltate. Pod oblong-lanceo-
late, flat and thin, indehiscent . . . . 78. PELTOPHORUM.
Leaves simply pinnate.
  Sepals and petals 5. Stamens 10; anthers all perfect, opening
  in terminal pores or short slits, or some minute and empty.
  Leaves abruptly pinnate . . . . : 79. CASSIA.
  Sepals and petals 5. Stamens 3 with perfect anthers opening in
  longitudinal slits, 2 small staminodia. Leaves unequally pin-
  nate . . . . 80. PETALOSTYLES.
  Sepals and petals 4 or 5. Stamens 2, opening in terminal
  pores. Leaves unequally pinnate or with sessile, digitate
  leaflets . . . . 81. LABICHEA.
  Sepals 4. Petals 3. Stamens 3 or 2 perfect, 4 or 5 minute
  staminodia. Leaves abruptly pinnate. Pod thick . . . . 82. TAMARINDUS.
  Sepals 4. Petals 5. Stamens 10, regular, with small anthers.
  Leaflets 1 or 2 pairs. Pod semiorbicular, turgid . . . . 84. CYNOMETRA.
Leaves of 2 leaflets or 2-lobed. Stamens 10 or fewer.
  Calyx-lobes or sepals valvate . . . . 83. BAUHINIA.
  Calyx-lobes or sepals imbricate . . . . 84. CYNOMETRA.

(73, *Barklya*, with 1-foliolate leaves, has the flowers nearly regular, but the upper petal
is outside.)
(85, *Erythrophlœum*, with bipinnate leaves, has the petals slightly imbricate, but the
flowers small, in dense spikes, as in the other *Mimoseæ*.)

SUBORDER III. **Mimoseæ.**—*Flowers 5-merous, 4-merous, rarely 3-
merous or 6-merous, small, regular, sessile in spikes or heads, or very rarely
shortly pedicellate. Sepals valvate, often united. Petals valvate (except in*
Erythrophlœum), *often united. Stamens equal to or double the number of
petals or indefinite. Radicle straight.*

Stamens twice as many as petals.
  Petals slightly imbricate . . . . 85. ERYTHROPHLŒUM.
  Petals strictly valvate.
    Anthers tipped with a gland.
      Pod large, coriaceous or woody, the sutures forming a per-
      sistent replum, the valves falling away in 1-seeded arti-
      cles. Tall woody climbers . . . . 86. ENTADA.
      Pod linear, 2-valved. Seeds thick. Flowers pedicellate.
      Tree . . . . 87. ADENANTHERA.
      Pod linear, twisted. Lower flowers of the spike with long,
      linear, coloured staminodia. Shrubs . . . . . . 88. DICHROSTACHYS.
    Anthers without any gland. Pod short, flat, falcate or
    oblique. Lower flowers of the spike often with long, linear,
    coloured staminodia. Herbs or undershrubs . . . . . 89. NEPTUNIA.
Stamens indefinite.
  Stamens, at least in the hermaphrodite flowers, all free . . . 90. ACACIA.
  Stamens monadelphous.
    Pod flat and thin, straight or scarcely falcate . . . . . . 91. ALBIZZIA.
    Pod curved or twisted, 2-valved and often reddish or pulpy
    inside, or separating into indehiscent articles . . . . . 92. PITHECOLOBIUM.

(*Achyronia villosa*, Wendl., and *Galega tricolor*, Hook. (*Callotropis*, Don ; *Accorombona*,
Endl.) are not Australian; the former is a *Priestleya*, the latter *Galega persica*.)

SUBORDER I. PAPILIONACEÆ.

Sepals united in a campanulate or tubular calyx, 5-toothed or cleft, or

4-toothed by the complete union of the 2 upper sepals, or 2-lobed, the upper lobe or lip entire or 2-toothed, the lower entire or 3-toothed, rarely irregularly split. Corolla very irregular, usually *papilionaceous*, that is of 5 petals, the upper one or *standard* (*vexillum*) outside in the bud, the 2 lateral ones or *wings* (*alæ*) intermediate, the 2 lowest ones more or less united along the lower edge or approximate, face to face, into a boat-shaped *keel* (*carina*), more or less enclosing the stamens and style. Stamens usually 10, either all free or all united in a tube or sheath enclosing the style, closed or open along the upper edge, or the upper stamen more or less free from the others, the filaments all free for some distance under the anthers. Ovules usually amphitropous, and the radicle of the embryo more or less curved over the edge of the cotyledons, rarely short and straight.

The subdivision of this large suborder into tribes is attended with very great difficulties, nor has any one character by which it has as yet been attempted proved constant. Those here adopted are such as have appeared the least objectionable, but there are connecting genera between all of them.

TRIBE I. PODALYRIEÆ.—Shrubs, rarely herbs or small trees. Leaves simple or digitately compound (except in a few species of *Gompholobium* and *Burtonia*). Stipellæ none. Stamens all free or scarcely united at the base. Pod not articulate.

This tribe was formerly united with *Sophoreæ*, and technically characterized as *Papilionaceæ* with free stamens ; but the affinity has always appeared to me much greater with *Genisteæ* with which it is connected by some S. African genera. The *Sophoreæ* seem rather to represent the *Galegeæ* and *Dalbergieæ*, connecting them by small gradations with *Cæsalpinieæ*. The connecting links between *Podalyrieæ* and *Sophoreæ* are supplied by the pinnate *Gompholobiums*, which, however, are very unlike *Sophoreæ*, and by a few almost herbaceous northern species of *Sophora* itself, which are certainly allied to the *Podalyrieæ* of the northern hemisphere.

1. JANSONIA, Kipp.
(Cryptosema, *Meissn.*)

Flowers 4 together, enclosed in the bud within an involucre of 4 bracts in 2 rows. Calyx very oblique, split on the upper side, with 2 upper minute teeth and 3 lower elongated lobes. Standard very small, recurved ; wings oblong; keel longer than the wings. Stamens free. Ovary sessile, with several ovules ; style filiform, with a small terminal stigma. Pod unknown. —Shrubby. Leaves opposite, simple, with stipules.

The genus is limited to a single species, endemic in Australia.

1. **J. formosa,** *Kipp. in Trans. Linn. Soc.* xx. 384. *t.* 16. A shrub, apparently of several feet, the young branches silky-pubescent. Leaves from ovate to lanceolate, obtuse and usually shortly mucronate, 1 to 2 in. long, rounded or cordate at the base on a short petiole, coriaceous, finely reticulate, glabrous above, silky-hairy or at length glabrous underneath. Stipules lanceolate-subulate, deciduous, or the upper ones shortly connate and persistent. Flower-heads 1 or 2, nearly sessile between the last leaves, recurved. Involucre at first globular, the 2 outer bracts nearly orbicular, about 5 lines long, thick, valvate, and closed over the inner rather thinner and smaller ones, all shorter than the open flowers, villous outside, glabrous within. Calyx about 9 lines long, densely hairy outside. Petals glabrous ; standard cordate, lanceolate, 1 to 1½ lines long, on a claw of nearly 2 lines ; lower petals on claws

of about 4 lines, the lamina of the keel about 6 lines. Ovary very villous. Ovules about 5.—*Cryptosema pimeleoides*, Meissn. in Pl. Preiss. ii. 207.

W. Australia, *Drummond, 3rd Coll. n.* 100; Scott's River, *Gilbert.*

2. BRACHYSEMA, R. Br.

(Leptosema, *Benth.*; Kaleniczenkia, *Turcz.*; Burgesia, *F. Muell.*)

Calyx-lobes nearly of equal length, the 2 upper ones often united higher up. Standard shorter and narrower than the wings, usually recurved; wings narrow; keel usually broader and longer than the wings, incurved. Stamens free. Ovary sessile or stipitate, with several ovules; style filiform, with a small terminal stigma. Pod ovoid or elongated, turgid, the valves usually coriaceous.—Shrubs or undershrubs. Leaves opposite or alternate and simple, or all reduced to small scales. Flowers red or rarely yellow-green or almost black, terminal or axillary, solitary or several together, or crowded on short radical scapes, the pedicels usually recurved so that the keel is turned uppermost. Bracteoles none, except in *B. bracteolosum.*

The genus is limited to Australia.

SECT. 1. **Eubrachysema.**—*Stems leafy. Ovary surrounded, within the stamens, by an inner cup-shaped or shortly sheathing disk.*

Leaves usually opposite, broad, truncate. Calyx deeply divided into
    lanceolate lobes . . . . 1. *B. præmorsum.*
Leaves mostly opposite, lanceolate or ovate-lanceolate. Calyx tubular-
    campanulate, the lobes shorter than the tube . . . . . . 2. *B. lanceolatum.*
Leaves various. Calyx shortly and broadly campanulate, the lobes
    shorter than the tube.
    Keel 3 times as long as the calyx. Bracteoles none . . . . . 3. *B. latifolium.*
    Keel not twice as long as the calyx. Bracteoles none.
       Leaves mostly alternate. Ovules about 20 . . . . 4. *B. undulatum.*
       Leaves mostly opposite or whorled. Ovules 4 . . . . . . 5. *B. subcordatum.*
    Keel not twice as long as the calyx. Bracteoles 2, orbicular . . 6. *B. bracteolosum.*

SECT. 2. **Leptosema.**—*Stems leafless, except small scales. No inner disk round the ovary.*

Stems winged, bearing the flowers at their notch-like nodes.
    Keel scarcely exceeding the calyx or shorter. Pod ovoid.
       Flowers mostly clustered. Bracts imbricate. Pod not exceed-
          ing the calyx . . . . 7. *B. bossiæoides.*
       Flowers solitary. Bracts minute or none. Pod more than twice
          as long as the calyx.
          Wings of the stem 2 to 3 lines broad on each side, striate.
             Calyx and pod hairy . . . . 8. *B. oxylobioides.*
          Wings of the stem not 1 line broad on each side. Calyx al-
             most and pod quite glabrous . . . . 9. *B. uniflorum.*
    Keel twice as long as the calyx. Pod long and cylindrical.
       Pedicels at the upper nodes of the branches . . . . 10. *B. aphyllum.*
       Pedicels 1 or 2 at the base of the stem . . . . 11. *B. macrocarpum.*
Flowers crowded on short radical scapes. Barren stems erect, dicho-
    tomous and leafless.
    Barren stems flat, softly silky, not spinous. Flowers scarcely
       1 in. long . . . . 12. *B. tomentosum.*
    Barren stems nearly terete, silky-pubescent, often spinescent.
       Flowers 1½ in. long . . . . 13. *B. Chambersii.*
    Barren stems terete, glabrous, spinescent. Flowers scarcely 1 in.
       long . . . . 14. *B. daviesioides.*

SECTION I. EUBRACHYSEMA.—Stems leafy. Ovary surrounded, within the stamens, by an inner cup-shaped or shortly sheathing disk.

1. **B. præmorsum,** *Meissn. in Pl. Preiss.* i. 25. A shrub of 2 or 3 ft., the young shoots pubescent or villous, glabrous with age. Leaves mostly opposite, shortly petiolate, obversely triangular, truncate or broadly sinuate–3-lobed at the top, mostly 1 to 1½ in. long and often as broad at the top, rounded at the base, thinly coriaceous, reticulate and glabrous when old, rarely a few smaller ones are obovate. Stipules setaceous, recurved, deciduous. Flowers red, usually 2, on short axillary leafy branches, or on peduncles bearing a pair of small leafy bracts. Calyx villous, ½ to nearly ¾ in. long, deeply divided into lanceolate segments, the 2 upper ones broader and more or less united. Standard on a rather slender claw, lanceolate, concave, reflexed, rather longer than the calyx if straightened; wings nearly as long as the keel, which is fully 1 in. long, broadly falcate. Inner disk short and cupular. Ovary short, densely villous, with 15 to 20 ovules. Pod ovoid, as long as the calyx. Seeds strophiolate.

W. Australia, *Drummond;* Preston river, *Preiss, n.* 824; Vasse river, *Mrs. Molloy;* Kalgan, Gordon, and Tweed rivers, *Oldfield;* Hay river, *Maxwell;* Hampden, *Clarke.*

2. **B. lanceolatum,** *Meissn. in Pl. Preiss.* i. 24. A shrub of 2 or 3 ft., the young branches silvery-white with a silky pubescence, at length glabrous. Leaves usually opposite, from broadly ovate-lanceolate to narrow-lanceolate, acute and often mucronate, 1½ to 3 or even 4 in. long; occasionally, however, they are all alternate, or some of them small and ovate, almost as in *B. latifolium;* all coriaceous, glabrous, and finely reticulate above, silvery-pubescent or at length glabrous underneath; stipules setaceous. Flowers red, axillary, solitary or clustered on short pedicels. Bracts very small. Calyx narrow campanulate, rather above ½ in. long, silvery-white with silky hairs, the lobes broad, acuminate, shorter than the tube. Standard lanceolate, about as long as the calyx; wings longer, with a broad basal auricle; keel broadly falcate, about 1 in. long. Ovary shortly stipitate, with 15 to 20 ovules, surrounded by a short inner disk. Pod oblong, loosely villous, as long as the calyx. Seeds rather large, strophiolate.—Bot. Mag. t. 4652, copied into Lemaire, Jard. Fleur. t. 301: *B. celsianum,* Lemaire?; Walp. Rep. v. 423.

W. Australia. From near Cape Riche, *Drummond, 5th Coll. n.* 20; *Preiss, n.* 815, 822, 823; to Point Henry, Salt river, and Vasse river, *Oldfield.*

3. **B. latifolium,** *R. Br. in Ait. Hort. Kew. ed.* 2. iii. 10. A diffuse, procumbent or half-climbing shrub, the young branches often silky-tomentose. Leaves alternate or very rarely opposite, shortly petiolate, ovate or almost orbicular, obtuse with a short often recurved point, 1 to 2 in. long, coriaceous, glabrous above, silky-tomentose or at length glabrous underneath. Flowers red, axillary, solitary or 2 or 3 together, and then often on an elongated, sometimes leafy peduncle. Calyx broadly campanulate, not above 4 lines long, silky-pubescent with short acute lobes. Standard obovate-oblong, nearly twice the length of the calyx; wings narrow, nearly as long as the keel; keel fully 1 in. long. Ovary stipitate, with about 10 or 12 ovules, surrounded by a cup-shaped inner disk. Pod nearly globular or shortly

oblong, about as long as the calyx or rather longer.—DC. Prod. ii. 105; Bot. Reg. t. 118; Bot. Mag. t. 2008; Lodd. Bot. Cab. t. 411.

W. Australia. King George's Sound, *R. Brown, Oldfield;* Hay district, *Preiss, n.* 807 ; S. coast ?, *Drummond, 3rd Coll. n.* 86 ; and eastward to Cape Arid and Cape Knob, *Maxwell.*

4. **B. undulatum,** *Ker, in Bot. Reg. t.* 642. An erect shrub, with weak diffuse or pendulous branches, silky-pubescent when young. Leaves usually alternate, from broadly ovate or almost orbicular to narrow oblong or almost linear, obtuse, with or without a small point, ¾ to 1½ or rarely 2 in. long, coriaceous, glabrous above, minutely silky-white underneath or at length glabrous, the margins generally undulate when the leaf is broad, recurved when it is narrow. Flowers yellowish-green or almost black, rarely red, axillary, solitary and pedicellate, or 2 or 3 together on a common, sometimes leafy peduncle. Calyx very broadly campanulate, silky-pubescent, about 4 lines long, divided to about the middle into broad obtuse or scarcely acute lobes. Standard ovate-lanceolate, rather longer than the calyx ; wings nearly as long as the keel; keel broad, not twice as long as the calyx. Ovary 15- to 20-ovulate, surrounded by a short inner disk. Pod ovoid, very hairy, usually exceeding the calyx.—DC. Prod. ii. 105 ; Lodd. Bot. Cab. t. 778 ; *Chorozema sericeum,* Sm. in Trans. Linn. Soc. ix. 253 ; *Podolobium (?) sericeum,* DC. Prod. ii. 103 ; *Oxylobium (?) sericeum,* Benth. in Ann. Wien. Mus. ii. 70 ; *Brachysema melanopetalum,* F. Muell. Fragm. iv. 11.

W. Australia. King George's Sound and adjoining districts, *Menzies, R. Brown, Drummond, 3rd Coll. n.* 87, *Preiss, n.* 808, 809, and others.

Var. *angustifolium.* Leaves mostly narrow-oblong.—Gordon, Tone, and Blackwood rivers, *Oldfield.* Dark-flowered and pale yellow-flowered specimens occur in both varieties.

5. **B. subcordatum,** *Benth.* A rigid shrub, apparently erect, with the habit of an *Oxylobium,* the young branches minutely silky-pubescent. Leaves rather crowded, opposite or in whorls of 3, very shortly petiolate, broadly cordate-ovate or almost orbicular, obtuse with a minute point, coriaceous with undulate margins, glabrous above, silky-pubescent underneath, or glabrous with age. Flowers apparently red, axillary, solitary or several on a short common peduncle, like those of *B. undulatum,* but much smaller. Calyx about 3 lines long, silky-tomentose, broadly campanulate, the lobes broad and shorter than the tube. Standard rather exceeding the calyx, broadly oblong, truncate, reflexed, on a broad erect claw ; wings as long as the keel ; keel broad, curved, half as long again as the calyx. Ovary very hirsute, with 4 to 6 ovules. Pod not seen.

W. Australia, *Drummond, 5th Coll. n.* 21.

6. **B. bracteolosum,** *F. Muell. Fragm.* iv. 10. Leaves alternate, oblong, linear or lanceolate, obtuse, 2 to 4 in. long, glabrous and coriaceous when full grown, with revolute margins, silky-hairy underneath when very young. Flowers red, axillary, solitary or 2 together on short slender pedicels, with 2 broadly orbicular bracteoles, nearly 3 lines long, close under the calyx. Calyx broadly campanulate, silky-villous, ¾ in. long, the lobes broad, obtuse, and shorter than the tube. Standard on a long claw, shortly ovate, concave, reflexed, shortly exceeding the calyx ; wings exceeding the standard, but

shorter and narrower than the keel; keel falcate, not twice the length of the calyx. Ovary very silky-hairy, with 6 ovules in the flower examined.

W. Australia. Cape Riche, *Herb. Mueller;* Kojonerup, *Herb. Oldfield.*

SECTION II. LEPTOSEMA.—Stems without any leaves, except small scales at the nodes. No prominent disk round the ovary within that which bears the stamens and lines the base of the calyx.

7. **B. bossiæoides,** *Benth.* Stems apparently erect, slightly branched, flat, with rather broad coriaceous wings descending from the nodes, and tapering both upwards and to the next lowest node on the same side, glabrous or silky-pubescent when young. Flowers clustered or rarely solitary, and mostly at the lower nodes, with several imbricate bracts on the very short peduncles. Calyx about 4 lines long, silky-hairy, deeply lobed, the 2 upper lobes connate to near the top. Standard lanceolate, about two-thirds the length of the calyx; wings narrow, from two-thirds to nearly the length of the keel; keel about as long as the calyx, broader and of a firmer consistence than the wings. Ovary villous, with 4 ovules. Pod broadly ovate, acuminate, rarely exceeding the calyx, hairy, turgid, especially towards the carinal suture, so that a transverse section is almost heart-shaped. Seeds not seen ripe.—*Leptolobium bossiæoides,* Benth. in Ann. Wien. Mus. ii. 84.

N. Australia. N. coast, *R. Brown;* Sims Island, *A. Cunningham.*

8. **B. oxylobioides,** *Benth.* Stems apparently several from the same stock, diffuse or suberect, 1 or more ft. long, flat, with coriaceous striate wings descending from the rather distant nodes, and about 2 or rarely 3 lines broad, pubescent when young but soon glabrous. Pedicels short, recurved, solitary at the upper nodes or 2 or 3 together in a short raceme. Calyx pubescent or villous, about 4 lines long, deeply lobed, the 2 upper segments united to the middle. Standard shorter than the calyx, the lamina oblong, concave, reflexed; wings scarcely exceeding the calyx; keel rather longer and broader. Ovary sessile, very villous, with about 20 crowded ovules; style rather short. Pod ovoid, inflated, acuminate, ¾ to 1 in. long, hirsute with long spreading hairs. Seeds as in *B. aphyllum.*—*Leptosema oxylobioides,* F. Muell. Rep. Burdek. Exped. 8.

Queensland. Shoalwater Bay, *R. Brown;* Repulse Bay, *A. Cunningham;* Port Sinclair, *Fitzalan;* Newcastle range, *F. Mueller.* There are also specimens in the Hookerian Herbarium, marked Victoria river, *Bynoe;* but there may be possibly some mistake. The specific name is unfortunately chosen, as there is no leafless species of *Oxylobium* known, and the flowers are very unlike.

9. **B. uniflorum,** *R. Br. Herb.* Habit of *B. oxylobioides,* but the wings of the stem very narrow, the whole breadth of the branch with the 2 wings rarely exceeding 2 lines, and scarcely striate. Flowers solitary at the nodes, on reflexed pedicels of 3 to 6 lines, rather smaller than in *B. oxylobioides,* but otherwise similar. Calyx only very slightly silky-pubescent. Ovary sessile, slightly villous. Pod ovoid, inflated, nearly 1 in. long, quite glabrous. Seeds, according to R. Brown's notes, about 15.

N. Australia. Islands of the Gulf of Carpentaria, *R. Brown (Herb. R. Br.)*

10. **B. aphyllum,** *Hook. Bot. Mag. t.* 4481. Stems erect or diffuse, 1 or several ft. long, flat, with rigid, coriaceous, and striate broad wings descending

from the nodes, truncate and forming an obtuse notch at the upper end, tapering downwards to the next node of the same side, glabrous and often somewhat glaucous. Flowers red, pedicellate and solitary in the upper notches. Calyx ¾ to nearly 1 in. long, glabrous or slightly silky-pubescent, deeply divided into narrow lanceolate segments, the 2 upper ones usually free or slightly connate. Standard ovate, very shortly clawed, scarcely half the length of the calyx, recurved between the upper lobes; wings nearly as long as the calyx; keel oblong-falcate, nearly twice the length of the calyx. Ovary silky-villous, with numerous crowded ovules. Pod nearly cylindric, turgid, furrowed at the sutures, 1½ to 2 in. long. Seeds small, without any strophiole; testa with an outer membranous coating, the inner coating thick and cartilaginous, especially at the back of the embryo, but not truly albuminous. —*Burgesia homaloclada,* F. Muell. Fragm. i. 222.

W. Australia, *Drummond, 4th Coll. n.* 37 ; White Peak, Murchison river, *Oldfield* and *Walcott.*

Cunningham's Herbarium contains specimens from York Sound, on the N.W. coast, of what, from the remains of a pedicel and calyx, I should suspect to be a *Brachysema* near *B. aphyllum* and *B. bossiœoides,* but distinct from both. There being, however, neither flowers nor fruit, it is useless to name or describe them.

11. **B. macrocarpum,** *Benth.* Stems numerous, from a woody, tufted stock, slightly branched, 1 ft. long or more, flat, with narrow, coriaceous, striate wings, more or less silky-hairy, or becoming glabrous with age, the nodes distant and the scales very minute. Flowers not seen. Fruiting pedicels solitary or very few near the base of the stems, although not on separate scapes. Pod 2 to 4 in. long, nearly cylindrical but tapering into a long point. Seeds not seen, but the persistent funicles show them to have been as least as numerous as in *B. aphyllum.*

W. Australia. Dirk Hartog's Island, plentiful, *Milne;* sand plains, Murchison river, *Oldfield* (the latter specimens too young to determine with certainty).

12. **B. tomentosum,** *Benth.* Barren stems erect and dichotomous, as in the following two species, but stouter, flattened to the breadth of 2 or 3 lines, densely and softly silky-pubescent, the ultimate branches flat and divergent, and not spinescent. Flowering scapes short and tufted, and flowers of *B. daviesioides.* Pod, also, as in that species, ovoid, turgid, acuminate, but larger, almost exceeding the calyx, and more silky.

W. Australia. Between Moore and Murchison rivers, *Drummond.*

13. **B. Chambersii,** *F. Muell.* Herb. Barren stems erect, dichotomous and spinescent as in *B. daviesioides,* but silky-pubescent, terete or slightly compressed. Flowering scapes short and tufted, but rather looser than in *B. daviesioides,* the flowers rather larger, attaining 1½ in., in one-sided racemes. Calyx softly villous, deeply divided into narrow lobes, the two upper ones united nearly to the top. Standard lanceolate, about two-thirds the length of the calyx; keel rather longer than the calyx; wings not quite so long. Ovary very densely villous, with numerous crowded ovules. Pod not seen.—*Leptosema Chambersii,* F. Muell. Rep. Burdek. Exped. 8.

N. Australia. Between the rivers Finke and Stephenson, *M'Douall Stuart.*

14. **B. daviesioides,** *Benth.* Rhizome thick and woody, emitting several

erect, rigid, leafless, barren stems, $\frac{1}{2}$ to $\frac{3}{4}$ ft. high, glabrous, nearly terete, striate, with numerous dichotomous corymbose branches, the ultimate branch-lets subulate and spinescent. Scales very minute at the base of the principal ramifications, and a few larger ones at the base of the stem. Flowers red, nearly sessile, in short unilateral or dichotomous racemes on very short scapes, forming very dense radical tufts, 2 or 3 in. high and often 3 to 4 in. diameter, with ovate or lanceolate, villous scales under the branches and very short pedicels. Calyx $\frac{3}{4}$ to nearly 1 in. long, reddish and villous, very deeply divided, the lobes narrow, the 2 upper ones united to near the top. Standard cordate-lanceolate, about two-thirds the length of the calyx ; wings narrow, nearly or quite as long as the calyx ; keel rather longer and broader, the petals shortly connate above the middle. Ovary sessile, short, villous, tapering into a long style ; ovules very numerous and crowded, on slender funicles. Pod ovoid, turgid, acuminate, shorter than the calyx. Seeds not seen.—*Kaleniczenkia daviesioides*, Turcz. in Bull. Mosc. 1853, i. 252.

W. Australia, *Drummond, 4th Coll. n.* 26.

3. OXYLOBIUM, Andr.

(Callistachys, *Vent.* ; Podolobium, *R. Br.*)

Calyx-lobes nearly of equal length, the 2 upper ones usually broader and united higher up. Petals clawed. Standard orbicular or reniform, emargi-nate, longer than the lower petals ; wings oblong ; keel broader than the wings and about the same length, straight or slightly curved, obtuse. Sta-mens free. Ovary sessile or stipitate, with several (4 to above 30) ovules, on straight, filiform funicles ; style incurved, filiform or thickened towards the base, with a small terminal stigma. Pod sessile or stalked, ovoid or ob-long, turgid, continuous inside or rarely with a cellular tissue forming irregular transverse half-dissepiments, or slightly lining the cavity ; valves usually co-riaceous. Seeds with or without a strophiole.—Shrubs or rarely undershrubs. Leaves on very short petioles, more or less distinctly verticillate or opposite, oc-casionally scattered or rarely all alternate, simple, entire or rarely with pungent lobes. Stipules setaceous, sometimes minute or none. Flowers yellow, or the keel and base of the standard, or rarely entirely, purple-red, in terminal or axillary racemes, either loose or contracted into corymbs or whorl-like clus-ters. Bracts and bracteoles very deciduous. Staminal disk usually very short. Ovary very villous, except in *O. staurophyllum*.

The genus is limited to Australia. It differs from *Chorizema*, chiefly in habit and in the proportions of the lower petals ; from *Gastrolobium* only in the number of ovules, 4 or more, not 2 only. Several species of this genus, as well as of *Gastrolobium*, are sent as the poison plant of W. Australia, especially *O. lineare, capitatum,* and *parviflorum.*

SERIES I. **Callistachyæ.**—*Leaves mostly irregularly verticillate. Inflorescence terminal, very dense. Ovules about 8. Strophiole none or minute.*

SERIES II. **Racemosæ.**—*Leaves mostly alternate (except in* O. obtusifolium). *Racemes terminal, loose. Ovules* 10 *to* 30. *Strophiole none.*

Leaves alternate, linear, 2 to 6 in. long. Ovules about 20 . . . 　4. *O. lineare.*
Leaves alternate or whorled, linear, ¾ to 1¼ in. long, hooked at the
　end, the margins revolute. Ovules above 20. . . . . . . 　7. *O. obtusifolium.*
Leaves mostly alternate, under 1 in. long, flat or nearly so. Ovules
　about 10.
　　Leaves oblong or lanceolate . . . . 　5. *O. carinatum.*
　　Leaves cuneate . . . . 　6. *O. spathulatum.*

SERIES III. **Ericoideæ.**—*Slender heath-like shrubs. Lea es small, mostly verticillate. Flowers axillary, or in short, terminal, umbel-like racemes. Ovules* 8 *to* 10. *Strophiole none (except in* O. Pulteneæ ?).

Flowers axillary, solitary. Leaves ovate, about 1 line long . . . 　9. *O. microphyllum.*
Flowers in short, terminal, umbel-like racemes.
　Leaves cordate-ovate or ovate-lanceolate, 1½ to 3 lines long . . 　8. *O. cordifolium.*
　Leaves lanceolate to linear, 2 to 4 lines long . . . . . . 10. *O. Pulteneæ.*
　Leaves linear-subulate, with hooked points, 2 to 4 lines long . 11. *O. hamulosum.*

SERIES IV. **Laxifloræ.**—*Shrubs or procumbent or trailing undershrubs. Leaves mostly opposite. Flowers in loose racemes, or few, axillary. Calyx glabrous or slightly pubescent. Ovules usually* 8. *Strophiole none.*

Stems diffuse procumbent or trailing.
　Leaves obovate, oval, oblong or lanceolate, obtuse or mucronulate 12. *O. scandens.*
　Leaves ovate, rigid, pungent-pointed . . . . 13. *O. procumbens.*
　Leaves crowded, rather small, truncate, 3-pointed. Flowers few,
　　solitary or scarcely racemose . . . . 14. *O. tricuspidatum.*
Stems erect. Leaves deeply cordate, rigid . . . . 15. *O. spectabile.*

SERIES V. **Gastrolobioideæ.**—*Rigid shrubs. Leaves mostly opposite or in threes, coriaceous. Flowers in axillary clusters, or terminal, short, corymbose racemes. Calyx villous. Ovules* 4 *or very rarely* 6. *Seeds strophiolate.*

Calyx 5 lines long, very villous. Petals purple, nearly twice as
　long. Leaves ovate or oblong. Bracts very broad . . . . 16. *O. atropurpureum.*
Calyx 3 to 4 lines long.
　Flowers mostly in a terminal corymb. Leaves ovate or oblong,
　　very obtuse or emarginate. Calyx very villous.
　　　Leaves broadly ovate . . . . 17. *O. retusum.*
　　　Leaves narrow oblong . . . . 18. *O. virgatum.*
　Flowers mostly in axillary clusters.
　　Leaves obovate or broadly oblong, strongly reticulate . . . 19. *O. reticulatum.*
　　Leaves, except the lowest, oblong, lanceolate or linear . . 20. *O. capitatum.*
　　Leaves cuneate-oblong, obovate-truncate or broadly spathulate,
　　　very obtuse or emarginate, faintly reticulate . . . . . 21. *O. cuneatum.*
　　Leaves small, elliptical-oblong, pungent, pointed . . . 22. *O. acutum.*

SERIES VI. **Podolobieæ.**—*Leaves mostly opposite. Flowers in loose or slender, axillary or terminal racemes. Calyx nearly glabrous. Ovules* 4 *or very rarely* 6. *Seeds (where known) not strophiolate.*

Leaves oblong or almost linear, obtuse, coriaceous. Racemes
　mostly terminal.
　Stipules inconspicuous. Leaves closely and minutely silky un-
　　derneath . . . . 23. *O. parviflorum.*
　Stipules setaceous. Leaves softly pubescent or villous underneath 24. *O. heterophyllum.*
Leaves pungent-pointed, entire or lobed. Racemes mostly axillary.
　Leaves lanceolate, entire, rounded or truncate at the base . . 25. *O. aciculiferum.*
　Leaves cuneate at the base, usually toothed or with divaricate,
　　pungent lobes.

Leaves ovate or lanceolate, toothed or shortly lobed . . . . 26. *O. trilobatum.*
Leaves deeply 3-lobed, the lateral lobes lanceolate and divaricate. . . . . 27. *O. staurophyllum.*

Callistachys linariæfolia, G. Don, Gen. Syst. ii. 117, is not recognizable from the very short diagnosis given, but is most probably the narrow-leaved variety of *O. ellipticum.*

SERIES I. CALLISTACHYÆ.—Leaves mostly irregularly verticillate. Inflorescence terminal, very dense. Ovules about 8. Seeds without any or with a very minute strophiole. The *Gastrolobioideæ* differ chiefly in their 4 ovules and strophiolate seed.

1. **O. Callistachys,** *Benth.* A tall shrub, the young branches often angular, more or less clothed with appressed silky hairs. Leaves mostly in irregular whorls of 3, from ovate-oblong and 1½ to 2 in. long, to lanceolate and 4 or 5 in. long, obtuse with a small callous point, coriaceous, glabrous and reticulate above, silky-pubescent underneath when young. Flowers yellow, in dense terminal racemes of 2 to 6 in. Pedicels short. Bracts and bracteoles setaceous, deciduous. Calyx about 4 lines long, hirsute with long silky hairs. Ovary very shortly stipitate with 6 to 8 ovules. Pod at least ½ in. long, acute, rigidly coriaceous, opening at the top only, more or less lined with cellular tissue which often dries up into partial transverse dissepiments. Seeds without any or with a scarcely perceptible strophiole.—*Callistachys lanceolata,* Vent. Jard. Malm. t. 115 ; DC. Prod. ii. 104 ; Bot. Reg. t. 216 ; Meissn. in Pl. Preiss. i. 26 ; *C. ovata,* Sims, Bot. Mag. t. 1925 (short broadleaved specimens) ; DC. Prod. ii. 104 ; Meissn. in Pl. Preiss. i. 26 ; *C. retusa,* Lodd. Bot. Cab. t. 1983 ? (from the fig., leaves very obtuse and marginate); *C. longifolia,* Paxt. Mag. viii. 31, with a fig. (long narrow-leaved specimens); *Chorizema Callistachys,* F. Muell. Fragm. iv. 18.

W. Australia. King George's Sound, chiefly in the sands by the sea, *Fraser* and others; King's river and Albany, *Preiss, n.* 849, 850 ; and eastward to Stokes Inlet, *Maxwell;* Gordon river, *Oldfield ;* Champion Bay, *Bowes.* The genus *Callistachys,* originally founded on this plant, has been subsequently distinguished from *Oxylobium* chiefly by the cellular tissue within the pod, which is usually abundant in this species, but occurs in several others, sometimes in the form of transverse raised lines, sometimes drying into hair-like papillæ or scarcely perceptible, and rarely furnishes even good specific characters. The incomplete dehiscence of the pod does however distinguish this species from all others.

2. **O. ellipticum,** *R. Br. in Ait. Hort. Kew. ed.* 2. iii. 10. An erect shrub, low and compact in mountain situations, tall, often straggling, sometimes above 10 ft. high when luxuriant, the branches silky-pubescent or tomentose. Leaves mostly in irregular whorls of 3, from oval-oblong or elliptical and under 1 in. long to oblong-linear or lanceolate and 2 or 3 in. long, mucronate, the margins recurved or revolute, coriaceous, glabrous and reticulate above, silky-pubescent or villous underneath. Stipules quite inconspicuous. Racemes densely corymbose, terminal or in the upper axils. Pedicels 1 to 4 lines long. Bracteoles linear, deciduous. Calyx softly villous, 3 lines long or rather more ; lobes as long as the tube, acuminate, the upper ones united much higher up. Standard much longer than the calyx ; wings and keel shorter. Ovary nearly sessile, with 8 to 10 ovules. Pod 4 to 6 lines long, acuminate, very villous, opening to the base, glabrous inside. Seeds without any strophiole.—DC. Prod. ii. 104 ; Hook. f. Fl. Tasm. i. 81 ; Bot.

Mag. t. 3249; *Gompholobium ellipticum,* Labill. Pl. Nov. Holl. i. 106. t. 135; *Callistachys elliptica,* Vent. Jard. Malm. under n. 115; *Chorizema ellipticum,* F. Muell. Pl. Vict. ii. 39; *Pleurandra (?) reticulata,* Hook. Journ. Bot. i. 245; *Oxylobium argenteum,* Kunze, in Linnæa, xx. 61; *O. Pulteneæ,* Lodd. Bot. Cab. t. 1947, not of DC.

N. S. Wales. Port Jackson, *R. Brown;* Hastings river, *Fraser;* Clarence river, *Beckler;* Argyle county, *A. Cunningham;* Illawarra, *Backhouse.*

Victoria. Sources of the Avon and Macalister rivers, *F. Mueller.*

Tasmania. Derwent river, *R. Brown;* abundant in heathy places in the southern part of the island, ascending in the mountains to 4000 ft., *J. D. Hooker.*

W. Australia. East River Flat, Stokes Inlet, *Maxwell.* This specimen (a small young one), and a few mountain ones, both from Victoria and from Tasmania, are silky-pubescent only on the young shoots and inflorescence, and come very near to *O. alpestre,* but the stipules are very small or quite inconspicuous, and the pod appears to be always quite glabrous and smooth inside.

Var. *angustifolium.* Leaves long and narrow. 1 find no other character; the distance of the bracteoles from the calyx is very variable.—*O. arborescens,* R. Br. in Ait. Hort. Kew. ed. 2. iii. 10; DC. Prod. ii. 104; Lodd. Bot. Cab. t. 163; Bot. Reg. t. 392; Bot. Mag. t. 2442; Hook. f. Fl. Tasm. i. 81; *O. Pulteneæ,* Paxt. Mag. ix. 149, with a fig., not of DC.; *O. angustifolium,* A. Cunn. Herb., erroneously referred in Ann. Wien. Mus. ii. 70, to *O. obtusifolium,* Sweet, a species which I had then mistaken. N. S. Wales and northern parts of Tasmania, including Port Dalrymple, *R. Brown.*

3. **O. alpestre,** *F. Muell. Trans. Phil. Inst. Vict.* i. 38. A bushy or diffuse shrub, attaining sometimes 3 or 4 ft., the young shoots silky-pubescent, at length nearly glabrous, much resembling the more glabrous varieties of *O. ellipticum.* Leaves mostly opposite, occasionally in whorls of 3, oblong or lanceolate, obtuse or mucronate, ¾ to 1½ in. long, the margins recurved, coriaceous, reticulate. Stipules narrow-lanceolate or linear-subulate, recurved, small but conspicuous at the upper leaves, at length deciduous. Racemes shortly corymbose, terminal or in the upper axils. Pedicels 1 to 3 lines long. Bracteoles linear, deciduous. Calyx silky-pubescent, about 3 lines long, the lobes acute, the 2 upper ones united above the middle. Standard half as long again as the calyx; wings and keel rather shorter. Ovary nearly sessile with about 6 ovules. Pod ovoid-oblong, acuminate, very villous, opening to the base, lined with a loose pithy substance or scale-like hairs. Seeds not strophiolate.

Victoria. Cobberas Mountains, Macalister river, Mount Butler, Timbertop, etc., *F. Mueller.* I was inclined to consider this as a variety of *O. ellipticum,* but, as observed by F. Mueller, the stipules are much more conspicuous, and in all the specimens I have seen the pods are slightly pithy inside. The whole plant is also usually much more glabrous.

SERIES II. RACEMOSÆ.—Leaves mostly alternate (more frequently verticillate in *O. obtusifolium*). Racemes terminal, loose. Ovules about 10 to 30. Seeds not strophiolate.

4. **O. lineare,** *Benth.* An erect shrub of 6 to 10 ft., with long and slender branches, minutely silky-pubescent when young. Leaves mostly alternate, linear or linear-lanceolate, 2 to 6 in. long, obtuse or mucronate, flat or with the margins recurved, reticulate, glabrous or silky underneath when very young. Flowers yellow or of a dull red, in rather loose terminal racemes or sometimes in the upper axils. Pedicels very short. Bracts narrow, very deciduous. Calyx 3½ to 4 lines long, silky-villous. Petals of *O. Callistachys*

or rather smaller. Ovary stipitate, with about 20 closely-packed ovules. Pod ovoid, acute, rarely above 5 lines long. Seeds minutely strophiolate.—*Callistachys linearis*, Benth. in Hueg. Enum. 27 ; Bot. Mag. t. 3882 ; Meissn. in Pl. Preiss. i. 27 ; *C. parviflora*, Benth. in Hueg. Enum. 26 ; Meissn. in Pl. Preiss. i. 26 ; *C. linariæfolia*, G. Don, Gen. Syst. ii. 117 ; *Chorizema lineare*, F. Muell. Fragm. iv. 17.

W. Australia. Swan River, *Fraser, Huegel, Preiss, n.* 853, and others ; northward to Murchison river, *Oldfield ;* southward to Preston river, Toby's Inlet, etc., *Preiss, n.* 851, 852, 854; Vasse, Gordon and Tone rivers, *Oldfield.* The breadth of the leaves, the size and colour of the flowers vary in almost all stations, without having any relation to each other. The short-leaved specimens sometimes much resemble *Gastrolobium Callistachys.*

5. **O. carinatum,** *Benth.* A low shrub, with ascending or erect simple or branched stems, often under 1 ft., but attaining 3 or 4 ft., more or less hirsute. Leaves scattered, oblong or lanceolate with a small rigid recurved point, often under ½ in. and rarely nearly 1 in. long, coriaceous, often undulate, glabrous or slightly hirsute, with a prominent midrib and transverse reticulations. Flowers bright yellow, in terminal racemes of 1 to 2 in. Pedicels short. Calyx 3 to 4 lines long, silky-hairy, the lobes narrow and acute, the 2 upper ones scarcely connate. Petals often scarcely exceeding the calyx-lobes. Ovary shortly stipitate, with about 10 ovules ; style rather thick with an oblique stigma. Pod ovoid, acuminate, 3 to 4 lines long without the stipes. Seeds shining black, without any strophiole.—*Callistachys carinata*, Meissn. in Pl. Preiss. i. 27 ; *Chorizema pubescens*, Turcz. in Bull. Mosc. 1853, i. 256.

W. Australia, *Drummond, 4th Coll. n.* 33 ; south side of Konkoberup hills, *Preiss, n.* 1068 ; Kalgan river, *Oldfield.* The species bears some resemblance to *Gastrolobium parvifolium.*

6. **O. spathulatum,** *Benth.* A shrub allied in appearance to *O. carinatum*, but evidently taller and more branched, the branches hirsute. Leaves alternate, linear-cuneate or almost oblong, with recurved points, mostly ¼ to ¾ in. long, rigid, with a prominent midrib, recurved margins and a few transverse reticulations. Racemes terminal, rather dense, 1 to 2 in. long. Pedicels short. Bracts linear-subulate, very deciduous. Calyx 4 lines long, silky-villous, the lobes lanceolate-subulate, the 2 upper ones rather broader and more connate. Petals larger than in *O. carinatum*, although scarcely exceeding the calyx. Ovary shortly stipitate, with about 10 ovules ; style rather thick, stigma oblique. Pod almost sessile, ovoid-oblong, acuminate. Seeds without any strophiole.—*Callistachys spathulata*, Meissn. in Pl. Preiss. ii. 208.

W. Australia, *Drummond, 2nd Coll. n.* 89.

7. **O. obtusifolium,** *Sweet, Fl. Austral. t.* 5. Stems 1 to 1½ ft. long, diffuse or procumbent, little branched, minutely silky-pubescent or at length glabrous. Leaves alternate or irregularly whorled in threes, linear, obtuse and recurved at the end, ¾ to 1 in. or rarely 1½ in. long, the margins much revolute, coriaceous and transversely reticulate above, the under side silky-hairy but usually almost concealed. Flowers orange-red, in terminal often 1-sided racemes of 1 to 2 in. on very short pedicels. Bracts and bracteoles linear, very deciduous. Calyx silky-pubescent, 4 to 5 lines long, not so broad as in other species and scarcely divided below the middle, the 2 upper

lobes united nearly to the top. Standard nearly twice as long as the calyx;
lower petals much shorter; keel obtuse, rather shorter than the wings. Ovary
on a rather long stipes with above 20 ovules. Pod not seen.

W. Australia. King George's Sound, *Baxter;* towards the Great Bight, *Maxwell.*
This species is in many respects allied to *Chorizema cytisoides*, but the keel is quite that of
Oxylobium.

SERIES III. ERICOIDEÆ.—Slender heath-like shrubs. Leaves small (1 to
3 or rarely 4 lines long), mostly verticillate. Flowers axillary or few in short
terminal umbel-like racemes. Ovules 8 to 10. Seeds not strophiolate (except perhaps in *O. Pulteneæ*).

8. **O. cordifolium,** *Andr. Bot. Reg. t.* 492. Branches slender, diffuse,
pubescent or hirsute when young. Leaves irregularly verticillate in threes,
ovate-cordate, 1½ to 2 lines or the upper ones 3 or rarely 4 lines long, the
margins revolute, often hirsute when young, nearly glabrous when full-grown.
Flowers orange-red, usually 3 or 4 together in little terminal heads or umbels,
the pedicels short and hirsute. Calyx villous, about 3 lines long, the lobes
longer than the tube, lanceolate, acuminate, the 2 upper ones shortly united.
Standard about 4 lines long and broad; keel equal to or rather longer
than the wings. Ovary almost sessile, with about 8 ovules. Pod ovoid,
acuminate, 4 or 5 lines long, scarcely coriaceous. Seeds without any strophiole.—DC. Prod. ii. 104; Bot. Mag. t. 1544; Lodd. Bot. Cab. t. 937;
Chorizema cordifolium, F. Muell. Fragm. iv. 17.

N. S. Wales. Botany Bay, *Banks and Solander;* Port Jackson, *R. Brown, Fraser,*
and others.

9. **O. microphyllum,** *Benth.* A much-branched diffuse or divaricate
shrub, the branches minutely tomentose-pubescent. Leaves scattered, about
1 line long, ovate, obtuse, with recurved margins. Flowers small, apparently
dark-coloured, axillary and solitary, shortly pedicellate. Calyx about 1½ lines
long, deeply divided into broad acute lobes, the 2 upper ones rather more
united. Petals nearly of equal length, shortly exceeding the calyx. Standard
very broad; keel broad, obtuse. Ovary on a rather long stipes, with about
8 ovules. Pod not seen.

W. Australia. Flats of E. Mount Barren, *Maxwell.* This has much the aspect of
Mirbelia oxylobioides, but the leaves are much smaller, the flowers all axillary, and I see no
trace of any dissepiment in the ovary slightly enlarged after flowering; there are, however,
none much advanced in the specimens I have seen.

10. **O. Pulteneæ,** *DC. Prod.* ii. 104. A heath-like shrub, with slender
branches, glabrous or minutely pubescent when young. Leaves alternate or
irregularly verticillate in threes, from ovate-lanceolate to linear, 2 to 4 lines
long, the margins much revolute, glabrous or scabrous above, minutely pubescent underneath. Flowers few, in short terminal racemes usually contracted
into umbels. Bracts very deciduous. Calyx 3 lines long or rather more,
pubescent with appressed hairs, divided to below the middle into lanceolate
acuminate lobes, the 2 upper ones shortly connate. Standard longer than the
calyx; wings and keel scarcely exceeding it. Ovary shortly stipitate, with 8
to 10 ovules. Pod nearly sessile, acuminate, 3 to 5 lines long. Seeds strophiolate according to De Candolle; I have not seen them.—*Pulteneæ sylvatica,*

Sieb. Pl. Exs.; *Callistachys sparsa,* A. Cunn.; Benth. in Ann. Wien. Mus. ii. 69.

N. S. Wales. Port Jackson, *Sieber, n.* 403; Hunter's River, *A. Cunningham ;* near Wollomby, Blue Mountains, *C. Moore.* The plant figured in ' Paxton's Magazine' as *O. Pultenææ* appears to be the narrow-leaved var. of *O. ellipticum.*

11. **O. hamulosum,** *Benth. in A. Gray, Bot. Amer. Expl. Exped.* i. 379. A heath-like shrub evidently very nearly allied to *O. Pulteneæ,* but the leaves are all narrow-linear, almost subulate, 3 to 6 lines long, hooked and pointed at the end, the margins much revolute as in the allied species. Flowers not seen. Fruiting racemes short, like those of *O. Pulteneæ,* and pod the same. Seeds not seen.

N. S. Wales. Hunter's River, *American Exploring Expedition ;* also in Herb. A. Cunningham, from Herb. Lambert. This will probably prove to be a variety of *O. Pulteneæ,* but the leaves are too distinct to unite it without having seen the flowers.

SERIES IV. LAXIFLORÆ.—Shrubs or procumbent or trailing undershrubs. Leaves mostly opposite. Flowers in loose racemes or few and axillary. Calyx glabrous or slightly pubescent. Ovules usually 8. Seeds not strophiolate.— This series differs from the *Podolobieæ* chiefly in the more numerous ovules.

12. **O. scandens,** *Benth. in Ann. Wien. Mus.* ii. 70. A shrub or under-shrub with weak procumbent or half climbing branches, pubescent when young. Leaves mostly opposite, from obovate or ovate-elliptical to ovate-lanceolate or narrow-oblong, obtuse or mucronulate, 1½ to 2 in. long or rarely more, the margins flat, narrowed at the base, reticulate, glabrous when full grown. Racemes terminal or in the upper axils, loose, with few yellow flowers. Pedicels sometimes as long as the calyx, with two small lanceolate-subulate bracteoles at a distance from the calyx. Calyx under 3 lines long, slightly pubescent with appressed hairs, the lobes broad, acute, about as long as the tube, and nearly equal. Standard about 5 lines diameter; wings shorter, obovate-oblong; keel rather smaller, the petals scarcely cohering. Ovary stipitate, with about 8 ovules. Pod very shortly stipitate, about ½ in. long, somewhat curved, acuminate, turgid, the valves transversely veined as in *O. Callistachys,* but the cellular tissue very scanty or scarcely any.—*Chorizema scandens,* Sm. in Trans. Linn. Soc. ix. 253; F. Muell. Pl. Vict. ii. 40; *Podolobium scandens,* DC. Prod. ii. 103; *Daviesia umbellata* and *D. humifusa,* Sieb. Pl. Exs. *Podolobium humifusum,* G. Don, Gen. Syst. ii. 116; *Mirbelia (?) Baxteri,* Lindl. Bot. Reg. t. 1434; *Chorizema Baxteri,* Grah. in Edinb. New Phil. Journ. 1830.

Queensland. Wide Bay, *Bidwill;* Moreton Bay, *W. Hill.*

N. S. Wales. Port Jackson, *R. Brown, Sieber, n.* 391, 392; Paramatta, *Woolls;* Blue Mountains, *Miss Atkinson,* Hastings river, *Beckler.*

Var. *obovatum.* Leaves all or almost all broadly obovate or orbicular. *Podolobium obovatum,* A. Gray, Bot. Amer. Expl. Exped. i. 379.—Hunter's River, *R. Brown* and others. I had long considered this as a distinct species, and had described it from the Paris Herbarium under the name of *O. diffusum,* but having now seen a considerable number of specimens from various sources, both in flower and fruit, I find that it only differs from *O. scandens* in the breadth of the leaves, and in this respect I have seen several interme-diate specimens.

13. **O. procumbens,** *F. Muell. in Trans. Phil. Inst. Vict.* i. 37. Rhi-

zome very thick and woody, emitting procumbent or ascending rigid stems of
¼ to 1 ft., pubescent when young. Leaves irregularly opposite or in threes,
ovate, with a fine pungent point, under 1 in. long, rigidly coriaceous, reticulate and glabrous when full grown, Flowers few, in short, loose, terminal,
pedunculate racemes, or in the upper axils. Bracts and bracteoles setaceous,
deciduous. Calyx about 4 lines long, slightly pubescent with appressed
hairs, the lobes about as long as the tube, all acuminate, the 2 upper ones
united above the middle. Standard ½ in. diameter; wings and keel shorter.
Ovary nearly sessile, with 6 to 10, usually 8, ovules. Pod very shortly stipitate, oblong, obtuse, about ½ in. long.

N. S. Wales. Maneroo, *S. Mossman.*
Victoria. Delatite river, Mount Macedon, ranges on the Upper Genoa river, etc., *F.
Mueller ;* ranges at Lexton, etc., **Whan.** F. Mueller now proposes to unite this with *O.
scandens,* but it appears to me constantly distinct in habit and foliage, in its longer deeper-
coloured flowers, and, in the only fruiting specimen seen, the pod is differently shaped.

14. **O. tricuspidatum,** *Meissn. in Pl. Preiss.* i. 30. A low diffuse or
procumbent glabrous shrub. Leaves mostly opposite, often crowded, obovate
or oblong-cuneate, truncate with 1 to 3 fine bristle-like or almost pungent
points, rarely above ¼ in. long, thin but rigid, reticulate. Flowers small and
few together in the upper axils or at the ends of the branches. Calyx 2 to 2½
lines long, glabrous or nearly so, the lower lobes narrow-lanceolate longer
than the tube, the upper ones broader, falcate, and united above the middle,
all usually with fine bristle-like points. Petals scarcely half as long again as
the calyx. Ovary nearly sessile, with 6 to 10 ovules. Pod sessile, ovoid, obtuse, slightly pubescent. Seeds not strophiolate.

W. Australia. Gravelly places, Hay district, *Preiss, n.* 1064, also *Drummond, n.*
266, *J. S. Roe.*

15. **O. spectabile,** *Endl. Nov. Stirp. Dec.* 2. Rigid, glabrous and
glaucous, or the young shoots slightly silky-pubescent. Leaves opposite,
deeply cordate, ovate or orbicular, mucronate, 1 to 1½ in. long, rigid, coriaceous, reticulate. Racemes terminal, loose, 1 to 2 in. long. Bracts lanceolate, the lower ones trifid, but all falling off before flowering ; bracteoles
none. Calyx broadly campanulate, 3 to 4 lines long, glabrous or nearly so,
the lobes rather shorter than the tube, the lower ones broad and rather obtuse,
the 2 upper ones broader, very obtuse, and united above the middle. Standard above ½ in. diameter; wings and keel rather shorter. Ovary stipitate,
with about 8 ovules. Pod not seen.—*Gastrolobium cordatum,* Benth. in
Lindl. Swan Riv. App. 13, t. 5 B.

W. Australia. Swan River, *Mangles ;* in the interior, *J. S. Roe.*

SERIES V. GASTROLOBIOIDEÆ.—Rigid shrubs. Leaves mostly opposite or in whorls of 3, coriaceous. Flowers in axillary clusters or short corymbose racemes. Calyx villous. Ovules 4 or very rarely 6. Seeds strophiolate.

These have the 4 ovules of the *Podolobieæ,* but differ from them, as from all other *Oxylobiums* (except perhaps *O. Pulteneæ*), in their strophiolate seeds. This character brings them
into close affinity with *Gastrolobium,* of several species of which they have the habit, differing only in the ovules, 4 instead of 2.

16. **O. atropurpureum,** *Turcz. in Bull. Mosc.* 1853, i. 250. A tall, stout shrub, the young branches angular and hoary or softly pubescent. Leaves mostly opposite, distinctly petiolate, ovate or elliptical-oblong, obtuse or emarginate, 2 to 4 in. long, rigidly coriaceous, penniveined and finely reticulate. Flowers much larger than in the following species, and apparently of a deep red, in dense axillary clusters or corymbose racemes. Bracts broadly orbicular, very deciduous. Calyx rather narrow, densely villous with long silky hairs, about ½ in. long, the broad lobes shorter than the tube, the 2 upper ones very shortly connate. Petals, including the claws, 9 to 10 lines long. Ovary shortly stipitate, with 4 ovules. Pod not seen.

W. Australia, Drummond, 5*th Coll. n.* 53, *Maxwell;* Champion Bay, *Bowen;* E. Mount Barren, *Maxwell?* (specimen in leaf only). This species bears much general resemblance to *Gastrolobium pyramidale.*

17. **O. retusum,** *R. Br. in Bot. Reg. t.* 913. A much-branched, rigid shrub, the young branches angular and hoary or pubescent. Leaves mostly opposite, petiolate, ovate or oblong-elliptical, obtuse, truncate or emarginate, usually 1 to 2 in. long, rigidly coriaceous, glabrous and reticulate above, silky-pubescent or rarely glabrous underneath. Flowers reddish-yellow, in dense, almost sessile, terminal clusters or corymbose racemes, or rarely also in the upper axils. Bracts ovate or oblong, very deciduous. Calyx very villous, about 3 or rarely nearly 4 lines long, divided to about the middle into broad-lanceolate lobes. Petals about half as long again as the calyx. Ovary very shortly stipitate, with 4 ovules. Pod ovoid, scarcely acuminate, about 4 lines long, very hairy. Seeds strophiolate.— *Chorizema coriaceum,* Sm. in Trans. Linn. Soc. ix. 254; *Podolobium (?) coriaceum,* DC. Prod. ii. 103; *Oxylobium ovalifolium,* Meissn. in Pl. Preiss. i. 28; *Callistachys tetragona,* Turcz. in Bull. Mosc. 1853, i. 249.

W. Australia. King George's Sound, *R. Brown, A. Cunningham,* and others; southern districts?, *Drummond, n.* 52, 3*rd Coll. n.* 83, 5*th Coll. n.* 56; stony places, Mount Manypeak and King George's Sound, *Preiss, n.* 813, 820; E. Mount Barren, *Maxwell.* The species often much resembles *Gastrolobium pyramidale,* but is more silky, and the ovary has always 4 ovules.

Var. *minus.* Leaves smaller. Flowers mostly terminal. Calyx less villous.— *Drummond, n.* 95, *and* 4*th Coll. n.* 20.—This passes almost into *O. reticulatum.* The differences indeed which separate *O. retusum, reticulatum, capitatum,* and *cuneatum,* are very slight, although the extreme forms are very different.

18. **O. virgatum,** *Hort. Kew.* An erect shrub, nearly allied to *O. retusum* and possibly a variety only, but the very much narrower leaves and smaller flowers give it a very different aspect. Leaves in threes or opposite, narrow-oblong or almost linear, rarely ovate-oblong, very obtuse and emarginate, ¾ to 1½ in. long, silky-pubescent underneath, much less rigid than in *O. retusum,* with the margins often recurved. Flowers in terminal, sessile, corymbose racemes or clusters, and occasionally also in the uppermost axils. Calyx nearly 3 lines long. Petals about half as long again. Ovary and pod of *O. retusum.*— *Gastrolobium retusum,* Lindl. Bot. Reg. t. 1647; Bot. Mag. t. 3328.

W. Australia. Only known in cultivation. The original specimens preserved from the Kew Garden correspond well with the figures given; one I have seen from the Edinburgh garden has the leaves rather broader.

19. **O. reticulatum,** *Meissn. in Pl. Preiss.* i. 29. Very nearly allied to *O. capitatum*, with the same indumentum and inflorescence. Leaves mostly opposite, obovate or broadly ovate-oblong, very obtuse or emarginate, with or without a minute recurved point, rarely above 1 in. long, very coriaceous, strongly reticulate and glabrous when full-grown. Flowers and pods rather smaller than in *O. capitatum*, but otherwise the same.

W. Australia. Sand-hills on the seashore and in the interior, *Preiss, n.* 840 *and* 831, *Drummond, n.* 95 *and* 205 (*or* 215 ?).

Var. *graċile.* Branches slender. Leaves rather smaller, often undulate on the edges, the reticulations not quite so coarse. Flowers and pods rather smaller.—*Gastrolobium axillare*, Meissn. in Bot. Zeit. 1855, 29.—*Drummond, 6th Coll. n.* 22.—These specimens agree perfectly with the diagnosis of Meissner, who refers to the same number of Drummond's, except that I have found in all the flowers that I have examined 4 ovules instead of 2.

20. **O. capitatum,** *Benth. in Hueg. Enum.* 28. A shrub or undershrub of 2 to 3 ft., with rigid, but not thick branches, minutely hoary or silky-pubescent when young. Lowest leaves sometimes obovate, all the others oblong, lanceolate or linear, 1 to 2 in. long, obtuse, with a short, usually recurved point, rigid, glabrous and reticulate above, minutely silky-pubescent or glabrous underneath, with a stout, prominent midrib. Stipules usually recurved. Flowers in axillary clusters or sometimes forming a terminal, compact, corymbose raceme or head. Bracts very deciduous. Calyx silky-villous, 3 or rarely nearly 4 lines long, rather broad, divided to about the middle into acuminate lobes, the 2 upper ones united at the base. Standard nearly twice as long as the calyx ; wings and keel much shorter. Ovary shortly stipitate, with 4 or rarely 6 ovules. Pod ovoid, scarcely acuminate, about 4 lines long, villous, often more or less lined with cellular tissue. Seeds with a rather large strophiole.—Bot. Reg. 1843, t. 16 ; Meissn. in Pl. Preiss. i. 30 ; *O. nervosum*, Meissn. in Bot. Zeit. 1855, 12 ; *Callistachys oxylobioides*, Meissn. in Pl. Preiss. i. 27.

W. Australia. Swan River, *Drummond, 1st Coll., Preiss, n.* 841, 842, 843, 844 ; stony places, foot of Mount Manypeak, *Preiss, n.* 805, 814 ; Vasse river, *Mrs. Molloy ;* Bunbury, Canning river, and Cape Leschenault, *Oldfield ;* between Moore and Murchison rivers, *Drummond, 6th Coll. n.* 21.

21. **O. cuneatum,** *Benth. in Lindl. Swan Riv. App.* 12. An erect, rigid shrub, attaining several ft., the specimens usually assuming a yellowish or a glaucous tint when dry, young branches angular, minutely tomentose-pubescent or almost silky. Leaves mostly opposite or in whorls of 3 or 4, from broadly obovate-triangular or spathulate to almost linear-cuneate, rounded, truncate or emarginate at the end, mostly 1 to 2 in. long, narrowed to the base, usually folded lengthwise, coriaceous, glabrous or slightly silky-pubescent underneath, the reticulate veins scarcely prominent except when the leaf is thin. Flowers yellow or the lower petals purple, in dense corymbose racemes or clusters, either all axillary or also terminal. Bracts narrow, very deciduous. Calyx about 3 lines long, silky-pubescent or villous, the lobes lanceolate, the upper ones slightly united and somewhat falcate. Standard nearly twice as long as the calyx. Ovary almost sessile, with 4 ovules. Pod ovoid, acuminate, rigid, 4 to 5 lines long.

W. Australia. From King George's Sound to Swan River and Murchison river, *Drummond* and others.

The following forms have been described by myself and others as distinct species, but different as some of them appear at first sight, we now find that they all pass into each other by insensible gradations, the only difference consisting really in the relative breadth of the leaves. All are closely allied to *O. reticulatum* and *O. capitatum*, but have the leaves more or less cuneate, not so coarsely reticulate, and a more dense inflorescence.

a. emarginatum. Leaves oblong-cuneate, mostly emarginate, rather small; axillary peduncles often growing out into leafy branches with terminal inflorescences.—*O. Drummondii*, Meissn. in Pl. Preiss. i. 30.—Swan River, *Drummond, n.* 72 *and* 210, *Preiss, n.* 800, in my set. This variety almost passes into *O. capitatum.*

b. cuneifolium. Leaves linear-cuneate, rounded at the end, mostly about 1½ to 2 in. long. —Swan River, *Drummond, 1st Coll., also n.* 71 *and* 207 (partly).

c. obovatum. Leaves very broadly cuneate, truncate, mostly about 1½ in. long.—*O. obovatum*, Benth. in Lindl. Swan Riv. App. 12; Bot. Reg. 1843, t. 36; Meissn. in Pl. Preiss. i. 29; Paxt. Mag. x. 243, with a fig.—Swan River, *Drummond, 1st Coll., also n.* 70 *and* 207 (partly).

d. dilatatum. Leaves very much dilated at the end, so as to be almost 2-lobed when truncate, or 3-lobed when also acuminate, and abruptly narrowed below the dilatation.—*O. dilatatum*, Benth. in Lindl. Swan Riv. App. 12; Meissn. in Pl. Preiss. i. 29.—Swan River, *Drummond, 1st Coll., also n.* 71.

22. **O. acutum,** *Benth.* Apparently a small shrub, our specimens, all with the root, from 1 to 1½ ft. high; branches few, erect, softly pubescent or villous. Leaves in whorls of 3 or scattered, ovate-elliptical or almost oblong, tapering into a pungent point, ½ to ¾ in. long, rigidly coriaceous, reticulate, silky-villous when young, glabrous when full-grown. Flowers all axillary, in loose clusters, shorter than the leaves. Calyx about 3 lines long, silky-villous, the 2 upper lobes shortly united. Standard about twice as long as the calyx; keel much curved, deeply coloured. Ovary nearly sessile, with 4 ovules. Pod acute, rather coriaceous, shortly exceeding the calyx.—*Gastrolobium acutum*, Benth. in Lindl. Swan Riv. App. 14; Bot. Mag. t. 4040.

W. Australia. Swan River, *Drummond, 1st Coll., also n.* 67 *and* 213. At first sight this much resembles *Gastrolobium epacridioides*, but the ovary is more sessile and always with 4 ovules.

SERIES VI. PODOLOBIEÆ.—Leaves mostly opposite. Flowers in loose or slender axillary or terminal racemes. Calyx nearly glabrous. Ovules 4 or very rarely 6. Seeds (where known) not strophiolate.

The first two species of this series have the habit of the racemose species of *Gastrolobium*, but the ovules are 4 and the seeds have no strophiole. The remaining three species are allied to the *Laxifloræ*, but with only 4 ovules and a narrow pod, which, with a somewhat peculiar habit, had induced the establishment of *Podolobium* as a distinct genus. But I find no character sufficiently distinct or consonant with habit to maintain it as such.

23. **O. parviflorum,** *Benth. in Lindl. Swan Riv. App.* 12. A tall spreading shrub, the young shoots hoary with a minute silky pubescence. Leaves alternate, opposite or in threes, narrow-oblong, slightly cuneate or linear, obtuse or emarginate, mostly about 1 in. long, coriaceous, glabrous above, minutely silky-pubescent underneath, the margins usually recurved. Flowers small, orange-yellow and purple, in slender racemes, terminal or in the upper axils, often 2 to 3 in. long. Calyx about 2 lines long, minutely pubescent, the lobes scarcely so long as the tube, acute, the 2 upper ones broader, falcate, and united nearly to the top into a truncate upper lip. Standard nearly 4 lines diameter, the lower petals rather shorter. Ovary on a rather long stipes, with 4 ovules. Pod stipitate, acuminate, 4 to 6 lines long, pubescent or

villous. Seeds often only 1 or 2, not strophiolate, embedded in a pithy substance lining the cell.—Meissn. in Pl. Preiss. i. 31.

W. Australia. Swan River, *Drummond, 1st Coll., J. S. Roe, Preiss, n.* 798 *and* 801, where it is said to be one of the worst of the poison plants; S. coast, at various points, from Phillips river to the Great Bight, *Maxwell.* Very nearly allied to *Gastrolobium crassifolium*, but I always find 4 ovules, besides that in that species the leaves are somewhat folded lengthwise, the margins never recurved.

24. **O. (?) heterophyllum,** *Benth.* Branches apparently diffuse, slender, minutely pubescent. Leaves mostly opposite, oblong-linear or lanceolate, or the lower ones obovate-oblong, obtuse, with a small point, $\frac{3}{4}$ to $1\frac{1}{2}$ in. long, the margins recurved, glabrous and reticulate above, loosely pubescent or villous underneath. Flowers not seen. Fruiting-racemes slender, 1 to 3 in. long. Pedicels 1 to 2 lines long. Calyx about 2 lines long, pubescent or hirsute, the lobes short, the 2 upper ones united to the middle. Standard about 3 lines diameter; keel nearly as long; wings rather shorter and narrow. Ovary stipitate, with 8 ovules. Pod shortly stipitate, oblong, 3 to 4 lines long, scarcely acute, hairy. Seeds not strophiolate.—*Chorizema heterophyllum,* Turcz. in Bull. Mosc. 1853, i. 255.

W. Australia, *Drummond, 5th Coll. n.* 27; gravelly soil, Oldfield river and mouth of Young river, *Maxwell.*

25. **O. aciculiferum,** *Benth.* Apparently a slender shrub; branches pubescent. Leaves lanceolate, with a fine pungent point, about 1 in. long, rounded or truncate at the base, quite entire, coriaceous, glabrous and shining above, with transverse reticulate veins, minutely pubescent and at length glabrous underneath. Stipules long and bristle-like. Racemes slender, axillary or terminating short leafy branches, the flowers few and distant. Calyx slightly silky-pubescent, about 2 lines long. Petals fully twice as long. Ovary stipitate, with 4 ovules. Pod not seen.—*Podolobium aciculiferum,* F. Muell. Fragm. i. 75.

Queensland. Brisbane river, *W. Hill,* a single specimen.

26. **O. trilobatum,** *Benth.* A shrub of several ft., sometimes almost glabrous, more frequently with pubescent branches. Leaves mostly opposite, from broadly ovate to lanceolate, 1 to 2 in. long, with pungent points and bordered by a few, distant, pungent teeth or lobes, of which 1 or 2 on each side near the base are usually larger than the others, coriaceous, glabrous, shining and reticulate above, pale and sometimes minutely pubescent underneath. Flowers yellow, in loose axillary or terminal racemes, often exceeding the leaves. Calyx slightly pubescent, about 2 lines long, the lobes shorter than the tube, the 2 upper ones united nearly to the top. Petals fully twice as long as the calyx. Ovary stipitate, silky-pubescent, with 4 or rarely 6 ovules. Pod stipitate, oblong, 3 to 4 lines long, straight or incurved, pubescent and turgid as in other *Oxylobiums*, but much narrower. Seeds not strophiolate. —*Pultenæa ilicifolia,* Andr. Bot. Rep. t. 320; *Chorizema trilobum,* Sm. in Trans. Linn. Soc. ix. 253; F. Muell. Fragm. iv. 19; *Podolobium trilobatum,* R. Br. in Ait. Hort. Kew. ed. 2, iii. 9; DC. Prod. ii. 103; Bot. Mag. t. 1477; Bot. Reg. t. 1333.

Queensland. Cabbage-tree Hills, Moreton Bay, *W. Hill.*

N. S. Wales. Port Jackson, *R. Brown, Sieber, n.* 395, *and Fl. Mixt. n.* 571, and others ; and northward to Clarence and Hastings rivers, *Beckler ;* Hunter's River, *Oldfield ;* New England, *C. Stuart ;* southward to Illawarra, *Shepherd ;* Twofold Bay, *F. Mueller.*

27. **O. staurophyllum,** *Benth.*   A divaricately branched, glabrous shrub, closely resembling *O. trilobatum,* and much better deserving that name. Leaves alternate or opposite, ¾ to 1½ in. long, with pungent points and a cuneate base, and deeply divided into 3 lanceolate, pungent lobes, the lateral ones divaricate and sometimes again 2-lobed, all coriaceous, shining, and strongly reticulate. Flowers yellow, in loose axillary racemes, rarely exceeding the leaves. Calyx glabrous, about 2 lines long, the lobes shorter than the tube, the 2 upper ones united, but not so high as in *O. trilobatum.* Petals fully twice as long as the calyx. Ovary on a long stipes, nearly glabrous, with 4 ovules. Pod narrow, like that of *O. trilobatum,* but glabrous. —*Podolobium staurophyllum,* DC. Prod. ii. 103 ; Bot. Reg. t. 959 ; Lodd. Bot. Cab. t. 1177 ; Paxt. Mag. iv. 171, with a fig.

N. S. Wales. Port Jackson to the Blue Mountains, *R. Brown, Sieber, n.* 393, and others ; between Emu Plains and Lachlan Depot, *A. Cunningham.*

4. CHORIZEMA, Labill.

(Orthotropis, *Benth.*)

Calyx-lobes nearly of equal length, the 2 upper ones usually broader and united higher up.   Petals clawed ; standard orbicular or reniform, emarginate, rather longer than the wings ; wings oblong ; keel much shorter than the wings, straight and obtuse or with an erect point, or rarely incurved. Stamens free.   Ovary sessile or stipitate, with numerous or rarely 8 to 10 ovules ; style usually short, incurved ; stigma terminal, frequently oblique. Pod ovoid, turgid or compressed, continuous inside.   Seeds not strophiolate. —Shrubs or undershrubs.   Leaves all alternate (except in *C. ericifolium*), simple, entire or prickly-toothed.   Stipules small, setaceous, sometimes wanting.   Flowers usually orange or red, in terminal racemes or rarely axillary. Pedicels short, with 2 small bracteoles, usually deciduous.   Ovary villous.

The genus is exclusively Australian. It differs from *Oxylobium* chiefly in habit and in the short or acuminate keel ; the pod is also usually less turgid.

Keel much curved, rostrate. Stigma fringed, very oblique. Erect
  rigid shrub. Leaves rigid, with pungent points . . . . .   1. *C. Dicksonii.*
Keel erect, obtuse or shortly pointed. Stigma not fringed. Leaves
  flat or the margins recurved.
  Leaves orbicular-cordate, rigid, pungent-pointed, quite entire, much
    undulate. Erect shrub . . . .   2. *C. nervosum.*
  Leaves cordate, all or almost all prickly-toothed.
    Tall erect shrub. Branches and under side of the leaves pubescent   3. *C. varium.*
    Branches and leaves glabrous.
      Tall shrub, with weak, slender branches. Leaves prickly-toothed   4. *C. cordatum.*
      Low or diffuse shrub or undershrub. Leaves prickly-lobed .   5. *C. ilicifolium.*
  Leaves quite entire or a few very loosely prickly-toothed.
    Low diffuse or ascending shrubs or undershrubs. Leaves rounded
      or almost cordate at the base, the lower ovate, the upper
      narrow. Style much curved.
      Leaves mostly or all ovate. Flowers few, large. Stigma
        very oblique . . . .   6. *C. rhombeum.*

Leaves mostly narrow, rarely all ovate. Flowers several, small.
 Stigma small, terminal.
   Leaves distant. Racemes loose, glabrous; pedicels slender   8. *C. angustifolium.*
   Leaves crowded. Racemes on long peduncles, hoary or
     silky-pubescent; pedicels short . . . .   9. *C. reticulatum.*
   Shrub, with slender, climbing, terete branches, Leaves nar-
     rowed at the base. Style nearly straight . . . . . .   7. *C. diversifolium.*
   Stems rigid, stout, suberect, very angular. Leaves few, oblong-
     linear, very thick. Style incurved . . . . 10. *C. trigonum.*
   Stems slender, ascending, angular-striate. Flowers small.
     Leaves obovate or cuneate. Keel acuminate . . . . . 11. *C. humile.*
     Leaves lanceolate, acute. Keel shortly acuminate . . . . 9. *C. reticulatum.*
     Leaves linear or oblong. Keel obtuse. . . . . 12. *C. parviflorum.*
 Keel erect, acutely acuminate. Leaves narrow-linear, with much-
   revolute margins (*Orthotropis*).
   Leaves obtuse. Racemes terminal, dense, and spike-like . . . 13. *C. cytisoides.*
   Leaves pungent-pointed. Racemes or clusters short, axillary . . 14. *C. Henchmanni.*
 Keel erect, obtuse or with a short recurved point. Stem erect.
   Leaves small, linear, with revolute margins . . . . 15. *C. ericifolium.*

(*Oxylobium carinatum* and *O. spathulatum* have nearly the habit of *Chorizema*, but the petals are nearly equal in length.)

1. **C. Dicksonii,** *Grah. in Maund. Botanist, t.* 106. An erect shrub of 1 to 3 ft., with numerous branches, pubescent when young. Leaves rather crowded, oblong-lanceolate or almost linear, tapering to a pungent point, often under ½ in. and rarely ¾ in. long, thick, rigid, glabrous and often shining, with a prominent midrib and transverse veins. Flowers red, rather large, in loose terminal racemes. Pedicels rather short. Calyx silky-villous, 4 lines long or rather more, the lobes all acuminate, the 2 upper ones broader, falcate, and united above the middle. Standard nearly twice as long as the calyx; wings scarcely exceeding the calyx, obliquely obovate; keel shorter, very much curved and rostrate. Ovary stipitate, with 8 to 10 ovules; style inflexed, with a very oblique stigma, fringed at the base on the upper side. Pod acuminate, 4 to 5 lines long. Seeds not seen.—Paxt. Mag. viii. 173, with a fig.; *C. costatum,* Meissn. in Pl. Preiss. i. 33.

W. Australia. Swan River, *Drummond, 1st Coll., and n.* 183; Darling Range, *Preiss, n.* 1036, *also n.* 1039 *and* 1040; Hampden, *W. Clarke.*

2. **C. nervosum,** *T. Moore, in Gard. Comp.* 1852, *with a fig. copied into Lemaire, Jard. Fleur. t.* 383. An erect rigid shrub with pubescent branches. Leaves broadly orbicular-cordate, with a pungent point, very much undulate, but not toothed, about ½ in. long, and often broader than long, very coriaceous, coarsely reticulate, usually glabrous. Racemes loose, few-flowered. Pedicels often as long as the calyx. Calyx 2 to 2½ lines long, nearly glabrous, the lobes rather acute, the 2 upper ones united to the middle. Standard about 3 lines long, but much broader; wings nearly as long; keel much shorter, broad with a short obtuse point. Ovary sessile, with 12 or more ovules; style glabrous, with a distinct capitate stigma. Pod about ½ in. long, resembling that of *C. ilicifolium.* Seeds shining.—*C. parvifolium,* Turcz. in Bull. Mosc. 1853, i. 253.

W. Australia, *Drummond, 5th Coll. n.* 25; gravelly plains of the range from E. to W. Moun Barren, *Maxwell.*

3. **C. varium,** *Benth. in Bot. Reg.* 1839, *t.* 49. An erect shrub of several ft., with pubescent branches. Leaves cordate-ovate, more or less prickly-toothed and undulate, 1 to 2 in. long, glabrous or slightly pubescent above and coarsely reticulate, pubescent or tomentose underneath. Racemes usually numerous, pubescent, the flowers not so distant as in *C. cordatum.* Pedicels short. Calyx about 3 lines long, like that of *C. cordatum,* but pubescent. Petals of *C. ilicifolium.* Ovary shortly stipitate, with numerous ovules. Pod stipitate, 6 to 8 lines long, often obtuse, Seeds smooth and shining.—Meissn. in Pl. Preiss. i. 32; Paxt. Mag. vi. 175, with a fig.

W. Australia. Swan River, *Drummond, 1st Coll. and n.* 184, *Preiss. n.* 1046, *Oldfield.*

4. **C. cordatum,** *Lindl. Bot. Reg.* 1838, *t.* 10. A glabrous shrub, with slender weak branches, very nearly allied to *C. ilicifolium,* but much larger, attaining several feet. Leaves cordate-ovate or ovate-lanceolate, 1 to 2 in. long, bordered by small prickly teeth or lobes, which are neither so deep nor so much undulate as in *C. ilicifolium,* and often very small. Flowers more numerous and larger than in that species, but otherwise similar. Pod also larger. Ovules 20 to 30.—Maund. Botanist, t. 89; Meissn. in Pl. Preiss. i. 32; Paxt. Mag. v. 97, with a fig.; *C. flavum,* Henfr. in Gard. Mag. i. 73, with a fig.; *C. superbum,* Lemaire, Illustr. Hortic. t. 29.

W. Australia. Swan River, *Drummond, 1st Coll.,* also *n.* 185, and 2*nd Coll. n.* 91, *Preiss. n.* 1042.

5. **C. ilicifolium,** *Labill. Voy.* i. 405, *t.* 21, *and Pl. Nov. Holl.* ii. 120. A small weak shrub, flowering often the first year so as to appear annual, with slender branches, sometimes erect and rigid, more frequently diffuse or almost filiform, glabrous or sprinkled with a few hairs when young. Leaves ovate to lanceolate, ¾ to 1 in. long, undulate and bordered with prickly teeth or lobes, often cordate at the base, glabrous, coriaceous and coarsely reticulate. Flowers orange-red, few and distant in axillary or terminal loose racemes. Calyx varying from under 2 to about 3 lines long, the lobes all acute, the 2 upper ones falcate and united to the middle. Standard broadly reniform, twice as long as the calyx; wings shorter than the standard; keel shorter than the calyx. Ovary nearly sessile, with 20 to 30 closely packed ovules; style short. Pod oblong, ½ in. long or shorter.—Bonpl. Jard. Malm. t. 35; DC. Prod. ii. 102; *Pultenæa nana,* Andr. Bot. Rep. t. 434; *Chorizema nanum,* Sims, Bot. Mag. t. 1032; DC. Prod. ii. 102; *C. triangulare,* Lindl. Bot. Reg. t. 1513; Meissn. in Pl. Preiss. i. 32, and ii. 208; Paxt. Mag. xiii. 73, with a fig.

W. Australia. King George's Sound and neighbourhood, *Labillardière, R. Brown, Preiss. n.* 1041, and others, and thence to the Great Bight, *Maxwell;* Swan River, *Drummond;* Flinders Bay, *Collie;* Blackwood river, *Oldfield.*

6. **C. rhombeum,** *R. Br. in Ait. Hort. Kew. ed.* 2, iii. 9. An undershrub (or herb?), with several ascending, simple or slightly branched stems, from ½ to 1½ ft. long, more or less angular or compressed, or at length terete and often pubescent. Lower leaves obovate or rhomboidal, passing into ovate or ovate-lanceolate, and under 1 in. long; upper ones often lanceolate and longer, flat or the margins slightly recurved, veined, glabrous or sprinkled

with a few hairs underneath. Flowers few and distant, on long terminal peduncles, forming loose racemes. Calyx 4 lines long or rather more, usually pubescent with appressed hairs, the upper lobes rather broader, united above the middle. Standard rather more than half as long again as the calyx; keel about the length of the calyx, erect and obtuse, but narrowed towards the top. Ovary shortly stipitate, with 20 or more ovules; style short, incurved, with a very oblique stigma. Pod more or less compressed, above ½ in. long, acuminate.—DC. Prod. ii. 103 ; *C. ovatum,* Lindl. Bot. Reg. t. 1528 ; Paxt. Mag. iv. 153, with a fig.; Reichb. Icon. Exot. t. 219 (an unusual garden form); Meissn. in Pl. Preiss. i. 32.

W. Australia. King George's Sound, *R. Brown,* and others, *Drummond, 2nd Coll. n.* 125 ; Cape Naturaliste, Gordon river, etc., *Oldfield ;* Mount Manypeak river, *Maxwell.*

7. **C. diversifolium,** *A. DC. Pl. Rar. Jard. Gen. 7e Not.* (1836), 44, *t.* 8. A tall shrub, with weak, slender, often climbing branches, slightly pubescent or at length glabrous. Leaves from ovate to narrow-lanceolate, 1 to 2 in. long or rather more, obtuse acute or mucronate, narrowed at the base, flat and not so rigid as in *C. rhombeum,* glabrous or slightly pubescent underneath. Flowers often numerous, in loose racemes. Calyx 4 lines long, glabrous or nearly so, the lobes acute, the 2 upper ones much falcate and united above the middle. Standard half as long again as the calyx; keel very obtuse, not exceeding the calyx. Ovary nearly sessile, with above 30 ovules ; style slightly curved, with a capitate slightly oblique stigma. Pod much flattened, about ¾ in. long, acuminate, transversely veined. Seeds numerous. —*C. spectabile,* Lindl. Bot. Reg. 1841, t. 45 ; Meissn. in Pl. Preiss. ii. 209 ; Bot. Mag. t. 3903 ; *C. rhombeum,* Lodd. Bot. Cab. t. 1619 (from the figure), not of R. Br.

W. Australia. Flinders Bay, *Collie ;* Swan River, *Drummond, Coll.* 2, *n.* 126 ; Cape Naturaliste, *Oldfield.*

8. **C. angustifolium,** *Benth. in Hueg. Enum.* 28, *and in Ann. Wien. Mus.* ii. 71. A low slender shrub or undershrub, with ascending branches of from ½ to 1½ ft., slightly pubescent when young. Lower leaves often ovate or lanceolate, acute, and almost pungent, like those of *C. rhombeum,* but more rigid, and sometimes bordered with a few small prickly teeth; upper leaves, and sometimes all linear or linear-lanceolate, 1 to 2 in. long, mucronate, the margins recurved, glabrous and reticulate above, usually silky-pubescent underneath. Flowers usually smaller than in *C. rhombeum,* in slender racemes. Calyx 2½ to nearly 3 lines long, glabrous or minutely pubescent, the lobes acute, the upper ones broader, but straight and separated below the middle. Standard nearly twice as long as the calyx; keel scarcely as long as the calyx, obtuse or with a short erect point. Ovary shortly stipitate, with 15 to 20 ovules ; style much incurved, with a small terminal stigma. Pod ½ to nearly ¾ in. long, slightly compressed.—Meissn. in Pl. Preiss. i. 33 ; *Dillwynia glycinifolia,* Sm. in Trans. Linn. Soc. ix. 264 ; DC. Prod. ii. 109 ; Bot. Reg. t. 1514 ; *Chorizema capillipes,* Turcz. in Bull. Mosc. 1853, i. 255 ; *C. denticulatum,* Turcz. l. c. 253 (specimens with nearly all the leaves broad and rigid).

W. Australia. King George's Sound, *Menzies, Baxter,* and others; *Drummond, 5th Coll. n.* 25 and 26 ; Stirling Terrace, *Preiss. n.* 1047, 1127, *Maxwell ;* Mount Clarence, *Oldfield ;* Bremer Bay and Cape Riche, *Maxwell.*

9. **C. reticulatum,** *Meissn. in Pl. Preiss.* i. 34. Stock woody, with erect or ascending slightly branched stems, of ½ to 1 ft., slightly silky-pubescent. Leaves rather crowded, lanceolate, acute, ½ to 1 in. long, rigid, reticulate, keeled underneath, the margin flat or slightly recurved. Flowers several, rather small, in terminal erect racemes, often 6 in. long, including the long peduncle; pedicels shorter than the calyx. Calyx about 2½ lines long, silky-pubescent; lobes acute or acuminate, nearly as long as the tube, the 2 upper ones broader and more united. Standard about twice as long as the calyx; wings nearly as long; keel scarcely exceeding the calyx, flattened towards the end with a short obtuse point. Ovary very shortly stipitate, with about 20 ovules; stigma capitate. Pod somewhat compressed, acuminate, 4 to 5 lines long.

W. Australia, *Drummond;* dense bushy places at the foot of Mount Wuljenup, *Preiss. n.* 1045 ; Vasse river, *Oldfield ;* Mount Manypeak river and Cape Riche, *Maxwell.*

The Vasse river specimens have the leaves rather more oblong or almost linear, with the margins slightly recurved, but I can see no difference between them and the southern ones. The short pedicels and long pedunculate racemes distinguish the species from all forms of *C. angustifolium,* independently of the foliage.

10. **C. trigonum,** *Turcz. in Bull. Mosc.* 1853, i. 254. A very rigid undershrub or shrub, the stems erect or ascending, not much branched, 1½ to above 2 ft. high, glabrous, very angular, the upper flowering branches often leafless. Leaves few, distant, erect, linear-oblong, 1½ to 2 in. long, with a short recurved point, narrowed at the base, thickly coriaceous, slightly folded lengthwise, glabrous, coarsely reticulate. Flowers on short pedicels, in terminal or lateral racemes. Calyx slightly pubescent, about 3 lines long, the lobes rather broad, the 2 upper ones more falcate and united above the middle. Standard twice as long as the calyx ; keel somewhat exceeding the calyx, shortly and obtusely acuminate. Ovary nearly sessile, with above 20 ovules; style short, incurved, with a capitate terminal stigma. Pod turgid, about ½ in. long.

W. Australia, *Drummond, 5th Coll. n.* 22 ; Phillips river and towards the Great Bight, *Maxwell.*

11. **C. humile,** *Turcz. in Bull. Mosc.* 1853, i. 254. A small diffuse branching shrub, more or less silky-pubescent, with erect or ascending angular branches, of ½ to 1 ft. Leaves from obovate or cuneate, and 2 to 4 lines long, to cuneate-oblong, and above ½ in., obtuse truncate or emarginate, with a small recurved point, pubescent on both sides. Flowers small, on short pedicels in terminal racemes. Calyx 2 to 2½ lines long, the lobes rather shorter than the tube, the 2 upper ones more obtuse, united above the middle. Standard nearly twice as long as the calyx ; keel slightly exceeding the calyx, flattened towards the end and shortly acuminate. Ovary shortly stipitate, with 15 to 20 or more ovules; styles slightly curved, with a small capitate stigma. Pod turgid, acuminate, 4 to 5 lines long.—*Oxylobium genistoides,* Meissn. in Bot. Zeit. 1855, 12.

W. Australia, *Drummond, 4th Coll. n.* 36, and between Moore and Murchison rivers, *Drummond, 6th Coll. n.* 9.

12. **C. parviflorum,** *Benth. in Ann. Wien. Mus.* ii. 71. An undershrub, with a thick rhizome, and numerous ascending, rather slender, angular,

striate stems, of about 1 ft., glabrous or slightly pubescent. Leaves not numerous, linear, and above 1 in. long, or shorter and oblong, obtuse or mucronate, the margins recurved. Flowers small, on very short pedicels, in rather long, slender, terminal racemes. Calyx 1½ lines long, slightly pubescent, the lobes shorter than the tube, the 2 upper ones broad, truncate and united nearly to the top. Standard very broad, twice as long as the calyx ; wings shorter ; keel much shorter, broad and obtuse. Ovary very shortly stipitate, with about 10 ovules ; style slightly incurved, with a capitate stigma. Pod oblique, often broader than long, very turgid, 3 to 4 lines diameter.—*C. Pulteneæ*, F. Muell. Fragm. iv. 19, but not the synonyms adduced.

Queensland. Keppel Bay, *R. Brown ;* E. coast, *A. Cunningham ;* Wide Bay, *Bidwill, Leichhardt ;* Stradbrooke Island, *Fraser ;* S. tributaries of Burnett river and Brisbane river, *F. Mueller.*

N. S. Wales. Port Jackson, on the Paramatta road, *R. Brown, Woolls ;* Hunters' river, *American Exploring Expedition ;* Hastings river, *Beckler ;* foot of Wacamurrum, *Leichhardt.*

13. **C. cytisoides,** *Turcz. in Bull. Mosc.* 1853, i. 256. Our specimens, with the root attached, all under 1 ft. high, nearly simple, slightly silky-pubescent towards the top. Leaves scattered, linear, obtuse, with a small point, ½ to ¾ in. long, the margins much revolute, glabrous above, silky pubescent underneath. Racemes terminal, oblong, dense and spike-like, 1 to 1½ in. long. Pedicels very short. Bracts and bracteoles larger and more persistent than in the other species, lanceolate, acuminate, silky-hairy. Calyx about 4 lines long, silky-hairy, deeply divided into lanceolate acuminate lobes, the 2 upper ones united to the middle. Standard about 6 lines long, including the claw, not emarginate, the sides reflexed ; wings rather shorter ; keel terminating in an erect recurved point, nearly as long as the wings. Ovary shortly stipitate, with about 15 ovules ; style short, curved, with a capitate stigma. Pod not seen.

W. Australia, *Drummond, 5th Coll. n.* 77 ; from King George's Sound to Bremer Inlet and Cape Riche, *Maxwell.*

14. **C. Henchmanni,** *R. Br. in Bot. Reg. t.* 986. An erect shrub or undershrub, 1 to 2 or 3 ft. high, with virgate branches, pubescent when young. Leaves linear, pungent, ½ to ¾ in. long, with smaller ones usually clustered in the axils, the margins revolute, glabrous above, the under side often pubescent, but usually concealed. Flowers red, in short racemes or clusters in the upper axils, often forming long terminal leafy raceme-like panicles. Pedicels 1 to 3 lines long. Calyx hirsute or silky-villous, about 3 lines long, the lobes narrow, longer than the tube, the 2 upper ones broader and united to the middle. Standard twice as long as the calyx ; wings shorter ; keel tapering into an erect point, nearly as long as the wings. Ovary shortly stipitate, with about 12 to 15 ovules ; style short, hooked, with a capitate stigma. Pod very turgid, rather obtuse, 4 to 5 lines long.—Bot. Mag. t. 3607 ; Meissn. in Pl. Preiss. i. 34, ii. 209 ; Lodd. Bot. Cab. t. 1233 ; Paxt. Mag. ii. 171, with a fig. ; *Podolobium (?) aciculare,* DC. Prod. ii. 103 ; *Chorizema Baueri,* Meissn. in Pl. Preiss. i. 34, ii. 209, not of Benth. ; *C. rhynchotropis,* Meissn. l. c. ii. 209 ; *Orthotropis pungens,* Benth. in Lindl. Swan Riv. App. 16.

W. Australia. King George's Sound, *Baxter*, Swan River, *Drummond*, 1*st Coll.* *and 2nd Coll. n.* 92, 93, 94; Harvey river and Mount Barker, *Oldfield;* S. coast, from King George's Sound to Cape Riche, *Maxwell.*

15. **C. ericifolium,** *Meissn. in Pl. Preiss.* ii. 209. Erect and much branched, 1 to 1½ ft. high, the branches rather slender, but rigid, striate, glabrous, often irregularly verticillate, the smaller ones sometimes spinescent. Leaves few and small, alternate or irregularly opposite or verticillate, linear, with revolute margins, 2 to 3 lines long. Flowers small, rather numerous, often verticillate, in terminal loose racemes, of 1 to 2 in. Pedicels very short. Calyx silky-pubescent, from 2 to nearly 3 lines long, the lobes shorter than the tube, the 2 upper ones broad, obtuse, and united nearly to the top. Standard scarcely twice as long as the calyx; wings shorter; keel much shorter, obtuse or with a very short recurved point. Ovary nearly sessile, with about 10 ovules; style incurved or hooked, with a capitate stigma. Pod ovoid or almost globular, about 2½ lines long, with a short flat point. Seeds usually few, not strophiolate.—*Dichosema racemosum*, Meissn. in Pl. Preiss. i. 78.

W. Australia, *Drummond*, 2*nd Coll. n.* 96; Murchison river and Champion Bay, *Oldfield;* near Albany, *Preiss. n.* 861. This species differs in habit from all others, approaching that of *Mirbelia daviesioides*, but there is no trace of any intrusion of the endocarp into the cavity of the pod. Some specimens from Bowes river, *Oldfield*, are remarkable for the size of the flowers, with the calyx-lobes narrower and more acute.

5. MIRBELIA, Sm.

(Dichosema, *Benth.*, Oxycladium, *F. Muell.*)

Calyx-lobes nearly of equal length, the 2 upper ones often broader and united higher up. Petals clawed; standard orbicular or reniform, emarginate or entire, longer than the lower petals; wings oblong; keel broader than the wings, and shorter or rarely of the same length. Stamens free. Ovary sessile or stipitate with 2 or several ovules; style usually short, incurved, with a terminal capitate stigma. Pod ovoid or oblong, turgid, divided longitudinally into 2 cells, by a false dissepiment projecting into the cavity from the lower suture and overlapped by or connate with the projecting placentas. Seeds without any strophiole.—Shrubs with the habit nearly of *Oxylobium* or of *Chorizema.* Leaves opposite verticillate or alternate, simple entire or prickly-toothed. Stipules small, setaceous or none. Flowers yellow, purple-red or blue, solitary or clustered in the axils of the leaves, or in axillary or terminal racemes. Bracts and bracteoles small or none. Ovary glabrous or villous. Endocarp of the pod separating from the epicarp in some species, but not in all.

The genus is limited to Australia. It is very nearly allied to *Oxylobium*, *Gastrolobium*, and *Chorizema*, differing chiefly in the remarkable 2-celled pod, the false dissepiment being already more or less apparent in the ovary at the time of flowering.

Plant not thorny. Leaves obtuse or pungent.
　Leaves dilated at the end into 3 to 7 prickly lobes or teeth . . . 1. *M. dilatata.*
　Leaves ovate, ovate-lanceolate or broadly oblong entire.
　　Keel much shorter than the wings.
　　　Leaves about 1 to 1½ in. long, quite glabrous. Flowers racemose. Ovary stipitate . . . . 2. *M. racemosa.*
　　　Leaves silky underneath. Flowers axillary or in terminal clusters. Ovary sessile . . . . 3. *M. grandiflora.*

Keel about as long as the wings.　Leaves usually ½ in. long or
　　less.　Calyx deeply lobed.
　　Leaves ovate-lanceolate, pointed . . . . 4. *M. subcordata.*
　　Leaves ovate, very obtuse, ¼ to ½ in. long . . . . 5. *M. ovata.*
　　Leaves ovate, very obtuse, 1 to 2 lines long . . . . . . 6. *M. oxylobioides.*
Leaves linear-oblong or narrow-linear, the margins recurved or re-
　　volute, entire.
　Ovules 2.
　　Leaves oblong, linear with recurved margins, mostly verticillate 7. *M. reticulata.*
　　Leaves narrow-linear, with revolute margins, all alternate . . 8. *M. aotoides.*
　Ovules 8 to 12.　Leaves narrow-linear, with revolute margins.
　　Leaves pungent.　Calyx-teeth very short, the upper lip broad
　　　and truncate . . . . 9. *M. pungens.*
　　Leaves obtuse.　Calyx-lobes acute, as long as the tube.
　　　Ovules about 12.　Pod broadly ovoid . . . . 10. *M. speciosa.*
　　　Ovules about 8.　Pod oblong . . . . 11. *M. floribunda.*
Plant thorny.
　　Thorns lateral, subulate, rather longer than the small leaves.
　　Leaves linear with revolute margins.　Ovary sessile . . . . 12. *M. spinosa.*
　　Leaves flat or folded lengthwise.　Ovary stipitate.
　　　Leaves 1 to 1½ lines long.　Ovary glabrous with 6 to 8 ovules 14. *M. microphylla.*
　　　Leaves 2 to 3 lines long.　Ovary pubescent with 12 to 15 ovules 13. *M. multicaulis.*
　　Branches leafless, spinescent at the end.
　　　Ovules above 12.　Pod oblong, deeply furrowed . . . . . . 15. *M. daviesioides.*
　　　Ovules 2 ?　Pod ovoid or almost globular, with prominent sutures 16. *M. oxyclados.*

1. **M. dilatata,** *R. Br. in Ait Hort. Kew. ed.* 2, iii. 21. An erect bushy
shrub, with angular pubescent branches, or quite glabrous. Leaves numerous,
scattered or irregularly verticillate, cuneate, much dilated and undulate at the
end with 3, 5 or 7 pungent-pointed lobes or teeth, ¾ to 1½ in. long, much
narrowed below the middle and sometimes petiolate, rigidly coriaceous,
strongly reticulate. Flowers of a bluish-purple, shortly pedicellate, solitary in the axils or forming terminal leafy racemes. Calyx slightly pubescent, 2 to 2½ lines long, the teeth or lobes short acute and nearly equal in
length, the 2 upper ones rather broader and more or less united. Standard
twice as long as the calyx ; wings nearly as long as the standard ; keel rather
shorter, obtuse. Ovary stipitate, villous in the originally described form,
with 6 to 8 ovules. Pod oblong, about ½ in. long.—DC. Prod. ii. 115 ;
Lindl. Bot. Reg. t. 1041 ; Lodd. Bot. Cab. t. 1367.

W. Australia. King George's Sound, *R. Brown, Baxter,* and others.
Var. *Meissneri.* Leaves shorter and more lobed than in the original form. Ovary gla-
brous.—*M. Meissneri,* Hook. Bot. Mag. t. 4419 ; *M. dilatata,* Meissn. in Pl. Preiss. i. 76.
Swan River, *Drummond,* 1st *Coll. Preiss, n.* 1049. Stirling range, *Maxwell;* Vasse river,
Oldfield. In some specimens the foliage of one form passes into that of the other, but I
have only observed the villous ovary in the King George's Sound form.

2. **M. racemosa,** *Turcz. in Bull. Mosc.* 1853, i. 282. A shrub apparently
loosely branched, perfectly glabrous. Leaves mostly opposite, petiolate, oval-
oblong, very obtuse, 1 to 1½ in. long, flat, coriaceous and shining, strongly
reticulate. Stipules persistent. Racemes loose, terminal, several-flowered.
Pedicels 2 to 3 lines long. Calyx quite glabrous, 2½ to nearly 3 lines long,
the lobes short and broad, the 2 upper ones very obtuse or truncate, united
nearly to the top. Standard ½ in. long, not emarginate, on a slender claw ;
wings nearly as long, keel short, shortly rostrate. Ovary stipitate, pubescent,

with 8 to 10 ovules. Pod about ½ in. long, oblong-elliptical, glabrous, coriaceous, the longitudinal dissepiment splitting as in other species but more pithy, the endocarp not separating.— *Chorizema magnifolium*, F. Muell. Fragm. iv. 18.

W. Australia, *Drummond, 5th Coll. n.* 59 ; Bremer Bay and Middle Mount Barren, *Maxwell.*

3. **M. grandiflora,** *Ait. in Hook. Bot. Mag. t.* 2771. A shrub or undershrub with diffuse or ascending branches of 1 to 2 ft., more or less silky-pubescent. Leaves alternate or opposite, ovate or ovate-lanceolate, acute or almost obtuse, ¾ to 1¼ in. long, coriaceous, glabrous shining and strongly reticulate above, silky-pubescent or villous underneath, the margins recurved. Flowers bright yellow and red, nearly sessile in axillary or terminal clusters, or rarely solitary. Calyx silky-villous, about 4 lines long, the lobes acute, as long as the tube, the 2 upper ones united to the middle. Standard large, deeply emarginate ; wings nearly as long ; keel much shorter, obtuse. Ovary sessile, very villous, with 10 to 15 ovules. Pod thickly oblong, hoary with appressed hairs, about ¾ in. long, very obtuse.—*Platylobium reticulatum*, Sieb. Pl. Exs. ; *Chorizema (?) platylobioides*, DC. Prod. ii. 103.

N. S. Wales. Port Jackson, *Sieber, n.* 371, 373, and *Fl. Mixt. n.* 607 ; Blue Mountains, *A. Cunningham ;* Illawarra, *Backhouse, M'Arthur.*

4. **M. subcordata,** *Turcz. in Bull. Mosc.* 1859, i. 282. A much-branched apparently small shrub, the branches terete, pubescent with very short spreading or reflexed hairs. Leaves mostly verticillate in threes, ovate-lanceolate with a small almost pungent point, rarely above ½ in. long, glabrous shining and reticulate above, pubescent underneath. Flowers very shortly pedicellate in the upper axils or in terminal clusters or short racemes. Calyx pubescent, 2½ to nearly 3 lines long, deeply divided into lanceolate acuminate lobes, the upper ones scarcely more united than the others. Standard rather longer than the calyx, emarginate ; wings and keel nearly as long, the latter somewhat curved, obtuse. Ovary nearly sessile, very villous, with 4 to 6 ovules. Pod oblong, villous, not seen full grown.

W. Australia, *Drummond, 5th Coll. n.* 60.

5. **M. ovata,** *Meissn. in Pl. Preiss.* i. 77. A diffuse or divaricate much-branched shrub, the young branches terete and tomentose or shortly villous. Leaves opposite or in threes, ovate, obtuse, under ½ in. long, with recurved or revolute margins, coriaceous, reticulate, glabrous above or scabrous-pubescent when young, more or less pubescent underneath. Flowers nearly sessile, 2 or 3 together at the ends of the branches, or in 2 or 3 distant pairs or threes in a terminal raceme. Calyx silky-pubescent, about 2½ lines long, deeply divided into lanceolate acuminate lobes as in *M. subcordata*, and petals nearly equal as in that species. Ovary nearly sessile, villous, with 6 to 8 ovules. Pod villous, thickly oblong, very obtuse, about 5 lines long.—*M. aspera*, Turcz. in Bull. Mosc. 1853, i. 281.

W. Australia, *Drummond, 5th Coll. n.* 28 ; Gordon river and Princess Royal Harbour, *Preiss, n.* 803 *and* 1202.

6. **M. oxylobioides,** *F. Muell. Fragm.* ii. 154, *and* iv. 12. A rigid divaricate shrub of several ft., with rather slender branches, the short branchlets

terete, tomentose-pubescent.   Leaves opposite or in whorls of 3 or scattered, ovate, 1 to 1½ or rarely 2 lines long, obtuse or with a minute recurved point, the margins recurved, scabrous above, silky-pubescent underneath.   Flowers few together at the ends of the short branchlets, on pedicels of about 1 line. Bracteoles small, narrow, deciduous.   Calyx silky-pubescent, about 2½ lines long.   Standard very broad; keel broad, deeply coloured.   Ovary nearly sessile, with about 12 to 14 ovules.   Pod ovoid, turgid, shortly acute, about ½ in. long, deeply furrowed on the upper side, divided by a dissepiment proceeding from the lower suture, but the endocarp not separating from the epicarp.

N. S. Wales.   Bushy rocks on the boundary of the marshy plains, Argyle county, *A. Cunningham.*

Victoria.   Deep rocky valleys, Haidinger range, Dandenong ranges, Snowy river, Mount Ligar, and adjoining ranges, generally at an elevation of 3000 to 4000 ft., *F. Mueller.*

7. **M. reticulata,** *Sm. in Ann. Bot.* i. 511, *and in Trans. Linn. Soc.* ix. 265.   A low shrub with slender but rigid angular branches quite glabrous or very minutely pubescent.   Leaves mostly verticillate in threes, oblong-linear with short pungent points, ½ to ¾ in. long, the margins recurved, glabrous, shining above with raised transverse reticulations, occasionally bordered by small tooth-like glands.   Flowers small, bluish-purple, shortly pedicellate in axillary clusters or short terminal corymbose racemes.   Calyx minutely pubescent, about 1½ lines long, the teeth much shorter than the tube, the 2 upper ones broad truncate and united nearly to the top.   Standard about 4 lines long, slightly emarginate; wings nearly as long; keel very short, obtuse. Ovary shortly stipitate, glabrous, with 2 ovules.   Pod ovoid, acute rarely above 2 lines long.—Vent. Jard. Malm. t. 119 ; DC. Prod. ii. 114 ; Bot. Mag. t. 1211 ; Lodd. Bot. Cab. t. 1371 ; *Pultenæa rubiæfolia,* Andr. Bot. Rep. t. 351 ; *Mirbelia rubiæfolia,* G. Don. Gen. Syst. ii. 126 ; *M. angustifolia,* Grah. in Edinb. N. Phil. Journ. 1838 (from the descr.).

N. S. Wales.   Port Jackson, *R. Brown, Sieber, n.* 368, and others.

8. **M. aotoides,** *F. Muell. in Trans. Phil. Inst. Vict.* iii. 53, *and Fragm.* iv. 11.   A rigid shrub with divaricate terete pubescent branches.   Leaves mostly alternate, narrow-linear, obtuse or with a small recurved or straight almost pungent point, the margins closely revolute, rarely above ⅓ in. long, glabrous or pubescent when young, and often silky underneath.   Flowers (yellow ?) nearly sessile, axillary and almost solitary or in short terminal corymbs.   Calyx about 2 lines long, pubescent with minute appressed hairs, the teeth very short, the upper ones truncate and united.   Standard very broad, not twice as long as the calyx, the claw short ; wings nearly as long, narrow ; keel rather shorter, obtuse.   Ovary sessile, glabrous, with 2 ovules. Pod broadly ovoid, very obtuse, shortly exceeding the calyx, separating into 2 closed hemicarpels, the'endocarp remaining attached to the epicarp.   Seeds ovoid.

Queensland.   Burnitt ranges, *F. Mueller.*

N. S. Wales.   Mount Mitchell, Clarence river, *Beckler.*   In one flower I found the pistil monstrous, with 3 carpels more or less developed.

9. **M. pungens,** *A. Cunn. in G. Don, Gen. Syst.* ii. 126.   A small shrub or undershrub, with ascending stems, often under 1 ft. high, the branches rather

slender, pubescent. Leaves alternate, narrow-linear, with pungent points and closely revolute margins, under ½ in. long, glabrous. Flowers (bluish-purple?) clustered in the upper axils on very short pedicels. Calyx slightly pubescent, rarely exceeding 2 lines and usually shorter, the lobes shorter than the tube, the 2 upper ones united in a truncate or shortly 2-lobed upper lip. Standard twice as long as the calyx; keel shorter than the wings but exceeding the calyx, obtuse or with a short erect point. Ovary glabrous or slightly hirsute, with 6 to 10 ovules. Pod ovoid, about 3 lines long, but not seen in a perfect state.—*Chorizema Baueri*, Benth. in Ann. Wien. Mus. ii. 71.

N. S. Wales. Paramatta and Richmond, *R. Brown;* rocky hills, Cox's river, *A. Cunningham;* Illawarra, *Backhouse;* New England, *C. Stuart;* also in *Leichhardt's* collection.

10. **M. speciosa,** *Sieb. in DC. Prod.* ii. 115. An erect shrub of 2 to 3 ft., with numerous virgate angular branches, slightly hoary-pubescent or glabrous. Leaves scattered or verticillate in threes, narrow-linear, obtuse with a small straight sometimes almost pungent point, ½ to ¾ in. long, the margins closely revolute, glabrous, scarcely reticulate. Flowers bluish-purple, almost sessile in the upper axils, the upper ones forming a terminal interrupted spike leafy at the base. Calyx fully 3 lines long, hoary-pubescent, the lobes acute, as long as the tube, the 2 upper ones united to the middle. Standard twice as long as the calyx, emarginate; wings nearly as long; keel very short, obtuse. Ovary sessile, glabrous, with about 12 ovules. Pod thickly ovoid, about 4 or 5 lines long.—Bot. Reg. 1841, t. 58; Reichb. Icon. Exot. t. 191.

N. S. Wales. Port Jackson, *Sieber, n.* 367; and *Fl. Mixt. n.* 570; abundant in arid rocky tracts of the Blue Mountains, *A. Cunningham;* New England, *C. Stuart;* Illawarra, *Fraser, Backhouse, Shepherd.* The figure of *M. floribunda,* Paxt. Mag. viii. 103, gives much more the idea of this plant than of the true western *M. floribunda.*

11. **M. floribunda,** *Benth. in Lindl. Swan Riv. App.* 12. A low, much-branched, divaricate or diffuse shrub or undershrub, the young branches hoary-pubescent. Leaves scattered or irregularly opposite or in threes, narrow-linear with a recurved point and closely revolute margins, usually under ½ in. long, glabrous and not reticulate above, slightly pubescent underneath. Flowers bluish-purple, numerous although solitary in each axil, scattered along the branches and not forming a terminal raceme as in *M. speciosa*, which this species closely resembles. Calyx under 3 lines long, silky-pubescent, the lobes equal to the tube, the 2 upper ones united above the middle. Standard fully twice as long as the calyx; wings shorter; keel shorter than the wings, but not so short nor so broad as in *M. speciosa*. Ovary sessile, glabrous; ovules usually 8. Pod oblong, 4 to 5 lines long, much narrower than in *M. speciosa.*—*M. speciosa*, Sweet, Fl. Austral. t. 34, not of Sieb.; *M. pulchella*, Meissn. in Pl. Preiss. ii. 221.

W. Australia. Swan River, *Drummond, 1st Coll. and 2nd Coll. n.* 92, *Burges;* Mount Yulagan, *Oldfield.* The flowers usually dry blue, but in Burges's specimens they look yellowish.

12. **M. spinosa,** *Benth.* A shrub of 2 or 3 ft., with ascending or virgate branches, glabrous or pubescent when young. Leaves narrow-linear, obtuse with revolute margins, under ½ in. long, usually clustered round a slender but rigid spreading thorn (an abortive branch), as long as or rather

longer than themselves, but in young shoots the leaves are solitary and alternate, without thorns.   Flowers pink or purple (yellowish when dry), axillary, sessile.   Calyx pubescent with appressed hairs, about 3 lines long, the lobes narrow, acuminate, nearly as long as the tube, the 2 upper ones united at least to the middle.   Standard nearly twice as long as the calyx, emarginate; wings shorter; keel still shorter, obtuse.   Ovary sessile, glabrous, with 6 to 8 ovules.   Pod not seen.—*Dichosema spinosum*, Benth. in Hueg. Enum. 35 ; Meissn. in Pl.Preiss. i. 77 ; *D. subinerme*, Meissn. l. c. i. 78 ; Moore, in Gard. Comp. i. 129, with a fig. copied into Lemaire, Jard. Fleur. t. 350.

W. Australia.   Swan River, *Huegel, Drummond, 1st Coll. and n.* 191, *Preiss, n.* 862; northward to Murchison river, and southward to Kalgan river, *Oldfield*, towards the Great Bight, *Maxwell*.   The spines in this species appear to be abortive branches proceeding from the centre of the tufts of leaves.   In *M. microphylla* they are mostly supra-axillary. In *M. multicaulis* they are lateral or quite away from the nodes yet they have no appearance of being epidermal productions.

13. **M. multicaulis,** *Benth.*   Our specimens show a thick rootstock, with numerous, erect, simple or slightly-branched stems of ½ to 1 ft., glabrous or slightly hoary.   Leaves alternate, mostly solitary, ovate or oblong, obtuse, 2 to 3 lines long, the margins flat, coriaceous and reticulate.   Spines subulate, often recurved, longer than the leaves, not usually arising from the nodes. Flowers axillary or clustered at the base of the spines, very shortly pedicellate, apparently purple.   Calyx minutely pubescent, about 2 lines long, the lobes shorter than the tube, the 2 upper ones truncate and united.   Standard twice as long as the calyx ; wings nearly as long ; keel much shorter, although exceeding the calyx, broad, incurved, obtuse.   Ovary on a long stipes, pubescent or villous, with 12 to 15 ovules.   Pod not seen.—*Dichosema multicaule*, Turcz. in Bull. Mosc. 1853, i. 283.

W. Australia, *Drummond, 4th Coll. n.* 34.

14. **M. microphylla,** *Benth.*   A diffuse or divaricate shrub, with slender glabrous or slightly hoary branches.   Leaves usually verticillate and clustered round or a little below subulate thorns, oblong or linear-cuneate, obtuse, not 2 lines long, flat or folded lengthwise.   Flowers small (yellow and purple ?), axillary and very shortly pedicellate.   Calyx minutely pubescent, 1 to nearly 1½ lines long, the teeth much shorter than the tube, the 2 upper ones broad, truncate, and united.   Standard twice as long as the calyx ; wings nearly as long ; keel short, obtuse.   Ovary stipitate, glabrous, with 6 to 8 ovules.   Pod not seen.—*Dichosema microphyllum*, Turcz. in Bull. Mosc. 1853, i. 283.

W. Australia, *Drummond, 5th Coll. n.* 85.

15. **M. daviesioides,** *Benth.*   An erect rigid leafless shrub, glabrous or the young shoots slightly pubescent, the branches numerous, terete, striate or sulcate, the smaller ones divaricate and almost all spinescent, sometimes small, fine, and phyllodineous.   Leaves replaced by minute scales.   Flowers small, shortly pedicellate, in racemes either terminal or close under the smaller phyllodineous branchlets.   Calyx about 2 lines long, the lobes much shorter than the tube, the 2 upper ones truncate and united.   Standard twice as long as the calyx ; wings rather shorter ; keel much shorter, broad, obtuse. Ovary stipitate, villous, with about 15 ovules (in Drummond's specimens).

Pod ovate, acuminate, about 4 lines long, the ventral or axillary face much flattened, the back more or less furrowed, divided longitudinally as in other *Mirbelias*, but the endocarp scarcely separating.—*Daviesia ramulosa*, Benth. in Lindl. Swan Riv. App. 14; *Chorizema daviesioides*, Meissn. in Pl. Preiss. i. 34; *Mirbelia aphylla*, F. Muell. Fragm. iv. 11.

W. Australia. Swan River, *Drummond, 1st Coll. and n.* 190; Gardner's River, *Maxwell;* Murchison river, *Oldfield.*

Var. (?) *rigida.* Specimens in fruit, with few stout but frequently spinescent branches. Pod oblong, scarcely furrowed on the back.—Dirk Hartog's Island, *Milne.*

16. **M. (?) oxyclada,** *F. Muell. Fragm.* iv. 12. A rigid leafless shrub, resembling *M. daviesioides*, but more slender, the branchlets usually spinescent. Leaves replaced by minute scales. Flowers unknown. Fruiting pedicels short, solitary or in pairs along the branches. Pod shortly stipitate, ovoid, turgid, but with prominent sutures and acuminate with the rigid persistent base of the style, 2 to 3 lines long, imperfectly divided into 2 cells by an incomplete dissepiment, the sutures persisting as a replum after the valves have fallen. Seed 1 in each cell, not strophiolate.—*Jacksonia viminalis*, A. Cunn.; Benth. in Ann. Wien. Mus. ii. 75; *Oxycladium semiseptatum*, F. Muell. in Hook. Kew Journ. ix. 20, and Fragm. i. 168.

N. Australia, *A. Cunningham;* sandstone table-land at the head of Victoria river, and in Arnhem's Land, *F. Mueller.*

6. ISOTROPIS, Benth.

Calyx deeply lobed, the 2 upper lobes united nearly to the top. Petals clawed. Standard orbicular, emarginate, longer than the wings; wings obovate, somewhat falcate; keel incurved, nearly as long as the wings. Stamens free. Ovary sessile, with numerous ovules; style incurved, filiform, with a minute terminal stigma. Pod oblong linear or lanceolate, acute, more or less turgid. Seeds not strophiolate.—Herbs or undershrubs, with diffuse or ascending stems. Leaves alternate, simple or unifoliolate, herbaceous. Stipules linear-falcate or minute. Flowers solitary, on axillary peduncles, or forming a loose terminal raceme. Ovary villous.

The genus is exclusively Australian. It is closely allied to *Oxylobium* and *Chorizema*, differing chiefly in habit, in the deeper-cleft calyx, and in the longer pod.

Leaves simple, sessile, continuous with the petiole.

Leaves obtuse, truncate or dilated and 2-lobed. Pedicels much longer than the calyx	1. *I. striata.*
Leaves few, cuneate acute or none. Flowers loosely racemose, the pedicels shorter or scarcely longer than the calyx	2. *I. Drummondii.*

Leaves of 1 leaflet, articulate on the petiole.

Calyx glabrous, 4 to 5 lines long. Racemes terminal, leafless. Stem-leaves cuneate-oblong or linear-obtuse	3. *I. juncea.*
Calyx tomentose, 4 lines long. Racemes terminal, leafless. Leaves ovate or oblong, very obtuse	4. *I. atropurpurea.*
Calyx pubescent, 3 lines long. Racemes mostly leaf-opposed, leafless. Stem-leaves lanceolate, acute	5. *I. filicaulis.*
Calyx pubescent, 2 lines long. Pedicels short, axillary or leaf-opposed. Stem-leaves linear, flat	6. *I. parviflora.*
Calyx tomentose, 3 lines long. Pedicels short, in a short terminal raceme or solitary. Stem-leaves terete, channelled above . .	7. *I. Wheeleri.*

1. **I. striata,** *Benth. in Hueg. Enum.* 28, *and in Ann. Wien. Mus.* ii. 71. A perennial or undershrub, more or less clothed with long hairs, silky and appressed on the upper branches, under side of the leaves, and calyxes, spreading on the lower part of the plant. Stems diffuse or ascending, ½ to 1½ ft. high. Leaves not numerous, the lower ones obovate or cuneate, very obtuse, truncate or broadly 2-lobed, sometimes exceeding 1 in., narrowed below the middle, either nearly sessile or tapering into a rather long petiole, but not articulate upon it, green on both sides, the upper ones narrower and sometimes acute, and some branches quite leafless and dichotomous. Stipules mostly falcate. Pedicels solitary in the upper axils, often 2 in. long or more, bearing a single large flower with a pair of linear bracteoles close under it. Calyx 4 to 5 lines long. Standard large, yellow with purple streaks; wings and keel purple. Ovary nearly sessile, with above 30 ovules. Pod ¾ to 1 in. long, much contracted towards the base, pubescent.—Meissn. in Pl. Preiss. i. 31; F. Muell. Fragm. iii. 16; *Callistachys cuneifolia*, Sm. in Trans. Linn. Soc. ix. 267; DC. Prod. ii. 104; *Chorizema spartioides*, Lodd. Bot. Cab. t. 1953; Paxt. Mag. x. 127, with a fig.; *Isotropis biloba*, Benth. in Hueg. Enum. 29, and in Ann. Wien. Mus. ii. 71.

W. Australia. King George's Sound, *R. Brown* and others, and thence to Swan River, *Huegel, Drummond,* and others; Vasse river, *Mrs. Molloy.* The leaves vary much in shape, from very much dilated and 2-lobed to narrow and scarcely obtuse, but we now find that all the different forms occur sometimes on the same specimen.

Var. (?) *parviflora.* Leaves narrow and flowers small, but the specimens insufficient for accurate determination.—Murchison river, *Oldfield.*

2. **I. Drummondii,** *Meissn. in Pl. Preiss.* i. 31. Stems several, from a perennial rootstock, ascending or erect, ½ to 1 ft. high, glabrous or hirsute towards the base and silky-hairy upwards, sometimes entirely leafless, sometimes with a few oblong-cuneate sessile leaves towards the base, rarely above ½ in. long. Stipules setaceous. Flowers like those of *I. striata,* but arranged in a loose terminal raceme, the pedicels rarely much longer than the calyx and usually shorter, each in the axil of a minute bract. Calyx 4 to 5 lines long. Standard veined, as in *I. striata.* Ovary stipitate, with 30 to 40 ovules or even more. Pod not seen.

W. Australia, *Drummond, n.* 277, *and 2nd Coll. n.* 95. F. Mueller unites this with *I. striata,* but I have not seen any specimens which appear to confirm the union.

3. **I. juncea,** *Turcz. in Bull. Mosc.* 1853, i. 251. Stems numerous from a perennial stock, ascending or erect, ½ to 1 ft. long, slender, wiry, slightly angular or compressed, glabrous. Leaves few, chiefly in the lower part of the stem, consisting of a single leaflet, always articulate on a rather long petiole, the lower ones very small, obovate or obcordate, the next cuneate-oblong, obtuse, mostly under ½ in., and often a few upper ones linear, and ¼ to nearly 1 in. long. Flowers as in *I. Drummondii,* in loose terminal racemes, but smaller. Pedicels almost filiform, shorter or rather longer than the calyx, each in the axil of a minute bract. Calyx glabrous, about 3 lines long, the lobes narrow, usually reflexed. Standard very broad, ½ in. long, more or less veined; wings and keel rather shorter. Ovary nearly sessile, with about 30 ovules. Young pod like that of *I. striata.*

W. Australia, *Drummond, 4th Coll. n.* 22.

4. **I. atropurpurea,** *F. Muell. Fragm.* iii. 16. A bushy shrub or undershrub of 1 to 2 ft., densely clothed with a short, velvety rusty or whitish tomentum. Leaves of a single leaflet, nearly orbicular, ovate or oblong, very obtuse, ½ to 1 in. long, always articulate on a petiole of 2 to 3 lines. Flowers in loose terminal racemes, on pedicels of 1 to 3 lines. Stipules, bracts, and bracteoles small and narrow. Calyx tomentose, nearly 4 lines long, the lobes not reflexed. Petals deep purple, the broad somewhat striate standard not much exceeding the others, all rather longer than the calyx. Ovary nearly sessile, with about 20 ovules; style usually short, but occasionally much longer and filiform. Young pod oblong, softly tomentose.

N. Australia. Hammersley Range, *F. Gregory's Expedition;* Attack Creek and between Mount Morphett and Bonney river, *M'Douall Stuart.*

5. **I. filicaulis,** *Benth. in Ann. Wien. Mus.* ii. 71. Stems erect or ascending from a perennial base, often above 1 ft. high, slender as in *I. juncea,* but more branched and terete, glabrous or pubescent with appressed hairs. Leaves consisting of a single leaflet, articulate on a rather short petiole, linear or lanceolate, acute, 1 to nearly 2 in. long, or rarely small and linear-cuneate. Flowers smaller than in *I. juncea,* in loose slender racemes, mostly leaf-opposed. Pedicels rather longer than the calyx. Bracteoles minute or none. Calyx nearly 3 lines long, pubescent, the lobes lanceolate, broader than in *I. juncea.* Standard very broad, shortly exceeding the calyx, slightly streaked ; wings and keel nearly as long. Ovary nearly sessile, with about 20 ovules. Pod linear, pubescent, often exceeding 1 in.—*Chorizema Leichhardtii,* F. Muell. Fragm. iv. 20.

Queensland. Shoalwater Bay, *R. Brown;* Port Curtis, *M'Gillivray;* Wide Bay, *Bidwill, Leichhardt.*

6. **I. parviflora,** *Benth. in Ann. Wien. Mus.* ii. 71. Very near *I. filicaulis,* and perhaps a variety. Stems slender, decumbent, branched, pubescent. Leaves of a single leaflet, articulate on a rather short petiole. Flowers smaller than in *I. filicaulis,* in all the specimens seen on short axillary or leaf-opposed pedicels. Calyx pubescent, about 2 lines long. Petals of *I. filicaulis,* but smaller. Pod linear, pubescent, about 1 in. long.

N. Australia, Islands of the Gulf of Carpentaria, *R. Brown.*

7. **I. Wheeleri,** *F. Muell. Herb.* A small bushy broom-like shrub, clothed with a soft but very short, close, hoary tomentum. Leaves few, nearly terete, channelled above, ¼ to ½ in. long or rarely longer, more or less distinctly articulate on a very short petiole. Flowers rather small and very few, in short terminal racemes, sometimes reduced to a single flower. Pedicels shorter than the calyx. Calyx tomentose, about 3 lines long. Petals rather longer and nearly equal in length, the standard broad and striate.

S. Australia. Between Stokes Range and Cooper's Creek, *Wheeler.*

7. GOMPHOLOBIUM, Sm.

Calyx deeply cleft, the tube very short, the lobes lanceolate, valvate, the 2 upper ones sometimes more falcate or slightly cohering, but not connate. Petals very shortly clawed. Standard orbicular or reniform, longer than the lower petals; wings oblong, more or less falcate; keel usually broader than the wings, obtuse. Stamens free. Ovary usually shortly stipitate or nearly sessile; style incurved, filiform or slightly thickened from the middle upwards; ovules several, usually 8 or more, rarely 4 or 6, the funicles long and thick, all curved or folded downwards. Pod broadly ovoid or nearly globular, usually oblique, inflated. Seeds small, without any strophiole.—Shrubs or rarely undershrubs, glabrous pubescent or hirsute with spreading hairs. Leaves simple or more frequently compound, the leaflets usually narrow, digitate or pinnate with the terminal leaflet sessile between the last pair. Stipules small, lanceolate or subulate, or none. Flowers yellow or red, terminal or rarely in the upper axils, solitary or 2 or 3 together or in short racemes. Bracts and bracteoles small, sometimes minute or none. Ovary glabrous in all except *G. Baxteri,* where the style is also exceptionally thickened at the base.

The genus is limited to Australia. It is readily distinguished from all except *Burtonia* by the calyx and pod, and is separated from that genus by the more numerous ovules, with the regularly-packed funicles all turned downwards. The ovules in both genera are usually scarcely larger than the breadth of the funicle.

Pedicels longer than the calyx, solitary or 2 or 3 together in a
 very loose raceme. Plant glabrous.
 Leaves all simple, sessile, coriaceous.
  Leaves broadly ovate or orbicular . . . . 1. *G. ovatum.*
  Leaves cordate-lanceolate . . . . 2. *G. amplexicaule.*
 Leaflets 3, very rarely 5 or more, digitate, the common petiole
  usually very short.
  Keel not ciliate (western species).
   Leaflets linear, or, if broad, truncate, mucronate, and veined
    above. Flowers large . . . . 5. *G. polymorphum.*
   Leaflets small, obcordate, not mucronate, the veins very ob-
    scure . . . . 6. *G. obcordatum.*
   Leaflets obovate or cuneate-oblong, mucronate with thickened
    margins, the veins obscure. Flowers small . . . . 7. *G. marginatum.*
  Keel ciliate or fringed at the edge (eastern species).
   Tall erect shrub. Leaflets 1 to 2 in. long. Standard above
    1 in. diameter. Keel densely fringed . . . . . . 3. *G. latifolium.*
   Diffuse or much branched. Leaflets under ¾ in. Standard
    ½ to ¾ in. Keel (often very shortly) ciliate . . . . 4. *G. Huegelii.*
 Leaves pinnate, with several leaflets . . . . 21. *G. pinnatum.*
Pedicels very short or not longer than the calyx, solitary or 2 or
 3 together in dense leafy corymbs or heads.
 Leaflets 3, digitate.
  Plant quite glabrous. Leaflets above ½ in. long.
   Leaflets with revolute margins. Calyx ½ in. long. Stan-
    dard ¾ in. . . . . 8. *G. grandiflorum.*
   Leaflets usually flat or recurved. Calyx 4 to 5 lines.
    Standard 6 or 7 lines long . . . . 9. *G. virgatum.*
  Branches pubescent or villous. Leaflets under ¼ in. long.
   Calyx glabrous.
    Stipules inconspicuous (eastern species).
     Flowers yellow, rather large. Ovules 12 to 20 . . 10. *G. minus.*
     Flowers red, small. Ovules about 6 . . . . . . 11. *G. uncinatum.*
    Stipules setaceous, persistent (western species) . . . 12. *G. Baxteri*
   Calyx villous. . . . . 13. *G. aristatum.*
 Leaves pinnate. Leaflets more than 3 (the petiole sometimes
  so short that they appear digitate).

Calyx pubescent or villous.
  Common petiole none.   Leaves linear, terete, with revolute
    margins . . . . 13. *G. aristatum.*
  Common petiole very short.   Leaflets linear, with recurved
    margins.   Flowers in dense leafy corymbs or heads . . 14. *G. burtonioides.*
  Common petiole elongated.
    Flowers in dense leafy corymbs or heads.   Leaflets linear,
      terete, revolute, usually fewer than 11.   Ovules 8   . 15. *G. capitatum.*
    Flowers few.   Leaflets linear-terete, revolute, usually
      fewer than 11.   Ovules 16 to 20 . . . . . . 16. *G. tomentosum.*
    Flowers few.   Leaflets numerous, often dilated at the
      end.   Ovules 8 . . . . 17. *G. Preissii.*
  Calyx glabrous.
    Viscid erect shrub (western species) . . . . 18. *G. viscidulum.*
    Shrubs, usually diffuse, not viscid (eastern species).
      Leaflets narrow-linear.   Common petiole very short . . 19. *G. glabratum.*
      Leaflets cuneate.   Common petiole elongated . . . . 20. *G. nitidum.*
      Leaflets narrow-linear.   Common petiole elongated . . 21. *G. pinnatum.*
Flowers in short corymbose racemes, pedunculate above the last
  leaves.   Leaves all or mostly pinnate.   Bracts persistent.
  Leaflets linear-terete, with revolute margins.
    Leaflets 3 to 7, under ½ in. long, with a very short common
      petiole . . . . 22. *G. Shuttleworthii.*
    Leaflets numerous, ¼ to ¾ in. long, on an elongated common
      petiole . . . . 23. *G. venustum.*
  Leaflets flat or nearly so, linear-lanceolate or oblong . . . . 24. *G. Knightianum.*

(The pinnate-leaved species with only 2 ovules to the ovary, are now transferred to *Burtonia.*)

1. **G. ovatum,** *Meissn. in Pl. Preiss.* i. 35. A glabrous undershrub, with simple or slightly branched ascending or erect stems, more or less compressed and prominently angular. Leaves simple, broadly ovate or orbicular, ¾ to 1 in. long, very obtuse or slightly pointed, coriaceous, penniveined. Flowers rather large, on long pedicels, solitary or 2 to 4 together in an irregular terminal raceme. Calyx 4 to 5 lines long, the 2 upper lobes slightly falcate. Standard 8 to 9 lines long, very broad, emarginate; wings and keel about 6 lines. Ovary almost sessile, with about 20 ovules; style rather attenuate at the base. Pod oblong, nearly twice as long as the calyx.

W. Australia, *Drummond ;* near Albany, *Preiss, n.* 1105 ; Ironstone hills, Blackwood river, *Oldfield.*

2. **G. amplexicaule,** *Meissn. in Pl. Preiss.* i. 36, *and* ii. 210. Closely allied to *G. ovatum,* and perhaps a variety. Stems decumbent or branched at the base, 1 to 2 ft. high, usually flattened, with prominent angles. Leaves simple, from ovate to lanceolate, acute, ½ to 1½ in. long, always cordate at the base, but not truly stem-clasping, coriaceous, obscurely veined. Flowers rather smaller than in *G. ovatum,* the pedicels not quite so long, and the lower ones occasionally axillary. Calyx about 4 lines long. Standard nearly twice as long. Ovules 15 to 20. Pod nearly globular, much longer than the calyx.

W. Australia, *Drummond, 2nd Coll. n.* 124 ; near Albany, *Preiss, n.* 1106 ; stony places, Tone river, *Oldfield.*

3. **G. latifolium,** *Sm. in Ann. Bot.* i. 505, *and* ii. 519, *and in Trans.*

Linn. Soc. ix. 249, *not of Labill.* A glabrous shrub, with erect virgate branches. Leaflets 3, on a very short common petiole, linear linear-lanceolate or linear-cuneate, acute or truncate and mucronate, 1 to 2 in. long, the margins flat or slightly recurved, the veins fine and almost longitudinal; stipules inconspicuous. Flowers large, yellow. Peduncles (or leafless flowering branches) solitary in the upper axils, nearly as long as or longer than the leaves, with a pair of small 3-foliolate bracts about the middle, or sometimes growing out into leafy branches with a terminal flower. Calyx ½ to ¾ in. long. Standard broad, usually above 1 in. long; lower petals nearly as long; keel incurved, very obtuse, densely fringed on the inner edge with short white woolly hairs. Ovary with about 20 ovules. Pod ovoid, ½ to ¾ in. long.—*G. fimbriatum*, Sm. Exot. Bot. t. 58; *G. psoraleæfolium*, Salisb. Parad. Lond. t. 6; *G. barbigerum*, DC. Prod. ii. 105; Bot. Mag. t. 4171; Paxt. Mag. xiv. 221, with a fig.

Queensland. Glasshouse mountains, *F. Mueller* (from a specimen in leaf only).
N. S. Wales. Port Jackson, *R. Brown, Sieber, n.* 361, *and Fl. Mixt. n.* 577, and others; Port Stevens, *Fraser, M'Arthur.*
Misled by Labillardière, most modern botanists have transferred the name of *G. latifolium* to the following species.

4. **G. Huegelii,** *Benth. in Hueg. Enum.* 29, *and in Ann. Wien. Mus.* ii. 72. A low, diffuse or much branched, glabrous shrub, rarely above 1 ft. high. Leaflets usually 3, on a very short common petiole, linear or the lower ones oblong-cuneate, or rarely almost obovate, mostly about ½ in. long, but occasionally very narrow, and ¾ or even 1 in. long, obtuse or mucronate, the margins flat, recurved or rarely revolute. Flowers yellow, solitary or 2 or 3 together; pedicels always longer than the calyx, attaining sometimes 1½ or 2 in. Calyx 4 to 5 lines long. Standard 6 or 7 lines long; wings rather shorter; keel as long as the wings and rather broader, very obtuse, usually bordered on the inner edge with minute white hairs, which sometimes almost disappear. Ovary with 15 to 20 ovules. Pod ovoid, longer than the calyx. — *G. latifolium*, Labill. Pl. Nov. Holl. i. 105, t. 133; DC. Prod. ii. 105; Hook. f. Fl. Tasm. i. 82, not of Sm.; *G. pedunculare*, Lodd. Bot. Cab. t. 1639? (from the figure).

N. S. Wales. Port Jackson, to the Blue Mountains, *R. Brown, A. Cunningham;* New England, *C. Stuart*, and to the southward, *Huegel, Backhouse.*
Victoria. King's Island, Port Phillip, *R. Brown;* Gipps Land, Dandenoug ranges, etc., *F. Mueller.*
Tasmania. Port Dalrymple and Derwent river, *R. Brown;* abundant in heathy places throughout the colony, *J. D. Hooker.*
Var. *leptophyllum.* Leaflets very fine, often ¾ in. long or more. Mountains of Victoria and in New England, also apparently Mount Mitchell, Clarence river, *Beckler*, but the latter specimens are past flower.
Some of the decumbent loose-flowered specimens of this species come very near to some forms of the western *G. polymorphum.*

5. **G. polymorphum,** *R. Br. in Ait. Hort. Kew. ed.* 2, iii. 11. A glabrous shrub or undershrub, truly polymorphous in aspect and leaflets. Stems usually numerous, slender, either decumbent and ½ to 1½ ft. long, or long and twining, or more or less erect and virgate. Leaflets 3, rarely 5 or even 7 or 9, digitate on a common petiole, usually short, but sometimes 3 or 4 lines

long, mostly linear, with recurved margins and short straight points, and ½ to 1 in. long, but sometimes the lower ones, or in luxuriant shoots, nearly all are broadly cuneate, spathulate or almost obovate, and then marked on the upper surface with raised oblique veins. Stipules setaceous. Pedicels long, terminal or leaf-opposed, solitary or 2 or 3 in a loose raceme. Bracteoles minute, at a distance from the calyx. Flowers varying from orange-yellow to a bright crimson. Calyx glabrous, about 4 lines long. Standard often twice as long as the calyx; wings considerably shorter; keel glabrous on the edge, rather shorter than the wings. Ovules 15 to 20. Pod ovoid-globular, much inflated, ½ to ¾ in. long.—Bot. Mag. t. 1533 ; DC. Prod. ii. 106 ; Meissn. in Pl. Preiss. i. 37 ; Paxt. Mag. vi. 151, with a fig. ; *G. grandiflorum*, Andr. Bot. Rep. t. 642, not of Sm. ; *G. pedunculare*, DC. Prod. ii. 105 ; Meissn. in Pl. Preiss. i. 36 ; *G. venulosum*, Lindl. in Bot. Reg. t. 1574 ; *G. tenue*, Lindl. Bot. Reg. t. ·1615 ; Meissn. in Pl. Preiss, i. 37 ; *G. versicolor*, Lindl. Bot. Reg. 1839, t. 43 ; Paxt. Mag. xii. 219, with a fig. ; Bot. Mag. t. 4179 (the latter a luxuriant form with long leaflets and large flowers).

W. Australia. King George's Sound, *R. Brown, Baxter*, and others, to Vasse and Swan rivers, *Drummond, Preiss, n.* 1108 (?), 1109, 1111, 1112, 1113, 1114, *Oldfield* and others, and extending eastward towards the Great Bight, *Maxwell*. The station, E. Australia, given by De Candolle for his *G. pedunculare*, was a mistake. The specimen in Sonder's herbarium, marked *Preiss, n.* 1108, is evidently *G. Huegelii*, with a strongly fringed keel, probably some mistake has occurred ; if it be really West Australian, and a form only of *G. polymorphum*, the distinction between the two species disappears.

6. **G. obcordatum,** *Turcz. in Bull. Mosc.* 1853, i. 258. An erect, much branched, glabrous shrub, with slender, angular, flexuose branches. Leaflets 3, on a very short common petiole, obovate or obcordate, very obtuse and not mucronate, 2 to 4 lines long, the margins usually reflexed, smooth or obscurely reticulate. Stipules minute or none. Pedicels terminal, solitary or 2 or 3 in a loose raceme, longer than the calyx. Flowers much smaller than in *G. polymorphum*. Calyx about 3 lines long, glabrous. Standard not twice as long as the calyx ; wings shorter, narrow ; keel about as long as the wings, much broader and incurved. Ovules about 8. Pod broader than long.

W. Australia, *Drummond, 5th Coll. n.* 42.

7. **G. marginatum,** *R. Br. in Ait. Hort. Kew. ed.* 2. iii. 11. A low, glabrous, and somewhat glaucous shrub, with slender, but rigid decumbent or ascending stems, under 1 ft. long. Leaflets 3 or rarely solitary, on a common petiole of ½ to 1½ lines, from obovate to linear oblong, ½ to ¾ in. long, with a short sharp point, coriaceous, bordered by a thickened nerve-like edge, the veins obscure. Stipules lanceolate-subulate or setaceous. Flowers small, yellow, few, in irregular loose terminal racemes or rarely solitary. Pedicels much longer than the calyx, bearing minute bracteoles below the middle. Calyx 2½ to 3 lines long. Standard about 4 lines long, deeply notched ; lower petals scarcely exceeding the calyx. Ovules 10 to 12. Pod much inflated, ¼ in. diameter.—DC. Prod. ii. 105 ; Bot. Reg. t. 1490 ; Meissn. in Pl. Preiss. i. 36.

W. Australia. King George's Sound, *Baxter ;* Swan River, *Drummond, n.* 219 ; Darling range, *Collie ;* Kalgan, Gordon, and Harvey rivers, *Oldfield*.

8. **G. grandiflorum,** *Sm. Exot. Bot. t.* 5, *and in Trans. Linn. Soc.* ix.

249. A glabrous, erect, branching shrub. Leaflets 3, on a very short common petiole, narrow-linear, with a short almost pungent point, $\frac{1}{2}$ to 1 in. long or rather more, the margins revolute, the veins inconspicuous. Flowers large, solitary or 2 or 3 together, shortly pedicellate, terminal or on very short axillary leafy branches. Calyx about $\frac{1}{2}$ in. long, glabrous, except minute hairs at the edge of the lobes. Standard broad, fully $\frac{3}{4}$ in. long; wings shorter; keel as long as the wings, broad, not fringed. Ovules varying from 8 to 14. Pod scarcely exceeding the calyx.—DC. Prod. ii. 105; Bot. Reg. t. 484? (the pedicels figured much longer than I have ever seen them); *G. maculatum*, Andr. Bot. Rep. t. 427.

N. S. Wales. Port Jackson, to the Blue Mountains. *R. Brown, Sieber, n.* 358, and others.
Victoria. A specimen from Latrobe river, *F. Mueller,* appears to be this species, but being in fruit only, it is doubtful.
Var. *setifolium*, DC. Leaflets narrower. Flowers rather smaller.— *G. setifolium*, Sieb. Pl. Exs.; *G. glaucescens*, A. Cunn. in Field. N. S. Wales, 346. Blue Mountains, *R. Brown, Sieber, n.* 363, *A. Cunningham*, and others.

9. **G. virgatum**, *Sieb. in DC. Prod.* ii. 105. An erect glabrous shrub, closely allied to *G. grandiflorum*, and differing chiefly in the shorter and less revolute leaflets and smaller flowers. Leaflets 3, the common petiole very short or scarcely any, linear, the margins always recurved, although rarely absolutely revolute, $\frac{1}{2}$ to $\frac{3}{4}$ in. long or rarely none. Flowers yellow, terminal, solitary or 2 or 3 together, the pedicels rarely so long as the calyx. Calyx 4 to 5 lines long. Standard 6 to 7 lines; keel shorter, not fringed. Ovary with about 8 ovules. Pod about as long as the calyx.—Reichb. Icon. Exot. t. 97.

Queensland. Sandy Cape, *R. Brown ;* islands of Moreton Bay, *F. Mueller.*
N. S. Wales. Port Jackson, *R. Brown, Sieber, n.* 360, *and Fl. Mixt. n.* 578, and others; Port Stephens, *M'Arthur.*
Var. *aspalathoides.* Leaflets narrower and more revolute, not above $\frac{1}{2}$ in. long.—*G. aspalathoides*, A. Cunn.; Benth. in Ann. Wien. Mus. ii. 72. Wellington valley.—This comes near to some forms of *G. minus*, but is quite glabrous, the leaves are longer and the ovules much fewer.

10. **G. minus**, *Sm. in Trans. Linn. Soc.* ix. 251. A much-branched shrub, sometimes low and decumbent, sometimes attaining 2 or 3 ft., the branches more or less pubescent, at least when young. Leaflets 3, on a very short common petiole, linear or almost subulate, obtuse, with a small straight or recurved point, under $\frac{1}{2}$ in. long, the margins revolute, glabrous. Flowers (yellow?) usually 2 or 3 together, the pedicels rarely exceeding the calyx and usually much shorter. Calyx $3\frac{1}{2}$ to 4 lines long. Standard 5 or rarely 6 lines long; wings and keel rather shorter, the keel not fringed. Ovary with 12 to 20 ovules. Pod shortly exceeding the calyx.—*Burtonia minor*, DC. Prod. ii. 106; *Gompholobium tetrathecoides*, Sieb. in DC. Prod. ii. 106.

N. S. Wales. Port Jackson, *R. Brown, Sieber, n.* 359, and others.
Victoria. Mount Abrupt, *F. Mueller ;* Wimmera, *Dallachy.*
S. Australia. Lofty Ranges, *F. Mueller ;* Penola, *J. Woods ;* mouth of the Glenelg, *Allitt ;* Kangaroo Island, *Waterhouse.*
Var. *grandiflora.* Flowers rather larger, pedicels rather longer; keel occasionally with a minute edging of white hairs. Ovules often above 20. To this variety belong most of the

southern specimens, and some seem almost to pass into *G. Huegelii,* although they have always shorter pedicels, smaller leaves, and more or less pubescent branches.

11. **G. uncinatum,** *A. Cunn.; Benth. in Ann. Wien. Mus.* ii. 72. An erect, much branched, heath-like shrub, the young branches usually minutely pubescent. Leaflets 3, on a very short common petiole or almost sessile, linear, obtuse, with a recurved point, rarely above 3 lines long, the margins revolute. Flowers red, much smaller than in the preceding species, solitary or 2 or 3 together, the pedicels shorter or rarely rather longer than the calyx. Calyx scarcely 3 lines long. Petals almost sessile ; standard very broad, about 4 lines long ; wings and keel rather shorter, the latter not fringed. Ovary with about 6 ovules. Pod nearly globular, rather longer than the calyx.

N. S. Wales. Blue Mountains, *A. Cunningham, Fraser ;* Paramatta, *Woolls ;* New England, *C. Stuart.*

12. **G. Baxteri,** *Benth.* An erect heath-like shrub, with numerous virgate pubescent branches. Leaflets usually 3, sessile or on a very short common petiole, narrow-linear, with revolute margins, mostly hooked at the ends, 3 or rarely 4 lines long. Stipules setaceous, persistent. Flowers terminal, solitary or 2 or 3 together, sessile within the last leaves. Calyx glabrous, apparently viscid, about 3 lines long, very angular in the bud. Standard 4 lines long or rather more ; wings shorter ; keel almost as long as the standard, straight, very obtuse. Ovary hirsute, almost sessile, with about 8 ovules ; style slightly thickened at the base. Pod ovoid-globular, slightly compressed, about as long as the calyx, more or less pubescent.

W. Australia. King George's Sound, *Baxter ;* Orleans Bay, Stokes Inlet, and Russell Ranges, *Maxwell.*

13. **G. aristatum,** *Benth. in Ann. Wien. Mus.* ii. 72. An erect shrub, with the habit and tomentose or villous branches of *G. tomentosum,* but the petiole of the leaves is never developed. Leaflets 3, 5, or rarely 7, sessile and digitate on a slight callosity of the branch, narrow-linear, with closely revolute margins, so as to be almost terete, with a fine almost pungent point, rarely more than ½ in. long, more or less pubescent. Stipules small, setaceous. Flowers terminal, nearly sessile, solitary or in compact leafy corymbs, but not so dense as in *G. capitatum.* Calyx villous, 3 to 3½ lines long, the buds tipped by the awn-like points of the sepals. Standard 5 to 6 lines long ; keel nearly as long, somewhat incurved, minutely ciliate on the edge. Ovary stipitate, with 10 to 12 ovules. Pod compressed-globular, scarcely exceeding the sepals.—Meissn. in Pl. Preiss. i. 38, and ii. 210.

W. Australia. Swan River, *Huegel, Drummond, Preiss, n.* 1199, and others. Var. (?) *laxum.* Leaflets usually 5, fine-pointed ; flowers 1 or 2, terminal, on pedicels nearly as long as the leaves ; calyx glabrous or slightly pubescent. Kalgan river, *Oldfield ;* Plantagenet and Stirling ranges, *Maxwell,* also one specimen in Herb. Sonder, with *Preiss's, n.* 1198.

Var. *muticum.* Leaflets usually 3, obtuse or with a short callous point ; keel rather more prominently ciliate ; ovules fewer, 8 according to Meissner, 6 in the flower I examined. *G. Drummondii,* Meissn. in Bot. Zeit. 1855, 25.—Between Moore and Murchison rivers, *Drummond, 6th Coll. n.* 10 ; Canning river, Mount Barker, and Kalgan river, *Oldfield.*

14. **G. burtonioides,** *Meissn. in Pl. Preiss.* i. 37, *and* ii. 210. An erect shrub, allied to *G. aristatum,* but at once distinguished by the broader

leaflets, with a more or less prominent common petiole, sometimes 1½ lines long. Young branches usually pubescent. Leaflets 3 or 5, the lateral ones usually inserted lower down than the 3 terminal ones, all linear, obtuse or mucronate, under ½ in. long, the margins recurved or revolute. Flowers yellow, terminal, solitary or several together, sessile or shortly pedicellate amongst the last leaves. Calyx hirsute or nearly glabrous, fully 4 lines long. Standard about ½ in. long; wings and keel scarcely exceeding the calyx. Ovary with about 8 ovules. Pod rarely exceeding the calyx.

W. Australia, *Drummond,* 3*rd Coll. n.* 74, *and* 5*th Coll. n.* 29 *and* 30 ; near Albany, *Preiss, n.* 1100, also 1193 ; heathy ground, Gordon river, *Oldfield ;* Salt river, *Maxwell.*

15. **G. capitatum,** *A. Cunn. in Lindl. Bot. Reg. t.* 1563. Very closely allied to *G. tomentosum,* and perhaps a variety. It is less tomentose, and sometimes glabrous or hirsute with a few long hairs. Foliage the same or the leaflets rather more slender. Flowers yellow, very shortly pedicellate, in dense, terminal, leafy corymbs, almost contracted into heads. Calyx usually very hirsute, about 4 lines long. Petals sometimes shorter than the calyx, and rarely exceeding it so much as in *G. tomentosum.* Ovules not more than 8 in any of the flowers I have opened. Pod usually shorter than the calyx. —Meissn. in Pl. Preiss. i. 38.

W. Australia. King George's Sound, *R. Brown* and others ; Swan River, *Drummond, n.* 129 ; Sussex district and Stirling Terrace, *Preiss, n.* 1200 *and* 1201 ; Blackwood and Canning rivers, *Oldfield.*

16. **G. tomentosum,** *Labill. in Pl. Nov. Holl.* i. 106, *t.* 134. An erect shrub of 1 to 3 ft., the young branches tomentose-villous. Leaves pinnate, the common petiole rarely above 2 or 3 lines long. Leaflets usually 5 or 7, but varying from 3 to 11, narrow-linear, the margins revolute so as to be almost terete, mucronate, about ½ in. or sometimes nearly ¾ in. long, more or less pubescent. Stipules subulate. Flowers yellow, terminal, few, in compact leafy corymbs or rarely solitary, the pedicels very short. Calyx villous, about 4 lines long, the lobes more or less fine-pointed. Standard about 6 lines long ; keel rather shorter, broad, somewhat curved, the edges minutely ciliate. Ovary with 16 to 20 ovules. Pod as long as or rather exceeding the calyx.—DC. Prod. ii. 106 ; Meissn. in Pl. Preiss. i. 40 ; Bot. Reg. t. 1474 ; *G. aciculare,* Reichb. Icon. Exot. t. 243 (from the fig.) ; *G. lanatum,* A. Cunn. in G. Don, Gen. Syst. ii. 118.

W. Australia. King George's Sound, *Labillardière, R. Brown,* and others ; Swan River, *Drummond, n.* 202, *Preiss, n.* 1198 ; Doubtful Island Bay, Champion Bay, and Murchison river, *Oldfield.*

17. **G. Preissii,** *Meissn. in Pl. Preiss.* i. 40. Very closely allied to *G. tomentosum,* and perhaps a variety. Branches and leaves hirsute with spreading hairs. Leaflets more numerous than in *G. tomentosum,* usually 11 to 19, but occasionally more, or on the lower leaves fewer, linear, with revolute margins, as in the allied species, but most of them more or less dilated at the end, with recurved points. Flowers few, in leafy terminal corymbs, as in *G. tomentosum,* but smaller. Calyx hirsute, 3 to 3½ lines long. Standard rather longer than the calyx ; keel shorter and straight, not ciliate. Ovules 8 in the flowers examined. Pod short, as in *G. capitatum.*

W. Australia, *Drummond, n.* 60 *and* 201 ; Stirling Terrace, *Preiss, n.* 1194.

18. **G. viscidulum,** *Meissn. in Pl. Preiss.* i. 39, *and* ii. 210. An erect, glabrous, more or less viscid, much branched shrub, of ½ to 1½ ft. Leaves pinnate, occasionally opposite, the common petiole rarely ½ in. long; leaflets 5 to 9, narrow-linear, with revolute margins, obtuse or nearly so, rarely ½ in. long, rather rigid. Stipules minute. Flowers terminal, solitary or 2 or 3 together, on very short pedicels within the last leaves. Calyx glabrous, about 3 lines long, the lobes broader and the buds more angular than in most species. Standard broad, 4 or 5 lines long; wings and keel shorter, the latter broad, slightly incurved, very minutely ciliate on the edge. Ovules usually 8.

W. Australia, *Drummond,* 3rd *Coll. n.* 75; Konkoberup Hills, *Preiss, n.* 1196; Fitzgerald river, Phillips Ranges, Stokes Inlet, Orleans Bay, etc., *Maxwell.*

19. **G. glabratum,** *DC. Prod.* ii. 106. A low decumbent or diffuse or rarely erect shrub, with the habit of *G. minus,* but readily known by the leaves really pinnate, although with the common petiole very short, rarely attaining 2 lines. Branches slender, virgate, minutely pubescent or rarely quite glabrous. Leaflets 5 or 7, very rarely 3, narrow-linear, with recurved or revolute margins, under ½ in. long, obtuse or minutely pointed, usually glabrous. Flowers terminal, solitary or 2 or 3 together, the pedicels very short. Calyx glabrous, 3 to nearly 4 lines long. Standard 4 to 5 lines; keel shorter, broad, somewhat incurved, not ciliate. Ovary with 8 to 10 ovules. Pod not seen.—*G. polymorphum,* Sieb. Pl. Exs. not of R. Br.

N. S. Wales. Port Jackson, *R. Brown, Sieber, n.* 362, and others; Double Bay, *Leichhardt.*

20. **G. nitidum,** *Soland. in Herb. Banks.* A much-branched, glabrous shrub. Leaves pinnate, with a common petiole of about ½ in.; leaflets usually about 7 to 11, oblong-cuneate or almost obovate, emarginate, 4 to 6 lines long, dark and shining above, glaucous underneath. Flowers terminal, solitary, rather large, on very short pedicels. Calyx glabrous, fully 5 lines long, the lobes of a thickish consistence and not separated so low down as in other species. Petals shortly exceeding the calyx, the keel not ciliate. Ovules (from R. Brown's notes) 4. Pod sessile, shorter than the calyx.

Queensland. Endeavour river, *Banks and Solander, R. Brown (Herb. Banks and R. Br.).*

21. **G. pinnatum,** *Sm. in Trans. Linn. Soc.* ix. 251. A glabrous undershrub, with slender, but rigid, ascending or erect stems, of 1 ft. or rather more, simple or little branched, and usually flexuose. Leaves pinnate; leaflets few in the lower leaves, on a short common petiole, often above 30 in the upper ones, with a common petiole of above 1 in., linear or almost subulate, 4 to 8 lines long, mucronate or acute, the margins revolute. Flowers few, in short, loose, terminal racemes, or rarely solitary, the pedicels longer than the calyx. Calyx about 3 lines long. Standard very broad, rather longer than the calyx, and the lower petals nearly as long. Ovary usually with 8 ovules. Pod ovoid-globular, rather longer than the calyx.— DC. Prod. ii. 106.

Queensland. Sandy Cape, *R. Brown;* Port Curtis, *M'Gillivray;* Wide Bay, *Bidwill;* Brisbane river and Moreton Bay, *F. Mueller.*
N. S. Wales. Port Jackson, *R. Brown, Woolls.*

22. **G. Shuttleworthii,** *Meissn. in Pl. Preiss.* i. 39. An erect, heath-like shrub, of ¼ to 1½ ft., the branches shortly pubescent. Leaves pin-nate, with a very short common petiole; leaflets 5 or 7, rarely 3, narrow-linear, with revolute margins, under ½ in. long, obtuse or with a short re-curved point, glabrous or scabrous-pubescent. Flowers purple-red, in short, corymbose, almost capitate racemes, pedunculate above the last leaves. Bracts and bracteoles linear-subulate, persistent. Pedicels much shorter than the calyx. Calyx 3 to 4 lines long, glabrous. Standard about 5 lines long, not so broad as in some species; keel shorter, slightly incurved, minutely or scarcely perceptibly ciliate on the edges. Ovules 6 in all the specimens I examined, but Meissner found only 4. Pod not seen.

W. Australia. Swan River, *Drummond, 1st Coll. and n.* 200; Darling Range, *Collie, Preiss, n.* 1178; Gordon river, *Oldjield.*

23. **G. venustum,** *R. Br. in Ait. Hort. Kew. ed.* 2, iii. 12. A gla-brous, often glaucous shrub or undershrub, with slender, flexuose stems, of 1 to 2 ft. Leaves pinnate, with a common petiole of ½ to 1 in.; leaflets usually 9 to 21, linear-terete, acute or mucronate, the margins closely revo-lute, ½ to ¾ in. long, transversely wrinkled when dry. Flowers pink or pur-ple, rather numerous, in a short corymbose raceme, on a long peduncle above the last leaves. Pedicels longer than the calyx, in the axils of subulate bracts. Calyx 3 to 4 lines long, glabrous. Standard very broad, rather longer than the calyx; wings and keel rather shorter. Ovules about 8. Pod very broad, as long as the calyx.—DC. Prod. ii. 106; Meissn. in Pl. Preiss. i. 40; Bot. Mag. t. 4258; Reichb. Icon. Exot. t. 76.

W. Australia. King George's Sound, *R. Brown* and others, *Drummond, 3rd Coll. n.* 72; near Albany, *Preiss, n.* 1102; Mount Barker, *Oldfield;* near Cape Paisley, *Maxwell.*

Var. (?) *læve.* Leaflets perfectly smooth, but revolute, as in the common form.—Middle Mount Barren, *Maxwell.*

24. **G. Knightianum,** *Lindl. in Bot. Reg. t.* 1468. A glabrous un-dershrub or small shrub, with slender, but rigid, ascending or erect stems, attaining 1 ft. or rather more. Leaves mostly pinnate, with 5 to 11 lanceo-late or linear leaflets, ½ to 1 in. long, obtuse or mucronate, flat or with slightly recurved margins, rigid and strongly reticulate on the upper side; the lower leaves, however, or nearly all those of the lower branches, some-times reduced to 3 short, ovate, obovate or oblong, digitate leaflets. Stipules subulate. Flowers pink or purple, in a short corymbose raceme, on a rather long peduncle above the last leaves. Bracts linear-subulate. Calyx scarcely 3 lines long. Standard broad, rather longer than the calyx; wings and keel rather shorter. Ovules about 8. Pod about 4 lines long.—Meissn. in Pl. Preiss. i. 40; *G. heterophyllum,* A. Cunn. in G. Don, Gen. Syst. ii. 118.

W. Australia. King George's Sound, *Baxter, A. Cunningham, Preiss, n.* 1103; Swan River, *Drummond, 1st Coll. and n.* 220; York District, *Preiss, n.* 1104; Gordon and Vasse rivers, *Oldfield;* Cape Knobb and Cape Le Grand, *Maxwell.*

8. BURTONIA, R. Br.

Calyx deeply cleft, the lobes longer than the tube, lanceolate, valvate, the 2 upper ones often broader or more obtuse, but not connate. Petals very shortly clawed ; standard orbicular or reniform, longer than the lower petals ; wings oblong or obovate, more or less falcate ; keel usually broader than the wings, obtuse. Stamens free. Ovary sessile or shortly stipitate ; style incurved, more or less dilated towards the base ; ovules 2, the funicles long and thick, one curved or folded upwards, the other downwards. Pod broadly ovoid or nearly globular, usually oblique, inflated. Seeds small, without any strophiole.—Shrubs or rarely undershrubs, glabrous or hirsute with spreading hairs. Leaves simple or compound, digitate, or pinnate with the terminal leaflet sessile between the last pair. Stipules minute or none. Flowers yellow, orange-red or bluish-purple, solitary in the axils of the upper leaves or forming terminal racemes. Bracts small ; bracteoles also small, usually below the middle of the pedicel. Ovary glabrous or villous.

The genus is limited to Australia. It is closely allied to *Gompholobium*, with the same diversity of foliage, valvate calyx, etc. ; differing chiefly in the ovules, always 2 only, with the funicles very long and thick, as in *Gompholobium*, but one always curved or folded upwards, the other downwards, not all downwards, as in that genus. The style is also much thicker at the base.

Leaves pinnate. Racemes terminal.
     Plant glabrous. Racemes 1- to 3-flowered.
         Leaflets few, subulate . . . . 1. *B. subulata.*
         Leaflets numerous, very small, obovate or obcordate . . . . 2. *B. foliolosa.*
     Plant very hirsute. Racemes elongated, many-flowered. Leaflets
         very numerous, small, ovate . . . . 3. *B. polyzyga.*
Leaflets 3, linear-revolute, sessile (the common petiole not produced).
     Pedicels in the upper axils.
         Calyx and ovary villous . . . . 4. *B. villosa.*
         Calyx and ovary glabrous.
            Branches tomentose. Leaflets rigid, often shorter than the internodes. Pedicels long . . . . 5. *B. Hendersonii.*
            Branches glabrous. Leaflets longer than the internodes and pedicels . . . . 6. *B. scabra.*
Leaves simple, linear or subulate. Racemes umbel-like, terminal . 7. *B. conferta.*

1. **B. subulata,** *Benth.* An erect, glabrous shrub, of 1 to 2 ft., with slender, rigid branches. Leaves pinnate, with a common petiole of 3 or 4 lines ; leaflets 5, 7, or rarely 9, linear-subulate, with revolute margins, mucronulate, ½ to ¾ in. long. Flowers few, in very short, terminal, almost corymbose racemes, or often quite solitary ; pedicels usually longer than the calyx. Calyx glabrous, about 4 lines long. Petals nearly equal in length, slightly exceeding the calyx. Style more slender than in most *Burtonias,* yet somewhat dilated at the base. Pod compressed-globular, scarcely exceeding the calyx.—*Gompholobium subulatum,* Benth. in Ann. Wien. Mus. ii. 72 ; *G. stenophyllum,* F. Muell. Fragm. iii. 30.

N. Australia. Regent's River, Brunswick Bay, N.W. coast, *A. Cunningham ;* islands of the Gulf of Carpentaria, *R. Brown ;* Port Essington, *Armstrong* ; Arnhem's Land, *F. Mueller.*

2. **B. foliolosa,** *Benth.* An elegant little shrub, quite glabrous and somewhat glaucous, with slender terete branches. Leaves pinnate, the common petiole rarely above ½ in. long ; leaflets 11 to 21 or even more, obovate or obcordate, very obtuse, 1 to 1½ lines long. Flowers small, few together, in loose, terminal, almost corymbose racemes, the pedicels rather longer than

the calyx. Calyx nearly 2 lines long, less deeply divided than in most other species, although the lobes are longer than the tube, the 2 upper ones more obtuse. Standard at least half as long again as the calyx; wings and keel shorter. Ovary sessile, glabrous; style slightly flattened towards the base. Pod not seen.—*Gompholobium foliolosum*, Benth. in Mitch. Trop. Austr. 348.

Queensland. Sandy forests, Warrego river, *Mitchell*; Dogwood Creek, *Leichhardt.*

3. **B. polyzyga,** *Benth.* An erect shrub, the branches, leaves, and inflorescence densely clothed with long, spreading, white, almost woolly hairs. Leaves pinnate, with a common petiole, of 1 to 2 in. or rather more; leaflets 29 to 51, oval obovate or oblong, rarely exceeding 2 lines. Flowers yellow, in long, loose, terminal racemes; the pedicels longer than the calyx. Bracts and bracteoles filiform. Calyx about 4 lines long. Petals rather longer, all of nearly equal length. Ovary almost sessile, hirsute with long white hairs. Pod not seen.—*Gompholobium polyzygum*, F. Muell. Fragm. iii. 29.

N. Australia. Between Mount Morphett and Bonny river, *M'Douall Stuart.*

4. **B. villosa,** *Meissn. in Pl. Preiss.* i. 41. A rigid, heath-like, erect shrub, with virgate branches, villous with short spreading hairs. Leaflets 3, sessile on the stem, narrow-linear, with revolute margins, obtuse or with a short, callous point, mostly about ¼ in. long and much longer than the internodes, glabrous or sprinkled with a few hairs. Flowers large, pale purple, solitary in the upper axils, on pedicels usually shorter than the leaves. Calyx hirsute, 4 to 5 lines long. Standard fully ¾ in. long; wings rather shorter; keel shorter than the wings, incurved. Ovary nearly sessile, villous; style flattened towards the base, ciliate on the inner edge. Pod somewhat compressed, shortly exceeding the calyx, as broad as long.—Bot. Mag. t. 4410.

W. Australia. King George's Sound, *R. Brown, Drummond;* sandy woods near Mount Wuljenup, *Preiss, n.* 1172; Mount Barker, *Oldfield.*

5. **B. Hendersonii,** *Benth.* A stout, rigid, divaricate shrub, the branches terete and tomentose when young. Leaflets 3, sessile on the stem, linear, with revolute margins, obtuse or with a hooked point, 2 to 3 lines long, and scarcely covering the internodes, rigid, glabrous, but appearing minutely glandular under a lens. Flowers orange-red, solitary in the upper axils, the pedicels longer than the leaves. Calyx glabrous outside, about 4 lines long, the lobes ciliate on the edge, whitish but scarcely tomentose inside. Standard very broad, above ½ in. long; wings scarcely exceeding the calyx; keel rather longer, deeply coloured, much incurved. Ovary nearly sessile, glabrous. Pod not seen.—*Gompholobium Hendersonii*, Paxt. Mag. xi. 103, with a fig. (from the figure and description).

W. Australia. From the interior, *J. S. Roe.*

6. **B. scabra,** *R. Br. in Ait. Hort. Kew. ed.* 2, iii. 12. A heath-like shrub, quite glabrous or the young branches minutely hoary. Leaflets 3, sessile on the stems, narrow-linear, with revolute margins, obtuse or with a hooked point, usually about ½ in. long, but varying from 4 to 8 lines. Flowers large, purple, solitary in the upper axils; pedicels shorter than the leaves. Calyx nearly 4 lines long, glabrous outside, sometimes slightly pubescent inside. Standard twice as long as the calyx; wings and keel shorter.

E 2

Ovary shortly stipitate, glabrous or pubescent; style flattened towards the base. Pod glabrous or shortly pubescent, scarcely exceeding the calyx.— DC. Prod. ii. 106 ; Meissn. in Pl. Preiss. i. 41 ; Bot. Mag. t. 5000 ; *Gompholobium scabrum*, Sm. in Trans. Linn. Soc. ix. 250 ; *B. sessilifolia*, DC. Prod. ii. 106 ; Deless. Ic. Sel. iii. t. 61 (with the leaflets rather shorter and slender) ; *B. pulchella*, Meissn. in Pl. Preiss. i. 41 ; Bot. Mag. t. 4392.

W. Australia. King George's Sound, *R. Brown* and others; near Albany, *Preiss, n.* 1177 : and eastward to Stokes Inlet, *Herb. F. Mueller ;* Swan River, *Drummond, n.* 199, *Preiss, n.* 1173, and others.

7. **B. conferta,** *DC. Prod.* ii. 106. A heath-like, glabrous shrub, with erect, virgate branches. Leaves crowded, all simple, narrow-linear or subulate, with closely reflexed margins, obtuse or shortly mucronate, rarely exceeding ½ in. in length. Flowers bluish-purple, usually numerous, in a short, dense, umbel-like, terminal raceme, the pedicels rarely longer than the calyx. Bracts small, subulate; bracteoles minute. Calyx 3 to 4 lines long, glabrous. Petals nearly equal in length and rarely exceeding the calyx; keel straighter than in the other species and shortly acuminate. Ovary shortly stipitate, glabrous; style flattened towards the base. Pod not exceeding the calyx, as broad as long.—Bot. Reg. t. 1600 ; Meissn. in Pl. Preiss. i. 42 ; Paxt. Mag. xii. 54, with a fig.

W. Australia. From King George's Sound, *R. Brown*, and Cape Riche to Swan River, *Drummond, 1st Coll. and n.* 197 *and* 198, *Preiss, n.* 1174, 1175, 1176, 1179, aud others; eastward to Bremer Bay and East Mount Barren, *Maxwell.*

It varies much in the leaves ; in some specimens rather thick, very obtuse, and about ¼ in. long ; in others slender, attaining sometimes above ½ in.

9. JACKSONIA, R. Br.

(Piptomeris, *Turcz.*)

Calyx deeply cleft, the tube usually very short, lobes valvate, the 2 upper ones broader, sometimes falcate, rarely connate. Petals shorter than the calyx or rarely exceeding it, nearly equal in length, the claws very short; standard orbicular or reniform, usually emarginate ; wings oblong ; keel nearly straight, obtuse, broader than the wings. Stamens free. Ovary sessile or stipitate ; style subulate, incurved, with a minute terminal stigma ; ovules 2 (except in *J. piptomeris*) attached by short funicles. Pod sessile or stipitate, ovate or oblong, flat or turgid. Seeds usually solitary, without any strophiole.—Shrubs or undershrubs, rigid and leafless, or rarely with a very few 1-foliolate leaves ; branches rigid, terete, angular or winged, the branchlets often phyllodineous or leaf-like, flat or terete or angular, very much branched and spinescent. Leaves replaced by very minute scales at the nodes. Flowers yellow or with an admixture of purple, either in terminal or lateral racemes or spikes, or scattered along the branches. Bracts small and scale-like. Bracteoles small, deciduous or persistent. Ovary villous.

The genus is limited to Australia. It is allied to *Gompholobium* and *Burtonia* in the deeply lobed valvate calyx, but very different in habit, in the short slender funicles of the ovules, in the pod, etc. In the two species (*J. vernicosa* and *J. thesioides*) where the calyx-tube is longer in proportion to the lobes, it is lined, at least halfway up, by the staminal disk.

Series I. **Phyllodineæ.**—*Phyllodineous branchlets flat, rigidly coriaceous, toothed or lobed, often prickly.*

Flowers in racemes or spikes terminating the phyllodia . . . . 1. *J. dilatata.*
Flowers in racemes or spikes lateral or terminating the normal
branches . . . . 2. *J. densiflora.*
Flowers in heads sessile amongst the phyllodia . . . . . . 3. *J. carduacea.*
Flowers mostly solitary on the teeth or lobes of the phyllodia . . 4. *J. floribunda.*

Series II. **Ramosissimæ.**—*Subphyllodineous branchlets crowded, linear, angular-striate, with projecting tooth-like nodes.*

Flowers in dense terminal spikes. Calyx very hirsute . . . . 5. *J. odontoclada.*
Flowers in short terminal racemes. Calyx sparingly silky-pubescent 6. *J. ramosissima.*

Series III. **Pungentes.**—*Branchlets more or less phyllodineous, spinescent, divaricate, terete angular or slightly compressed.*

Undershrubs, occasionally with a few leaves at the base, panicu-
lately branched upwards.
  Flowers on the pungent branches . . . . 7. *J. foliosa.*
  Flowers on virgate unarmed branches; pungent branches sterile
  (see below, *Scopariæ*).
Branching shrubs.
  Branchlets terete sulcate or slightly angular.
    Flowers 2 or 3 below the summits of the pungent branchlets,
      which are usually divaricately trichotomous . . . . 8. *J. spinosa.*
    Flowers in clusters or very short racemes, terminal or at the
      base of the pungent branchlets.
      Pungent branchlets mostly simple. Calyx minutely pubes-
      cent . . . . 9. *J. stricta.*
      Pungent branchlets mostly branched. Calyx silky-pubes-
      cent or villous . . . . 10. *J. hakeoides.*
    Flowers in distinct terminal racemes. Branchlets crowded . 11. *J. furcellata.*
  Branchlets very angular or compressed, usually silky-pubescent.
    Pungent branchlets short, crowded and forked or trichotomous.
      Pungent branchlets slender. Flowers distinctly racemose.
      Pod distinctly though shortly stipitate . . . . . . 11. *J. furcellata.*
      Pungent branchlets stout. Flowers in irregular racemes
      mixed with phyllodia. Pod nearly sessile . . . . 12. *J. horrida.*
    Pungent branchlets mostly simple and not crowded.
      Pod ovate, nearly sessile. Flowers shortly or loosely race-
      mose . . . . 13. *J. sericea.*
      Pod oblong on a stipes of about 2 lines. Flowers rather
      large, few, distant . . . . 14. *J. Sternbergiana.*
(*J. Lehmanni, racemosa, macrocalyx,* and perhaps some others among *Scopariæ,* have occasionally a few pungent barren branchlets near the base of the stem.)

Series IV. **Scopariæ.**—*Flowering branches virgate or rush-like without pungent branchlets. Barren stems or branches usually similar, although occasionally, in the lower part of the stem, much-branched flexuose or pungent.*

Calyx-tube half as long as the lobes or longer, turbinate, 10-nerved.
  Calyx 4 lines long, the tube half as long as the lobes . . . . 15. *J. vernicosa.*
  Calyx 2 lines long, the tube nearly as long as the lobes . . . 16. *J. thesioides.*
Calyx-tube very short, without prominent nerves.
  Calyx-lobes deciduous (or persistent in *J. restioides, umbellata,*
    and *capitata*). Buds not angular.
    Branches flat or angular.
      Pod sessile.
        Pod oblong-lanceolate, flat, ¾ in. long. Branches flat . 17. *J. compressa.*

Pod broad, turgid, 3 to 4 lines long. Branches usually
    angular . . . . 13. *J. sericea*.
Pod on a long stipes, about ½ in. long.
    Flowers racemose. Calyx 2½ to 3 lines long . . . . . 18. *J. scoparia*.
    Flowers few. Calyx 5 to 6 lines long . . . . 14. *J. Sternbergiana*.
Branches terete or sulcate.
    Flowers distant or racemose.
        Calyx villous. Standard large, exceeding the calyx.
            Flowers few, distant. Calyx 5 lines long. Keel much
                shorter than the wings . . . . 19. *J. restioides*.
            Flowers in lateral clusters. Calyx about 4 lines. Keel
                as long as the wings . . . . 20. *J. velutina*.
        Calyx silky-pubescent. Standard shorter than the calyx ;
            keel as long as the standard.
            Ovules 2.
                Calyx 4 to 5 lines long. Pod acuminate, on a stipes
                  of 1 line . . . . 21. *J. Lehmanni*.
                Calyx 3 lines long. Pod scarcely acute, almost sessile 22. *J. racemosa*.
            Ovules 4 to 6 . . . . 26. *J. piptomeris*.
    Flowers small, in terminal heads or umbels. Calyx-lobes
        persistent.
        Flowers distinctly pedicellate. Keel long . . . . . 23. *J. umbellata*.
        Flowers almost sessile. Keel short . . . . 24. *J. capitata*.
Calyx-lobes persistent. Buds prominently angled.
    Branches flat. Calyx 2 to 2½ lines long . . . . 25. *J. alata*.
    Branches terete, sulcate.
        Calyx about 4 lines long.
            Ovules 2. Keel very short . . . . 27. *J. angulata*.
            Ovules 4 to 6. Keel as long as the standard . . . . 26. *J. piptomeris*.
        Calyx about 6 lines long. Keel as long as the standard . 28. *J. macrocalyx*.

SERIES I. PHYLLODINEÆ.—Branchlets, either barren or flower-bearing,
phyllodineous, flat, rigidly coriaceous, toothed or lobed, often pungent.

1. **J. dilatata,** *Benth. in Ann. Wien. Mus.* ii. 74. An erect shrub, silky-
pubescent or tomentose, more or less rust-coloured under the inflorescence.
Branchlets leaf-like, flat, lanceolate, 2 to 4 in. long, the nodes forming tooth-
like notches on the edges with an arched nerve from the midrib to each node.
Flowers sessile in oblong spikes or heads at the ends of some of the leaf-like
branchlets, which taper more to the end than the barren ones. Bracts ovate,
scale-like, 1 to 1½ lines long, bracteoles often longer and lanceolate, all very
deciduous. Calyx villous, about 3 lines long. Standard about as long as
the calyx, lower petals shorter. Ovary very shortly stipitate. Pod ovate,
shorter than the calyx.

N. Australia. Melville Island, *Fraser;* Victoria river, *Bynoe;* Islands of the Gulf of
Carpentaria, *R. Brown;* Port Essington, *Armstrong;* Arnhem's Land, *F. Mueller.*

2. **J. densiflora,** *Benth. in Lindl. Swan Riv. App.* 13. A stout erect
or procumbent shrub, the young shoots and inflorescence densely rusty-villous,
older branches terete or slightly compressed, glabrous or minutely pubescent.
Barren leaf-like branchlets flat, obovate oblong or lanceolate, 1½ to 3 in. long,
usually rounded at the end and narrowed into a kind of petiole at the base,
the margins undulate, sinuate, with pungent teeth or short lobes. Racemes
or flowering branches either terminal or in the axil of a small scale under the
phyllodia, always very villous, 2 to 4 in. long. Flowers dense or distant,

sometimes intermixed with a few small phyllodia either barren or bearing 1 or 2 flowers. Bracts and bracteoles linear-subulate, deciduous. Pedicels very short. Calyx villous, about ½ in. long, the angles very prominent in the bud. Standard nearly as long as the sepals, lower petals shorter. Ovary very shortly stipitate. Pod ovate, included in the calyx.—*J. floribunda,* Meissn. in Pl. Preiss. i. 43, not of Endl.

W. Australia. Swan River, *Drummond, 1st Coll.*

Var. *laxiflora.* Racemes loose and rather less villous, the flowers distant.—*Drummond, 4th Coll. n.* 24.

3. **J. carduacea,** *Meissn. in Bot. Zeit.* 1855, 25. An erect shrub with virgate clustered branches, angular or somewhat compressed, silky or softly villous. Barren phyllodineous branchlets numerous, flat, sessile, cuneate-oblong, ½ to 1 in. long, prickly-toothed or sometimes forked, thick, usually 3-nerved, glabrous or silky-pubescent. Flowers in terminal heads within the uppermost phyllodineous branchlets and shorter than them. Pedicels very short. Bracts and bracteoles subulate-acuminate. Calyx about ½ in. long, loosely clothed with long silky hairs, the lobes with long subulate points. Petals nearly equal in length, shorter than the calyx. Ovary very shortly stipitate. Pod not seen.

W. Australia. Between Moore and Murchison rivers, *Drummond, 6th Coll. n.* 14.

4. **J. floribunda,** *Endl. Stirp. Decad. in Ann. Wien. Mus.* ii. 197 (*from the description*). A rigid erect shrub with forked branches, more or less silky-pubescent, or at length glabrous, more or less flattened or at length terete. Phyllodineous branchlets flat, cuneate oblong or linear, irregularly divided into deep divaricate rigid teeth or lobes, which are obtuse or terminate in a small scale, or the terminal ones almost pungent; most of these branchlets are narrowed as it were into a petiole, and when broad are strongly reticulate, the upper ones passing sometimes into forked racemes. Flowers mostly solitary on some of the tooth-like lobes of the branchlets. Pedicels very short. Bracts and bracteoles minute. Calyx 5 to 7 lines long, silky-pubescent. Standard and wings much shorter than the calyx; keel longer than the standard but shorter than the calyx. Pod nearly sessile, ovate, shorter than the calyx.—*J. grevilleoides,* Turcz. in Bull. Mosc. 1853, i. 259.

W. Australia, *Drummond, 4th Coll. n.* 32; between King George's Sound and Cape Riche, *Harvey;* Point Henry and Stirling Range, *Oldfield.* Endlicher describes the flowers as rather larger than I find them in our specimens, with the keel not exceeding the other petals; but the characteristic reticulate phyllodia and inflorescence are not found in any other species.

Series II. RAMOSISSIMÆ.—Barren branchlets subphyllodineous, but neither flat nor pungent, crowded, linear, angular-striate, with projecting tooth-like nodes.

5. **J. odontoclada,** *F. Muell. Herb.* Pubescent or villous with short loose hairs and densely branched. Barren branchlets short, crowded, linear, angular or somewhat flattened, the minute scales confluent with tooth-like projections at the nodes. Flowers sessile or nearly so in dense terminal spikes or heads. Bracteoles lanceolate, adnate to the base of the calyx, and longer than its tube. Calyx densely hirsute, 4 to nearly 5 lines long, the tube under

1 line, the 2 upper lobes often connate to the middle. Petals shorter than the calyx; keel shortly acuminate, broader than the wings and longer than the standard. Ovary sessile. Young pod densely villous, turgid, but not seen ripe.

N. Australia. Gulf of Carpentaria, *F. Mueller,* also *M'Douall Stuart's* expedition, lat. 17° 58

6. **J. ramosissima,** *Benth. in Mitch. Trop. Austr.* 258. Glabrous, very densely branched. Barren branchlets crowded, linear, angular or somewhat flattened, the small scales at the nodes often rigid and spreading. Flowers in short terminal racemes, on pedicels scarcely exceeding the subtending scales. Bracteoles ovate, adnate to the base of the calyx and often as long as its tube. Calyx sparingly pubescent, about 4 lines long, membranous, the 2 upper lobes broader, falcate and often cohering above the middle. Petals shorter than the calyx. Ovary sessile. Pod ovoid-oblong, acuminate, hirsute, shorter than the calyx.

Queensland. Sutton river, *F. Mueller*, Belyando river, *Mitchell.*

SERIES III. PUNGENTES.—Branchlets either barren or flower-bearing, spinescent, subphyllodineous but not flat, divaricate, simple or branched, terete or angular, sometimes resembling short lateral prickles, sometimes terminating all the branches.

7. **J. foliosa,** *Turcz. in Bull. Mosc.* 1853, i. 260. Rhizome thick, with several ascending or erect rigid stems of ½ to 1 ft., simple in the lower part and usually bearing a few petiolate obovate or orbicular toothed leaves, the upper portion leafless with numerous rigid divaricate paniculate branches, often slightly flattened and striate or angular, the whole plant minutely silky-pubescent, the ultimate branchlets divaricate and often pungent. Flowers few, shortly pedicellate, below the summits of the branchlets. Calyx 3 to 4 lines long. Petals and ovary of *J. spinosa.* Pod not seen.

W. Australia, *Drummond, 4th Coll. n.* 25. The species is nearly allied to *J. spinosa,* but with a different habit.

8. **J. spinosa,** *R. Br. in Ait. Hort. Kew. ed.* 2, iii. 13. A rigid much-branched shrub of 2 to 4 ft. or sometimes more slender and twice as high, glabrous or very minutely hoary, the branches angular or striate, the smaller ones trichotomous, divaricate rigid and pungent-pointed. Flowers rather small, usually 2 or 3 below the spinous point of some of the smaller branchlets. Pedicels short. Bracteoles minute. Calyx minutely silky-pubescent, about 3 lines long, the buds not angular. Petals about as long as the calyx and all of nearly equal length; keel deeply coloured. Ovary slightly contracted at the base. Pod sessile, oblong, usually falcate, 3 to 4 or even 5 lines long.— DC. Prod. ii. 107; Meissn. in Pl. Preiss. i. 45; *Gompholobium spinosum,* Labill. Pl. Nov. Holl. i. 107. t. 136.

W. Australia. King George's Sound, *Labillardière, R. Brown,* and others; towards Cape Riche, *Harvey, Drummond, 3rd Coll. n.* 81 *and* 103; and eastward to Espérance Bay, *Maxwell;* Two-peopled Bay and Gordon river, *Preiss, n.* 1074; Point Henry and Wilson's Inlet, *Oldfield;* Mount Barker, *Oldfield.* In the latter specimen the flowering branches are less spinous and the flowers more abundant, almost paniculate.

9. **J. stricta,** *Meissn. in Bot. Zeit.* 1855, 27. Possibly a small-flowered

variety of *J. hakeoides*, with smaller spines. Branches elongated, glabrous, striate and furrowed but without very prominent angles; phyllodineous branchlets terete, rigid, pungent, divaricate, rarely ½ in. long, simple or rarely with 1 or 2 forks. Flowers clustered under the phyllodineous branchlets as in *J. hakeoides*, but rather smaller and the calyx less pubescent.

W. Australia. Between Moore and Murchison rivers, *Drummond, 6th Coll. n.* 12; Murchison river, *Oldfield*.

10. **J. hakeoides,** *Meissn. in Pl. Preiss.* i. 45. Very nearly allied to some forms of *J. furcellata* and *J. horrida*, but less silky and sometimes almost glabrous. Branches striate, without very prominent angles and scarcely flattened; phyllodineous branchlets terete or very slightly flattened, simple or once or twice forked, very rigid, striate, usually but not always pungent. Flowers of *J. horrida*, in short racemes, often reduced to almost sessile clusters at the base of the pungent phyllodineous branchlets, rarely terminal. Pedicels short, with small lanceolate bracteoles about the middle. Calyx about 5 lines long, velvety or silky-pubescent with shining hairs often of a golden colour. Petals and ovary of *J. horrida*. Pod not seen.—*J. fasciculata*, Meissn. in Pl. Preiss. ii. 212; *J. ulicina*, Meissn. in Bot. Zeit. 1855, 26.

W. Australia, *Drummond, n.* 261, *2nd Coll. n.* 113, *3rd Coll. n.* 88, *and 6th Coll. n.* 13. This species varies much in the phyllodineous branchlets, simple or once or twice forked, very stout or slender, in some specimens altogether 2 or 3 lines long, in others the ultimate forks ½ to ¾ in., but the several forms described by Meissner often pass insensibly one into the other even on the same specimen.

11. **J. furcellata,** *DC. Prod.* ii. 107. A tall shrub, more or less silky-pubescent or sometimes shortly villous, attaining sometimes 15 ft. or more in height, the branches then usually pendulous. Smaller phyllodineous branchlets crowded, compressed and narrow-linear or angular, branched or forked, usually but not always pungent, short and divaricate or long and erect. Flowers in terminal racemes of 2 to 4 or even 5 in., without any barren branchlets intermixed except quite at the base. Pedicels short. Bracteoles very small, at a distance from the calyx. Calyx 4 to 5 lines long. Petals shorter, nearly equal in length. Pod stipitate, ovate, turgid, almost acute, about as long as the calyx.—*Gompholobium furcellatum*, Bonpl. Jard. Malm. 30. t. 11; *J. dumosa*, Meissn. in Pl. Preiss. i. 44, and ii. 212.

W. Australia, *Baudin's Expedition;* King George's Sound to Swan River, *Fraser, Drummond, Preiss, n.* 1069, 1070, 1071, *and* 1081; *Drummond, 3rd Coll. n.* 80 (branchlets short slender divaricate), *2nd Coll. n.* 112 (branchlets softly villous, short, slender, much-branched and crowded, terminal racemes long and many-flowered), *Maxwell, Oldfield,* and others (both forms). The species is very near *J. horrida,* differing in the more slender phyllodia, more racemose flowers, shorter more stipitate pod, etc.

12. **J. horrida,** *DC. Prod.* ii. 107. A tall much-branched shrub, more or less silky-pubescent. Branches virgate, very angular or flattened when young; phyllodineous branchlets crowded, often 2 or 3 times forked, the ultimate forks linear, pungent, rigid, flat with a raised nerve, usually ½ to ¾ in. long and quite entire, sometimes long and slender, or short and broad, rarely with 1 or 2 notch-like teeth. Flowers either 1, 2 or 3, at the base of some of the smaller phyllodineous branchlets, or forming irregular racemes intermixed with phyllodia. Pedicels shorter than the calyx with minute bracteoles.

Calyx fully 4 lines long, silky-pubescent. Standard nearly as long as the calyx, wings and keel shorter. Pod shortly stipitate, oblong, turgid, somewhat falcate, shortly acuminate, 3 to 5 lines long.—Meissn. in Pl. Preiss. i. 45; ii. 212.

W. Australia. King George's Sound and adjoining districts, *R. Brown, Fraser, Drummond, 3rd Coll. n.* 79, *Preiss, n.* 1080, and others; Point Possession, *Collie;* Tone river.

Var. (?) *tenuis.* Branches much more slender and inflorescence nearly that of *J. hakeoides,* but the branchlets much more angular.—Gardner and Salt rivers, *Maxwell.* The habit is almost that of *J. spinosa,* but the flowers are larger, the calyx more silky, etc.

Var. (?) *racemosa.* Racemes elongated, many-flowered. Champion Bay, *Oldfield,* a single small specimen.

13. **J. sericea,** *Benth. in Hueg. Enum.* 31, *and in Ann. Wien. Mus.* ii. 74. A large shrub, decumbent ascending or tall with pendulous branches, angular striate or flattened, either unarmed or the pungent phyllodineous branchlets linear, spreading, 2 to 4 lines long, simple or rarely with 1 or 2 short forks. Flowers either solitary near the base of some of the smaller branches or in loose terminal irregular racemes. Pedicels shorter than the calyx with very minute bracteoles. Calyx $3\frac{1}{2}$ to $4\frac{1}{2}$ lines long, silky-pubescent. Petals all shorter than the calyx. Pod nearly sessile, broadly ovate, more or less turgid, very villous, 3 to 4 lines long.—Meissn. in Pl. Preiss. i. 44, and ii. 211; *J. gracilis,* Meissn. l. c. i. 44, and ii. 211.

W. Australia. King George's Sound, *Huegel, Baxter;* Swan River, *Fraser, Drummond, 3rd Coll. n.* 92; *Preiss, n.* 1073 and 1158; southern districts, *Oldfield.*

Var. *robusta.* Branches thick, with very prominent angles, *Drummond, 3rd Coll. n.* 85. The species is very nearly allied to *J. horrida.*

14. **J. Sternbergiana,** *Hueg. Bot. Archiv. t.* 3. An erect shrub, attaining many ft. in height, with pendulous branches prominently angled or flattened, glabrous or rarely minutely pubescent, the smaller branchlets rigid and pungent, but usually rather slender and divaricate. Flowers few, shortly pedicellate along some of the longer branchlets, but scarcely racemose. Calyx nearly glabrous, 5 to 6 lines long. Standard nearly as long : lower petals shorter. Ovary stipitate. Pod oblong, often slightly falcate, fully $\frac{1}{2}$ in. long, pubescent, on a stipes of about 2 lines.—Meissn. in Pl. Preiss. i. 43, and ii. 211.

W. Australia. Swan River, *Fraser, Huegel, Drummond, 1st Coll. and n.* 260 (very slender), also *2nd Coll. n.* 111, *Preiss, n.* 1076, *Oldfield,* and others. The pungent branchlets in this, as in *J. sericea,* are occasionally wanting.

SERIES IV. SCOPARIÆ.—Flowering branches usually virgate or rushlike, without pungent branchlets. Barren stems or branches usually similar, although occasionally in the lower part of the stem much-branched, flexuose or pungent.

15. **J. vernicosa,** *F. Muell. Herb.* Branches very numerous, slender, striate but scarcely angled, usually glabrous, the young shoots apparently somewhat glutinous. Flowers not numerous, in terminal racemes, on very short pedicels. Bracteoles minute, adnate to the base of the calyx. Calyx 4 lines long, slightly pubescent, the lobes not twice as long as the tube, the 2 upper ones broader and falcate, the tube campanulate, 10-ribbed, lined nearly to the

top by the staminal disk. Petals shorter than the calyx. Ovary sessile, short, tapering into the style. Pod sessile, villous, turgid, shorter than the calyx, but exposed, owing to the calyx-lobes being reflexed or deciduous.

N. Australia. Gulf of Carpentaria, *F. Mueller.*

16. **J. thesioides,** *A. Cunn.; Benth. in Ann. Wien. Mus.* ii. 74. A tall shrub, with the habit of *J. scoparia.* Branches numerous, elongated, not spinescent, with 2 or 3 very prominent angles, minutely silky-hoary or quite glabrous. Flowers yellow, smaller than in *J. scoparia,* sessile or shortly pedicellate in terminal one-sided racemes. Bracteoles minute, just below the calyx. Calyx about 2 lines long, minutely silky-hoary, the lobes a little longer than the tube, which is 10-nerved and lined at least halfway up by the staminal disk. Standard and wings about as long as the calyx ; keel a little shorter. Ovary sessile, tapering into a short style. Pod oblong, somewhat incurved, turgid, 2 to 2½ lines long.

N. Australia. Prince of Wales's Islands, etc., *R. Brown;* Victoria river and Macadam range, *F. Mueller.*
Queensland. Endeavour river, *Banks and Solander, A. Cunningham, W. Hill;* Cape Flinders, *A. Cunningham;* Lizard Island, *M'Gillivray.*

17. **J. compressa,** *Turcz. in Bull. Mosc.* 1853, i. 260. Stem erect, virgate ; branches very flat, minutely silky-pubescent, without pungent branchlets. Flowers few, very shortly pedicellate and distant along the upper branchlets. Bracteoles very small, ovate, under the calyx. Calyx about 4 to 5 lines long, shortly silky-villous, the lobes deciduous. Petals shorter, the keel at least as long as the standard. Ovary sessile. Pod sessile, oblong-lanceolate, much flattened, about ¾ in. long, minutely glandular and pubescent with very short close hairs. Funicles slender.

W. Australia, *Drummond, 5th Coll. n.* 35 and 36.

18. **J. scoparia,** *R. Br. in Ait. Hort. Kew. ed.* 2, iii. 13. A tall shrub or small tree, usually entirely leafless, but occasionally the young plants or the base of the branches have a few petiolate, oblong or oval-elliptical, herbaceous leaves, ¾ to 2 in. long. Branches numerous, erect or pendulous, elongated, not spinescent, angular, glabrous or minutely hoary-pubescent. Flowers yellow, in one-sided racemes, either terminal or from the upper nodes. Pedicels rarely as long as the calyx, without bracteoles. Calyx membranous, minutely silky-hoary, 2½ to 3 lines long, divided nearly to the base. Standard rather longer than the calyx, lower petals rather shorter. Ovary stipitate. Pod flat, oblong, usually 4 to 6 lines long, on a stipes of 1 to 2 lines, tipped by the persistent style.—DC. Prod. ii. 107 ; Lodd. Bot. Cab. t. 427 ; *J. macrocarpa,* Benth. in Ann. Wien. Mus. ii. 74 (the filaments erroneously described as toothed) ; *Viminaria lateriflora,* Link, Enum. Hort. Berol. i. 403 ; DC. Prod. ii. 107 (from the short description given).

Queensland. Sandy Cape, Broad Sound, *R. Brown;* Port Curtis, *M'Gillivray;* Burnitt river, *F. Mueller;* Barcoo river, *Mitchell;* Brisbane river and Moreton Bay islands, *F. Mueller;* Pine river, *Fitzalan;* Rockhampton, *Dallachy.*
N. S. Wales. Port Jackson to the Blue Mountains, *R. Brown, Sieber, n.* 372, and others ; northwards to the Hastings and Macleay rivers, *Beckler;* New England, *C. Stuart.*
Var. *parviflora.* Flowers, especially the calyx, much smaller.—Macleay river, *Beckler.*

and Clarence river, *Wilcox;* also in the collection of Sydney woods, of the Paris Exhibition, n.º 117.

Var. *macrocarpa.* Pod rather longer, at length nearly glabrous. I can find no other difference.—*J. cupulifera,* Meissn. in Bot. Zeit. 1855, 27.

W. Australia. Between Moore and Murchison rivers, *Drummond,* 6*th Coll. n.* 11; Murchison river, *Oldfield.*

19. **J. restioides,** *Meissn. in Pl. Preiss.* i. 46. Stems erect, branched, virgate, rush-like, terete or nearly so, sulcate, silky-pubescent when young, at length glabrous, 1 to 2 ft. high, the barren branches rigid but not spinescent. Flowers few, on very short pedicels along the upper branches or in terminal racemes. Bracteoles small, lanceolate, below the calyx. Calyx silky-villous, about 5 lines long, the tube nearly one-third as long as the lobes but without prominent ribs, the lobes apparently persistent but not seen in fruit. Standard large, shortly exceeding the calyx ; wings shorter ; keel much shorter. Ovary stipitate. Pod not seen.

W. Australia. Swan River, *Drummond,* 1*st Coll. and n.* 263; *Preiss, n.* 1079; Darling range, *Collie.*

20. **J. velutina,** *Benth.* Branches in our specimens elongated, simple, rather thick, sulcate but scarcely angular, shortly velvety-pubescent, without barren branchlets. Flowers in lateral clusters or very short racemes. Bracteoles small, subulate. Calyx pubescent or shortly villous, nearly 4 lines long, the tube very short, the lobes deciduous. Standard exceeding- the calyx, but not so large as in *J. restioides ;* wings and keel shorter and nearly equal to each other. Ovary stipitate, densely covered with long hairs. Pod not seen.

W. Australia. Swan River, *Oldfield.* The specimens insufficient for a full description, but not referable to any other species.

21. **J. Lehmanni,** *Meissn. in Pl. Preiss.* i. 46. A shrub or undershrub of ½ to 1½ ft., with erect branching stems, silky-pubescent when young, at length glabrous, terete or slightly compressed, striate, sulcate or almost angular, all leafless and unarmed, or rarely with one or two short, simple or divaricately-forked, spinescent, barren branchlets near the base. Flowers few, along the upper slender branches. Pedicels as long as the calyx or rather shorter, with very minute bracteoles about the middle. Calyx minutely silky-pubescent, 4 to 5 lines long, the lobes deciduous and the buds not prominently angular. Petals shorter than the calyx, the keel about as long as the standard. Ovary stipitate. Pod ovate, acuminate, 5 to 6 lines long, on a stipes of about 1 line.

W. Australia. King George's Sound, *R. Brown* (with short pedicels) ; Swan River, *Drummond,* 1*st Coll. and n.* 89 ; Canning river, *Preiss, n.* 1077 ; towards the Great Bight, *Maxwell.* This plant seems to connect *J. racemosa* with *J. angulata,* differing from the former in its larger flowers, scattered along the branches rather than truly, racemose, and larger pods ; and from *J. angulata* in the calyx not nearly so angular in the bud, with deciduous lobes, and in the keel as long as the standard, as in *J. racemosa.* The southern specimens have the branches slightly striate, in the Swan River ones they are more sulcate or almost angular.

22. **J. racemosa,** *Meissn. in Pl. Preiss.* ii. 212. An undershrub or shrub of ½ to 1 ft., with erect virgate branching stems, silky-pubescent when young, at length glabrous, terete or slightly compressed, sometimes scarcely

striate, more frequently more or less sulcate, all leafless and unarmed, or with a few simple or branched, spinescent, barren branchlets near the base. Flowers usually numerous on short pedicels, in terminal racemes either short and dense or long and interrupted. Calyx about 3 lines long, silky-pubescent, with deciduous lobes, the bud terete. Petals rather shorter than the calyx, the deeply-coloured keel as long as the yellow standard. Ovary very shortly stipitate. Pod ovate, less turgid than in many species, though not so flat as in *J. scoparia,* scarcely acuminate, 3 to 4 lines long, on a very short stipes.

W. Australia, *Drummond, 3rd Coll. n.* 78; Stirling range, Phillips river, Cape Arid, and W. Mount Barren, *Maxwell;* Tone river, *Oldfield.*

Var. *pubiflora.* Pubescence of the young branches, inflorescence, and calyx less appressed, and more dense. Barren branchlets more spinescent.—*Drummond.*

23. **J. umbellata,** *Turcz. in Bull. Mosc.* 1853, i. 261. A shrub or undershrub, with erect or ascending, rush-like stems, in some specimens under 1 ft. and much branched, in others more simple and 1½ to 2 ft. high, silky-pubescent when young, terete or slightly compressed and rarely sulcate. Barren branchlets sometimes none or few, small, and simple, sometimes short, much branched, bushy, and almost spinescent or much incurved. Flowers several, in terminal umbels, with rarely 1 lower down. Pedicels shorter than the calyx. Bracteoles very small. Calyx silky-pubescent, 2½ to nearly 4 lines long, the buds acuminate, scarcely angular, the lobes persistent. Petals rather shorter than the calyx, the keel deeply coloured and at least as long as the standard. Ovary shortly stipitate. Pod nearly sessile, ovate, acute, about as long as the calyx.—*J. capitata,* Benth. in Ann. Wien. Mus. ii. 74, not of Meissn. (with rather shorter pedicels); *J. juncea,* Turcz. in Bull. Mosc. 1853, i. 261 (flowering branches elongated, barren branches often wanting.

W. Australia. S. coast, *R. Brown,* also *Drummond, 5th Coll. n.* 33 *and* 34; Cape Riche to Kalgan river, *Oldfield;* Stokes Inlet and Cape Legrand, *Maxwell.* It varies much in the size of the flowers.

24. **J. capitata,** *Meissn. in Pl. Preiss.* i. 45 *and* ii. 212, *not of Benth.* An erect shrub or undershrub, with virgate rush-like or broom-like stems, ¾ to 1½ ft. high, silky-pubescent when young, the barren branches rigid and often recurved, but rarely spinescent. Flowers nearly sessile, in very short terminal heads or spikes, with very rarely 1 or 2 lower down. Bracts and bracteoles ovate or ovate-lanceolate, sometimes nearly 1 line long. Calyx about 3 lines long, pubescent or almost villous, the buds not angular and less slender than in most species, the lobes persistent. Standard about as long as the calyx; keel shorter. Ovary sessile. Pod sessile, ovoid, shorter than the calyx.

W. Australia, *Drummond, 3rd Coll. n.* 88; Peel district, *Preiss, n.* 1078; Gardner ranges, *Herb. F. Mueller;* between Eyre ranges and Oldfield ranges, *Oldfield.*

25. **J. alata,** *Benth. in Hueg. Enum.* 30, *and in Ann. Wien. Mus.* ii. 74. An erect much-branched shrub or undershrub, of ½ to 1 ft., glabrous or the young parts slightly silky-pubescent, the branches flattened with 2 narrow wings descending from the distant tooth-like nodes; barren

branches not spinescent. Flowers small, nearly sessile, in short terminal racemes. Bracteoles minute. Calyx slightly silky-pubescent, about 3 lines long, the buds prominently angular, the lobes persistent. Standard and wings nearly as long as the calyx; keel rather shorter. Ovary very shortly stipitate, with a short incurved style. Pod almost sessile, ovate, much compressed, shorter than the calyx.

W. Australia. Swan River, *Huegel, Drummond,* 1*st Coll., also n.* 193, *and* 3*rd Coll. n.* 32, and others; in the interior, *Preiss, n.* 1119.

26. **J. piptomeris,** *Benth.* A rigid erect broom-like shrub, with virgate branches, silky-pubescent when young, usually terete and striate, without spinescent branchlets in any specimens seen. Flowers scattered along the upper branches. Pedicels much shorter than the calyx, with minute bracteoles. Calyx minutely silky-pubescent, 3 to 4 lines long, slightly angular in the bud, the lobes deciduous. Petals nearly as long as the calyx, the deeply-coloured keel as long as the yellow standard. Ovary stipitate, with 4 to 6 ovules; style slender. Pod on a rather long stipes when young, but not seen fully formed.—*Piptomeris aphylla,* Turcz. in Bull. Mosc. 1853, i. 258.

W. Australia. Towards Cape Riche, *Drummond,* 5*th Coll. n.* 32. This species closely resembles *J. racemosa* in aspect, but the calyx is rather more angular in the bud, like the smaller forms of *J. angulata.* The lobes are deciduous and the keel as long as the standard, as in *J. racemosa;* and the increased number of ovules distinguishes it from all other *Jacksonias.*

27. **J. angulata,** *Benth.* An undershrub, with numerous ascending or erect stems of ½ to 1 ft., silky-pubescent when young, sometimes nearly terete or sulcate, more frequently angular or somewhat flattened, nearly simple or branching, or very flexuose. Flowers few, very shortly pedicellate on some of the lower or intermediate nodes of the more simple and straighter branches. Bracteoles very small. Calyx silky-pubescent, about 4 lines long, the buds very prominently angled, the lobes apparently persistent. Petals shorter than the calyx, the keel shorter than the others. Ovary nearly sessile. Pod not seen fully formed.—*J. Lehmanni,* var., Meissn. in Pl. Preiss. i. 46 (as to Drummond's specimens).

W. Australia, *Drummond, n.* 262, *and* 5*th Coll. n.* 143 (the latter with smaller flowers).

28. **J. macrocalyx,** *Meissn. in Bot. Zeit.* 1855, 26. Very near *J. angulata,* but larger in all its dimensions, the rigid ascending stems often above 1 ft. high, the shorter ones divaricately dichotomous, with the ultimate branchlets spinescent, all sulcate or striate but scarcely angular or compressed, the lower part of the stem occasionally bearing a few obovate-oblong leaves of ½ to 1 in. Flowers few, lateral. Pedicels very short, with 2 small bracteoles. Calyx pubescent, ½ in. long, the angles very prominent in the bud, the lobes persistent. Petals not seen perfect, but the standard appears to be at least as long as the calyx. Pod ovate-acuminate, rather turgid, nearly sessile, shorter than the calyx.

W. Australia. Between Moore and Murchison rivers, *Drummond,* 6*th Coll. n.* 15.

10. SPHÆROLOBIUM, Sm.

(Roea, *Hueg.*)

Calyx-lobes imbricate, the two upper ones larger, falcate, united into an upper lip. Petals with short claws; standard orbicular or reniform, emarginate; wings rather shorter, oblong, usually falcate; keel longer or rather shorter than the wings, straight or curved. Stamens free. Ovary stipitate; style much incurved, subulate or dilated at the base, usually with a longitudinal membrane or a ring of hairs under the stigma; ovules 2, with short thick funicles. Pod small, stipitate, oblique, globular or compressed. Seeds 1 or 2, not strophiolate.—Glabrous shrubs or undershrubs, with rush-like stems, óften leafless. Leaves, when present, narrow, entire, alternate or irregularly opposite or whorled. Flowers yellow or red, in terminal racemes or in lateral racemes or clusters. Filaments of the outer stamens often somewhat dilated below the middle. Ovary always glabrous.

The genus is limited to Australia. It is readily known by its habit, by the small stipitate nearly globular pod, and, in all species except *S. medium* and *S. euchilus*, by the appendages of the style; *S. euchilus* has also a somewhat different habit, but is nearer to this than to any other genus.

SECT. I. **Roea.**—*Calyx-tube narrow-turbinate, longer than the lobes. Stigma surrounded by a ring of white hairs. Flowering stems more or less leafy, at least at the base. Racemes terminal.*

Calyx 3 to 4 lines long.
  Leaves narrow-linear. Bracts persistent, at least the lower ones . 1. *S. linophyllum.*
  Leaves lanceolate or cuneate. Bracts none . . . . 2. *S. nudiflorum.*
Calyx scarcely 2 lines long. Leaves very few and small. Bracts
none . . . . 3. *S. grácile.*

SECT. II. **Eusphærolobium.**—*Style with a longitudinal wing under the stigma, or without any appendage. Flowering stems leafless.*

Calyx-tube narrow-turbinate, longer than the lobes. Racemes lateral 4. *S. racemulosum.*
Calyx-tube about as long as the upper lip. Flowers lateral, in clusters
  or solitary.
  Stems winged . . . . 5. *S. alatum.*
  Stems terete.
    Keel broad and obtuse, slightly curved.
      Calyx not 2 lines long. Keel about as long as the wings . . 6. *S. vimineum.*
      Calyx 3 lines long. Keel longer than the wings . . . . 7. *S. grandiflorum.*
      Keel very much curved. Calyx 1½ lines long . . . . . . 8. *S. fornicatum.*
    Calyx-tube not above half as long as the upper lip.
    Keel longer than the standard.
      Keel narrow, curved, more or less acuminate, about 3 lines long 9. *S. medium.*
      Keel broad, straight, very obtuse, nearly 6 lines long . . . . 10. *S. scabriusculum.*
    Keel broad and obtuse, not longer than the standard.
      Branches erect, rigid, sometimes spinescent, without lateral
      branchlets . . . . 11. *S. macranthum.*
      Branchlets divaricate or recurved, spinescent . . . . . . 12. *S. daviesioides.*

SECT. III. **Euchiloides.**—*Calyx-tube very short. Style subulate, without any appendage. Flowering stems leafy.*

Leaves narrow-linear. Flowers axillary . . . . 13. *S. euchilus.*

SECT. I. **Roea.**—Calyx-tube narrow-turbinate, longer than the lobes. Stigma surrounded at the base by a membrane or fringe of hairs. Flowering stems leafy or leafless.—*Roea*, Hueg.; Benth. in Hueg. Enum. 34, and in Ann. Wien. Mus. ii. 77.

1. **S. linophyllum,** *Benth.* Stems numerous, from a thick stock, ascending or erect, ¾ to 1½ ft. high, not much branched, terete or slightly angular. Leaves narrow-linear, not numerous and sometimes very few, the larger ones ½ to 1 in. long and rather thick, the upper ones passing into the bracts. Flowers (yellow?) in a loose terminal raceme, each one in the axil of a leafy lanceolate bract of 1 to 3 lines; pedicels rather shorter than the calyx. Bracteoles small, deciduous. Calyx fully 3 lines long, the narrow turbinate tube longer than the lobes. Standard half as long again as the calyx; wings nearly as long, much falcate; keel not exceeding the calyx, very obtuse, the edges fringed with white hairs. Outer filaments flattened. Style subulate, folded inwards near the end, bearded with a ring of white hairs under the stigma. Pod ovoid, inflated, very oblique, nearly 3 lines long.—*Roea linophylla*, Hueg. Enum. 34; Benth. in Ann. Wien. Mus. ii. 77; Meissn. in Pl. Preiss. i. 58.

W. Australia. King George's Sound, *R. Brown, Huegel, Preiss, n.* 1115 *a, and* 1121; Swan River, *Huegel, Drummond,* 1*st Coll. and n.* 252, *Preiss, n.* 880.

2. **S. nudiflorum,** *Benth.* A glabrous undershrub, with the habit of *S. linophyllum.* Stems several from a thick stock, ½ to 1 ft. high, simple or little branched, sulcate or slightly angular. Leaves more numerous than in *S. linophyllum,* lanceolate or linear-cuneate, acute or rather obtuse, rarely above ½ in. long. Flowers in a terminal raceme, without bracts or bracteoles. Pedicels very short. Calyx about 5 lines long, the narrow turbinate tube longer than the lobes. Standard much longer than the calyx; keel about as long as the calyx, obtuse, not fringed, the petals soon separating. Outer filaments slightly flattened. Style folded near the end, bearded with a ring of white hairs under the stigma. Pod nearly globular, 2 to 3 lines diameter. —*Roea nudiflora,* Meissn. in Pl. Preiss. i. 59; *Sphærolobium foliosum,* F. Muell. Fragm. i. 166.

W. Australia. King George's Sound, *R. Brown,* and thence to Cape Riche, *Drummond,* 3*rd Coll. n.* 82, *Preiss, n.* 1018, and to Cape Arid and Cape Legrand, *Maxwell.*

3. **S. gracile,** *Benth.* Stems from a thick rootstock, numerous, slender, not above 6 in. high in our specimens, with a very few short linear leaves near the base, or some quite leafless. Flowers small, distant, in a slender raceme occupying more than half the stem, without any bracts. Pedicels recurved, shorter than the calyx. Calyx scarcely 2 lines long, the lobes much shorter than the narrow turbinate tube, and less unequal than in the other species. Standard very broad, half as long again as the calyx; wings shorter; keel still shorter, broad and almost truncate. Style folded in at the end, bearded with a ring of white hairs round the stigma. Pod ovoid, inflated, not very oblique, scarcely 2 lines long.

W. Australia. Sand plains N. of Murchison river, *Oldfield.*

SECT. II. EUSPHÆROLOBIUM.—Calyx-tube short or rarely longer than the upper lip. Style either with a longitudinal membranous wing or appendage below the stigma on the inner edge, or without any appendage. Flowering stems leafless.

4. **S. racemulosum,** *Benth.* Stems apparently herbaceous, slender, wiry, terete, above 1 ft. long, all quite leafless in our specimens. Racemes lateral, slender, 1 to 2 in. long, with occasionally a few minute bracts near the base. Pedicels short, slender, solitary at each scar or bract. Calyx rather above 2 lines long, the tube, including the narrow turbinate base, longer than the lobes. Standard fully twice as long as the calyx; lower petals shorter than the standard, about equal in length, the keel much incurved, broad and very obtuse. Style much incurved, folded in at the end as in *Roea,* but with a very broad longitudinal appendage under the stigma, as in *Eusphærolobium.* Pod stipitate, ovoid or nearly globular, 2 lines long.

W. Australia, *Drummond,* 3rd *Coll. n.* 76; Phillips river, *Maxwell.* This species seems very closely to connect the *Roeas* with the true *Sphærolobiums,* having the calyx and petals of the former with the style nearly of the latter, and an inflorescence different from both.

5. **S. alatum,** *Benth. in Hueg. Enum.* 32, *and in Ann. Wien. Mus.* ii. 76. Stems erect, slightly branched, 2 ft. high or more, very angular or winged, the wings of the lower portion occasionally 4 or 5 lines broad, the others very narrow or reduced to prominent angles. Leaves none on any of our specimens. Racemes lateral towards the ends of the branches, sometimes very short and 3- or 4-flowered, sometimes ¾ in. long, with 8 to 10 flowers. Calyx rather more than 2 lines long, usually marked with black streaks or blotches, the tube rather shorter than the upper lip. Standard ½ in. broad; keel obtuse, as long as the wings. Style curved, with a short, broad, membranous vertical wing under the stigma. Pod about 2 lines diameter.—*S. stenopterum,* Meissn. in Pl. Preiss. i. 57.

W. Australia. King George's Sound, *Fraser, Huegel, Oldfield;* southern districts, *Drummond,* 5th *Coll. n.* 48; near Albany, *Preiss, n.* 1117; Hay and Phillips rivers, *Maxwell.* In my former description the character of the wingless style had by some error been transferred from *S. medium* to this species, which misled Meissner in distinguishing his *S. stenopterum.*

6. **S. vimineum,** *Sm. in Ann. Bot.* i. 509, *and in Trans. Linn. Soc.* ix. 261. Stems ascending or erect, from a few inches to above 2 ft. high, with slender, terete, wiry branches, all leafless or the barren branches bearing a few, scattered, linear or narrow-lanceolate leaves, rarely exceeding ¼ in. in length. Flowers numerous, usually clustered 2 or 3 together along the smaller branches, forming dense or interrupted terminal racemes. Pedicels very short. Calyx 1½ to nearly 2 lines long, the tube about as long as the upper lip. Petals about twice as long as the calyx; keel somewhat incurved, very obtuse, as long as the wings. Style much curved from near the base, with a long, narrow wing along the inner edge. Pod scarcely 2 lines diameter.—Bot. Mag. t. 969; DC. Prod. ii. 108; Lodd. Bot. Cab. t. 1753; Hook. f. Fl. Tasm. i. 84; *S. minus,* Labill. Pl. Nov. Holl. i. 108, t. 138.

N. S. Wales. Port Jackson, *R. Brown* and others; Blue Mountains, *Miss Atkinson;* and northward to Hastings river, *Beckler.*

Victoria. Heathy places in the southern districts, *Adamson, Robertson;* moist places, Dandenong Mountains, *F. Mueller;* Wimmera, *Dallachy.*

Tasmania, *R. Brown;* abundant, usually in marshy or grassy places, in many parts of the colony, *J. D. Hooker.*

S. Australia. Encounter Bay, *Whittaker;* Lofty Ranges, *F. Mueller.*

7. **S. grandiflorum,** *R. Br. Herb.; Benth. in Hueg. Enum.* 32, *and in Ann. Wien. Mus.* ii. 76. Stems 1 to 3 ft. high, terete, rather thick, not striate, all leafless in our specimens. Flowers red, usually in pairs, in the axils of truncate, scale-like bracts, in rather dense, terminal racemes. Calyx 3 lines long, the tube nearly as long as the lobes, which are often bordered with black. Standard nearly twice as long as the calyx, very broad and deeply emarginate; wings much shorter; keel longer than the wings, broad, very obtuse, not much curved. Style much curved, with a rather short and broad longitudinal wing immediately under the stigma, on the inner edge. Pod rather broader than long.—Meissn. in Pl. Preiss. i. 57.

W. Australia. King George's Sound, *Menzies, Huegel,* and others; near Albany, *Preiss, n.* 1116, *also Drummond, 2nd Coll. n.* 116; bogs on the Vasse river, *Oldfield; Phillips river, Maxwell.*

8. **S. fornicatum,** *Benth. in Hueg. Enum.* 32, *and in Ann. Wien. Mus.* ii. 76. Stems erect, terete, scarcely striate, usually more slender than in *S. vimineum,* and in our specimens all leafless. Flowers rather small, in terminal racemes, solitary or in clusters of 2 or 3, on very short pedicels. Calyx 1½ lines long, the tube about as long as the upper lip. Petals half as long again as the calyx, the standard rather longer than the others; keel very broad, much incurved, forming a prominent obtuse angle on the back, the front very broad and obtuse. Style abruptly incurved above the base, with a longitudinal membrane under the stigma on the inner edge, usually rather short and broad. Pod not seen.—Meissn. in Pl. Preiss. i. 58.

W. Australia. Swan River, *Huegel, Mangles, Drummond, 1st Coll., Preiss,n.* 1115 *and* 1124; Mount Melville, Plantagenet district, *Preiss, n.* 1122; Robertson's Brook, *Maxwell. S. medium,* Meissn. in Pl. Preiss. i. 58, not of R. Br., appears to be only a slight variety with rather smaller flowers.

9. **S. medium,** *R. Br. in Ait. Hort. Kew. ed.* 2. iii. 14. Stems erect, 1 to 2 ft. high, usually more striate than in *S. vimineum.* Leaves on the barren branches small, subulate, often opposite or in whorls of 3, the flowering stems leafless. Flowers usually numerous, densely clustered, in terminal racemes. Calyx about 2 lines long, the tube not half so long as the upper lip. Standard orbicular, rather longer than the calyx; wings at least as long; keel rather longer than the wings, slightly curved, more or less acuminate. Style slightly curved, tapering from the dilated base to the end, but not winged. Pod fully 2 lines diameter. Seeds mottled.—DC. Prod. ii. 108; *S. acuminatum,* Benth. in Hueg. Enum. 32, and in Ann. Wien. Mus. ii. 76; Meissn. in Pl. Preiss. i. 58.

W. Australia. King George's Sound, *Menzies, R. Brown,* and others; and thence to Swan River, *Huegel, Drummond, 1st Coll., and n.* 216 *and* 218, *Preiss, n.* 1111, 1120, 1123, *and* 1126, and others.

10. **S. scabriusculum,** *Meissn. in Pl. Preiss.* ii. 214. Stems rather firm, terete, striate or slightly sulcate, and often scabrous with minute raised dots, leafless in our specimens. Flowers in rather loose terminal racemes, usually solitary within the minute, scale-like, truncate bracts, pendulous, and remarkable for their large keel, usually bordered by a pale colour. Calyx about 4 lines long, the tube about half as long as the upper lip. Standard and wings rather longer than the calyx; keel usually about 6 lines

long, broadly obovate. Style slender at the base, much curved upwards, with a very narrow longitudinal membrane on the inner edge.

W. Australia, *Drummond, 2nd Coll. n.* 114.

11. **S. macranthum,** *Meissn. in Pl. Preiss.* ii. 213. Stems more rigid and thicker than in *S. vimineum*, erect, terete, scarcely striate, occasionally spinescent at the end, but without the lateral branchlets of *S. daviesioides*. Leaves on a very few barren branches only, small, subulate, and usually verticillate. Flowers clustered, forming rather dense racemes, and often, but not always, larger than in *S. vimineum*. Calyx about 2 lines long, the tube scarcely half so long as the upper lip. Standard broad, rather longer than the other petals; keel broad and rounded at the top, not so incurved as in *S. fornicatum*. Style incurved, with a narrow membranous wing under the stigma. Pod broader than long.—*S. Drummondii,* Turcz. in Bull. Mosc. 1853, i. 267.

W. Australia. Swan River, *Drummond, 1st Coll, 2nd Coll. n.* 115, *and 5th Coll. n.* 47; from King George's Sound to Murchison river, *Oldfield.*

S. crassirameum, Meissn. in Bot. Zeit. 1855, 28,—from between Moore and Murchison rivers, *Drummond, 6th Coll. n.* 20,—differs slightly in the branches rather thicker than usual.

Var. *pulchellum.* Of smaller stature, with smaller flowers, appearing more red in the dried state.—*S. pulchellum,* Meissn. in Bot. Zeit. 1855, 28.—Between Moore and Murchison rivers, *Drummond, 6th Coll. n.* 19.

Var. *parviflorum.* Flowers smaller, all yellow.—Clay plains, near M'Callum Inlet, *Maxwell.*

12. **S. daviesioides,** *Turcz. in Bull. Mosc.* 1853, i. 266. A low shrub, with terete, divaricate branches, numerous short, divaricate or recurved, spinescent branchlets, all leafless, and occasionally a few longer barren branches, with a few, small, distant, linear-lanceolate leaves. Flowers solitary or 2 or 3 together, along some of the smaller spinescent branchlets; pedicels rather slender. Calyx rather more than 2 lines long, the tube about half as long as the upper lip. Standard nearly 4 lines long, very broad; lower petals rather shorter, the keel broad, curved, very obtuse, yellow like the standard, or deeply coloured. Style much curved, with a rather broad longitudinal wing under the stigma on the inner edge.

W. Australia, *Drummond, 5th Coll. n.* 46; Phillips Ranges, Cape Arid, and Cape Le Grand, *Maxwell.* This may prove to be a variety of *S. macranthum.*

Section III. Euchiloides.—Calyx-tube very short. Style subulate, without any appendage. Flowering-stems leafy.

13. **S. euchilus,** *Benth.* Stems from a thick rootstock, erect or ascending, 1 to 1½ ft. high, slender, nearly terete, minutely silky-pubescent, all leafy. Leaves alternate or rarely irregularly opposite, narrow-linear, obtuse or with a callous point, ½ to 1½ in. long, the margins closely revolute, minutely silky-pubescent or almost silvery. Pedicels axillary, slender, often ½ in. long, with a pair of minute bracteoles above the middle. Calyx glabrous, about 3 lines long, the tube very short, the 2 upper lobes broadly obovate, almost orbicular, very obtuse, the 3 lower lobes very small. Standard 4 lines long; keel shorter, broad, incurved, acute, almost rostrate. Ovary stipitate, villous. Style subulate, without any appendage. Pod ovate, very turgid,

villous, about 4 lines long, on a stipes of 1 line. Seeds rather large, without any strophiole.—*Euchilus linearis*, Benth. in Hueg. Enum. 35, and in Ann. Wien. Mus. ii. 80 ; Meissn. in Pl. Preiss. i. 72.

W. Australia, *Huegel ;* Swan River, *Drummond, 1st Coll. and. n.* 295, *Preiss, n.* 879 *and* 1110 ; Vasse river, *Oldfield ;* base of Mount Melville, *Maxwell.*

11. VIMINARIA, Sm.

Calyx-teeth short, equal. Petals on rather long claws. Standard orbicular ; wings oblong, shorter than the standard ; keel slightly curved, about as long as the wings. Stamens free. Ovary nearly sessile ; style filiform, with a small terminal stigma ; ovules 2, with short funicles. Pod sessile, ovoid-oblong, usually indehiscent, the pericarp thickly membranous. Seed usually solitary, filling the cavity, with a very small annular strophiole.— Shrub with rush-like stems. · Leaves alternate, mostly reduced to a long filiform petiole. Flowers small, in terminal racemes.

The genus is limited to a single species, with the flowers nearly of a *Daviesia*, but very distinct in the fruit, which is almost that of a *Melilotus*.

1. **V. denudata**, *Sm. Exot. Bot.* 51, *t.* 27, *and in Ann. Bot.* i. 507, *and Trans. Linn. Soc.* ix. 261. A glabrous shrub, sometimes erect, attaining 10 to 20 ft., with long, wiry, pendulous branches, more rarely low and decumbent. Leaves reduced to filiform petioles, of from 3 to 8 or even 9 in., the lower ones or those of luxuriant branches occasionally bearing at the extremity 1 to 3 oval-oblong or lanceolate, herbaceous leaflets, of ½ to 1½ in. Flowers small, orange-yellow, in long terminal racemes. Pedicels rarely as long as the calyx, in the axils of small scale-like bracts, without bracteoles. Calyx nearly 2 lines long, including the short, turbinate, disk-bearing base. Petals about twice as long. Pod 2 to 3 lines long. Albumen rather thicker than in the other *Podalyrieæ* where it has been observed.—DC. Prod. ii. 107 ; Bot. Mag. t. 1190 ; Meissn. in Pl. Preiss. i. 57 ; Paxt. Mag. xiv. 123, with a fig ; *Sophora juncea*, Schrad. Sert. Hannov. t. 3 ; *Pultenæa juncea*, Willd. Spec. ii. 506 ; *Daviesia denudata*, Vent. Choix, t. 6 ; *D. juncea*, Pers. Syn. i. 454, not of Sm.

N. S. Wales. Port Jackson, *R. Brown, Sieber, n.* 369, *and Fl. Mixt. n.* 553, and others ; and northward to Port Macquarrie, *Fraser, A. Cunningham ;* Hastings river, *Beckler.*

Victoria. Port Phillip, *R. Brown ;* from Gipps' Land to the Glenelg, *F. Mueller* and others.

Tasmania? Although the station is given by De Candolle, the plant has never been found in the island by Gunn or any other of its most zealous explorers. In F. Mueller's herbarium are some fragments marked Tasmania, *Fitzalan*, but without any precise locality, and there may be some mistake.

S. Australia. Torrens river, *Whittaker ;* Lofty Ranges, *F. Mueller.*

W. Australia. King George's Sound, *R. Brown ;* to Cape Leeuwin, *Collie ;* Swan River, *Preiss, n.* 1023 : and Murchison river, *Oldfield.*

V. (?) Preissii, Meissn. in Pl. Preiss. i. 57,—from sandy places, inundated in winter, on the Canning river, *Preiss, n.* 1024,—which I have not seen, is not sufficiently described to be recognizable. Meissner himself doubts whether it may not be a *Sphærolobium.*

12. DAVIESIA, Sm.

Calyx-teeth short, either all equal or the 2 upper ones united in a truncate upper lip, the disk-bearing base either shortly turbinate or elongated and stalk-like. Petals on a slender claw; standard orbicular or reniform, emarginate; wings falcate-oblong or obovate, not longer than the standard; keel more or less incurved, obtuse or almost acute, not exceeding the wings. Stamens free, the 5 outer filaments often flattened and sometimes cohering in a tube, although readily separable. Ovary shortly stipitate, tapering into a subulate style, with a small terminal stigma; ovules 2, with short funicles. Pod nearly sessile or stipitate, more or less flattened, acute, triangular, the upper suture nearly straight, the dorsal or lower suture much curved, forming almost a right angle. Seeds solitary or rarely 2, with a rather large strophiole.—Shrubs or undershrubs. Leaves alternate, simple, entire, coriaceous or rigid, either flat and horizontal or vertical, or terete and spinescent, sometimes decurrent along the stem, or reduced to short prickles or teeth, or entirely wanting. Stipules none or very minute. Flowers usually small, yellow orange or red, in axillary or lateral racemes or pedunculate umbels, occasionally reduced to short clusters or rarely solitary or terminal. Bracts at the base of the racemes small, dry and scale-like, those under the pedicels similar or a few of them, in a few species, much enlarged over the fruit. Bracteoles none. Ovary glabrous.

The genus is limited to Australia. The short calyx-teeth usually distinguish it from almost all *Podalyrieæ* except *Viminaria* and *Latrobea*, but cannot be absolutely relied upon; the pod, however, is quite peculiar. It is also in most cases readily known by the habit, and is indeed so natural that, numerous as are the species, I have been unable to distribute them into distinct sections. The following *series* are founded chiefly on the foliage and on the degree of development of the inflorescence. In the first five series the leaves are horizontally flat or very rarely terete and then obtuse, and never vertically compressed; in the last four they are terete or vertically flattened and pungent or altogether wanting. Some species of this genus, when not in flower, have been occasionally mistaken for phyllodineous Acacias.

SERIES I. **Involucratæ.**—*Flowers umbellate, at the ends of the peduncles. Upper bracts orbicular, often small at the time of flowering, but much enlarged afterwards, and enclosing the fruits. Leaves flat, horizontal.*

Leaves very much reticulate.
Leaves 3 to 4 in. long, deeply cordate, with rounded auricles . 1. *D. cordata.*
Leaves ovate, about 2 in. long, narrowed at the base . . . . 2. *D. ovata.*
Leaves ½ to ¾ in. long, orbicular, slightly cordate, pungent-
pointed . . . . 3. *D. crenulata.*
Leaves thick, the reticulations few or not very prominent.
Leaves elliptical-oblong, obtuse. Bracts very concave. Calyx
glabrous; teeth very short . . . . 4. *D. oppositifolia.*
Leaves oblong-linear, mostly pointed. Bracts flat. Calyx pu-
bescent: teeth acuminate . . . . 5. *D. alternifolia.*
Leaves linear. (Bracts narrow?) Calyx glabrous; teeth very
short . . . . 6. *D. elongata.*

SERIES II. **Umbellatæ.**—*Flowers umbellate, at the ends of the peduncles. Bracts all small and not enlarging. Leaves flat, horizontal, usually small, rigid, with pungent points.*

(*D. umbellulata* and *D. corymbosa*, amongst the *Racemosæ*, have the flowers occasionally almost umbellate.)

Leaves oblong, narrowed at both ends or almost linear.
Pedicels articulate, at some distance from the calyx, and not
enlarged at the articulation . . . . 7. *D. pedunculata.*

Pedicels articulate close to the calyx and there expanded into a ring　8. *D. mollis,* var.
Leaves obovate or orbicular . . . .　8. *D. mollis.*
Leaves broadly ovate-cordate, tapering into a pungent point . .　9. *D. concinna.*

SERIES III. **Racemosæ.**—*Flowers racemose, the common rhachis elongated, either flowering from the base, or the pedicels crowded towards the end. Bracts small. Leaves flat, horizontal.*

Leaves (under 1 in.) linear or linear-lanceolate, pungent, not reticulate. Racemes flowering above the middle. Bracts very small 10. *D. umbellulata.*
Leaves broad, obtuse or acute, young buds very obtuse. Pods
under ½ in.
　Leaves (under 1 in.) ovate or orbicular, cordate or very obtuse at
　the base, sessile, not reticulate. Racemes flowering from
　the middle. Bracts small . . . .　11. *D. buxifolia.*
　Leaves (1 to 3 in.) ovate or ovate-lanceolate, reticulate. Ra-
　cemes flowering from the base. Bracts ovate or oblong,
　nearly as long as the pedicels . . . .　12. *D. latifolia.*
Leaves narrow, elongated, the lower ones rarely ovate-oblong, not
　pungent.
Branches slightly angular. Flowers numerous. Calyx 1 to 1¼
　line long.
　Flowering branches all leafy, never spinescent (Eastern species)　13. *D. corymbosa.*
　Upper flowering branches often spinescent and leafless (Western
　species) . . . .　14. *D. horrida.*
Branches very angular. Flowers few, distant. Calyx above 2 lines 15. *D. reclinata.*
Leaves rounded at the end or emarginate. . Young buds shortly
　acuminate. Pod nearly 1 in. (Western species).
　Leaves oblong. Calyx-teeth obtuse, the 2 upper united and rounded 16. *D. obtusifolia.*
　Leaves obovate. Calyx-teeth all acuminate and distinct . . . 17. *D. obovata.*

SERIES IV. **Calamiformes.**—*Leaves long, very narrow or terete, obtuse or with cal-lous or hooked points.*
Racemes loose. Leaves striate.
　Leaves flat . . . . 18. *D. longifolia.*
　Leaves terete . . . . 19. *D. chordophylla.*
Racemes or clusters very short. Leaves not striate, mostly terete 20. *D. nematophylla.*

SERIES V. **Fasciculatæ.**—*Flowers in axillary clusters or very short racemes. Leaves flat, horizontal, pungent-pointed.*

Leaves above 1 in. long, oblong, lanceolate or almost ovate, nar-
　rowed at the base (Western species).
　Flowers axillary. Calyx-base narrow-turbinate. Keel obtuse . 21. *D. daphnoides.*
　Flowering nodes often leafless. Calyx-base very short. Keel rostrate 22. *D. nudiflora.*
Leaves under 1 in. long (except *D. rhombifolia*), obovoid, rhomboid
　or broadly cordate-ovate or oval-oblong. Calyx with a trun-
　cate upper lip (Western species).
　Branches glabrous, very rigid.
　　Leaves rhomboidal-acuminate, much reticulate . . . . . 23. *D. rhombifolia.*
　　Leaves broadly cordate, shortly acuminate, not reticulate . . 24. *D. cardiophylla.*
　　Leaves obovate, with recurved points, not reticulate . . . 25. *D. Drummondii.*
　Branches hirsute. Leaves oval or oval-oblong, not reticulate,
　　straight-pointed . . . . 26. *D. filipes.*
Leaves under 1 in. long, cordate-ovate, lanceolate or linear. Calyx
　5-toothed (Eastern species).
　Branches not spinescent.
　　Leaves cordate or ovate, much acuminate, usually under ½ in.
　　long. Pedicels filiform . . . . 27. *D. squarrosa.*

Leaves linear, with revolute margins, about 1 in. long. Pedicels very short . . . . 29. *D. acicularis.*
Branches spinescent. Leaves ovate, lanceolate or linear, pungent-pointed, but scarcely acuminate . . . . 28. *D. ulicina.*

SERIES VI. **Teretifoliæ.**—*Leaves terete or slightly compressed or rarely vertically dilated towards the top, at length articulate on the stem, usually short or pungent-pointed. Flowers solitary, clustered or rarely racemose.*

(See also *Calamiformes.*)

Leaves crowded, glaucous, thickly conical, pithy, divaricate . . 30. *D. pachyphylla.*
Leaves cylindrical, erect, the pungent point very short . . . . 31. *D. teretifolia.*
Leaves divaricate, very pungent.
   Upper calyx-teeth truncate. Filaments slender.
     Bracts imbricate, longer than the pedicels . . . . . . 33. *D. hakeoides.*
   Bracts very small.
     Keel not much curved, obtuse (Eastern species) . . . . 32. *D. genistifolia.*
     Keel much curved, almost rostrate (Western or Southern species).
       Calyx nearly 2 lines long, with a narrow base.
         Branches straight or scarcely flexuose. Leaves incurved or spreading . . . . 34. *D. colletioides.*
         Branches very flexuose. Leaves short, reflexed . . 35. *D. reversifolia.*
       Calyx about 1¼ line long, the base very short.
         Leaves mostly ½ in. long or more, often dilated upwards, at length articulate on the stem . . . . 36. *D. incrassata.*
         Leaves few, rarely above ¼ in. long, appearing like prickles, continuous with the stem . . . . . . 37. *D. brevifolia.*
Upper calyx-teeth small, distinct. Filaments much dilated and cohering.
   Leaves not crowded, mostly ¾ to 1 in. long, slender or flat . 38. *D. Preissii.*
   Leaves crowded, about ½ in. long, thick and scarcely flattened 39. *D. spinosissima.*
Upper calyx-teeth acuminate. Filaments free. Racemes loose 40. *D. pachylina.*

SERIES VII. **Verticales.**—*Leaves vertically flattened, often attached by a broad base, but not decurrent, usually pungent-pointed. Flowers solitary or clustered or umbellate-racemose.*

Branchlets not spinescent.
   Flowers in a pedunculate umbel or short raceme. Leaves almost rhomboidal . . . . 41. *D. quadrilatera.*
   Flowers in sessile clusters or very short racemes.
     Leaves crowded, at least half as broad as long, attached by a very broad base . . . . 42. *D. striata.*
     Leaves narrow, dilated upwards. Filaments free . . . 36. *D. incrassata.*
     Leaves linear, compressed. Filaments cohering . . . . 38. *D. Preissii.*
     Leaves lanceolate, or, if ovate, contracted at the base.
       Branches angular . . . . 43. *D. polyphylla.*
Branchlets spinescent. Flowers solitary. Leaves small . . 44. *D. microphylla.*

SERIES VIII. **Decurrentes.**—*Leaves terete or vertically compressed or with a vertical dorsal projection, usually pungent, decurrent at the base into raised angles on the branches. Flowers clustered or shortly racemose.*

Leaves slightly decurrent.
   Leaves few, terete or conical, not above ¼ in. long, resembling prickles . . . . 37. *D. brevifolia.*
   Leaves crowded, nearly terete, almost ½ in. long . . . . 39. *D. spinosissima.*
Leaves very decurrent.
   Leaves terete or slightly compressed. Branchlets very flexuose. 45. *D. flexuosa.*
   Leaves much compressed, lanceolate. Branchlets straight . . 46. *D. pectinata.*

Leaves horizontally dilated or channelled on the upper edge, the
    dorsal midrib or wing decurrent . . . . 47. *D. trigonophylla*.
Leaves resembling pinnate lobes of very glaucous flat phyllodine-
    ous branchlets.   Inflorescence on the face of the branchlets . 48. *D. epiphylla*.

SERIES IX. **Aphyllæ.**—*Leaves none* (*except sometimes in* D. divaricata). *Flowers
solitary clustered or rarely racemose.*

Branches terete. Bracts very small.
    Branches very thick, cylindrical and pithy . . . . 49. *D. euphorbioides*.
    Branches rigid, but scarcely pungent. Pedicels very short . . 37. *D. brevifolia*, var.
    Branches elongated, not pungent. Pedicels slender, short . . 52. *D. aphylla*.
    Branchlets divaricate, pungent. Pedicels longer than the calyx 50. *D. divaricata*.
    Branchlets divaricate, slender, 2–3-chotomous, the lower ones
        spinescent. Flowers terminal . . . . 51. *D. paniculata*.
Branches terete, elongated. Bracts imbricate, covering the rhachis 53. *D. juncea*.
Branches flat. Calyx-teeth long, acute.
    Branches winged. Racemes short, dense, with ovate bracts . . 54. *D. alata*.
    Branches flat, but not winged. Racemes loose, few-flowered.
        Bracts minute or none . . . . 55. *D. anceps*.

SERIES I. INVOLUCRATÆ.—Flowers umbellate at the ends of the peduncles.   Upper bracts orbicular, often small at the time of flowering, but much
enlarged afterwards, and exceeding or enclosing the fruits.   Leaves flat, horizontal.

1. **D. cordata,** *Sm. in Trans. Linn. Soc.* ix. 259. A glabrous erect shrub,
of 2 to 3 ft., the branches elongated, with 2 or 3 prominent angles.   Leaves
ovate or ovate-lanceolate, acute, 3 to 4 in. long, deeply cordate and embracing
the stem by their large rounded auricles, rigidly coriaceous, strongly reticulate on both sides.   Peduncles from 1 to 4 or 5 in each axil, usually shorter
than the leaves, bearing each an umbel of 8 to 12 flowers on short slender
pedicels.   Bracts under the umbel 2 to 4, orbicular, small when the flowers
first open, but soon enlarging, and when the fruit is ripe, 1 to 1½ in. diameter, thin, scarious and elegantly veined.   Calyx about 2 lines long, the 2
upper teeth broad, truncate and connate.   Standard yellow, about 4 lines
diameter; lower petals shorter, falcate, purple.   Pod above ½ in. long.—DC.
Prod. ii. 114 ; Bot. Reg. t. 1005 ; Meissn. in Pl. Preiss. i. 56.

W. Australia.   King George's Sound, *Menzies, A. Cunningham* ; Swan River, *Drummond, n.* 223 ; Darling Range and Mount Wuljenup, *Preiss, n.* 1137 and 1136 ; Vasse and
Blackwood rivers, *Oldfield*.

2. **D. ovata,** *Benth.* A glabrous shrub, with the habit, angular stems,
and large bracts of *D. cordata.*   Leaves ovate or elliptical, with a small
callous point, mostly about 2 in. long, narrowed at the base, rigidly coriaceous and strongly reticulate as in *D. cordata.*   Peduncles axillary, but often
growing out into leafy branches, with a terminal umbel, surrounded when in
fruit by 2 or 3 large orbicular bracts, rigidly scarious and elegantly veined as
in *D. cordata*, and of about the same size.   Calyx and pod of *D. cordata.*
Flowers not seen.

W. Australia, *Drummond, n.* 23.

3. **D. crenulata,** *Turcz. in Bull. Mosc.* 1853, i. 265. A rigid shrub,
the branches more slender than in the two preceding species, and usually pu-

hescent with short stiff hairs, the rest of the plant glabrous. Leaves often opposite, orbicular-cordate, with a short pungent point, ½ to ¾ in. diameter, the margin undulate and slightly crenulate, rigid, shining and strongly reticulate. Peduncles axillary, longer than the leaves, bearing an umbel of 3 to 5 or rarely 6 rather small flowers, subtended by 2 or rarely 3 or 4 orbicular bracts, small at the time of flowering, but afterwards much enlarged, scarious and much veined, attaining often 1 in. diameter, and closing over the fruit; inner bracts small and narrow. Calyx not 2 lines long, the upper teeth broad and truncate. Standard nearly 5 lines diameter; keel much shorter. Filaments as in several allied species, folded inwards above the middle. Pod about ½ in. long and nearly as broad.—*D. calystegia,* Turcz. in Bull. Mosc. 1853, i. 264; *D. parifolia,* F. Muell. Fragm. iv. 16.

W. Australia. Between Swan River and Cape Riche, *Drummond, 5th Coll. n.* 40 (in flower), and *4th Coll. n.* 30 (in fruit); Kojonerup hills, *Maxwell.*

4. **D. oppositifolia,** *Endl. in Ann. Wien. Mus.* ii. 199. A glabrous shrub of several feet, the branches stout, very prominently 3- or 4-angled. Leaves often irregularly opposite or in whorls of 3, oblong-elliptical, obtuse or rarely with a minute callous point, in most specimens 1 to 2 in. long, but occasionally attaining 3 or even 4 in., thickly coriaceous, with a thickened margin, the veins few and only conspicuous on the young leaf. Peduncles about as long as the leaves, bearing an umbel of 3 to 6 flowers, subtended by 2 or 3 orbicular very concave rigidly membranous slightly veined bracts, which are at first as long as the flowers and enlarge often to above 1 in. diameter, completely enclosing the fruits. Calyx under 2 lines long, the teeth very short, the 2 upper ones broad and truncate. Standard about 4 lines diameter; wings and keel rather shorter. Five outer filaments much broader than the others. Pod ½ in. long.—Meissn. in Pl. Preiss. i. 55.

W. Australia. King George's Sound, *R. Brown, A. Cunningham, Huegel, Drummond,* and others; Mount Wuljenup, *Preiss, n.* 856.

5. **D. alternifolia,** *Endl. in Ann. Wien. Mus.* ii. 199. A shrub or undershrub, with decumbent or ascending stems of 1 to 2 ft., slightly angular and often minutely pubescent. Leaves glabrous, alternate or rarely opposite or in threes, linear-oblong or oblong-cuneate, rarely obovate-oblong, 1 to 2 in. long, with a short rigid point or rarely obtuse, much narrowed below the middle, with a pair of minute teeth (stipules?) near the base, thickly coriaceous, the veins not numerous nor very prominent. Peduncles rather longer than the leaves, bearing an umbel usually of 3 flowers. Bracts usually 3, orbicular, attaining at length nearly 1 in. diameter, flat or scarcely concave, rigid and more reticulate than in *D. oppositifolia.* Calyx pubescent, above 2 lines long, the teeth acuminate, the 2 upper ones broad, truncate and united. Standard (according to Preiss) flame-coloured; wings and keel rather shorter, the latter purple. Pod about ½ in. long.—Meissn. in Pl. Preiss. i. 55.

W. Australia. King George's Sound, *Huegel, Drummond, 4th Coll. n.* 31, and others; near Albany, *Preiss, n.* 855; Kalgan and King rivers, *Oldfield.* The lower leaves of the main branches are sometimes reduced to small scales, which Endlicher considers as the only true leaves, designating the others as phyllodia; they all however appear to me to be true leaves.

D. ternata, Endl. l. c., from the same locality, appears to me, from his description, to be the same species, in which the leaves of the side branches are occasionally ternately verticillate.

6. **D. elongata,** *Benth.* A glabrous shrub or undershrub, with ascending angular stems, of 2 ft. or more. Leaves alternate, the lowest sometimes oblong-cuneate, the others linear or linear-oblong, obtuse or with a short callous point, 2 to 4 or even 5 in. long, thickly coriaceous, obscurely veined, narrowed at the base, but without stipular teeth. Peduncles shorter than the leaves, bearing an umbel of 2 or 3 flowers, with 2 narrow-oblong or linear-cuneate bracts, not so long as the flower, but probably enlarging afterwards. Flowers apparently like those of *D. oppositifolia*, the calyx quite glabrous, with very short teeth. Fruit not seen.

W. Australia, *Drummond, 2nd Coll. n.* 136. The species is evidently allied to, but distinct from *D. oppositifolia*, although the specimens are not good enough for a full description.

SERIES II. UMBELLATÆ.—Flowers umbellate at the ends of the peduncles. Bracts all small and not enlarging. Leaves flat, horizontal, usually small and rigid with pungent points.

7. **D. pedunculata,** *Benth. in Lindl. Swan Riv. App.* 14. A low shrub, the short slender terete branches occasionally pubescent. Leaves glabrous, oblong or almost linear, narrowed at both ends with a fine pungent point, ½ to ¾ in. long, rather thick, rigid, scarcely veined besides the midrib. Peduncles slender, glabrous, much longer than the leaves or rarely about their length, bearing a terminal umbel of 3 to 6 flowers, and rarely a single flower lower down. Bracts very small. Pedicels longer than the calyx, articulate at some distance from it, and thickened above the articulation. Calyx 1½ lines long, without the narrow-turbinate stipitiform base, the teeth short and obtuse, the 2 upper ones truncate and united. Standard twice as long as the calyx, lower petals rather shorter. Pod only seen young.—Meissn. in Pl. Preiss. i. 53.

W. Australia. Swan River, *Drummond, 1st Coll. and n.* 229 ; Darling Range, *Preiss, n.* 1043.

Var. *minor.* Leaves linear, not pungent, Konkoberup hills, *Preiss, n.* 1154 (I have not seen the specimens).

8. **D. mollis,** *Turcz. in Bull. Mosc.* 1853, i. 263. A shrub, apparently decumbent, the branches leaves and peduncles in their original form hirsute, with soft spreading hairs. Leaves rather crowded, obovate, obtuse, with a short often pungent point, ½ to ¾ in. long, thick, coriaceous, not much veined besides the midrib. Peduncles usually exceeding the leaves, bearing an umbel of 3 or rarely 4 or 5 flowers. Bracts minute. Pedicels about 2 lines long, articulate a little below the calyx and there dilated into a ring. Calyx 2 lines long, pubescent, the teeth nearly as long as the tube, acuminate, the 2 upper ones broad, truncate and united. Standard nearly twice as long as the calyx ; keel shorter, much incurved, almost rostrate. Filaments less unequal than in the preceding species. Pod about ½ in. long and broad.

W. Australia, *Drummond, 5th Coll. n.* 39 ; Rocky Hills, inland from Cape Legrand, *Maxwell.*

Var. *minor.* Softly pubescent or glabrous. Leaves under ¼ in. long, from broadly obovate or orbicular to narrow-obovate or oblong, or in some specimens all oblong ; narrowed at both ends, or almost lanceolate. Flowers smaller, with the calyx-teeth less acuminate than in the ordinary form.—*D. lancifolia*, Turcz. in Bull. Mosc. 1853, i. 263. *Drummond, 5th Coll. n.* 28, and *suppl. n.* 23 ; W. Mount Barren, *Maxwell.* Besides the shorter leaves, this may be always distinguished from *D. pedunculata* by the very short base of the calyx and by the truncate dilatation of the pedicel.

9. **D. concinna,** *R. Br. Herb.* Branches elongated, slender, slightly angular-pubescent. Leaves broadly ovate-cordate, tapering into a pungent point, not exceeding ¼ in., almost veinless except the midrib. Flowers small, in umbels of 3 or 4, on a common peduncle about as long as the leaves. Bracts very small. Pedicels slender, 2 to 4 lines long. Calyx about 1 line long. Petals and pod of *D. umbellulata,* from which the species differs chiefly in the broad, less coriaceous leaves, and more umbellate inflorescence.

Queensland. Rock hills, Pine Port, *R. Brown.*
N. S. Wales. Hastings river, *Beckler (Herb. R. Br. and F. Muell.).*

Series III. Racemosæ.—Flowers racemose, the common rhachis elongated, either flowering from the base, or the pedicels crowded towards the end, but not so distinctly umbellate as in the 1st and 2nd series. Bracts small, under the pedicels, or the lower ones at the base of the peduncle without flowers. Leaves flat, horizontal, obtuse or more or less pointed in the first 2 or 3 species.

10. **D. umbellulata,** *Sm. in Ann. Bot.* i. 507, *and in Trans. Linn. Soc.* ix. 258. A slender much-branched shrub, glabrous or sparingly pubescent, the branches sulcate. Leaves lanceolate or linear-lanceolate, ½ to ¾ in. long, rigid with a pungent point, 1-nerved, flat, not reticulate. Racemes in some specimens shorter than the leaves, in others twice as long, flowering from the middle upwards or at the end only. Pedicels usually 2 to 3 lines. Bracts about ⅓ line long. Calyx about 1 line long, the teeth short and obtuse, the two upper ones broad, truncate and united nearly to the top. Petals twice as long as the calyx. Pod about 5 lines long.—*D. racemulosa,* DC. Prod. ii. 114 (from the character given).

Queensland. Peele's Island, Moreton Bay, *Fraser, A. Cunningham;* Wide Bay, *Bidwill.*
N. S. Wales. Port Jackson, *R. Brown;* Sandstone Ranges around Biroa, *Leichhardt.*
This species has considerable affinity on the one hand with the western *D. pedunculata,* and on the other with those specimens of *D. ulicina* in which the inflorescence is slightly elongated.
Var. *pubigera.* Branches virgate. Leaves small, mostly 3 to 4 lines long. Racemes few-flowered, the rhachis short.—*D. pubigera,* A. Cunn. ; Benth. in Ann. Wien. Mus. ii. 75. Bushy forest lands and grassy banks of the Cugeegong river, N.° of Bathurst, *A. Cunningham.* This form has the habit of *D. squarrosa,* with the leaves of *D. ulicina,* but the upper teeth of the calyx are more united and truncate, and the inflorescence is evidently a reduced form of that of *D. umbellulata.*

11. **D. buxifolia,** *Benth.* A glabrous shrub, with numerous angular branches. Leaves broadly ovate or orbicular and cordate at the base, or rarely oval-oblong, obtuse or with a minute point, ¼ to ¾ in. long, quite sessile, coriaceous, shining, not reticulate. Racemes usually longer than the leaves, flowering from above the middle. Bracts minute. Pedicels about 2

lines long. Calyx about 1 line long, with short teeth, the 2 upper ones
broad, truncate and united. Standard fully twice as long as the calyx.
Lower petals rather shorter. Pod about 5 lines long.

N. S. Wales. Between Wombim river and False Bay, *Mossman.*
Victoria. Avon Ranges, Macalister and Genoa rivers, *F. Mueller.*
W. Australia. " King George's Sound " (probably to the eastward), *Baxter.*

F. Mueller had formerly proposed this as a species, under the name of *D. cordifolia,*
which I have not adopted, on account of the older *D. cordata,* Sm. He now considers it as
a variety of *D. latifolia.* If so, the specimens show it to be a very well-marked and distinct
form.

12. **D. latifolia,** *R. Br. in Ait. Hort. Kew. ed.* 2, iii. 20. A glabrous
shrub, of 2 to 5 ft. Leaves ovate-elliptical or ovate-lanceolate, usually terminating in a callous point, rarely almost pungent or quite obtuse, mostly 2 to
3 in. long, or smaller only on elongated side-branches, narrowed into a petiole,
rigid, but strongly reticulate. Flowers small, orange-yellow, numerous, in
racemes of 1 to 2 in. flowering often from near the base. Bracts ovate or
oblong, 1 to 2 lines long, densely imbricate before the flowers are full-grown.
Pedicels rarely exceeding the bracts till after flowering. Calyx 1 line long, the
teeth very short, the two upper ones broad, truncate and united. Standard
fully twice as long as the calyx, the lower petals nearly as long. Pod about
5 lines long.—Andr. Bot. Rep. t. 638 ; Bot. Mag. t. 1757 ; DC. Prod. ii.
113 ; Hook. f. Fl. Tasm. i. 83 ; Paxt. Mag. iv. 223, with a fig.

N. S. Wales. Blue Mountains, *Fraser, A. Cunningham, Sieber, n.* 349, and others,
and northward to New England, *C. Stuart;* Clarence river, *Beckler.*
Victoria. Port Phillip, *R. Brown;* common in the wet forest valleys, often forming
an impenetrable jungle, *F. Mueller,* and called " native Hop," *Mossman* and others.
Tasmania. Derwent river and Port Dalrymple, *R. Brown;* common throughout the
colony, *J. D. Hooker.*

Var. *parvifolia.* Leaves oval-oblong, often under 1 in. long, more rounded at the base
and less veined. To this variety belong some of the most northern as well as of the southern
specimens ; they may be only lateral branches of large-leaved shrubs. They may at first
sight appear to connect the species with *D. buxifolia,* but in the latter, the leaves on the
main stems are always orbicular-cordate, and quite sessile, whilst the larger leaves of
D. latifolia are always narrowed at the base into a petiole.

13. **D. corymbosa,** *Sm. in Ann. Bot.* i. 507, *and in Trans. Linn. Soc.*
ix. 258. A glabrous shrub, of 2 to 4 or 5 ft., the branches slightly angular.
Leaves usually lanceolate or linear, rarely broader and oblong, with a short
callous point or rarely quite obtuse, 1½ to 3 in. long or sometimes almost 4
in., rigid, 1-nerved, and when broad more or less reticulate. Racemes usually
shorter than the leaves, and flowering from above the middle or from the end
only, with long slender pedicels, the bracts small, obovate, spreading under
the pedicels, with numerous others crowded at the base of the peduncle
without flowers, but occasionally the racemes flower more regularly from below the middle. Calyx about 1 line long, the teeth short, the 2 upper ones
broad, truncate and united. Standard 3 times as long as the calyx ; keel
rather shorter. Pod nearly ½ in. long.—DC. Prod. ii. 113 ; Andr. Bot.
Rep. t. 611 ; *D. mimosoides,* Bot. Mag. t. 1957 ; *D. glauca,* Lodd. Bot. Cab.
t. 43 (from the figure) ; *D. macrophylla,* Endl. Nov. Stirp. Dec. 15 (a luxuriant garden specimen with the lower leaves broad, above 4 in. long).

N. S. Wales. Port Jackson, to the Blue Mountains, *R. Brown, Sieber, n.* 350, and others; northward to New England, *C. Stuart;* Macleay, Hastings, and Clarence rivers, *Beckler*, and southward to Illawarra, *Shepherd.*

Victoria. Port Phillip, *R. Brown* and others: frequent from Gipps Land to Melbourne and the Grampians, *F. Mueller* and others.

S. Australia. Lofty, Flinders, and Bugle ranges, *F. Mueller* and others.

Var. *mimosoides.* Leaves usually narrow, with more pinnate and less reticulate veins. Flowers smaller, the racemes flowering from below the middle; bracts smaller and less spreading.—*D. mimosoides*, R. Br. in Ait. Hort. Kew. ed. 2, iii. 20; DC. Prod. ii. 114. *D. virgata*, A. Cunn. in Bot. Mag. t. 3196. *D. linearis*, Lodd. Bot. Cab. t. 1615. *D. leptophylla*, A. Cunn. in G. Don, Gen. Syst. ii. 125.—This appears to bé the most common form over the whole range, from the Blue Mountains to Victoria and S. Australia. The original *D. corymbosa* is probably limited to Port Jackson and the Blue Mountains. *D. virgata, linearis,* and *leptophylla* represent a remarkably narrow-leaved form, from the barren parts of the Blue Mountains, which appears, however, in our numerous specimens to pass very gradually into the common *mimosoides* variety.

14. **D. horrida,** *Meissn. in Pl. Preiss.* i. 54. An erect, glabrous, often glaucous, rigid shrub of several feet, the smaller branches usually leafless, paniculate, divaricate and spinescent, as in *D. divaricata*, but scarcely striate. Leaves chiefly on the main branches, linear or linear-lanceolate, obtuse or shortly pointed, rarely pungent, 1½ to 4 or even 5 in. long, very rigid and phyllodia-like, the midrib scarcely prominent, and the lateral veins inconspicuous. Racemes almost always on leafless spinescent branchlets, rarely axillary, usually loose, the rhachis often nearly 1 in. long, but sometimes very short. Bracts small. Pedicels slender, as long as the calyx or longer. Calyx about 2 lines long, including the stalk-like narrow-turbinate base, the teeth short, but all acute and distinct. Standard twice as long as the calyx; keel small, obtuse. Pod only seen young.

W. Australia. Swan River, *Drummond, n.* 230, *Preiss, n.* 1142 and 1171, and others; Geographe Bay, Gordon, and Kalgan rivers, etc., *Oldfield.* When leafless or nearly so, this species closely resembles *D. divaricata*, but it may be readily known by the acute distinct calyx-teeth.

15. **D. reclinata,** *A. Cunn. Herb.* Quite glabrous. Branches elongated, acutely angular. Leaves linear, obtuse or mucronate, 2 to 3 or rarely nearly 4 in. long. Racemes not so long as the leaves, but with few distant pedicels of 2 or 3 lines, each in the axil of a minute bract. Calyx nearly 3 lines long, including the stalk-like turbinate base, the teeth acuminate, nearly as long as the tube, the 2 upper ones broad, truncate and united. Standard not twice as long as the calyx; wings and keel shorter. Filaments alternately much dilated. Pod not seen.

N. Australia. Arnhem N. Bay, *R. Brown;* Sims Island, *A. Cunningham.*

16. **D. obtusifolia,** *F. Muell. Fragm.* ii. 104. A glabrous shrub, with virgate, somewhat angular branches. Leaves broadly or narrow-oblong, very obtuse and rounded or emarginate at the end, narrowed into a petiole at the base, 1½ to 3 in. long, thickly coriaceous, veinless or with a very faintly prominent midrib. Racemes axillary, very short and few-flowered. Bracts very small. Pedicels short, thickened under the calyx. Calyx very smooth, nearly 1½ lines long, including the slender, turbinate, stalk-like base, the young bud obtusely acuminate, the lower teeth very small, the 2 upper ones longer, very

broad, rounded and united. Standard twice as long as the calyx; keel considerably shorter. Pod very coriaceous, nearly 1 in. long.

W. Australia, *Drummond;* King George's Sound, *Collie, Baxter;* near Cape Riche *Harvey.* This sometimes resembles *D. corymbosa,* var. *mimosoides,* in aspect, but its affinity seems rather with *D. obovata,* especially in the shape of the bud, the thick consistence of the calyx, the short keel, and large pod.

17. **D. obovata,** *Turcz. in Bull. Mosc.* 1853, i. 261. A stout glabrous shrub, the young branches sometimes flattened, but not angular. Leaves from broadly obovate to almost oblong, rounded at the end, and very obtuse or emarginate, 2 to 4 in. long, narrowed into a petiole, very thickly coriaceous, veinless or more or less penniveined. Flowers the largest in the genus, few, in short racemes. Bracts oblong, 1 to 2 lines long; pedicels often not longer. Calyx nearly 3 lines when in flower, thick, with the teeth all narrow and acuminate, enlarging much when in fruit, often attaining ½ in. diameter, the bud acuminate. Standard ½ in. long; keel much shorter. Pod 1 in. long or even more.

W. Australia, *Drummond,* 5*th Coll. n.* 41.

SERIES IV. CALAMIFORMES.—Leaves long, very narrow or terete, obtuse or with callous or hooked points.

The three species here collected have a common aspect, differing from the two preceding series in their long, narrow leaves, and from the next following one in their leaves never pungent. In inflorescence they pass from the *Racemosæ* to the *Fasciculatæ.*

18. **D. longifolia,** *Benth. in Lindl. Swan Riv. App.* 14. Glabrous with long, rather slender, angular or deeply sulcate branches, becoming, however, at length nearly terete. Leaves rigid, narrow-linear, obtuse or with a callous point, 2 to 6 in. long, striate, with 3 to 5 prominent parallel ribs. Racemes 1 to 2 in. long, loose, and few-flowered. Bracts minute. Pedicels slender. Calyx 1½ lines long, including the narrow, stalk-like, turbinate base, the teeth short, the 2 upper ones truncate and united. Petals not twice as long as the calyx. Pod only seen young.—Meissn. in Pl. Preiss. i. 55.

W. Australia. Swan River, *Drummond,* 1*st Coll., Preiss, n.* 1184; East Shoal Pass and M'Callum's Inlet, *Maxwell.*

19. **D. chordophylla,** *Meissn. in Pl. Preiss.* i. 48. A glabrous shrub, with long, slender, sulcate branches, agreeing in every respect with *D. longifolia,* except that the leaves are more slender and terete, varying from 3 to 9 in. in length, striate, terminating in innocuous deciduous black points. Flowers in loose racemes, as in *D. longifolia,* and of the same size, with similar short calyx-teeth. Pod about 4 lines long.

W. Australia. Swan River, *Drummond, n.* 240.

20. **D. nematophylla,** *F. Muell. Herb.* A glabrous shrub, with numerous, erect, slender, but rigid, terete or slightly angular branchlets. Leaves very narrow-linear, terete or slightly flattened, 1 to 2 in. long, obtuse or with a short, hooked, innocuous point, not sulcate. Flowers small, shortly pedicellate, on a very short common peduncle. Bracts very small. Calyx 1¼ lines long, including the narrow, stalk-like, turbinate base, the teeth very short,

the 2 upper ones truncate and united.   Petals twice as long as the calyx ;
keel obtuse, not much curved.   Pod only seen young.

W. Australia, *Drummond, 4th Coll. n.* 27 ; Phillips Ranges, *Maxwell.*

SERIES V.   FASCICULATÆ.—Flowers in axillary clusters or very short
compact racemes.   Leaves flat, horizontal, pungent-pointed.

21. **D. daphnoides,** *Meissn. in Pl. Preiss.* i. 54.   A rigid, glabrous,
somewhat glaucous shrub, of 2 or 3 ft., the branches prominently angled.
Leaves oblong-lanceolate, pungent-pointed, 1 to 3 in. long, narrowed into a
petiole, very thick and rigid, obscurely veined.   Flowers few, small, in axillary clusters or exceedingly short racemes.   Bracts very small.   Pedicels very
short, besides the narrow, stalk-like, turbinate base of the calyx.   Broad tube
of the calyx scarcely 1 line long, the teeth very short, the 2 upper ones broad,
truncate and united.   Standard about twice the length of the calyx ; keel
rather shorter, obtuse, and little curved.   Pod not seen.

W. Australia.   Swan River and to the northward, *Drummond, n.* 225, *and 6th Coll.
n.* 17 ; sandy plains of Quangen, *Preiss, n.* 1144.   Allied on the one hand to *D. obtusifolia*
and on the other to *D. nudiflora,* it has the stalk-like base of the calyx of the former, with
the pungent leaves and truncate upper calyx-teeth of the latter.

22. **D. nudiflora,** *Meissn. in Pl. Preiss.* i. 53.   An erect glabrous
shrub, of 2 to 3 ft., the branches sulcate, but scarcely angular.   Leaves
oblong-lanceolate or sometimes almost ovate, tapering into a long pungent
point, 1 to 2 in. long, narrowed at the base, rigidly coriaceous, with a prominent midrib and pinnate veins.   Flowers clustered or very shortly racemose
at the leafless lower nodes of the upper branches or rarely axillary.   Bracts
small.   Pedicels 2 lines long or more, conspicuously thickened below the articulation.   Calyx 1½ lines long, the turbinate base exceedingly short, the
teeth short, somewhat acute, the 2 upper ones truncate and united.   Standard
fully twice as long as the calyx ; keel rather shorter, much incurved, distinctly rostrate.   Ovary longer than in most species.   Pod 6 to 7 lines long.

W. Australia.   Swan River, *Preiss, n.* 1143, *Drummond, n.* 226, *Oldfield.*
Var. *lanceolata.*   Branches elongated.   Leaves narrow-lanceolate, almost veinless.
Flowering nodes less destitute of leaves.—*Drummond, n.* 133.

23. **D. rhombifolia,** *Meissn. in Pl. Preiss.* i. 56.   A glabrous shrub,
of 1½ to 2 ft., with long, rigid, apparently divaricate or ascending branches,
more or less angular and sulcate.   Leaves broadly ovate or rhomboidal,
tapering into a pungent point and narrowed at the base, about ¾ in. long on
the flowering branches, but occasionally twice that on the main stems, rigid,
with a nerve-like margin, prominent midrib, and reticulate veins.   Flowers
few, in axillary clusters, on slender pedicels of 2 to 3 lines.   Calyx about 1½
lines long, including the narrow turbinate base, the lower teeth very small,
the 2 upper ones larger, broad, truncate and united.   Petals about twice as
long as the calyx, the keel not much shorter than the others and very obtuse.
Pod not seen.

W. Australia.   Swan River, *Drummond, n.* 224 ; Sussex district and Darling Range,
Preiss, n. 1145 *and* 1146 ; Mount Yulagan, *Oldfield.*

24. **D. cardiophylla,** *F. Muell. Fragm.* ii. 105.   Apparently a low,
divaricate, much-branched shrub, quite glabrous, the branches scarcely angu-

lar. Leaves sessile, broadly cordate, tapering into a pungent point, under ½ in. long, thick, with a prominent midrib, but otherwise veinless. Flowers 1 to 3 in the axils, on slender pedicels of 1 to 3 lines. Bracts very small. Calyx nearly 2 lines long, the turbinate base short; teeth very short, the 2 upper ones broad, truncate, and united. Petals twice as long as the calyx, the keel nearly as long as the others, much incurved, obtuse. Pod not seen.

W. Australia. Between Swan River and Cape Riche, *Harvey ;* sandy plains near Belgarup, *Oldfield.* Allied to the eastern *D. squarrosa*, but much more rigid, with larger, broader, less acuminate leaves, and the flowers twice the size, with a much more distinct upper lip to the calyx.

25. **D. Drummondii,** *Meissn. in Pl. Preiss.* i. 53. A rigid, glabrous shrub, the young branches prominently angled. Leaves numerous, rarely exceeding ½ in., obovate or obovate-oblong, tapering into a recurved pungent point, narrowed at the base, but sessile, very rigid, with nerve-like margins and a prominent midrib, but otherwise veinless, often folded lengthwise. Flowers few together in axillary clusters, on pedicels of 2 to 3 lines, slightly thickened at the top. Calyx nearly 1½ lines long, the turbinate base very short, the teeth rather short, the 2 upper ones truncate and united. Petals twice as long as the calyx, the keel nearly as long as the others, much curved rather acute, but not beaked. Pod not seen.

W. Australia. Swan River, *Drummond, n.* 227 ; near Kojonerup, *Oldfield.*

26. **D. filipes,** *Benth. in Mitch. Trop. Austr.* 363. Branches slender, terete, virgate, softly hirsute in our specimens as well as the leaves. Leaves oblong or oval-oblong, shortly pungent-pointed, under ½ in. long, veinless except the midrib. Flowers solitary or 2 together, on filiform pedicels about as long as the leaves. Bracts very small. Calyx 1¼ to 1½ lines long, the turbinate base rather short, the teeth short, the 2 upper ones broad, truncate, and united. Standard twice as long as the calyx ; keel rather shorter, incurved, obtuse. Pod only seen young.

Queensland. On the Maranoa river, *Mitchell.*

27. **D. squarrosa,** *Sm. in Ann. Bot.* i. 507, *and in Trans. Linn. Soc.* ix. 257. A glabrous or pubescent shrub, with slender terete or slightly angular branches. Leaves numerous, sessile, spreading or reflexed, cordate or ovate-lanceolate, tapering into a pungent point, mostly 3 to 4 lines long, veinless except the prominent midrib. Flowers small, solitary or 2 together, on pedicels of 2 to 3 lines, with a few minute bracts at their base. Calyx about 1 line long, the turbinate base very short, the teeth rather short, the 2 upper ones broader and shortly united, but not forming a truncate upper lip. Standard twice as long as the calyx ; keel rather shorter, much incurved, obtuse. Pod about 5 lines long.—DC. Prod. ii. 114.

N. S. Wales. Port Jackson to the Blue Mountains, *R. Brown, Sieber, n.* 348, and others. The calyx is that of *D. ulicina*, from which this species is chiefly distinguished by the very acuminate leaves, the slender pedicels, and the absence of thorny branches.

Var. *villifera.* Branches and younger leaves pubescent or villous.—*D. villifera*, A. Cunn.; Benth. in Ann. Wien. Mus. ii. 76.

Queensland. Brisbane river, Moreton Bay, *A. Cunningham, Fraser, F. Mueller*, and others.

28. **D. ulicina,** *Sm. in Ann. Bot.* i. 506, *and in Trans. Linn. Soc.* ix. 256. A rigid, bushy shrub, attaining sometimes several feet, the branches more or less angular, the smaller ones generally ending in short thorns, glabrous or hirsute with spreading hairs. Leaves from broadly ovate to lanceolate or linear, ending in a pungent point, usually under ½ in., but occasionally attaining 1 in. when narrow and luxuriant, very rigid, veinless, except the midrib. Flowers solitary or clustered or rarely in umbels of 3 or 4, on a short common peduncle. Bracts very small. Pedicels short. Calyx 1 to 1¼ lines long, the turbinate base very short, the 2 upper teeth rather broader than the others, but scarcely united at the base. Standard very broad, more than twice as long as the calyx; keel shorter, much incurved, obtuse. Pod 4 to 5 lines long.—DC. Prod. ii. 114; Lodd. Bot. Cab. t. 44; Paxt. Mag. iv. 29, with a fig.; *D. ulicifolia,* Andr. Bot. Rep. t. 304; *D. umbellulata,* DC. Prod. ii. 114 (partly); Hook. f. Fl. Tasm. i. 82, not of Sm.; *D. genistoides,* Lodd. Bot. Cab. t. 1552.

Queensland. Moreton Bay and Burnett river, *F. Mueller.*

N. S. Wales. Port Jackson, *R. Brown, Sieber, n.* 353, and others; northward to New England, *C. Stuart;* and southward to Twofold Bay, *F. Mueller.*

Victoria. From Gipps' Land and Watson's Promontory to the Murray and the Glenelg, *F. Mueller, Robertson,* and others.

Tasmania. Derwent river, *R. Brown;* most abundant throughout the island, *J. D. Hooker.*

S. Australia. Lynedoch Valley and Crystal Brook, *F. Mueller;* Encounter Bay, *Whittaker;* Mount Serle, *Warburton.*

The following forms appear in their extremes very distinct, but are connected by many gradations :—

a. *subumbellata.* Glabrous and luxuriant. Leaves narrow, ½ to 1 in. long. Common peduncle often 1 or even 2 lines long.—*D. umbellata,* Labill. Pl. Nov. Holl. i. 107, t. 137.—Victoria and Tasmania.

b. *communis.* Glabrous or hirsute. Leaves lanceolate, mostly under ½ in. Pedicels very short, clustered.—From N. S. Wales to Tasmania.

c. *ruscifolia.* Glabrous or hirsute, with the inflorescence of the common form, but the leaves broader, mostly ovate, always pungent and under ¼ in. long.—*D. ruscifolia,* A. Cunn.; Benth. in Ann. Wien. Mus. ii. 75; Schlecht. Linnæa, xx. 665.—Victoria, Tasmania, and S. Australia.

d. *angustifolia.* Glabrous. Leaves linear, but flat, not terete nor laterally compressed, as in *D. genistifolia,* and the calyx quite that of *D. ulicina.*—Queensland and N. S. Wales.

29. **D. acicularis,** *Sm. in Ann. Bot.* i. 506, *and in Trans. Linn. Soc.* ix. 255. An erect shrub, with virgate branches, glabrous or pubescent. Leaves crowded, linear, pungent-pointed, mostly about 1 in. long, the margins revolute, the midrib conspicuous at the base only. Flowers solitary or clustered, the pedicels very short. Calyx 1½ to nearly 2 lines long, the turbinate base very short, the 5 teeth nearly equal, lanceolate, and scarcely shorter than the tube. Standard twice as long as the calyx; keel shorter, obtuse. Pod 4 to 5 lines long, acuminate.—DC. Prod. ii. 114; Bot. Mag. t. 2679; Lodd. Bot. Cab. t. 1234 (leaves much less crowded, but probably the same species).

N. S. Wales. Port Jackson, *R. Brown, Sieber, n.* 347, and others; arid bushy skirts of Liverpool Plains, *A. Cunningham.* In young specimens, 1 or 2 of the lowest leaves are occasionally oblong-lanceolate, 1½ to 2 in. long, and not pungent.

SERIES VI. TERETIFOLIÆ.—Leaves terete or slightly compressed or rarely vertically dilated towards the top, at length articulate on the stem and not decurrent, usually short or pungent-pointed, and never horizontally flattened. Flowers solitary, clustered, or shortly racemose.

30. **D. pachyphylla,** *F. Muell. Fragm.* iv. 15. Very rigid, glabrous, and often becoming very glaucous, the branches thick and terete. Leaves numerous, divaricate, thickly oblong-conical, tapering into a pungent point, ½ to ¾ in. long, nearly three lines diameter at the base, terete or slightly compressed laterally, quite smooth, without ribs or veins, of a dense pithy consistence. Flowers several, in loose racemes, shorter than the leaves. Pedicels nearly as long as the calyx. Bracts minute. Calyx 1½ lines long, including the narrow turbinate base, the teeth minute. Standard half as long again as the calyx; keel very obtuse. Pod about ½ in. long, of the same blue-glaucous colour as the rest of the fruiting specimens.

W. Australia, *Drummond, 5th Coll. n.* 45; mountains near Gardner and Phillips rivers, *Maxwell.*

31. **D. teretifolia,** *R. Br. Herb.* A glabrous shrub, of 2 to 3 ft., with numerous, erect, terete branches. Leaves erect, cylindrical, ¾ to 1½ in. long, about 1 line diameter, with a short pungent point, smooth or obscurely ciliate. Racemes loose, few-flowered, not exceeding the leaves. Pedicels longer than the calyx. Bracts very small. Calyx above 2 lines long, including the narrow turbinate base, the teeth short, the 2 upper ones connate. Standard nearly twice as long as the calyx; lower petals shorter, the keel much incurved, almost rostrate. Pod ¾ in. long or rather more, very coriaceous.

W. Australia. King George's Sound, *Baxter;* Phillips Ranges and Cape Arid, *Maxwell.* The species is evidently allied to the eastern *D. genistifolia,* but readily distinguished as well by the foliage as by the loose inflorescence, larger flowers, and large thick pod.

32. **D. genistifolia,** *A. Cunn.; Benth. in Ann. Wien. Mus.* ii. 75. A glabrous shrub, with slender, slightly sulcate branches. Leaves linear-terete or very slightly laterally compressed, divaricate, pungent-pointed, mostly ½ to 1 in. long, smooth or sulcate. Pedicels slender, 1½ to 2 lines long, in clusters or exceedingly short racemes. Bracts small, obovate. Calyx 1½ lines long, including the narrow, almost stalk-like, turbinate base; the teeth very short, the 2 upper ones broad, truncate, and united. Petals twice as long as the calyx, of nearly equal length, the keel obtuse. Pod about 4 lines long.

Queensland. Moreton Bay, *Fraser.*
N. S. Wales. Port Jackson, *R. Brown;* Paramatta, *Woolls;* Williams river, *Backhouse;* Hunter's River to the south-west of Mount Cunningham and open forest land at Illawarra, *A. Cunningham;* New England, *C. Stuart.*
Victoria. Wimmera, *Dallachy.*
S. Australia. Crystal Brook and towards Mount Remarkable, *F. Mueller.*
Var. *colletioides.* Leaves rather shorter, terete.—*D. colletioides,* A. Cunn.; Benth. in Ann. Wien. Mus. ii. 76.—Forest land near Bathurst, *A. Cunningham.* Both forms of this species may be readily distinguished from *D. ulicina* by the calyx, as well as by the leaves never flattened horizontally; the keel is much more obtuse and the pod much smaller than in the western *D. colletioides,* Meissn.

33. **D. hakeoides,** *Meissn. in Pl. Preiss.* i. 47. A glabrous, erect,

rigid shrub, with terete or slightly compressed sulcate branches, allied in foliage to *D. genistifolia* and *D. incrassata*, but with the inflorescence and bracts of *D. juncea*. Leaves terete or very slightly compressed, rigid and pungent, in some specimens the lower ones 4 to 8 in. long and almost erect, the upper ones 1 to 1½ in., slender and divergent, in others all stout, 1 to 1½ in. long, or in one variety very few, short, and divaricate or recurved, almost as in *D. brevifolia*. Flowers usually small, in very short, sessile racemes. Bracts imbricate, concave, the outer ones short, the inner ones 1½ in. long, concealing the rhachis and pedicels at the time of flowering, often fallen off from the fruiting raceme. Calyx 1 line long, with a very short turbinate base and small teeth, the 2 upper ones truncate and united. Petals twice as long as the calyx, nearly equal in length, the keel much curved, rather acute or almost acuminate. Pod 4 to 5 lines long.

W. Australia. Swan River and to the northward, *Drummond, n.* 238, 4*th Coll. n.* 136, *and* 6*th Coll. n.* 16, *Preiss, n.* 1156 *and* 1157 ; Murchison river, *Oldfield.*

Var. *subnuda.* Leaves few, under ½ in. long, very divaricate or recurved, the lower ones of each branch reduced to small scales.—*Drummond, n.* 42.

Var. *major.* Bracts and flowers considerably larger, but I find no other difference.— Granite hills north from Cape Paisley, *Maxwell.*

34. **D. colletioides,** *Meissn. in Pl. Preiss.* i. 48, *not of A. Cunn.* Closely allied to the terete-leaved forms of *D. incrassata* and to the var. *colletioides* of *D. genistifolia*, having the narrow base to the calyx of the latter species, but the flowers and pods are much larger than in either. The calyx, including the base, is 2 lines long when fully out, the keel twice as long as the calyx, much incurved, and almost rostrate, as in *D. incrassata.* Pod ¾ in. long.

W. Australia. Swan River, *Fraser ;* south districts ?, *Drummond, 2nd Coll. n.* 104 ; near Albany, *Preiss, n.* 1180, *partly (and* 1163 ?) ; King George's Sound, *Maxwell ;* Geographe Bay, *Oldfield.* It is doubtful whether this may not be a large-flowered variety of *D. incrassata.* In some of Drummond's specimens the base of the calyx is much broader than in others. Under *n.* 1180 of Preiss, I have generally found this and the terete-leaved form of *D. incrassata* mixed.

35. **D. reversifolia,** *F. Muell. Fragm.* i. 145. A bushy rigid intricately-branched shrub, glabrous and somewhat glaucous, very nearly allied to *D. colletioides*, Meissn., with the same inflorescence flowers and fruit, but the branches are very flexuose, and the leaves numerous, scarcely ½ in. long, very rigid and pungent and remarkably reflexed.

W. Australia. Fitzgerald ranges, *Maxwell.* This plant, notwithstanding its singular aspect, may very probably be only an accidental form of *D. colletioides.*

36. **D. incrassata,** *Sm. in Trans. Linn. Soc.* ix. 255. A rigid much branched glabrous shrub, with terete or slightly compressed branches. Leaves divaricate, always very rigid and pungent-pointed, often appearing continuous with the stem, but at length articulate and not really decurrent, usually ½ to 1 in. long, either terete and tapering to a point or vertically dilated towards the end, gradually narrowing to the base. Pedicels 1 to 2 lines long, usually several together on a very short common peduncle. Bracts very small. Calyx 1¼ or rarely 1½ lines long with a very short turbinate base, the teeth very short, the 2 upper ones broad, truncate and united. Petals fully twice as long as the calyx, the keel much incurved, almost rostrate. Filaments free

G 2

as in most *Daviesias*. Pod 6 to 7 lines long, rather turgid.—DC. Prod. ii.
114 ; *Acacia dolabriformis*, Wendl. Comm. Acac. 55 ; *D. physodes*, A. Cunn.
in G. Don, Gard. Dict. ii. 125 ; Bot. Mag. t. 4244 (a cultivated form with
remarkably dilated leaves); Meissn. in Pl. Preiss. i. 49 ; *D. Benthamii*, Meissn.
in Pl. Preiss. i. 48 (a slender-leaved form with small flowers) ; *D. brachyphylla*, Meissn. l. c. i. 49 (a short-leaved form).

S. Australia. Port Lincoln, *Wilhelmi ;* Kangaroo Island, *Waterhouse*.
W. Australia. King George's Sound, *R. Brown, Menzies*, and others; and thence to
Vasse and Swan rivers, *Huegel, Drummond, n.* 236, 241, 242, etc., *Preiss, n.* 1161, 1162,
1164, 1165, 1168, 1169, 1170, and eastward in the plate above quoted, but all the
forms are so frequently intermixed as to prevent the characterizing any distinct varieties.
The species is exceedingly variable in its leaves, sometimes all terete and either mostly
under ½ in. and recurved, or those of the principal branches 1 in. long, straight or almost
incurved ; sometimes on the same specimen a few more or less dilated upwards, very rarely
all dilated and scarcely ever so much so as represented in the plate above quoted, but all the
forms are so frequently intermixed as to prevent the characterizing any distinct varieties.
The slender-leaved forms come very near in appearance to *D. genistifolia*, but the turbinate
base of the calyx is much shorter and the pod much larger and more turgid. In Drummond's specimens, 5th Coll. n. 37, the leaves are very few, from ¼ to ½ in. long, almost connecting the species with *D. brevifolia*.

37. **D. brevifolia,** *Lindl. in Mitch. Three Exped.* ii. 201. An erect
shrub with broom-like, rigid, terete, somewhat flexuose branches. Leaves
few, distant, linear-conical, rigid and pungent, 1 to 3 lines long, the thick
base continuous with the stem but not decurrent. Flowers usually several
together on very short pedicels, the common peduncle rarely 1 line long.
Bracts very small. Calyx about 1¼ lines long, the turbinate base very short,
the teeth very short, especially the 2 upper ones, which are very obtuse, truncate or scarcely prominent. Keel twice as long as the calyx, rostrate. Pod
½ in. long, turgid.

Victoria. Glenelg river, *Mitchell, Robertson ;* Scrub of Concorooa, *Schulzer ;* Wimmera, *Dallachy ;* Grampians, *F. Mueller*.
S. Australia. Encounter Bay, *F. Mueller ;* Mount Lofty, *Whittaker*. The species
is nearly allied to *D. incrassata*, but the leaves, reduced to short spines, are more continuous
with the stem.
Var. (?) *ephedroides*. Branches often clustered, sometimes quite leafless, but usually with
a very few small spine-like leaves.
W. Australia, *Drummond, n.* 137. The specimens are very bad, and may possibly
belong to a form of *D. aphylla*.

38. **D. Preissii,** *Meissn. in Pl. Preiss.* i. 50. A glabrous much branched
rigid shrub, resembling *D. incrassata* and *D. colletioides*, but readily distinguished by the calyx and stamens. Leaves linear-falcate, vertically compressed especially towards the base or rarely terete, straight or falcate, tapering into a pungent point, ½ to 1 in. long, attached by a broad base and, when
flat, usually striate. Flowers 2 or 3 together, on pedicels of 2 or 3 lines on
a short common peduncle. Bracts minute. Calyx rigid, turbinate-campanulate, about 1½ lines long, the teeth erect, short, somewhat acute, the 2 upper
ones distinct, rather smaller and not truncate. Keel more than twice as long
as the calyx, much incurved but scarcely rostrate. Filaments much dilated,
especially the outer ones, and cohering in a tube but readily separable. Pod
probably large, but only seen young.

W. Australia. King George's Sound, *A. Cunningham, Baxter ?, Drummond, 5th*

Coll. n. 38; near Hassell, Hay district, *Preiss, n.* 1153; Vasse river and Gordon ranges, *Oldfield.* In Cunningham's specimens the leaves are vertically 1 to 2 lines broad; in others (from Baxter?) they are thick and almost terete; in Drummond's the lower ones are flattened and striate, the upper ones terete as in *D. incrassata.*

39. **D. spinosissima,** *Meissn. in Pl. Preiss.* i. 51.   A rigid glabrous shrub, with thick terete sulcate branches.   Leaves crowded, linear, almost terete or laterally compressed, thick and rigid with pungent points, mostly nearly ½ in. long, divergent and often somewhat recurved, the base broad, apparently almost decurrent when young, but at length articulate.   Flowers solitary or rarely 2 together, the pedicels 2 or 3 lines long.   Calyx with a striate turbinate base, nearly 2 lines long, the teeth broad, obtuse or almost acute, the 2 upper ones shorter but not truncate.   Standard fully 5 lines diameter; keel twice as long as the calyx, incurved but scarcely rostrate.   Filaments much dilated, cohering in a tube, but readily separable.   Pod about ½ in. long, thickly turgid.

W. Australia. King George's Sound, *Baxter, Harvey;* near Mount Wuljenup, *Preiss, n.* 1152; Kalgan river, *Oldfield.* The species differs chiefly from *D. Preissii* in its short crowded leaves.

40. **D. pachylina,** *Turcz. in Bull. Mosc.* 1853, i. 263.   A low glabrous or minutely pubescent shrub, with numerous slender somewhat angular and often flexuose branches.   Leaves vertically compressed, narrow-linear, pungent-pointed, mostly ¾ to 1 in. long, the edges much thickened.   Flowers few together in loose pedunculate racemes or almost solitary.   Bracts few, very small.   Pedicels short.   Calyx 1½ lines long, the teeth all narrow-acuminate nearly as long as the tube, the 2 upper ones united to the middle.   Standard twice as long as the calyx and not so broad as in some species; lower petals as long, keel incurved.   Stamens free.   Pod only seen young.

W. Australia, *Drummond, 5th Coll. n.* 43.   This species bears much resemblance to *D. anceps* in its flowers and peculiar calyx, but the habit and foliage are quite different.

SERIES VII. VERTICALES.—Leaves vertically flattened, often attached by a broad base but scarcely decurrent, usually pungent-pointed.   Flowers solitary, clustered, or umbellate-racemose.

41. **D. quadrilatera,** *Benth. in Lindl. Swan Riv. App.* 14.   A glabrous glaucous shrub, with rigid branches, terete or nearly so.   Leaves vertical, erect, oblong-rhomboidal, from about 4 lines to nearly 1 in. long, and about half as broad, rounded at the upper end towards the stem with an erect point at the outer angle, truncate at the base with a reflexed point at the outer angle, either sessile or slightly decurrent, thickly coriaceous, obscurely severalnerved.   Peduncles about as long as the leaves, bearing an umbel of 3 to 6 flowers which rarely breaks out into a very short raceme.   Bracts minute.   Calyx 1½ lines long, including the narrow turbinate base, the teeth very short and nearly equal.   Petals nearly equal, twice as long as the calyx, the keel much curved.   Pod above ½ in. long.—*Meissn. in Pl. Preiss.* i. 52.

W. Australia. Swan River, *Drummond, 1st Coll. and n.* 228, *Preiss, n.* 1139; Murchison river, *Oldfield.*

42. **D. striata,** *Turcz. in Bull. Mosc.* 1853, i. 264.   Glabrous, with long thick rigid branches, terete and striate or slightly angular.   Leaves crowded,

rigid, vertically flat, falcate ovate or almost rhomboidal, pungent-pointed, attached by their broad base, mostly about $\frac{1}{2}$ in. long and $\frac{1}{4}$ in. broad, but sometimes nearly as broad as long, more or less striate. Flowers usually several, clustered on a very short common peduncle with minute bracts. Calyx about $1\frac{1}{2}$ lines long including the turbinate base, the teeth very short, the 2 upper ones united and almost truncate. Keel much incurved, almost rostrate. Outer filaments dilated but scarcely cohering. Pod not seen.—*D. adnata,* F. Muell. Fragm, ii. 105.

W. Australia, *Drummond, 4th Coll. n.* 29 ; South-West Bay, *Maxwell.*

43. **D. polyphylla,** *Benth. in Lindl. Swan Riv. App.* 14. Glabrous and much-branched, the young branches angular or sulcate. Leaves vertically flattened but thick and rigid, linear or lanceolate, usually falcate, with pungent or rarely almost callous points, $\frac{1}{4}$ to 1 in. long, the edges thickened, usually narrowed towards the base. Flowers small, usually clustered on a very short peduncle, with minute bracts. Pedicels slender, 1 to 2 in. long. Calyx scarcely 1 line long, the turbinate base very short, the teeth small, the 2 upper ones very obtuse or truncate. Petals more than twice as long as the calyx, the keel much incurved, almost rostrate. Filaments slender. Pod 5 to 6 lines long, turgid.—Meissn. in Pl. Preiss. i. 50.

W. Australia. S. coast, *R. Brown ;* Swan River, *Fraser, Drummond,* 1*st Coll. and n.* 231, 232, 233 ; *Preiss, n.* 1149.

D. angulata, Benth. in Lindl. Swan Riv. App. 14; Meissn. in Pl. Preiss. i. 50, proves to be only a luxuriant state of *D. polyphylla* passing sometimes into the common form on the same specimen.

44. **D. microphylla,** *Benth.* Glabrous. Branches striate with raised lines, the smaller branchlets ending in stout thorns. Leaves vertically flat, thick and rigid, ovate or lanceolate, pungent-pointed, 1 to 2 or rarely 3 lines long, the edges thickened, the base broad, but usually narrower than the middle of the leaf. Pedicels in our specimens always solitary, slender, 1 to 2 lines long, with minute bracts at the base. Calyx about 1 line long, the turbinate base very short, the teeth small, the 2 upper ones very obtuse or truncate. Petals and pod of *D. polyphylla.*—*D. incrassata,* Meissn. in Pl. Preiss. i. 49, not of Sm.

W. Australia. Swan River and Darling range, *Preiss, n.* 1150 *and* 1155 ; *Drummond, n.* 32. The spinescent branchlets as well as the broad vertical base of the leaves readily distinguish this from all the forms I have seen of the true *D. incrassata,* besides the solitary flowers, which may not prove constant.

Series VIII. Decurrentes.—Leaves terete, or vertically compressed or with a prominent wing-like dorsal midrib, usually pungent-pointed, decurrent at the base into raised angles along the branches. Flowers clustered or shortly racemose.

45. **D. flexuosa,** *Benth. in Hueg. Enum.* 32, *and in Ann. Wien. Mus.* ii. 75. Glabrous and very rigid ; branches very angular with the decurrent bases of the leaves, the lower portion straight, with linear, vertically flattened leaves of 1 to 2 in. or more, the flowering branches very flexuose, with small shorter leaves often nearly terete, all tapering into a pungent point. Flowers small, clustered on a very short common peduncle with very small bracts.

Pedicels rarely as long as the calyx. Calyx about 1¼ lines long, shortly turbinate at the base, the teeth short, the upper ones truncate and united. Filaments rather flat, but free. Keel much incurved but obtuse. Pod about ¾ in. long.—Meissn. in Pl. Preiss. i. 51.

W. Australia. King George's Sound, *Huegel* and others; *Drummond, 5th Coll. n.* 43: Sterling Terrace, *Preiss, n.* 1180 (*in part*).

46. **D. pectinata,** *Lindl. in Mitch. Three Exped.* ii. 151. Glabrous and very rigid. Branches all very prominently angled with the decurrent bases of the leaves and not flexuose. Leaves vertically flat, lanceolate or linear-lanceolate, tapering from the broad decurrent continuous base to the pungent point, the lower ones often above 1 in. long, the upper ones under ½ in., horizontally divaricate and straight or falcate and recurved, more rarely incurved, varying in breadth at the base from 1 to nearly 3 lines. Flowers very small, in dense axillary clusters or very short racemes. Bracts ovate, concave, longer than in the preceding species but not imbricate. Pedicels exceedingly short. Calyx about 1 line long, obtuse at the base, the teeth very short, the 2 upper ones truncate. Keel scarcely twice as long as the calyx, incurved, obtuse. Pod 5 to 6 lines long.—*D. decurrens* and *D. prionoides,* Meissn. in Pl. Preiss. i. 52; *D. latipes,* F. Muell. in Linnæa, xxv. 390.

Victoria. Near Mount Hope, *Mitchell;* Wimmera, *Dallachy.*
S. Australia. Dombey Bay, *Wilhelmi.*
W. Australia. From King George's Sound to Swan River, *Drummond, n. 234 and 235, Preiss, n.* 1141, 1147, *and* 1148, *and* others; northward to Murchison river, *Oldfield;* and eastward to Cape Knobb and Cape Le Grand, *Maxwell.*
In the majority of specimens the leaves are recurved-falcate, but in some they are all or mostly straight or incurved. In a few of Baxter's, from King George's Sound, they are occasionally dilated upwards or oblong-falcate.

47. **D. trigonophylla,** *Meissn. in Pl. Preiss.* ii. 213. A rigid shrub, glabrous or minutely and densely pubescent, the branches broadly winged by the decurrent bases of the leaves. Leaves under ½ in. long, continuously decurrent, divaricate, recurved and tapering into a pungent point as in *D. pectinata,* but the upper edge dilated into a horizontal concave or channelled lamina, the midrib on the back forming the prominent decurrent wing. Flowers few, in little axillary clusters or very short racemes, with very small bracts. Calyx 1¼ lines long, the upper teeth truncate and united. Keel much curved. Pod fully ½ in. long, very turgid.

W. Australia, *Drummond, 3rd Coll. n.* 77, *and 5th Coll. n.* 42.

48. **D. epiphylla,** *Meissn. in Bot. Zeit.* 1855, 27. A glabrous, very glaucous shrub, with broad thick flat phyllodineous branchlets, pinnately lobed, the lobes (or decurrent vertical leaves) triangular or lanceolate, ¼ to ½ in. long, tapering to a pungent point, and occasionally branchlets proceed from one of the faces instead of the edges of the branches. Flowers not seen. Fruiting pedicels 3 to 4 lines long, solitary or 2 or 3 together on a very short common peduncle, usually inserted in the centre of one of the faces of the phyllodineous branch. Bracts several, small, the lower ones imbricate. Fruiting calyx broad and oblique, about 4 lines long, including a turbinate base of about 1 line, the teeth short and broad, the 2 upper ones distinct but very obtuse. Pod coriaceous, more than 1 in. long.

W. Australia. Gardener's Range, between Moore and Murchison rivers, *Drummond,* *6th Coll. n.* 18.

SERIES IX. APHYLLÆ.—Leaves none.   Flowers solitary, clustered or rarely racemose.

49. **D. euphorbioides,** *Benth.* Erect, glabrous and glaucous. Branches cylindrical, not sulcate, very thick, of a pithy texture inside, the small branchlets erect, several inches long, 3 to 5 lines diameter, contracted at the base. Leaves replaced by minute scattered prickly conical scales, rarely 1 line long. Flowers several in very short racemes or clusters.   Pedicels about as long as the calyx.   Bracts minute.   Calyx broad, 1 line long or rather more, the teeth very short, the 2 upper ones truncate.   Petals and pod only seen very imperfect, yet evidently showing the genus.

W. Australia, *Drummond,* 3rd *Coll. n.* 76.

50. **D. divaricata,** *Benth. in Hueg. Enum.* 31, *and in Ann. Wien. Mus,* ii. 75.   A glabrous tall erect shrub, quite leafless, paniculately branched, with divaricate sulcate spinescent branchlets, the leaves replaced by minute obtuse or mucronate scales.   Racemes short and few-flowered, inserted on the smaller spinescent branchlets.   Bracts very minute.   Pedicels slender, usually longer than the calyx.   Calyx about 2 lines long, including the narrow turbinate stalk-like base, the teeth very short broad and obtuse, the 2 upper ones almost truncate.   Standard about twice as long as the calyx; keel rather shorter, obtuse.   Pod only seen young.—Meissn. in Pl. Preiss. i. 47.

W. Australia. Swan River, *Huegel, Drummond,* 1st *Coll. and n.* 110 ; *Preiss, n.* 1166 *and* 1167 ; Vasse river, Point Gregory, and Murchison river, *Oldfield.* Some almost leafless specimens of *D. horrida* much resemble this species, but may always be known by the calyx.

51. **D. paniculata,** *Benth. in Hueg. Enum.* 31, *and in Ann. Wien. Mus.* ii. 75.   A glabrous erect leafless shrub, the branches slender, terete, paniculate, the lower barren ones often spinescent, the flowering ones unarmed. Leaves replaced by minute scales.   Flowers on slender pedicels, irregularly racemose, forming a loose terminal dichotomous or trichotomous panicle. Bracts minute.   Calyx about 1 line long, besides a narrow stalk-like base scarcely distinguishable from the pedicel, but at length articulate upon it, the broad tube truncate or obscurely toothed.   Standard twice as long as the calyx ; keel rather shorter, obtuse.   Pod only seen young.

W. Australia. Swan River, *Huegel.* This species, which I have seen in no other collection, differs widely from the rest of the genus in inflorescence.

52. **D. aphylla,** *F. Muell. Herb.* A glabrous shrub of 2 to 3 ft., with leafless terete broom-like branches, neither furrowed nor spinescent. Leaves replaced by minute, often scarcely perceptible scales.   Racemes lateral, very short.   Bracts very small.   Pedicels slender, shorter than the calyx.   Calyx 1½ lines long, with a short turbinate base, the teeth short, the 2 upper ones broad, more or less united.   Standard about twice as long as the calyx ; keel shorter, broad, curved, very obtuse.   Pod only seen young.

W. Australia. Oldfield river, *Maxwell.* Allied to some forms of *D. brevifolia,* but there are no leaves whatever, the calyx is much smaller, and the keel not at all rostrate.

53. **D. juncea,** *Sm. in Trans. Linn. Soc.* ix. 260.   A glabrous shrub or

undershrub, with long erect leafless rush-like slightly branched stems, terete and smooth or slightly sulcate and not spinescent. Leaves replaced by minute scales, very rarely forming short pungent points. Racemes lateral, distant, very short and few-flowered, the rhachis and pedicels concealed at the time of flowering by rigid chaffy imbricate bracts, the outer ones broad and short, the inner ones narrower and often 2 lines long, all very obtuse and striate. Calyx 1 to 1½ lines long, with short teeth, the 2 upper ones truncate, united or distinct. Standard about twice as long as the calyx; keel shorter, curved and almost acute. Pod ¾ in. long, very acute.—DC. Prod. ii. 114; Meissn. in Pl. Preiss. i. 47.

W. Australia. King George's Sound, *Menzies, R. Brown*, and others, and thence to Swan River, *Drummond, Preiss, n.* 1159, 1160, *and* 1181, and others. I do not feel certain that there are not two species here confounded. In the King George's Sound specimens I find the calyx usually 5-toothed, as described by Smith; those from Swan River are numerous, all in flower, and one also in fruit; they belong to two varieties, both with the 2 upper teeth of the calyx united in a truncate upper lip. In one of these forms the flowers are very small and the calyx-teeth very short, and they only differ from *D. hakeoides* in the absence of leaves, except very rarely a few very small ones on barren branches; in the other form the flowers are rather larger and the upper lip of the calyx is very prominent.

54. D. alata, *Sm. in Trans. Linn. Soc.* ix. 259. Branches from a short woody base, long and virgate, leafless, flat or 3-angled, with the angles more or less winged and quite glabrous. Leaves replaced by minute scales. Racemes very short, almost capitate, the pedicels very short. Bracts almost imbricate, the inner ones often 2 lines long, but not so rigid as in *D. juncea* and *D. hakeoides*, and often fringed at the edge. Calyx about 2 lines long, the teeth lanceolate, as long as the tube, the 2 upper ones broader but distinct. Standard not twice as long as the calyx, lower petals shorter. Pod 4 to 5 lines long.—Bot. Reg. t. 728; DC. Prod. ii. 114.

N. S. Wales. Port Jackson, *R. Brown, Sieber, n.* 356, and others.

55. D. anceps, *Turcz. in Bull. Mosc.* 1853, i. 266. A glabrous shrub, about 2 ft. high, with slender leafless branches, flat but not winged. Leaves replaced by minute scales. Flowers solitary or 2 or 3 together in a short raceme at the ends of small axillary branches, the pedicels each in the axil of an acute scale-like bract. Calyx about 2 lines long, the teeth subulate-acuminate, as long as the tube, the 2 upper ones united to the middle. Standard twice as long as the calyx; keel nearly as long, incurved, obtuse. Pod only seen young.

W. Australia, *Drummond, 5th Coll. n.* 86; Phillips river, *Maxwell.*

13. AOTUS, Sm.

Calyx, 2 upper lobes broader and more or less united in an upper lip. Petals rather long-clawed; standard nearly orbicular, longer than the lower petals; wings oblong; keel incurved. Stamens free. Ovary sessile or stipitate, with 2 ovules on short straight funicles; style filiform, with a minute terminal stigma. Pod ovate, flat or turgid, 2-valved. Seed reniform, without any strophiole (except in *A. gracillima*).—Shrubs, with branches often virgate. Leaves simple, scattered or ternately whorled, the margins recurved or revolute. Stipules none. Flowers in axillary clusters, often in threes, on short pedicels, or rarely in short terminal racemes. Bracts small and very deciduous; bracteoles none. Ovary villous.

The genus is limited to Australia. It differs from *Pultenæa* chiefly in the want of stipules and bracteoles, and in most cases in the want of any strophiole to the seeds. It is, in most cases, readily distinguished from *Dillwynia* and *Latrobea* by the recurved, not incurved, margins of the leaves, independently of the seeds.

Leaves scattered or imperfectly whorled, narrow, with much revolute
    margins. Calyx usually above 1½ lines long.
Leaves linear, obtuse or with recurved points. Keel purple.
    Calyx under 2 lines, the upper lobes falcate or truncate, united
        above the middle. Plant tomentose, hoary or nearly gla-
        brous.
        Seeds strophiolate (western species) . . . . 1. *A. gracillima.*
        Seeds not strophiolate (eastern species) . . . . 2. *A. villosa.*
    Calyx nearly 2 lines, the lobes nearly equal. Leaves softly to-
        mentose . . . . 3. *A. mollis.*
    Calyx above 2 lines, the upper lobes united above the middle.
        Flowers axillary . . . . 4. *A. Preissii.*
        Flowers crowded, in short, terminal, leafless racemes . . . 5. *A. phylicoides.*
Leaves mostly lanceolate and almost acute.
    Branches villous. Flowers large. Keel yellow. Ovary sti-
        pitate . . . . 6. *A. lanigera.*
    Branches minutely hoary. Keel purple. Ovary sessile . . 7. *A. genistoides.*
Leaves all or almost all in whorls of 3, folded lengthwise or broad
    with recurved margins. Calyx small, membranous. Pedicels
    recurved.
Leaves lanceolate, folded lengthwise and prominently keeled . . 8. *A. carinata.*
Leaves oval-oblong or broadly lanceolate, not reticulate, silky-
    villous when young. Ovary sessile . . . . 9. *A. passerinoides.*
Leaves cordate-ovate, sharp pointed. Stipules none. Ovary sti-
    pitate . . . . 10. *A. cordifolia.*

Sphærolobium euchilus has almost the technical characters of *Aotus*, but the lax foliage and long-pedicellate flowers give it a different habit, and the calyx is quite distinct.

Aotus Würthii, Regel, in Bot. Zeit. 1851, 596, is described with the leaves channelled above and convex underneath, and is therefore probably a *Dillwynia*, perhaps *D. floribunda.*

1. **A. gracillima,** *Meissn. in Pl. Preiss.* i. 59. A tall shrub, with elongated branches, closely resembling the more slender forms of *A. villosa*, the branches hoary or slightly tomentose. Leaves narrow-linear, obtuse, 3 to 6 lines long, with closely revolute margins, nearly glabrous above, hoary or tomentose underneath. Flowers rather smaller than in *A. villosa*, bright-coloured and very numerous, forming long dense leafy racemes below the ends of the branches. Calyx tomentose, scarcely 1½ lines long, the lobes rather shorter than the tube, the 2 upper ones broader and more united. Petals fully twice as long as the calyx. Pod rather smaller than in *A. villosa*. Seeds (only seen in A. Cunningham's specimens) like those of *A. villosa*, except that they have a deeply-lobed membranous strophiole.—Bot. Mag. t. 4146 ; *A. intermedia*, Meissn. in Pl. Preiss. i. 60.

W. Australia. King George's Sound and adjoining districts, *A. Cunningham, Baxter*, and others, *Preiss, n.* 863, 864, and 871 ; Swan River, *Drummond, 1st Coll. n.* 246. Were it not for the strophiole of the seeds, which remains to be verified on other specimens, I should have considered this as a slender variety of *A. villosa.*

2. **A. villosa,** *Sm. in Ann. Bot.* i. 504, *and in Trans. Linn. Soc.* ix. 249.

A bushy heath-like shrub, the branches terete, often long and virgate, usually densely tomentose or softly villous, rarely hoary or almost glabrous. Leaves narrow-linear or rarely oblong, obtuse or with recurved points, 3 to 6 lines long, the margins closely revolute, glabrous or pubescent above when young, the under surface pubescent, but usually concealed. Flowers yellow, with a purple or dark-coloured keel, axillary, solitary or in clusters of 2 or 3, often forming long leafy spikes or racemes below the ends of the branches. Pedicels short, without bracts or bracteoles. Calyx pubescent or villous, 1½ lines long or rather more, the lobes as long as the tube, the 2 upper ones broader, falcate and united to the middle. Standard twice as long as the calyx, emarginate; lower petals nearly as long, the keel incurved, very obtuse. Ovary stipitate. Pod 2 to 2½ lines long, somewhat turgid. Seeds not strophiolate.—Bot. Mag. t. 949; DC. Prod. ii. 108; Lodd. Bot. Cab. t. 1353; Hook. f. Fl. Tasm. i. 83; *Pultenæa villosa*, Andr. Bot. Rep. t. 309; *Pultenæa ericoides*, Vent. Jard. Malm. t. 35; *Daviesia ericoides*, Pers. Syn. i. 454; *Aotus ferruginea*, Labill. Pl. Nov. Holl. i. 104, t. 132; *Aotus ericoides*, G. Don, Gen. Syst. ii. 120; *Pultenæa rosmarinifolia* and *P. virgata*, Sieb. Pl. Exs.; *Aotus virgata*, DC. Prod. ii. 108.

Queensland. Moreton Island, *F. Mueller.*

N. S. Wales. Port Jackson, *R. Brown*, *Sieber*, n. 387, 389; *Fl. Mixt.* n. 581, and others; Port Stephens and Illawarra, *M'Arthur.*

Victoria. Common in wooded valleys and heath ground, Gipps' Land, near Brighton, etc., *F. Mueller.*

Tasmania. Most abundant throughout the island, occasionally covering many acres of ground, *J. D. Hooker.*

S. Australia, *Sturt.*

In a few (imperfect) specimens the leaves are all short and broad, almost ovate, in others they are fully ¼ in. long, thick and nearly glabrous, but obtuse, as in other forms of *A. villosa*, and with the calyx of the species. Minute truncate bracts may occasionally be seen, when the buds are as yet only ¼ line long.

Var. *subspinescens.* Closely tomentose. Branches short, divaricate and often spinescent. Leaves short and narrow.—In the Murray Desert, on the Wimmera, etc., *F. Mueller, Dallachy,* and others.

3. **A. mollis,** *Benth. in Mitch. Trop. Austr.* 236. Nearly allied to *A. villosa.* Branches densely velvety-tomentose. Leaves linear-oblong, obtuse, 4 to 8 lines long, the margins revolute, softly pubescent above, densely rusty-tomentose underneath. Flowers clustered in the axils, often arranged in irregular but distinct whorls, nearly sessile. Bracts about 1 line long, truncate, very deciduous. Calyx villous, nearly 2 lines long, the lobes nearly equal. Petals not twice as long as the calyx, the keel dark and very much incurved. Ovary shortly stipitate. Pod rather larger than in *A. villosa.* Seeds not strophiolate.

Queensland. From the Mantuan Downs to the Maranoa, *Mitchell.*

N. S. Wales. New England, *C. Stuart;* Clarence river, *Beckler.* It is possible that this may prove to be a variety of *A. villosa*, but, besides the indumentum and other minor characters, the more regular calyx appears to be constant.

4. **A. Preissii,** *Meissn. in Pl. Preiss.* ii. 214. A low shrub, with ascending or erect stems, apparently not exceeding 1 ft., pubescent or hirsute. Leaves linear, obtuse, rarely above ¼ in. long, the margins much revolute, scabrous or hirsute. Flowers axillary, solitary or clustered, as in *A. villosa.*

Bracts 1 line long, very obtuse, very deciduous. Calyx very villous, above 2 lines long, the lobes longer than the tube, all acute, the 2 upper ones falcate and united to the middle. Standard nearly $\frac{1}{2}$ in. long; lower petals shorter, the keel broad, much incurved, purple. Ovary distinctly stipitate. Pod not seen.

W. Australia. Swan River, *Drummond, 2nd (and 3rd?) Coll. n.* 95.

Var. *leiophylla,* Meissn. in Pl. Preiss. ii. 215. Leaves rather broader, glabrous and shining above; *A. procumbens,* Meissn. in Pl. Preiss. i. 60.—Swan River and southern interior, *Preiss, n.* 845 *and* 882; Robertson's Brook, *Maxwell.*

The species is very near *A. villosa,* differing chiefly in the longer and more deeply-cleft calyx.

5. **A. phylicoides,** *F. Muell. Herb.* Apparently a straggling shrub, the branches clothed with a whitish tomentum, usually close and dense, but sometimes loose and woolly. Leaves linear or slightly lanceolate-linear, obtuse or with recurved points, $\frac{1}{4}$ to $\frac{1}{2}$ in. long, with recurved or revolute margins, nearly glabrous and shining above when full-grown, tomentose underneath. Flowers in short, dense, terminal, leafless racemes. Pedicels rather shorter than the calyx. Bracts, as in other species, very deciduous. Calyx about 3 lines long, the lobes as long as the tube, the 2 upper ones much broader, very much falcate, and usually united to the middle. Standard about half as long again, wings and very broad keel not much shorter. Ovary nearly sessile. Pod broadly ovate, almost orbicular, about 3 lines diameter, the valves slightly convex. Seeds reniform, very strongly pitted, not strophiolate.

W. Australia. Murchison river and Port Gregory, *Oldfield.* The inflorescence is quite exceptional in the genus, and the calyx shows an approach to that of *Sphærolobium,* but the other characters, as well as the foliage, are quite those of *Aotus.*

6. **A. lanigera,** *A. Cunn.; Benth. in Ann. Wien. Mus.* ii. 78. A stouter shrub than *A. villosa,* the branches tomentose and villous, with soft spreading hairs. Leaves oblong-lanceolate or linear, mostly acute, $\frac{1}{2}$ to $\frac{3}{4}$ in. long, the margins revolute, hairy when young, at length glabrous, smooth and shining above, hoary underneath. Flowers axillary, as in *A. villosa,* but much longer, and all yellow. Bracts above 1 line long, very deciduous, though not so very early as in *A. villosa.* Calyx villous, $2\frac{1}{2}$ lines long, the lobes acuminate, the 2 upper ones rather broader and slightly falcate. Standard $\frac{1}{2}$ in. long; lower petals rather shorter. Ovary stipitate. Pod very villous, much flattened, very obtuse, above 3 lines long. Seeds not strophiolate.

Queensland. Islands of Moreton Bay, *A. Cunningham, Fraser, F. Mueller;* Wide Bay, *Bidwill;* also *Leichhardt.*

N. S. Wales. Port Macquarrie, *Backhouse.*

7. **A. genistoides,** *Turcz. in Bull. Mosc.* 1853, i. 268. Branches erect, virgate, minutely hoary-pubescent. Leaves mostly irregularly verticillate in threes, lanceolate and almost acute, 4 to 8 lines long, the margins revolute, coriaceous, glabrous or scabrous above, hoary-tomentose underneath. Flowers axillary, as in *A. villosa,* but larger. Bracts exceedingly deciduous. Calyx silky-pubescent, 2 lines long, the lobes shorter than the tube, all acute, the 2 upper ones broader, very falcate, and united to the middle. Standard fully twice as long as the calyx; lower petals shorter, the keel purple. Ovary sessile or scarcely contracted at the base, and much shorter than in the other

species. Style long and very slender. Pod ovate, very obtuse, turgid, about
2 lines long. Seeds not strophiolate.

W. Australia, *Drummond, 5th Coll. n.* 61 *and* 63.

8. **A. carinata,** *Meissn. in Pl. Preiss.* ii. 215. An elegant shrub, the
branches and foliage densely clothed with long, soft, silky hairs. Leaves
regularly verticillate in threes, and crowded into 6 rows, spreading, lanceolate, acute, under ½ in. long, folded lengthwise and prominently keeled underneath. Flowers solitary or 2 or 3 together in each axil, on exceedingly short
pedicels. Bracts minute, ovate. Calyx not 1 line long, membranous, the lobes
much shorter than the tube, the 2 upper ones rather broader and more united.
Standard fully 4 lines long; wings shorter; keel still shorter, purple, incurved. Ovary small and sessile; style very slender. Pod not seen.

W. Australia, *Drummond, 2nd Coll. n.* 102, *also (3rd Coll.?) n.* 86.

9. **A. passerinoides,** *Meissn. in Pl. Preiss.* i. 61. Stems erect,
simple or slightly branched, 1 to 1½ ft. high, densely and softly villous.
Leaves almost all in whorls of 3, oval-oblong or lanceolate, obtuse or
almost acute, under ½ in. long, the margins recurved or slightly revolute,
softly silky-villous on both sides, or at length glabrous above. Flowers axillary, clustered, the pedicels very villous, about ½ line long. Calyx membranous, about 1 line long, the lobes somewhat obtuse, the 2 upper ones broad,
falcate and united above the middle. Standard about 3 lines long; keel
shorter, purple. Ovary sessile; style filiform, hirsute. Pod not seen.

W. Australia. Near Albany, *Preiss, n.* 868.

10. **A. cordifolia,** *Benth. in Hueg. Enum.* 33, *and in Ann. Wien. Mus.*
ii. 78. An erect shrub of several ft., the branches terete, loosely pubescent, and
villous with fine spreading hairs. Leaves in whorls of 3, ovate-cordate, acute
and sometimes almost pungent, 4 to 8 lines long, herbaceous, the margins
slightly recurved, undulate and almost denticulate, finely reticulate, hirsute or
at length glabrous. Flowers axillary, solitary or 2 or 3 together, the pedicels very short. Bracts small, ovate, concave. Calyx slightly pubescent,
membranous, 1 to 1¼ lines long, the lobes rather shorter than the tube, the
2 upper ones truncate and united above the middle. Standard 3 to 3½ lines
long; lower petals rather shorter, the keel incurved, obtuse. Ovary stipitate. Pod small.—Meissn. in Pl. Preiss. i. 61.

W. Australia. Swan River, *Huegel, Drummond, 1st Coll. and n.* 251, *Preiss,
n.* 1050, and others.

14. PHYLLOTA, DC.

Calyx, 2 upper lobes broader, and sometimes united into an upper lip.
Petals clawed; standard nearly orbicular, longer than the lower petals; wings
oblong; keel much incurved. Stamens either all, or at least the 5 outer
ones, more or less adnate to the petals at the base, and sometimes all united
with them in a ring or short tube. Ovary sessile, with 2 ovules on short
funicles; style dilated or thickened at the base, incurved and subulate upwards; stigma small, terminal. Pod ovate, somewhat turgid, 2-valved.
Seed reniform, without any strophiole.—Shrubs, usually heath-like. Leaves
scattered, linear, with revolute margins. Stipules none, or very minute in

P. humifusa. Flowers axillary or terminal. Bracteoles often leaf-like, in-serted under the calyx and usually closely pressed to it. Ovary small, very villous.

The genus is limited to Australia. It differs from *Dillwynia*, which it resembles in habit, in the revolute. not involute, margins of the leaves, and in the absence of any stro-phiole; from *Aotus* in the presence of bracteoles; and from both, as well as from all other allied genera, in the tendency to a union of the filaments with the petals.

Flowers axillary, either along the branches or forming apparently terminal leafy heads or spikes.
  Keel acute. Style bearded upwards on the inner side.
    Flowers ½ in. long. Pedicels very short . . . . 1. *P. barbata.*
    Flowers ¼ in. long. Pedicels filiform, 2 or 3 times as long
    as the sinuate leaves . . . . 2. *P. gracilis.*
  Keel rather obtuse, as long as the standard ; wings much smaller.
    Style slender, not bearded, but hairy to the middle . . . . 3. *P. Sturtii.*
  Keel obtuse, not longer than the wings. Style glabrous, much
  dilated below the middle.
    Erect shrub. Flowers usually crowded towards or at the ends
    of the branches . . . . 4. *P. phylicoides.*
    Slender procumbent shrub. Flowers few . . . . . . 5. *P. humifusa.*
Flowers sessile, in clusters of leaves, terminating the branches or
very short axillary branchlets . . . . 6. *P. pleurandroides.*

1. **P. barbata,** *Benth. in Hueg. Enum.* 33, *and in Ann. Wien. Mus.* ii. 78. A heath-like shrub of several ft., with pubescent or villous branches. Leaves linear, obtuse, rarely exceeding ½ in., the margins closely revolute, glabrous, scabrous or sparingly hairy. Flowers axillary, solitary, nearly sessile, usually longer than the leaves. Bracteoles leafy, dilated at the base, as long as or longer than the calyx. Calyx about 3 lines long, the lobes acu-minate, rather longer than the tube, the 2 upper ones united to the middle. Petal-claws rather short; standard more than ½ in. long, acuminate; keel nearly as long, deeply coloured, rather narrow, incurved and acute or acumi-nate ; wings shorter and narrower. Five at least of the filaments adnate to the petal-claws at the base. Style dilated downwards, longitudinally fringed from the middle upwards, and on the inner side with dense white woolly hairs. Pod ovoid-oblong, turgid, 2 to 3 lines long. Seeds not seen.—Meissn. in Pl. Preiss. i. 59.

W. Australia. King George's Sound to Cape Riche, *Huegel, A. Cunningham, Preiss, n.* 846, and others; Wilson's Inlet, *Oldfield.*
P. villosa, Turcz. in Bull. Mosc. 1853, i. 267, from the same district, *Gilbert, n.* 255, which I have not seen, would appear from the character given, to be a rather more hairy form of *P. barbata.*

2. **P. gracilis,** *Turcz. in. Bull. Mosc.* 1853, i. 267. Branches long and very slender, hoary with a close pubescence. Leaves linear, obtuse, rarely above 1 line long, the margins closely revolute, hoary-pubescent. Flowers few, on filiform axillary pedicels, 2 to 3 times as long as the leaves. Bracteoles ovate, obtuse, concave, keeled, close under the calyx and shorter than its tube. Calyx minutely hoary, about 1½ lines long, the lobes rather longer than the tube, acute, the 2 upper ones united into a truncate upper lip. Standard 2½ lines long; wings shorter; keel nearly as long as the standard, incurved and acute. Five of the filaments adnate to the petals at their base. Style thickened downwards, incurved attenuate and pubescent

or ciliate along the inner side above the middle.   Pod ovate, obtuse, about 2
lines long, minutely pubescent.   Seed without any strophiole.

W. Australia, *Drummond, 4th Coll. n.* 91.

3. **P. Sturtii,** *Benth.*   A shrub, with the habit apparently of *P. barbata*
and *P. phylicoides*, the branches usually minutely tomentose-pubescent.
Leaves 3 to 4 lines long, obtuse or mucronulate, the margins closely revolute,
more or less tuberculate and sprinkled with rigid hairs when young. Flowers
crowded in short leafy spikes at the ends of the branchlets.   Bracteoles leafy,
keeled, very acute, almost pungent, as long as the calyx.   Calyx about 3
lines long, the lobes acuminate, almost pungent, the 2 upper ones rather
broader, the lowest one very narrow.   Standard ovate, complicate ; keel much
incurved, almost acute, but scarcely rostrate, nearly as long as the standard ;
wings shorter and narrower.   Stamens adnate to the petals at the base. Style
slender, hairy to the middle, but not bearded.   Ovary sessile, slightly hairy.

S. Australia, *C. Sturt.*   The shape of the flowers is as it were intermediate between
that of *P. barbata* and *P. phylicoides*, the style is rather different from either.

4. **P. phylicoides,** *Benth. in Ann. Wien. Mus.* ii. 77.   An erect
heath-like shrub, of several ft., the branches terete, glabrous pubescent or
hirsute.   Leaves numerous, narrow-linear, mostly about ½ in. long, but in
some specimens nearly ¾ in., in others not above 4 lines, obtuse or with
callous usually recurved points, the margins revolute, more or less tuberculate,
scabrous, and sometimes sprinkled with erect hairs, rarely quite smooth and
glabrous.   Flowers almost sessile in the upper axils, forming terminal leafy
heads or spikes, or becoming lateral by the elongation of the terminal shoot.
Bracteoles leafy, lanceolate, acuminate, longer than the calyx-tube, and often
exceeding the lobes.   Calyx 2½ to 3½ lines long, glabrous or villous, the
lobes about as long as the tube, the 2 upper ones broad and shortly united,
the lowest rather longer than the lateral ones.   Standard 4 to 6 lines long ;
lower petals rather shorter, the keel broader than the wings, much incurved,
but obtuse.   Filaments and petal-claws all united at the base in a ring or
short tube.   Ovary tapering into the style, which is much dilated below the
middle, and quite glabrous.   Pod ovate or shortly oblong, included in the
calyx.   Seeds without any strophiole.—*Pultenæa phylicoides, P. aspera, P.
comosa*, and *P. squarrosa*, Sieb. in DC. Prod. ii. 113 ; *Phyllota pilosa, P.
aspera, P. comosa, P. Billardieri, P. grandiflora, P. squarrosa*, and *P. Baueri*,
Benth. in Ann. Wien. Mus. ii. 77.

Queensland.   Sandy Cape, *R. Brown ;* Moreton Island, *M'Gillivray, F. Mueller.*
N. S. Wales.   Port Jackson to the Blue Mountains, *R. Brown, Sieber, n.* 405, 406,
407, 408, *and Fl. Mixt. n.* 583, and others, Illawarra, *Shepherd ;* near Goulburn, *C. Moore.*
The characters upon which, after De Candolle, I had endeavoured to distinguish several
species, entirely break down when applied to the large number of specimens I have now had
before me. I am unable to distribute them even into marked varieties, much as they differ
in the size of the flowers, the erect spreading or recurved leaves, etc. The supposed differences in inflorescence depend often on the period of development.

5. **P. humifusa,** *A. Cunn. Herb.* (under *Dillwynia*).   Stems prostrate,
with slender ascending branches, glabrous or nearly so.   Leaves narrow-
linear, obtuse or with a recurved point, 2 to 3 lines long, the margins revolute, scabrous or glabrous.   Stipules very minute or quite inconspicuous.

Flowers few, axillary, below the ends of the branches, on very short pedicels. Bracteoles leafy, linear, longer than the calyx-tube. Calyx about 2 lines long, the lobes short, with subulate points, the 2 upper ones broader at the base. Standard nearly twice as long as the calyx; lower petals rather shorter, the keel purple and much incurved, but obtuse. Filaments mostly adnate to the petals at the base. Ovary tapering into a thickish style, attenuate and curved upwards. Pod not seen.

N. S. Wales. Wombal Brush, Argyle county, *A. Cunningham.*

6. **P. pleurandroides,** *F. Muell. in Trans. Phil. Inst. Vict.* i. 38. Branches virgate or diffuse, pubescent or villous. Leaves scattered, few along the branches, but often crowded at the ends, narrow-linear, obtuse or with a short recurved point, under ½ in. long, the margins closely revolute, glabrous scabrous or hirsute when young, the broad midrib alone appearing underneath. Flowers small terminal or on very short axillary shoots, sessile in a dense tuft of floral leaves, ciliate and imbricate at the base. Bracteoles broad, obtuse, shorter than the calyx-tube. Calyx pubescent, 1½ to nearly 2 lines long, the lobes short, the 2 upper ones truncate and more united. Petal-claws nearly as long as the calyx-tube, more or less adnate to the stamens at their base; standard fully twice as long as the calyx; lower petals rather shorter, the keel much incurved, but obtuse. Ovary tapering into the style, which is dilated downwards and slightly pubescent, but without any longitudinal row of hairs. Pod broadly ovate, shorter than the calyx.

Victoria. In the Grampians, *F. Mueller.*
S. Australia. Mount Barker, *Whittaker;* Kangaroo Island, *F. Mueller;* Spencer's Gulf, *Wilhelmi.*

15. GASTROLOBIUM, R. Br.

Calyx 5-lobed, the 2 upper lobes usually broader and united higher up. Petals clawed. Standard orbicular or reniform, emarginate, longer than the lower petals; wings oblong; keel broader than the wings and usually shorter. Stamens free. Ovary stipitate or rarely sessile, with 2 ovules on straight and filiform funicles. Style incurved, filiform, with a small terminal stigma. Pod ovoid or nearly globular, turgid, continuous inside, the valves coriaceous. Seeds (where known) strophiolate.—Shrubs. Leaves on very short petioles, more or less distinctly verticillate or opposite, or occasionally scattered, simple and entire, usually rigid. Stipules setaceous, rarely wanting. Flowers yellow or the keel and base of the standard purple-red, in terminal or axillary racemes, either loose or contracted into corymbs or whorl-like clusters. Bracts and bracteoles usually very deciduous, in a few species the brown rigid bracts persist nearly till the flowers open. Staminal disk usually very short. Ovary very villous.

The genus is limited to West Australia. It is closely allied on the one hand to the strophiolate species of *Oxylobium*, only differing from them in the number of ovules constantly 2, and on the other to *Pultenæa*, from which it is distinguished by the habit, the coriaceous leaves, the bracteoles either deciduous or inconspicuous and the more coriaceous turgid pod. Several of the species are sent as the Poison-plant of W. Australia, especially *G. bilobum* and *G. Callistachys.*

SERIES I. **Axillares.**—*Racemes contracted into clusters or heads, all or mostly axillary.*

Leaves ovate or oblong, obtuse truncate or equally rounded at both ends, coriaceous.

　Leaves mostly above 1 in. long.　Calyx-lobes nearly equal.

　　Leaves ovate or oblong, glabrous.　Flowers large.　Calyx very

　　　villous . . . .　1. *G. pyramidale.*

　　Leaves oblong, tomentose underneath.　Flowers rather small.

　　　Calyx silky . . . .　2. *G. Lehmanni.*

　　Leaves mostly under 1 in. long.　Calyx silky, the 2 upper lobes

　　　connate above the middle.　Flowers small.

　　Branchlets very angular, minutely silky-hairy.　Leaves ½ to 1

　　　in. long . . . .　3. *G. pulchellum.*

　　Branchlets terete, loosely pubescent.　Leaves mostly under

　　　½ in. . . . .　5. *G. Brownii.*

　Leaves under ½ in. long.　Calyx villous, the lobes nearly equal .　6. *G. reticulatum.*

Leaves thin, cordate-orbicular or broadly obovate, truncate, much

　undulate.　Flowers and fruits small . . . .　7. *G. truncatum.*

Leaves cuneate or spathulate, obtuse or truncate, or with a very

　short point.

　Calyx silky, about 2 lines long.

　　Leaves oblong, slightly cuneate, mostly under ½ in. long . .　5. *G. Brownii.*

　　Leaves spathulate, ½ to 1 in. long, or obovate under ½ in. . .　8. *G. spathulatum.*

　Calyx loosely villous, 3 to 4 lines long.

　　Leaves obovate, truncate, with a small point.　Ovary stipitate　9. *G. plicatum.*

　　Leaves cuneate, 3-pointed.　Ovary sessile . . . .　10. *G. tricuspidatum.*

Leaves narrow-linear.　Stipules very long . . . .　4. *G. stipulare.*

Leaves narrow-oblong or cuneate, with a short pungent point . .　5. *G. Brownii.*

Leaves tapering into a pungent point.

　Leaves obovate-rhomboidal . . . .　11. *G. obovatum.*

　Leaves ovate-acuminate . . . .　12. *G. epacridioides.*

Leaves with lateral pungent lobes or teeth.

　Leaves 3-lobed, the lateral lobes divaricate.　Racemes loose, but

　　short and axillary . . . .　13. *G. trilobum.*

　Leaves with pungent teeth or lobes above the middle.　Flowers

　　densely clustered in the axils . . . .　14. *G. ilicifolium.*

SERIES II. **Racemosæ.**—*Racemes terminal or axillary, elongated, cylindrical one-sided or with a few distant pairs of flowers or rarely short and dense.*

Leaves broad, obtuse truncate or emarginate, with or without a

　small deciduous point.

　Leaves much undulate.　Ovary-stipes rather long.

　　Racemes long, pedunculate.　Calyx 3 lines or more.　Standard

　　　large.　Style thick.　Pod as broad as long . . . . .　15. *G. villosum.*

　　Racemes sessile, 1 in. long.　Flowers small.　Calyx barely 2

　　　lines long.　Style slender.　Pod longer than broad . . .　16. *G. polystachyum.*

　Leaves flat, ovate.　Ovary nearly sessile.　Style slender.　Pod

　　longer than broad . . . .　17. *G. ovalifolium.*

　Leaves flat, oval or oblong.　Racemes loose.　Ovary on a long

　　stipes.　Flowers large.　Calyx 4 or 5 lines long . . . . .　18. *G. grandiflorum.*

　Leaves orbicular-cordate.　Racemes dense, oblong.　Ovary stipi-

　　tate.　Calyx 3 lines . . . .　19. *G. pycnostachyum.*

　Leaves flat, cuneate . . . .　30. *G. velutinum.*

Leaves tapering to a pungent point.

　Leaves broadly cordate or triangular, entire or prickly-toothed .　20. *G. spinosum.*

　Leaves broadly ovate or orbicular, entire . . . .　21. *G. rotundifolium.*

　Leaves oblong-elliptical, entire.

　　Calyx villous, scarcely 2 lines long.　Bracts brown scarious .　22. *G. microcarpum.*

　　Calyx silky-pubescent, fully 3 lines long.　Bracts brown, sca-

　　　rious . . . .　23. *G. oxylobioides.*

Calyx glabrous, fully 5 lines long. Bracts large membranous . 24. *G. calycinum*.
Leaves with lateral pungent lobes or angles . . . . 13. *G. trilobum*.
Leaves narrow or cuneate, obtuse or emarginate.
  Leaves linear, 1 to 2 in. long, scarcely verticillate.
    Leaves flat or the margins recurved. Racemes long . . . 25. *G. Callistachys*.
    Leaves folded lengthwise or involute. Racemes short and
      dense . . . . 26. *G. stenophyllum*.
  Leaves oblong.
    Leaves opposite or verticillate, strongly keeled, about 1 in. long 27. *G. crassifolium*.
    Leaves under ½ in. long, the midrib scarcely prominent.
      Leaves crowded, scarcely verticillate. Calyx glabrous . . 28. *G. parvifolium*.
      Leaves verticillate. Calyx villous. Racemes very short . 29. *G. hamulosum*.
  Leaves, emarginate.
    Racemes cylindrical.
      Leaves mostly ½ in. long, the margins slightly recurved . . 30. *G. velutinum*.
      Leaves mostly 1 in. long, the margins revolute . . . . 31. *G. bidens*.
    Racemes contracted almost into an umbel . . . . 32. *G. bilobum*.

SERIES I. AXILLARES.—Racemes contracted into clusters or heads, all or mostly axillary, assuming the appearance of whorls.

1. **G. pyramidale,** *T. Moore, in Gard. Comp.* i. 81, *with a fig.* A tall stout handsome shrub, the young branches softly pubescent or loosely villous. Leaves mostly verticillate in threes, ovate or rarely broadly obovate-oblong, very obtuse truncate or emarginate, with or without a small point, 1 to 1½ or in cultivation 2 in. long, rounded or broadly cordate at the base, rigidly coriaceous, reticulate, glabrous, the margins often thickened and nerve-like. Stipules usually long. Flowers bright yellow with a red keel, in dense short umbels or heads, on short peduncles, axillary or terminal. Calyx 3 to 4 lines long, very villous, the lobes nearly equal. Ovary very shortly stipitate. Pod not seen.—*Oxylobium ovalifolium*, Lindl. and Paxt. Fl. Gard. ii. 63, t. 85, not of Meissn.; *Gastrolobium polycephalum*, Turcz. in Bull. Mosc. 1853, i. 274 ; *G. crenulatum*, l. c. 273 (with rather narrower leaves and smaller flowers).

W. Australia, *Drummond, 5th Coll. n. 54 and* 55 ; Cheynes beach, S. coast, *Maxwell.* The species very closely resembles some specimens of *Oxylobium retusum*, but the indumentum is not silky, and the ovary 2-ovulate.

2. **G. Lehmanni,** *Meissn. in Pl. Preiss.* i. 70, *and* ii. 217. An erect shrub, probably of several ft., the branches softly tomentose-pubescent. Leaves oblong, very obtuse or emarginate, mostly 1 to 2 in. long, or smaller on the side-branches, rounded at the base, very coriaceous, the thickened nerve-like margins often minutely crenulate, glabrous above, the reticulations scarcely visible there, and quite concealed underneath by a soft whitish tomentum. Flowers rather small, in axillary clusters. Calyx silky-tomentose, the lobes rather narrow, about equal to the tube, the 2 upper ones a little more connate than the others. Ovary shortly stipitate. Pod acuminate, rather longer than the calyx.

W. Australia. In the interior, *Preiss, n.* 806, *Drummond, 3rd Coll. n.* 95.

3. **G. pulchellum,** *Turcz. in Bull. Mosc.* 1853, i. 274. Branches silky-pubescent, much more slender than in *G. Lehmanni*, which this species resembles in some respects. Leaves oblong, obtuse or emarginate, with a minute recurved point, under 1 in. long, rounded at the base, coriaceous, reticulate, glabrous. Flowers rather small, clustered in the upper axils or in a

terminal head, not usually exceeding the leaves. Calyx silky-pubescent, about 2½ lines long, the lobes rather shorter than the tube, the 2 upper ones united above the middle. Keel deeply coloured. Ovary on a rather long stipes. Pod not seen.

W. Australia, *Drummond, 5th Coll. n.* 57.

4. **G. stipulare,** *Meissn. in Pl. Preiss.* ii. 218. Branches erect, densely villous as well as the young leaves with long soft hairs. Leaves crowded, scarcely or irregularly whorled, narrow-linear, mucronate, 1 to 1½ in. long, rigid, at length nearly glabrous, the margins revolute, the midrib and transverse veins very prominent underneath. Stipules setaceous, remarkably long, often attaining ½ in. Flowers not seen. Fruiting pedicels short, axillary. Calyx silky-villous, about 2½ lines long, the 2 upper lobes rather broader than the others. Pod ovoid, turgid, acute, about 3 lines long. Seeds 2, strophiolate.

W. Australia, *Drummond, 3rd Coll. n.* 93. This species has no immediate affinity with any other one of the genus, and has more the aspect of some *Oxylobiums,* but the young pods show that there are only 2 ovules.

5. **G. Brownii,** *Meissn. in. Pl. Preiss.* i. 71. A shrub, attaining several feet, the branches softly pubescent or villous. Leaves obovate or oblong, but usually broader above than below the middle, and sometimes almost linear-cuneate, rounded or truncate, with a short rigid or pungent point, in some specimens not exceeding ½ in., in others ¾ to 1 in. long, obtuse at the base, coriaceous, reticulate, glabrous, often slightly undulate. Flowers in axillary clusters, often crowded in the upper part of the branches, but rarely exceeding the leaves. Calyx softly villous or nearly glabrous, 2 to 2½ lines long, the upper lobes united above the middle. Ovary on a long stipes. Pod not seen ripe.

W. Australia. Rocky summit of Mount Wuljenup, Plantagenet district, *Preiss, n.* 802 (*Herb. Sonder*).

G. Hookeri, Meissn. in Pl. Preiss. i. 71, from *Drummond, n.* 73, and, according to Meissner, *n.* 209, does not appear to me to differ from *G. Brownii,* except in its shorter, less mucronate leaves.

6. **G. reticulatum,** *Benth.* A low shrub, with erect, rather thick, virgate, loosely tomentose branches. Leaves in irregular whorls of 3, crowded, oval or oblong, 2 to 3 or rarely 4 lines long, obtuse, erect and often almost imbricate at the base, spreading or recurved upwards, thickly coriaceous, glabrous and densely reticulate. Stipules none. Flowers axillary, usually forming irregular whorls below the ends of the branches. Bracts concave, but falling off long before flowering. Pedicels short, villous. Bracteoles none. Calyx very villous, nearly 3 lines long, the lobes broad, nearly equal, about as long as the tube. Petals about half as long again as the calyx; keel deeply coloured, broad, obtuse. Ovary nearly sessile. Pod nearly globular, often slightly acute, coriaceous, villous, about 3 lines long. Seeds strophiolate.—*Eutaxia reticulata,* Meissn. in. Pl. Preiss. i. 65 ; *E. punctata,* Turcz. in. Bull. Mosc. 1853, i. 272.

W. Australia, *Drummond, 5th Coll. n.* 69 ; in the interior, *Preiss, n.* 870. I have not seen Preiss's specimen, but Meissner's description leaves no doubt of the identity with Drummond's plant.

7. **G. truncatum,** *Benth.* Branches numerous, but short and slender,

hirsute with long spreading hairs. Leaves nearly orbicular, mostly about ½ in. long, truncate at the end, the midrib produced into a long fine point, broad and often cordate at the base, the margins much undulate, scarcely coriaceous, glabrous and reticulate above, pale underneath and sprinkled with a few long hairs. Stipules setaceous, deciduous. Flowers small, in axillary clusters of 3, sometimes borne on a short common peduncle, the pedicels about 1 line long, recurved, each in the axil of a small ovate or lanceolate very deciduous bract. Calyx membranous, slightly hairy, about 1½ lines long, the lobes about as long as the tube, the 2 upper ones rather more united than the others. Standard twice as long as the calyx. Ovary shortly stipitate. Pod nearly 3 lines long, shortly acuminate.

W. Australia, *Drummond* (5*th Coll. ?*), *n.* 30. The leaves of this species are much thinner than usual in the genus, and recall in some measure the foliage of *Aotus cordifolius.* The seeds of the first pod I opened also had no strophiole, but I found the strophiole perfect in the seeds of several other pods. and the general habit and stipules are those of *Gastrolobium.*

8. **G. spathulatum,** *Benth. in. Lindl. Swan Riv. App.* 14. A small shrub, with erect virgate nearly simple stems, often under 1 ft. high, slightly hoary or silky-tomentose. Leaves spathulate, ¼ to 1 in. long, 3 or 4 lines broad, rounded truncate or emarginate at the end, with a minute point, tapering at first rapidly, and afterwards gradually to the base, folded lengthwise, coriaceous, much reticulate and glabrous. Flowers small, in axillary clusters or short racemes, never exceeding the leaves. Calyx silky-pubescent, about 2 lines long, the two upper lobes united above the middle. Keel deeply coloured. Ovary on a long stipes. Pod not seen ripe.—Meissn. in Pl. Preiss. i. 71.

W. Australia, 1*st Coll. and n.* 72, also (according to Meissner) *n.* 208 ; Mount Bakewell, *Preiss, n.* 800, mixed with *Oxylobium cuneatum,* which this species much resembles, but may be distinguished by the more spathulate and much more reticulate leaves, besides the constant difference in the number of ovules.

Var. (?) *latifolium.* Stems loosely villous. Leaves broader and not above ½ 'n. long, loosely tomentose underneath when young. Calyx more villous.—W. Australia, *Drummond ;* Phillips Ranges, *Maxwell.* The specimens are scarcely sufficient to determine whether they may not rather be a variety of *G. Brownii.*

9. **G. plicatum,** *Turcz. in Bull. Mosc.* 1853, i. 274. A stout, rigid shrub, the young branches tomentose-villous. Leaves obovate-cuneate, truncate, with a small recurved point, the angles rounded, mostly about 1 in. long, folded lengthwise, very coriaceous, glabrous and often somewhat glaucous, the reticulations fine and not very conspicuous. Stipules long. Flowers axillary, loosely clustered. Calyx very villous, about 3 lines long, the lobes nearly equal. Standard twice as long as the calyx ; keel deeply coloured. Ovary on a long stipes. Pod stipitate, very hirsute, exceeding the calyx.

W. Australia, *Drummond,* 5*th Coll. n.* 50.

10. **G. tricuspidatum,** *Meissn. in Pl. Preiss.* i. 66. A stout, rigid shrub, the branches scarcely angular, softly villous. Leaves rather crowded, cuneate, truncate, or shortly 3-lobed, with 3 short rigid points, ¾ to 1½ in. long, very coriaceous, more or less folded lengthwise, villous when young, at length glabrous, reticulate underneath, and drying usually of a yellowish tinge. Flowers axillary, clustered. Calyx loosely villous, about 3 lines long, the lobes rather broad, acuminate, the 2 upper ones shortly united. Ovary ses-

sile, with the 2 ovules at the base of the cavity. Style nearly straight. Pod sessile, acute, enclosed in the calyx. Seeds strophiolate.

W. Australia, *Drummond, n.* 212; in the interior, *Preiss, n.* 839.

11. **G. obovatum,** *Benth. in Lindl. Swan Riv. App.* 14. Branches rather slender, tomentose-pubescent. Leaves more scattered than in most species, from obovate to rhomboidal, under 1 in. long, tapering more or less into a pungent point, narrowed below the middle, folded lengthwise, coriaceous, reticulate, glabrous. Flowers axillary, in rather loose clusters, like those of *G. spathulatum.* Calyx about 2 lines long, silky-villous, the 2 upper lobes united above the middle. Standard twice as long as the calyx; keel-petals often free. Ovary on a rather long stipes. Pod not seen.

W. Australia. Swan River, *Drummond,* 1*st Coll.,* also *n.* 74 *and* 206; Mount Bakewell, *Preiss, n.* 874 *and* (according to Meissner) *n.* 800 *in part.*

12. **G. epacridioides,** *Meissn. in Pl. Preiss.* i. 72. A tall shrub, with slender, virgate, loosely villous branches. Leaves numerous, very spreading, ovate, acuminate, with a pungent point, 4 to 8 lines long, coriaceous, very rigid, shining above, glabrous, coarsely reticulate. Stipules none. Flowers axillary, loosely clustered. Calyx silky-villous, nearly 3 lines long, the lobes acute or acuminate, the 2 upper ones slightly united at the base. Standard about twice as long as the calyx; keel deeply coloured. Ovary shortly, but distinctly stipitate. Pod ovoid, acute, about 3 lines long, transversely veined, loosely hairy. Seeds not seen.

W. Australia, *Drummond, n.* 196; Darling Range, *Fraser, Preiss, n.* 837; King George's Sound, *Maclean.* The species is in many respects allied to *Oxylobium acutum.*

13. **G. trilobum,** *Benth. in Lindl. Swan Riv. App.* 13. Much-branched, not very stout, and quite glabrous. Leaves rhomboidal or 3-lobed, ¾ to 1 in. long, tapering into a pungent point, the lateral lobes or angles very divaricate, sometimes lanceolate, almost as in *Oxylobium staurophyllum,* sometimes very broad and short, always ending in pungent points, the leaf usually folded lengthwise, very coriaceous, often glaucous, the fine reticulations scarcely prominent. Flowers few, in loose axillary racemes, not usually exceeding the leaves. Calyx quite glabrous or minutely pubescent, 2 to 2½ lines long, the lower lobes acute, the 2 upper ones broad, united above the middle into a truncate upper lip. Standard about twice as long as the calyx; the lower petals nearly as long, the keel not so deeply coloured as in some species. Ovary stipitate. Pod not seen.—Meissn. in Pl. Preiss. i. 66.

W. Australia. Swan River, *Drummond,* 1*st Coll. and n.* 87 (*or* 187 ?); sterile places near Williams, *Preiss, n.* 825. The inflorescence is sometimes almost that of the *Racemosæ,* but the racemes are short and rarely more than 4-flowered.

14. **G. ilicifolium,** *Meissn. in Pl. Preiss.* i. 67. A tall, rather stout shrub, the young branches softly pubescent or villous. Leaves from narrow-cuneate to broadly elliptical, 1½ to 2½ in. long, obtuse acute or truncate, bordered, especially above the middle, by a few pungent teeth, sometimes very prominent and undulate, the leaf entire and tapering at the base, glabrous, coriaceous, smooth above, penniveined and finely reticulate underneath. Stipules rather long. Flowers axillary, densely clustered. Calyx silky-villous, 2 to 2½ lines long, the 2 upper lobes slightly united at the base.

Standard twice as long as the calyx; keel deeply coloured. Ovary stipitate.
Pod broadly ovoid, rather longer than the calyx.

W. Australia, *Drummond, n.* 76 *and* 211; Mount Bakewell, *Preiss, n.* 821. The
species approaches sometimes in foliage *Oxylobium tricuspidatum*, but the leaves are more
toothed and the ovary different.

Var. *lobatum.* Leaves narrow, with revolute margins, divided above the middle into 1,
2 or 3 pairs of opposite, divaricate, short, broad, divaricate lobes, all ending in pungent
points, and there are often small pungent teeth in the undulate sinuses.—*G. verticillatum,*
Meissn. in Bot. Zeit. 1855, 28.—Between Moore and Murchison rivers, *Drummond, 6th
Coll. n.* 24. Different as this appears at first sight, there are specimens in which the leaves
pass from the one form to the other, and the inflorescence and flowers are the same in all.

SERIES II. RACEMOSÆ.—Racemes terminal or occasionally axillary, elongated, cylindrical or 1-sided, or reduced to a few distant pairs of flowers.

15. **G. villosum,** *Benth. in Lindl. Swan Riv. App.* 13. A decumbent
shrub, with ascending stems, the branches in our specimens above 1 ft. long,
softly tomentose, and hirsute with spreading hairs. Leaves opposite, ovate,
broadly oblong or almost ovate-lanceolate, very obtuse, truncate or emarginate, with or without a small point, 1 to 2 in. long, the margins undulate,
the base broad, truncate or slightly cordate, coriaceous, glabrous or slightly
hirsute underneath. Racemes terminal, pedunculate, often 3 or 4 in. long,
Bracts brown and rigid, lanceolate-subulate, often persisting till the flowers
expand. Calyx about 3 lines long, villous, the 2 upper lobes connate about
to the middle. Standard twice as long as the calyx, orange-red, lower petals
shorter, purple-red, the keel shorter than the wings. Ovary stipitate. Style
short, incurved. Pod broadly ovoid, stipitate, about 4 lines long.—Bot. Reg.
1847, t. 45; Meissn. in Pl. Preiss. i. 68.

W. Australia. Swan River, *Drummond, 1st Coll. and n.* 194; Darling Range,
Preiss, n. 810.

16. **G. polystachyum,** *Meissn. in Pl. Preiss.* ii. 217. An erect or
spreading shrub, the branches tomentose-villous. Leaves mostly opposite.
broadly oblong, often more or less cuneate, truncate at the end, ¾ to 1 in.
long or rarely more, the margins undulate, the base rounded, coriaceous, glabrous above, softly pubescent or villous or rarely at length glabrous underneath. Racemes axillary or terminal, scarcely pedunculate, and rarely above
1 in. long. Bracts brown and rigid, ovate, concave, rather acute. Flowers
much smaller than in *G. villosum.* Calyx villous, about 2 lines long, the
upper lobes scarcely united at the base. Standard twice as long as the calyx;
wings as in *G. villosum,* shorter than the standard and longer than the keel.
Pod shortly stipitate, ovoid, acute, scarcely 3 lines long.—*Oxylobium batillum,*
Hook. Ic. Pl. t. 612.

W. Australia, *Drummond, 2nd Coll. n.* 97 *and Suppl. n.* 32.

17. **G. ovalifolium,** *Henfr. in Gard. Comp.* i. 41, *with a fig., copied in
Lemaire, Jard. Fl. t.* 247. Apparently a low, diffuse or procumbent shrub,
the young branches villous. Leaves mostly opposite, ovate, orbicular or
broadly oblong, rounded at both ends, and often emarginate, ½ to 1 in. long,
the margins thickened and nerve-like, not undulate, coriaceous, glabrous
above, reticulate and villous or at length glabrous underneath. Racemes
nearly sessile, rather slender, 1 to 3 in. long. Bracts brown, rigid, acumi-

nate. Flowers nearly sessile, smaller than in *G. villosum.* Calyx villous, 2¼ to nearly 3 lines long. Petals less unequal in size than in *G. villosum*, the lower ones deeply coloured. Ovary very shortly stipitate; style filiform, rather long. Pod ovoid, acute, about 3 lines long.

W. Australia, *Drummond, n.* 31, 39, *and Suppl. n.* 27. The description and figure agree in every respect with our plant, except that the racemes are usually longer in the wild specimens.

18. **G. grandiflorum,** *F. Muell. Fragm.* iii. 17. Apparently a tall shrub, with something of the aspect of *Oxylobium Callistachys*, the young branches and inflorescence softly silky-pubescent, the full-grown foliage nearly glabrous and glaucous. Leaves opposite or the upper ones alternate, from ovate to oblong, obtuse or emarginate, 1½ to 3 in. long, flat, coriaceous. Racemes short, loose, axillary and terminal, with few, large, pedicellate flowers. Calyx softly pubescent, 4 to 5 lines long, the lobes much shorter than the tube, the 2 upper ones broad, falcate, and united nearly to the middle. Standard fully ¾ in. diameter, lower petals rather shorter, the keel much incurved and deeply coloured. Ovary very hairy, on a long glabrous stipes. Pod short, but not seen ripe.

N. Australia. Whittington Range, *M'Douall Stuart;* Purdie's Ponds, *Waterhouse.*

19. **G. pycnostachyum,** *Benth.* A rigid shrub, apparently with the habit of *G. ovalifolium*, but nearly glabrous or with a close hoary or almost silvery tomentum on the branches and under side of the leaves. Leaves mostly opposite, broadly cordate-ovate or orbicular, 1 to 1½ in. long, very obtuse, flat, rigidly coriaceous, of a pale silvery or yellowish colour. Racemes terminal, oblong, very dense, about 1 in. long. Flowers crowded, the pedicels short. Calyx softly villous, about 3 lines long, the lobes shorter than the tube, the 2 upper scarcely broader than the others, but rather more united. Standard 5 to 6 lines diameter; lower petals shorter; keel broad and much incurved. Ovary very hairy, on a rather long glabrous stipes. Pod not seen.

W. Australia. East Mount Barren, *Maxwell.*

20. **G. spinosum,** *Benth. in Lindl. Swan Riv. App.* 13. A shrub, of 2 to 4 ft., usually quite glabrous, but sometimes the young shoots clothed with a very evanescent wool, and the calyx and pedicels with a more persistent down. Leaves mostly opposite, broadly ovate-cordate, ending in a pungent point and bordered with pungent teeth, or rarely almost or quite entire, ¾ to 1½ in. long, often as broad as long, rigidly coriaceous, and often glaucous. Racemes loose, pedunculate, 1 to 1½ in. long. Calyx broad, about 2 lines long, the lobes much shorter than the tube, the 2 upper ones united nearly to the top. Standard striate, ½ in. diameter; wings rather shorter and scarcely exceeding the keel, which is broad and deeply coloured. Ovary on a rather long stipes; style rigid. Pod glabrous, ovoid-falcate, acuminate, 4 to 5 lines long.—Meissn. in Pl. Preiss. i. 68; Paxt. Mag. xi. 171, with a fig.; *G. Preissii*, Meissn. in Pl. Preiss. i. 68.

W. Australia. Swan River, *Drummond, 1st Coll. and n.* 48 *and* 186; Kalgan river, Cape Naturaliste and Freemantle, *Oldfield.* Drummond's specimens, 2nd Coll. n. 130, referred here in Pl. Preiss. ii. 216, belong in our sets to *Bossiæa Aquifolium.*

Var. *triangulare.* Leaves triangular-cordate, quite entire, with pungent points at the angles. Flowers smaller, the racemes looser and more pedunculate.—Stony places, Port Gregory, *Oldfield.*

21. **G. rotundifolium,** *Meissn. in Pl. Preiss.* ii. 216. An erect, rigid shrub, of about 1 ft., the young branches loosely villous or woolly, at length glabrous. Leaves mostly opposite, broadly ovate or orbicular, tapering into a pungent point, ¾ to 1 in. long or rarely more, the margins somewhat undulate in our specimens, the base rounded, coriaceous, glabrous or loosely villous underneath. Racemes short and sessile. Bracts brown and rigid, broad, concave, rather obtuse, imbricate and persistent till the flower opens. Calyx softly villous, under 3 lines long, the lobes all acute, the 2 upper ones rather broader, but not more united. Standard not twice the length of the calyx; keel as long as the wings. Ovary almost sessile; style rather dilated. Pod not seen.

W. Australia, *Drummond, 2nd Coll. n.* 99.

22. **G. microcarpum,** *Meissn. in Pl. Preiss.* i. 70 (as a var. of *G. oxylobioides*). A rigid shrub, the branches minutely silky-hoary when young. Leaves mostly in whorls of 3 or 4, elliptical-oblong, tapering into a pungent point and narrowed at the base, ¾ to 1½ in. long, very rigid, coriaceous, and reticulate, glaucous or hoary underneath. Racemes loose, 1 to 3 in. long, pubescent or villous. Bracts brown and rigid, but very deciduous. Calyx slightly villous, about 2 lines long, the upper lobes broad, truncate, and united nearly to the top. Keel shorter than the wings. Ovary on a long stipes. Pod (according to Meissner) scarcely 2 lines long, on a stipes as long as the calyx.

W. Australia, *Drummond, n.* 205, *Preiss, n.* 816, 817. This may be, as suggested by Meissner, a variety of *G. oxylobioides,* but the flowers are much smaller and more numerous, and the stipes of the pod and ovary much longer.

23. **G. oxylobioides,** *Benth. in Lindl. Swan Riv. App.* 13. An erect shrub, of 1 to 2 ft., not much branched, glabrous or the young shoots and racemes slightly silky-hoary. Leaves opposite or in threes, elliptical-oblong, broad or narrow, tapering to a pungent point, 1 to 1½ or rarely 2 in. long, coriaceous, rigid, reticulate, and often folded lengthwise. Racemes terminal or in the upper axils, consisting of few flowers in distant pairs or whorls of 3. Bracts ovate, acuminate, brown and rigid, but very deciduous. Pedicels short. Calyx about 3 lines long, silky-pubescent, the lobes broad, the 2 upper ones united above the middle. Standard about twice as long as the calyx; keel rather shorter than the wings. Ovary rather shortly stipitate. Pod about 3 lines long, on a stipes very much shorter than the calyx.

W. Australia. Swan River, *Drummond, 1st Coll., Oldfield;* S. Hutt river and Murchison river, *Oldfield.* One of the poison-plants.

G. Drummondii, Meissn. in Pl. Preiss. i. 69, appears to me to be referable to the form originally described of *G. oxylobioides,* and the var. *microcarpum,* Meissn. l. c. 70, to be a narrow-leaved form, sufficiently constant to be considered as a distinct species, as far as can be judged from our specimens.

24. **G. calycinum,** *Benth. in Lindl. Swan Riv. App.* 13. An erect shrub, nearly allied to *G. oxylobioides,* but quite glabrous. Leaves opposite or in threes, oblong-elliptical or more frequently from ovate-lanceolate to lanceolate, with a pungent point, 1 to 2 in. long, coriaceous, rigid, reticulate,

and often glaucous. Racemes terminal or in the upper axils, with few large flowers, in distant pairs or whorls of 3. Bracts larger and more membranous than in any other species, ovate, concave, often 4 or 5 lines long, besides a long point. Calyx 5 to nearly 6 lines long, the lobes rather longer than the tube, the 2 upper ones broad, rounded at the end, and united above the middle. Standard about ¾ in. diameter; keel deeply coloured, rather shorter than the wings. Ovary on a very short stipes.—Meissn. in Pl. Preiss. i. 69.

W. Australia. Swan River, *Drummond*, 1*st Coll. and n.* 203 ; Mount Bakewell and Goderich district, *Preiss, n.* 835 *and* 836 ; Blackwood river, *Oldfield*, where it is known as the York-Road Poison.

25. **G. Callistachys,** *Meissn. in Pl. Preiss.* ii. 216. An erect shrub, of 2 or 3 ft., with virgate branches, minutely and closely silky-pubescent. Leaves like those of *Oxylobium lineare*, alternate or irregularly verticillate, linear, obtuse or sometimes retuse, and minutely mucronulate, 1 to 2 in. long, flat or with recurved margins, glabrous or silky-pubescent underneath. Flowers rather large, in terminal racemes of 3 to 4 in. Bracts not seen. Calyx silky-pubescent, fully 3 lines long, the 2 upper lobes broadly falcate, but scarcely connate. Standard nearly twice as long as the calyx ; wings and keel scarcely shorter. Ovary stipitate, with a rather thick style. Pod about 4 lines long, broadly ovoid, scarcely acute, on a stipes of 1½ lines.— *G. lineare*, Meissn. in Bot. Zeit. 1855, 30 (leaves scarcely narrower, calyx rather less pubescent).

W. Australia. Swan River, *Drummond, 3rd Coll. n.* 90, *Cook ;* between Moore and Murchison rivers, *Drummond, 6th Coll. n.* 25 ; Kalgan woods, *Oldfield.* This is sent as one of the Swan River poison-plants.

26. **G. stenophyllum,** *Turcz. in Bull. Mosc.* 1853, i. 275. An erect, leafy shrub, less virgate than *G. Callistachys*, the branches silky-pubescent. Leaves alternate or irregularly verticillate, linear, obtuse, with a minute point. 1 to 1½ in. long, folded lengthwise or the margins involute, silky-pubescent when young, at length glabrous, transversely reticulate. Racemes cylindrical, dense, about 1 in. long. Bracts small, narrow. Calyx silky-pubescent, 2 to 2½ lines long, the upper lobes broader, obtuse, united to the middle. Standard not twice as long as the calyx ; lower petals not much shorter than the standard. Ovary shortly stipitate. Pod not seen.

W. Australia, *Drummond, 5th Coll. n.* 52.

27. **G. crassifolium,** *Benth.* An erect, rigid shrub, very much resembling *Oxylobium parviflorum*, but with thicker and more rigid leaves. Branches angular, very minutely silky-pubescent. Leaves opposite or in threes, narrow-oblong, obtuse, with a minute point, ¾ to 1½ in. long, very rigid, somewhat folded lengthwise, glabrous or minutely silky-pubescent when young, the veins often scarcely conspicuous. Flowers small, in terminal racemes, of ½ to 1 in. Calyx sparingly pubescent, about 2 lines long, the 2 upper lobes united nearly to the top. Standard about twice as long as the calyx ; lower petals shorter, the keel deeply coloured. Ovary-stipes rather long. Pod broadly ovoid, obtuse, 2 to 2½ lines long, on a stipes of about 1 line.

W. Australia, *Drummond, n.* 32.

28. **G. parvifolium,** *Benth. in Lindl. Swan Riv. App.* 13. A rigid, spreading, Epacris-like shrub, the branches pubescent. Leaves crowded, irregularly verticillate, narrow-oblong, obtuse, with a minute point, under ½ in. long, thickly coriaceous, convex underneath, glabrous and reticulate, the midrib scarcely conspicuous. Racemes terminal, rather dense, rarely exceeding 1 in. when in flower, often 2 in. in fruit, the rhachis and short pedicels softly and densely pubescent. Calyx glabrous and veinless, broadly campanulate, about 2 lines long, the 2 upper lobes almost completely united into a truncate upper lip, the 3 lower ones much shorter. Standard twice as long as the calyx; wings shorter, oblong; keel still shorter, almost as broad as long, all on rather long claws. Ovary-stipes long; style short. Pod compressed-globular, oblique, very obtuse, glabrous.—Meissn. in Pl. Preiss. i. 69.

W. Australia. Swan River, *Drummond, 1st Coll.;* Mount Bakewell, *Preiss, n.* 1017.

29. **G. hamulosum,** *Meissn. in Pl. Preiss.* ii. 218. Branches numerous, rather slender, hoary-tomentose. Leaves mostly verticillate in threes, obtuse, with a small often recurved point, about 3 or 4 lines long, rigid, glabrous when full-grown, strongly reticulate, the midrib scarcely prominent. Racemes terminal, short, consisting of 2 or 3 pairs of flowers or whorls of 3 each. Bracts lanceolate. Pedicels very short. Calyx villous with spreading hairs, about 3 lines long, the lobes acuminate, the 2 upper ones shortly united. Standard not twice as long as the calyx, the lower petals not much shorter; keel deeply coloured. Ovary very shortly stipitate; style rather thick.

W. Australia, *Drummond, 2nd Coll. n.* 106.

30. **G. velutinum,** *Lindl. in Paxt. Fl. Gard.* iii. 76, *with a woodcut.* An elegant shrub, the branches rather stout, angular, minutely silky-pubescent. Leaves verticillate in threes or fours, from obovate or obcordate to linear-cuneate, very obtuse or truncate, emarginate, about ½ in. or rarely ¾ in. long, the margins recurved, coriaceous, reticulate, glabrous above, usually pubescent underneath. Flowers orange-red, in terminal rather dense racemes, of 1 to 1½ in., the rhachis and pedicels softly villous. Bracts ovate, very deciduous. Calyx pubescent or nearly glabrous, broad, fully 2 lines long, the 2 upper lobes broad and united nearly to the top. Standard twice as long as the calyx; wings shorter but considerably exceeding the deeply-coloured keel. Ovary stipitate; style short. Pod ovoid, scarcely acute, about 3 lines long, on a very short stipes.—*G. emarginatum,* Turcz. in Bull. Mosc. 1853, i. 273.

W. Australia, *Drummond, 5th Coll. n.* 51, *and Suppl. n.* 27, *Maxwell;* Kalgan river, *Oldfield.* Turczaninow's name is much the most appropriate, but Lindley's has the right of priority. *G. cuneatum,* Henfr. in Gard. Comp. i. 49, with a fig. copied into Lemaire, Jard. Fl. t. 258, may be the same species, although the raceme is figured as much longer.

31. **G. bidens,** *Meissn. in Bot. Zeit.* 1855, 29. A much stouter shrub than *G. velutinum,* the branches terete or slightly compressed, softly tomentose or villous when young. Leaves mostly opposite, linear-cuneate, truncate and often emarginate with a minute point, or sometimes 3-pointed, about 1 in. long, the margins much revolute, coriaceous, glabrous, and scarcely reticulate above, softly villous, almost woolly underneath. Racemes slender, ses-

sile, villous, 1 to 1½ in. long, many-flowered. Bracts concave, acuminate, persisting almost to the opening of the flowers. Calyx not 2 lines long, villous, the lobes all acute, the 2 upper ones rather broader and shortly united at the base. Ovary on a long stipes; style short. Pod not seen.

W. Australia. Between Moore and Murchison rivers, *Drummond, 6th Coll. n.* 23.

32. **G. bilobum,** *R. Br. in Ait. Hort. Kew. ed.* 2, iii. 16. A tall shrub, the young branches angular and usually silky-pubescent. Leaves mostly verticillate in threes or fours, from obovate to narrow-oblong, always more or less cuneate, truncate or emarginate with 2 short rounded lobes, and minutely mucronate, ¾ to 1½ in. long or rarely smaller, thinly coriaceous, glabrous and veined above, pale and often minutely silky-pubescent underneath. Flowers numerous, in very short, almost umbel-like, terminal racemes rarely exceeding the leaves. Pedicels much longer than in the preceding species. Calyx silky-pubescent, 2 to 4 lines long, the 2 upper lobes broader, more obtuse, and united to about the middle. Standard about twice as long as the calyx; wings and keel rather shorter, the latter deeply coloured. Ovary on a long stipes; style slender. Pod stipitate, ovoid or oblong, rather acute, 2, 3, and even 4 lines long.—DC. Prod. ii. 110; Bot. Reg. t. 411; Bot. Mag. t. 2212; Lodd. Bot. Cab. t. 70; Meissn. in Pl. Preiss. i. 66, ii. 216.

W. Australia. King George's Sound and adjacent districts, *R. Brown, Preiss, n.* 799, *Drummond, 2nd Coll. n.* 137, and others; common from thence to Blackwood and Murray rivers, *Oldfield;* and eastward to Cape Le Grand, *Maxwell.* Said to be the worst of the poison-plants.

Var. *angustifolium.* Leaves linear-cuneate, but not otherwise differing from the common form.—*G. corymbosum,* Turcz. in Bull. Mosc. 1853, i. 272.—*Drummond, 5th Coll. n.* 58.

16. PULTENÆA, Sm.

(Euchilus, *R. Br.;* Spadostyles, *Benth.;* Urodon, *Turcz.*)

Calyx : 2 upper lobes more or less united into an upper lip, and sometimes much larger than the lower ones, rarely all nearly equal. Petals on rather long claws; standard nearly orbicular, longer than the lower petals; wings oblong; keel incurved. Stamens free. Ovary sessile or rarely shortly stipitate, with 2 ovules on short funicles; style subulate, often more or less dilated downwards; stigma small, terminal. Pod ovate, flat or turgid, 2-valved. Seed reniform, strophiolate.—Shrubs. Leaves alternate or rarely opposite or in verticils of 3, simple. Stipules linear-lanceolate or setaceous, brown and scarious, closely pressed on the branch, and more or less united in the axil of the leaf, the points or sometimes nearly the whole stipule free and spreading, or the stipules minute and free, rarely quite deficient, those of the floral leaves often much enlarged whilst the lamina is reduced. Flowers yellow orange or mixed with purple, rarely pink, either axillary and solitary and then frequently collected in leafy heads or tufts near the ends of the branches, or crowded in terminal heads and surrounded within the floral leaves by imbricate, scarious, brown bracts or enlarged stipules without any lamina. Bracts under each flower usually small. Bracteoles persistent (except in *P. pinifolia*), either close under the calyx or adnate with its tube. Ovary villous or rarely glabrous.

The genus is limited to Australia, presenting considerable diversity in foliage, inflorescence, and calyx, and closely allied in character to the other *Podalyrieæ* with 2 ovules
and strophiolate seeds, yet not generally difficult to distinguish. From *Gastrolobium* it
differs chiefly in habit, in the persistent bracteoles, and the more sessile, less turgid pod;
from *Latrobea, Eutaxia,* and *Dillwynia,* which are even closer connected with it, the bracteoles close to or upon the calyx afford the most prominent distinction, although accompanied
usually by other slight differences in habit or in flower. In the following distribution of its
numerous species, the sections proposed, founded on foliage and inflorescence, are perhaps
scarcely worthy of ranking higher than so many series.

SECT. I. **Eupultenæa.**—*Leaves alternate, with recurved or revolute margins. Ovary
quite sessile (except in* P. conferta).

Bracteoles adnate to (apparently inserted on) the calyx-tube (except
 in *P. pycnocephala ?*).
Flowers in dense, terminal, sessile heads, surrounded by imbricate
 stipular bracts, the inner ones longer than the pedicels.
 Stipules minute.
  Leaves cuneate-oblong or broadly-cuneate, rounded or trun-
   cate at the end, flat or nearly so, glabrous or silvery-
   white underneath.
   Leaves ½ to 1½ in. long. Flower-heads rather large . . 1. *P. daphnoides.*
   Leaves under ½ in. long. Flower-heads small . . . . 2. *P. stricta.*
   Leaves linear, obtuse or retuse . . . . 3. *P. retusa.*
   Leaves linear or lanceolate, acute or almost pungent . . . 4. *P. Benthamii.*
  Stipules conspicuous, often 1 line long or more.
   Leaves shortly obovate, coriaceous, shining above, densely
    silky-white underneath. . . . . 5. *P. pycnocephala.*
   Leaves narrow-oblong, obtuse. Flower-heads large, silky-
    hairy. Petals persistent . . . . 6. *P. myrtoides.*
   Leaves oblong or elliptical, aristate. Petals deciduous.
    Whole plant softly villous . . . . 7. *P. mucronata.*
   Leaves linear.
    Leaves nearly sessile, obtuse, or with a fine point. Flower-
     heads softly villous. Bracts deciduous . . . . . 8. *P. polifolia.*
    Leaves on petioles of 2 to 4 lines. Bracts few . . . . 9. *P. petiolaris.*
    Leaves linear-acute or rarely oblong. Bracts persistent and
     completely covering the calyxes, glabrous or rarely
     slightly silky . . . . 10. *P. paleacea.*
Flowers few or in small heads, the bracts shorter than the pedi-
 cels. Leaves under ½ in. long.
 Leaves ovate lanceolate or oblong . . . . 11. *P. Gunnii.*
 Leaves obovate or broadly cuneate, usually emarginate or 2-
  lobed . . . . 12. *P. scabra.*
 Leaves linear-cuneate, obtuse or truncate . . . . 13. *P. microphylla.*
Bracteoles close under the calyx, but free from it or scarcely adnate.
 Leaves small, obovate, very silky underneath, glabrous above . . 5. *P. pycnocephala.*
 Leaves linear.
  Erect shrubs.
   Flowers sessile, in small heads. Leaves under ½ in. long.
    Calyx with a large upper lip . . . . 14. *P. Drummondii.*
   Flowers in terminal umbels. Leaves about 1 in. long. Calyx-
    lobes nearly equal . . . . 15. *P. pinifolia.*
  Prostrate shrubs. Leaves small.
   Flowers terminal, on filiform pedicels. Stipules distinct . . 16. *P. pedunculata.*
   Flowers axillary. Pedicels shorter than the calyx.
    Stipules distinct. Calyx about 3 lines long, the large
     upper lobes free . . . . 17. *P. conferta.*
    Stipules none. Calyx under 2 lines, the large upper lobes
     united . . . . 18. *P. diffusa.*

SECT. II. **Aciphyllum.**—*Leaves alternate, rigid, concave, keeled, transversely reticulate. Flowers terminal, solitary or 2 or 3 together.*

Leaves lanceolate, pungent . . . . 19. *P. reticulata.*
Leaves obovate or oblong, obtuse . . . . : . . 20. *P. ochreata.*
Leaves narrow-linear, acute, crowded . . . . : . . 21. *P. aspalathoides.*

SECT. III. **Euchilus.**—*Leaves all or mostly opposite or in whorls of three, flat, concave or the margins slightly recurved, often 1- or 3-nerved or penniveined, rarely reticulate. Flowers axillary or crowded at the ends of the branches. Two upper lobes of the calyx much larger than the others. Ovary often contracted at the base into a very short stipes.*

Leaves flat or with recurved margins. Bracteoles close under the
　calyx.
　Leaves broadly obovate or obcordate. Flowers nearly sessile in
　　the upper axils, or in a terminal leafy head . . . . . . 22. *P. obcordata.*
　Leaves very small, orbicular. Flowers axillary, on slender pedi-
　　cels . . . . 23. *P. rotundifolia.*
Leaves concave or with incurved margins. Bracteoles close under
　the calyx.
　Leaves oblong-linear or cuneate, obtuse. Flowers nearly sessile,
　　crowded at the ends of the branches . . . . 24. *P. calycina.*
　Leaves opposite, oblong or lanceolate, fine- or pungent-pointed,
　　rigid, penniveined. Flowers in the upper axils. Pedicels
　　short . . . . 25. *P. spinulosa.*
　Leaves in threes, oblong-linear, obtuse. Pedicels nearly or quite
　　as long as the leaves. . . . . 26. *P. tenella.*
Leaves flat or concave, with fine or pungent points. Bracteoles
　on the calyx-tube. Flowers pedicellate.
　Glabrous or pubescent. Leaves broad, flat or slightly concave,
　　all in threes . . . . 27. *P. ternata.*
　Very villous. Leaves small, ovate or lanceolate, concave, oppo-
　　site, in threes or irregularly scattered . . . . 28. *P. styphelioides.*

(A very few species of the following section have the leaves occasionally irregularly whorled, but with the calyx-lobes much less dissimilar. A few species again have the calyx of *Euchilus*, but with the leaves all alternate.)

SECT. IV. **Cœlophyllum.**—*Leaves all alternate, either flattened but more or less concave, or with incurved margins, or darker-coloured underneath than above, or, if linear-terete or trigonous, channelled above, 1- or 3-nerved or quite nerveless, without transverse veins or reticulations.*

Leaves nearly flat or very concave, but open on the upper side in A, B, F, and G, and sometimes in E, in *P. neurocalyx* under D, and rarely in *P. tenuifolia* under H.
Leaves terete or trigonous and channelled above, the involute margins closed in C, D, and H, and sometimes in E, in *P. setulosa* and rarely in *P. humilis* under F.
Flowers in dense terminal heads, not intermixed with leaves, and not growing out into leafy shoots in A, B, C, and most species of D.
Flowers axillary or in leafy heads growing out into leafy shoots, or solitary and terminal in E, F, G, and H, and in *P. adunca* and *P. neurocalyx* under D.

A. *Leaves concave or nearly flat, obtuse or nearly so. Stipules none or minute and quite free. Flowers in terminal heads or umbels. Bracteoles close under the calyx.* (*Eastern species except* P. urodon.)
Flowers in small terminal umbels or umbel-like racemes, quite gla-
　brous.
　Leaves linear-cuneate . . . . 29. *P. altissima.*
　Leaves small, obovate . . . . 30. *P. obovata.*
Flowers in dense terminal heads. Calyx villous.
　Leaves incurved at the end. Stipules minute, setaceous . . . 31. *P. incurvata.*

Leaves not or scarcely incurved. Stipules none.
  Calyx shortly villous, the 2 upper lobes rather large . . . . 32. *P. subumbellata.*
  Calyx plumose with long hairs, the 2 upper lobes very large . 33. *P. urodon.*

B. *Leaves concave or nearly flat, obtuse acute or mucronate but not pungent, often narrow-linear but not terete. Stipules conspicuous, united at the base within the leaf, at least on the young shoots. Flowers in dense terminal heads. (Eastern species.)*
Bracteoles inserted on or adnate to the calyx-tube.
  Leaves linear, acute, almost flat.
    Stipules long, imbricate. Leaves 1 to 1½ in. long. Calyx
    villous . . . . 34. *P. stipularis.*
    Stipules not imbricate. Leaves under ¾ in. Calyx and whole
    plant glabrous . . . . 35. *P. glabra.*
  Leaves very concave, linear-obtuse, mucronate or scarcely acute,
  rarely above ⅓ in. long.
    Leaves not aristate. Bracts broad, imbricate. Calyx hirsute
    with short hairs . . . . 36. *P. dentata.*
    Upper leaves and calyx-lobes aristate. Bracts few. Calyx
    villous with long hairs . . . . 37. *P. aristata.*
Bracteoles close under the calyx-tube, but free from it.
  Leaves narrow-oblong or almost linear, obtuse. Flowers nearly
  sessile . . . . 38. *P. plumosa.*
  Leaves linear or linear-lanceolate, acute. Flowers shortly pedicel-
  late in the head . . . . 39. *P. viscosa.*

C. *Leaves terete or trigonous, not pungent, channelled above. Stipules united at the base within the leaf, at least on the young shoots. Flowers in dense terminal heads. (Eastern species.)*
Bracteoles close under the calyx, but free from it.
  Leaves crowded, incurved, with fine points . . . . 40. *P. echinula.*
  Leaves obtuse acute or shortly mucronate.
    Bracts broad. Flowers yellow . . . . 41. *P. hibbertioides.*
    Bracts narrow. Flowers pink . . . . 42. *P. rosea.*
Bracteoles inserted on or adnate to the calyx-tube . . . . . . 43. *P. mollis.*

D. *Leaves terete or trigonous, channelled above. Stipules small and distinct or minute or none. Flowers in terminal, leafless, or rarely leafy heads, or (in P. neurocalyx) axillary. Bracteoles close under the calyx, but free from it. (Western species.)*
Flower-heads dense. Bracts imbricate, completely covering the
  calyx.
    Heads ovoid-globular. Bracts appressed, shortly toothed . . . 44. *P. strobilifera.*
    Heads broadly globular. Bracts spreading at the top, deeply lobed 45. *P. ericifolia.*
Flower-heads dense. Bracts imbricate but few and small, or much
  shorter than the calyx.
    Heads many-flowered. Leaves straight or incurved. Stipules
    minute, setaceous. Bracts entire or 2-lobed . . . . . . 46. *P. verruculosa.*
    Heads small. Leaves small, mostly recurved. Stipules minute
    but broad. Bracts with a small leaf-like middle lobe . . . 47. *P. empetrifolia.*
Flowers few, in a head or cluster, becoming lateral by the growing
  out of the shoot. Bracts none besides the floral leaves.
    Calyx silky-villous, the 2 upper lobes much larger than the others 48. *P. adunca.*
    Calyx glabrous, rigid, strongly striate as well as the bracteoles,
    the lobes nearly equal . . . . 49. *P. neurocalyx.*

E. *Leaves rigid, pungent, nearly flat, concave or terete and channelled above. Stipules conspicuous, usually setaceous. Flowers in leafy heads or clusters, at length axillary and lateral. (Eastern species.)*
Leaves lanceolate, concave or conduplicate. Flowers distinctly pe-
  dicellate . . . . 50. *P. rigida.*

Leaves cordate-lanceolate or linear, nearly flat, concave or with involute margins. Flowers almost or quite sessile . . . . . . 51. *P. juniperina.*
Leaves linear-terete or trigonous, channelled above. Flowers almost
or quite sessile . . . . 52. *P. acerosa.*

F. *Leaves concave or nearly flat, rarely terete and then rigid, but not pungent. Flowers axillary, or, if terminal, solitary or in small leafy heads growing out into leafy shoots. Bracteoles inserted on or adnate to the calyx-tube. (Eastern species.)*

Flowers sessile or nearly so, all axillary.
Calyx 2 upper lobes much longer than the others, broad, falcate
and united to the middle.
Leaves elliptical-oblong or linear, usually open above . . . 53. *P. humilis.*
Leaves linear-terete or trigonous, channelled above . . . . 55. *P. setulosa.*
Calyx 2 upper lobes but little longer than the others.
Leaves small, oblong-cuneate, obtuse . . . . 54. *P. parviflora.*
Leaves narrow-linear, mucronate. Stipules imbricate . . . 56. *P. vestita.*
Leaves small, obovate or lanceolate, mucronate and recurved . 57. *P. procumbens.*
Flowers in small, terminal, leafy heads, or solitary and terminal.
Stipules small.
Pubescence rust-coloured, loose. Leaves linear, obtuse . . . . 58. *P. hispidula.*
Pubescence silky or hoary. Leaves from obovate to narrow-linear.
Flowers sessile or nearly so . . . . 60. *P. largiflorens.*
Flowers distinctly pedicellate . . . . 59. *P. laxiflora.*
Flowers all axillary, pedicellate. Stipules small.
Plant more or less villous. Pedicels short. Leaves small.
Leaves small, mucronate or fine-pointed and recurved at the end 57. *P. procumbens.*
Leaves mostly obtuse, 2 to 4 lines long or rather more, broad
or narrow . . . . 61. *P. villosa.*
Leaves numerous, about 1 line long, broad, obtuse, and recurved
at the end . . . . 62. *P. foliolosa.*
Plant glabrous. Leaves ¼ to 1 in. long . . . . 63. *P. flexilis.*

G. *Leaves concave or nearly flat, not pungent. Flowers axillary, or, if terminal, solitary. Bracteoles inserted close under the calyx, but free from it. (Eastern species.)*

Plant quite glabrous. Flowers all axillary.
Leaves ½ in. long or more. Flowers pedicellate. Calyx upper
lobes large . . . . 64. *P. euchila.*
Leaves small, almost imbricate. Flowers usually nearly sessile.
Calyx-lobes nearly equal. . . . . , ... 65. *P. selaginoides.*
Branches pubescent or villous. Flowers all axillary.
Leaves small, broad, rigid, squarrose . . . . 66. *P. densifolia.*
Leaves elliptical, oblong or obovate, obtuse. Bracteoles narrow . 67. *P. elliptica.*
Leaves linear or linear-oblong. Bracteoles broad . . . . . 68. *P. subspicata.*
Leaves acute, mostly 3-nerved, hirsute or ciliate with long hairs . 69. *P. villifera.*
Flowers terminal, solitary, surrounded by imbricate bracts and a
tuft of floral leaves. Leaves acute, mostly 3-nerved.
Plant villous . . . . 70. *P. involucrata.*
Plant glabrous . . . . 71. *P. Muelleri.*

H. *Leaves linear-terete, channelled above, slender or small, not pungent, and rarely rigid. Flowers solitary sessile axillary or terminal.*

Flowers terminal, surrounded by broad imbricate bracts . . . . 72. *P. prostrata.*
Flowers axillary or surrounded by a tuft of floral leaves. Bracts
small and few or none.
Leaves mostly about ½ in. long. Bracteoles under the calyx.
Calyx about 3 lines long . . . . 73. *P. canaliculata.*
Leaves not exceeding 4 lines or very slender.

Calyx about 2 lines. Bracteoles inserted on the tube at its
base . . . . 74. *P. fasciculata.*
Calyx under 1½ lines long. Bracteoles close under the calyx . 75. *P. tenuifolia.*

(*P. laxa*, Kunze, in Linnæa, xvi. 319, from a specimen raised by Lehmann from Australian seed, is insufficiently described to be recognizable. It is said to be near *P. flexilis*, but
the characters given are very different.)

(*P. crassifolia*, Lodd., *P. eriophora*, Lodd., and *P. incarnata*, Mackay, enumerated in
Steud. Nomencl. ed. 2, are unpublished garden names, referring probably to some of the
above species.)

Sect. I. EUPULTENÆA.—Leaves alternate, with recurved or revolute margins. Ovary quite sessile, except in *P. conferta*.

This section may be compared in foliage to *Aotus* and others of the preceding genera, in
which the tendency of the margin is always to be recurved, instead of incurved as in the
fourth and fifth sections and in the three succeeding genera. Where the leaf is nearly flat,
the difference is more observable in the withered leaf; but even when fresh or pressed quite
flat it is generally indicated by the under surface being paler or more hoary, instead of being
deeper coloured than the upper one. When the leaf is terete, it is grooved on the under
instead of the upper side. The summit of the leaf is sometimes incurved although the margins may be recurved.

1. **P. daphnoides,** *Wendl. Bot. Beob.* (1798), 49, *and in Hort. Herrenh. t.* 17. An erect shrub of 3 to 6 ft., the branches virgate, slightly angular, minutely silky-pubescent or hoary. Leaves cuneate-oblong, rounded
or rarely truncate at the end, with a minute point, ¾ to 1½ in. long, or in the
southern varieties shorter and broader, flat, glabrous or with a few hairs along
the midrib above, pale or silvery underneath. Stipules minute. Flowers
shortly pedicellate, in dense terminal sessile heads, usually shorter than the
last leaves, which form an involucre round it. Bracts ovate, imbricate, the
outer ones short and persistent, the inner ones often 3 lines long and deciduous. Bracteoles small, linear, inserted about the middle of the calyx-tube.
Calyx silky-hairy, 2½ to nearly 3 lines long, the lobes lanceolate, shorter than
the tube, the 2 upper ones broader. Petals deciduous; standard nearly
twice as long as the calyx; lower petals shorter, the keel obtuse. Ovary
villous. Style slightly thickened towards the base. Pod obliquely ovate,
acuminate or mucronate, flat, about 3 lines long.—Andr. Bot. Rep. t. 98;
Sm. in Trans. Linn. Soc. ix. 247; DC. Prod. ii. 110; Bot. Mag. t. 1394;
Lodd. Bot. Cab. t. 1143; Hook. f. Fl. Tasm. i. 86.

N. S. Wales. Port Jackson, *R. Brown, Sieber, n.* 419, *Fl. Mixt. n.* 557, and others;
Twofold Bay, *F. Mueller.*

Victoria. Common on barren ranges, *F. Mueller.*

Tasmania. Port Dalrymple and Kent's Island, *R. Brown;* not uncommon in various
parts of the island, *J. D. Hooker.*

S. Australia. Along river banks, *Whittaker.*

Var. *obcordata.* Leaves shorter (mostly ½ to ¾ in. long) and broader, more truncate, with
a more prominent point. *P. obcordata*, Andr. Bot. Rep. t. 574; DC. Prod. ii. 110. To
this form belong the majority of the Tasmanian specimens.

2. **P. stricta,** *Sims, in Bot. Mag. t.* 1588. An erect or decumbent
shrub of 1 to 2 ft., with slender scarcely angular branches more or less silky-
pubescent. Leaves obovate oblong or cuneate, obtuse with a small usually
recurved point, 3 to 4 lines or rarely ½ in. long, flat or nearly so, glabrous
above, pale and often silky-pubescent underneath, especially when young.

Stipules minute. Flowers not numerous, nearly sessile in small dense heads within the last leaves. Bracts imbricate, usually deciduous, the inner ones often at least 2 lines long. Bracteoles linear or oblong, concave, inserted on the middle of the calyx-tube. Calyx about 2 lines long, silky-villous ; lobes lanceolate, shorter than the tube, the 2 upper ones broader and united to the middle. Petals deciduous ; standard nearly twice as long as the calyx ; lower petals shorter, the keel obtuse. Ovary villous ; style slightly thickened to-wards the base. Pod triangular-ovate, acute, about 3 lines long, silky-pubescent.—DC. Prod. ii. 111; Lodd. Bot. Cab. t. 974; Hook. f. Fl. Tasm. i. 86 ; *P. capitellata,* Sieb. in DC. Prod. ii. 112.

N. S. Wales. Port Jackson, *Sieb. n.* 413 ; Bargo Brush, *Backhouse.*
Victoria. Creswick Creek and Dandenong ranges, and Grampian Hills, *F. Mueller ;* mouth of the Glenelg, *Allitt.*
Tasmania. Port Dalrymple, *R. Brown;* abundant in various localities, on moist peaty soil, *J. D. Hooker.*
S. Australia. S. E. interior, *F. Mueller.*

3. **P. retusa,** *Sm. in Ann. Bot.* i. 502, *and in Trans. Linn. Soc.* ix. 247. A shrub with the slender virgate branches of *P. stricta,* often angular and usually silky-pubescent. Leaves linear or linear-cuneate, very obtuse or more frequently emarginate, and sometimes dilated at the end, 2 to 4 lines or rarely ½ in. long in the normal form, flat, pale underneath, but usually glabrous on both sides. Stipules very small. Flowers few, in small terminal heads sessile within the last leaves. Bracts imbricate, the inner ones often 1½ lines long, but very deciduous, the outer ones smaller. Bracteoles broadly linear, inserted on the calyx-tube. Calyx about 2 lines long, silky-hairy, the 2 upper lobes broader, but scarcely more united. Standard not twice as long as the calyx. Ovary villous ; style filiform almost from the base. Pod broadly ovate, almost triangular, about 3 lines long, rather flat.—DC. Prod. ii. 112 ; Bot. Mag. t. 2081 ; Bot. Reg. t. 378.

Queensland. Shoalwater Bay, *R. Brown ;* Glasshouse mountains and Brisbane river, *F. Mueller.*
N. S. Wales. Port Jackson to the Blue Mountains, *R. Brown, Sieber, n.* 415, *and Fl. Mixt. n.* 558 ; northward to Hastings and Clarence rivers, *Beckler ;* and southward to Twofold Bay, *F. Mueller.*
Victoria. Snowy and Broadribb rivers, *F. Mueller.*
Var. *linophylla.* Branches more pubescent, and scarcely angular or quite terete, leaves rather longer, usually about ½ in., rarely nearly 1 in., flowers larger, the calyx often 2½ lines long, and rather more numerous in larger heads.—*P. linophylla* (*P. bracteata* on the plate), Schrad. Sert. Hannov. t. 18 ; Sm. in Trans. Linn. Soc. ix. 247 ; DC. Prod. ii. 112 ; *P. glaucescens,* Sieb. Pl. Exs. n. 417, and Fl. Mixt. n. 561.—Port Jackson. Sieber's specimens of *P. amœna,* Pl. Exs. n. 414, and Fl. Mixt. n. 559, appear to be intermediate between the two forms.
Some specimens without flowers, from Burnett river, Queensland, *F. Mueller,* appear to be nearly allied to *P. retusa,* but with very narrow leaves, almost terete, with closely revolute margins.

4. **P. Benthamii,** *F. Muell. in Trans. Phil. Inst. Vict.* i. 38. A rigid erect shrub, the young branches slightly angular and silky-pubescent. Leaves lanceolate or oblong-linear, rigid, acute or pungent-pointed, mostly about ½ in. long and rarely exceeding 8 lines, flat or with recurved margins, glabrous above, pale or silky-pubescent underneath. Stipules small. Flowers few to-gether, rather large, in terminal heads, sessile within the last leaves. **Bracts**

imbricate, the inner deciduous ones 2 lines long, 2-lobed, but otherwise entire, ciliate. Bracteoles inserted on the calyx-tube, oblong, very concave or keeled. Calyx silky-villous, 3 lines long; lobes shorter than the tube, the 2 upper ones much broader and united above the middle. Standard twice as long as the calyx; lower petals shorter. Ovary villous, tapering into the style. Pod only seen young.

Victoria. Along springs and rivulets in the Grampians, and amongst rocks on Mount Abrupt, *F. Mueller.*

Var. *elatior,* F. Muell. Tall, with elongated slender branches. Leaves narrow.—Yowaka river, and foot of Mount William, *F. Mueller.*

5. **P. pycnocephala,** *F. Muell. Herb.* Branches and under side of the leaves silvery-white with a soft dense silky-pubescence. Leaves broadly obovate, obtuse or with a short recurved point, 3 to 4 lines long, coriaceous, glabrous smooth and shining above, the margins slightly recurved. Stipules appressed, nearly 1 line long. Flowers in dense globular heads, sessile above the last leaves. Bracts imbricate, broad, densely covered with silky hairs, except a narrow margin. Calyx sessile, nearly 4 lines long, very silky, the lobes narrow, acute. Bracteoles narrow-cuneate, shortly 3-toothed, very silky outside, inserted at the very base of the calyx or close under it and nearly as long. Petals not half as long again as the calyx, mostly persistent after flowering. Pod sessile, acuminate, oblique, much flattened, very silky, about as long as the calyx.

N. S. Wales. Bluff Mountain, New England, *C. Stuart.* An elegant species allied to *P. myrtoides,* but distinct in indumentum, as well as in the shape of the leaf.

6. **P. myrtoides,** *A. Cunn.; Benth. in Ann. Wien. Mus.* ii. 81. A tall shrub, with virgate terete branches, usually silky-pubescent. Leaves narrow-oblong, often more or less cuneate, obtuse, with a very minute point, $\frac{1}{2}$ to $\frac{3}{4}$ in. long, flat or the margins recurved, glabrous above, pale and sometimes silky-pubescent underneath. Stipules lanceolate, acuminate, closely pressed, about 1 line long. Flowers numerous, nearly sessile in dense globular terminal heads, sessile within the last leaves, which are however rarely so long as the flowers. Bracts imbricate, ovate or lanceolate, silky-hairy at the edges. Bracteoles inserted on the calyx-tube, broadly oblong or ovate, very concave and keeled, almost boat-shaped, 1 to $1\frac{1}{2}$ lines long. Calyx silky-pubescent, $2\frac{1}{2}$ lines long, the lobes shorter than the tube, the 2 upper ones united above the middle. Petals persistent till the fruit is ripe. Standard nearly twice as long as the calyx; lower petals nearly as long, the keel obtuse. Ovary villous; style flattened at the base. Pod ovate-lanceolate, acuminate, about 4 lines long.

Queensland. Islands of Moreton Bay, *A. Cunningham, Fraser;* in the Cypress-Pine country, *Leichhardt.*

7. **P. mucronata,** *F. Muell. Fragm.* i. 8. An erect or diffuse shrub, of 2 to 6 ft., loosely villous with few spreading hairs. Leaves oblong or narrow-elliptical, mucronate, mostly about $\frac{1}{2}$ in. long, flat or with recurved margins, not coriaceous, green but loosely hairy above, hoary-pubescent and loosely villous underneath. Stipules rather long and appressed. Flower-heads dense and sessile above the last leaves, but all past flower in our specimens.

Bracts imbricate, broad, silky-villous, 2-lobed. Bracteoles linear-setaceous, inserted on the calyx-tube. Calyx about 2½ lines long, the lobes fine-pointed, the 2 upper ones rather broader and more united. Petals not seen. Pod acuminate, rather longer the calyx.

Victoria. Granite hills, Futter's Range, and May-Day Hills, *F. Mueller.*

S. Australia. Lofty Range, *Wurth,* the specimen small, and iu leaf only, and therefore doubtful.

The species may possibly prove to be a broad-leaved variety of *P. polifolia.*

8. **P. polifolia,** *A. Cunn. in Field. N. S. Wales,* 346. A shrub with terete virgate branches, more or less villous when young, with soft spreading hairs. Leaves linear, obtuse, with a fine straight or recurved point, the margins recurved or revolute, in some specimens all under ½ in., in others 1½ in. long, glabrous or rarely hairy above, hoary underneath, and often hirsute with long hairs, especially on the midrib. Stipules rather long, appressed. Flowers numerous, in dense terminal heads, sessile within the last leaves. Bracts broad, imbricate, softly villous, the inner ones 2 lines long and bifid. Bracteoles very concave, keeled, inserted on the calyx-tube. Calyx 3 lines long, softly villous; lobes finely acuminate, shorter than the tube, the 2 upper ones broad and united at the base. Standard not twice as long as the calyx; lower petals shorter. Ovary hirsute, tapering into the style. Pod very oblique, acuminate, longer than the calyx.—*P. rosmarinifolia,* Lindl. Bot. Reg. t. 1584 (the West Australian origin a mistake); *P. mucronata,* Lodd. Bot. Cab. t. 1711? (from the figure); *P. rosmarinifolia,* Endl. Nov. Stirp. Dec. 4 (with the pubescence rather more silky).

N. S. Wales. Port Jackson, *R. Brown, Woolls;* brushy hills, Blue Mountains, *A. Cunningham;* Dividing Range, between Nangatta and Bondi, *E. B. Sharpe.*

9. **P. petiolaris,** *A. Cunn.; Benth. in Ann. Wien. Mus.* ii. 82. A procumbent or straggling shrub, with numerous ascending branches, hirsute with short spreading hairs. Leaves on remarkably long petioles, those of the floral ones often attaining 3 or 4 lines, linear, obtuse, with a short recurved point, ½ to ¾ in. long, the margins revolute, often sprinkled with a few hairs above, the under side hirsute, especially the midrib. Stipules with spreading or recurved fine points. Flowers in dense terminal heads, sessile within the last leaves. Bracts few, besides the stipules of the floral leaves. Bracteoles inserted above the middle of the calyx-tube, linear-subulate, ciliate. Calyx about 3 lines long, hirsute; lobes acuminate, rather longer than the tube, the 2 upper ones united to the middle. Petals not much longer than the calyx-lobes, the keel dark-coloured. Ovary very silky-villous, tapering into the subulate style. Pod not seen.

Queensland. Brisbane river, *A. Cunningham, F. Mueller;* Burnett river, *F. Mueller.*

10. **P. paleacea,** *Willd. Spec. Pl.* ii. 506. A shrub, with slender diffuse or divaricate branches, silky-pubescent when young. Leaves linear, with fine straight or recurved points and revolute margins, ½ to ¾ in. long, glabrous above, pale and usually silky-hairy underneath. Stipules appressed, often 2 lines long. Flowers in dense, but not large, terminal heads, sessile within the last leaves. Bracts imbricate, glabrous, scarcely ciliate, completely covering the calyxes, the inner ones 3 to 4 lines long. Bracteoles inserted on the

calyx-tube, concave, carinate. Calyx silky-hairy, about 3 lines long, the lobes lanceolate, much shorter than the tube, the 2 upper ones united above the middle. Standard nearly twice as long as the calyx; lower petals shorter. Ovary villous, gradually tapering into a long style. Pod compressed, silky, longer than the calyx, and tapering into the long persistent silky base of the style.—Sm. in Trans. Linn. Soc. ix. 246; DC. Prod. ii. 112; Lodd. Bot. Cab. t. 291.

N. S. Wales. Port Jackson to the Blue Mountains, _R. Brown, Sieber, n._ 416, _Fl. Mixt. n._ 560, _Fraser, R. Cunningham, Woolls,_ and others; New England, _C. Stuart._

Victoria. Upper Genoa river, _F. Mueller,_ the specimen in young bud only, and therefore in some measure doubtful.

Var. _obtusata._ Leaves rather broader and less acute.—Hunter's River, _R. Brown;_ between Suggerah and Lake Macquoy, _Leichhardt._

Var. _sericea._ Branches, under side of the leaves, and bracts, closely silky, but leaves acute and bracts covering the calyxes, as in the normal form.—Marshy places, near Brighton, Victoria, _F. Mueller;_ near Melbourne, _Adamson._

11. **P. Gunnii,** _Benth. in Ann. Wien. Mus._ ii. 82. An erect diffuse or spreading shrub, of 1 to 3 ft., the numerous slender branches pubescent or hirsute with spreading hairs or at length glabrous. Leaves varying from ovate and 1 to 2 lines long, to oblong or almost linear and 3 to 5 lines long, obtuse, always convex or with recurved margins, often shining above, pale and sometimes hairy underneath. Stipules small, usually spreading. Flowers from 2 or 3 to about 8, in small terminal heads. Bracts imbricate, but usually shorter than the very short pedicels. Bracteoles inserted on the calyx-tube, small, lanceolate or linear. Calyx pubescent or villous, about 2 lines long or rather more, the lobes lanceolate, not acuminate, as long as the tube, the 2 upper ones more falcate and united to the middle. Standard twice as long as the calyx; lower petals shorter, the keel deeply coloured. Ovary villous; style subulate. Pod obliquely ovate, acute, flat, about 3 lines long, the fruiting pedicels often nearly as long as the calyx.—Hook. f. Fl. Tasm. i. 88, t. 13; _P. bæckeoides,_ A. Cunn.; Benth. in Ann. Wien. Mus. ii. 83 (described from a very imperfect fragment).

Victoria. Port Phillip, _R. Brown,_ and thence to Gipps' Land and Australia Felix; common in the Stringy-Bark and other ranges, _F. Mueller._

Tasmania. Port Dalrymple and Derwent river, _R. Brown;_ abundant throughout the colony, ascending to from 2000 to 3000 ft., _J. D. Hooker._

The larger-leaved forms can always be distinguished from _P. striata_ by the very much smaller bracts.

12. **P. scabra,** _R. Br. in Ait. Hort. Kew._ ed. 2. iii. 18. A shrub of 3 to 4 ft., with divaricate branches, terete and softly pubescent or villous, the down often rust-coloured. Leaves from obovate to narrow-cuneate, under ½ in. long, truncate, emarginate or shortly 2-lobed, and often mucronate, the margins revolute, scabrous above, tomentose or hirsute underneath. Stipules spreading or recurved. Flowers sessile in the upper axils or 3 or 4 together at the ends of the branches. Bracts very small or none besides the stipules of the floral leaves. Bracteoles inserted on the calyx-tube, linear or lanceolate. Calyx broad, about 2 lines long, the lobes as long as the tube, the 2 upper ones broader and united to the middle. Standard about twice as long as the calyx; lower petals rather shorter, the keel deeply coloured. Ovary

villous, tapering into a subulate style. Pod ovate, pubescent, nearly 3 lines long. Seeds ovate, the hilum at the broad end with a fringed strophiole.

N. S. Wales. Port Jackson to the Blue Mountains, *R. Brown, Sieber, n.* 386, *Fl. Mixt.* 563, 564, *and* 592, and others.

Victoria. In the Grampians, *Mitchell;* Mount Disappointment, the Grampians, etc., *F. Mueller;* Wimmera, *Dallachy.*

Var. *montana.* More tomentose. Leaves frequently obovate, very retuse, scarcely mucronate.—*P. montana,* Lindl. in Mitch. Three Exped. ii. 178.—To this form belong the Victorian specimens.

Var. *biloba.* Tomentum short and often hoary or even silvery on the under side of the leaves. Leaves narrow-cuneate, dilated and 2-lobed at the end, with a short recurved point. —*P. biloba,* R. Br. in Bot. Mag. t. 2091 (a starved specimen); DC. Prod. ii. 110; Lodd. Bot. Cab. t. 550; *P. deltoidea,* Sieb. Pl. Exs.—Port Jackson, *R. Brown, Sieber, n.* 388, etc.

13. **P. microphylla,** *Sieb. in DC. Prod.* ii. 112. A dwarf, diffuse, much branched shrub, the branches slender, hoary silky-pubescent or villous. Leaves linear-cuneate, usually narrow and 3 to 4 lines long, truncate or retuse, with a recurved point and revolute margins; sometimes longer and flatter, rounded at the end, with a recurved point, but never exceeding ½ in., usually glabrous above and hoary tomentose underneath, rarely softly villous when young. Stipules very small, spreading or recurved. Flowers in the upper axils on very short pedicels, or 2 or 3 together at the ends of the short branchlets. Bracts very small or none besides the stipules of the floral leaves. Bracteoles inserted on the calyx-tube, small, linear. Calyx scarcely 2 lines long, hoary or silky-villous, the lobes nearly as long as the tube, the 2 upper ones broad and united above middle. Standard about twice as long as the calyx; lower petals shorter. Ovary villous. Pod broadly and obliquely ovate, not acuminate, about 3 lines long.—*P. stenophylla,* A. Cunn. in G. Don, Gen. Syst. ii. 124; *P. uncinata,* A. Cunn.; Benth. in Ann. Wien. Mus. ii. 88.

Queensland. Brisbane river, *Fraser.*

N. S. Wales. Port Jackson to the Blue Mountains, *R. Brown, Sieber, n.* 418, *Fl. Mixt. n.* 562, and others; Goulburn, *Backhouse;* Lachlan river, *Fraser.*

Var. *cuneata.* Leaves broadly cuneate-truncate, 3 to 4 lines long.—*P. cuneata,* Benth. in Ann. Wien. Mus. ii. 83.—To this belong the Queensland specimens, and some from New England, *C. Stuart.* When the leaves are nearly flat, it has some resemblance to *P. largiflorens,* which, however, besides the usually incurved margins of the leaves, may be readily distinguished by the large upper lip of the calyx.

14. **P. Drummondii,** *Meissn. in Pl. Preiss.* ii. 219. Branches slender, elongated, pubescent when young with rather rigid appressed hairs. Leaves linear, with revolute margins, obtuse or with recurved points, under ½ in. long, very scabrous above, pale and more or less hairy underneath. Stipules appressed. Flowers 3 to 6 together, sessile in heads either terminal or on very short lateral shoots, so as to appear axillary, surrounded by a few broad, bifid bracts. Bracteoles ovate, concave, inserted immediately under the calyx and as long as the tube. Calyx pubescent, 3 lines long, the 2 upper lobes rounded and united into a broad emarginate upper lip, the 3 lower ones much shorter, acute. Standard twice as long as the calyx; keel much shorter, straight. Ovary villous; style scarcely thickened at the base, hooked at the top.

W. Australia, *Drummond, 2nd Coll. n.* 127; Vasse and Murchison rivers, *Oldfield.*

15. **P. pinifolia,** *Meissn. in Pl. Preiss.* ii. 220. An erect shrub, with virgate loosely pubescent or villous branches. Leaves narrow-linear, with very short callous or recurved points and revolute margins, 1 to 1½ in. long, glabrous or sprinkled with a few hairs above, pale or hoary underneath. Stipules spreading. Flowers rather large and numerous, in terminal umbels or heads, sessile within the last leaves, although each flower is distinctly pedicellate. Bracts apparently imbricate, rather narrow and bifid, but mostly fallen off from our specimens. Bracteoles linear, inserted under the calyx and very deciduous. Calyx silky-pubescent, broad, 3 lines long; lobes broad, acute, about as long as the tube, the 2 upper ones slightly larger and more united. Standard twice as long as the calyx ; keel much incurved, but very obtuse. Ovary villous, shortly tapering into the style. Pod very villous, obliquely and broadly ovate, turgid, 4 to 5 lines long.

W. Australia, *Drummond, 2nd Coll. n.* 109. This species differs from the rest of the genus in the very deciduous bracts, but the habit, inflorescence, strophiolate seeds, and other characters, are quite those of *Pultenæa.*

16. **P. pedunculata,** *Hook. Bot. Mag. t.* 2859. A prostrate much-branched shrub, often forming large patches, the branchlets slender, but rigid, terete, loosely pubescent or villous. Leaves linear or oblong-lanceolate, narrowed at both ends, the margins recurved or revolute, 2 to 3 lines or rarely ¼ in. long, rigid, with fine almost pungent points, which at length wear off, glabrous or scabrous above, pale and often hairy underneath. Stipules erect. Flowers small, solitary or 2 together at the ends of the branchlets, on filiform pedicels, longer than the leaves. Bracteoles linear, inserted at the base of the calyx, free or scarcely adnate. Calyx 1½ lines long, very open; lobes all acuminate, spreading, the 2 upper ones broader and shortly united. Standard twice as long as the calyx; lower petals rather shorter, the keel incurved, dark coloured. Ovary villous ; style scarcely thickened at the base. Pod obliquely ovate, obtuse or with a recurved point, turgid, about 3 lines long. —Hook. f. Fl. Tasm. i. 91; *P. diemenica,* Turcz. in Bull. Mosc. 1853, i. 277.

N. S. Wales. Port Jackson, *R. Brown.*
Victoria. Port Phillip, *R. Brown ;* Glenelg river, *Mitchell ;* Windu Valley, *Robertson ;* Forest Creek, *F. Mueller ;* Wimmera, *Dallachy.*
Tasmania. Derwent river, *R. Brown ;* common in sandy plains, near Hobarton, and probably throughout the island, *J. D. Hooker.*
S. Australia. Mount Lofty, *Whittaker ;* Port Lincoln, *Wilhelmi ;* Lofty and Bugle ranges, *F. Mueller.*

17. **P. conferta,** *Benth.* Much branched and apparently diffuse, the short ascending branchlets glabrous or nearly so. Leaves crowded, but not opposite, linear, obtuse or with recurved points, 2, 3 or rarely 4 lines long, rather thick, the margins closely revolute, glabrous or the upper ones slightly hairy. Stipules lanceolate-subulate, imbricate on the branchlets. Flowers axillary on short pedicels. Bracteoles lanceolate-subulate, inserted immediately under the calyx and free from it. Calyx pubescent, with appressed hairs, about 3 lines long, the 2 upper lobes large, oblong, scarcely acute, free; the lower ones shorter, narrow-lanceolate, falcate. Petals all reddish when dry ; standard not half as long again as the calyx ; lower petals rather

shorter. Ovary villous, slightly contracted at the base, as in most species of the section *Euchilus;* style subulate. Pod not seen.

W. Australia. *Drummond, 5th Coll. n.* 70. From the number quoted this should be *Euchilus purpureus,* Turcz. in Bull. Mosc. 1853, i. 276, but there may be some mistake as the character given does not agree with our specimens.

18. **P. diffusa,** *Hook. f. Fl. Tasm.* i. 91. *t.* 14. A low diffuse or procumbent shrub, with short ascending or erect branches, somewhat angular, minutely hoary or glabrous. Leaves linear, 2 to 3 lines long, obtuse or shortly mucronate, flat or convex and glabrous above, the margins folded back and adnate to the lower surface leaving exposed only the broad midrib, often sprinkled with a few short hairs. Stipules minute or none. Flowers small, in the upper axils, resembling those of *P. pedunculata,* usually 2 opposite to each other, each in the axil of a small bract at the base of a young axillary shoot, and often 2 such axillary shoots each with 2 flowers are opposite to each other. Pedicels slender, about 1 line long. Bracteoles immediately under the calyx and shorter than its tube, ovate-lanceolate, slightly glandular-toothed. Calyx slightly strigose-pubescent, 1½ lines long or rather more; lobes shorter than the tube, the 2 upper ones rounded or truncate, united above the middle into a broad upper lip. Standard about twice as long as the calyx; lower petals shorter; the keel much curved. Ovary silky-pubescent; style much dilated at the base. Pod not seen.—*Phyllota diffusa,* F. Muell. Fragm. i. 8.

Tasmania. Sandy plains in various localities, *J. D. Hooker.* The seeds not being as yet known, the genus of this species cannot be determined with certainty. It was removed by F. Mueller to *Phyllota* on account of the want of stipules, but that occurs in a few other undoubted *Pultenæas,* and the free filaments and other characters are much more those of *Pultenæa* than of *Phyllota.* It appears to be nearly allied to *P. pedunculata.*

SECTION II. ACIPHYLLUM.—Leaves alternate, rigid, concave and keeled but the margins flat, transversely reticulate. Flowers terminal, solitary or 2 or 3 together. Ovary quite sessile.

The rigid coriaceous leaves recall those of *Gastrolobium* and of the multi-ovulate genera allied to it; but the inflorescence and flowers are quite those of *Pultenæa.*

19. **P. reticulata,** *Benth.* An erect rigid shrub of several feet; branches virgate, minutely silky-pubescent when young. Leaves from ovate-lanceolate to almost linear, rigid, tapering into a pungent point ⅛ to ¾ in. long, concave above, glabrous and strongly reticulate on both sides, the midrib prominent underneath. Stipules small. Flowers terminal or in the upper axils, solitary or rarely 2 or 3 together, surrounded by 2 or 3 broad almost orbicular bracts nearly 2 lines long, often very deciduous. Bracteoles immediately under the calyx, short and broad. Calyx silky-pubescent or nearly glabrous, broad, 3 to 3½ lines long; lobes lanceolate, almost pungent, much longer than the tube, the 2 upper ones shortly united. Standard twice as long as the calyx; lower petals much shorter, the keel slightly curved, deeply coloured. Ovary villous; style filiform. Pod villous, about 3 lines long, broadly ovate, obtuse, turgid, the valves hard.—*Daviesia reticulata,* Sm. in Trans. Linn. Soc. ix. 256; *Jacksonia reticulata,* DC. Prod. ii. 107; *Pultenæa aciphylla,* Benth. in Hueg. Enum. 35, and in Ann. Wien. Mus. i. 81; Meissn. in Pl. Preiss. i. 74, and ii. 219.

W. Australia. King George's Sound, *Menzies, R. Brown, Huegel, Drummond,* 2nd *Coll. n.* 108; Sussex district, Stirling Terrace and near Albany, *Preiss, n.* 833, 847, and 848; Kalgan and Vasse rivers, *Oldfield.*

20. **P. ochreata,** *Meissn. in Pl. Preiss.* i. 75, *and* ii. 219. A tall erect shrub allied to *P. reticulata,* the branches much more slender, glabrous or minutely pubescent. Leaves obovate oblong or almost linear, very obtuse, rarely above ½ in. long, rigid, concave with flat margins, glabrous and strongly reticulate on both sides, the midrib prominent underneath. Stipules rather broad. Flowers of *P. reticulata,* but rather smaller, solitary and terminal. Bracts not 1 line long. Bracteoles immediately under the calyx, short, nearly orbicular. Calyx broad not above 2 lines long, the lobes very acute but not pungent. Ovary villous; style filiform. Pod villous, broadly ovate, obtuse, 3 to 4 lines long, the valves coriaceous and turgid.

W. Australia, *Drummond,* 2nd *Coll. n.* 107; Wellington district, *Preiss, n.* 1038.

21. **P. aspalathoides,** *Meissn. in Pl. Preiss.* i. 73, *and* ii. 219. A shrub of 2 or 3 ft., the branches terete, silky-pubescent when young. Leaves distant along the branches, crowded on the smaller branchlets and round the flowers, narrow-linear, ⅓ to ¾ in. long, rigid, tapering to a fine point, but scarcely pungent, concave, glabrous or hirsute with soft fine hairs, strongly reticulate on both sides, the midrib prominent underneath. Stipules narrow. Flowers terminal, solitary or 2 or 3 together, almost sessile within a dense tuft of floral leaves, with a very few ovate concave bracts. Bracteoles immediately under the calyx, oblong, concave. Calyx silky-villous, nearly 2 lines long; lobes acute, nearly equal, about as long as the tube. Standard more than twice as long as the calyx; lower petals rather shorter, the keel much incurved. Ovary villous; style filiform. Pod broadly ovate, almost acute, about 2 lines long, much flatter than in *P. reticulata.*

W. Australia. King George's Sound, *R. Brown,* and others, *Drummond,* 3rd *Coll. n.* 96; near Albany, *Preiss, n.* 838, *and* 1195; near Mount Barker, *Maxwell;* Wilson's Inlet, *Oldfield.*

SECTION III. EUCHILUS.—Leaves all or mostly opposite or in whorls of three, flat or concave, or in some species the margins slightly recurved, often 1- or 3-nerved or penniveined, somewhat reticulate in a few species. Flowers axillary or crowded at the ends of the branches. Two upper lobes of the calyx much larger than the others. Ovary often contracted at the base into a very short stipes. Style often broad at the base.

The opposite or verticillate leaves, the remarkable development of the upper lobes of the calyx, the much dilated base of the style and shortly stipitate ovary, all characters more prominent in this than in any other section, might have justified the retaining it as a distinct genus, did they all generally accompany each other in the same species, but each one is most prominent in a different species, and each one may be traced in almost as great a degree in some one or more species belonging to other sections.

22. **P. obcordata,** *Benth.* An erect much-branched shrub, the young branches tomentose-pubescent. Leaves opposite or in whorls of three or scattered, broadly obovate or obcordate, 2 to 4 or rarely 5 lines long, very obtuse truncate or emarginate, coriaceous, rigid, softly pubescent when young, at length nearly glabrous and obscurely reticulate above, tomentose-pubescent underneath with the margins slightly recurved. Stipules minute. Flowers

in the upper axils or forming a short terminal leafy head. Bracteoles immediately under the calyx, small, linear. Calyx pubescent, 2 to 2½ lines long, the tube very short, the 2 upper lobes large, obovate, obtuse, the lower ones much shorter, linear-lanceolate, ciliate. Standard half as long again as the calyx; lower petals rather shorter, the keel deeply coloured. Ovary villous, contracted at the base·but scarcely stipitate. Pod ovate, flattened, as long as the calyx.—*Euchilus obcordatus,* R. Br. in Ait. Hort. Kew. ed. 2. iii. 17; Lodd. Bot. Bab. t. 60; Bot. Reg. t. 403; Meissn. in Pl. Preiss. i. 72.

W. Australia. King George's Sound, *R. Brown, Baxter, Drummond,* and others, *Preiss, n.* 804; and from thence along the coast eastward to the Great Bight, *Maxwell.*

23. **P. rotundifolia,** *Benth.* A diffuse shrub, with slender terete pubescent branches. Leaves small but not crowded, opposite, broadly obovate or orbicular, very obtuse, 1 or rarely 1½ lines long, the margins recurved, glabrous above when full-grown, strigose or hirsute underneath with a prominent midrib. Stipules small. Flowers small, on filiform pedicels often ½ in. long. Bracteoles immediately under the calyx, linear-subulate. Calyx about 2 lines long, slightly pubescent, the tube very short, 2 upper lobes large, obovate or oblong, somewhat falcate, the lower ones much shorter, narrow-lanceolate. Standard not twice as long as the calyx; lower petals shorter, the keel deeply coloured. Ovary shortly stipitate. Pod rather longer than the calyx, obliquely ovate, the valves thin.—*Euchilus rotundifolius,* Turcz. in Bull. Mosc. 1853, i. 277; *Euchilus crinipodus,* F. Muell. Fragm. i. 145.

W. Australia, *Drummond, 5th Coll. n.* 78; E. Mount Barren, Phillips and Fitzgerald ranges, *Maxwell.*

24. **P. calycina,** *Benth.* Branches minutely hoary-pubescent. Leaves irregularly opposite or in whorls of three, oblong-linear or slightly cuneate, obtuse, 4 to 6 lines long, rigid, slightly concave, the midrib prominent and the veins more or less reticulate underneath, glabrous or with a few silky hairs when young. Stipules small. Flowers in the upper axils or in a short terminal leafy head, on pedicels of about 1 line, with small stipular bracts. Bracteoles immediately under the calyx, narrow, concave. Calyx hairy, 3 to 3½ lines long, the 2 upper lobes large, obovate-oblong, very obtuse, the lower ones much shorter, lanceolate-subulate, ciliate. Standard broad, rather longer than the calyx; lower petals nearly as long, the keel deeply coloured. Ovary villous, narrowed into a short stipes, and tapering into a subulate style. Pod not seen.—*Euchilus calycinus,* Turcz. in Bull. Mosc. 1853, i. 276.

W. Australia, *Drummond, 5th Coll. n.* 75.

25. **P. spinulosa,** *Benth.* Much branched and apparently diffuse, the branchlets slightly hairy. Leaves opposite, crowded, lanceolate, with a fine almost pungent point, 3 to 4 lines long, rigid, concave, keeled and transversely reticulate, glabrous or with a few long hairs. Stipules often 2 lines long. Flowers in the upper axils on very short pedicels. Bracteoles close under the calyx, long, linear-subulate, ciliate. Calyx nearly 3 lines long, loosely hairy, the tube very short; 2 upper lobes large, oblong, with fine points, lower ones much shorter, lanceolate-subulate, ciliate. Petals scarcely exceeding the calyx; the keel dark-coloured. Ovary villous, contracted into

a very short stipes. Young pod oblong-falcate, acûminate, but not seen ripe.
—*Euchilus spinulosus*, Turcz. in Bull. Mosc. 1853, i. 275.

W. Australia, *Drummond, 5th Coll. n.* 71.

26. **P. tenella,** *Benth.* Stems slender, much-branched, diffuse or procumbent, the smaller branches silky-pubescent. Leaves in whorls of 3, oblong-
linear, obtuse, narrowed at the base, 2 or rarely 3 lines long, coriaceous, the
margins involute, glabrous above, convex and sprinkled with a few silky hairs
underneath. Stipules very small. Pedicels axillary, often longer than the
leaves. Bracts minute. Bracteoles small, linear, inserted close under the
calyx but scarcely adnate to it. Calyx about 3 lines long, slightly silky-pubescent or glabrous, the 2 upper lobes broad and falcate, the lower ones narrow, but nearly as long. Standard nearly twice as long as the calyx, lower
petals shorter, all apparently yellow. Pod sessile, flat, nearly orbicular, about
3 lines long.

Victoria. Haidinger range at an elevation of 5000 ft., *F. Mueller.*

27. **P. ternata,** *F. Muell. Fragm.* i. 8, *and* iv. 21. An erect, usually
glabrous, often glaucous shrub, the branches terete. Leaves all in whorls of
three, in the original form broadly rhomboidal, truncate, or shortly tapering,
the midrib produced into a more or less pungent point, from 2 or 3 lines to
¾ in. long, and usually rather broader than long, flat or concave, often 3- or
5-nerved at the base. Stipules small. Flowers in the upper axils on pedicels of 1 to 2 lines. Bracteoles subulate, inserted on the base of the calyx.
Calyx above 3 lines long, the lobes longer than the tube, the 2 upper ones
broad, falcate, acute, united above the middle, the lower ones lanceolate-subulate. Petals nearly of equal length, twice as long as the calyx ; keel large,
almost hood-shaped, very obtuse. Ovary sessile, glabrous, tapering into the
flattened style. Pod ovate, turgid, about 3 lines long.—*Spadostyles Cunninghamii*, Benth. in Ann. Wien. Mus. ii. 81 ; *Gastrolobium Huegelii*, Henfr. in
Gard. Mag. i. with a fig. ; *Aotus cordifolius*, Lindl. and Paxt. Fl. Gard. i. 76,
not of Benth. ; *Spadostyles ternata*, F. Muell. First Gen. Rep. 12 ; *Pullenæa
oxalidifolia*, A. Cunn. in Steud. Nom. Bot. ed. 2.

N. S. Wales. Williams river, *R. Brown ;* Blue Mountains, *A. Cunningham ;* Nar-
gas, *M'Arthur.*
Victoria. Stony scrubby hills, Buffalo ranges, *F. Mueller.*
Var. *pubescens.* Branches more or less pubescent. Leaves broad but with long pungent
points.—Newcastle and Ruined-Castle Creek, *Leichhardt ;* Hunter's
river ?, *Vicary ;* Clarence river, *Beckler.*
Var. *cuspidata.* Branches slender, pubescent. Leaves small, tapering into pungent
points. Pedicels slender, as long as or sometimes longer than the calyx.—*Oxylobium spinosum*, DC. Prod. ii. 104; *Euchilus cuspidatus*, F. Muell. in Trans. Phil. Inst. Vict. ii. 68.
Queensland. Burnett and Brisbane rivers, *F. Mueller ;* Wide Bay, *Bidwill ;* Ipswich,
Nernst.

28. **P. styphelioides** (misspelt *staphyleoides*), *A. Cunn. in G. Don,
Gen. Syst.* ii. 124. A tall much-branched shrub, softly pubescent or hirsute
with spreading hairs. Leaves often irregularly opposite or in threes, or all
alternate, ovate or lanceolate, tapering into a rigid almost pungent point, 2 to
3 or rarely 4 lines long, concave or with involute margins, sometimes recurved
at the end, the midrib and sometimes also oblique lateral veins prominent.
Stipules small. Flowers axillary on slender pedicels of 1 to 2 lines. Brac-

teoles inserted on the calyx-tube, linear or subulate, shorter than the lobes. Calyx villous, 3 to 3½ lines long, the lobes subulate-acuminate all longer than the tube; the 2 upper ones broad falcate and united at the base, the lower ones narrow. Petals rather longer than the calyx. Ovary sessile, villous with a few long hairs, tapering into a subulate style. Pod ovate, rather obtuse, turgid, shorter than the calyx.—*P. epacridea,* F. Muell. Fragm. iv. 21.

N. S. Wales. High ranges of the interior, *Fraser;* Argyle county, *M'Arthur;* Murray river, *A. Cunningham.*

Victoria. Scrubby and stony ridges between the Broken and Ovens rivers, Mount Pleasant and Mount Hunter, *F. Mueller.*

This species is one of those which connect *Euchilus* with *Cœlophyllum,* some specimens of *P. procumbens* closely resemble it in habit, but the calyx is very different. It is also nearly allied to *P. humilis.*

SECTION IV. CŒLOPHYLLUM.—Leaves all alternate, either flattened but more or less concave in withering, or with involute margins, or darker-coloured underneath than above, or linear-terete and channelled on the upper side, the margins never recurved, although the end of the leaf may be so, 1- or 3-nerved or quite nerveless, without transverse veins or reticulations. Calyx-lobes nearly equal or the 2 upper ones large. Ovary sessile.

Some species of this section (especially *P. urodon, P. euchila, P. humilis,* etc.) have the calyx of *Euchilus,* but the leaves are all alternate. A few species are nearly allied in habit and character to *Latrobea diosmifolia,* but have the calyx less regular, and the bracteoles close under the calyx. Those with linear-terete or subulate leaves approach *Dillwynia* in foliage, but differ in the presence of stipules, and in the bracteoles close under or adnate to the calyx, besides that the standard is not so broad as it usually is in that genus.

29. **P. altissima,** *F. Muell. Herb.* A tall glabrous shrub, almost growing into a small tree. Leaves linear-cuneate, obtuse, rarely above ½ in. long, much narrowed at the base, concave above, faintly 1-nerved and often darker coloured underneath. Flowers small, several together in little terminal umbel-like racemes, rarely exceeding the last leaves. Bracts minute. Pedicels rather shorter than the calyx. Bracteoles very small, ovate, close under the calyx, but scarcely adnate to it. Calyx about 2 lines long, glabrous, the lobes as long as the tube, rather obtuse, the 2 upper ones a little broader. Standard not twice as long. Ovary glabrous.

N. S. Wales. Twofold Bay and Upper Genoa river, *F. Mueller.*

30. **P. obovata,** *Benth.* A glabrous, much-branched, rather slender shrub, the young branches slightly angular. Leaves on rather long petioles, obovate or broadly cuneate, obtuse, 2 to 3 lines long, concave, darker-coloured underneath, without any prominent midrib. Stipules very small. Flowers in terminal heads or umbels. Bracts very small. Pedicels rarely 1 line long. Bracteoles under the calyx or rather below it, small, ovate, obtuse. Calyx 1½ to 2 lines long, glabrous or minutely ciliate ; lobes all obtuse, rather longer than the tube, the 2 upper ones rather broader and united at the base. Ovary glabrous, tapering into a subulate style. Pod not seen.

N. S. Wales. Bargo Brush, Argyle county, *Backhouse.*

In all the flowers I examined I uniformly found 2 ovaries or distinct carpels, but that may not be constant in the species The bracteoles in this as in *P. urodon* are not so close to the calyx as in other Pultenæas, yet not so distant from it nor so small as in *Eutaxia* and *Dillwynia.*

31. **P. incurvata,** *A. Cunn. in Field, N. S. Wales,* 346. A low shrub with elongated slender branches, terete and softly pubescent or villous when young. Leaves lanceolate, obtuse or acute, incurved, otherwise flat or concave, mostly 3 to 4 lines long, rather thick, darker coloured on the under side, without any midrib. Stipules minute or none. Flowers in terminal heads, sessile within the last leaves. Bracts few, rather narrow, 3-fid. Bracteoles inserted under the calyx, linear, villous with long hairs. Calyx silky-villous, about 3 lines long; lobes lanceolate, longer than the tube, the 2 upper ones united to the middle. Standard nearly twice as long as the calyx; keel scarcely exceeding the calyx-lobes. Ovary villous; style erect, hooked at the end. Pod obtuse, not 2 lines long, the valves very convex.

N. S. Wales. Marshes near Sydney, *R. Brown;* margins of peaty bogs, King's Tableland, Blue Mountains, *A. Cunningham.* The species is very nearly allied to *P. subumbellata,* and perhaps a variety, but the stipules are more conspicuous, the flowers smaller, and the bracts and bracteoles different, at least in R. Brown's specimens.

32. **P. subumbellata,** *Hook. Bot. Mag. t.* 3254. A shrub either low and erect or taller and straggling, the branches virgate, rather slender, terete, pubescent when young. Leaves narrow-oblong or almost linear, obtuse, under ¼ in. and usually 3 to 4 lines long, rather incurved than recurved at the end, otherwise flat or concave, darker-coloured underneath, without any midrib, usually glabrous. Stipules none. Flowers all yellow, in dense terminal heads, sessile within the last leaves. Bracts few, short, broad, ciliate. Bracteoles inserted under the calyx, linear or oblong. Calyx softly villous, about 2 lines long; lobes lanceolate, scarcely so long as the tube, the 2 upper ones united to the middle. Standard more than twice as long as the calyx; lower petals shorter. Ovary villous, tapering into an erect, rather thick style, hooked at the top. Pod about 2 lines long, obliquely and broadly ovate, obtuse, turgid.—Bot. Reg. t. 1632; Hook. f. Fl. Tasm. i. 87.

N. S. Wales. Near Mount Imlay, *F. Mueller.*
Victoria. Australian Alps, at an elevation of 5000 ft., *F. Mueller.*
Tasmania. Port Dalrymple, *R. Brown;* abundant in moist situations, ascending to 4000 ft., *J. D. Hooker.*

The alpine specimens are small and slender, with small leaves and flowers; but they pass gradually into the larger forms. In the absence of stipules, as well as in general habit, this species connects *Pultenæa* with *Latrobea diosmifolia.* The name *subumbellata* was unfortunately chosen, for the flower-heads are as compact as in most capitate species.

33. **P. urodon,** *Benth.* A low shrub, with erect or ascending stems, in our specimens simple or not much branched, 1 to 1½ ft. high, glabrous and glaucous or more or less villous with long loose hairs. Leaves numerous, narrow-oblong, obtuse or scarcely acute, 2 to 4 lines or rarely nearly ¼ in. long, flat or concave, rather glaucous on both sides or darker-coloured underneath, the midrib slender or inconspicuous, glabrous or villous with long spreading hairs. Stipules none. Flowers in rather large dense very villous heads, sessile within the last leaves, which are larger, broader, thinner, and more acute than the others. Bracteoles inserted a little below the calyx, linear-subulate, plumose-hairy. Calyx membranous, about ½ in. long, densely clothed with very long soft hairs; lobes all much longer than the tube, the 2 upper ones broadly oblong with short fine points, the lower ones rather

shorter, very narrow, tapering into long fine plumose points. Standard rather longer than the calyx, emarginate; lower petals rather shorter; keel purple, obtuse. Ovary very villous, sessile or slightly contracted at the base; style subulate. Pod not seen ripe.—*Urodon capitatus*, Turcz. in Bull. Mosc. 1849, ii. 17; *U. dasyphyllus*, Turcz. l. c. 1853, i. 268.

W. Australia, *Drummond, n.* 21, 23, 24, 98, 267, *and 5th Coll. Suppl. n.* 47.
The larger hairy and more acute leaves by which *U. dasyphyllus* was distinguished may be sometimes found on the same specimen as the smaller obtuse glabrous ones of *U. capitatus.* The species has the foliage and inflorescence of *P. subumbellata*, differing in the large very hairy flower-heads, and especially in the calyx, which is that of some species of the section *Euchilus.*

34. **P. stipularis,** *Sm. Bot. N. Holl.* 35, *t.* 12, *and in Trans. Linn. Soc.* ix. 245. A tall shrub with erect virgate terete branches, usually glabrous, but the surface almost concealed by the crowded leaves and appressed stipules. Leaves linear, acute, with a short fine but scarcely pungent point, 1 to 1½ in. long, flat above, with slightly prominent margins, usually darker-coloured underneath with a scarcely prominent midrib, ciliate with a few long hairs, but otherwise glabrous. Stipules narrow, often above 3 lines long. Flowers numerous in dense heads, sessile within the last leaves. Bracts imbricate, but not numerous, narrow, bifid, acuminate. Bracteoles inserted on the calyx-tube and as long as its lobes, linear, ciliate. Calyx fully 4 lines long, ciliate or hirsute with long hairs; lobes lanceolate, subulate, much longer than the tube, the 2 upper ones broader, and united at the base. Standard scarcely half as long again as the calyx; lower petals shorter. Ovary with a few long hairs, tapering into the style. Pod not seen.—Bot. Mag. t. 475; DC. Prod. ii. 112; Reichb. Icon. Exot. t. 192; *P. proteoides* and *P. psoraleoides*, Sieb. Pl. Exs.

N. S. Wales. Port Jackson to the Blue Mountains, *R. Brown, Sieber, n.* 382, and others.

35. **P. glabra,** *Benth.* Allied in habit to *P. stipularis* and to *P. aristata*, but readily known by the peculiar calyx and the absence of all hairs. Branches terete, virgate. Leaves linear, acute, rigid, ½ to ¾ in. long, flat or concave or the margins slightly involute, the midrib slightly prominent, the under surface usually darker-coloured. Stipules subulate, acuminate, often spreading. Flowers rather smaller than in *P. stipularis*, in dense heads, sessile within the last leaves. Bracts few besides the stipules of the floral leaves, and these usually with a few setæ in their axils. Bracteoles inserted on the calyx-tube and as long as its lobes, broadly lanceolate, with 2 or 3 brown setæ or acuminate scales in their axils. Calyx quite glabrous, 2½ lines long, the broad lanceolate very acute lobes nearly equal and spreading, as long as the tube. Standard twice as long as the calyx; lower petals nearly as long; keel nearly straight, obtuse. Ovary glabrous, tapering into a flattened style. Pod not seen.

N. S. Wales. Blue Mountains, *R. Cunningham.*

36. **P. dentata,** *Labill. Pl. Nov. Holl.* i. 103, *t.* 131. A rigid heath-like shrub of 1 to 3 ft.; branches rigid, virgate, minutely silky-pubescent or villous when young, rarely quite glabrous. Leaves linear, linear-oblong or narrow-lanceolate, usually narrowed at both ends, but scarcely acute, under

½ in. and sometimes only ¼ in. long, flat or concave, rigid, glabrous, darker-coloured or somewhat silvery underneath, the midrib rarely visible. Stipules small. Flowers in dense terminal heads, sessile within the last leaves. Bracts very broad, imbricate, the inner ones 1½ lines long. Bracteoles inserted on the calyx-tube, ovate or oblong, bifid with a third central subulate lobe. Calyx silky-villous, 2 to 2½ lines long, the lobes rather shorter than the tube, the 2 upper ones united to the middle. Standard about twice as long·as the calyx; lower petals rather shorter. Ovary villous, tapering into a slender style. Pod 2 lines long or rather more, ovate, acute or rarely obtuse.—DC. Prod. ii. 112 ; Hook. f. Fl. Tasm. i. 88 ; *P. argentea*, A. Cunn. in Field. N. S. Wales, 347 ; *P. pimeleoides*, Hook. f. Fl. Tasm. i. 88.

N. S. Wales. Port Jackson, *R. Brown;* barren rocky hills, Cox's river,· *A. Cunningham ;* Illawarra, *M'Arthur.*

Victoria. Moist grassy places, Australia Felix, *F. Mueller.*

Tasmania. Port Dalrymple, *R. Brown;* northern parts of the island, Rocky Cape, Woolnorth, Hampshire Hills, etc., *Gunn, J. D. Hooker.*

This species varies much in the size of the flowers, those of Cunningham's specimens are the smallest, those described as *P. pimeleoides* the largest, but I have been unable to mark out distinct varieties. In M'Arthur's specimens they are as large as in the common Tasmanian ones.

37. **P. aristata,** *Sieb. in DC. Prod.* ii. 112. Branches erect, virgate, glabrous. Leaves crowded, linear, rarely above ½ in. long, obtuse or acute with a long bristle-like point, which wears off on the lower leaves, concave or with involute margins, darker-coloured underneath, the midrib inconspicuous, glabrous or ciliate with a few long hairs. Stipules narrow, rather long, appressed or with spreading points. Flowers in dense terminal heads, sessile within the last leaves and usually not exceeding them. Bracts few, narrow, bifid. Bracteoles inserted on the calyx-tube and as long as its lobes, linear, ciliate with long hairs and occasionally with a brown seta in the axil. Calyx about 4 lines long ; lobes nearly equal, subulate-acuminate, longer than the tube, ciliate or hirsute with long hairs. Standard not half as long again as the calyx ; keel rather shorter, incurved. Ovary hirsute with long hairs ; style slightly dilated downwards. Pod not seen.—Reichb. Icon. Exot. t. 195.

N. S. Wales. Port Jackson, *Sieber, n.* 383, *Fl. Mixt. n.* 555 ; Illawarra, *M'Arthur.*

38. **P. plumosa,** *Sieb. in DC. Prod.* ii. 111. Branches erect, virgate, glabrous or silky-hairy. Leaves crowded, narrow-oblong or slightly cuneate, under ½ in. long, obtuse or scarcely mucronate, concave and glabrous above, darker-coloured underneath, the midrib scarcely conspicuous, usually, especially the upper ones, hirsute with long silky hairs. Stipules long and often ciliate. Flowers in dense terminal heads, sessile within the last leaves, without any or with very few bracts besides the stipules of the floral leaves. Bracteoles inserted under the calyx, linear, ciliate, with 2 broad stipules. Calyx hirsute or ciliate with soft hairs, 3 lines long or rather more; lobes acuminate, longer than the tube, the 2 upper ones much broader and more united. Standard about half as long again as the calyx ; keel shorter. Ovary with a few long hairs, tapering into a subulate style. Pod not seen.—Reichb. Icon. Exot. t. 193 ; *P. canescens*, A. Cunn. in Field, N. S. Wales, 346.

N. S. Wales. Port Jackson to the Blue Mountains, *Sieber, A. Cunningham.*

39. **P. viscosa,** *R. Br. Herb.* An erect shrub of 3 to 4 ft.; branches
virgate, terete, pubescent or villous. Leaves linear, acute or nearly so, ¼ to
¾ in. long, concave, glabrous or silky-pubescent underneath, the midrib obtuse or slightly prominent. Stipules subulate, with recurved points. Flowers
in terminal heads, sessile within the last leaves, though each flower is shortly
pedicellate. Bracts scarcely longer than the pedicels. Bracteoles inserted
under the calyx, ovate-lanceolate, rather large. Calyx villous, nearly 3 lines
long, lobes lanceolate, nearly as long as the tube, the 2 upper ones united to
the middle. Standard half as long again as the calyx; lower petals rather
shorter. Ovary villous; style subulate. Pod ovate, acuminate, about 3 lines
long.

N. S. Wales. Paramatta, *R. Brown, Woolls;* southern districts ?, *Mossman;* Wombaya Ranges, *F. Mueller.*
Victoria. Mount Sturgeon, *Robertson.*
From R. Brown's name, it would appear that the plant is more or less viscid when fresh.
This character does not show in the dried specimens seen either in his or other herbaria.
The species is closely allied to *P. hibbertioides,* but the leaves appear to be constantly open
on the upper side, not slender and terete as in the latter species.

40. **P. echinula,** *Sieb. in DC. Prod.* ii. 112 (spelt *echinata* in Spreng.
Syst. Cur. Post. 173). Apparently a straggling shrub, the older branches
denuded of leaves and tuberculate or echinate with the remains of their petioles. Leaves crowded on the younger branches, often incurved, linear-terete,
almost subulate, mucronate, rarely exceeding ½ in., channelled above by the
involute margins, often tuberculate outside and sometimes hirsute with soft
hairs. Stipules rather long. Flowers in dense heads, sessile within the last
leaves, with few bracts besides the stipules of the floral leaves. Bracteoles
inserted under the calyx, oblong or lanceolate. Calyx 2½ lines long, glabrous
or hirsute; lobes lanceolate, nearly equal, about as long as the tube. Standard not twice as long as the calyx; lower petals rather shorter. Ovary villous; style subulate. Pod not seen.—Reichb. Icon. Exot. t. 196.

Queensland. Brisbane river, *Fraser ?*
N. S. Wales. Port Jackson to the Blue Mountains, *Sieber, n.* 384, *R. Cunningham.*
Sieber's specimens have the upper leaves more hairy than R. Cunningham's, whilst in the
latter the calyx is villous, which in Sieber's is glabrous. Fraser's specimen is past flower
and the leaves are nearly smooth, but all appear to belong to the same species.

41. **P. hibbertioides,** *Hook. f. Fl. Tasm.* i. 89. A much-branched
diffuse shrub, forming large bushes of 1 to 2 ft., more or less clothed with
long soft spreading hairs. Leaves linear-terete, obtuse or shortly acute, mostly
about ½ in. long, channelled above by the involute margins, glabrous, pubescent or softly villous. Stipules with long subulate points. Flowers not
numerous, in dense heads, sessile amongst the last leaves, but each flower on
a pedicel of 1 line or rather more. Bracts imbricate, bifid, the inner ones
above 2 lines long, usually striate. Bracteoles inserted under the calyx,
concave and keeled, at least as long as the calyx. Calyx 2 to 2½ lines long;
lobes all acute or acuminate, the 2 upper ones rather broader. Standard
nearly twice as long as the calyx; lower petals rather shorter. Ovary villous, tapering into a subulate style. Pod rather narrow, acute or acuminate,
2 to 3 lines long.

Victoria. Buffalo ranges, *F. Mueller.*

Tasmania. Between Launceston and George Town, *Gunn.*
Var. *conferta.* Pedicels short. Bracts and bracteoles smaller.—Australia Felix, *F. Mueller.*
The species is nearly allied to *P. mollis,* but the bracteoles are quite distinct from the calyx. It differs from *P. viscosa* in its much narrower terete leaves, larger bracts, etc.

42. **P. rosea,** *F. Muell. Fragm.* ii. 15. An erect heath-like shrub, the branches virgate, glabrous or sprinkled with a few hairs. Leaves linear-terete, obtuse or with short callous points, under ½ in. long, channelled above by the involute margins, slightly scabrous. Stipules subulate-pointed. Flowers pink, in terminal heads, sessile within the last leaves. Bracts few, narrow, trifid. Bracteoles inserted under the calyx, linear-lanceolate. Calyx silky-pubescent, 2½ lines long ; lobes lanceolate, as long as the tube, the 2 upper ones more united. Petals nearly of equal length, not twice as long as the calyx. Ovary villous ; style subulate. Pod 2 lines long, acuminate.—*Burtonia subalpina,* F. Muell. in Trans. Phil. Inst. Vict. i. 39.

Victoria. In the Grampians, Mount William, at an elevation of 5000 ft., *F. Mueller.*
The species is chiefly distinguished from the preceding by the unusual colour of the flowers.

43. **P. mollis,** *Lindl. in Mitch. Three Exped.* ii. 260. A bushy shrub, the branches clothed with soft spreading hairs. Leaves narrow-linear, almost terete, acute or obtuse, mostly about ½ in. long, concave or channelled above, usually incurved, softly pubescent or villous. Stipules spreading. Flowers in dense terminal heads, sessile amongst the last leaves. Bracts short, few besides the broad bract-like stipules of the floral leaves. Bracteoles inserted on the calyx-tube near its base, narrow, keeled. Calyx about 3 lines long, villous ; lobes broad, scarcely so long as the tube and all nearly equal. Standard not twice as long as the calyx ; lower petals rather shorter. Ovary villous, tapering into a subulate style. Pod not seen.

Victoria. Wannon river, at the foot of the Grampians, *Mitchell ;* in the Grampians and Mount Macedon, *F. Mueller.*
Var. (?) *canescens.* Bracts small. Calyx hoary-pubescent, the bracteoles small and narrow.
S. Australia. Marble range, *Wilhelmi ;* near Spencer's Gulf, *F. Mueller.*

44. **P. strobilifera,** *Meissn. in Pl. Preiss.* i. 75, *and* ii. 220. An erect heath-like shrub of ½ to 1½ ft., closely resembling *P. ericifolia,* except in the shape of the flower-heads and bracts. Branches virgate, terete, minutely hoary. Leaves linear-terete, obtuse, rarely above 3 lines long and often much shorter, channelled above, rather thick, glabrous, smooth or slightly wrinkled. Stipules small. Flowers in dense terminal ovoid or scarcely globose heads. Bracts numerous and closely imbricate, broad, shortly toothed, pubescent or ciliate, the inner ones about 3 lines long. Bracteoles inserted close under the calyx, linear or subulate, hirsute with long hairs. Calyx under 4 lines long, the lobes narrow, acuminate, longer than the tube, the 2 upper ones united above the middle. Standard not half as long again as the calyx ; keel not exceeding the calyx-lobes. Ovary villous, tapering into an erect style, hooked at the top. Pod ovoid, turgid, about 3 lines long.—*P. ptero-nioides,* Turcz. in Bull. Mosc. 1853, i. 280.

W. Australia. In the interior, *Drummond, n.* 247, *and* 5*th Coll. n.* 62 (*or* 67 ?), *Preiss, n.* 1185 ; Minto district, *Preiss, n.* 1190 ; Stirling and Plantagenet ranges, *Maxwell.*

45. **P. ericifolia,** *Benth. in Lindl. Swan Riv. App.* 13. An erect heath-like shrub, with terete minutely pubescent branches. Leaves linear-terete, spreading, obtuse or mucronate, under ½ in. long, channelled on the upper side, glabrous, usually tuberculate or irregularly wrinkled. Stipules small. Flowers in dense terminal broadly globose heads, sessile within the last leaves. Bracts numerous, imbricate in several rows, but spreading at the top, deeply trifid, with a subulate central, and broad lateral lobes, the inner ones 3 to 4 lines long, all fringed with long hairs. Bracteoles inserted under the calyx, linear or subulate, hirsute with long hairs. Calyx fully 4 lines long, the lobes narrow, acuminate, longer than the tube, the 2 upper ones united above the middle. Standard not half as long again as the calyx; keel not exceeding the calyx-lobes. Ovary villous, tapering into an erect style, hooked at the top. Pod not seen.—Meissn. in Pl. Preiss. i. 75, and ii. 219.

W. Australia. King George's Sound, *R. Brown;* Swan River, *Drummond, 1st Coll. and n.* 248, *Preiss, n.* 1189; Konkoberup Hills and near Albany, *Preiss, n.* 1186 *and* 1187.

46. **P. verruculosa,** *Turcz. in Bull. Mosc.* 1853, i. 278. An erect heath-like shrub, the young branches loosely pubescent. Leaves linear-terete or trigonous, obtuse or scarcely pointed, mostly 3 to 4 lines or rarely 5 lines long, channelled above, glabrous, smooth or tuberculate, incurved, spreading or rarely recurved. Stipules subulate. Flowers in terminal heads, sessile amongst the last leaves, occasionally proliferous; buds acute. Bracts few, broadly ovate, not above 2 lines long, entire or shortly 2-lobed. Bracteoles inserted under the calyx, short, lanceolate. Calyx pubescent, or villous at the top with long soft hairs, about 3 lines long; lobes shorter than the tube, acuminate, the 2 upper ones broader and united nearly to the middle. Standard half as long again as the calyx, dark with a yellow edge; keel shorter, almost acuminate. Ovary villous, tapering into an erect rather thick style, hooked at the end. Pod not seen.

W. Australia. King George's Sound, *Collie, Drummond, 5th Coll. n.* 65; Kalgan river, *Oldfield;* Stirling Range and eastward to the Great Bight, *Maxwell.*
Var. *pilosa.* Sprinkled with spreading hairs. Bracts rather longer and more numerous, connecting this species with *P. ericifolia.*—Cheynes Beach, *Oldfield.*
Var. *brachyphylla.* Leaves shorter, thicker, and often very shortly mucronate. Flowers rather smaller, the heads often proliferous with few leaf-like bracts.—*P. brachyphylla,* Turcz. in Bull. Mosc. 1853, i. 279.—*Drummond, 5th Coll. n.* 68, *Maxwell.*
Var. *recurva.* Leaves more spreading, mostly recurved at the end, as in *P. empetrifolia,* but stipules, bracteoles, etc., of *P. verruculosa.* Flowers as in the var. *brachyphylla.*—King George's Sound, *Collie.*

47. **P. empetrifolia,** *Meissn. in Pl. Preiss.* i. 76. A diffuse or divari-cate shrub, rarely above 1 ft. high, with numerous slender hoary-pubescent branches. Leaves numerous, linear-terete, obtuse or shortly mucronate and mostly recurved at the end, in some specimens all under 2 lines long and often scarcely 1 line and thick, in others more slender and 2 to 3 lines long, all channelled above, glabrous and smooth or tubercular-scabrous. Stipules lanceolate, rather broad. Flowers in small dense terminal heads, occasionally proliferous; buds acute. Bracts (or floral leaves) few, shorter than the calyx, with 2 broad stipular lobes, and usually a central green leaf-like lobe.

Bracteoles inserted under the calyx, oblong, acute. Calyx $2\frac{1}{2}$ lines long, glabrous or with a few long hairs; lobes lanceolate, acute, rigid but not striate, as long as the tube, the 2 upper ones broader and slightly connate, the lower ones usually with a dark spot in each sinus. Standard about half as long again as the calyx; keel shorter, deeply coloured. Ovary villous; style subulate. Pod shorter than the calyx in our specimens.—*P. verticillata,* Turcz. in Bull. Mosc. 1853, i. 279.

W. Australia, *Drummond,* 5*th Coll. n.* 62 *and* 64. In inflorescence, this species connects the preceding with the following two species, the bracts are peculiar, and in a young state the acutely acuminate buds are very prominent.

48. **P. adunca,** *Turcz. in Bull. Mosc.* 1853, i. 279. A slender heath-like erect shrub, the young branches silky-pubescent. Leaves linear-terete, obtuse or with a hooked point, under $\frac{1}{2}$ in. long, channelled above, glabrous or scabrous-pubescent. Stipules small. Flowers at first in terminal heads, which soon grow out into leafy branches, leaving the flowers axillary near their base, without any other bracts than small floral leaves. Bracteoles inserted under the calyx, linear. Calyx nearly 3 lines long, silky-villous; lobes longer than the tube, the 2 upper ones broad, falcate, united to above the middle; the lower ones short and narrow. Standard half as long again as the calyx, dark-coloured with a yellow edge; lower petals shorter, dark-coloured. Ovary villous, tapering into a thick erect style, hooked at the top. Pod not seen.

W. Australia, *Drummond,* 5*th Coll. n.* 66. The calyx-lobes are more unequal in this species than in any other one of the section.

49. **P. neurocalyx,** *Turcz. in Bull. Mosc.* 1853, i. 281. A slender but rigid, heath-like, diffuse or divaricate shrub, with the habit, inflorescence, and rigid calyx-lobes of *P. empetrifolia,* but at once known by the striate bracteoles and calyx. Leaves linear or lanceolate, 1 or rarely 2 lines long, obtuse, concave or channelled above, 1- or 3-nerved underneath, usually glabrous. Stipules none or very small. Flowers at first in small, terminal, sessile, leafy heads, which soon grow out into leafy branches. Bracts or floral leaves ovate, striate, with scarious ciliate margins. Bracteoles inserted under the calyx, ovate or oblong, 3-nerved. Calyx glabrous or slightly pubescent, about $2\frac{1}{2}$ lines long; lobes all broadly lanceolate, scarcely acute, rigid, 3- or 5-nerved. Standard half as long again as the calyx or rather more; keel much shorter. Ovary villous, tapering into a thick erect style, hooked at the top. Pod not seen.

W. Australia, *Drummond,* 5*th Coll. n.* 63.
Var. *major.* Leaves 2 to 3 lines long. Flowers larger, the calyx 3 lines long.—Phillips ranges, Mount Bland, Robertson's Brook, etc., *Maxwell.*

50. **P. rigida,** *R. Br. Herb.* A much-branched, rigid shrub, the young branches terete, hoary-pubescent. Leaves lanceolate, very rigid, tapering into pungent points, concave or conduplicate, glabrous or obscurely penni-veined. Stipules with long subulate points. Flowers not seen. Fruiting pedicels axillary, solitary, 1 to 2 lines long. Bracteoles inserted close under the calyx, and as long as its tube, ovate-lanceolate. Calyx nearly 2 lines

long, the lobes subulate-acuminate. Pod very turgid, ovoid, 2 to 3 lines long.

S. Australia. Memory Cove, *R. Brown* (*Hb. Br.*).

51. **P. juniperina,** *Labill. Pl. Nov. Holl.* i. 102, *t.* 130. A much-branched prickly shrub, attaining several feet, the young branches terete and usually villous. Leaves linear or linear-lanceolate, spreading, rigid and pungent-pointed, under ½ in. long, concave or with involute margins, darker-coloured underneath, usually glabrous. Stipules subulate or those of the floral leaves broader. Flowers in the uppermost axils, usually 2 or 3 together at the ends of the smaller branches, with occasionally 1 or 2 leafless stipular bracts, shorter than the very short pedicels. Bracteoles inserted under the calyx, lanceolate. Calyx glabrous or pubescent, about 2 lines long or rather more; lobes lanceolate, rather shorter than the tube, the 2 upper ones broader, falcate, and united to the middle. Standard fully twice as long as the calyx; lower petals rather shorter. Ovary pubescent, tapering into a subulate style. Pod obliquely ovate, shortly acuminate, often above 3 lines long.—DC. Prod. ii. 112; Hook. f. Fl. Tasm. i. 90.

Victoria. Frequent in rocky places in the Australian Alps, also in the Grampians, *F. Mueller.*

Tasmania. Port Dalrymple and Derwent river, *R. Brown;* abundant throughout the island, ascending to 4000 ft., *J. D. Hooker.*

Var. *latifolia.* Leaves lanceolate, rounded or sometimes almost cordate at the base, tapering into a rigid pungent point.—*P. cordata,* Grah. in Edinb. Phil. Journ. 1836; Bot. Mag. t. 3443; Hook. f. Fl. Tasm. i. 90; by a clerical error, *P. cordifolia,* Benth. in Ann. Wien. Mus. ii. 82.—Tasmania, *R. Brown* and others.

Some Tasmanian specimens, as well as others from the Grampians, are quite intermediate between the two forms; and in the true *P. juniperina* the breadth of the leaves is very variable.

52. **P. acerosa,** *R. Br. Herb.* A rigid much-branched shrub of 1½ to 2 ft., the branches divaricate or spreading, loosely pubescent or tomentose. Leaves linear-subulate or trigonous, channelled above, rigid with more or less pungent points, crowded on the smaller branches and 3 to 5 lines long, divaricate on the longer branches, and nearly ½ in. Stipules appressed, subulate-pointed. Flowers sessile or nearly so, crowded in short leafy heads growing out into leafy shoots. Bracteoles inserted close under the calyx, oblong with subulate points. Calyx about 2½ lines long, glabrous or ciliate, the lobes broad with rigid subulate points, the 2 upper ones rather broader and more united. Standard half as long again as the calyx. Ovary villous; style subulate. Pod ovoid, very turgid, about as long as the calyx.

S. Australia. Memory Cove, *R. Brown;* Lofty Range, *Whittaker, F. Mueller;* Encounter Bay, *C. Stuart;* Port Lincoln, *Wilhelmi.*

53. **P. humilis,** *Benth. in Hook. f Fl. Tasm.* i. 91. Either a low shrub, with a thick stock and decumbent ascending or scarcely erect branches of from ½ to 1 ft., or erect with spreading branches and several feet high, pubescent or hirsute with spreading hairs. Leaves linear- or elliptical-oblong, obtuse or rather acute, rarely exceeding ½ in., rather thick, concave or with involute margins, often darker coloured outside, the midrib scarcely conspicuous, hirsute with long soft hairs when young, at length glabrous. Stipules narrow, except those of the floral leaves. Flowers axillary, but crowded into

short leafy spikes at or near the ends of the branches, often with a few broad stipular leafless bracts. Bracteoles inserted on the calyx near its base, linear-subulate, ciliate, often 2-stipulate. Calyx villous or ciliate with long hairs, 3 to 4 or even 5 lines long ; lobes acuminate, much longer than the tube, the 2 upper ones broad and united to the middle, the lower ones very narrow. Petals not much longer than the calyx-lobes ; keel dark-coloured. Ovary with a few long hairs; style much dilated downwards. Pod not seen.—*Spadostyles Huegelii* and *S. Benthamii,* Endl. Nov. Stirp. Dec. 3.

N. S. Wales. On the Murrumbidgee, *M'Arthur.*

Victoria. In the Grampians, *F. Mueller ;* Wimmera, *Dallachy ;* mouth of the Glenelg, *Allitt.*

Tasmania. Epping Forest, Launceston and Hobarton road, *Gunn.* The aspect is sometimes nearly that of *P. elliptica,* but differs in the calyx, the insertion of the bracteoles, etc., characters which bring it nearer to *P. villosa.*

54. **P. parviflora,** *Sieb. in DC. Prod.* ii. 111. Branches numerous, slender, pubescent with greyish appressed hairs. Leaves oblong-cuneate, obtuse, 1 to 2 or rarely 3 lines long, concave or with incurved margins, darker-coloured underneath with a slender midrib, glabrous or sprinkled with a few hairs when young. Stipules appressed. Flowers few, small, in the upper axils. Bracteoles inserted on the calyx-tube near its base, linear-subulate, ciliate, 2-stipulate. Calyx about 2 lines long; lobes acuminate, longer than the tube, the 2 upper ones broad, falcate and united at the base. Ovary hairy at the top; style dilated downwards.

N. S. Wales. Port Jackson, *R. Brown, Sieber, n.* 399 ; *Fl. Mixt. n.* 589 ; near Penrith, *Backhouse.*

Var. *angustifolia.* Leaves narrower and more acute or mucronate, sometimes almost terete and channelled above. Bracteoles usually shorter.—To this belong the Murrumbidgee specimens, and some from Stringy Bark Forest, *F. Mueller ;* Creswick diggings, *W. S. Whan.*

The aspect of the broader-leaved forms is sometimes nearly that of *P. elliptica,* but they always differ in the calyx, the insertion of the bracteoles, etc. As the leaves become narrower they are at the same time more rigid, acute, and spreading ; but in some specimens we see a passage from the one form to the other.

55. **P. setulosa,** *Benth.* Apparently procumbent, with silky-pubescent branchlets. Leaves linear, terete or trigonous, channelled above, mucronate, 2 to 3 or rarely 4 lines long, glabrous or silky-pubescent. Stipules appressed, with long, fine, erect or spreading points. Flowers axillary, nearly sessile, forming leafy heads or clusters at or below the ends of the branches. Bracts none besides the floral leaves. Bracteoles linear, inserted on the base of the calyx-tube, often 2-stipulate. Calyx 3 to 3½ lines long, slightly silky-pubescent, the lobes all tapering to fine points, the 2 upper ones broad, fal-cate and united above the middle, the lower ones shorter and much narrower. Standard scarcely twice as long as the calyx. Ovary very villous. Style glabrous, flattened at the base. Pod ovate, shorter than the calyx.

Queensland. Broad Sound, *Bowman.* Certainly allied to *P. humilis,* with nearly the same calyx ; but the small leaves not at all opened out, and the fine points to the stipules, leaves, and calyx-lobes give it a peculiar aspect.

56. **P. vestita,** *R. Br. in Ait. Hort. Kew. ed.* 2, iii. 19. A shrub of 1½ to 2 ft., with spreading diffuse or virgate branches, completely covered when young by the long closely-imbricate stipules, each pair united almost to the

top. Leaves narrow-linear or rarely lanceolate, shortly mucronate, under ½ in. long, concave or with involute margins, rigid, glabrous or minutely silky-pubescent, quite nerveless. Flowers in dense leafy terminal heads, occasionally growing out into short leafy proliferous spikes. Floral leaves or bracts with large broad stipules and a small lamina. Bracteoles inserted close under the calyx, stipular, scarious, lanceolate, usually with 2 pointed lobes and a central long ciliate point. Calyx 3 to 4 lines long, the tube glabrous or silky-hairy; lobes lanceolate, with rigid subulate points, hirsute or ciliate, the 2 upper ones broader at the base. Petals scarcely exceeding the calyx-lobes; keel almost as long as the standard. Ovary silky-villous; style subulate. Pod flattened, enclosed in the calyx.—DC. Prod. ii. 112; *P. fuscata,* F. Muell. in Trans. Vict. Inst. 119.

S. Australia. Port Lincoln, *R. Brown;* Salt's Creek, *F. Mueller.*
W. Australia. Point Sherratt, Esperance Bay, *Maxwell.*

57. **P. procumbens,** *A. Cunn. in Field, N. S. Wales,* 347. A diffuse or prostrate much-branched rigid shrub, the young branches pubescent. Leaves obovate-oblong or lanceolate, 1 to 2 or rarely 3 lines long, mucronate or fine-pointed, recurved at the end, but concave or with involute margins, darker-coloured and glabrous or slightly hairy outside, the midrib scarcely prominent. Stipules rather broad. Flowers solitary in the upper axils, forming sometimes short, leafy, terminal spikes. Bracteoles inserted on the middle of the calyx-tube, small, acute. Calyx glabrous or ciliate, about 2 lines long; lobes lanceolate, fine-pointed, about as long as the tube, the 2 upper ones broader and rather more united. Petals about twice as long as the calyx, the keel deeply coloured. Ovary bearing a few long hairs; style dilated downwards.—*P. setigera,* A. Cunn.; Benth. in Ann. Wien. Mus. ii. 82.

N. S. Wales. Blue Mountains, Cox's river, and N. of Bathurst, *A. Cunningham;* Argyle county, *M'Arthur;* rocky ridges near Mundoora, *Herb. F. Mueller.*

58. **P. hispidula,** *R. Br. Herb.* An erect much-branched shrub of 3 to 4 ft.; branches slender, terete, hirsute with rust-coloured hairs. Leaves linear, obtuse or with a minute recurved point, about 3 lines long, concave above, scabrous or hirsute underneath, the midrib scarcely conspicuous. Stipules small. Flowers small, in small terminal sessile heads. Bracts few, scarcely exceeding the short pedicels. Bracteoles inserted on the base of the calyx, oval-oblong. Calyx scarcely more than 1 line long, the lobes or teeth short, the 2 upper ones rather broader. Standard more than twice as long as the calyx, lower petals rather shorter. Ovary villous; style filiform. Pod not seen.

N. S. Wales. St. George's river, *R. Brown.* In habit this approaches at first sight *P. villosa,* but the flowers are much smaller, and the inflorescence and bracteoles quite different (*Herb. R. Br.*)

59. **P. laxiflora,** *Benth.* A prostrate shrub, with slender, terete branches, silky-pubescent when young. Leaves narrow-linear, obtuse or scarcely acute, mostly 3 to 4 lines long, concave or channelled above, convex underneath, without any prominent midrib, minutely pubescent or glabrous. Stipules narrow, appressed. Flowers in terminal clusters, at first surrounded by short imbricate bracts, which soon fall off, and the cluster grows out into a leafy shoot, with 3 to 6 flowers at its base, on silky-pubescent pedicels of 2

to 3 lines. Bracteoles close under the calyx, linear-lanceolate, with 2 stipular lobes. Calyx slightly silky-pubescent, about 2 to 2½ lines long; lobes all acute or acuminate, the 2 upper ones broader and united at the base. Standard half as long again as the calyx; lower petals rather shorter, the keel deeply coloured. Ovary villous, tapering into a subulate style. Pod not seen.

Victoria. Near the Western frontier, *Robertson ;* Grampians, *F. Mueller.*
S. Australia. Onkaparinga river and Encounter Bay, *F. Mueller.*

60. **P. largiflorens,** *F. Muell. Herb.* A rigid apparently divaricate shrub, the young branches minutely hoary or almost silky-pubescent. Leaves from broadly obovate and 2 or 3 lines long, to linear-cuneate and ¼ in. long, obtuse or truncate, more or less concave and glabrous above, silky-pubescent or at length glabrous underneath, with a prominent midrib. Stipules very small. Flowers nearly sessile, in axillary or terminal clusters, surrounded when young by a few short broad imbricate bracts, which usually fall off before the flowers expand. Bracteoles inserted near the top of the calyx-tube, small, lanceolate. Calyx silky-pubescent, 2 lines long or rather more; lobes scarcely so long as the tube, the 2 upper ones broad and united nearly to the middle. Standard twice as long as the calyx; lower petals nearly as long, the keel almost acuminate, Ovary villous; style filiform. Pod obliquely ovate, acute, silky, scarcely exceeding the calyx, more or less flattened.

Victoria. Forest Creek and Mount M'Ivor, *F. Mueller.*
S. Australia. Encounter Bay, *Whittaker ;* Lofty Range, *F. Mueller.*
The Victorian specimens have generally much narrower leaves than the S. Australian ones, although some branches of the latter have also sometimes the narrow leaves of the former.

61. **P. villosa,** *Willd. Spec. Pl.* ii. 507. A low or spreading much-branched shrub, pubescent or villous, rust-coloured when dry. Leaves usually oblong or somewhat cuneate, but varying from linear to obovate, obtuse or scarcely pointed, 2 to 3 or rarely when narrow nearly 4 lines long, concave or with incurved margins, tubercular or hirsute underneath, the midrib slender. Stipules small, narrow or broad. Flowers usually entirely yellow, solitary in each axil, but sometimes forming short terminal leafy racemes. Pedicels short, but slender. Bracteoles inserted on the calyx-tube, but sometimes very near its base, linear, with occasionally 1 or 2 setæ in their axil. Calyx from 1½ to above 2 lines long ; lobes acuminate, longer than the tube, the 2 upper ones broad, falcate and united to the middle, the lower ones narrow. Petals nearly equal in length, twice as long as the calyx. Ovary more or less hairy ; style subulate. Pod scarcely exceeding the calyx.—Sm. in Ann. Bot. i. 503, and in Trans. Linn. Soc. ix. 248; Bot. Mag. t. 967; DC. Prod. ii. 113; *P. polygalifolia,* Rudge, in Trans. Linn. Soc. xi. 303, t. 25 ; DC. Prod. ii. 111 ; *P. lanata,* A. Cunn.; Benth. in Ann. Wien. Mus. ii. 83 (a small-leaved form).

Queensland. Brisbane river, Moreton Bay, *A. Cunningham, Fraser, F. Mueller.*
N. S. Wales. Port Jackson to the Blue Mountains, *R. Brown, Sieber, n.* 421, *Fl. Mixt. n.* 588, and others ; northward to Hastings and Clarence rivers, *Beckler.*
Victoria. Australia Felix, near Mount Zero, *F. Mueller.*
Var. *latifolia.* Leaves small, very pubescent, from narrow-cuneate to broadly obovate. Flowers rather large.—*P. ferruginea,* Rudge, in Trans. Linn. Soc. xi. 300, t. 23 ; DC.

Prod. ii. 111; *P. lanata*, Sieb. Pl. Exs.; *Spadostyles ramulosa*, Endl. Nov. Stirp. Dec. 21 (from the description, no specimen having been preserved).—Port Jackson, *R. Brown, Sieber, n.* 420, etc.

Var. *glabrescens.* Leaves linear or narrow-oblong, 2 to 4 lines long, glabrous or slightly pubescent. Flowers of the common variety.—Port Jackson, *R. Brown. Spadostyles concolor*, Endl. Nov. Stirp. Dec. 20, of which no specimen has been preserved, is probably, from the description, this variety.

P. racemulosa, Sieb. Pl. Exs. n. 594; DC. Prod. ii. 111, is not in any of the sets of Sieber's plants which I have seen; from the character given it is probably one of the numerous forms of *P. villosa.*

62. P. foliolosa, *A. Cunn.; Benth. in Ann. Wien. Mus.* ii. 83. A spreading shrub, with elongated branches and very numerous short slender branchlets, rusty-pubescent or villous. Leaves numerous and very small, rarely 1 line long, ovate or oblong, obtuse, recurved at the end but concave above or with involute margins, darker-coloured and hirsute or scabrous underneath, the midrib not conspicuous. Stipules small. Flowers nearly sessile in the upper axils. Bracteoles inserted on the calyx, small, leafy with hair-like stipules. Calyx pubescent or villous, 2 lines long or rather more; lobes broad, not longer than the tube, the 2 upper ones much falcate and united to the middle. Standard fully twice as long as the calyx; lower petals rather shorter, apparently of the same colour. Ovary with a few long hairs; style somewhat dilated downwards. Pod turgid, obtuse, shorter than the calyx, which is somewhat enlarged after flowering.

N. S. Wales. Westward of Wellington valley, *A. Cunningham;* Lachlan river, *Fraser.*
Victoria. Barren hills between Meadow Creek and King river, *F. Mueller.*

63. P. flexilis, *Sm. in Ann. Bot.* i. 502, *and in Trans. Linn. Soc.* ix. 248. A shrub either quite glabrous or with a few appressed hairs on the young shoots and backs of the leaves. Leaves linear or linear-oblong, often slightly cuneate, obtuse or mucronate, ½ to 1 in. long, flat or concave, the under side darker-coloured, with a prominent midrib. Stipules very small. Flowers solitary in the upper axils, shortly pedicellate. Bracteoles inserted on the calyx near its base, small, lanceolate. Calyx glabrous or slightly ciliate, about 2 lines long; lobes rather broad, acute, shorter than the tube, the 2 upper ones broader and falcate. Standard fully twice as long as the calyx; lower petals not much shorter. Ovary with a few long hairs at the top; style dilated downwards. Pod obliquely ovate or ovate-oblong, turgid, about 3 lines long.—DC. Prod. ii. 111; Bot. Reg. t. 1694; *Dillwynia teucrioides*, Sieb. Pl. Exs.; *P. Sweetii*, Don, in Steud. Nomencl. ed. 2.

N. S. Wales. Port Jackson to the Blue Mountains, *R. Brown, Sieber, n.* 423, and others. Allied in habit to *P. euchila*, but the calyx-lobes are less disproportioned, the bracteoles are not under the calyx, the style less dilated at the base, etc.

Var. *mucronata.* Leaves narrow, with a fine pungent point.—Clarence river, *Beckler.*

64. P. euchila, *DC. Prod.* ii. 112. A glabrous shrub, with rather slender branches, nearly resembling at first sight *P. flexilis*, but differing in bracteoles, calyx, and style. Leaves linear-cuneate, obtuse, 5 to 9 lines long, flat or slightly concave, of a darker or a more silvery colour underneath than above, the midrib slender. Stipules minute. Flowers axillary, on pedicels of 2 to 3 lines. Bracteoles linear-subulate, inserted close under the calyx and shorter than its tube. Calyx 3 to $3\frac{1}{2}$ lines long, the lobes much longer than the tube, the 2 upper ones large, falcate and united above the middle, the lower ones narrow-lanceolate. Petals nearly equal, half as long again as the calyx, the keel slightly incurved. Ovary glabrous, tapering into the much dilated style. Pod longer than the calyx, coriaceous, turgid when ripe, with a flat point.—*Dillwynia cuneata*, Sieb. Pl. Exs. ; *Spadostyles Sieberi*, Benth. in Ann. Wien. Mus. ii. 81.

Queensland. Near the Brisbane, *Leichhardt ;* Ipswich, *Nernst.*

N. S. Wales. Port Jackson, *Sieber, n.* 422, and others ; Hunter's River, *R. Brown ;* Clarence river, *Beckler.*

65. **P. selaginoides,** *Hook. f. Fl. Tasm.* i. 87. An erect glabrous shrub, with the habit of a *Diosma.* Leaves numerous, obovate- or oblong-cuneate, almost imbricate, obtuse or with a short thick point, rarely above 2 lines long, thick, concave above, very convex and almost nerveless underneath. Stipules reduced to minute tubercles. Flowers axillary, forming a leafy head or tuft, at or below the ends of the branchlets. Bracts small, oblong. Pedicels short. Bracteoles lanceolate, concave, immediately under the calyx, but free from it. Calyx about 2 lines long, the lobes as long as the tube, obtuse, the 2 upper ones rather broader and united to the middle. Standard fully twice as long as the calyx. Ovary sessile, silky-villous. Style slender. Pod not seen.

Tasmania. Eastern parts of the island, St. Paul's River, Avoca, *Gunn, C. Stuart.* In foliage this is certainly allied to *P. subumbellata, obovata,* etc., but the inflorescence is quite different. Although the flowers are sometimes apparently in terminal heads, the heads are always interspersed with leaves, and the new shoots continue the axis, instead of growing out from under the heads, as in *P. subumbellata.*

66. **P. densifolia,** *F. Muell. in Trans. Vict. Inst.* 119. A rigid shrub, the branches diffuse or divaricate, tomentose-pubescent, but long concealed by the closely-appressed imbricate stipules. Leaves numerous, broadly obovate, obtuse or scarcely acute, rarely above 2 lines long, rigidly coriaceous, concave or conduplicate, but with recurved squarrose ends, glabrous on both sides, faintly penniveined underneath. Flowers axillary, sessile, forming leafy tufts below the summits of the branches, the stipules of the floral leaves large and imbricate, and with the broad scarious bracts and bracteoles concealing the calyx, the bracteoles inserted close under the calyx, but free from it. Calyx 2 to $2\frac{1}{2}$ lines long, slightly-pubescent or ciliate, the lobes broad with short pungent points, the 2 upper ones rather more united. Petals not much longer than the calyx. Style villous and slightly thickened below the middle. Pod obliquely ovate, shortly acute, silky-pubescent, scarcely exceeding the calyx.

Victoria. In the Murray desert, *F. Mueller.*

S. Australia. Port Lincoln, *Wilhelmi ;* Encounter Bay, *C. Stuart.*

67. **P. elliptica,** *Sm. in Trans. Linn. Soc.* ix. 246. Branches virgate, terete, pubescent, softly villous or at length glabrous. Leaves crowded, elliptical-oblong or when small often oblong-cuneate or almost obovate, rarely above $\frac{1}{2}$ in. long except when very luxuriant, and in some specimens not $\frac{1}{4}$ in., obtuse or minutely mucronate, concave and glabrous above, darker coloured underneath, the slender midrib sometimes quite inconspicuous, the upper ones often ciliate with long hairs. Stipules closely appressed, often 2

lines long. Flowers axillary towards the ends of the branches, forming at first an oblong leafy head, the floral leaves distinctly petiolate with broad bract-like stipules. Bracteoles inserted close under the calyx, linear or lanceolate. Calyx broad and membranous, about $2\frac{1}{2}$ lines long, lobes all lanceolate, rather shorter than the tube, ciliate with long hairs. Standard above $\frac{1}{2}$ in. long, lower petals rather shorter. Ovary mixed with a few long hairs, tapering into a subulate style. Pod ovate, rather turgid, scarcely exceeding the calyx.—Rudge, in Trans. Linn. Soc. xi. 302, t. 24; DC. Prod. ii. 111; *P. tuberculata*, Pers. Syn. i. 434 (from the short diagnosis); *P. hypolampra*, Sieb. in DC. Prod. ii. 111; Reichb. Icon. Exot. t. 194.

N. S. Wales. Port Jackson, *R. Brown, Sieber, n.* 394, 396, *and* 397, and *Fl. Mixt. n.* 591, and others.

Var. *thymifolia.* Leaves and flowers smaller, but not otherwise differing from the common form.—*P. thymifolia*, Sieb. in DC. Prod. ii. 111.—Port Jackson, *Sieber, n.* 398, and *Fl. Mixt.* n. 590, and others.

68. **P. subspicata,** *Benth.* A low procumbent or diffuse much-branched shrub, the branches glabrous or hirsute with long hairs. Leaves linear or linear-oblong, obtuse or with a minute point, mostly 3 to 4 lines long, rather rigid, concave, glabrous or ciliate with a few long hairs. Stipules appressed or slightly spreading, those of the floral leaves large and broad. Flowers almost sessile in the upper axils, often forming ovoid leafy heads or short spikes. Bracteoles inserted close under the calyx, broad, 2-lobed, with or without a central point or lobe. Calyx about $2\frac{1}{2}$ lines long, the lobes longer than the tube, lanceolate-subulate, the 2 upper ones broad at the base. Standard twice as long as the calyx; lower petals shorter. Ovary with a few hairs; style dilated downwards. Pod not seen.

N. S. Wales, *Vicary, Clowes, Bynoe;* near Yass, *Backhouse.* This species is allied to *P. villifera* and *P. elliptica*, with the habit of *P. humilis*, and bracteoles different from any.

69. **P. villifera,** *Sieb. in DC. Prod.* ii. 111. A large erect shrub, the branches loosely villous. Leaves often crowded or clustered on the smaller branches, oval-elliptical or lanceolate, $\frac{1}{3}$ to $\frac{3}{4}$ in. long, acute or tapering into an almost pungent point, rather rigid, flat or concave, more or less hirsute or ciliate with long hairs, usually 3-nerved underneath, the petioles more distinct than in most species, often above 1 line long. Stipules lanceolate, loose, those of the floral leaves much longer and bract-like. Flowers nearly sessile in the upper axils or clustered on the short branches, sometimes with one or two leafless stipular bracts. Bracteoles inserted close under the calyx, lanceolate. Calyx 2 lines long; lobes acuminate, rather longer than the tube, the 2 upper ones rather broader at the base and often shortly united. Standard twice as long as the calyx, lower petals rather shorter, all of the same colour. Ovary villous; style filiform. Pod ovate or oval-oblong, shortly acuminate, about 3 lines long.

N. S. Wales. Port Jackson to the Blue Mountains, *Sieber, n.* 390, and *Fl. Mixt. n.* 556, *A. Cunningham,* and others.

Var. (?) *australis.* Leaves usually under $\frac{1}{2}$ in., broadly ovate in some specimens, narrow-lanceolate in others. Calyx shorter than in the northern specimens, with shorter broader lobes.

S. Australia. Port Lincoln, *R. Brown, F. Mueller;* Encounter Bay, *Wilhelmi.*

70. **P. involucrata,** *Benth.* A much-branched shrub, loosely villous with spreading hairs. Leaves linear or lanceolate, mostly acute, rarely exceeding ½ in. long, narrowed at the base but the petioles very much shorter than in *P. villifera,* or scarcely any, concave and glabrous above, very convex loosely hairy and 1-nerved underneath, usually clustered on the smaller branchlets. Stipules rather long, with recurved points. Flowers solitary on very short lateral leafy branchlets, sessile in a tuft of leaves and closely surrounded within the floral leaves by a few short imbricate bracts. Bracteoles broad, concave, as long as the calyx and inserted close under it. Calyx not 2 lines long, the lobes broad and obtuse or the lower ones scarcely acute. Standard more than twice as long as the calyx. Pod not seen.

S. Australia. Lofty Ranges, *F. Mueller.* The species seems to be allied on the one hand to *P. villifera,* differing in inflorescence, in the almost sessile leaves, etc., on the other hand to *P. Muelleri,* differing in indumentum, shorter calyx, broader leaves, etc.

71. **P. Muelleri,** *Benth.* An erect much-branched shrub of several feet, the young shoots loosely silky-pubescent. Leaves linear or lanceolate, tapering into a short almost pungent point, narrowed at the base, mostly under ½ in. long, concave and glabrous above, more or less silky-hairy underneath, more or less prominently 1- or 3-nerved. Stipules rather long, appressed. Flowers terminal solitary, almost sessile above the last leaves, and surrounded by short broad imbricate bracts. Bracteoles ovate-oblong, concave, nearly as long as the calyx and inserted close under it. Calyx 2½ lines long, slightly silky or nearly glabrous, the lobes acute or mucronate, the 2 upper ones rather broader and more united. Standard nearly twice as long as the calyx, keel deeply-coloured. Pod sessile, ovate, about 3 lines long, somewhat turgid.

Victoria. Abundant in some of the Australian Alps at an elevation of 4000 or 5000 ft., Mount Useful, Baw Baw, Bunieye Creek, etc., *F. Mueller;* near Shipton, *W. S. Whan.* In the latter specimens the stipules and bracts have fine setaceous points.

72. **P. prostrata,** *Benth. in Hook. f. Fl. Tasm.* i. 89. A prostrate or diffuse shrub, forming depressed patches of 1 ft. or more, of a silvery-grey from the short appressed silky pubescence of the young shoots and inflorescence. Leaves linear-terete, obtuse, 2 to 3 or rarely 4 lines long, channelled above by the involute margins, becoming glabrous with age. Stipules small. Flowers terminal, but often appearing axillary from the shortness of the flowering branchlets, solitary and sessile within 4 to 6 broad closely imbricated bracts covering the calyx-tube, the inner ones often 2 lines long, obtuse or jagged. Bracteoles inserted on the calyx-tube at its base, oval-oblong, scarious. Calyx silky-pubescent, 2½ or nearly 3 lines long; lobes shorter than the tube, the 2 upper ones broader and more united. Standard twice as long as the calyx; keel considerably shorter. Ovary villous; style subulate. Pod ovoid, obtuse, about as long as the calyx.

Victoria. Heaths on the Glenelg, *Robertson;* near Portland, *Allitt;* in the Murray scrub, *F. Mueller;* Wimmera, *Dallachy.*
Tasmania. Plains near Ross, *Gunn.*
S. Australia. In the Tattiara country, *Woods.*
The species is readily distinguished from all others except *P. Muelleri* and *P. involucrata,* by the flowers singly surrounded by imbricate bracts.

73. **P. canaliculata,** *F. Muell. in Trans. Vict. Inst.* 119. A much-branched shrub, with the habit nearly of *P. mollis,* but the pubescence closer and more hoary or almost silky, and the flowers not capitate. Leaves about ½ in. long, narrow-linear, terete or slightly dilated upwards, obtuse, concave or channelled above by the involute margins, softly hoary-pubescent or almost silky-pubescent on both sides. Stipules narrow, pubescent. Flowers sessile in the upper axils, forming short leafy spikes or heads, without any bracts except the stipules of the floral leaves. Bracteoles inserted close under the calyx, linear or filiform. Calyx villous, about 3 lines long; lobes all rather broad, subulate-acuminate, rather longer than the tube. Petals dark-coloured; standard not twice as long as the calyx; wings and keel rather shorter. Ovary hirsute, tapering into a subulate style. Pod ovate-oblong, shorter than the calyx.

Victoria. Cape Otway, *F. Mueller;* Cape Nelson, *Allitt;* Corner Inlet, *Wilhelmi.*
S. Australia. Mount Lofty, *Whittaker;* Encounter Bay, *F. Mueller.*
The species differs from *P. tenuifolia* in its longer and thicker leaves and much larger flowers.

74. **P. fasciculata,** *Benth. in Ann. Wien. Mus.* ii. 82. A prostrate or diffuse much-branched shrub, with the silvery-grey appressed pubescence of *P. prostrata,* but a very different inflorescence. Leaves linear-terete with short fine points, 2 to 3 or rarely 4 lines long, channelled above with the involute margins, usually sprinkled with short silky hairs, but sometimes glabrous or nearly so. Stipules subulate-acuminate. Flowers axillary, solitary, nearly sessile, with a single small broad lobed bract at their base. Bracteoles inserted on the calyx at its base, subulate-acuminate. Calyx silky-pubescent, about 2 lines long; lobes subulate-acuminate, as long as the tube, the 2 upper ones broader and more united. Standard fully twice as long as the calyx; lower petals shorter. Ovary villous; style subulate. Pod ovate, acute, rather flat, not longer than the calyx.—Hook. f. Fl. Tasm. i. 92.

Victoria. Cobberas mountains, at an elevation of 5000 ft., *F. Mueller.*
Tasmania. Summit of the Western Mountains and Arthur's Lake, at an elevation of 2 to 3000 ft., *Lawrence, Gunn.*
The species is very nearly allied to *P. tenuifolia,* but the bracteoles appear to be constantly aduate to the base of the calyx.

75. **P. tenuifolia,** *R. Br. in Bot. Mag. t.* 2086. A small prostrate shrub, the slender branches and foliage softly pubescent or villous. Leaves narrow-linear or terete, obtuse or scarcely acute, 2 to 4 lines long, concave or channelled above by the involute margins. Stipules acuminate. Flowers solitary or 2 together, sessile on the smaller branchlets and often shorter than the surrounding leaves. Bracts broad, bifid, acute, twice as long as the calyx. Bracteoles inserted under the calyx, ovate or oblong, or sometimes there are no bracts besides the floral leaves and then the bracteoles are leafy and 2-stipulate. Calyx pubescent, 1¼ lines long; lobes nearly equal, subulate-acuminate, longer than the tube. Standard about 2 lines long; keel short, obtusely acuminate. Ovary villous, very short; style slender. Pod very obliquely ovate, not 2 lines long, the valves thin.—DC. Prod. ii. 113 ; Meissn. in Pl. Preiss. i. 76, ii. 200 ; Hook. f. Fl. Tasm. i. 92 ; *P. candida,* Lodd. Bot. Cab. t. 1236 ? from the figure.

Victoria. Port Phillip, *R. Brown ;* Wilson's Promontory and Murray river, *F. Mueller.*
Tasmania. Sandy lands near the sea, *J. D. Hooker.*
S. Australia. All along the coast, *R. Brown, F. Mueller,* and others.
W. Australia. King George's Sound, *R. Brown, Preiss, n.* 1203, and others.
Var. *glabra.* Almost entirely glabrous, leaves small. Wimmera river, *Dallachy ;* Venus Bay, *Warburton.*
Var. *recurvifolia.* Leaves crowded, short, linear or almost linear-cuneate, recurved, chan-nelled above. Calyx-lobes short and broad.—Near Portland, *Allitt.*
P. filifolia, F. Muell. Fragm. i. 9, from Kangaroo Island, *Bannier,* appears to me to be a luxuriant form of *P. tenuifolia* with remarkably long slender leaves, but F. Mueller's her-barium only contains a single specimen past flower, scarcely sufficient for identification.

17. LATROBEA, Meissn.

(Leptocytisus, *Meissn.*)

Calyx 5-lobed, the lobes nearly equal, ribbed, or very short. Petals shortly clawed ; standard ovate or nearly orbicular, obtuse or acuminate, longer than the lower petals ; wings narrow, keel straight or slightly incurved, as long as the wings or rather longer. Stamens free. Ovary sessile or stipitate, with 2 ovules on short funicles ; style filiform or slightly thickened at the base with a small terminal stigma. Pod flattened, ovate or lanceolate. Seeds reniform, strophiolate.—Heath-like shrubs with usually virgate branches. Leaves al-ternate or scattered, simple, linear, concave or channelled above. Stipules none. Flowers yellow (or purplish ?) terminal, or rarely apparently axillary from the shortness of the flowering branch, solitary or in corymbs or heads. Bracts and bracteoles none or small and inserted at a distance from the calyx.

The genus is entirely Australian. It is nearly allied to *Aotus,* differing in the strophio-late seeds and the leaves concave or with involute not revolute margins ; and to the latter sections of *Pultenæa,* from which it is only distinguished by the more regular calyx and the usual absence of bracteoles. *L. diosmifolia* closely connects *Latrobea* with *Pultenæa sub-umbellata* and *P. urodon.*

SECT. I. **Eulatrobea.**—*Calyx-teeth shorter than the ribless tube. Flowers solitary or two together.*

Plant pubescent. Leaves rigid, pungent-pointed . . . . 1. *L. pungens.*
Plant glabrous. Leaves obtuse.
Flowers 5 to 6 lines long. Ovary on a long stipes . . . . . . 2. *L. genistoides.*
Flowers 3 to 4 lines long. Ovary almost sessile . . . . . . 3. *L. Brunonis.*

SECT. II. **Leptocytisus.**—*Calyx-lobes longer than the ribbed tube. Flowers solitary or few or many in terminal corymbs or heads.*

Calyx glabrous. Flowers solitary or few, almost racemose.
Pod lanceolate or ovate-lanceolate . . . . 4. *L. tenella.*
Calyx silky-pubescent. Corymbs few-flowered . . . . 5. *L. hirtella.*
Calyx villous. Corymbs dense, many-flowered. Pod ovate . . . . 6. *L. diosmifolia.*

SECTION I. EULATROBEA.—Calyx-teeth shorter than the tube which is campanulate, membranous, without prominent ribs. Flowers solitary or 2 together.—Shrubs, glabrous in every part.

1. **L. pungens,** *Benth.* Branches rigid, apparently procumbent or spreading, very softly pubescent or shortly villous. Leaves sessile, lanceolate, pungent-pointed, mostly about $\frac{1}{2}$ in. long or rather more, concave, rigid, 1-nerved, softly pubescent. Stipules none, but a small scarious stipule-like scale on each side at the base of each branchlet. Flowers axillary, solitary,

much shorter than the leaves. Pedicels much shorter than the calyx, with a very small ovate obtuse bract at their base, and 2 rather larger bracteoles a little higher up. Calyx scarcely 1½ lines long, thin, glabrous, the teeth short and equal. Petals twice as long as the calyx; standard obovate-oblong; keel slightly incurved; wings scarcely so long as the other petals. Ovary elongated, glabrous, with 2 ovules; style subulate, incurved, with a small stigma. Pod not seen.—*Daviesia abnormis*, F. Muell. Fragm. ii. 106.

W. Australia. South-west Bay, *Maxwell.*

2. **L. genistoides,** *Meissn. in Pl. Preiss.* ii. 219. An erect, glabrous often glaucous shrub, closely allied to *L. Brunonis*, but with larger flowers and broader leaves. Leaves cuneate-oblong or elliptical, sometimes almost oval, 4 to 8 lines long, obtuse or with a small callous point, concave, rather thick and nerveless. Flowers on short axillary leafy shoots, with the lower leaves reduced to scales. Calyx thin and membranous, not larger than in *L. Brunonis*, the teeth very short. Standard ½ in. long; keel nearly as long. Ovary slightly hairy, on a rather long stipes. Pod not seen.—*Pultenæa genistoides*, Meissn. in Pl. Preiss. i. 73.

W. Australia. King George's Sound, *Collie, Preiss, n.* 1021 *and* 1101, *Harvey, Oldfield,* and others.

3. **L. Brunonis,** *Meissn. in Pl. Preiss.* ii. 219. A glabrous shrub of 2 to 3 ft., the young branches angular. Leaves linear or linear-cuneate, obtuse, under ½ in. long, rather thick, concave above, nerveless underneath. Stipules none. Flowers apparently axillary, but really terminating very short axillary shoots bearing a few small leaves of which the lower ones or sometimes all are reduced to short scales. Bracteoles none. Calyx broadly campanulate, about 1 line long, very obtuse at the base, thin and membranous, the teeth short and broad. Petals on very short claws; standard nearly 4 lines diameter; keel nearly as long. Ovary elongated, almost sessile, villous; the ovules distant on slender funicles; style subulate. Pod ovate, acuminate, much longer than the calyx.—*Pultenæa Brunonis*, Benth. in Ann. Wien. Mus. ii. 81; Meissn. in Pl. Preiss. i. 73.

W. Australia. King George's Sound, *R. Brown;* near Albany, *Preiss, n.* 876, also *Drummond.*

Section II. Leptocytisus, *Meissn.* (as a genus).—Calyx-tube 10-ribbed, the lobes rigid, as long as or longer than the tube, each with a prominent midrib. Flowers solitary, few or many in a dense corymb or head.

R. Brown, in his MS., had already indicated *L. Brunonis* and *L. diosmifolia* as congeners under the name of *Acarpha,* and *L. tenella* is too closely connected with the latter through *L. hirtella* to be generically separated from it. The imbricate æstivation of the calyx-lobes, the short funicles, etc., remove them far from *Burtonia,* to which I had formerly referred *L. diosmifolia.*

4. **L. tenella,** *Benth.* A heath-like shrub, with long slender virgate branches, quite glabrous or slightly pubescent when young. Leaves narrow-linear or subulate, obtuse or acute, 2 to 3 or rarely 4 lines long. Flowers small, terminal, solitary or 2 or 3 together in short racemes, on slender pedicels, shorter or scarcely longer than the calyx. Bracts and bracteoles none. Calyx 1 to nearly 2 lines long, glabrous; tube exceedingly short; lobes

rigid, narrow-linear, acute with a prominent midrib. Petals rather longer than the calyx-lobes. Ovary shortly stipitate, pubescent; style filiform. Pod flattened, oblong-lanceolate, acute, 3 to 4 lines long. Seeds usually 2, on very short funicles.—*Burtonia (?) tenella*, Meissn. in Pl. Preiss. i. 42; *Leptocytisus tenellus*, Meissn. l. c. ii. 211.

W. Australia. Swan River, *Preiss, n.* 178, *also Drummond, 2nd Coll. n.* 133, *3rd Coll. n.* 94.

Var. *platycarpa.* More rigid. Leaves broader and more obtuse. Flowers rather larger. Pod obliquely ovate-lanceolate, about twice as long as broad.—King George's Sound, *Harvey, Drummond.*

Var. *grandiflora.* Foliage of the preceding variety. Flowers still larger, deeply coloured. Calyx-lobes often not much longer than the tube, sometimes slightly ciliate. Standard about 4 lines long. Pod not seen.—*Drummond, 4th Coll. n.* 79, *5th Coll. n.* 74 Rocky Ranges, Middle Mount Barren, and towards Cape Riche, *Maxwell.*

5. **L. hirtella,** *Benth.* An erect shrub of 1 to 2 ft., very closely allied to *L. diosmifolia*, but with fewer less villous flowers and much less conspicuous bracteoles. Branches virgate, shortly pubescent when young. Leaves oblong-linear or slightly cuneate, obtuse, 2 to 3 lines long, thick, but flatter and rather broader than in *L. diosmifolia*, glandular-papillose and sprinkled with a few hairs, which soon wear off. Flowers few, in dense terminal corymbs, similar to those of *L. diosmifolia* in structure and proportions, but rather smaller. Bracts and bracteoles very small and deciduous. Calyx rather under 2 lines long, silky-pubescent. Standard not above 3 lines long. Pod only seen young, when it is not so broad as in *L. diosmifolia.*—*Leptocytisus hirtellus*, Turcz. in Bull. Mosc. 1853, i. 258.

W. Australia, *Drummond, 5th Coll. n.* 72.

6. **L. diosmifolia,** *Benth.* Stems or branches virgate, glabrous or clothed with rather long silky hairs. Leaves numerous, linear or linear-lanceolate, obtuse or with a short callous point, about 3 or rarely nearly 4 lines long, glabrous, rather thick. Flowers in dense terminal corymbs or heads on very short villous pedicels. Bracteoles rather small, linear. Calyx about 2 lines long or rather more, hirsute, the tube prominently 10-nerved; lobes lanceolate, rather longer than the tube, each with a prominent midrib. Standard about 4 lines diameter; wings considerably shorter; keel almost as long as the standard. Ovary nearly sessile or very shortly stipitate, densely villous with long hairs; style subulate. Pod broadly ovate, longer than the calyx, very hirsute. Seeds shining, on very short funicles.—*Burtonia diosmifolia*, Benth. in Hueg. Enum. and in Ann. Wien. Mus. ii. 73; Meissn. in Pl. Preiss. i. 42, ii. 211.

W. Australia. King George's Sound, *R. Brown, Huegel,* and others; near Albany, *Preiss, n.* 857; Avon river, *Preiss, n.* 858, *also Drummond, 2nd Coll. n.* 18 *and* 134; Kalgan and Gordon rivers, *Oldfield.*

Var. *glabrescens.* Calyx and pedicels glabrous or nearly so.—King George's Sound, *Preiss, n.* 883 (*Meissner*).

18. EUTAXIA, R. Br.

(Sclerothamnus, *R. Br.*)

Calyx 5-lobed, the 2 upper lobes more or less united into an upper lip. Petals on rather long claws; standard orbicular, longer than the lower petals, entire or nearly so; wings oblong; keel nearly straight, obtuse, shorter than the wings. Stamens free. Ovary narrowed at the base or stipitate, with 2 ovules on short or slender funicles; style filiform and incurved or thicker and hooked at the end; stigma small, terminal. Pod ovate, flattened or turgid, 2-valved. Seeds reniform, strophiolate.—Shrubs, usually glabrous or nearly so. Leaves small, opposite, decussate, entire, concave or with involute margins, 1- or 3-nerved underneath, not reticulate. Stipules minute or none. Flowers axillary, solitary or 2 to 4 together, sometimes crowded at the ends of the branches. Bracteoles on the short pedicel not close to the calyx, often very small. Ovary villous. Strophiole usually 2-lobed.

The genus is entirely Australian. It is closely allied to *Pultenæa*, differing only in the decussate leaves and in the bracteoles neither close to nor adnate to the calyx, although sometimes very near it. F. Mueller unites it with *Dillwynia*, but the peculiar habit seems to justify the retaining both these natural groups as distinct genera, although not very strictly limited by floral characters.

SECT. I. **Eutaxia.**—*Ovary sessile or nearly so. Style rather thick, abruptly curved or hooked at the end. Pod flat or the valves convex, usually very oblique.*

Calyx upper lip truncate or emarginate.
Leaves obtuse. Flowers solitary . . . . 1. *E. cuneata.*
Leaves (½ to ¾ in. long) acuminate, often pungent. Flowers 2 to 4   2. *E. myrtifolia.*
Calyx-lobes all erect, acuminate, the 2 upper ones united to the middle. Flowers solitary.
Calyx 10-nerved. Leaves ½ to ¾ in. long . . . . 3. *E. epacridioides.*
Calyx 5- or 6-nerved. Leaves under ½ in. long.
Leaves flat or concave, narrow-linear or slightly cuneate.
Leaves distant . . . . 4. *E. virgata.*
Leaves crowded, acute . . . . 5. *E. densifolia.*
Leaves linear-terete, channelled above, rather obtuse, crowded . 6. *E. dillwynioides.*
Leaves obovate or oblong, 2 to 3 lines long, obtuse . . . . 7. *E. parvifolia.*

SECT. II. **Sclerothamnus.**—*Ovary stipitate. Style subulate, elongated, incurved. Leaves small.*
Pod turgid . . . . 8. *E. empetrifolia.*

E. Strangeana, Turcz. in Bull. Mosc. 1853, i. 270, said to come from New Zealand, involves some mistake. No such plant is known in New Zealand. The description applies in many respects, but not in all, to the Tasmanian *Pultenæa prostrata.*

E. Baxteri, Knowl. and Westc., attributed by mistake to me in Walp. Rep. i. 573, is unknown to me. From the description, it is probably either *Oxylobium scandens* or some *Pultenæa* of the section *Euchilus.*

SECTION I. EUTAXIA.—Pod flat or the valves convex. Style rather thick, erect, abruptly incurved or hooked at the end.

1. **E. cuneata,** *Meissn. in Pl. Preiss.* i. 65. A glabrous shrub of about 3 ft. Leaves obovate-oblong, cuneate, rounded at the end or rather obtuse, 3 to 4 lines long, concave above, convex underneath, with the midrib slightly prominent. Flowers axillary, solitary, pendulous, rather larger than in *E. myrtifolia.* Calyx smooth or faintly ribbed, the 2 upper lobes united in a broad truncate emarginate upper lip, the lower lobes lanceolate, acute. Petals orange or red, the keel dark-purple, nearly straight. Pod sessile, lanceolate, rather turgid, slightly hairy.

W. Australia. Rocky places in the Konkoberup Hills, *Preiss, n.* 1022. I have not seen this species, and have taken the above characters from Meissner. It appears to have the calyx of *E. myrtifolia,* but rather larger solitary flowers, and the foliage of *E. parvifolia.*

2. **E. myrtifolia,** *R. Br. in Ait. Hort. Kew. ed.* 2, iii. 16. A glabrous shrub of 2 to 3 ft., with rather slender branchlets, angular when young. Leaves from obovate-oblong or elliptical to linear, mostly ½ to ¾ in. long, tapering into a sharp usually pungent point, more or less concave above, darker-coloured underneath, with a prominent midrib and sometimes also 2 lateral nerves. Flowers yellow with a dark-orange keel, axillary, 2 to 4 together or rarely solitary. Pedicels short, with very small bracteoles above the middle. Calyx 2 to 2½ lines long, the tube 6-ribbed, the sixth rib between the 2 upper lobes; lobes shorter than the tube, the 2 upper ones broadly falcate and united into a truncate emarginate upper lip. Standard nearly twice as long as the calyx; keel much shorter. Ovary villous, slightly contracted at the base, tapering into a rather thick erect style, hooked near the end. Pod ovate, 2 to 3 lines long, the valves convex.—Bot. Mag. t. 1274; DC. Prod. ii. 109; Meissn. in Pl. Preiss. i. 66, ii. 216; *Dillwynia myrtifolia,* Sm. in Trans. Linn. Soc. ix. 263; *D. obovata,* Labill. Pl. Nov. Holl. i. 110, t. 140.

W. Australia. King George's Sound, *Menzies, R. Brown, Preiss, n.* 1019, *Drummond, 2nd Coll. n.* 110, and others; and eastward towards the Great Bight, *Maxwell.*

A specimen, without flowers, with larger almost obovate leaves, from Rottenest Island, *A. Cunningham,* may possibly belong to this species.

3. **E. epacridioides,** *Meissn. in Pl. Preiss.* i. 64, ii. 215. A glabrous shrub, with the habit and foliage nearly of *E. myrtifolia,* but in our specimens the branches are more rigid and virgate. Leaves from narrow-oblong to linear, ½ to ¾ in. long, tapering into a short sometimes pungent point, concave above, dark-coloured or very glaucous underneath, the midrib less conspicuous than in *E. myrtifolia.* Flowers usually 2 together in each axil, apparently of the size of those of *E. myrtifolia,* but only seen faded. Calyx 3 lines long or rather more, the tube 10-ribbed; lobes all acuminate, the 2 upper ones broader and united to the middle. Petals, according to Preiss, red. Pod almost sessile, flat, obliquely ovate, slightly hairy, about 3 lines long.

W. Australia, *Drummond, 3rd Coll. n.* 128; near Mounts Melville and Elphinstone, *Preiss, n.* 412 *and* 867.

4. **E. virgata,** *Benth. in Hueg. Enum.* 34, *and in Ann. Wien. Mus.* ii. 80. A glabrous shrub of 2 to 3 ft., with long slender virgate or rarely divaricate branches. Leaves oblong-linear or linear-cuneate and about ½ in. long on the principal branches, narrow-linear and much smaller on the smaller ones, obtuse or acute, but not pungent, concave above, the midrib slightly prominent underneath or inconspicuous. Flowers solitary in each axil, but sometimes crowded near the ends of the branches or in pairs terminating short axillary shoots, rather larger than in *E. myrtifolia.* Pedicels short, with linear-lanceolate bracteoles about the middle. Calyx 2 to 2½ lines long, 6-ribbed, the sixth rib between the 2 upper lobes; lobes all acute, the 2 upper ones united to about the middle. Standard fully twice as long as the calyx; keel short and narrow. Ovary very shortly stipitate, villous, tapering into a rather thick style, hooked at the top. Pod flat, about 3 lines long.— *E. ericoides,* Meissn. in Pl. Preiss. i. 63.

W. Australia. Swan River, *Huegel, Drummond,* 1*st Coll. and n.* 245, *Preiss, n.* 877, *Oldfield;* King George's Sound and Mount Manypeak, *Maxwell.* The southern specimens have the calyx-lobes rather less acuminate, but do not otherwise differ.

5. **E. densifolia,** *Turcz. in Bull. Mosc.* 1853, i. 271. Closely allied on the one hand to *E. parvifolia* and on the other to *E. dillwynioides.* Leaves crowded, narrow-linear oblong-linear or linear-cuneate, acute, mostly 3 to 4 lines long, concave above, the midrib prominent underneath. Flowers solitary in the upper axils, but often crowded at or near the ends of the branches, rather larger than in *E. parvifolia,* the calyx-lobes much more acuminate, and the linear bracteoles often nearly as long as the calyx and more or less herbaceous. Pod not seen.

W. Australia, *Drummond,* 5*th Coll. n.* 76; Russell Range and Mount Bland, *Maxwell.*

6. **E. dillwynioides,** *Meissn. in Pl. Preiss.* i. 63. A much-branched heath-like glabrous shrub, with the habit of *E. parvifolia,* but differing from all others of the genus in the narrow-linear leaves with the margins involute, so as to be usually terete and only channelled above, obtuse or with a short callous point, 2 to 4 lines or rarely nearly ½ in. long. Flowers solitary in the upper axils, but often crowded towards the ends of the branches. Pedicels about 1 line long, recurved after flowering, with linear bracteoles above the middle. Calyx about 3 lines long, obscurely 5-nerved and more or less reticulate; lobes all acuminate, longer than the tube, the 2 upper ones united above the middle. Standard not twice as long as the calyx; keel considerably shorter. Ovary nearly sessile, villous; style short, erect, hooked at the top. Pod nearly flat, obliquely oval-oblong, about 3 lines long.

W. Australia, *Drummond*; in the interior, *Preiss, n.* 1191.

7. **E. parvifolia,** *Benth. in Hueg. Enum.* 34, *and.in Ann. Wien. Mus.* ii. 80. A much-branched shrub with the habit of many *Epacrideæ,* glabrous or the young shoots very slightly pubescent. Leaves crowded, obovate or oblong, mostly obtuse and 1 to 2 lines long, rarely on luxuriant branches narrower almost acute and 3 lines long, concave above, either nerveless or 1- or 3-nerved underneath. Flowers solitary in the upper axils, but often crowded at or near the ends of the branches. Pedicels short or rather long, with oblong or linear bracts above the middle. Calyx about 2 lines long, finely or obscurely 5-nerved; lobes all acute or acuminate, longer than the tube, the 2 upper ones united to about the middle. Standard not twice as long as the calyx; keel not exceeding the calyx-lobes. Ovary villous, nearly sessile; style short, much-curved or hooked above the middle. Pod flat, obliquely oval-oblong, almost falcate, about 3 lines long.—Meissn. in Pl. Preiss. i. 65; *E. obovata,* Turcz. in Bull. Mosc. 1853, i. 271.

W. Australia. King George's Sound, *R. Brown, A. Cunningham, Huegel, Baxter,* and others, *Drummond,* 5*th Coll. n.* 73, *and Suppl. n.* 46, *Preiss, n.* 1020.

SECTION II. SCLEROTHAMNUS.—Pod turgid. Style subulate, elongated, incurved. Leaves small, rarely above 3 lines long.

8. **E. empetrifolia,** *Schlecht. Linnæa,* xx. 667. A glabrous divaricate or diffuse shrub with rigid branches, sometimes short ending in slender spines, sometimes elongated slender and erect. Leaves usually elliptical-oblong or

linear, 1 to 3 lines long, rigid, concave, obtuse or almost acute, without any dorsal midrib, rarely broadly oblong or almost ovate or obovate. Flowers small, on axillary pedicels of 1 to 2 lines, with a pair of leafy linear obtuse bracteoles a little below the calyx. Calyx glabrous, 1½ to nearly 2 lines long; lobes acute or acuminate, rather longer than the tube, the 2 upper ones more or less united. Standard about 3 lines long or rather more; lower petals shorter, the keel deeply coloured. Ovary stipitate, silky-villous; style subulate, incurved. Pod ovoid or nearly globular, very turgid, varying from 1½ to nearly 3 lines long, and the stipes from ½ to 1 line. Seeds black.— *Sclerothamnus microphyllus*, R. Br. in Ait. Hort. Kew. ed. 2, iii. 16 ; *E. leptophylla*, Turcz. in Bull. Mosc. 1853, i. 268 ; *E. diffusa*, F. Muell. Fragm. i. 7.

Victoria. Sandy, stony or rocky hills in Australia Felix and the Grampians, and abundant in the Murray scrub, *F. Mueller* and others.

S. Australia. Islands off the coast, *R. Brown ;* from the Murray to the western limits, *F. Mueller* and others.

W. Australia. King George's Sound, *Baxter ;* Swan River to King George's Sound, *Drummond, 1st Coll., also 4th Coll. n.* 142 *and Suppl. n.* 35; Kalgan river, Gardner Ranges and eastward to the Great Bight, *Maxwell.*

The extreme forms of this plant are so different in aspect that it is difficult, at first sight, to consider them all as varieties of one species. Most of those from the mountain districts of Victoria are unarmed, with broadly cuneate-oblong leaves of 2 or 3 lines or even longer; many from the Murray scrub and South Australia are stunted, spinescent, with very numerous narrow leaves, from ½ to 1 line long ; Drummond's W. Australian ones are heath-like, unarmed, with fine slender leaves of 1 to 2 or even 3 lines ; others again, from near the Great Bight or from the scrub of S. Australia and Victoria, have short broad leaves or in some instances two or three of the above forms are on the same specimen. The size of the pod and the length of the stipes also vary much and not in any relation to the differences in the foliage.

19. DILLWYNIA, Sm.

Calyx-lobes short or as long as the tube, the 2 upper ones more or less united in an upper lip. Petals clawed ; standard broader than long; wings narrow ; keel shorter, straight or scarcely incurved. Stamens free. Ovary shortly stipitate, with 2 ovules on short funicles ; style erect, rather thick, hooked below the top, with a truncate or thick stigma. Pod nearly sessile, ovate or rounded, turgid, 2-valved. Seeds reniform, strophiolate.—Heath-like shrubs. Leaves alternate or scattered, simple, narrow-linear or terete, channelled above. Stipules none. Flowers yellow or orange-red, few together in axillary or terminal racemes or corymbs, rarely solitary. Bracts small, brown, very deciduous ; bracteoles small on the short pedicels.

The genus is entirely Australian. It differs from *Aotus* in the strophiolate seeds and in the leaves channelled above and not underneath, from *Pultenæa* in the bracteoles at a distance from the calyx and usually deciduous, from *Latrobea* in the calyx style and pod, but is closely connected with the two latter genera through *D. brunioides.*

SECT. I. **Dillwyniastrum,** *DC.—Calyx distinctly turbinate at the base, the 2 upper lobes broad, falcate, and united to the middle. Petals deciduous ; standard on a long claw, the lamina above twice as broad as long.*

Keel acuminate, nearly as long as the wings. Racemes on long terminal peduncles . . . . 1. *D. hispida.*
Keel obtuse, much shorter than the wings.
  Racemes terminal, sessile, corymbose or pedunculate . . . . . 2. *D. ericifolia.*
  Flowers all axillary, solitary or in short racemes or clusters . . . 3. *D. floribunda.*

Sect. II. **Xeropetalum,** *R. Br.—Calyx obtuse or very shortly turbinate at the base. Petals persistent; standard-claw shorter than the calyx, the lamina rather broader than long or rarely twice as broad as long.*

Calyx-lobes nearly as long as the tube, all acute, the 2 upper ones united
to the middle only or a little above it.

Flowers axillary . . . . 4. *D. Preissii.*
Flowers in terminal sessile heads or corymbs . . . . 5. *D. brunioides.*

Calyx-lobes short, the 2 upper ones united in a broad upper lip, scarcely
emarginate (except in *D. uncinata*), and longer than the lower
ones.

Leaves rigid and pungent.
Leaves keeled. Flowers mostly corymbose (Eastern species) . . 6. *D. juniperina.*
Leaves not keeled. Flowers in the upper axils (Western species) . 7. *D. pungens.*

Leaves not pungent, mostly ¼ to ½ in. long, and rather slender.
Glabrous or ashy-pubescent. Flowers in small terminal corymbs,
exceeding the last leaves or in the upper axils . . . . . . 8. *D.cinerascens.*
Branches and calyx tomentose-villous. Flowers solitary or few in
terminal heads, shorter than the last leaves . . . . 9. *D. divaricata.*

Leaves mostly under ¼ in. long, rather thick, obtuse, and often recurved. Flowers few, terminal, longer than the last leaves. Calyx
upper lip often shortly 2-lobed . . . . 10. *D. patula.*

D. subaphylla, Colla, Hort. Ripul. App. ii. 347, is so imperfectly described, without
flowers or fruit, as to be absolutely unrecognizable, but is probably no *Dillwynia.*

Section I. Dillwyniastrum.—Calyx with a distinct turbinate base below the insertion of the petals and stamens, and from ¼ to ⅓ the length of
the whole calyx, the lobes short, the 2 upper ones broad, rounded-falcate,
and united to about the middle. Petals deciduous; standard on a claw
nearly as long as the calyx, the lamina above twice as broad as long.

1. **D. hispida,** *Lindl. in Mitch. Three Exped.* ii. 251. A shrub, attaining several feet, but showing in several specimens erect stems of 6 in. clustered
on a thick rootstock; branches and foliage scabrous-pubescent or hirsute,
rarely quite glabrous. Leaves ¼ to ½ in. long, obtuse or scarcely pointed, not
twisted and without any prominent keel. Flowers rather large, of a deep
orange mixed with purple-red, in clusters or short racemes, on terminal peduncles much longer than the leaves. Bracteoles very small or none. Calyx
hirsute or rarely glabrous, 3 to 3½ lines long, the turbinate base rather long,
the 2 upper lobes broadly falcate acute and shortly united. Petals deciduous;
standard with a long claw, the lamina more than twice as broad as long;
wings much shorter; keel tapering into a recurved point, nearly as long as
the wings. Pod nearly globular.—*D. scabra,* Schlecht. Linnæa, xx. 666;
Henfr. in Gard. Comp. i. 25, with a fig. copied into Lemaire, Jard. Fleur.
t. 296.

Victoria. Near Mount Napier, *Mitchell;* in the Murray scrub, and on the Murrum-
bidgee, *F. Mueller;* Wendu Valley, *Robertson;* Wimmera, *Dallachy.*
S. Australia. Encounter Bay, *Whittaker;* Barossa Range, *Behr;* Bugle Range,
Port Lincoln, Guichen Bay, *F. Mueller.*
The more glabrous specimens often resemble *D. ericifolia,* var. *peduncularis,* but the
flowers are rather larger, and always readily known by the peculiar keel.

2. **D. ericifolia,** *Sm. Ann. Bot.* i. 510; *Exot. Bot. t.* 25, *and in Trans.
Linn. Soc.* ix. 262. An erect heath-like shrub, usually attaining several feet,
but sometimes dwarf and stunted; the branches erect and virgate, or short

L 2

and divaricate, glabrous or pubescent. Leaves numerous, rather slender, usually ¼ to ½ in. long, but sometimes nearly ¾ in. or under 2 lines, terete or scarcely keeled, straight or spirally twisted when dry, obtuse, with a very short recurved or straight, but scarcely pungent point, rarely quite obtuse. Flowers yellow, in very short racemes or clusters, sometimes several together, almost sessile in a terminal leafy corymb, sometimes each one on ·a terminal or rarely axillary long or short peduncle Calyx glabrous, silky-pubescent, or shortly scabrous-hirsute, 2 to 3½ lines long, distinctly turbinate at the base, the lobes shorter than the tube, the 2 upper ones broadly rounded and falcate, united to the middle. Petals deciduous ; standard with a claw usually as long as the calyx, the lamina more than twice as broad as long ; wings much shorter ; keel still shorter, obtuse. Pod ovate or nearly globular, slightly exceeding the calyx.—*Pultenæa retorta*, Wendl. Hort. Herrenh. t. 9.

Queensland. Moreton Island, *F. Mueller.*

N. S. Wales. Port Jackson to the Blue Mountains, *R. Brown* and others; and northward to Hastings and Clarence rivers, *Beckler.*

Victoria. Low stony scrubby hills, Buffalo Range.

Tasmania. Common in poor wet sandy soils, especially in the northern parts of the island, *J. D. Hooker.*

Various forms assumed by this plant have been generally recognized as species, but the differences are so slight, depending chiefly on indumentum, length, and degree of twisting of the leaves, or length of peduncles, and the passages from the one to the other so gradual, that it is often very difficult to separate them even as varieties. The following are the most prominent :—

a. *normalis.* Branches pubescent. Leaves mostly 4 to 6 lines long, spreading, twisted, with straight or slightly recurved points. Flowers rather large, usually rather numerous, in sessile terminal leafy corymbs.—*D. ericifolia*, Sm., as above; DC. Prod. ii. 108 ; Lodd. Bot. Cab. t. 1277; Benth. in Ann. Wien. Mus. ii. 78. The commonest Port Jackson form, including *D. ericoides*, Sieb. Pl. Exs. n. 412, and Fl. Mixt. n. 585, *D. pinifolia*, Sieb. n. 424, *D. seriphioides*, Endl. Nov. Stirp. Dec. 14, and probably *Aotus ericoides*, Paxt. Mag. v. 51, with a fig.

b. *phylicoides.* Branches foliage and calyx scabrous, pubescent with short rigid hairs. Leaves mostly 2 to 3 lines long, spreading, twisted, with straight or slightly recurved points and less slender than in other forms. Flowers nearly sessile, but not so numerous as in the normal form.—*D. phylicoides*, A. Cunn. in Field, N. S. Wales, 347 ; Benth. in Ann. Wien. Mus. ii. 78. Rocky hills in the Blue Mountains, *A. Cunningham, Fraser ;* Mount Mitchell, *Beckler.*—*D. speciosa*, Paxt. Mag. vii. 27, with a fig., raised from Baron Huegel's seeds, is probably this variety or very near it.

c. *parvifolia.* Glabrous or nearly so. Leaves mostly 2 lines long or under, spreading, often twisted, with straight or slightly recurved points. Flowers rather small, usually few, the clusters sessile or shortly pedunculate.—*D. parvifolia*, R. Br. in Bot. Mag. t. 1527 ; Lodd. Bot. Cab. t. 559 ; DC. Prod. ii. 108 ; Benth. in Ann. Wien. Mus. ii. 79 ; *D. microphylla*, Sieb. Pl. Exs. n. 410, and 553, and Fl. Mixt. n. 586. Port Jackson, *R. Brown ;* Blue Mountains and southward to the Murray river and Victoria.

d. *tenuifolia.* Branches slightly pubescent. Leaves 2 to 4 lines long or rarely more, spreading or erect, usually straight with straight or recurved points, and more slender than in other forms. Flowers few, middle-sized, the clusters sessile or nearly so. Calyx glabrous or silky-pubescent.—*D. tenuifolia*, Sieb. in DC. Prod. ii. 109 ; Benth. in Ann. Wien. Mus. ii. 79 ; *D. ramosissima*, Benth. l. c. Chiefly in the Blue Mountains, *R. Brown, Sieber, n.* 409, and *Fl. Mixt. n.* 587, and others.

e. *peduncularis.* Branches glabrous or slightly pubescent. Leaves 3 to 6 lines long, usually slender. Flowers middle-sized, in loose clusters of 2 or 3, on peduncles usually exceeding the leaves and sometimes several times as long. Calyx usually glabrous or nearly so.—*D. peduncularis*, Benth. in Ann. Wien. Mus. ii. 78 ; *D. filifolia*, Endl. Nov. Stirp. Dec. 13.—Port Jackson, *Sieber, n.* 553, in part, and others, and northward to Moreton

Island. This variety usually appears very distinct in inflorescence, assuming the aspect of *D. hispida*, but with the flowers of *D. ericifolia*, and when the peduncles are shorter, it passes gradually into the vars. *tenuifolia* or *glaberrima*.

f. *glaberrima*. Quite glabrous. Leaves usually crowded, rarely very spreading, ¼ to ½ in. long or often more, rather slender, not twisted, the point recurved or rarely straight. Flowers rather large, in dense terminal corymbs, sessile or shortly pedunculate.—*D. glaberrima*, Sm. in Ann. Bot. i. 510, and in Trans. Linn. Soc. ix. 263 ; Bot. Mag. t. 944 ; Lodd. Bot. Cab. t. 582; Labill. Pl. Nov. Holl. i. 109, t. 139 ; DC. Prod. ii. 108 ; Benth. in Ann. Wien. Mus. ii. 79 ; Hook. f. Fl. Tasm. i. 85.—Tasmania and southern districts of Victoria, also Port Jackson, *R. Brown*, a form passing into the vars. *peduncularis* or *tenuifolia*.

3. **D. floribunda,** *Sm. in Ann. Bot.* i. 510 ; *Exot. Bot. t.* 26, *and in Trans. Linn. Soc.* ix. 262. A tall erect heath-like shrub, either quite glabrous or more or less pubescent, or the branches, foliage and calyxes densely hirsute. Leaves usually crowded, ¼ to ½ in. long or rather more, obtuse or with a minute point, not keeled, straight, sometimes as slender as in *D. ericifolia*, but usually thicker. Flowers on very short pedicels, solitary or 2 or 3 together, all axillary, but often crowded into leafy racemes below or very near the ends of the branches. Bracts often broad and above 1 line long, but so deciduous as to be rarely seen. Calyx 2½ to 3 lines long, with a distinct turbinate base, the lobes short, often tipped with a small gland, the 2 upper ones broad falcate and united to the middle. Petals deciduous ; standard with the broad claw as long as the calyx, the lamina more than twice as broad as long ; wings much shorter ; keel still shorter, obtuse. Pod scarcely exceeding the calyx.—DC. Prod. ii. 108 ; Lodd. Bot. Cab. t. 305 ; Benth. in Ann. Wien. Mus. ii. 79 ; Hook. f. Fl. Tasm. i. 85 ; *D. ericifolia*, Sims, Bot. Mag. t. 1545, not of Sm. ; *D. rudis*, Sieb. in DC. Prod. ii. 109 ; *D. hispidula* and *D. teretifolia*, Sieb. Pl. Exs. ; *D. elegans*, Endl. Nov. Stirp. Dec. 13 ; *D. clavata*, Paxt. Mag. vii. 117.

Queensland. Wide Bay, *Bidwill*.

N. S. Wales. Port Jackson to the Blue Mountains, *R. Brown, Sieber, n.* 400, 402, 404, and others ; Macquarrie river, *A. Cunningham* ; Hastings river, *Beckler*.

Victoria. Sandy and stony ridges on the Broken River, *F. Mueller* ; in the Grampians, *Wilhelmi, Dallachy*, and others ; Glenelg river, *Robertson*.

Tasmania. Common in dry stony places in various parts of the colony, *J. D. Hooker*.

S. Australia. Mount Lofty, *Whittaker* ; Forest Creek, *F. Mueller* ; Kangaroo Island, *Waterhouse*.

The flowers of this species are not to be distinguished from those of *D. ericifolia*, but the inflorescence appears to be constant.

Var. *sericea*. Branches foliage and calyxes hoary-pubescent or hirsute.—*D. sericea*, A. Cunn. in Field, N. S. Wales, 347 ; *D. adenophora*, Endl. Nov. Stirp. Dec. 24. Chiefly in the Blue Mountains.

Aotus Wurthii, Regel, in Bot. Zeit. 1851, 596, raised from seeds collected near Adelaide, appears from the description to be a form of *D. floribunda*. No *Aotus* has the leaves channelled above and convex underneath.

SECTION II. XEROPETALUM, *R. Br.*—Calyx obtuse or very shortly and obscurely turbinate at the base. Petals persistent ; standard-claw shorter than the calyx, the lamina rather broader than long or rarely twice as broad as long.

4. **D. Preissii,** *Benth.* A shrub of 4 or 5 ft., with erect elongated branches, glabrous or hoary-pubescent. Leaves rigid, obtuse or with a minute recurved point, mostly ½ to ¾ in. long, not keeled, usually glabrous. Flowers

1 to 3 in each axil along the branches. Pedicels short. Calyx villous, above 3 lines long, obtuse at the base; lobes lanceolate, all acute, as long as the tube, the 2 upper ones united above the middle. Petals persistent; standard not twice as broad as long, not above half as long again as the calyx; wings nearly as long; keel much shorter, obtuse or scarcely acuminate. Pod not seen.—*Aotus (?) dillwynioides,* Meissn. in Pl. Preiss. i. 60, and ii. 215.

W. Australia, *Drummond,* 2nd *Coll. n.* 103; Canning river, *Preiss, n.* 872; Harvey river, *Oldfield.* The bracts, foliage, petals and style are those of *Dillwynia,* and the seeds may well have a strophiole, for in other species it cannot be seen at the time of flowering.

5. **D. brunioides,** *Meissn. in Pl. Preiss.* i. 62. A heath-like shrub, with much of the aspect of *Pultenæa subumbellata* or of *Latrobea diosmifolia,* but with the foliage and style of *Dillwynia.* Branches slightly hoary-pubescent. Leaves rather crowded, mostly about ¼ in. long, spreading or recurved, thicker than in most *Dillwynias,* obscurely keeled, obtuse or very shortly pointed, glabrous, tubercular-scabrous, or the upper ones slightly hirsute. Flowers 3 to 10 or 12 together, in dense terminal heads or corymbs. Bracteoles occasionally rather long, but mostly small and deciduous. Calyx villous, about 3 lines long, scarcely turbinate at the base, the lobes all acute, nearly as long as the tube, the 2 upper ones united to the middle. Petals apparently persistent; standard reniform, the lamina nearly twice as broad as long; wings almost as long; keel shorter and obtusely acuminate. Pod not seen.

N. S. Wales. Blue Mountains, *R. Cunningham.*

6. **D. juniperina,** *Sieb.; Benth. in Hueg. Enum.* 33. A rigid shrub with divaricate pubescent or loosely villous branches. Leaves ¼ to ½ in. long, very straight, strongly keeled, rigid, with strong pungent points. Flowers nearly sessile, several together in terminal clusters, or rarely 2 or 3 apparently axillary. Calyx pubescent, 2 to 2½ lines long, scarcely turbinate at the base; lobes short, the 2 upper ones united into a broad upper lip, either quite entire or minutely emarginate. Petals persistent; standard scarcely twice as broad as long; wings nearly as long; keel much shorter and obtuse. Pod about as long as the calyx.—Lodd. Bot. Cab. t. 401; *D. cinerascens,* DC. Prod. ii. 109, not of R. Br.

Queensland. Moreton Bay, *C. Stuart;* near Dalby, *E. G. Moberly.*
N. S. Wales. Port Jackson, *R. Brown, Sieber, n.* 411, *Woolls,* and others; Blue Mountains, *Miss Atkinson;* Campden Range brushes, *A. Cunningham;* Glen Finlass, *M'Arthur;* New England, *C. Stuart, Beckler.*
Victoria. Rocky Mountains on Macalister river and Futter's Range, *F. Mueller.*
Loddiges says that he received his seeds from Tasmania in 1818, but I have seen no specimens from thence.

7. **D. pungens,** *Mackay; Benth. in Ann. Wien. Mus.* ii. 79. A glabrous or slightly pubescent shrub, the branches often divaricate or elongated and pendulous. Leaves rather crowded, mostly above ½ in. long, rigid, with a strong pungent point, as in *D. juniperina,* but not keeled. Flowers in short axillary racemes or clusters crowded at the ends of the branches into an oblong leafy raceme-like panicle, or almost corymbose. Calyx glabrous or silky-pubescent, about 2 lines long, the base obtuse or scarcely turbinate; lobes short and broad, the 2 upper ones united into a broad truncate entire or slightly emarginate upper lip. Petals persistent; standard not twice as broad

as long, the claw shorter than the calyx; wings nearly as long as the standard; keel much shorter, obtuse. Pod ovoid-oblong, exceeding the calyx. —Lodd. Bot. Cab. t. 1502; *Eutaxia pungens,* Sweet, Fl. Austral. t. 28; *Daviesia condensata,* Turcz. in Bull. Mosc. 1853, i. 265.

W. Australia. King George's Sound, *R. Brown, Baxter;* southern districts, *Drummond,* 5*th Coll. n.* 50; Phillips Ranges, *Maxwell;* Canning river, *Oldfield.* The upper lip of the calyx is longer than the lower lobes, but not so much so as represented in Sweet's figure.

8. D. cinerascens, *R. Br. in Bot. Mag. t.* 2247. A heath-like shrub, slightly hoary or nearly glabrous, resembling some forms of *D. ericifolia,* but usually more slender, and in some western specimens the branchlets often end in slender thorns. Leaves ¼ to ½ in. long or even more, rather slender, not keeled nor twisted, obtuse or with a short point, and usually more or less recurved at the extremity. Flowers in small terminal almost sessile corymbs or short racemes or rarely also in the upper axils. Calyx 2 to 2½ lines long, slightly pubescent, the turbinate base exceedingly short; lobes short, the 2 upper ones united into a broad slightly emarginate upper lip, longer than the lower ones. Petals persistent; standard not twice as broad as long; wings nearly as long; keel much shorter, obtuse. Pod ovate, very obtuse, exceeding the calyx.—Lodd. Bot. Cab. t. 527; Benth. in. Ann. Wien. Mus. ii. 79; Hook. f. Fl. Tasm. i. 85; *D. acicularis,* Meissn. in Pl. Preiss. i. 62, not of Sieb.

N. S. Wales. Hunter's River, *Oldfield;* Bendinine, *M'Arthur.*
Victoria. Buffalo Range, Wilson's Promontory, vicinity of Melbourne, etc., *F. Mueller;* Wimmera, *Dallachy;* mouth of Glenelg river, *Robertson, Allitt.*
Tasmania. Derwent river, *R. Brown;* common in grassy and heathy places throughout the colony, *J. D. Hooker.*
S. Australia. Forest Creek, *F. Mueller.*
W. Australia. Swan River, *Huegel, Drummond, Coll.* 1 *and n.* 243 *and* 244, *Preiss, n.* 875, 881; Darling Range, *Collie, Preiss, n.* 873; Blackwood river, Victoria Plains, etc., *Oldfield.*
Var. (?) *laxiflora.* Leaves crowded, rather long and erect. Racemes rather longer, with more flowers.—*D. laxiflora,* Benth. in Hueg. Bot. Arch. t. 9, and in Ann. Wien. Mus. ii. 79. Hunter's River, *Oldfield.—D. acicularis,* Sieb. in DC. Prod. ii. 109; Benth. in Ann. Wien. Mus. ii. 79, is probably the same variety, but the flowers in Sieber's specimens are still young, and there is a variety of *D. ericifolia,* which is near it in foliage.

9. D. divaricata, *Benth.* Branches divaricate, softly tomentose or villous. Leaves scattered or rather crowded towards the ends of the branches, about 3 to 8 lines long, obtuse or with a minute callous point, not keeled, glabrous or tomentose. Flowers solitary or few together, almost sessile, terminal, and shorter than the last leaves. Calyx 2½ lines long, silky-villous, obtuse at the base, lobes as long as the tube, the 2 upper ones united into a broad truncate emarginate upper lip. Petals persistent; standard not twice as broad as long; wings nearly as long; keel shorter, obtuse. Pod ovoid, about 3 lines long.—*Eutaxia divaricata,* Turcz. in Bull. Mosc. 1853, i. 270.

W. Australia, *Drummond, 4th Coll. n.* 23; sandy flats, Phillips river, *Maxwell.*

10. D. patula, *F. Muell. Fragm.* iv. 16. Branches divaricate, loosely hoary-pubescent. Leaves loosely scattered, erect, spreading or recurved, 2 to 3 lines long, rather thick, obtuse. Flowers solitary or several together on short pedicels at the ends of the branches. Calyx silky-villous, about 3 lines long, the tube very shortly turbinate at the base; lobes short, the 2 upper ones united into a broad truncate emarginate upper lip, longer than the lower ones. Petals persistent; standard nearly twice as long as the calyx and rather broader than long, with a short claw; wings nearly as long as the standard; keel much shorter, obtuse or very shortly and broadly acuminate. Pod only seen young.—*Eutaxia uncinata,* Turcz. in Bull. Mosc. 1853, i. 269; *E. spar-sifolia,* F. Muell. in Trans. Vict. Inst. i. 118; *E. patula,* F. Muell. in Dietr. Fl. Univ. N. Ser. t. 17.

S. Australia. Near the Murray river, *F. Mueller.*

W. Australia, *Drummond,* 5*th Coll. n.* 49, *J. S. Roe ;* Vasse river, *Oldfield ;* Phillips river and to the eastward, *Maxwell.*

TRIBE II. GENISTEÆ.—Shrubs or herbs, very rarely small trees. Leaves simple or with 1, 3 or more digitate leaves (pinnately 3-foliolate in *Goodia*). Stamens all united in a sheath, open on the upper side in all the Australian genera (except in one species of *Hovea,* where the upper stamen is free), or in a closed tube in several European and African genera. Pod dehiscent, not articulate.

This tribe is closely allied to *Podalyrieæ,* from which it is technically distinguished by the monadelphous stamens, and in the Australian genera there is no difficulty in separating them, for in those species of *Daviesia* and *Phyllota* where the stamens are united at the base, the union is too short to be confounded with the sheath of *Genisteæ.* On the other hand, *Goodia,* by its foliage, connects *Genisteæ* with *Galegeæ,* and may be equally well classed in either tribe.

20. PLATYLOBIUM, Sm.

Calyx: 2 upper lobes very large, free or shortly united;· the lower ones small and narrow. Petals clawed; standard orbicular or reniform, wings oblong-obovate, much shorter; keel obovate, nearly as long as the wings. Stamens all united in a sheath open on the upper side; anthers uniform: Ovary sessile or stipitate, with several ovules; style subulate, incurved, with a small terminal stigma. Pod sessile or stipitate, very flat, winged along the upper suture, opening elastically in 2 valves, rolled back but not separating from the wing. Seeds strophiolate.—Slender shrubs. Leaves opposite, entire or with pungent angles, reticulate. Flowers yellow, solitary, in opposite axils. Bracts brown and scarious, imbricate, in 2 or 3 pairs at the base of the pedicels; bracteoles similar but longer, under the calyx.

The genus is limited to Eastern Australia. It is closely allied to the opposite-leaved *Bossiæas,* differing chiefly in the pod, and generally in the proportion of the petals.

Leaves all or mostly triangular, with pungent-pointed angles.
Pedicels much longer than the bracts. Pod stipitate, and several
times longer than calyx . . . . 1. *P. triangulare.*
Pedicels completely concealed by the bracts. Pod sessile, not
twice as long as the calyx . . . . 2. *P. obtusangulum.*
Leaves all ovate, cordate or lanceolate, the lateral angles or auricles
obtuse.
Pedicels completely concealed by the bracts. Pod sessile . . . 2. *P. obtusangulum.*
Pedicels longer than the bracts. Pod stipitate . . . . . . 3. *P. formosum.*

1. **P. triangulare,** *R. Br. in Ait. Hort. Kew. ed.* 2, iv. 266, *not of Sims.* A straggling or procumbent shrub, with slender glabrous or villous

stems, rarely exceeding 1½ ft. Leaves broadly triangul   cordate-hastate, the angles terminating in short pungent points, or the lower leaves rarely broadly cordate, with the lateral angles rounded, mostly ¾ to 1 in. long, veined and glabrous or scabrous above, glabrous or pubescent underneath. Pedicels in the upper axils ¼ to ½ in. long, the bracts at the base obtuse, striate, ½ to 1 line long, the bracteoles under the calyx rather longer and narrower. Calyx clothed with long appressed hairs, 4 or rarely 5 lines long. Standard reniform, deeply emarginate, about twice as long as the calyx ; lower petals not exceeding the calyx. Ovary stipitate, villous or ciliate, with 6 to 8 ovules. Pod above 1 in. long, besides a stipes of 1 to 2 lines, usually glabrous or slightly ciliate when ripe.—*P. Murrayanum,* Hook. Bot. Mag. t. 3259 ; Hook. f. Fl. Tasm. i. 96.

Victoria. Forest land near Portland, *Robertson ;* mouth of the Glenelg, *Allitt.*
Tasmania. Port Dalrymple, *R. Brown ;* light sandy soil near Rocky Cape and George Town, *Gunn.*

2. **P. obtusangulum,** *Hook. Bot. Mag. t.* 3258. A slender shrub, but less straggling and more erect than *P. triangulare,* the stems and leaves glabrous or nearly so. Leaves from broadly triangular to ovate-cordate, hastate or cordate-lanceolate, mostly ¾ to 1 in. long, with a small terminal pungent point, the lateral angles either acute and pungent, as in *P. triangulare,* or rounded and obtuse, as in *P. formosum.* Pedicels very short and completely concealed by the imbricate bracts at their base, of which the inner ones are fully 2 lines long, overlapping the bracteoles under the calyx, which often attain 3 lines. Calyx about ½ in. long, very hairy. Standard shortly exceeding the calyx, lower petals shorter. Ovary sessile, hairy, with about 4 ovules. Pod sessile, rarely 1 in. long, hairy all over.—*P. triangulare,* Sims, Bot. Mag. t. 1508 ; Hook. f. Fl. Tasm. i. 96, but not of R. Br. ; *P. macrocalyx,* Meissn. in Pl. Preiss. i. 80.

Victoria. Port Phillip, *R. Brown ;* common about Melbourne and to the Glenelg, *Robertson ;* not rare in sterile ranges and heath ground, *F. Mueller ;* Wimmera, *Dallachy.*
Tasmania. Common in many parts of the island, *J. D. Hooker.* R. Brown's herbarium contains no Tasmanian specimens of this species, but it appears from the ' Hortus Kewensis ' that it was raised from seeds gathered by him in the island.
S. Australia. Encounter Bay and about Adelaide, *Whittaker ;* Barossa and Lofty ranges, *F. Mueller ;* Kangaroo Island, *Waterhouse.*

3. **P. formosum,** *Sm. in Trans. Linn. Soc.* ii. 350, *and Bot. Nov. Holl.* 17, *t.* 6. A handsome shrub, attaining often 4 or 5 ft., the branches more robust than in the preceding species, glabrous or slightly pubescent. Leaves from broadly heart-shaped to ovate or rarely ovate-lanceolate, acute, with a small rigid point but without lateral angles, 1 to 2 in. long, strongly reticulate and rather coriaceous, glabrous or slightly pubescent underneath. Pedicels hairy, often fully ½ in. long, always exserted from the bracts at their base, which, as well as the bracteoles, are usually as large as in *P. obtusangulum,* glabrous or more or less hairy. Calyx 4 to 5 lines long, very hairy. Standard nearly twice as long ; wings and keel shorter. Ovary stipitate, villous all over or near the sutures only, with about 8 ovules. Pod 1 to 1½ in. long, on a stipes of from one-fourth as long as to longer than the calyx,

loosely hairy or at length glabrous.—Vent. Jard. Malm. t. 31; Bot. Mag. t. 469; DC. Prod. ii. 116; Paxt. Mag. xiii. 195, with a fig.

Queensland. Glasshouses, *W. Hill, F. Mueller.*

N. S. Wales. Port Jackson to the Blue Mountains, *Banks and Solander, Sieber, n.* 373, and others; Bathurst plains, *Fraser;* northward to Hastings river, *Beckler,* and southward to Twofold Bay, *F. Mueller.*

Victoria. Forest Creek, Dandenong and Disappointment mountains, and generally in wooded hills, Gipps' Land, *F. Mueller.*

Tasmania. Plentiful about Mount Direction, N. E. of Launceston, *Gunn.*

Var. *parviflora.* Usually distinguished from the larger N. S. Wales form by the narrow leaves, shorter pedicels, smaller flowers, more glabrous bracts, and by the ovary villous near the sutures only and not all over; but I do not find one of these characters constant, and some of the southern specimens are in all respects intermediate between the extreme Port Jackson forms.—*P. parviflorum,* Sm. Bot. Nov. Holl. 18; Bot. Mag. t. 1520; DC. Prod. ii. 116; Lodd. Bot. Cab. t. 1241; Paxt. Mag. xi. 219, with a fig.; *P. ovatum,* Sieb. in DC. Prod. ii. 116.—The best characterized specimens from Port Jackson, *R. Brown, Sieber, n.* 374, and others.

The synonym of " *Cheilococca apocynifolia,* Salisb. Prod. 412," given by Smith under *Platylobium formosum,* in Bot. Nov. Holl. 17, and copied from him by De Candolle, Endlicher, and many others, appears to be entirely a mistake. There is no such name in Salisbury's work, and the page quoted is one of those of the index.

21. BOSSIÆA, Vent.

(Scottea, *R. Br.;* Lalage, *Lindl.*)

Calyx: 2 upper lobes or teeth broader and usually much larger than the others, distinct or united in an upper lip, 3 lower ones equal. Petals clawed; standard orbicular or reniform, usually reflexed; wings narrow; keel broader and usually shorter than the wings, rarely longer or exceeding the standard. Stamens all united in a sheath open on the upper side; anthers uniform, ovate or oblong, versatile. Ovary stipitate or nearly sessile, with several ovules; style subulate, incurved; stigma small, terminal. Pod sessile or stipitate, flat, not winged; valves completely separating, thin, with the edges nerviform or thickened. Seeds strophiolate.—Shrubs or rarely undershrubs, occasionally leafless; branches terete or flattened, very rarely angular and not sulcate. Leaves alternate or opposite, simple, entire or rarely toothed, often articulate on a very short petiole. Stipules small, brown, lanceolate or setaceous. Flowers axillary, solitary or in clusters of 2 or 3, yellow orange or red. Bracts at the base of the pedicel, 2, 3, or more, imbricate, the outermost very small and persistent, the inner ones often much longer and very deciduous; bracteoles on the pedicel very small and persistent, or longer and deciduous.

The genus is limited to Australia, and, with *Platylobium,* is distinguished from other *Genisteæ* by the anthers all perfectly uniform, attached by the middle, with a more perceptible connectivum.

SERIES I. **Oppositifoliæ.**—*Leaves opposite. Calyx upper lobes obtuse. Pod glabrous, exserted, on a long stipes.*

Calyx upper lobes not longer than the lower. Keel and wings much
  longer than the standard. Leaves denticulate . . . . . . 1. *B. dentata.*
Calyx upper lobes very large. Keel and wings shorter than the
  standard.
Leaves reniform, sinuate and prickly-toothed . . . . . . 2. *B. Aquifolium.*

Leaves entire.
    Leaves 1 to 3 lines long and broad.
        Pubescence rigid. Leaves broadly obovate . . . . . . 3. *B. strigillosa.*
        Glabrous or nearly so. Leaves broadly cordate acute or
           mucronate . . . . 4. *B. cordigera.*
        Glabrous. Leaves orbicular, very obtuse . . . . . . 5. *B. lenticularis.*
        Leaves oblong-lanceolate, ½ to ¾ in. long . . . . 6. *B. Kiamensis.*

SERIES II. **Eriocarpæ.**—*Leaves alternate. Calyx upper lobes acuminate acute or mucronate. Ovary very hairy. Pod nearly sessile, hairy.*

Leaves mostly above ½ in. long, or, if small, narrow. Pod much
    longer than broad.
    Leaves ovate to lanceolate, acute, ¾ to 1½ in. long. Calyx-lobes
      longer than the tube . . . . 7. *B. ornata.*
    Leaves narrow-oblong or linear, obtuse, ½ to ¾ in. long. Calyx-
      lobes longer than the tube . . . . 8. *B. eriocarpa.*
    Leaves narrow cordate-lanceolate, rarely above ½ in. long. Calyx-
      lobes scarcely so long as the tube. Branchlets often spinescent 9. *B. divaricata.*
Leaves under ¼ in. long and broad. Pod about as broad as long.
    Leaves ovate, flat. Calyx upper lobes very large, obovate, almost
      covering the pod. Branches spinescent . . . . 10. *B. calycina.*
    Leaves very small, orbicular, with revolute margins. Calyx upper
      lobes not much larger than the lower. Pod exserted . . . 11. *B. foliosa.*

SERIES III. **Normales.**—*Leaves alternate. Calyx upper lobes rounded or truncate. Ovary glabrous or ciliate on the edge. Pod sessile or stipitate, glabrous.*

Branches terete or angular. Leaves not distichous.
    Branches pubescent or villous, not spinescent.
      Leaves cordate lanceolate or linear, pungent-pointed . . . . 12. *B. cinerea.*
      Leaves linear-cuneate, obtuse retuse or 2-lobed . . . . . 13. *B. biloba.*
    Whole plant glabrous. Branchlets spinescent.
      Calyx 2½ to 3 lines long. Keel nearly as long as the standard.
        Pod on a long stipes . . . . 14. *B. Preissii.*
      Calyx 2 to 2½ lines long. Lower petals much shorter than the
        standard. Pod nearly sessile . . . . 15. *B. concinna.*
Branches terete or slightly compressed. Leaves distichous, usually
    small and rigid (except *B. linophylla*).
    Keel much longer than the standard.
      Leaves ovate-cordate or cordate-lanceolate . . . . . . 16. *B. carinalis.*
      Leaves linear-lanceolate, rounded or narrowed at the base . 17. *B. rupicola.*
    Keel shorter than the standard.
      Leaves mostly above ¼ in. long.
        Leaves narrow-linear, crowded, ½ to 1 in. long . . . . . 18. *B. linophylla.*
        Leaves ovate or oblong, very obtuse, not coriaceous.
          Branches elongated, weak and straggling, villous . . . 19. *B. disticha.*
          Stems prostrate, pubescent, rarely above 1 ft. long . . 20. *B. prostrata.*
      Leaves mostly under ¼ in. long, rigid.
        No thorns.
          Branches pubescent. Leaves ovate or cordate.
            Pedicels longer than the leaves. Pod sessile, thin with
             nerve-like margins. Ovules 6 or more . . . . 21. *B. buxifolia.*
           Pedicels short. Pod on a long stipes, with much
             thickened margins. Ovules 2 or 3 . . . . . 21. *B. Brownii.*
          Plant quite glabrous. Pedicels very short. Pod on a
           long stipes. Ovules 2 or 3.
            Leaves obovate or rhomboidal. Pod with thick mar-
             gins . . . . 23. *B. rhombifolia.*
            Leaves cordate, acute . . . . 24. *B. pulchella.*

Branches spinescent.
  Glabrous. Leaves broadly obovate or obcordate. Pedicels
    short, axillary . . . . 25. *B. microphylla.*
  Minutely strigose-pubescent. Pedicels slender, extra-
    axillary . . . . 26. *B. peduncularis.*
Branches much flattened or winged. Leaves distichous or none.
  Branches leafy.
    Keel glabrous. Pod coriaceous, the stipes much longer than
      the calyx . . . . 27. *P. heterophylla.*
    Keel edged with a short wool. Pod thin, the stipes not ex-
      ceeding the calyx . . . . 28. *P. rufa.*
  Branches leafless, winged.
    Keel not exceeding the standard.
      Keel edged with a short wool (Western species) . . . . 28. *B. rufa.*
      Keel glabrous (Eastern or Northern species).
        Calyx with 5 nearly equal lobes. Flowers nearly sessile,
          with imbricate bracts . . . . 29. *B. bracteosa.*
        Calyx 2 upper lobes broad and united. Pedicels longer
          than the bracts.
          Pod thin, with nerve-like margins . . . . 30. *B. riparia.*
          Pod broad with thickened margins.
            Flowers small. Pod distinctly stipitate . . . . 31. *B. ensata.*
            Flowers rather large. Pod almost sessile . . . . 32. *B. scolopendria.*
    Keel more or less exceeding the standard. Flowers large.
      Stem-wings usually narrow and slightly indented at the nodes.
        Keel much longer than the standard . . . . 33. *B. Walkeri.*
      Stem-wings broad, with very projecting lobes or angles under
        the nodes. Keel rather longer than the standard . . . 34. *B. phylloclada.*

SERIES I. OPPOSITIFOLIÆ.—Leaves opposite. Calyx : 2 upper lobes or
teeth obtuse, broad, free or united. Pod glabrous, on a long stipes.

1. **B. dentata,** *Benth.* An erect glabrous shrub, attaining 7 or 8 ft.,
although sometimes much smaller, the branches terete and often glandular-
scabrous. Leaves opposite, from broadly ovate-cordate or triangular to hastate-lanceolate or almost linear, ½ to 1 in. long, acute or obtuse, irregularly
denticulate. Flowers orange-red or yellow, more or less tinged with green.
Pedicels solitary. Bracts broad, rigid, striate, 1 or 2 inner ones and bracteoles often 2 to 3 lines long, but very deciduous. Calyx 3 to 4 lines long,
the lobes or teeth short, obtuse, all of equal length but the 2 upper ones
broader. Standard twice as long as the calyx ; keel and wings nearly three
times the calyx. Ovary on a long stipes, glabrous, with about 4 ovules.
Pod 1 in. long or more and about 4 lines broad, on a stipes exceeding the
calyx ; valves coriaceous, with thick margins.—*Scottea dentata,* R. Br. in
Ait. Hort. Kew. ed. 2, iv. 269 ; DC. Prod. ii. 118; Lodd. Bot. Cab. t.
1458 ; Meissn. in Pl. Preiss. i. 87.

W. Australia. King George's Sound to Stirling range, and eastward to the great
Australian Bight, *R. Brown, Baxter, Drummond, Oldfield, Maxwell,* and others.

The leaves vary so much in aspect as to appear at first to characterize distinct species, but
the real difference is only in breadth, and the following varieties pass insensibly one into the
other :—

  a. latifolia. Leaves nearly triangular, flat, ½ in. long and broad.—*Drummond, n.* 88,
and other collections.

  b. hastata. Leaves ovate-hastate or hastate-lanceolate, ¾ to above 1 in. long, ¼ to ½
in. broad.—*Scottea dentata,* Bot. Reg. t. 1233 ; Maund, Botanist, t. 134 ; *S. lævis,* Lindl.
Bot. Reg. t. 1652.—*Preiss, n.* 1034, and other collections.

c. angustifolia. Leaves hastate-lanceolate to almost linear, the margins revolute.—*Scottea angustifolia,* Lindl. in Bot. Reg. t. 1266 ; Meissn. in Pl. Preiss. i. 87.—*Preiss, n.* 1035, and other collections.

2. **B. Aquifolium,** *Benth.* A glabrous shrub, with slender terete branches. Leaves opposite, broadly cordate, almost reniform, pungent-pointed, deeply sinuate and prickly-toothed, $\frac{1}{2}$ to $\frac{3}{4}$ in. broad and nearly as long. Flowers solitary, yellow, on pedicels shorter than the calyx. Inner bracts and bracteoles 2 to 3 lines long, rigid, but very deciduous. Calyx about $2\frac{1}{2}$ lines long ; lobes short, the 2 upper ones broad, rounded-truncate, lower ones narrow, not much shorter. Standard nearly three times as long as the calyx ; lower petals rather shorter. Ovary stipitate, glabrous, with about 4 ovules. Pod in our specimen $\frac{3}{4}$ in. long and about half as broad, thickened at the upper suture.

W. Australia, *Drummond, 2nd Coll. n.* 130; Harvey river, *Clarke.* Evidently nearly allied to *B. dentata,* although the proportion of the petals is so different.

3. **B. strigillosa,** *Benth.* Branches divaricate, rigid, rather slender, clothed when young, as well as the under side of the leaves, with short rigid almost appressed hairs. Leaves opposite, broadly obovate-orbicular, with a recurved point, mostly about 2 lines long and broad, very rigid, with recurved margins, obtuse at the base, glabrous and veined above. Stipules subulate, recurved. Pedicels short. Bracteoles lanceolate-subulate. Calyx pubescent, the upper lobes large and rounded, the lower ones small and narrow.

W. Australia, *Drummond, 5th Coll. ?, n.* 81. I have seen neither full-blown flowers nor fruit, but the petals and stamens in the young bud, and the remains of fruiting pedicels, are sufficient to indicate the genus.

4. **B. cordigera,** *Benth. in Hook. f. Fl. Tasm.* i. 95, *t.* 16. An elegant straggling or almost trailing shrub, with numerous slender terete branches, glabrous or minutely pubescent. Leaves opposite, on short slender petioles, broadly orbicular-cordate, acute or mucronulate, $1\frac{1}{2}$ to 3 lines diameter, glabrous or minutely pubescent underneath. Pedicels filiform, usually several times as long as the leaves. Bracteoles small, a little below the calyx. Calyx about 3 lines long, the 2 upper lobes broadly orbicular, a little longer than the tube, lower ones very small. Standard fully twice as long as the calyx ; wings and keel rather shorter. Ovary stipitate, glabrous, with 6 to 8 ovules. Pod $\frac{1}{2}$ to $\frac{3}{4}$ in. long, about $\frac{1}{4}$ in. broad, on a stipes longer than the calyx, the margins nerve-like.—F. Muell. Fragm. iii. 100.

Victoria. Mount Macedon, *F. Mueller.*
Tasmania. Widely distributed over the northern parts of the island, from the sea-level to 4000 ft. elevation, *J. D. Hooker.*

5. **B. lenticularis,** *Sieb. in DC. Prod.* ii. 117, *not of Lodd.* A straggling diffuse glabrous shrub, with numerous slender terete branchlets. Leaves opposite, orbicular, very obtuse, $1\frac{1}{2}$ to 3 lines diameter, rather rigid. Stipules minute. Pedicels rarely above twice as long as the calyx, often articulate and bearing the usual bracts a little above the base. Bracteoles small, above the middle. Calyx about 3 lines long, the 2 upper lobes longer than the tube, broadly orbicular ; lower lobes much shorter and narrow. Standard very broad, twice as long as the calyx ; wings and keel rather shorter. Ovary stipitate, glabrous, with about 4 ovules. Pod $\frac{1}{2}$ to $\frac{3}{4}$ in. long, about

¼ in. broad, on a stipes fully twice as long as the calyx, the margins nerve-like.

N. S. Wales. Port Jackson, *R. Brown, Sieber, n.* 425; Mount Tomah, *R. Cunningham.*

6. **B. Kiamensis,** *Benth.* Branches rather slender, divaricate, minutely hoary. Leaves opposite, oblong-lanceolate or elliptical, entire, obtuse with a short rigid point, ½ to ¾ in. long, coriaceous, veined, glabrous or minutely pubescent underneath. Stipules setaceous. Pedicels 2 to 3 lines long, with minute bracteoles a little below the calyx. Calyx 2 to 2½ lines long, glabrous or nearly so, the 2 upper lobes very broadly orbicular-falcate, as long as the tube, the lower ones shorter. Standard not twice as long as the calyx; wings rather shorter; keel as long as the wings, broad and deeply coloured. Ovary glabrous, on a rather long stipes, with 2 or 3 ovules. Pod not seen.

N. S. Wales. Near Kiama, Illawarra, *Backhouse.*

SERIES II. ERIOCARPÆ.—Leaves alternate. Calyx upper lobes acuminate acute or mucronate. Ovary very hairy. Pod nearly sessile, hairy.

In *B. ornata*, the upper lobes of the calyx are so much narrower and longer than in the generality of the genus, that the character was considered sufficient to separate it under the name of *Lalage*, they are however broader than the lower lobes, and, through the other species of this series, they pass so gradually into the more usual shape observed in *Bossiæa*, that I have felt that *Lalage* could not be maintained even as a section.

7. **B. ornata,** *Benth.* An erect shrub or undershrub of 1 to 2 ft., the branches terete or slightly flattened, pubescent densely villous or rarely nearly glabrous. Leaves varying from broadly ovate almost cordate to narrow-lanceolate, acute or scarcely obtuse, in some specimens ¾ to 1 in., in others 1 to 2 in. long, flat or with recurved almost revolute margins, thinly coriaceous, reticulate, glabrous or scabrous above, loosely pubescent underneath. Stipules from lanceolate to setaceous, rather long or very small. Pedicels usually 2 or 3 together, 1 to 3 or even 4 lines long, shortly pubescent or villous with long hairs. Inner bracts and bracteoles oblong or linear, often 2 or 3 lines long, but so deciduous as to be rarely seen. Calyx 4 to 5 lines long, very hairy, the lobes all acuminate and longer than the tube, the 2 upper ones broader and somewhat falcate, the lower ones narrow, but not shorter. Standard broad, often ¾ in. diameter, but variable in size; keel rather shorter, wings still shorter and much narrower. Ovary sessile, very hairy, with 10 to 12 ovules. Pod sessile, about 1 in. long and ¼ in. broad, clothed or sprinkled with long hairs.

W. Australia. From Stirling Range to Swan River, *Drummond, 1st Coll. and n.* 253 and 254; *Preiss, n.* 999, 1001, 1002, 1007, 1010, *Oldfield,* and others.

There are three principal forms: *Lalage ornata,* Lindl. Bot. Reg. t. 1722, Maund, Botanist, t. 141, Meissn. in Pl. Preiss. i. 85, with ovate flat leaves; *L. hoveæfolia,* Benth. in Lindl. Swan Riv. App. 15, Meissn. in Pl. Preiss. i. 86, Paxt. Mag. ix. 171, with a fig., with oblong-lanceolate leaves with somewhat recurved margins; and *L. angustifolia,* Meissn. in Pl. Preiss. i. 86, with linear or narrow-lanceolate leaves, the margins often almost revolute. These all appear quite distinct in some specimens, but in others they pass very gradually one into the other, or two forms may be observed on the same specimen so as to prevent their separation. The degree of hairiness is variable in all, as is the breadth of the inner bracts. The bracteoles are usually narrow but often concealed by the inner bracts and falling off with them so as to be quite overlooked.

Lalage acuminata, Meissn. in Pl. Preiss. i. 86, and *L. stipularis*, Meissn. l. c. 87, both described from specimens of Preiss's, without flowers or fruit, n. 1003 and 1006, neither of which I have seen, appear from the descriptions given to come within the limits of *B. ornata*.

8. **B. eriocarpa,** *Benth. in Hueg. Enum.* 36. A rigid shrub of 1 to 2 ft., with ash-coloured but scarcely pubescent branches, terete or slightly flattened. Leaves narrow-oblong, lanceolate or linear, very obtuse but often with a setaceous recurved point, mostly ½ to ¾ in. long, the margins recurved, veined and glabrous or scabrous above, glaucous underneath or the upper ones slightly hairy. Stipules small, lanceolate or setaceous. Pedicels slender but very variable in length. Inner bracts and bracteoles oblong or linear but very deciduous. Calyx 3 to 5 lines long, glabrous pubescent or silky-villous, the lobes much longer than the tube, all acuminate, the 2 upper ones much broader, slightly falcate and united above the middle. Petals red ; standard twice as long as the calyx ; wings and keel shorter. Ovary shortly stipitate, very villous, with 4 to 6 ovules. Pod rather small, very villous, the valves flat with nerve-like margins.—Meissn. in Pl. Preiss. i. 83 ; *B. ovalifolia*, Endl. Nov. Stirp. Dec. 21 ; *B. Endlicheri*, Meissn. in Pl. Preiss. i. 83 ; *B. nervosa*, Meissn. in Bot. Zeit. 1855, 31.

W. Australia. King George's Sound, *Huegel;* Swan River, *Drummond, 1st Coll., and n.* 255 *and* 256 ; *Preiss, n.* 1000, 1004, 1005, 1009 ; between Moore and Murchison rivers, *Drummond, 6th Coll. n.* 29.

The extreme forms of this variable species may be characterized as follows, but they are connected by many intermediates.

a. *normalis*. Flowers large; calyx glabrous or pubescent, bracteoles usually broad.

b. *eriocalyx*. Flowers smaller, pedicels shorter ; calyx villous, bracteoles narrow.—*B. Endlicheri*, var. *angustifolia*, Meissn. in Pl. Preiss. i. 83 ; *B. Gilberti*, Turcz. in Bull. Mosc. 1853, i. 285, from Gilbert's collection, n. 313, which I have not seen, must, from the character given, be the same variety.

9. **B. divaricata,** *Turcz. in Bull. Mosc.* 1853, i. 285. A rigid shrub with distichous divaricate branches, hoary as in *B. eriocarpa*, but frequently ending in thorns. Leaves ½ to ¾ in. long, lanceolate or almost linear, acute with short straight points, otherwise veined and scabrous above and hoary or pubescent underneath as in *B. eriocarpa*. Flowers smaller than in that species, but otherwise similar. Calyx-lobes all acuminate and longer than the tube, the 2 upper ones much broader than the others. Ovary very villous, with 4 to 6 ovules. Pod not seen.—*B. lalagoides*, F. Muell. Fragm. iv. 12.

W. Australia, *Drummond, 5th Coll. n.* 83 ; Gardner river, *Maxwell.*

10. **B. calycina,** *Benth.* A low rigid divaricate shrub with hoary branches, the smaller ones ending in fine thorns. Leaves ovate or elliptical, acute or acuminate, 2 to 3 lines long, flat, rigid, transversely veined, glabrous. Pedicels solitary, short, with small oblong or linear bracteoles about the middle. Calyx nearly glabrous, the tube about 1 line long, the 2 upper lobes obovate, mucronate, fully 3 lines long and almost scarious, the lower lobes about half as long, narrow and acute. Petals shorter than the calyx, nearly equal in length. Ovary nearly sessile, villous, with 4 or 5 ovules. Pod scarcely exceeding the calyx and almost as broad as long, very hairy ; valves flat.—*Platylobium ? spinosum*, Turcz. in Bull. Mosc. 1853, i. 284.

W. Australia, *Drummond, 5th Coll. n.* 84 or 85.

11. **B. foliosa,** *A. Cunn. in Field, N. S. Wales,* 347. An erect shrub, with numerous often distichous branches, terete and hoary with a minute tomentum. Leaves broadly orbicular, rarely 2 lines diameter and often only 1 line, with revolute margins, glabrous or scabrous above, hoary underneath. Stipules small, broad. Pedicels very short, with very small bracts. Calyx pubescent, 1½ lines long, the upper lobes falcate, acute, not much larger than the lower ones. Standard twice as long as the calyx; keel nearly as long. Ovary stipitate, very villous, with about 3 ovules. Pod nearly orbicular, about 3 lines diameter, on a short stipes, rusty-villous, valves slightly convex. —F. Muell. Fragm. iii. 100; *B. distichoclada,* F. Muell. in Trans. Phil. Soc. Vict. i. 39, and in Hook. Kew Journ. viii. 13.

N. S. Wales. Brushy forest land near Bathurst, *A. Cunningham.*
Victoria. In the Australian Alps from the Mitta-Mitta to the tributaries of the Snowy River at an elevation of 4 to 5000 ft., as well on rocks as on the peaty margins of rivulets, *F. Mueller.*

SERIES III. NORMALES.—Leaves alternate. Calyx 2 upper lobes rounded or truncate, free or united in an upper lip. Ovary glabrous or rarely ciliate on the edge. Pod sessile or stipitate, glabrous.

12. **B. cinerea,** *R. Br. in Ait. Hort. Kew. ed.* 2, iv. 268. An erect or rarely straggling much-branched shrub of 2 to 4 or 5 ft., the branches terete or slightly angular, pubescent or villous. Leaves from broadly ovate-lanceolate to linear lanceolate, tapering into a pungent point, about ½ in. long or shorter, rarely ¾ in., often almost cordate at the base, rigid, with recurved margins, glabrous or sprinkled with a few hairs above, more pubescent or villous underneath. Pedicels slender with very small bracteoles. Calyx glabrous, 2 to 2½ lines long, the lobes much shorter than the tube, 2 upper ones very broad and rounded, the lower ones small. Standard broad, fully twice as long as the calyx; keel about as long, deeply coloured. Ovary stipitate, glabrous, w.th 4 to 6 ovules. Pod ½ to ¾ in. long, ¼ in broad, on a stipes about as long as the calyx; valves thin, with nerve-like margins.—Bot. Reg. t. 306; DC. Prod. ii. 117; Hook. f. Fl. Tasm. i. 95; *B. coccinea,* Bonpl. Jard. Malm. 128, t. 52; *B. cordifolia,* Sweet, Fl. Austral. t. 20; *B. tenuicaulis,* Grah. in Edinb. New Phil. Journ. 1840; Bot. Mag. t. 3895.

N. S. Wales. Near Twofold Bay, *F. Mueller.*
Victoria. From Brighton to about Melbourne, *F. Mueller;* and thence to the Glenelg, *Robertson.*
Tasmania. Port Dalrymple, *R. Brown;* abundant in dry situations throughout the island, *J. D. Hooker.*
Var. (?) *rosmarinifolia.* Leaves crowded, linear-lanceolate or linear, with short pungent points, ¾ to 1 in. long, the margins revolute. Flowers rather small.—*B. rosmarinifolia,* Lindl. in Mitch. Three Exped. ii. 178.—Mount William in Victoria, *Mitchell, F. Mueller.* I follow F. Mueller in uniting this as a variety with *B. cinerea,* but with some hesitation, as its aspect is somewhat different, and I have seen no regular chain of intermediates.

13. **B. biloba,** *Benth. in Hueg. Enum.* 36. Stems from a thick stock, erect, little-branched, ¾ to 1½ ft. high, rigid, angular, loosely pubescent or villous. Leaves linear-cuneate, very obtuse retuse or 2-lobed, sometimes all under 1 in. and rather broad, sometimes narrow and 1 to 2 in. long, rigid with revolute margins, glabrous and green, or slightly hairy underneath. Pedicels short, with softly setaceous bracteoles. Calyx silky-villous, 4 to 5 lines

long, the lobes longer than the tube, the 2 upper ones united into a broad truncate emarginate upper lip, the lower ones lanceolate, nearly as long, all silky inside. Standard broad, nearly twice as long as the calyx; wings and keel shorter. Ovary stipitate, glabrous, with about 6 ovules. Pod ½ to ¾ in. long, rather broad, on a short stipes, but not seen perfect.—Meissn. in Pl. Preiss. i. 85.

W. Australia. King George's Sound, *Huegel, Oldfield;* Swan River, *Drummond,* 1st *Coll. and n.* 264, *Preiss, n.* 1061; S. Hutt and Murchison rivers, *Oldfield.*

14. **B. Preissii,** *Meissn. in Pl. Preiss.* i. 82. A glabrous much-branched shrub, the branches terete or angular, the smaller ones spinescent. Leaves oblong-cuneate or lanceolate, obtuse or acute, 3 to 4 or rarely 5 lines long, flat, rigid, prominently or obscurely veined. Pedicels nearly as long as the leaves, with small deciduous bracteoles. Calyx rather thick, 2½ to 3 lines long, the lobes or teeth very short, the 2 upper ones united in a broad truncate emarginate upper lip. Standard more than twice as long as the calyx, orbicular; wings and keel as long as the standard. Ovary stipitate, glabrous, with 12 to 15 ovules. Pod elongated, rather broad, on a stipes longer than the calyx.—*B. rigida,* Turcz. in Bull. Mosc. 1853, i. 285.

W. Australia, *Drummond,* 5th *Coll. n.* 79, *J. S. Roe;* King George's Sound, *Baxter;* rocky sterile places near Cape Riche, *Preiss, n.* 986; upper Kalgan river, *Oldfield;* Kojonerup and grassy flats on Salt river, and eastward to Cape Arid and Cape Le Grand, *Maxwell.* I have not seen Preiss's original wild specimen, but a cultivated one from the Hamburg garden is certainly conspecific with the wild ones examined from other collectors.

15. **B. concinna,** *Benth.* A glabrous spinescent shrub allied to *B. Preissii,* but with smaller flowers, different in the proportion of the petals and an almost sessile pod. Branches angular, the smaller ones often reduced to short leafless thorns. Leaves oblong or cuneate, obtuse, 1½ to 3 lines long, the margins often slightly recurved. Pedicels about as long as the leaves, with small deciduous bracteoles. Calyx 2 to 2½ lines long, the 2 upper lobes united in a broad emarginate upper lip nearly as long as the tube, the lower lobes scarcely shorter. Standard twice as long as the calyx; lower petals considerably shorter, the keel slightly fringed on the edge. Ovary shortly stipitate, glabrous, with 5 to 7 ovules. Pod ¾ in. long, narrow, almost sessile, but not quite ripe in our specimens.

W. Australia, *Drummond,* 5th *Coll. n.* 81, *and Suppl. n.* 41 (very spinescent specimens with dark-coloured flowers); Grass-tree plains between M'Callum and Stokes Inlets, *Maxwell* (more leafy and less spinescent, with apparently bright yellow flowers).

16. **B. carinalis,** *Benth. in Mitch. Trop. Austr.* 290. A shrub with the habit almost of *B. rhombifolia,* but with flowers near those of *B. dentata.* Branches terete, softly pubescent. Leaves distichous, ovate or broadly ovate-lanceolate, mostly cordate, mucronulate or scarcely obtuse, 4 to 6 lines long, often oblique at the base, coriaceous, prominently veined, minutely hoary or glabrous. Pedicels short, with small bracteoles. Calyx 4 to 5 lines long, quite glabrous and rather thick, the lobes much shorter than the tube, the 2 upper ones united in a broad emarginate upper lip, the lower ones narrow and rather shorter. Standard very broad, half as long again as the calyx; wings twice as long, and keel incurved, rather narrow, three times as long as the calyx. Ovary on a long stipes, glabrous, with 8 to 10 ovules. Pod not seen.

Queensland. Sandstone gullies of the Mantuan Downs, *Mitchell;* also in *Leich-hardt's* collection.

17. **B. rupicola,** *A. Cunn. Herb.* A shrub with the habit and flowers of *B. carinalis,* but very different leaves. Branches terete, pubescent. Leaves distichous, narrow-lanceolate, acute, ½ to 1 in. long, rarely 2 lines broad, narrowed or scarcely obtuse at the base, coriaceous, veinless except the midrib. Pedicels short, the small bracteoles near the base. Calyx 3 to 3½ lines long, the large upper lip nearly as long as the tube, the narrow lower lobes scarcely shorter. Standard broad, reflexed, rather longer than the calyx; wings longer, obovate-falcate; keel broad incurved, exceeding the calyx by 5 or 6 lines. Ovary on a long stipes, glabrous, with 8 to 10 ovules. Pod with broad thick margins when young, not seen ripe.

Queensland. Brisbane river, *Fraser;* Mount Lindsay at an elevation of 5700 ft., *A. Cunningham.*

18. **B. linophylla,** *R. Br. in Ait. Hort. Kew. ed.* 2, iv. 268. An erect much-branched shrub of 2 to 4 ft., the branches rather slender, flattened when young, glabrous or sparingly pubescent. Leaves numerous, narrow-linear, obtuse or shortly pointed, ½ to 1 in. long or rarely more, the margins recurved, glabrous or sprinkled when young with a few short hairs. Pedicels solitary or 2 together, filiform, shorter than the leaves, with small narrow bracteoles near the calyx. Calyx glabrous or sprinkled with a few hairs, about 1½ lines long, the lobes much shorter than the tube, the 2 upper ones united in a broad truncate emarginate upper lip, the lower ones shorter. Standard fully twice as long as the calyx; wings nearly as long; keel shorter. Anthers small, almost globular. Ovary stipitate, glabrous, with about 6 ovules. Pod ½ to ¾ in. long, the margins nerviform, the stipes much longer than the calyx. —Bot. Mag. t. 2491; DC. Prod. ii. 117; Lodd. Bot. Cab. t. 174; Meissn. in Pl. Preiss. i. 83; ii. 221; Maund, Botanist, t. 68.

W. Australia. King George's Sound and adjoining districts, *R. Brown, Drummond, 2nd Coll. n.* 121, *Preiss, n.* 1011, 1012, 1013, and others; to Stirling range and Mount Gairdner, *Maxwell;* Blackwood river, *Oldfield;* Vasse river, *Mrs. Molloy.*

19. **B. disticha,** *Lindl. Bot. Reg.* 1841, *t.* 55. A shrub with weak straggling slender terete or slightly flattened branches, softly pubescent or villous. Leaves distichous, ovate or oblong, very obtuse, but often minutely mucronate, mostly under ½ in. long or on luxuriant branches ¾ in., not coriaceous, sprinkled with appressed hairs on both sides. Pedicels slender, solitary or 2 together, the small narrow bracteoles above the middle. Calyx softly pubescent, 2 to 2½ lines long, the lobes rather shorter than the tube, the 2 upper ones broad rounded-falcate, the lower ones nearly as long, lanceolate. Standard twice as long as the calyx; wings and keel shorter. Ovary stipitate, usually ciliate on the sutures, with 6 to 8 ovules. Pod glabrous, under 1 in. long, rather broad, the stipes shorter than the calyx.—Meissn. in Pl. Preiss. ii. 221.

W. Australia, *Drummond, 2nd Coll., n.* 124, in our sets, or *n.* 122, according to Meissner.

20. **B. prostrata,** *R. Br. in Ait. Hort. Kew. ed.* 2, iv. 268. A small shrub or undershrub with a thick woody stock and slender prostrate stems

from a few inches to about 1 ft. long, terete or slightly flattened, usually pubescent. Leaves distichous, ovate or oblong, obtuse or scarcely acute, ¼ to ½ in. long or rarely more, glabrous or sprinkled with a few hairs, the petioles frequently rather long and slender. Pedicels usually much longer than the leaves, pubescent, with small deciduous bracteoles. Calyx about 2 lines long. the 2 upper lobes broadly falcate and united above the middle, the lower ones short and narrow. Standard fully twice as long as the calyx; wings and keel considerably shorter. Ovary very shortly stipitate, glabrous or ciliate, with 6 to 10 ovules. Pod nearly sessile, glabrous, ¾ to 1 in. long, rarely 2½ lines broad.—Bot. Mag. t. 1493; DC. Prod. ii. 117; Hook. f. Fl. Tasm. i. 94; *B. ovata*, Sm. in Trans. Linn. Soc. ix. 303; *B. linnæoides*, G. Don, Gen. Syst. ii. 129; *B. nummularia*, Endl. Nov. Stirp. Dec. 22; *B. humilis*, Mcissn. in Pl. Preiss. i. 85.

Queensland. Wide Bay, *Bidwill.*
N. S. Wales. Port Jackson to the Blue Mountains, *R. Brown, Sieber, n.* 351, and others; northward to New England, *C. Stuart;* and southward to Yowaka river, *F. Mueller.*
Victoria. Port Phillip, *R. Brown;* Australia Felix and Wimmera river, *F. Mueller.*
Tasmania. Abundant in dry soil throughout the island, *J. D. Hooker.*
S. Australia. Onkaparinga, Stringy Bark, Lofty and Bugle ranges, Rivoli Bay, etc., *F. Mueller.*

In the southern specimens the stems are usually short, very slender, almost filiform, in several of the northern ones they are longer firmer and more branched. The size of the flowers is variable, and here and there a few flowers, possibly imperfect ones, may be found abnormally almost sessile.

21. **B. buxifolia,** *A. Cunn. in Field, N. S. Wales,* 348. A procumbent or diffuse shrub with numerous slender terete or scarcely flattened branches, minutely but softly pubescent. Leaves broadly ovate or almost cordate, acute, 1½ to nearly 3 lines long, nearly flat, coriaceous, transversely wrinkled and sprinkled with a few hairs above, loosely pubescent underneath. Pedicels much longer than the leaves, with small broad deciduous bracteoles above the middle. Calyx minutely pubescent, scarcely 2 lines long, the upper lobes broad truncate as long as the tube and more or less united, the lower ones lanceolate and much shorter. Standard broad, twice as long as the calyx; wings and keel shorter. Ovary nearly sessile, glabrous or with ciliate edges, with about 6 ovules. Pod usually ¾ to 1 in. long, ¼ in. broad, almost sessile, the margins nerviform.—*B. decumbens*, F. Muell. Fragm. i. 9.

N. S. Wales. Near George's river, *R. Brown;* rocky brushy hills in the Blue Mountains, *A. Cunningham,* and others; northward to Clarence river, *Beckler;* and southward to Argyle County, *Lhotsky.*
Victoria. In the higher Australian Alps on the Genoa and Delatite rivers, *F. Mueller;* Mount Macedon, *Dallachy.*

22. **B. Brownii,** *Benth.* An erect apparently stout much-branched shrub of 3 or 4 ft., with the habit of *B. rhombifolia,* the branches terete or slightly compressed, softly pubescent. Leaves distichous, broadly ovate, almost cordate, obtuse or mucronulate, 2 to 4 lines long, mostly oblique at the base, flat, coriaceous, loosely pubescent or hairy. Pedicels mostly shorter than the calyx, with deciduous bracteoles near the base. Calyx about 2 lines long; lobes much shorter than the tube, the 2 upper ones broad, rounded-falcate. Petals fully twice as long as the calyx, the keel nearly as long as the standard. Ovary glabrous, on a long stipes, with usually 3 ovules. Pod ¾

M 2

to 1 in. long, about 5 lines broad, the margins broad, the upper one much thickened; the stipes longer than the calyx.

Queensland. Port Bowen, *R. Brown, also in Leichhardt's Collection.* Allied in foliage to *B. buxifolia,* but the pod is that of *B. rhombifolia.*

23. **B. rhombifolia,** *Sieb. in DC. Prod.* ii. 117. A tall much-branched shrub, quite glabrous and often glaucous, the young branches often flattened. Leaves distichous, from obovate to broadly rhomboidal, usually mucronulate, 2 to 3 or rarely 4 lines long and broad, coriaceous, flat, with a prominent midrib. Pedicels shorter than the calyx, with small broad bracteoles below the middle. Calyx about 4 lines long, the lobes much shorter than the tube, the 2 upper ones very broad and falcate, the lower ones small but nearly as long. Petals twice as long as the calyx, nearly equal in length. Ovary on a long stipes, quite glabrous, with 2 or 3 ovules. Pod ¾ to nearly 1 in. long, about 4 lines broad, the margins, especially the upper one, broad and thick, the stipes longer than the calyx, the seeds often separated by a cellular substance as in *B. heterophylla.*—*B. lenticularis,* Lodd. Bot. Cab. t. 1238, from the fig., not of Sieb.

Queensland. In the gullies of the Mantuan Downs, *Mitchell;* Dogwood Creek, *Leichhardt.*

N. S. Wales. Port Jackson to the Blue Mountains, *R. Brown, Sieber, n.* 354, *and Fl. Mixt. n.* 568, and others.

B. rotundifolia, DC. Prod. ii. 117, from "eastern New Holland," must, from the character given, be closely allied to the above, perhaps a luxuriant variety, with the leaves 4 to 5 lines long and 5 to 6 lines broad.

24. **B. pulchella,** *Meissn. in Pl. Preiss.* i. 84. A much-branched shrub of 2 or 3 ft., with rather slender terete or scarcely compressed branches, minutely pubescent. Leaves broadly heart-shaped, mostly acute, 2 to 3 lines long, coriaceous, glabrous. Pedicels very short, the small broad bracts coriaceous and imbricate; bracteoles 2½ to 3 lines long, falling off before the flower expands. Calyx 2½ lines long, glabrous or nearly so; lobes or teeth very short, the 2 upper ones united in a broad emarginate upper lip. Standard broad, more than twice as long as the calyx; keel nearly as long; wings very narrow. Ovary on a long stipes, glabrous, with 2 ovules. Young pod with the upper margin much thickened.

W. Australia. Swan River, *Drummond,* 1*st Coll. and n.* 250; Darling Range, *Preiss, n.* 1032.

25. **B. microphylla,** *Sm. in Trans. Linn. Soc.* ix. 303. A spreading much-branched shrub, attaining several feet, the branches terete or slightly flattened, hoary with a minute tomentum, the smaller ones ending in subulate thorns. Leaves broadly obovate or obcordate, truncate, emarginate, or with a small recurved point, 1 to 3 lines long, coriaceous, glabrous above, sprinkled with a few hairs underneath. Pedicels shorter or scarcely longer than the leaves, usually without bracteoles. Calyx glabrous, 2 to 2½ lines long, the lobes shorter than the tube, the 2 upper ones very broadly falcate, the lower ones small but scarcely shorter. Standard and keel twice the length of the calyx; wings narrow and shorter. Ovary stipitate, glabrous, with about 6 ovules. Pod rarely above ½ in. long, 2 to 3 lines broad, on a stipes about as long as the calyx, the margins nerviform.—DC. Prod. ii. 117; Lodd. Bot.

Cab. t. 656 ; F. Muell. Fragm. iii. 99 ; *Platylobium microphyllum*, Sims, Bot. Mag. t. 863 ; *P. obcordatum*, Vent. Jard. Malm. under n. 31.

N. S. Wales. Port Jackson to the Blue Mountains, *R. Brown, Sieber, n.* 355, *and Fl. Mixt. n.* 569, and others ; northward to Clarence river, *Beckler ;* and southward to Illawarra, *Shepherd.*

Victoria. Wooded or bushy hills, chiefly granitic, on the Tumbo and at the mouth of Snowy River, *F. Mueller.*

26. **B. peduncularis,** *Turcz. in Bull. Mosc.* 1853, i. 287. A rigid spreading shrub with terete or scarcely flattened branches, the short divaricate branchlets ending in slender thorns, pubescent with appressed hairs. Leaves oblong, very obtuse, 1½ to 3 lines long, coriaceous, with recurved margins, sprinkled on both sides with appressed hairs. Pedicels filiform, ¼ to ½ in. long, all inserted at some distance above the axils, with minute bracteoles above the middle. Calyx scarcely 2 lines long, sprinkled with appressed hairs, the 2 upper lobes broadly falcate, as long as the tube, the lower ones narrow and rather shorter. Standard and keel twice as long as the calyx. Ovary stipitate, glabrous, with about 4 ovules. Pod not seen.

W. Australia, *Drummond, 5th Coll. n.* 80.

27. **B. heterophylla,** *Vent. Jard. Cels. t.* 7. A low glabrous and often glaucous shrub or undershrub ; branches erect or ascending, from under 1 ft. to 2 ft. high, often much flattened. Leaves distichous, often distant, the lower ones ovate, obtuse or nearly orbicular, the upper ones gradually narrower or sometimes nearly all oblong or linear, the larger ones ¾ to 1 in. long, rather thick and nearly veinless. Pedicels shorter than the leaves, with small bracteoles below the middle. Calyx 2½ to nearly 3 lines long, the lobes all short, the upper ones very broadly falcate. Standard very broad, twice as long as the calyx ; keel rather shorter, deeply coloured. Ovary stipitate, glabrous, with about 6 ovules. Pod ¾ to 1 in. long, on a stipes longer than the calyx, the margins much thickened, the seeds separated by cellular tissue.— DC. Prod. ii. 117 ; Lodd. Bot. Cab. t. 271 ; *Platylobium lanceolatum*, Andr. Bot. Rep. t. 205 ; *P. ovatum*, Andr. Bot. Rep. t. 266 ; *Bossiæa lanceolata*, Bot. Mag. t. 1144 ; *B. ovata*, G. Don, Gen. Syst. ii. 128.

Queensland. Wide Bay, *Bidwill, in Hook. Herb.,* but possibly some error.

N. S. Wales. Port Jackson, *Banks and Solander, R. Brown, Sieber, n.* 352, and others ; Mount Imlay, near Twofold Bay, *F. Mueller.*

Victoria. Near Lake Victoria, Gipps' Land, *F. Mueller.*

28. **B. rufa,** *R. Br. in Ait. Hort. Kew. ed.* 2, iv. 267. A tall erect glabrous shrub with flattened branches, sometimes spinescent, bordered by narrow wings, sometimes disappearing in leafy specimens, more prominent on the leafless ones, with slightly indented nodes. Leaves when present obovate or oblong, ¼ to ¾ in. long, very obtuse, rather thin, on slender petioles. Pedicels solitary or clustered, proceeding from leafless nodes even on leafy specimens, filiform, with very small bracteoles near the calyx. Calyx about 2 lines long, the 2 upper lobes united in a broad truncate emarginate upper lip, the lower ones small and acute. Standard broad, more than twice as long as the calyx ; wings rather shorter ; keel much shorter, edged with a close woolly fringe. Ovary stipitate, with 8 to 10 ovules. Pod ¾ to above 1 in. long, about ¼ in. broad, the margins nerviform, the stipes rather shorter or scarcely

longer than the calyx.—DC. Prod. ii. 117; *B. ensata,* Meissn. in Pl. Preiss. i. 81, not of Sieb.

W. Australia, *R. Brown, Drummond, Preiss,* and others.

The following forms, different as they look, pass much into each other; all have similar flowers and fruit, differing from *B. riparia* in the comparatively larger standard and in the small fringed keel, and from *B. ensata* by the same characters and in the narrower pod without the thick margins of that species.

     a. *normalis.* Branches elongated, leafless or nearly so, not spinescent.—Lodd. Bot. Cab. t. 1119, appears to represent this form.—King George's Sound, *R. Brown, Drummond, 5th Coll. n. 87, and in another Coll. n.* 84 ; Phillips river, *Maxwell.*

     b. *oxyclada.* Branches numerous, divaricate, narrow or rarely very broad, leafless, mostly spinescent.—*B. oxyclada,* Turcz. in Bull. Mosc. 1853, i. 284.—*Drummond, 5th Coll. n.* 82.

     c. *virgata.* Branches elongated, more or less leafy, not spinescent.—*B. virgata,* Hook. Bot. Mag. t. 3986 ; *B. paucifolia,* Benth. in Bot. Reg. 1841, Misc. 53 ; Meissn. in Pl. Preiss. i. 81.—*Drummond, n.* 258 ; Tone, Gordon, and Blackwood rivers, *Oldfield.*

     d. *foliosa.* Leaves rather numerous, branches often spinescent.—*B. paucifolia,* Lindl. Bot. Reg. 1843, t. 63 ; *B. spinescens,* Meissn. in Pl. Preiss. i. 82.—Swan River, *Drummond, 1st Coll. and n.* 259 ; stony hills, York district, *Preiss, n.* 1030, 1031; stony places, S. Hutt river, *Oldfield.*

Closely allied to this is a stunted shrub from Dirk Hartog's Island, *Milne,* with very spinescent flat branches, and small coriaceous broadly obovate leaves, near those of *B. rhombifolia* or *B. microphylla,* but without flowers to determine its characters.

29. **B. bracteosa,** *F. Muell. Herb.* A glabrous leafless shrub, nearly allied to *B. ensata,* but more rigid, with flattened branches and broad thickly coriaceous wings, indented at the nodes so as to form prominent obtuse angles, with erect brown lanceolate scales, often 1 to 2 lines long, but very deciduous. Flowers solitary at the nodes, small and almost sessile. Bracts imbricate, the inner ones and bracteoles above 1 line long, but often very deciduous. Calyx about 2 lines long, rather rigid, the lobes shorter than the tube, obtuse, and all 5 nearly equal. Standard twice as long as the calyx ; keel scarcely shorter than the standard, deeply coloured, quite glabrous. Ovary stipitate, with about 8 ovules. Pod not seen.

Victoria. In the Australian Alps, on the Mitta-Mitta and Macalister rivers, at an elevation of 3000 to 4000 ft., and on Mount Latrobe, *F. Mueller.* F. Mueller is disposed to include this among the forms of *B. scolopendria,* but independently of the more rigid habit, the differences in the bracts and calyx are generally very constant in *Bossiæas.*

30. **B. riparia,** *A. Cunn. Herb.* A glabrous leafless shrub, usually procumbent or diffuse, resembling some of the small-flowered specimens of *B. ensata,* but different in the proportion of the petals and in the pod. It is also usually more branched, the wings very narrow and less indented at the nodes. Flowers small, on short pedicels, with very small bracteoles. Calyx under 2 lines long, the lobes very short, the 2 upper ones united in a broad truncate upper lip as in *B. ensata,* the lower ones narrow, but scarcely shorter. Petals fully twice as long as the calyx, the keel almost as long as the standard, the wings scarcely so long. Ovary stipitate, glabrous, with 4 to 6 ovules. Pod ¾ in. long or rather more, scarcely more than 2 lines broad, on a stipes usually longer than the calyx ; valves thin, with nerve-like margins.—*B. ensata,* Hook. f. Fl. Tasm. i. 94, but scarcely of Sieber.

N. S. Wales. Downs of Mineri, on the upper branches of the Lachlan river, *A. Cunningham* (specimens imperfect).

Victoria. Maneroa, *F. Mueller*, and possibly some specimens from other localities, not determinable for want of flowers or fruit.
Tasmania. Derwent river, *R. Brown;* abundant throughout the island in various soils and situations, ascending to 4000 ft., *J. D. Hooker.*
S. Australia. Port Lincoln, *F. Mueller* (doubtful specimens, without flowers or fruit).

31. **B. ensata,** *Sieb. in DC. Prod.* ii. 117. An erect or procumbent glabrous leafless shrub, very variable in aspect, and not always easy to distinguish from *B. scolopendria*, with which F. Mueller unites it, and of which it may very likely prove a small-flowered variety, but the fruits of the different forms are as yet insufficiently known. Branches flat and winged as in that species, but usually narrower. Flowers smaller, the pedicels shorter and consequently less covered by the small bracts and bracteoles. Calyx rarely above 2 lines long, the keel much shorter than the standard, and glabrous as in *B. scolopendria*. Ovary distinctly stipitate, glabrous, with about 6 ovules. Pod not seen fully ripe, but apparently not so broad and more stipitate than in *B. scolopendria*, although with much more of the character of that species than of *B. riparia.*—Sweet, Fl. Austral. t. 51 ; *B. rufa*, Maund, Botanist, t. 81, not of R. Br.

Queensland. Moreton Bay, *F. Mueller.*
N. S. Wales. Port Jackson to the Blue Mountains, *R. Brown, Sieber, n.* 434, and others; and in the scrub in the interior, *Fraser ;* Port Macquarrie, *Backhouse ;* and southward to Twofold Bay, *F. Mueller.*
Victoria. Snowy River, *F. Mueller.*

32. **B. scolopendria,** *Sm. in Trans. Linn. Soc.* ix. 303. An erect glabrous often somewhat glaucous leafless shrub, the branches flattened, with coriaceous distichous wings descending from the nodes, the 2 wings with the branch varying from 1 to 6 lines broad, the nodes scarcely indented. Leaves only on young seedlings and 1 or 2 occasionally at the base of the branches. Flowers usually solitary at the nodes on very short pedicels, the bracteoles under 1 line long, but almost covering the pedicel. Calyx about $2\frac{1}{2}$ lines long or rather more ; lobes shorter than the tube, the 2 upper ones very broad, united in a broad emarginate upper lip, the lower ones nearly as long, lanceolate, acute. Standard fully $\frac{1}{2}$ in. broad; wings shorter; keel still shorter. Ovary almost sessile, glabrous, with about 6 ovules. Pod almost sessile, 1 in. long or more, about 5 lines broad, the valves flat but with broad very much thickened margins, more prominent even than in *B. heterophylla*, but without any cellular matter between the seeds inside.—Bot. Mag. t. 1235 ; DC. Prod. ii. 116 ; Lodd. Bot. Cab. t. 1747; *Platylobium scolopendrium,* Vent. Jard. Malm. t. 55 ; Andr. Bot. Rep. t. 191.

N. S. Wales. Port Jackson, *R. Brown, Sieber, n.* 357, and many others. The other stations usually given belong either to *B. ensata*, which, as above stated, may be only a small-flowered variety, or to *B. riparia*, differing more essentially both in flowers and fruit.

33. **B. Walkeri,** *F. Muell. Fragm.* ii. 120, iii. 97, 166. A glabrous leafless shrub with flattened winged branches, usually narrow and slightly indented as in *B. ensata*, but thickly coriaceous as in *B. bracteosa*. Pedicels usually solitary, $\frac{1}{4}$ to $\frac{1}{2}$ in. long, the bracteoles all fallen from our specimens. Calyx about 5 lines long, very oblique, the lobes shorter than the tube, the 2 upper ones rounded and united to the middle, the lower ones narrower and

shorter. Standard scarcely twice as long as the calyx, wings about as long as the standard ; keel considerably longer. Ovary stipitate, with about 20 ovules. Pod above 2 in. long, about 4 lines broad, on a stipes shorter than the calyx ; valves apparently flat, with slightly thickened margins, but not seen quite ripe.

N. S. Wales. Peel Range, between the Lachlan and Murrumbidgee rivers, *Walker ;* and thence to the Barrier Range, *Victorian Expedition.*

34. **B. phylloclada,** *F. Muell. in Trans. Phil. Inst. Vict.* iii. 52, *Fragm.* ii. 120. A tall glabrous glaucous leafless shrub, the branches flattened and very broadly winged, the wings forming angles projecting under the nodes from $\frac{1}{4}$ to $\frac{1}{2}$ in., sometimes acute and pungent-pointed, sometimes shorter and obtuse. Pedicels solitary or more frequently clustered, rather slender, with small narrow bracteoles about the middle. Calyx 4 to 5 lines long, the 2 upper lobes as long as the tube, broadly obovate and distinct, the lower ones very small. Standard not twice as long as the calyx ; wings shorter and narrow ; keel as long as the standard or rather longer, woolly-ciliate on the edge. Ovary stipitate, glabrous, with 10 to 12 ovules. Pod above 1 in. long, $\frac{1}{4}$ in. wide, very flat, with slightly thickened margins.

N. Australia. Islands of the Gulf of Carpentaria, *R. Brown ;* Fitzmaurice river, a chief component of the scrub between Van Alphen and Nicholson rivers, *F. Mueller.*

22. TEMPLETONIA, R. Br.

(Nematophyllum, *F. Muell.*)

Calyx 2 upper lobes or teeth completely united or rarely distinct, 2 lateral ones often shorter, lowest one the longest. Standard orbicular or obovate, usually reflexed ; wings narrow, usually shorter than the standard ; keel as long as the standard or shorter, the petals slightly united. Stamens all united in a sheath open on the upper side ; anthers alternately long and erect and short and versatile. Ovary sessile or stipitate, with several ovules or rarely only 2 or 3 ; style incurved, filiform ; stigma small, terminal. Pod sessile or stipitate, much flattened, ovate-oblong or linear, often oblique, completely dehiscent, the valves coriaceous, without thickened sutures. Seeds strophiolate.—Glabrous shrubs or rarely undershrubs, occasionally leafless, the branches angular or sulcate-striate. Leaves when present alternate, simple, .entire. Stipules minute or spinescent. Flowers axillary, solitary or 2 or 3 together, red or yellow. Bracts 2 or 3 at the base of the pedicel as in *Bossiæa,* but usually very minute ; bracteoles at or above the middle.

The genus is limited to Australia. It has as much variety in habit as *Bossiæa,* from which it differs in the calyx, anthers, and pod, as well as in the striate-sulcate branches.

Stems leafy. Stipules minute or inconspicuous.
  Leaves cuneate or obovate, obtuse or emarginate. Tall shrub with
    large red flowers . . . . 1. *T. retusa.*
  Leaves ovate or elliptical, acute. Low undershrub . . . . . 2. *T. Drummondii.*
  Leaves narrow-linear.
    Calyx small, the lobes scarcely acute, shorter than the tube . . 3. *T. Muelleri.*
    Calyx-lobes acuminate, much longer than the tube . . . . . 4. *T. Hookeri.*
Stems leafy or leafless. Stipules spinescent, recurved . . . . . 5. *T. aculeata.*
Stems leafless. Stipules minute, inconspicuous. Flowers small.
  Stems terete . . . . 6. *T. egena.*
  Stems flattened . . . . `. . 7. *T. sulcata.*

1. **T. retusa,** *R. Br. in Ait. Hort. Kew. ed.* 2, iv. 269. A tall, glabrous, often somewhat glaucous shrub, the branches angular and more or less sulcate. Leaves from broadly obovate to narrow cuneate-oblong, sometimes all under ¾ in., sometimes all above 1 in. long, obtuse emarginate or minutely mucronate, thickly coriaceous, nearly sessile or articulate on a short thick petiole. Pedicels shorter or longer than the calyx, rigid, with obtuse bracteoles near the middle, thickened and sulcate under the calyx after flowering. Calyx 3 to nearly 4 lines long, with 4 very short broad teeth, the lowest rather more prominent. Petals red or rarely white, 1 to 1½ in. long, all narrow, on short claws, and of nearly equal length. Ovary stipitate, glabrous, with 10 to 15 ovules. Pod 1½ to nearly 2 in. long and about 5 lines broad, oblique, with a rigid hooked point, the stipes longer than the calyx ; valves very coriaceous.—DC. Prod. ii. 118 ; Bot. Reg. t. 383 ; Bot. Mag. t. 2334 ; Lodd. Bot. Cab. t. 526 ; Meissn. in Pl. Preiss. i. 88 ; *Rafnia retusa,* Vent. Jard. Malm. t. 53 ; *T. glauca,* Sims, Bot. Mag. t. 2088 ; Bot. Reg. t. 859 ; Lodd. Bot. Cab. t. 644 ; DC. Prod. ii. 118.

S. Australia. Memory Cove, *R. Brown ;* from Port Lincoln around Spencer's Gulf to Lake Torrens and Flinders Range, *F. Mueller, Whittaker, Warburton,* and others.

W. Australia. From the Great Australian Bight to King George's Sound and Stirling Range and thence to Geographe Bay and Swan River, *Fraser, Collie, Drummond, n.* 278, *Preiss, n.* 1082 *and* 1083, *Oldfield,* and others ; Dirk Hartog's Island, *Milne.*

The dissimilarity of the anthers is not so striking as in other species, yet on examination 5 will always be found longer and attached by the base, and 5 attached near the middle, all are much narrower than in *Bossiæa.*

2. **T. Drummondii,** *Benth.* Stems in our specimens ascending from a thick rhizome, not 6 in. high, simple or branched, glabrous, angular-striate. Leaves ovate or elliptical-oblong, mucronate, ¾ to 1 in. long, rigid, glabrous, glaucous, articulate on a petiole of 1 to 4 lines. Stipules minute. Pedicels 3 to 4 lines long, the bracteoles rather below the middle. Calyx about 2 lines long, the 4 teeth or lobes nearly equal, much shorter than the tube, the uppermost broad and obtuse, the lateral ones rather smaller and almost acute, the lowest rather longer. Flowers not seen. Pod above 1 in. long, 4 to 5 lines broad, on a very short stipes ; valves coriaceous, very convex.

W. Australia, *Drummond.*

3. **T. Muelleri,** *Benth.* A glabrous shrub or undershrub, with a thick stock and ascending or erect virgate stems of 1 to 2 ft., more or less sulcate-striate. Leaves few, the lower ones narrow-oblong, the upper ones linear, 1 to 1½ in. long or in some specimens attaining 3 in., obtuse or with short recurved points, flat or concave, coriaceous, continuous or rarely when narrow showing a tendency to an articulation. Stipules minute. Pedicels solitary or 2 together, the bracteoles above the middle ½ to 1 line long. Calyx 2 to 2½ lines long, the 4 lobes nearly equal and shorter than the tube, the uppermost broad, the lowest rather longer. Standard broad, reflexed, more than twice as long as the calyx ; keel broad and nearly as long as the standard ; wings much narrower and shorter. Ovary stipitate, with about 6 ovules. Pod oblong, oblique, about ¾ in. long and 5 lines broad, the stipes longer than the calyx ; valves convex, almost turgid.—*Bossiæa stenophylla,* F. Muell. Fragm. i. 9.

Queensland. Wide Bay, *Bidwill, Leichhardt.*
N. S. Wales. Hawkesbury river, *R. Brown;* Cugeegong river, *A. Cunningham;*
New England, near Tenterfield, *C. Stuart.*
Victoria. Murray river, *Prince Paul Wilhelm;* Wimmera river and Mount Arapiles,
near Lake Hindmarsh, *Dallachy;* Milton, near Port Phillip, *Weidenbach.*

4. **T. Hookeri,** *Benth.* A tall slender shrub with erect branches, glabrous or slightly pubescent. Leaves rather crowded, linear-terete or almost filiform with a short recurved point, 1 to 3 in. long or even more, glabrous, usually
articulate near the middle, showing that the lower portion is a petiole, with
which the leaflet is occasionally continuous. Stipules minute. Pedicels filiform, often 1 in. long, with small bracteoles near the top. Calyx 5 to 6 lines
long, the 4 lobes acuminate, longer than the tube, the uppermost with an inflexed point, the lateral ones shorter, the lowest considerably the longest.
Petal-claws short; standard broad, 6 to 7 lines long; keel about as long;
wings much smaller. Ovary stipitate, with about 6 ovules. Pod ¾ to above
1 in. long, about 4 lines broad, on a stipes longer than the calyx, oblique
when young, but scarcely so when full grown; valves coriaceous, slightly
convex.—*Nematophyllum Hookeri,* F. Muell. in Hook. Kew. Journ. ix. 20.

N. Australia. N.W. coast, *Bynoe;* Islands of the Gulf of Carpentaria, *R. Brown;*
Hooker's and Sturt's Creeks, *F. Mueller.*

5. **T. aculeata,** *Benth.* A low rigid shrub or undershrub, with a thick
rhizome and numerous simple or branched often flexuose stems, rarely exceeding 1 ft. in height, sulcate-striate and glabrous or nearly so. Leaves few or
sometimes none, the lower ones obovate or oblong, the uppermost linear, ½ to
1 in. long, rigid and glabrous. Stipules rigid, recurved, prickly, and often 2
lines long. Pedicels short, the bracteoles about the middle. Calyx about 3
lines long, the 4 lobes of nearly equal length, rather shorter than the tube,
the uppermost rather broader and the lowest rather longer than the lateral
ones. Standard broad, reflexed, twice as long as the calyx; keel about as
long as the standard; wings shorter. Ovary stipitate, with about 6 ovules.
Pod about ¾ in. long, on a stipes longer than the calyx; valves coriaceous,
but nearly flat.—*Bossiæa aculeata,* F. Muell. Fragm. ii. 120.

W. Australia, *Drummond, n. 141, and 2nd Coll. n. 101;* Culjong river, *Oldfield.*

6. **T. egena,** *Benth.* A tall glabrous leafless shrub, with numerous
erect terete sulcate branches, the nodes bearing only minute protuberances.
Pedicels solitary or 2 together, rarely 1 line long, with small orbicular bracteoles close under the calyx. Calyx 1½ lines long or rather more, with 5
nearly equal broad obtuse teeth, much shorter than the tube, the lowest rather
the longest. Petals on rather long claws, scarcely twice as long as the calyx,
the standard rather longer than the others. Ovary shortly stipitate, with 6
to 8 ovules. Pod nearly sessile, obliquely oblong, 6 to 8 lines long and about
4 broad, the valves very coriaceous and slightly convex.—*Daviesia egena,* F.
Muell. in Trans. Vict. Inst. 118; *Bossiæa egena,* F. Muell. in Hook. Kew
Journ. viii. 43; Fragm. iii. 94.

N. Australia. Hooker and Sturt's Creeks, *F. Mueller.*
N. S. Wales. Deserts of the Murray and Darling, *Victorian Expedition;* northward
to Mount Aiton, *A. Cunningham.*
Victoria. Clayey and sandy, somewhat saline deserts of the Murray and Murrum-
bidgee, *F. Mueller.*
S. Australia. Barren bushy places along Spencer's Gulf to Lake Torrens, *F. Mueller.*

7. **T. sulcata,** *Benth.*   A tall rigid glabrous leafless shrub, with numerous divaricate branches much flattened, sulcate, striate, and often ending in stiff thorns, notched at the nodes.   Pedicels very short; bracteoles under the calyx, concave, sometimes nearly 1 line long.   Calyx 1½ lines long, somewhat rigid, the 4 lobes broad, obtuse, nearly as long as the tube, the lateral ones rather smaller than the others.   Standard scarcely twice as long as the calyx; lower petals shorter.   Ovary almost sessile, with 2 to 4 ovules.   Pod sessile, very obliquely ovate, ½ to ¾ in. long, with a hooked point, valves thickly coriaceous and convex.—*Bossiæa sulcata,* Meissn. in Pl. Preiss. i. 81; *B. Rossii,* F. Muell. Fragm. iii. 94, 168.

Victoria. Mallee scrub, from the junction of the Murrumbidgee and Murray to Lake Hindmarsh, *F. Mueller.*

W. Australia, *Drummond, n.* 107, 108, *and* 144, *J. S. Roe;* sandy plains of the Avon, *Preiss, n.* 1028; clayey places, Phillips river, *Maxwell.*

23. HOVEA, R. Br.

(Poiretia, *Sm.;* Plagiolobium, *Sweet;* Platychilum, *Delaun.*)

Calyx upper lobes united into a broad truncate upper lip, entire or slightly emarginate, the 3 lower ones much smaller, lanceolate. Petals clawed; standard nearly orbicular, emarginate; wings shorter, obliquely obovate, auriculate on the inner side at the base; keel much shorter, slightly incurved, obtuse, the petals slightly cohering.   Stamens all united in a sheath open on the upper side and sometimes split also on the lower side, or rarely the uppermost stamens, and very rarely the lowest free; anthers alternately long and erect and short and versatile.   Ovary sessile or stipitate with 2 or rarely more ovules; style incurved, rather thick; stigma terminal.   Pod sessile or stipitate, turgid, very obliquely globular or ovoid, the valves at length entirely separating.   Seeds reniform on short funicles, strophiolate.— Shrubs.   Leaves alternate, simple, entire or prickly-toothed, glabrous above, often tomentose underneath.   Stipules setaceous, minute or none.   Flowers blue or purple, in axillary clusters or very short racemes or rarely solitary.

The genus is entirely Australian, and easily recognized by the habit, the calyx, the colour of the flower and short turgid pod, although it is closely connected with *Templetonia,* through *H. longipes.*

Ovary and pod sessile (Eastern species).
 Pod glabrous or slightly pubescent.
  Stems decumbent. Lower leaves ovate, upper ones lanceolate
   or linear . . . . 2. *H. heterophylla.*
  Stems erect. Leaves numerous, long, mostly narrow-linear . 1. *H. linearis.*
 Pod tomentose or villous.
  Leaves oblong, lanceolate or linear, obtuse at both ends . . 3. *H. longifolia.*
  Leaves narrowed at both ends . . . . 4. *H. acutifolia.*
Ovary and pod stipitate, always glabrous.
 Leaves with numerous oblique parallel veins. Calyx lower lobes
  nearly as long as the upper (East tropical species) . . . 5. *H. longipes.*
 Leaves smooth or reticulate the primary veins distant (when con-
  spicuous) transverse or arcuate. Calyx lower lobes much
  shorter than the upper (Western species).
  Leaves under ¼ in. long, narrow, entire. Branches spines-
   cent . . . . 6. *H. acanthoclada.*

Leaves mostly above 1 in. long. No thorns.
Leaves ovate or lanceolate, prickly-toothed. Upper stamen
   free . . . . 7. *H. chorizemifolia.*
Leaves entire or slightly crisped, the margins flat or recurved.
   Upper stamen united with the rest.
   Tall shrub. Leaves elliptical to narrow-lanceolate. Calyx
     lower lobes at least ½ as long as the upper. Ovules 2. 8. *H. elliptica.*
   Stems low or not much branched. Leaves ovate to
     linear. Calyx lower lobes very small. Ovules usually
     3 or more . . . . 9. *H. trisperma.*
Leaves ½ to 1 in. long, narrow, rigid, the margins much re-
   volute.
   Leaves obtuse with a small scarcely pungent point . . . 10. *H. stricta.*
   Leaves pungent-pointed . . . . 11. *H. pungens.*

1. **H. linearis,** *R. Br. in Ait. Hort. Kew. ed.* 2, iv. 275. Apparently a
low shrub, with erect not much-branched stems, of 1 to 2 ft., closely tomentose
or pubescent or at length nearly glabrous. Leaves nearly all narrow-linear,
1½ to 3 in. long, obtuse with a small point, coriaceous with recurved margins,
more or less reticulate, quite glabrous or pubescent underneath, the lower
ones occasionally lanceolate or oblong-elliptical. Flowers rather small, solitary or 2 or 3 in each axil, on very short pedicels. Calyx about 2 lines long,
more or less silky-hairy. Staminal sheath open on the upper side only. Ovary
glabrous. Pod sessile, glabrous, about 4 lines broad and long.—DC. Prod.
ii. 115 ; Lodd. Bot. Cab. t. 1222 ? ; *Poiretia linearis,* Sm. in Trans. Linn.
Soc. ix. 304.

N. S. Wales. Port Jackson, *R. Brown, Sieber, n.* 375, and others ; and northward
to Clarence river, *Beckler.*

This very much resembles the long linear-leaved varieties of *H. longifolia.* It appears to
be a smaller plant, the leaves are not so coriaceous, the flowers smaller and the pod shorter,
and always quite glabrous. The specimens figured, Bot. Reg. t. 463, and in Paxt. Mag. xii.
75, seem to connect this also with the following.

2. **H. heterophylla,** *A. Cunn. in Hook. f. Fl. Tasm.* i. 93, *t.* 15. Very
closely allied to *H. linearis,* and perhaps a variety only, although usually very
different in aspect. Stems decumbent or prostrate at the base, with ascending or erect slender branches, clothed with a short close tomentum. Lower
leaves ovate, intermediate ones lanceolate, the uppermost narrow, or sometimes
quite linear, and rarely above 1½ in. long, obtuse or almost acute, the margins
often recurved, reticulate and glabrous or slightly hairy underneath. Flowers
of *H. longifolia* or rather larger. Calyx 2 to 3 lines long, the lower lobes
sometimes nearly as long as the upper lip, but very narrow. Ovary and pod
of *H. longifolia,* but quite glabrous or shortly pubescent.

Queensland. Stradbrooke Island, *Fraser, A. Cunningham ;* Sandstone Hills, towards
Brisbane, *Leichhardt.*

N. S. Wales. Paramatta, *Woolls ;* grassy valleys, near Macquarrie river, *A. Cunningham ;* Twofold Bay, *F. Mueller.*

Victoria. Glenelg river, *Robertson ;* Mounts Dandenong and Disappointment, *F.
Mueller ;* in the Tattiara country, *J. E. Woods.*

Tasmania. Abundant in dry and stony places in various parts of the island, *J. D.
Hooker.*

3. **H. longifolia,** *R. Br. in Ait. Hort. Kew. ed.* 2, iv. 275. A stout erect
shrub, attaining 8 to 10 ft. ; branches usually erect, softly tomentose or the

smaller varieties bushy and stunted. Leaves oblong-lanceolate or linear, obtuse, with or without a small callous point, all under ¾ in. long in some varieties, in others all above 2 in., thickly coriaceous, with flat recurved or revolute margins, glabrous above and smooth and shining or densely reticulate, the primary veins when conspicuous transverse or arcuate, more or less rusty-tomentose underneath. Flowers very shortly pedicellate in axillary clusters, which sometimes grow out into interrupted spikes or racemes, or rarely solitary. Bracts and bracteoles small, usually obtuse. Calyx tomentose, 2 to 3 lines long; lobes all short, the upper broad truncate lip not much longer than the lower lobes, which are usually more obtuse than in *H. linearis.* Standard twice as long as the calyx. Staminal tube open on the upper side only. Ovary tomentose. Pod sessile, 4 to 6 lines broad, softly rusty-tomentose or almost villous.

N. Australia. Port Essington, *A. Cunningham* (specimens imperfect).
Queensland. Near Mount Owen, *Mitchell;* Newcastle Range and Suttor river, *F. Mueller;* Shoalwater Bay passages, *R. Brown;* Moreton Bay, *A. Cunningham;* near Warwick, *Beckler;* Ipswich, *Nernst.*
N. S. Wales. Port Jackson to the Blue Mountains, *R. Brown, Sieber,* and others ; Macquarrie and Lachlan rivers, *A. Cunningham;* northward to Clarence river, *Beckler,* and New England, *C. Stuart,* and southward to Twofold Bay, *F. Mueller.*
Victoria. Gipps' Land, Mitta Mitta river, Mount Butler, etc., ascending to 5000 ft., *F. Mueller.*
Tasmania. Port Dalrymple and Bass's Straits, *R. Brown;* abundant throughout the island, *J. D. Hooker.*
S. Australia. Rocky sides of Mount Remarkable and adjacent hills, *F. Mueller.*
The following forms, usually considered as distinct species, pass into each other by such insensible gradations, that I am unable to distinguish them otherwise than as varieties.
a. *normalis.* Leaves linear with revolute margins, usually 1½ to 3 in. long, but in some specimens shorter. Flower-clusters often shortly racemose.—*H. longifolia,* Bot. Reg. t. 614; Lodd. Bot. Cab. t. 994 ; *H. racemulosa,* Benth. in Bot. Reg. 1843, t. 4 (the supposed Swan River origin probably a garden mistake).—Port Jackson, *R. Brown, Sieber, n.* 376, and others, also *R. Brown's* Queensland specimens. Some specimens are very difficult to distinguish from *H. linearis.*
b. *rosmarinifolia.* Leaves linear, very obtuse and much revolute, but much shorter than in the normal form, and very much reticulate, otherwise closely connecting the short-leaved specimens of *a* and *c.*—*H. rosmarinifolia,* A. Cunn. in Field, N. S. Wales, 348.—Blue Mountains.
c. *lanceolata.* Leaves oblong, or lanceolate, or broadly linear, with flat or recurved margins, ¾ to 3 in. long, often rather thick, closely, but often densely tomentose underneath. —*H. lanceolata,* Sims, Bot. Mag. t. 1624 ; Bot. Reg. t. 1427 (a weak slender form ?) ; DC. Prod. ii. 115 ; *H. apiculata,* A. Cunn., and *H. mucronata,* A. Cunn. in G. Don, Gen. Syst. ii. 126 ; *H. purpurea,* Lodd. Bot. Cab. t. 1457 ; Maund, Botanist, t. 72 ; Hook. f. Fl. Tasm. i. 93, but scarcely of Sweet ; *H. Beckeri,* F. Muell. in Linnæa, xxv. 391.—Extends over the whole range of the species and the most common form. The more northern specimens have often elongated interrupted inflorescences, and the tomentum of the under side of the leaves thin and pale-coloured ; the Tasmanian and Victorian ones, especially those from high elevations, have usually smaller much more revolute leaves, and the tomentum more rusty.
d. *pannosa.* Leaves linear or oblong, rather large and very coriaceous, the tomentum soft and dense, often almost woolly. Flowers rather large, in close clusters, the calyx densely hirsute, the lower lobes narrow and acute. Pod often rusty-woolly.—*H. purpurea,* Sweet, Fl. Austral. t. 13 ; Bot. Reg. t. 1423 ; *H. villosa,* Lindl. in Bot. Reg. t. 1512 ; *H. pannosa,* A. Cunn. in Bot. Mag. t. 3053 ; *H. lanigera,* Lodd. in Steud. Nom. Bot. ed. 2 ; *H. ramulosa,* A. Cunn. in Bot. Reg., under n. 4 (a narrow-leaved form connecting it with the normal variety).—Blue Mountains and Queensland. The Port Essington specimens are also nearest to this variety.

4. **H. acutifolia,** *A. Cunn. in G. Don, Gen. Syst.* ii. 126. A tall shrub, allied to the var. *pannosa* of *H. longifolia*, but with the leaves always narrowed at both ends. Branches densely tomentose-villous. Leaves elliptical-oblong or lanceolate, acuminate or acute, narrowed at the base, mostly 2 to 3 in. long, and the larger ones 1 in. broad in the middle, the margins slightly recurved, densely but minutely reticulate above, loosely tomentose-villous underneath, the primary veins few, nearly transverse or arcuate. Flowers in clusters of 2 or 3, like those of *H. longifolia*, var. *pannosa* or rather larger. Ovary sessile, tomentose-villous. Pod not seen.

Queensland. Brisbane river, *Fraser, A. Cunningham, F. Mueller*, and others; Pine river, *Fitzalan*.

5. **H. longipes,** *Benth. in Hueg. Enum.* 37. A tall shrub, resembling at first sight some forms of *H. longifolia*, but readily distinguished by the venation of the leaf as well as by the flower and fruit. Branches, under side of the leaves and calyxes hoary or slightly rusty, with a close or soft tomentum. Leaves from oval-elliptical to oblong or lanceolate, obtuse with a minute callous point, ¾ to 1½ in. long, coriaceous with slightly recurved margins, glabrous above and marked with numerous oblique parallel slightly reticulate primary veins. Flowers usually 2 or 3 together, each on a pedicel longer than the calyx, with 2 minute bracteoles near the end. Calyx very broadly campanulate, scarcely 2 lines long, the lobes or teeth all very short, the upper lip broad and truncate, but scarcely exceeding the lower lobes. Standard very broad, twice as long as the calyx. Ovary quite glabrous. Pod 4 or 5 lines broad and long, very coriaceous, quite glabrous, on a stipes from the length of the calyx to twice as long.—*H. leiocarpa*, Benth. Mitch. Trop. Austr. 289.

Queensland. Keppel Bay, *R. Brown;* dry forest and sheltered valleys, Mantuan Downs and Maranoa river, *Mitchell;* Burdekin river, *F. Mueller;* edge of the scrub, near Rockhampton, *Thozet;* Fitzroy river, *Bowman*.

N. S. Wales. Clarence river, *Beckler*.

6. **H. acanthoclada,** *F. Muell. Fragm.* iv. 15. A rigid spreading divaricately branched shrub, with the habit of some of the spinescent *Bossiæas;* branches closely rusty-tomentose, the smaller ones ending in slender thorns. Leaves scattered or clustered, rarely 3 lines long, linear-oblong, obtuse, with revolute margins, glabrous above, rusty-tomentose underneath. Flowers usually solitary, on short pedicels. Bracteoles minute. Calyx minutely tomentose, turbinate-campanulate, about 2 lines long, the upper lip truncate, with acuminate angles, the lower lobes or teeth much shorter. Petals and ovary not seen. Staminal sheath open on the upper side. Pod obliquely ovoid, glabrous, on a stipes as long as the calyx-tube.—*Daviesia acanthoclada*, Turcz. in Bull. Mosc. 1853, i. 262.

W. Australia, *Drummond, 5th Coll. n.* 96; Phillips river, *Maxwell*. This species connects in some measure *Hovea* with *Bossiæa*, but the pod is entirely that of the former genus.

7. **H. chorizemifolia,** *DC. Prod.* ii. 116. Usually a stout branching shrub of several ft., although occasionally the stems are nearly simple from a thick stock; branches rusty-tomentose, villous or nearly glabrous. Leaves from ovate to lanceolate, pungent-pointed, sinuate and prickly-toothed, often

undulate, 1½ to nearly 3 in. long, coriaceous, reticulate and usually glabrous. Flowers 2 to 6 together, rather small, on short pedicels. Bracteoles subulate. Calyx silky-villous, 2 to 3 lines long; upper lip very large and broad, lower lobes very small. Ovary glabrous, with 2 ovules. Staminal sheath open on the upper side, and sometimes also on the under side, the upper stamens usually quite free. Pod glabrous, about as broad as long, on a short stipes. —Bot. Reg. t. 1524 , Maund, Botanist, t. 130 ; *Plagiolobium chorizemifolium,* Sweet, Fl. Austral. t. 2 ; Meissn. in Pl. Preiss. i. 80 ; *P. ilicifolium,* Sweet, l. c. in a note ; Meissn. l. c.; *H. ilicifolia,* A. Cunn. in Bot. Reg. 1844, t. 58 (a narrow-leaved branch).

W. Australia. Common from King George's Sound to Swan River, *R. Brown, Fraser, Drummond,* 1st *Coll. and n.* 181, *Preiss, n.* 1052, 1058, 1060, and others. I have in vain endeavoured to distribute our numerous specimens into appreciable varieties.

8. **H. elliptica,** *DC. Prod.* ii. 115. A tall shrub, attaining sometimes 8 to 10 ft., the branches rather slender, rusty with a minute tomentum. Leaves from ovate-elliptical to narrow-lanceolate, usually narrowed at both ends, obtuse or emarginate, with a minute point or rarely almost acute, 2 to 3 in. long on the larger branches, 1 to 2 in. on the smaller ones, not very coriaceous, reticulate with the primary veins nearly transverse, glabrous above, pale underneath with an appressed pubescence, the petioles often rather long. Flowers blue, in axillary clusters or short racemes, the pedicels often as long as or longer than the calyx. Calyx 2½ or rarely 3 lines long ; upper lip as long as the tube and very broad, lower lobes about half as long. Standard fully twice as long as the calyx. Staminal sheath open on the upper side and scarcely splitting on the lower side. Ovary glabrous with 2 ovules. Pod stipitate, glabrous, about as long as broad.— Lodd. Bot. Cab. t. 1450 ? (a starved specimen?) ; *Poiretia elliptica,* Sm. in Trans. Linn. Soc. ix. 305 ; *Platychilum Celsianum,* Herb. Amat. t. 187, according to G. Don, Gen. Syst. ii. 127 ; DC. Prod. ii. 116 ; *Goodia simplicifolia,* Spreng. Syst. Cur. Post. 267 ; *Hovea Celsi,* Bonpl. Jard. Malm. t. 51 ; DC. Prod. ii. 115 ; Bot. Reg. t. 280 ; Bot. Mag. t. 2005 ; Maund, Botanist, t. 40 ; Lodd. Bot. Cab. t. 1488 ; Meissn. in Pl. Preiss. i. 79, and ii. 221 ; Paxt. Mag. iii. 241 with a fig. ; *H. latifolia,* Lodd. Bot. Cab. t. 30 (?) (from the figure).

W. Australia. King George's Sound, *R. Brown, Menzies, Drummond,* and others ; Sussex and Plantagenet districts, *Preiss, n.* 1053 *and* 1055 ; Vasse river, *Oldfield.*

9. **H. trisperma,** *Benth. in Hueg. Enum.* 37. An undershrub or little-branched shrub, with ascending or almost erect stems, of 1 to 2 ft., glabrous or pubescent. Lower leaves ovate or elliptical, ¾ to 1½ in. long, upper ones lanceolate or linear, 1 to 3 in. long, obtuse with a short point or acute, glabrous and reticulate above, slightly tomentose or loosely pubescent underneath. Flowers rather large, usually 2 or 3 together on short pedicels. Bracts and bracteoles subulate. Calyx silky-villous, 3 to 4 lines long, the tube very short, the upper lip very large and rounded-truncate, the lower lobes not ⅓ as long. Staminal sheath usually open on the lower as well as on the upper side, leaving the lowest stamen free, and rarely the uppermost one also. Ovary glabrous with 2 or more ovules, usually 3 in the normal form. Pod fully as broad as long, glabrous, on a stipes from half as long to

fully as long as the calyx.—Meissn. in Pl. Preiss. i. 79 ; *H. Manglesii*, Lindl. Bot. Reg. 1838, t. 62 (said to differ in the sessile ovary, but it is figured as stipitate).

W. Australia. King George's Sound to Swan River, *R. Brown, Huegel, Drummond, 1st Coll. and n.* 179 *and* 180, *Preiss, n.* 1051, and others.

Var. *crispa.* More slender. Leaves from ovate to lanceolate, often slightly cordate, the margins crisped and sometimes almost crenulate. Calyx rather more hairy. Ovules 2 to 4. —*H. crispa*, Lindl. Bot. Reg. 1839, Misc. 19 ; Meissn. in Pl. Preiss. i. 79.—Swan River, *Drummond, 1st Coll. ;* Darling range, *Preiss, n.* 1063. *H. splendens,* Paxt. Mag. x. 103, from the figure given, must be very near this variety.

Var. *grandiflora.* Larger and nearly glabrous. Leaves more coriaceous, quite entire, strongly reticulate. Flowers large. Calyx often near 5 lines long. Ovules often 6.— *H. elliptica*, Meissn. in Pl. Preiss. i. 79, not of DC.—Swan River, *Drummond, 1st Coll. and n.* 178.

10. **H. stricta,** *Meissn. in Pl. Preiss.* i. 79, *and in Bot. Zeit.* 1855, 30. Stems little-branched, erect, rigid, 1 to 1½ ft. high, loosely tomentose-villous. Leaves lanceolate or linear, or the lower ones ovate-lanceolate, often erect, ½ to 1 in. long, obtuse with a short rigid point, the margins revolute, rounded or cordate at the base, rather rigid, reticulate and glabrous above, hoary-to-mentose or loosely villous underneath. Flowers 2 or 3 together on short pedicels. Bracts and bracteoles subulate. Calyx 2 to 3 lines long, silky or loosely villous, upper lip large and broad, lower lobes much shorter narrow and acute, but not so small as in *H. trisperma.* Standard ½ in. diameter. Staminal sheath often splitting on the lower side as well as the upper, but I have never seen any stamen free. Ovary stipitate, glabrous, with 2 ovules. Pod not seen.

W. Australia. Sandy woods, Sussex district, *Preiss, n.* 1057 ; Vasse river, *Oldfield ;* Swan River and between Moore and Murchison rivers, *Drummond, 6th Coll. n.* 27.

11. **H. pungens,** *Benth. in Hueg. Enum.* 37, *and in Hueg. Arch. Bot. t.* 7. An erect rigid not much-branched shrub of 1 to 2 ft., the branches softly pubescent or villous. Leaves linear or lanceolate, ½ to 1 in. long, very spreading, rigidly coriaceous with pungent points, the margins much revolute, reticulate above, glabrous or the upper ones sprinkled with a few hairs. Stipules setaceous, often more conspicuous than in other species, but sometimes minute or deciduous. Flowers 1 to 3 together on short pedicels. Bracteoles subulate. Calyx 2½ to 3 lines long, silky-villous or hirsute, the upper lip very broad, the lower lobes rather shorter. Standard about ½ in. diameter. Staminal sheath open on the upper side only. Ovary glabrous, with 2 or very rarely 3 ovules. Pod very oblique, rather longer than broad, the stipes very nearly as long as the calyx.—Meissn. in Pl. Preiss. i. 78 ; Maund, Botanist, t. 164 ; Paxt. Mag. vi. 101, and x. 51 with figs.

W. Australia. Swan River, *Drummond, 1st Coll. and n.* 177, *Preiss, n.* 1054, *Oldfield* and others ; Gordon river, *Oldfield ;* along the coast to East Mount Barren, *Maxwell.* Var. *ulicina.* Lower leaves rather broader, calyx rather more hairy.—*H. ulicina*, Meissn. in Bot. Zeit. 1855, 30.—Between Moore and Murchison rivers, *Drummond, 6th Coll. n.* 26. Some specimens are, however, scarcely distinguishable from the common *H. pungens*, even as a variety.

24. GOODIA, Salisb.

Calyx 2 upper lobes united in a 2-toothed upper lip, 3 lower ones equal.

Petals clawed; standard orbicular; wings narrow; keel **broader**, incurved, obtuse.   Stamens all united in a sheath open on the upper side; anthers all versatile, alternately smaller.   Disk annular between the stamens and ovary. Ovary stipitate, with 2 to 4 ovules; style subulate, incurved; stigma small, terminal.   Pod stipitate, flat, valves thin with a nerviform edge.   Seeds strophiolate.—Shrubs.   Leaves pinnately 3-foliolate, with entire leaflets.   Flowers yellow mixed with purple, in terminal or leaf-opposed racemes.   Stipules bracts and bracteoles membranous, but so deciduous as to be rarely seen but in very young branches or racemes.

The genus is limited to Australia, and although nearly allied to *Bossiæa* in its flowers and fruit, has the inflorescence of *Crotalaria*, and differs from all other *Genisteæ* in its pinnately trifoliolate leaves.   It would therefore be equally well placed under *Galegeæ*, next to *Ptychosema*.

Glabrous or the young shoots minutely pubescent . . . . 1. *G. lotifolia.*
Softly pubescent all over . . . . 2. *G. pubescens.*

G. (?) polysperma, DC. Prod. ii. 117, is *Argyrolobium Andrewsianum*, Steud., a South African, not an Australian plant.

G. retusa, Mackay, and *G. subpubescens*, Sweet, in Steud. Nom. Bot. ed. 2, are unpublished garden names, probably of some varieties of *G. lotifolia*.

1. **G. lotifolia,** *Salisb. Parad. Lond. t.* 41.   A tall much-branched shrub, either quite glabrous or the young shoots minutely pubescent and often glaucous.   Leaflets ovate or obovate, very obtuse, $\frac{1}{2}$ to $\frac{3}{4}$ in. long, the lateral ones usually at a considerable distance from the terminal one, the petiole slender. Racemes loose, many-flowered, 2 to 4 in. long.   Calyx $2\frac{1}{2}$ to 3 lines long, the lower lobes linear-lanceolate nearly as long as the tube, the upper lip very broad, more or less 2-toothed at the top.   Standard about twice as long as the calyx, notched, yellow with a purple base; lower petals rather shorter. Pod varying from $\frac{1}{2}$ to nearly 1 in. long and 3 to 4 lines broad, on a stipes much longer than the calyx, the upper suture often dilated, the valves thin with transverse reticulations sometimes very prominent, sometimes scarcely perceptible.—DC. Prod. ii. 117 ; Bot. Mag. t. 958 ; Lodd. Bot. Cab. t. 696 ; Hook. f. Fl. Tasm. i. 97 ; Meissn. in Pl. Preiss. i. 88.

N. S. Wales.   Hastings river, *Fraser.*
Victoria.   In the scrub of the interior, *F. Mueller ;* near Portland, *Robertson.*
Tasmania.   Common in various parts of the island, *J. D. Hooker.*
S. Australia.   Flinders range, Guichen and Rivoli Bays, Crystal Brook, etc., *F. Mueller ;* Onkaparinga range, *Whittaker.*
W. Australia, *Drummond, 5th Coll. n.* 89 ; Hay district, *Preiss, n.* 1014,
G. medicaginea, F. Muell. Fragm. i. 10, the prevailing Continental form, with shorter and smoother pods, passes gradually into the more common Tasmanian form.

2. **G. pubescens,** *Sims, Bot. Mag. t.* 1310.   Most probably a variety of *G. lotifolia*, differing in being softly pubescent all over, the leaflets usually narrower and more approximate at the end of the shorter petiole, the flowers rather smaller.—DC. Prod. ii. 117 ; Hook. f. Fl. Tasm. i. 97.

Victoria.   Dandenong mountains, *F. Mueller.*
Tasmania.   Common throughout the island, *J. D. Hooker.*   It is united with *G. lotifolia* by F. Muell. Fragm. i. 10.

25. CROTALARIA, Linn.

Calyx-lobes nearly equal, or the 2 upper ones and the 3 lower ones more or less united. Standard orbicular or ovate; wings shorter; keel incurved or angled, terminating inwards in a straight or incurved beak. Stamens all united in a sheath, open along the upper side; anthers alternately long and erect and short and versatile. Ovary sessile or stipitate, with 2 or more ovules; style much incurved or suddenly bent inwards, with a longitudinal line of hairs above the middle on the inner side (sometimes very small); stigma terminal. Pod turgid or inflated, continuous inside. Seeds not strophiolate, on slender funicles.—Herbs or shrubs. Leaves simple or digitately compound with 1, 3 or (in species not Australian) 5 or 7 leaflets often marked with pellucid dots. Stipules free from the petiole, occasionally decurrent along the stem, frequently small or wanting. Flowers yellow or blue, in simple terminal racemes, becoming sometimes leaf-opposed, with a bract, often very small, under each pedicel and minute bracteoles adnate to the calyx-tube or just below it.

A very large and well-marked genus, widely dispersed over the tropical and warm regions both of the New and the Old world. Of the 14 Australian species 8 are East Indian, and 2 of these are also abundant in many parts of Africa and America, the remaining 6 are endemic, 3 of them belonging to a group peculiarly Australian.

SERIES I. **Simplicifoliæ.**—*Leaves simple, continuous with the short petiole, the Australian species all herbs or undershrubs.*

Ovary and pod pubescent or villous.
  Leaves ovate. Stipules leafy, semilunar or falcate. Flowers
    blue . . . . 1. *C. verrucosa.*
  Leaves oblong, linear, or rarely obovate. Stipules setaceous or
    none. Flowers yellow.
    Ovules 2. Pod usually 1-seeded, not exceeding the calyx.
      Flowers small . . . . 2. *C. crispata.*
    Ovules numerous. Pod many-seeded, above 1 in. long.
      Flowers rather large . . . . 3. *C. juncea.*
Ovary and pod quite glabrous.
  Upper leaves usually linear. Petals and pod not exceeding the
    calyx.
    Calyx 3 to 4 lines long, silky-pubescent or shortly villous,
      the 2 upper lobes united . . . . 4. *C. linifolia.*
    Calyx 1 in. long, densely hirsute with long spreading hairs,
      the upper lobes free . . . . 5. *C. calycina.*
  Upper leaves broad, oblong-cuneate or rarely almost linear.
    Petals and pod much longer than the calyx.
    Leaves oblong cuneate. Flowers large in loose racemes.
      Calyx 4 to 6 lines; pod 1½ in. long . . . . . . 6. *C. retusa.*
    Leaves oval-elliptical or oval-lanceolate. Flowers numerous
      in dense racemes. Calyx not above 3 lines, pod, under ¾
      in. long . . . . 7. *C. Mitchelli.*

SECT. II. **Unifoliolatæ.**—*Leaves simple, the petiole articulate or geniculate above the middle. Stem shrubby.*

Flowers under ¾ in. long; standard obtuse.
  Leaves pubescent or villous, at least underneath. Stipules none
    or not decurrent . . . . 8. *C. Novæ-Hollandiæ.*
  Whole plant quite glabrous. Stipules decurrent . . . . . 9. *C. crassipes.*
Flowers 1½ in. long or more; standard acute or acuminate . . 10. *C. Cunninghamii.*

SERIES III. **Digitatæ.**—*Leaves all or mostly compound with* 3 *rarely* 5 *digitate leaflets. Herbs or shrubs.*

Ovules 2. Pod small, as broad as long. Herb with small flowers 11. *C. trifoliastrum.*
Ovules many. Pod oblong, much longer than the calyx.
  Ovary and pod sessile or nearly so.
    Calyx deeply lobed. Standard almost acute, slightly exceed-
      ing the calyx. Pod hirsute with spreading hairs . . . 12. *C. incana.*
    Calyx divided to the middle. Standard broad and obtuse,
      much longer than the calyx. Pod pubescent, tomentose
      or almost glabrous . . . . 13. *C. dissitiflora.*
  Ovary and pod on a long stipes. Flowers large.
    Leaflets 3. Standard acute . . . . 14. *C. laburnifolia.*
    Leaflets usually 5. Standard very obtuse . . . . . . 15. *C. quinquefolia.*

SERIES I. SIMPLICIFOLIÆ.—Leaves simple, continuous with the short petiole, the Australian species all herbaceous or undershrubs.

1. **C. verrucosa,** *Linn.; DC. Prod.* ii. 125. A stout erect minutely pubescent annual of 1½ to 3 ft.; branches divaricate with prominent angles almost winged. Leaves usually ovate-rhomboidal 2 to 4 in. long, but passing sometimes into ovate-acuminate or almost lanceolate and 5 or 6 in. long, always very obtuse. Stipules semilunar or falcate, horizontally spreading. Flowers pale-blue, in loose terminal or leaf-opposed racemes. Calyx about 4 lines long, the lobes acuminate, longer than the tube, all free, the lowest rather the narrowest. Standard broad, above ½ in. diameter. Ovary sessile, very villous all over or on the inner side, with above 20 ovules. Pod oblong, villous, 1½ to 2 in. long.—Wight, Ic. t. 200; F. Muell. Fragm. iii. 54.

N. Australia. Victoria river and stony hills and grassy banks on the Wickham river, *F. Mueller.*

Queensland. Endeavour river, *R. Brown;* Cape Upstart, *M'Gillivray;* Bowen river, *Bowman* Rockhampton, *Thozet;* Edgecombe Bay, *Dallachy.*

The species is common in East India, and is now spread over many parts of tropical Africa and America.

2. **C. crispata,** *F. Muell. Herb.* A low much-branched softly villous herb, the stems diffuse or ascending and not exceeding 1 ft. Leaves from obovate-oblong to narrow oblong-cuneate, or broadly linear, very obtuse, ½ to 1 in. long, villous on both sides. Flowers small, few, in short loose terminal racemes. Bracts and bracteoles minute, ovate-acute or lanceolate, villous outside, glabrous inside. Calyx about 3 lines long, deeply cleft, the 3 lower lobes shortly united, the 2 upper ones broader, all lanceolate, very glabrous and sometimes viscous inside, villous outside, the margins often recurved and crisped after flowering. Petals scarcely exceeding the calyx. Ovary sessile, very villous, with 2 ovules. Pod ovoid, villous, scarcely exceeding the calyx. Seed usually solitary, black and shining.

N. Australia. Islands of the Gulf of Carpentaria, *R. Brown;* Victoria, Fitzmaurice and Baines rivers, *F. Mueller.*

This plant is referred by F. Mueller, Fragm. iii. 55, to *C. ramosissima,* Roxb., which it resembles in many respects, but which, in its large flowers and broad reflexed viscous bracts, is nearer to *C. lunulata,* Heyne. Both these species are allied to *C. paniculata,* and *C. crispata* is undoubtedly connected with them, although rather more distinct from all, than they are from each other.

3. **C. juncea,** *Linn.; DC. Prod.* ii. 125. An erect annual, attaining
N 2

many feet, with few, erect, sulcate-striate, silky-pubescent branches. Leaves simple, nearly sessile, oblong or linear, obtuse, 1 to 3 in., or when narrow, 4 in. long or more, glabrous above or nearly so, pubescent underneath. Flowers rather large, yellow, not numerous, in a long terminal raceme. Calyx tomentose, 4 to 6 or even 7 lines long, deeply divided into narrow-lanceolate nearly equal lobes, the 2 upper ones truncate or hooked at the top. Petals slightly exceeding the calyx, the standard usually pubescent. Ovary sessile, villous, with about 20 ovules. Pod above 1 in. long, densely clothed with a rusty tomentum.—Andr. Bot. Rep. t. 422; Bot. Mag. t. 490; W. and Arn. Prod. Fl. Ind. 185 (with the synonyms given); F. Muell. Fragm. iii. 51.

N. Australia. Upper Victoria river, Hooker's and Sturt's Creeks, *F. Mueller.*
Queensland. Broad Sound, *R. Brown;* Logan river, *Fraser;* Port Denison, *Bowman;* Rockhampton, *Thozet, Dallachy;* Wide Bay, *Leichhardt.*
The species is common in East India, where it is much cultivated for the fibrous bark, used as a substitute for hemp under the name of *Sunn.*

4. **C. linifolia,** *Linn. f.; DC. Prod.* ii. 128. A perennial with a thick rhizome, or sometimes annual, exceedingly variable in aspect, usually silky-pubescent or villous, sometimes clothed with long spreading hairs or with a close or woolly white tomentum, often drying black. Stems erect or ascending, from a few inches to 1½ ft. high. Leaves simple, the lower ones, or nearly all in the smaller forms, obovate or oblong, obtuse and mostly under ½ in. long, the upper ones, or nearly all in the elongated varieties, narrow-oblong or linear, 1 to 2 in. long, obtuse or almost acute. Flowers small, yellow, in loose terminal racemes. Calyx 3 to 4 lines long, divided nearly to the base into 2 lips, the upper one 2-toothed, the lower one 3-lobed to the middle. Petals not exceeding the calyx. Ovary sessile, glabrous, with 10 to 20 ovules. Pod ovoid-globular, scarcely exceeding the calyx.—W. and Arn. Prod. Fl. Ind. 190; Benth. in Hook. Lond. Journ. ii. 569; F. Muell. Fragm. iii. 55; *C. stenophylla,* Vog.; Benth. l. c.; *C. melanocarpa,* Wall.; Benth. l. c.

N. Australia. Islands of the Gulf of Carpentaria, *R. Brown;* N. W. Coast, *Bynoe;* Goulburn Island, *A. Cunningham;* Victoria river and Macadam ringe, *F. Mueller.*
Queensland. Endeavour river, Keppel Bay, Shoalwater Bay, etc., *R. Brown;* Cape York, *M'Gillivray;* Percy Island, *A. Cunningham;* Rockhampton, *Thozet* and others; Port Denison, *Fitzalan;* Moreton Bay, *M'Gillivray, F. Mueller,* and others.
N. S. Wales. Clarence river, *Beckler.*
R. Brown's herbarium comprises a particularly instructive series of specimens connecting the different forms, which have at first sight the appearance of distinct species. Amidst all these varieties the species is easily recognized by the calyx and pod.

5. **C. calycina,** *Schranck, Pl. Rar. Hort. Monac. t.* 12. A decumbent or nearly erect annual, 1 to 1½ ft. high, not much branched, villous with appressed or scarcely spreading hairs. Leaves simple, nearly sessile, from short and oblong, to lanceolate or linear and 2 to 6 in. long, glabrous or nearly so above, villous underneath. Flowers in terminal racemes, remarkable for their large pendulous calyx, often fully 1 in. long, thickly covered with long spreading rusty hairs, deeply divided into nearly equal lobes the 2 upper ones rather broader. Petals pale yellow, shorter than the calyx. Ovary sessile, glabrous, with above 30 ovules. Pod oblong, not exceeding the calyx.—Benth. in Hook. Lond. Journ. ii. 564; *C. anthylloides,* D. Don; W. and Arn. Prod. Fl. Ind. 181, and of some others, not of Lam.

N. Australia. Arnhem S. Bay, *R. Brown ;* S. Goulburn Island, *A. Cunningham.*
Queensland. Endeavour river and Broad Sound, *R. Brown ;* Rockhampton, *Thozet ;* Fitzroy river, *Dallachy ;* Port Denison, *Fitzalan.*

6. **C. retusa,** *Linn.; DC. Prod.* ii. 125. An erect perennial or undershrub of 1½ to 3 ft., with few stiff erect branches, hoary with a short pubescence. Leaves simple, cuneate-oblong, very obtuse or retuse, 1½ to 3 in. long, glabrous above, hoary or silky-pubescent underneath. Flowers yellow, rather large, pendulous. Bracts and bracteoles as in the other Australian species small and narrow. Calyx 4 to 6 lines long, slightly pubescent, the tube broad, the lobes longer than the tube, the 2 upper ones rather broader, the lateral ones shortly united with the narrow lowest one. Standard broadly orbicular, ¾ to 1 in. diameter. Ovary sessile, glabrous, with 18 to 20 ovules. Pod glabrous, much inflated, often attaining 1½ in. in length.—Bot. Reg. t. 253; Bot. Mag. t. 2561; W. and Arn. Prod. 187; F. Muell. Fragm. iii. 51.

N. Australia. Sandy rocky situations on Victoria river, and Sea Range, Arnhem's Land, *F. Mueller ;* Albert river, *Henne.*
Queensland, *Bowman ;* Edgecombe Bay, *Dallachy.*
The species is widely spread over the warmer regions of the globe, both in the New and the Old World, but chiefly near the sea.

7. **C. Mitchelli,** *Benth. in Mitch. Trop. Austr.* 120. A perennial with a thick stock and erect branching stems of 1½ to 3 ft., more or less pubescent or tomentose. Leaves ovate-elliptical, ovate-lanceolate or rarely almost obovate or narrow-oblong, obtuse but usually less so than in *C. retusa,* 2 to 3 or rarely 4 in. long, glabrous above, hoary or loosely pubescent underneath. Flowers much smaller and more numerous than in *C. retusa,* in a dense terminal raceme often attaining 4 to 6 in. Calyx rarely 3 lines long, slightly pubescent, the lobes scarcely longer than the tube. Standard 5 to 6 lines diameter. Ovary sessile, glabrous, with 8 to 10 ovules. Pod under 1 in. long.

N. Australia. Wills's Creek, *Howitt's Expedition.*
Queensland. Bed of the Balonne river, *Mitchell ;* Dawson and Brisbane rivers, *F. Mueller ;* Wide Bay, *Bidwill ;* Rockhampton, *Thozet, Dallachy ;* Port Denison, *Fitzalan.*
N. S. Wales. Hunter's River, *R. Brown ;* Clarence river, *Beckler ;* head of the Gwydir, *Leichhardt.*
Much as the shape of the leaves varies, they are always broader and less cuneate than in *C. retusa,* and the pellucid dots are much less conspicuous.

SERIES II. UNIFOLIOLATÆ.—Leaves simple, the petiole articulate or geniculate above the middle. Stem shrubby.

The three following species appear to be anomalous in the development of their flowers. Most of the specimens of *C. Novæ-Hollandiæ* have the lower buds of the raceme still unopened whilst the upper ones are fully out, and I have observed it also in one or two racemes of *C. Cunninghamii,* in another I see undeveloped buds irregularly mixed. The only 2 specimens of *C. crassipes* are not in a state to show the order of development. It remains, however, as yet doubtful whether the inflorescence is really centrifugal, or whether the development of the lower buds has been from some cause retarded after their first appearance.

8. **C. Novæ-Hollandiæ,** *DC. Prod.* ii. 127. An erect shrub, of 2 or 3 ft., with terete or angular closely tomentose branches. Leaves oval-elliptical or oblong, very obtuse, 2 to 3 in. long, glabrous or pubescent above, silky-pubescent, tomentose or villous underneath, the petiole from ¼ to ¾ in.

long, more or less distinctly articulate or geniculate above the middle. Flowers yellow, rather numerous, in terminal racemes, variable in size. Bracts small and narrow. Calyx about 3 to $3\frac{1}{2}$ lines long, the lobes all acuminate, nearly equal and scarcely longer than the tube. Standard 6 to 8 lines diameter, glabrous. Ovary sessile, pubescent or villous, with 15 to 20 ovules or even more. Pod 1 to $1\frac{1}{2}$ in. long, tomentose-pubescent, or at length nearly glabrous.—*C. oblongifolia*, Hook. Ic. Pl. under n. 830; *C. Mitchelli*, F. Muell. Fragm. iii. 50, not of Benth.

N. Australia. N.W. coast, *Bynoe;* Nichol Bay, *F. Gregory's Expedition;* Upper Victoria river and Alligator Point, *F. Mueller ;* Gulf of Carpentaria, *Henne, Landsborough ;* near Mount Humphries, *M'Douall Stuart.*

The above specimens comprise the 3 following rather marked forms:—

a. *parviflora.* Leaves glabrous above. Flowers small. Pod oblong, about $\frac{3}{4}$ in. long.

b. *oblongifolia.* Leaves glabrous above or slightly pubescent. Flowers rather large. Pod above 1 in. long, much inflated.

c. *lasiophylla.* Leaves softly pubescent on both sides. Flowers rather large. Bracts closely reflexed.

9. **C. crassipes,** *Hook. Ic. Pl. t.* 830. Apparently a tall and erect plant, closely allied to *C. Novæ-Hollandiæ,* with the same oblong or elliptical obtuse leaves on articulate or geniculate petioles, but the whole plant is perfectly glabrous, and the subulate stipules, and the back of the petiole are continued below the insertion of the leaf into raised angles shortly decurrent on the stem. The inflorescence is that of *C. Novæ-Hollandiæ,* the flowers rather larger and the calyx-lobes rather longer. The ovary is as glabrous as the rest of the plant.

N. Australia. N.W. coast, *Bynoe.*

10. **C. Cunninghamii,** *R. Br. in. App. Sturt Exped.* 8. A shrub, of 2 to 3 ft., with softly tomentose terete or slightly angular branches. Leaves ovate, usually broad, very obtuse, $1\frac{1}{2}$ to 3 in. long, densely and softly tomentose-pubescent or villous on both sides, the petiole $\frac{1}{4}$ to $\frac{3}{4}$ in. long, articulate or geniculate above the middle. Stipules and bracts softly subulate, sometimes rather long, but very deciduous. Racemes terminal, usually short and dense, sometimes reduced to a sessile cluster, rarely 4 to 5 in. long. Flowers very large, of a yellowish-green colour, more or less streaked with dark lines. Calyx tomentose, the tube about 3 lines long, the lobes varying from that length to twice as long, all nearly equal. Standard ovate, acuminate, about $1\frac{1}{2}$ in. long when fully developed; keel rather longer; wings shorter. Ovary shortly stipitate, villous, with 20 or more ovules. Pod coriaceous, tomentose, $1\frac{1}{2}$ in. long.—Hook. Ic. Pl. t. 829; F. Muell. Fragm. iii. 52.

N. Australia. Common on the sandy shores of the N.W. coast, *Bynoe ;* from Cygnet Bay, *A. Cunningham,* to Victoria river, and the Gulf of Carpentaria, *F. Mueller, Leichhardt ;* sandy ridges of the Hammersley Range, *F. Gregory's Expedition ;* Nichol Bay and De Grey river, *Ridley's Expedition ;* Mount Humphries, *M'Douall Stuart.*

S. Australia. Towards Spencer's Gulf, *Warburton* ; near Cooper's Creek, *Wheeler, Howitt's Expedition.*

W. Australia. Sharks Bay, *M. Brown.*

C. Sturtii, R. Br. in App. Sturt Exped. 7, gathered by D. Sturt between latitudes 28° and 26°, which I have not seen, is believed both by Hooker and by F. Mueller to be the same as *C. Cunninghamii,* the specimen of the latter seen by R. Brown having been imperfect as to inflorescence, and there is nothing in R. Brown's diagnoses of *C. Sturtii,* which does not agree perfectly with the common state of *C. Cunninghamii*, but in the description the tenth stamen is said to be free, which I have never found in *C. Cunninghamii.*

SERIES III. DIGITATÆ.—Leaves all or mostly digitately compound, with 3 leaflets in the Australian species, very rarely reduced to 1 in a few leaves of one species, 5 or 7 in some East Indian species. Herbs or shrubs.

11. **C. trifoliastrum,** *Willd.; W. and Arn. Prod.* 191. A perennial with rather slender, erect ascending or decumbent branching stems, usually 1 to 2 ft. high, more or less pubescent. Leaflets 3, usually oblong-cuneate, but varying from obovate and under ½ in. long, to linear-cuneate and about 1 in. long, very obtuse or retuse, glabrous above, hoary or pubescent underneath, the petiole slender. Flowers usually small, but variable in size, in terminal racemes of 1 to 3 in. Calyx pubescent, about 2 lines long, the lobes narrow and much longer than the tube, the 2 upper ones rather smaller than the others. Standard broad, exceeding the calyx, but usually shorter than the straight beak of the keel. Ovary sessile, pubescent, with 2 ovules. Pod about 2 lines broad and not longer, tapering into a short hooked point, pubescent or nearly glabrous.—Wight, lc. t. 421.

N. Australia. N.W. coast, *Bynoe;* Victoria river, *F. Mueller;* islands of the Gulf of Carpentaria, *R. Brown, Henne;* Port Essington, *Armstrong.*

Queensland. Bay of Inlets and Shoalwater Bay, *R. Brown;* Wide Bay, *Bidwill;* Port Curtis, *M'Gillivray;* Rockhampton, *Thozet, Dallachy;* Port Denison, *Fitzalan.*

F. Mueller, Fragm. iii. 56, unites this with *C. medicaginea*, Lam., but the latter appears to be always prostrate, with small broad leaflets, unless when drawn up in luxuriant grass, the racemes much shorter, the flowers smaller, the standard larger in proportion to the keel, etc.

F. Mueller's herbarium contains also a single imperfect specimen from the Gulf of Carpentaria, Landsborough, allied to *C. trifoliastrum*, but evidently shrubby, with woolly tomentose branches and larger flowers, closely resembling *C. Notonii,*.W. and Arn. Prod. ii. 192 (the same as *C. rostrata*, W. and Arn. l. c. 191), but the materials are insufficient for ascertaining whether it be a distinct species.

12. **C. incana,** *Linn.; DC. Prod.* ii. 132. An erect herb, usually annual, attaining 2 or 3 ft., the branches tomentose, pubescent or rusty-villous. Leaflets 3, obovate or orbicular, very obtuse, usually ½ to 1 in. long, glabrous above, more or less ciliate on the edge and sometimes hairy underneath, on a long common petiole. Flowers small, yellow, in short terminal or leaf-opposed racemes. Calyx 3 to 4 lines long or rarely rather more, the lobes finely acuminate, several times longer than the small tube. Standard as long as or rather longer than the calyx, broad, but almost acuminate; wings narrow; keel nearly as long as the standard; anthers smaller than in most species and rather less disproportioned. Ovary sessile, villous, with numerous densely crowded ovules. Pod sessile, 1 to 1½ in. long, usually much inflated and hirsute with spreading hairs.—Benth. in Mart. Fl. Bras. Leg. 27; F. Muell. Fragm. iii. 53; *C. affinis*, DC. Prod. ii. 132; *C. herbacea*, Schweigg. in Schranck, Syll. Pl. Ratisb. ii. 77; *C. cubensis*, DC. Prod. ii. 131; *C. Schimperi*, A. Rich. Fl. Abyss. i. 151.

Queensland. Keppel Bay, *R. Brown;* Moreton island and Gilbert river, *F. Mueller;* Rockhampton and Bowen river, *Bowman, Dallachy.* The species is widely dispersed over the tropical and subtropical regions of the New and the Old World. The figure in Bot. Reg. t. 377, usually quoted for it, represents rather *C. striata*, DC. Prod. ii. 131, another common tropical species, but not as yet found in Australia.

13. **C. dissitiflora,** *Benth. in Mitch. Trop. Austr.* 386. An erect perennial, of 1 to 2 ft., the branches hoary or silky-tomentose. Leaflets usually 3, broadly obovate, oblong, or rarely almost linear, very obtuse, rarely above 1 in. long, and often much smaller, usually glabrous above and hoary-tomentose or silky underneath, rarely glabrous or villous on both sides. Flowers yellow, in a rather loose terminal raceme, often elongating to 5 or 6 in. Calyx 2 to 3 lines long, the lobes rarely longer than the tube, and all nearly equal. Standard broad, twice as long as the calyx; keel rather shorter. Ovary shortly stipitate, more or less pubescent or villous, with 10 to 12 ovules. Pod pubescent, tomentose, or nearly glabrous.

N. Australia. Gulf of Carpentaria, *Landsborough.*
Queensland. Balonne river, *Mitchell ;* Suttor river, *F. Mueller ,* Broad Sound, *Bowman ;* Rockhampton, *Thozet, Dallachy ;* heads of the Isaacs and Bowen rivers, *Bowman.*

Var. *eremæa.* Leaflets narrow, the lateral ones often small or wanting, softly tomentose hoary or white.—*C. eremæa,* F. Muell. Rep. Greg. Pl. 5 ; referred to *C. dissitiflora,* F. Muell. Fragm. iii. 56.—Flooded border of Wills's Creek, *Murray,* also—
N. S. Wales. Between the Darling and Cooper's Creek, *Neilson.*
S. Australia. Cooper's Creek, *C. A. Gregory ;* towards Spencer's Gulf, *Warburton.*
Var. *rugosa.* Leaves soft, silky-villous on both sides. Flowers small.—Sturt's Creek and Newcastle Water, *F. Mueller.*
Var. (?) *grandiflora.* Very silky-villous. Flowers much larger.—Hammersley Range, *F. Gregory's Expedition ;* Port Nichol and De Grey River, *Ridley's Expedition.*

14. **C. laburnifolia,** *Linn.; DC. Prod.* ii. 130. An erect glabrous shrub of several feet, with rather slender terete branches. Leaflets 3, petiolulate, ovate, mostly acute, 1 to 2 in. long, on a rather long common petiole. Flowers large, yellow, in loose terminal or leaf-opposed racemes. Calyx 5 to 6 lines long, the lobes acuminate, much longer than the broad tube. Standard broadly ovate, shortly acuminate, fully ¾ in. long ; wings not half so long, broad, with the transverse folds particularly prominent ; keel with a long straight beak, as long as the standard. Ovary on a long stipes, glabrous with 20 to 30 or more ovules. Pod 1¼ in. long or more, on a stipes much longer than the calyx.—W. and Arn. Prod. 193 ; F. Muell. Fragm. iii. 53.

Queensland. Cape Cleveland, *A. Cunningham ;* Cape Upstart, *M'Gillivray ;* sandy shores of the Burdekin, Dawson and Burnett rivers, *F. Mueller ;* Port Denison, *Fitzalan ;* Burdekin and Bowen rivers, *Bowman.*
The species is common on the sandy coasts of East India.

15. **C. quinquefolia,** *Linn. ; DC. Prod.* ii. 135. An erect annual, attaining 3 or 4 ft., with a hollow stem, glabrous or silky-pubescent. Leaflets usually 5, lanceolate or linear, obtuse, 1½ to 3 in. long or the central one longer, almost sessile on a rather long common petiole. Flowers large, yellow, in loose terminal or leaf-opposed racemes. Bracts lanceolate, acuminate, reflexed. Calyx 5 to 6 lines long, the lobes broad, acuminate, scarcely longer than the tube. Standard broad, very obtuse, about ¾ in. diameter ; wings rather shorter ; keel with an acute curved beak. Pod glabrous, about 2 in. long, on a stipes equal to or longer than the calyx.—W. and Arn. Prod. 194.

Queensland. On the lower Burdekin river, *Bowman.* The species is generally dispersed over E. India and the Archipelago.

26. PENTADYNAMIS, R. Br.

Calyx-lobes nearly equal. Standard broad, without auricles, but with callosities decurrent on the claw; keel obtuse, as long as the wings. Upper stamen free, the others united; anthers alternately long and erect, and short and versatile. Ovary with several ovules; style incurved, bearded upwards along the inner side; stigma terminal. Pod flattened.—Herbs. Leaves 3-foliolate. Leaflets sessile (digitate?). Flowers yellow, in axillary racemes.

The genus is limited to the single Australian species.

1. **P. incana,** *R. Br. in App. Sturt Exped.* 76. An erect branching hoary-white herb or perhaps undershrub, of about 2 ft. Leaflets linear, obtuse, the central one the longest and scarcely 1 in. long. Racemes many-flowered, the pedicels as long as the calyx. Calyx-lobes acute, as long as the tube. Petals more than twice as long as the calyx. Ovary pubescent. Unripe pod hoary-white, acuminate by the incurved base of the style.

S. Australia. On sandhills with *Crotalaria Sturtii, Sturt.* I have seen no specimens answering at all to the above character, which I have taken from R. Brown. The affinity of the genus must therefore remain uncertain till the plant has been again seen and examined. The author suspects that one of my *Vignas* may be another species, but I have seen no plant, allied to *Vigna*, which has any tendency to the dimorphous anthers, hitherto, among *Phaseoleæ*, only observed in *Mucuna, Teramnus,* and *Dioclea.* F. Mueller, Fragm. iii. 56, refers the plant, without hesitation, to *Crotalaria dissitiflora*, Benth., var. *eremæa,* but R. Brown describes the keel as obtuse, the stamens diadelphous, and the pod flattened, all of them characters incompatible with *Crotalaria.*

27. ROTHIA, Pers.

(Westonia, *Spreng.*; Xerocarpus, *Guillem. and Perr.*)

Calyx narrow, the lobes nearly of equal length, the 2 upper ones rather broader. Standard ovate or oblong; wings narrow; keel-petals like the wings and scarcely cohering. Stamens all united in a sheath open on the upper side; anthers small, uniform. Ovary sessile, with several ovules; style straight, not bearded, with a terminal stigma. Pod linear or linear-lanceolate, acute, not divided inside, opening when ripe on the upper side as a follicle. Seeds without any strophiole.—Annuals. Leaves digitate, with 3 leaflets. Stipules free. Flowers very small, leaf-opposed.

Besides the Australian species, which is also E. Indian, there is only one other species from tropical Africa.

1. **R. trifoliata,** *Pers.; DC. Prod.* ii. 382. A diffuse or prostrate annual, attaining 1 to 1½ ft., softly hairy in all its parts. Leaflets from almost obovate to narrow-oblong, quite entire, ¼ to ½ in. or rarely ¾ or even 1 in. long, on a short common petiole. Stipules ovate and leaf-like, but small. Flowers rarely above 2 lines long, solitary or 2 together on very short pedicels opposite the leaves. Bracts and bracteoles small, setaceous. Pod narrow-linear, 1 to 2 in. long.—W. and Arn. Prod. 195; Wight, Ic. t. 199.

N. Australia. Upper Victoria river, *F. Mueller.* Not uncommon in E. India.

TRIBE III. TRIFOLIEÆ.—Herbs, very rarely shrubs. Leaves usually pinnately or rarely digitately 3-foliolate, the veinlets of the leaflets extending to the edge and often produced into minute teeth. Peduncles, racemes or flower-heads axillary, or apparently terminal by the reduction of the upper floral leaves, never leaf-opposed. Upper stamens free, except in *Ononis*, the others united in a sheath. Ovules 2 or more (except in *Medicago lupulina*). Pod not articulate.

This tribe, consisting chiefly of European and North Asiatic plants, with a few American or tropical species, is represented in Australia by only one indigenous species, and that one closely allied to some east Mediterranean forms. Several European species have however become more or less established in waste or cultivated places in the settled colonies, especially the following :—

Ononis Natrix, Linn. A low much-branched perennial, more or less viscid-pubescent. Leaves pinnately trifoliolate or the upper ones 1-foliolate, with denticulate leaflets. Flowers rather large, yellow, often streaked with purple, solitary on axillary peduncles. Stamens all united in a closed tube. Pod straight, oblong, turgid.—Campbell's Creek, *Herb. F. Mueller.*

Medicago sativa, Linn., or cultivated *Lucern.* A perennial with ascending or erect stems, of 1 to 2 ft. Leaves pinnately 3-foliolate. Peduncles axillary, bearing a short close raceme of violet or blue flowers. Upper stamen free. Pod spirally twisted so as to form 2 or rarely 3 complete coils, without tubercles or prickles.—Rocky pastures, Victoria and S. Australia, *F. Mueller.*—*M. lupulina*, Linn., a softly pubescent or hairy annual. Leaves pinnately 3-foliolate. Peduncles axillary, bearing a compact raceme or head of very small yellow flowers. Ovary 1-ovulate. Pod small, kidney-shaped, marked with veins curved almost into a spire, the minute base of the style completing the spire.—Gabo island, *Maplestone*, and about Port Jackson.—*M. denticulata*, Willd. An annual, glabrous or nearly so. Leaves pinnately 3-foliolate. Stipules finely-toothed. Flowers very small, yellow, few, in little heads on axillary peduncles. Pod spirally-twisted, forming 2 or 3 loose flat coils veined on the surface and edged with 2 rows of hooked or curved prickles.—Waste places, Queensland, N. S. Wales, Victoria, and S. Australia.

Melilotus parviflora, Desf. A slender branching annual. Leaves pinnately 3-foliolate with narrow leaflets. Flowers small, yellow, in slender axillary racemes. Pod of 1 or 2 seeds, straight, thick, small, but longer than the calyx, indehiscent.—About Port Jackson, in Tasmania and in S. Australia.

Trifolium pratense, Linn., or *Red Clover.* A more or less hairy decumbent or erect perennial, of 1 to 2 ft. Leaves digitately 3-foliolate. Flowers of a purplish-red, in ovoid showy heads, apparently terminal, with 2 sessile 3-foliolate leaves close under the head. Claws of the lower petals adhering to the staminal tube, as in all the following species of *Trifolium.* Pod small, enclosed in the calyx.— Pastures on the Snowy River, in the Australian Alps, *F. Mueller.*—*T. repens*, Linn., or *White* or *Dutch Clover.* A glabrous or slightly hairy perennial, the stems creeping and rooting at the nodes. Leaves digitately 3-foliolate, with obovate leaflets. Peduncles axillary, long and erect, bearing a globular dense umbel of white flowers. Pod 2- to 4-seeded, usually protruding from the calyx, but enclosed in the withered corolla.—Victoria and S. Australia.—*T. agrarium*, Linn., or *Hop Clover.* A slender branching annual, glabrous or nearly so. Leaves pinnately 3-foliolate, with obovate or obcordate leaflets. Flowers small, yellow, 30 to 50 together, in loosely globular or ovoid heads, turning a pale-brown in fading, persistent, reflexed, with a strongly striate standard concealing the small pod.—Victoria and S. Australia.—*T. procumbens*, Linn. (*T. minus*, Sm), differing from *T. agrarium*, in being more slender and procumbent, with smaller flowers, less than 20 in the head, and the faded standard scarcely striate. Victoria and Tasmania.

28. TRIGONELLA, Linn.

Calyx-teeth nearly equal. Petals free from the staminal tube; standard obovate or oblong, narrowed at the base but scarcely clawed; wings and keel shorter, obtuse. Upper stamen free or at first united with the others; filaments not dilated; anthers uniform. Ovary sessile or shortly stipitate, with several ovules; style filiform. Pod either linear straight or curved, or in species not Australian, flat and falcate, or short with a long beak, 2-valved or indehiscent. Seeds not strophiolate.—Herbs, often strong-scented. Leaves pinnately 3-foliolate, the leaflets usually denticulate. Stipules adnate to the petiole. Flowers yellow white or blue, in axillary heads umbels or short racemes.

The genus is rather numerous in species in the warmer extratropical regions of the northern hemisphere in the Old World, one of the common ones being also found in South Africa. The only Australian species is very nearly allied to an Egyptian one, although not quite identical with any form hitherto observed there.

1. **T. suavissima,** *Lindl. in Mitch. Three Exped.* i. 255. An annual, either quite glabrous or sprinkled with a few hairs on the under side of the leaves and on the calyxes, the stems prostrate or ascending, from ½ to 2 or 3 ft. in length. Leaflets broadly obovate or obcordate, rarely above ½ in. long, more or less denticulate, on a long slender petiole. Stipules semisagittate, deeply toothed. Flowers small, yellow, in sessile clusters. Calyx about 2 lines long, the lobes lanceolate-subulate, rather rigid, fully as long as the tube. Standard longer than the calyx; wings and keel scarcely shorter. Upper stamen free. Pod linear, curved, almost obtuse, ½ to ¾ in. long, and about 1 line broad, opening in 2 thin reticulate valves, either flat or undulate.

N. S. Wales. On the Darling river, *Mitchell*, also *Victorian Expedition;* Molle's Plains, *A. Cunningham.*

Victoria. Wimmera, *Dallachy.*

S. Australia. Central Australia, *M'Kinlay's Expedition;* grassy and saline plains towards Cudnaka, *F. Mueller.*

W. Australia. Between Moore and Murchison rivers, *Drummond, 6th Coll. n.* 30.

The species is closely allied to *T. hamosa* of the northern hemisphere, which is also found in S. Africa, and to the E. Mediterranean *T. microcarpa,* Poir., and *T. anguina,* Delile, but not quite identical with either.

TRIBE IV. EULOTEÆ.—Herbs, rarely shrubs. Leaves pinnate or sometimes apparently digitately 3-foliolate, the lowest pair of leaflets taking the place of stipules; leaflets entire. Flowers capitate or umbellate, on axillary peduncles. Upper stamen usually free, at least at the base, the others united in a sheath; filaments either all or 5 only dilated towards the end. Pod not articulate.

29. LOTUS, Linn.

Calyx-lobes nearly equal or the lowest longer. Standard obovate or orbicular; keel much incurved, beaked. Upper stamen free, the rest united in a sheath; filaments, above the sheath, alternately dilated near the top; anthers uniform. Ovary sessile, with several ovules; style bent above the ovary, glabrous, with a terminal stigma. Pod usually linear, terete, with cellular partitions between the seeds. Seeds not strophiolate.—Herbs, or, in species not Australian, undershrubs. Leaves of 4 or 5 leaflets, 3 almost digitate at the end of the petiole, 1 or 2 close to the stem, taking the place of stipules. Real stipules reduced to minute tubercles or dark spots, or entirely wanting. Flowers yellow pink or white, usually several together in an umbel, on an axillary peduncle, with a leaf-like bract under the umbel.

The genus is widely spread over the temperate regions of the northern hemisphere in the Old World, the mountains of tropical Asia, and extratropical South Africa. Of the Australian species, one has a very wide European and Asiatic range, the other is endemic.

Flowers yellow.   Calyx-lobes about as long as the tube . . . . . 1. *L. corniculatus.*
Flowers pink or white.   Calyx-lobes usually longer than the tube . . 2. *L. australis.*

L. tetragonolobus, Linn., an annual with deep purple-red flowers, belonging to the section *Tetragonolobus,* with the pods winged, a native of Southern Europe, has been introduced as a weed of cultivation in the Bugle Range, S. Australia, *F. Mueller.*

1. **L. corniculatus,** *Linn.; Ser. in DC. Prod.* ii. 214. A perennial, with prostrate, decumbent, ascending or almost erect stems, from a few in. to nearly 2 ft. high, the Australian specimens usually glabrous or somewhat glaucous, but often hairy in other countries. Leaflets usually obovate or ovate, acute, and rarely much above ½ in. long, the 2 stipular ones broader and very oblique, but sometimes all are narrow. Flowers yellow, often tinged with b.ight red, from about 5 to near 10 in the umbel. Calyx 3 to 3½ lines long, usually slightly hairy, the lobes narrow and subulate-acuminate, about as long as the tube. Standard fully 5 lines diameter; wings nearly as long; keel with a long straight beak. Pod linear, terete, straight, rather slender, 1 to 1½ in. long. Seeds globular, separated by thin transverse partitions.—Hook. f. Fl. Tasm. i. 98.

N. S. Wales. Port Jackson to the Blue Mountains, *R. Brown* and others; New England and Clarence river, *Beckler.*
Victoria. Wimmera, *Dallachy.*
Tasmania. Port Dalrymple, *R. Brown;* abundant in rich soils and marshy places, affording good pasturage and ascending to 4000 ft., *J. D. Hooker.*
S. Australia. Near Bethanie, *F. Mueller.*
The species is widely spread over Europe, temperate Asia, and the mountainous districts of East India.

2. **L. australis,** *Andr. Bot. Rep. t.* 624. A perennial, sometimes almost shrubby at the base, with diffuse ascending or erect stems, either glabrous and glaucous or more frequently pubescent on the younger branches and peduncles, and in some Queensland specimens softly villous all over. Leaflets usually narrower than in *L. corniculatus,* and the stipulary ones less dissimilar, but varying from obovate and all under ½ in. long, to linear and 1 to 1½ in. long. Inflorescence and pod of *L. corniculatus,* and the flowers scarcely to be distinguished except by the colour, which is usually pink, but varies from white to a purple-red; they are also very variable in size, in some forms much smaller, in others much larger than in *L. corniculatus;* the tube of the calyx is also shorter, and the lobes longer than in that species.— Ser. in DC. Prod. ii. 212; Bot. Mag. t. 1365; Hook. f. Fl. Tasm. i. 98; *L. lævigatus,* Benth. in Mitch. Trop. Austr. 62; *L. albidus,* Lodd. Bot. Cab. t. 1063; Maund, Botanist, t. 211.

N. Australia. E. tributaries of Victoria river, *F. Mueller;* Nichol Bay and De Grey river, *Ridley's Expedition.*
Queensland. Keppel Bay, *R. Brown;* Port Curtis, *M'Gillivray;* Moreton Bay, *A. Cunningham;* near Mount Faraday, *Mitchell;* Edgecombe Bay, Rockhampton, etc., *Dallachy.*
N. S. Wales. Hunter's River, *R. Brown, Oldfield;* Macquarrie river, *Mitchell;* northwards to Clarence river, *Beckler;* New England, *C. Stuart;* and in the interior to the Murray.
Victoria. Pasture land, frequent on the coast and on several ranges in the interior, *F. Mueller;* Wimmera, *Dallachy.*
Tasmania. Port Dalrymple, *R. Brown;* sandy shores of the N. coast, *J. D. Hooker.*
S. Australia. Near Adelaide, *Whittaker;* Lofty and Bugle ranges, Mount Remarkable, etc., *F. Mueller;* Kangaroo Island, *Waterhouse.*

W. Australia. Flinders Bay, *Collie ;* between Moore and Murchison rivers, *Drummond, 6th Coll. n.* 32, *Oldfield.*

Var. *parviflorus.* Leaves small, usually broad. Flowers often solitary or only 2 or 3 together on the peduncle, very much smaller, and often but not always deeply coloured, the calyx-lobes very fine and scarcely so long in proportion to the tube as usual in *L. australis.*— *L. coccineus,* Schlecht. Linnæa, xxi. 452.— Paterson's River, *R. Brown ;* Peel's Range, *A. Cunningham ;* in the Murray desert and numerous S. Australian localities, *Behr, F. Mueller,* and others; between Moore and Murchison rivers, *Drummond, 6th Coll. n.* 31, *Oldfield ;* Nichol Bay and De Grey river, *Ridley's Expedition.* Most of the specimens have a very different aspect from those of the normal *L. australis,* but there are also too many intermediate forms to admit of characterizing it as a distinct species.

TRIBE V. GALEGEÆ.—Herbs not twining, shrubs or rarely tall trees or woody climbers. Leaves pinnate, rarely reduced to 1 or 3 leaflets. Stipellæ none, or setaceous in a few pinnate genera. Upper stamen usually free, at least at the base, the others united in a sheath, very rarely all united ; filaments filiform. Ovules usually 2 or more. Pod not articulate, 2-valved, except in *Psoralea.*

This tribe comprises a large number of genera from all parts of the world, generally distinguished by their stems not twining, pinnate leaves without tendrils, diadelphous stamens, and 2-valved pod ; but to all these characters there are exceptions, connecting them with almost all the other tribes of Papilionaceæ. Thus, amongst Australian genera, *Psoralea* is connected with the 1-seeded *Hedysareæ* by its indehiscent 1-seeded pod ; *Ptychosema* and *Lamprolobium* have the stamens of *Genisteæ ;* and *Millettia* has the habit of *Dalbergieæ.* *Goodia,* placed in *Genisteæ,* might be nearly as well inserted next to *Ptychosema,* among *Galegeæ ;* and in several genera of *Phaseoleæ,* there are species which, by their erect stems, 5- or 7-foliolate leaves, or evanescent stipellæ, connect that tribe with *Galegeæ.*

30. **PSORALEA,** Linn.

(Meladenia, *Turcz.*)

Calyx-lobes nearly equal or the lowest the largest, or the 2 upper ones united. Standard ovate or orbicular ; wings slightly adhering to the keel, which is slightly incurved, obtuse, and shorter than the other petals. Upper stamen free or more or less adhering to the others ; anthers uniform. Ovary with a single ovule ; style filiform or dilated at the base. Pod small, ovate, not dehiscent, the pericarp usually adhering to the seed.—Herbs undershrubs or rarely shrubs, dotted with black or transparent glands. Leaves of 3, 5, or 7 digitate entire leaflets, or of 1 or 3 pinnately arranged entire or toothed leaflets, or in species not Australian pinnate with several leaflets. Stipules attached by a broad base. Flowers purple pink blue or white, usually small, and in the Australian species, in axillary spikes or racemes. Bracts membranous, deciduous, each usually with 2 or 3 flowers in its axil.

A large genus, widely distributed over various parts of the globe, but most abundant in S. Africa and N. America. Of the Australian species, one is also a native of the Indian Archipelago, the remainder are endemic, but some are very nearly allied to some of the E. Mediterranean species.

Leaves all 1-foliolate. Leaflets entire or toothed.
  Calyx lower lobe much longer than the others.
    Plant softly pubescent or silky-villous. Leaflets entire . . . 1. *P. badocana.*
    Plant hispid. Leaflets toothed . . . . 2. *P. Archeri.*
  Calyx-lobes nearly equal in length. Plant very dark and rough,
    with glandular dots.

Plant pubescent or villous. Flowers in nearly globular short
racemes . . . . 3. *P. balsamica.*
Plant glabrous or slightly hoary. Flowers small, in loose elon-
gated racemes . . . . 10. *P. leucantha.*
Leaves all pinnately 3-foliolate, or the lower ones rarely 1-foliolate.
Calyx lower lobe much longer than the lateral ones. Leaflets
entire.
Flowers in dense heads. Calyx very hispid, the lower lobe
long-lanceolate. Petals shorter than the calyx . . . . 4. *P. plumosa.*
Flowers in interrupted spikes. Petals longer than the calyx.
Calyx pubescent . . . . 5. *P. pustulata.*
Calyx densely clothed with soft white woolly hairs . . . 6. *P. lachnostachys.*
Calyx lower lobe scarcely exceeding the upper ones. Leaflets
usually toothed.
Calyx softly silky-villous or black, 2 to 4 lines long, com-
pletely concealing the pod.
Calyx 3 to 4 lines long, the lateral lobes short. Plant
usually white-tomentose . . . . 7. *P. eriantha.*
Calyx about 2 lines long, the lobes nearly equal. Plant
hoary or pubescent . . . . 8. *P. patens.*
Calyx hoary-tomentose or slightly pubescent, 1 to 1½ lines
long, open when in fruit, and scarcely exceeding or shorter
than the pod.
Leaflets ovate or elliptical, mostly ¾ to 1 in. long . . . . 9. *P. cinerea.*
Leaflets oblong or lanceolate, 1½ to 3 in. long . . . . . 10. *P. leucantha.*
Leaves digitately 3- to 7-foliolate. Leaflets entire.
Spikes dense. Calyx 2 to 2½ lines long, black or silky-villous,
much longer than the pod . . . . 12. *P. adscendens.*
Racemes slender. Calyx 1 to 1½ lines long, slightly pubescent,
about as long as the pod . . . . 11. *P. tenax.*

1. **P. badocana,** *Benth.* An erect stout undershrub or shrub, of 2 to
3 ft., softly tomentose or silky-villous all over and strongly scented, the black
dots mostly concealed by the indumentum. Leaflet single, on a petiole articulate near the top, ovate to lanceolate, obtuse or scarcely acute, 2 to 3 in. long,
entire, softly villous on both sides and usually silky underneath. Stipules
linear-subulate or broad and subulate-acuminate, often ½ in. long. Flowers in
dense heads or short spikes, all axillary or sessile or very shortly pedunculate.
Calyx softly villous, fully 5 lines long in the normal forms, including the
lower lobe, which is much longer than the others and boat-shaped. Petals
shorter than the lower calyx-lobe. Pod small, reticulate, glandular.—*Liparia
badocana,* Blanco, Fl. Filip. 597 ; *Meladenia densiflora,* Turcz. in Bull. Mosc.
1848, i. 576.

N. Australia. Gulf of Carpentaria, *R. Brown ;* N. coast, *A. Cunningham ;* Port
Essington, *Armstrong.* Also in New Guinea, *M'Gillivray,* and the Philippine Islands,
Cuming, n. 1149 *and* 1649.

Var. *grandiflora.* Spikes crowded and often 1 to 1½ in. long. Calyx fully 7 lines long.
Pod tomentose.—*P. Archeri,* F. Muell. Fragm. iv. 21, partly.—Regent's River, N. W.
coast, *A. Cunningham ;* Upper Victoria river, *F. Mueller.*

Var. ? *cephalantha.* Rather less villous. Stipules broad, triangular, very shortly acuminate. Spikes or flower-heads globular. Calyx rather smaller than in the normal form, the
lowest lobe scarcely so long as the petals.—*P. cephalantha,* F. Muell. Fragm. iv. 35.

Queensland. Mount Elliott, *Dallachy.*

2. **P. Archeri,** *F. Muell. Fragm.* iv. 21 (*partly*). Very nearly allied
to *P. badocana,* with which F. Mueller unites it, but apparently an erect

coarse annual of 1 to 2 ft., hirsute with much more rigid and spreading hairs than in that species. Leaves similarly 1-foliolate, the leaflet ovate or oblong, obtuse, 2 to 3 in. long, but always toothed. Flower-heads sessile and very hispid. Flowers of *P. badocana*, but the calyx-lobes more subulate. Pod ovate, almost acute, very hispid.

N. Australia. Upper Victoria river, *F. Mueller.*

3. **P. balsamica,** *F. Muell. in Trans. Vict. Inst.* iii. 55. A tall undershrub, softly pubescent or villous, or at length nearly glabrous, the branches very rough with numerous prominent tubercular glands. Leaflets single, on a petiole articulate at the top, elliptical-oblong obtuse or mucronate, denticulate, 1½ to 3 in. long, slightly pubescent, the black or brown glandular dots numerous. Flowers shortly pedicellate, in nearly globular short racemes, in the upper axils or at the ends of the branches. Calyx shortly villous, 2 to 2½ lines long, open when in fruit, the lobes shorter than the tube, the lowest broader but not longer than the others. Petals rather longer than the calyx; keel shorter than the others and obtuse. Pod obovate, very obtuse, pubescent, nearly as long as the calyx.

N. Australia. Van Alphen river, *F. Mueller;* Attack Creek, *M'Douall Stuart.*

4. **P. plumosa,** *F. Muell. Fragm.* iv. 22. Apparently an annual, erect, nearly simple, 6 to 9 in. high, very hispid with long spreading rigid hairs. Leaves pinnately 3-foliolate; leaflets obovate or elliptical, acute or mucronulate, quite entire, ¾ to 1 in. long, the lateral veins few. Flower-spikes dense, short, shortly pedunculate, very hispid. Bracts broad. Calyx-tube very short, the lowest lobe lanceolate, acuminate, nearly 4 lines long, the upper ones scarcely half as long and narrow. Petals shorter than the lower calyx-lobe; keel with a short erect point. Pod ovate, slightly hirsute.

N. Australia. Hooker's Creek, *F. Mueller.*

5. **P. pustulata,** *F. Muell. in Trans. Vict. Inst.* iii. 54. An undershrub with stout rigid erect branches attaining 5 to 10 ft. in height, loosely pubescent and sprinkled with large almost scale-like glands. Leaves pinnately 3-foliolate; leaflets obovate or oblong, very obtuse, 1 to 2 in. long, entire, softly pubescent. Stipules broad, rigid, striate. Flowers shortly pedicellate, in rather loose axillary racemes of 2 to 4 in., flowering almost from the base. Calyx about 3 lines long, slightly pubescent and sprinkled with prominent glands; the lobes rather broad, the 4 upper ones very short, the lowest nearly twice as long. Petals half as long again as the calyx, the standard rather broad. Pod enclosed in the somewhat inflated calyx, ovate-oblong, slightly hairy, very glandular.

N. Australia. Victoria river, and sources of Nicholson river, Gulf of Carpentaria, *F. Mueller;* Nichol Bay, *F. Gregory's Expedition.*

6. **P. lachnostachys,** *F. Muell. Fragm.* iii. 105. From the single fragment of this plant in F. Mueller's herbarium, it would appear to have the habit, foliage, inflorescence, and floral characters of *P. pustulata*, except that the flowers are rather larger, and the rhachis of the racemes, the pedicels, and the calyxes are densely clothed with soft, very white, woolly hairs.

N. Australia. Nichol Bay, *F. Gregory's Expedition.*

7. **P. eriantha,** *Benth. in Mitch. Trop. Austr.* 131. A perennial with a woody rhizome and prostrate or ascending stems of 1 to 2 ft., hoary or white with a short or soft tomentum. Leaves pinnately 3-foliolate; leaflets ovate obovate or almost orbicular, toothed, hoary or white-tomentose, the terminal one usually ¾ to 1 in. or rarely 1½ in. long, the lateral ones smaller. Stipules short. Spikes pedunculate, sometimes dense and 1 to 2 in. long, more frequently rather loose and 2 to 3 or even 4 in. long. Flowers bluish, almost sessile. Calyx 3 to nearly 4 lines long, clothed with a soft white tomentum or pubescence, the lowest lobe rather longer than the 2 uppermost, the lateral ones much shorter. Petals rather longer than the calyx. Pod ovoid, obtuse, tomentose or villous, shorter than the calyx.

Queensland. In the bed of the Balonne river, near St. George's Bridge, *Mitchell.*
N. S. Wales. Near the Darling river, *Dallachy and Goodwin;* sandhills near Menindee, *Victorian Expedition.*
Victoria. Along the Murray from the junction of the Murrumbidgee to the western limits of the colony, *F. Mueller.*
S. Australia. Along the Murray, *F. Mueller;* Mount Kingston, *M'Douall Stuart.*
W. Australia, *Drummond, 4th Coll. n.* 158, *5th Coll. n.* 96.
The species is nearly allied to, although not identical with, *P. Jaubertiana,* Fenzl, from the E. Mediterranean region.

8. **P. patens,** *Lindl. in Mitch. Three Exped.* ii. 9. A perennial with a woody rhizome and erect or ascending branches of 1 to 2 ft., hoary-tomentose or pubescent. Leaves pinnately 3-foliolate. Leaflets from ovate-rhomboid to broadly lanceolate, obtuse, usually rounded at the base, mostly 1 to 1½ in. long, denticulate, green or minutely hoary-tomentose. Spikes at first dense, but afterwards elongated and interrupted, on very long peduncles. Flowers nearly sessile, the bracts small. Calyx softly silky with white or black hairs, about 2 lines long, the lobes about as long as the tube, the lowest much broader but not longer than the others. Standard and wings half as long again as the calyx; keel shorter, obtuse. Pod tomentose, much shorter than the calyx.—*P. australasica,* Schlecht. Linnæa, xx. 668.

N. Australia. Victoria river, *F. Mueller;* Attack Creek, *M'Douall Stuart's Expedition.*
Queensland. On the Burdekin and near Port Denison, *Bowman, Dallachy.*
N. S. Wales. On and near the Lachlan river, *A. Cunningham, Fraser, Mitchell;* near Wellington, *C. Moore.*
Victoria. On the Murray river, *F. Mueller.*
S. Australia. S. coast, *R. Brown;* common about Bethanie, *Behr;* Lofty Ranges, *F. Mueller;* about Adelaide, *Whittaker;* Spencer's Gulf, *Warburton.*
This plant, *P. eriantha* and *P. cinerea,* are all considered by F. Mueller as forms of one species; they belong to the same group as *P. plicata,* Delile, from Africa and the E. Mediterranean region, and are all nearly allied to that species, although none are quite identical with it.

9. **P. cinerea,** *Lindl. in Mitch. Three Exped.* ii. 65. A perennial with ascending or erect branching stems as in *P. patens,* but more slender, minutely hoary as well as the leaves. Leaves pinnately 3-foliolate; leaflets ovate or elliptical, cuneate at the base, mostly ¾ to 1 or 1½ in. long, mucronate and irregularly denticulate. Racemes pedunculate, slender, loose and much longer than the leaves. Flowers very small, shortly pedicellate. Calyx scarcely above 1 line long, rather open, the teeth short broad and nearly equal. Petals but little longer than the calyx, the keel very obtuse. Pod about as long as the calyx, slightly hairy.—*P. Drummondii,* Meissn. in Bot. Zeit. 1855, 31.

N. Australia. Victoria river, *Bynoe, F. Mueller;* Gulf of Carpentaria, *Landsborough.*
N. S. Wales. Lachlan and Darling rivers, *A. Cunningham, Mitchell, Dallachy and Goodwin.*
W. Australia. Between Moore and Murchison rivers, *Drummond, 6th Coll. n.* 33.

10. **P. leucantha,** *F. Muell. in Trans. Vict. Inst.* iii. 54. A tall undershrub or shrub with spreading branches, minutely hoary-tomentose or glabrous. Leaves pinnately 3-foliolate or some of them 1-foliolate; leaflets oblong or lanceolate, mucronate but otherwise obtuse, entire or bordered by small crenatures or prominent glands, 1½ to 3 or even 4 in. long, hoary-tomentose or nearly glabrous. Flowers small, white with a blue keel, pedicellate in rather loose pedunculate racemes of 2 to 3 in. or sometimes longer, often clustered along the rhachis. Calyx about 1½ lines long, the lobes of nearly equal length, the 2 upper ones united, the lowest broader than the others. Petals twice as long as the calyx; standard obovate; keel rather shorter. Pod nearly glabrous, exceeding the calyx, very oblique and much wrinkled.

N. Australia. Hammersley range, Nichol Bay, *F. Gregory's Expedition;* gravelly banks of the Victoria river, *F. Mueller, Bynoe;* Attack Creek, *M'Douall Stuart;* Arnhem's Land, *F. Mueller;* Gulf of Carpentaria, *Landsborough.*
Queensland. Bowen river, *Bowman;* Bogie river, and Edgecombe Bay, *Dallachy.*

11. **P. tenax,** *Lindl. in Mitch. Three Exped.* ii. 10. A perennial with decumbent or ascending rather slender branching-stems of 1 to 2 ft., glabrous or minutely pubescent with appressed hairs. Leaves digitately 5- or 7-foliolate with linear-lanceolate or oblong-elliptical acute leaflets of ¾ to 1½ in., or the lower leaves with 3 broader and more obtuse leaflets, all quite entire, glabrous or sprinkled underneath with a few appressed hairs. Flowers small, blue (or purple?), very shortly pedicellate in dense or interrupted racemes of 1 to 4 in., on very long peduncles. Calyx 1 to 1½ lines long, the lobes acute, as long as the tube, the 2 upper ones united to the middle, the lowest one broad. Standard usually not half as long again as the calyx; keel shorter, obtuse. Pod ovoid, slightly pubescent, about as long as the open fruiting calyx.

Queensland, *Bidwill;* Dawson river and Peak Downs, *F. Mueller.*
N. S. Wales. Grassy banks of rivers towards the Lachlan and Darling, *A. Cunningham, Dallachy and Goodwin;* open forest land, Argyle county, *A. Cunningham;* Puen Buen, *American Exploring Expedition;* New England, *C. Stuart;* head of the Gwydir and plains of the Condamine, *Leichhardt;* Clarence river, *Beckler.* In *R. Brown's* collection without the precise station.
Var. (?) *major.* Leaves all 3-foliolate with broader leaflets. Standard nearly twice as long as the calyx.—N. S. Wales, *A. Cunningham;* Moreton Bay, *C. Stuart.*

12. **P. adscendens,** *F. Muell. in Trans. Vict. Inst.* i. 40. A perennial with a woody rhizome and straggling or ascending branching stems of 1 to 2 ft. or more, glabrous or sparingly pubescent. Leaves digitately 3-foliolate; leaflets mostly lanceolate or oblong-elliptical, quite entire, 1 to 2 or even 3 in. long, glabrous or sprinkled with a few hairs, those of the lower leaves smaller and broader. Flowers purple pink or white, nearly sessile in dense ovoid or cylindrical spikes of ¾ to 3 in., rarely fewer-flowered and almost interrupted, on very long peduncles. Calyx 2 to 2½ lines long, clothed with soft silky white or black hairs, lobes scarcely so long as the tube and nearly equal. Standard about half as long again as the calyx; keel short. Pod

ovate, densely granular-glandular, much shorter than the calyx.—*P. Gunnii,* Hook. f. Fl. Tasm. i. 99.

N. S. Wales. Near Naugas, *Backhouse.*

Victoria. Grassy moist banks of rivers and along the torrents in the Australian Alps, *F. Mueller.*

Tasmania. Near Woolnorth, *Gunn.*

S. Australia. Torrens and Gawler rivers, Barossa range, near Villunga, etc., *F. Mueller.*

Var. *parva.* Flowers rather smaller, few, in less dense spikes, calyx hoary-pubescent.—*P. parva,* F. Muell. in Trans. Vict. Inst. i. 40.—Dry pastures, Thompson and Latrobe rivers, *F. Mueller,* also most of the S. Australian specimens.

The works in which Dr. Hooker and Dr. Mueller respectively described this species were published at about the same time, but the latter may have been before the Australian world a few months before the former was issued in England.

31. INDIGOFERA, Linn.

(Sphæridiophorum, *Desv.*)

Calyx-tube short broad and oblique, the teeth or lobes nearly equal or the lowest longest. Standard ovate or orbicular, sessile or narrowed into a short claw; keel erect, obtuse or acuminate, with a hollow protuberance or spur on each side. Upper stamen free from the base, the others united in a sheath open on the upper side; anthers uniform, tipped by the point of the connectivum resembling a small gland. Ovary sessile or nearly so, with several or rarely 1 or 2 ovules; style incurved at the top, with a terminal stigma. Pod oblong, linear or rarely globular, terete or rarely flattened, straight or incurved, 2-valved, divided transversely between the seeds by cellular tissue. Seeds globular, or truncate at each end, or flattened, not strophiolate.—Herbs undershrubs or shrubs, more or less clothed or sprinkled with appressed hairs attached by the centre, sometimes mixed with loose hairs or tomentum. Leaves in the Australian species 1-foliolate or pinnate with 3 or more leaflets, occasionally stipellate. Stipules small, setaceous. Flowers usually red or purple, in axillary spikes or racemes. Bracts usually small and deciduous. Bracteoles none. Standard usually silky-pubescent outside.

A very large and distinct genus widely spread over the warmer regions of the globe, especially numerous in tropical and southern Africa. Of the Australian species 9 out of the 11 herbaceous ones are common in India, the remaining 2 herbaceous ones and the 6 shrubby ones are all endemic.

Calyx-lobes very much longer than the very short tube. Herbs or undershrubs (except *I. rugosa*).
  Leaves simple, nearly sessile.
    Leaves linear or narrow-oblong.
      Flowers in short sessile spikes. Pod globular, 1-seeded . .   1. *I. linifolia.*
      Flowers in long pedunculate racemes. Pod linear, several-seeded . . . .   5. *I. haplophylla.*
    Leaves cordate-ovate. Pod short, usually 2-seeded . . . .   2. *I. cordifolia.*
  Leaflets single, ovate, rugose, very white, on a petiole of 3 to 4 lines . . . . 13. *I. rugosa.*
  Leaves pinnately 3-foliolate. Flowers scarcely 2 lines long, in very short sessile spikes.
    Plant conspicuously glandular-dotted. Ovules 2. Pod ovoid-oblong, 2 lines long, reflexed, pubescent and glandular . .   4. *I. glandulosa.*

Plant with very small glandular dots. Ovules 8 or more.
   Pod slender, reflexed, glabrous, nearly ½ in. long . . . .   6. *I. trifoliata.*
Plant pale or hoary. Ovules many. Pod spreading, slightly
   incurved, 1 to 1½ in. long, obtusely 4-angled . . . . .   7. *I. trita.*
Leaves pinnate, with several pairs of leaflets.
  Pod short. Ovules and seeds 2. Spikes short dense and sessile   3. *I. enneaphylla.*
  Pod linear. Ovules and seeds several.
    Calyx much shorter than the petals.
     Flowers rather crowded in sessile racemes. Pod 1 to 1½
      in. long, with an incurved point . . . .   8. *I. parviflora.*
     Flowers very small, distant, in slender racemes. Pod ½ to
      ¾ in. long, straight, very slender, often viscid . . .   9. *I. viscosa.*
    Calyx-lobes about as long as the petals. Racemes loose.
     Plant hirsute with spreading hairs . . . . 10. *I. hirsuta.*
Calyx-teeth all very short, the lower ones rarely rather longer than
  the tube. Shrubs (except *I. pratensis*).
  Leaflet 1, articulate on the petiole.
   Leaflet broadly obovate, with parallel prominent pinnate veins . 12. *I. monophylla.*
   Leaflet ovate or oval-oblong, reticulate and very rugose . . . 13. *I. rugosa.*
  Leaflets usually 5, obovate or orbicular, the veins scarcely con-
   spicuous . . . . 14. *I. saxicola.*
  Leaflets 9 or more.
   Herb or undershrub. Leaflets mucronate and distinctly veined.
    Stipellæ setaceous. Flowers 5 to 6 lines long . . . . . 11. *I. pratensis.*
   Shrubs. Leaflets obscurely veined. Stipellæ none or replaced
    by small glands.
    Calyx truncate, the teeth scarcely prominent. Plant nearly
     glabrous. Flowers 3 to 4 lines long, on rather long pedi-
     cels. Pod quite glabrous . . . . 15. *I. australis.*
    Calyx-teeth distinct, the lowest as long as the tube or nearly
     so.
     Whole plant slightly canescent. Pods pubescent, at least
      when young. Leaflets contracted at the base or petiolu-
      late . . . . 16. *I. brevidens.*
     Branches densely tomentose. Leaflets sessile, orbicular or
      nearly so, thick, glabrous on both sides . . . . . 17. *I. coronillæfolia.*

I. atropurpurea, Ham. (Wight, Ic. t. 369), a Himalayan shrubby species which is occasionally grown in our plant-houses, has been sent by C. Moore from Richmond river, and
is in Leichhardt's collection from Glendon, but is probably not indigenous to Australia.
I. decora, Lindl. Bot. Reg. 1846, t. 22, a Chinese shrubby species, has also escaped from
gardens in the neighbourhood of Moreton Bay.

§ 1. *Herbaceous species.*

1. **I. linifolia,** *Retz ; DC. Prod.* ii. 222. A slender much-branched
diffuse or procumbent annual or perennial of ½ to 1½ ft., more or less hoary
or white. Leaves simple, almost sessile, linear or rarely oblong-lanceolate,
½ to 1½ in. long. Flowers very small, in sessile spikes, very short when in
flower, and rarely lengthening to ½ in. when in fruit. Calyx-teeth subulate-
pointed, much longer than the tube, the lower ones as long as the petals.
Standard sessile, about 1½ lines long. Ovary sessile, with 1 ovule. Pod
nearly globular, white-tomentose, about 1 line diameter.—W. and Arn. Prod.
198 ; F. Muell. Fragm. iii. 101 ; *Sphæridiophorum linifolium*, Desv., and *S.
abyssinicum*, Spach, in Jaub. and Spach, Ill. Pl. Or. v. 103, t. 494.

N. Australia. N. Goulburn Island, *A. Cunningham ;* Victoria river, *F. Mueller ;*
Port Essington, *Armstrong ;* Sweers island, *Henne.*

Queensland. Endeavour river, *Banks and Solander;* Shoalwater Bay, *R. Brown;* Dawson river, *F. Mueller;* Rockhampton, *Dallachy;* Bremer river, *Fraser;* also in *Leichhardt's* collection.

N. S. Wales. Sandy banks of Hunter's River, *A. Cunningham.*

The species is common in tropical Asia and some parts of Africa. Wight, Ic. t. 313, re-presents a short broad-leaved variety not found in Australia.

2. **I. cordifolia,** *Heyne; DC. Prod.* ii. 222. A diffuse or prostrate branching annual or perennial of a few inches or rarely extending to nearly 1 ft., hoary or white with forked hairs, not so appressed as usual in the genus. Leaves simple, nearly sessile, broadly ovate-cordate, mostly about ½ in. long. Flowers very minute, in short sessile spikes or clusters. Calyx-tube scarcely any; lobes subulate, about 1 line long. Petals scarcely exceeding the calyx; standard narrowed into a long claw; keel acuminate. Ovary with 2 ovules. Pod about 2 lines long, pubescent, usually 2-seeded.

N. Australia. Port Essington, *Armstrong.* Found also in tropical Africa, E. India and Timor.

3. **I. enneaphylla,** *Linn.; DC. Prod.* ii. 229. A prostrate straggling or rarely erect perennial of 1 to 1½ ft., hoary or almost silky-pubescent, the branches angular with the hairs sometimes spreading. Leaflets 5 to 9, mostly alternate or scarcely opposite, obovate or oblong-cuneate, 3 to 4 or rarely 5 lines long. Flowers very small, in sessile spikes which are short and dense when in flower, rarely ½ in. long in fruit. Calyx-teeth much longer than the tube and shorter than the petals. Standard about 3 lines long, narrowed into a broad claw; keel narrow, almost acuminate. Ovary with 2 or rarely 3 ovules. Pod terete, about 3 lines long, usually 2-seeded.—W. and Arn. Prod. 199; Wight, Ic. t. 403; F. Muell. Fragm. iii. 102.

N. Australia. Nichol Bay, *F. Gregory's Expedition;* Depuech Island, *Bynoe;* stony, chiefly basaltic, plains and hills, Upper Victoria river, *F. Mueller.*

Queensland. Bay of Inlets, *Banks and Solander;* Keppel Bay, *R. Brown;* Dawson river, *F. Mueller;* Fitzroy river, *Dallachy;* Port Denison, *Fitzalan;* Connor's River, *Bowman.*

N. S. Wales. Between Darling river and Cooper's Creek, *Neilson.*

The species is common in the plains of India.

4. **I. glandulosa,** *Willd.; DC. Prod.* ii. 223. An annual or perennial with the habit nearly of *I. trifoliata,* but usually rather stouter and taller, more pubescent, and conspicuously marked with glandular dots, especially on the under side of the leaflets. Leaflets 3, from obovate to narrow-oblong, obtuse, ¾ to 1 in. long, more hoary than in *I. trifoliata.* Flowers very small, in sessile racemes very short at first, but lengthening out to nearly ½ in. Calyx glandular and hirsute, about 1 line long, the lobes subulate, much longer than the tube. Standard about 2 lines long, narrowed at the base, pubescent and glandular outside. Ovules 2. Pod reflexed, ovoid-oblong, almost 4-angled, about 2 lines long, glandular and pubescent.—W. and Arn. Prod. 199; Wight, Ic. t. 330; *Psoralea Leichhardtii,* F. Muell. Fragm. iv. 22.

Queensland. Comet river, *Leichhardt.* The species is widely spread over E. India. It has the habit and foliage of *I. trifoliata,* with the pod of *I. enneaphylla.*

5. **I. haplophylla,** *F. Muell. Fragm.* iii. 102. An erect or rarely diffuse branching herb of 1 ft. or rather more, pale or hoary with a slight pubescence. Leaves simple, almost sessile, linear or very narrow-oblong, mostly 1 to 2 in.

long. Racemes slender, pedunculate, scarcely exceeding the leaves. Calyx-lobes much longer than the tube, but much shorter than the petals. Standard narrowed at the base, but not clawed; keel obtuse. Pod straight, rather slender, cylindrical, spreading, often 1 in. long or rather more.

N. Australia. Rocky springs and torrents on the Upper Victoria river, *F. Mueller ;* islands of the Gulf of Carpentaria, *R. Brown.*

6. **I. trifoliata,** *Linn. ; DC. Prod.* ii. 223. Stock perennial, with seve-ral ascending or erect rather slender stems of ½ to 1 or 1½ ft., the pubescence very short and scarcely hoary. Leaflets 3, on a slender petiole, obovate-ob-long cuneate or narrow-oblong, mostly ½ to 1 in. long, green or slightly hoary underneath. Racemes sessile, exceedingly short. Flowers very small. Calyx-teeth much longer than the tube, but scarcely so long as the claws of the lower petals. Standard about 2 lines long, narrowed into a short broad claw ; keel obtuse, pubescent and glandular as well as the standard. Pod slender, reflexed, about ½ in. long, with 4 prominent angles or narrow longitudinal wings, many-seeded.—W. and Arn. Prod. 201 ; Wight, Ic. t. 314 ; F. Muell. Fragm. iii. 104.

Queensland. Bustard Bay, *Banks and Solander ;* Northumberland island, *R. Brown ;* Glasshouse mountains, *F. Mueller ;* along the coast and adjoining islands, *A. Cunningham, M'Gillivray, Henne, Dallachy,* also in *Leichhardt's* collection. The species is widely spread over E. India and the Archipelago. It is always much more slender than *I. trita,* with more closely sessile short spikes or clusters of much smaller flowers.

7. **I. trita,** *Linn. f. ; DC. Prod.* ii. 232. A decumbent or suberect per-ennial of ½ to 1½ ft., pale or hoary with a minute pubescence. Leaflets 3, or very rarely 5, on a rather rigid petiole, from broadly obovate and 3 or 4 lines long, to elliptical-oblong and above 1 in. long. Flowers small, very nearly sessile ; the racemes sometimes short, dense, and nearly sessile, some-times pedunculate, interrupted, and attaining several inches. Calyx-teeth much longer than the tube, but not exceeding the claws of the lower petals. Standard nearly 3 lines long, narrowed at the base but scarcely clawed ; keel almost acu-minate. Pod rather rigid, usually incurved, obscurely quadrangular, with thickened sutures, 1 to 1½ in. long or even more.—Hook. Comp. Bot. Mag. i. t. 16 ; W. and Arn. Prod. 203 ; Wight, Ic. t. 315, 386 ; F. Muell. Fragm. iii. 103.

N. Australia. Gravelly plains, Upper Victoria river, *F. Mueller ;* islands of the Gulf of Carpentaria, *R. Brown, Henne ;* and adjoining mainland, *Landsborough.*
Queensland. In the interior, *Mitchell ;* Peak Downs, *F. Mueller ;* Bowen river, *Bowman.*
N. S. Wales. Between Darling river and Cooper's Creek, *Neilson.*
The species is widely spread over E. India and the Archipelago. *I. Leschenaultii* and *I. timoriensis,* DC. Prod. ii. 223, which have been referred to *I. trifoliata,* both belong to *I. trita.*

8. **I. parviflora,** *Heyne ; W. and Arn. Prod.* 201. An erect herb of 1 to 2 ft., pale or hoary with a minute pubescence. Leaflets 9 to 13 or rarely fewer, linear or rarely oblong, mostly ½ to 1 in. long. Racemes usually short, rather loose, rarely lengthening out to 1 or 2 in., and flowering almost from the base. Calyx small, the lobes much longer than the tube but much shorter than the petals. Standard 2 to 2½ lines long, narrowed into a short claw, glabrous or nearly so ; keel terminating in a linear obtuse point protruding beyond the wings. Anthers small, tipped with a minute point. Pod nearly

glabrous, linear, with thickened sutures, 1 to 1½ in. long, straight except an incurved or hooked end.—*I. deflexa*, Hochst. in A. Rich. Fl. Abyssin. i. 178; *I. oxycarpa*, F. Muell. Fragm. iii. 103.

N. Australia. Nichol Bay and De Grey river, *Ridley's Expedition ;* stony hills and gravelly banks, Victoria river and Sturt's Creek, *F. Mueller.*
Queensland. Walloon, *Bowman.*
The species is common in the E. Indian peninsula, also in Abyssinia and Cordofan.

9. **I. viscosa,** *Lam. ; DC. Prod.* ii. 227. A slender wiry annual or perennial, with much-branched decumbent or erect stems of ½ to 1 ft., more or less clothed with spreading glandular viscid hairs, mixed with the ordinary pubescence of the genus. Leaflets 9 to 15, ovate or oblong, sometimes all under 2 lines, sometimes 3 to 4 lines long. Flowers very small, distant, in slender racemes rather shorter than the leaves. Calyx-lobes much longer than the tube, but not exceeding the claws of the lower petals. Standard almost sessile, about 1½ lines long ; keel obtuse, the lateral spurs very short. Pod slender, straight, spreading or pendulous, ½ to ¾ in. long, torulose, with viscid hairs mixed with the ordinary pubescence.—W. and Arn. Prod. 200 ; Wight, Ic. t. 404 ; F. Muell. Fragm. iii. 104.

N. Australia. Victoria river, *F. Mueller ;* Brinkley's Bluff, *M'Douall Stuart;* islands of the Gulf of Carpentaria, *R. Brown ;* Port Essington, *Armstrong.*
Queensland. Endeavour river, *Banks and Solander, R. Brown ;* Port Curtis, *M'Gillivray ;* E. coast, *A. Cunningham ;* Port Denison, *Fitzalan ;* Rockhampton, *Bowman ;* Comet river, *Leichhardt.*
The species is widely spread over tropical Asia and Africa.

10. **I. hirsuta,** *Linn. ; DC. Prod.* ii. 228. A decumbent or ascending branching annual, 1 to 2 ft. high, remarkable for the spreading hairs which clothe the branches, petioles, inflorescence, and calyx. Leaflets 7 to 11, obovate or oblong, ¼ to 1 in. long, with stiff appressed hairs. Racemes usually dense, shortly pedunculate, 1 to 4 in. long. Calyx with scarcely any tube, the subulate lobes often nearly as long as the petals. Standard fully 3 lines long, narrowed into a distinct claw. Pod about ½ in. long, straight, quadrangular, reflexed on the peduncle, very hirsute.—W. and Arn. Prod. 204 ; Hook. Comp. Bot. Mag. ii. t. 24 ; Benth. Fl. Hongk. 76 ; F. Muell. Fragm. iii. 105.

N. Australia. N. coast, *R. Brown ;* Victoria river and Arnhem's Land, *F. Mueller ;* Port Essington, *Armstrong ;* islands off the coast, *A. Cunningham, Henne.*
Queensland. Bay of Inlets, *Banks and Solander ;* Keppel Bay, *R. Brown ;* Port Denison, *Fitzalan ;* Rockhampton, *Bowman ;* Taylor's Range, *Fraser ;* Brisbane river, *F. Mueller.*
N. S. Wales. Hunter's River, *Leichhardt.*
The species is widely distributed over tropical Asia and Africa, and now introduced also into some parts of tropical America.

11. **I. pratensis,** *F. Muell. Rep. Burdek. Exped.* 10. A diffuse perennial, pale or hoary with the ordinary pubescence of the genus, the branches angular, ascending to 1 or 2 ft. Leaflets about 13 to 21, from broadly oval-oblong to narrow-oblong, obtuse with a fine straight point, ½ to 1 in. long, the pinnate veins usually conspicuous underneath. Stipules setaceous, often 3 to 4 lines long, and small setaceous stipellæ usually present. Flowers rather large, in pedunculate racemes longer than the leaves, the pedicels 2 to

3 lines long. Calyx above 1 line long, the teeth shorter than the tube as in the shrubby species. Standard nearly 6 lines long; keel almost acuminate. Pod cylindrical, straight, rather thick, 1 to 1½ in. long.

Queensland. Bay of Inlets, *Banks and Solander;* Broad Sound and Keppel Bay, *R. Brown;* along various points of the E. coast and adjoining islands, *A. Cunningham, M'Gillivray, Henne,* and others; in the interior, *Mitchell;* on the Burdekin, *F. Mueller;* Brisbane river, *Fraser, F. Mueller,* etc.; Mackenzie Hill, *Leichhardt.*

§ 2. *Shrubby species.*

12. **I. monophylla,** *DC. Prod.* ii. 222. A straggling shrub of 2 to 3 ft., the branches and foliage very hoary or white with minute hairs. Leaflets single, articulate on the petiole with a pair of minute stipellæ, broadly obovate, ¾ to 1 in. long, the parallel primary veins very prominent underneath. Racemes shortly pedunculate, at length exceeding the leaves. Calyx-teeth shorter than the tube, the upper ones broad and distant. Standard sessile, about 3 lines long; keel almost acute. Pod straight, above 1 in. long, softly tomentose.—F. Muell. Fragm. iii. 44.

N. Australia. Baudin's Expedition; Depuech Island, N.W. coast, *Bynoe;* Nichol Bay and Fortescue river, *F. Gregory's Expedition.*

13. **I. rugosa,** *Benth.* A shrub of 2 or 3 ft., the branches white with a dense soft tomentum. Leaflets solitary, ovate or oval-oblong, very obtuse, about ¾ in. long in the specimen before me, reticulately veined and very rugose, soft and white on both sides with a minute almost silky pubescence; the petiole articulate above the middle. Flowers in short dense nearly sessile racemes, the petals as well as the calyx white-tomentose. Calyx about 2 lines long, the subulate lobes longer than the tube. Standard nearly 4 lines long, broad, acute, sessile; lower petals narrow and rather shorter. Anthers very prominently apiculate. Ovary glabrous with about 10 ovules. Pod not seen.

N. Australia. Bed of the Fortescue river, N.W. coast, *F. Gregory's Expedition* (*Herb. F. Mueller*).

14. **I. saxicola,** *F. Muell. Herb.* A shrub of 3 or 4 ft. with spreading branches, slightly hoary with a minute pubescence. Leaflets 5 or rarely 7, obovate or orbicular and very obtuse, mostly about 1 in. long, on petiolules of 1 to 2 lines, the pinnate veins scarcely prominent. Racemes pedunculate, rather slender, longer than the leaves. Calyx-teeth shorter than the tube, the upper ones broad and distant. Standard sessile, 3 lines long or rather more; keel acute. Pod terete, spreading, ½ to ¾ in. long, straight.

N. Australia. Port Essington, *A. Cunningham;* grassy stony plains, Sea Range, *F. Mueller.*

15. **I. australis,** *Willd.; DC. Prod.* ii. 226. An erect branching shrub of 2 to 4 ft., assuming occasionally the appearance of a low undershrub, either glabrous or slightly sprinkled with the small hairs of the genus. Leaflets usually 9 to 17, oblong, obtuse or retuse, ½ to ¾ in. long, but varying to broadly ovate, almost orbicular in some specimens or nearly linear in others. Stipules small; stipellæ none except minute glands. Flowers red and showy, in dense or loose racemes shorter or rather longer than the leaves, the pedicels usually longer than the calyx. Calyx about 1 line long, broad

and obliquely truncate, the teeth either inconspicuous or the lower ones especially slightly prominent, but always much shorter than the tube. Standard truncate at the base, with an exceedingly short claw, 3 to 4 lines long. Pod spreading, terete, straight or nearly so, 1 to 1½ in. long.—Vent. Jard. Malm. t. 45 ; Bot. Reg. t. 386 ; Lodd. Bot. Cab. t. 149 ; Hook. f. Fl. Tasm. i. 99 ; *I. angulata,* Lindl. Bot. Reg. t. 991 ; *I. sylvatica,* Sieb. in Hook. Bot. Mag. t. 3000 ; *I. ervoides,* Meissn. in Pl. Preiss. i. 88.

Queensland. Brisbane river, Moreton Bay, *Fraser, A. Cunningham, F. Mueller,* and others.

N. S. Wales. Port Jackson to the Blue Mountains, *R. Brown, Sieber, n.* 379, 380, and others; Hunter's River, *Backhouse;* northward to Hastings and Clarence rivers, *Beckler;* New England, *C. Stuart;* and southward to Argyle county, *M'Arthur;* Twofold Bay, *F. Mueller;* and in the interior to the Darling and Lachlan rivers, *A. Cunningham, Neilson,* etc.

Victoria. Grassy places near Melbourne and Port Phillip to Gipps' Land, etc., *F. Mueller,* and others.

Tasmania. Derwent river, *R. Brown :* abundant in many places throughout the island, *J. D. Hooker.*

S. Australia. Lofty Ranges, *F.Mueller.*

W. Australia, *Drummond, 5th Coll. Suppl. n.* 44 ; rocks of Mount Matilda, *Preiss, n.* 1967.

Amidst all its variations, this species may be known by its glabrous, not hoary, aspect, notwithstanding the minute hairs often visible under a lens, by the very short or quite obsolete teeth of the calyx, and by the pod glabrous even when quite young. The following are the principal forms it assumes, which, although they often pass one into another, are nevertheless sometimes considered as distinct species.

a. *angulata.* Tall, with angular branches. Flowers large and showy. The most common form in our gardens, and received from Port Jackson, Port Phillip, and other maritime districts, the only one in Tasmania, and includes Drummond and Preiss's W. Australian specimens.

b. *gracilis,* DC. Branches terete and as well as the petioles and racemes more slender. Flowers rather smaller. Apparently common in Victoria, the Blue Mountains, and in the northern districts of N. S. Wales to the Brisbane in Queensland.

c. *minor.* More scrubby and branched, of a pale colour, the branchlets short and somewhat angular. Leaflets small, with small stipellary glands. Flowers small, in short racemes. Chiefly in the interior of N. S. Wales and S. Australia, on the Lachlan and Darling, etc., and northward to Clarence river.

d. *signata,* F. Muell. Rigid, very glabrous, apparently almost leafless, the numerous rigid petioles bearing very small obcordate obovate or cuneate leaflets in distant pairs, with very prominent dark-coloured stipellary glands. Flowers as in the var. *minor.*—Between Ovens river and Mayday Hills in Victoria, *F. Mueller;* Port Jackson, *Herb. Mueller;* and Queensland, *Bowman.*

e. *platypoda.* With the same rigid aspect and few small leaflets with prominent stipellary glands as the var. *signata,* but the common petioles very rigid and flattened, often above 1 line broad.—New England, *C. Stuart;* Arne river, *Beckler.*

16. **I. brevidens,** *Benth. in Mitch. Trop. Austr.* 385. A slender shrub, very nearly allied to *I. australis,* of which F. Mueller considers it a variety, but always hoary or silvery with the appressed forked pubescence of the genus or white with a denser tomentum. Leaflets from about 9 to 21, obovate or oblong, obtuse or mucronate, rarely ½ in. long, usually firmer than in *I. australis,* and hoary or white on both sides, more or less petiolulate. Stipules rather short and deciduous or rarely more persistent and recurved. Flowers rather smaller than in *I. australis,* the calyx-teeth much more prominent although still very short, the lowest occasionally as long as the tube.

Standard densely silky-pubescent. Pod always pubescent or tomentose, at least when young.—*I. lasiantha*, F. Muell. in Rep. Greg. Exped. 6.

N. Australia. M'Donnell Range and up to lat. 20° 20′, *M'Douall Stuart's Expedition.*

Queensland. St. George's Bridge on the Balonne, *Mitchell ;* Peak Downs, *F. Mueller ;* in the scrub north of Expedition Range, *Leichhardt ;* also in *Bowman's* and other collections.

N. S. Wales. On the Darling, *Neilson* and others.

S. Australia. Cooper's Creek, *A. C. Gregory ;* towards Spencer's Gulf, *Warburton.*

W. Australia. Murchison river, *Oldfield.*

Var. *uncinata.* Stipules persistent, broader at the base, recurved and sometimes spinescent.

Var. (?) *galegoides*, R. Br. Branches softly tomentose ; leaflets numerous, very white underneath. Pod small, loosely pubescent.—Cumberland Islands, *R. Brown.*

17. **I. coronillæfolia,** *A. Cunn. Herb.* Shrubby, the branches very white with a dense soft tomentum. Leaflets 15 to 21 or more, sessile, broadly ovate or orbicular, very obtuse, 2 to 4 lines long, thick, veinless, and almost glabrous, the common petiole white-tomentose. Racemes about as long as the leaves. Calyx distinctly but shortly toothed. Petals hoary-pubescent. Pod straight, nearly terete, about 1 in. long, glabrous.

N. S. Wales. In the barren rocky country west of Wellington Valley, *A. Cunningham ;* Castlereagh, *C. Moore.* The species requires further investigation from more perfect specimens, but the foliage gives it a different aspect from any forms of *I. australis* or of *I. brevidens.*

A fragment from M'Donnell's Range, M'Douall Stuart's Expedition, with the leaflets white tomentose on both sides, has some affinity to *I. coronillæfolia*, but is too imperfect for determination.

32. PTYCHOSEMA, Benth.

Calyx turbinate, the 2 upper lobes united in a truncate emarginate upper lip. Petals on rather long claws ; standard nearly orbicular, emarginate ; wings falcate-oblong, free ; keel shorter, nearly straight, obtuse. Stamens all united in a sheath open on the upper side ; anthers uniform. Ovary sessile, with several ovules ; style short, inflexed, the stigma oblique outwards. Pod . . .—Herb. Leaves unequally pinnate. Peduncles terminal, 1-flowered.

The genus is limited to the single Australian species, apparently allied on the one hand to *Goodia*, on the other to *Tephrosia*.

1. **P. pusillum,** *Benth. in Lindl. Swan Riv. App.* 16. A small, slender, nearly glabrous perennial, the diffuse stems mostly 2 to 3 in. long. Leaflets 7 to 11, from obovate to linear-cuneate, obtuse or acute, mostly 2 to 3 lines long, much narrowed at the base, green on both sides, the primary veins very prominent underneath. Peduncles slender, terminal, 1 to 2 in. long, articulate and bearing 1 or 2 small bracts at about two-thirds of their length, the upper portion or pedicel bearing 2 bracteoles at some distance from the calyx. Calyx slightly hirsute, about 3 lines long, the lobes about as long as the tube, the 3 lower ones lanceolate, acute. Standard about 4 lines diameter, strongly striate-veined outside, without callosities inside. Ovules 4 or 5. Pod unknown.

W. Australia. Swan River, *Drummond, 1st Coll.,* and *n.* 251.

33. LAMPROLOBIUM, Benth.

Calyx deeply cleft, the 2 upper lobes united nearly to the top. Standard orbicular, narrowed into a short claw; wings obliquely oblong, free; keel much curved, obtuse. Stamens all united in a sheath open on the upper side; anthers uniform. Ovary shortly stipitate, with several ovules; style filiform, incurved, with a terminal stigma. Pod stipitate, oblong-linear, very flat, 2-valved, with transverse partitions between the seeds, the valves coriaceous. Seeds oblong, with a fleshy strophiole. Radicle short, quite straight. —Shrub. Leaves pinnate, without stipellæ. Stipules minute. Flowers yellow, small, solitary (or 2 or 3 ?) on terminal or lateral peduncles. Bracts and bracteoles minute and very deciduous.

The species is limited to a single species endemic in Australia. In the structure of the seeds, with a straight embryo, it differs from · all *Galegeæ* except the S. American genera *Brongniartia* and *Harpalyce*.

1. **L. fruticosum,** *Benth.* An erect shrub of a man's height, the branches softly pubescent. Leaflets 3, 5 or 7, or rarely solitary in the upper leaves, oblong, obtuse or mucronate, 1 to 2 in. long, coriaceous, glabrous or sprinkled with appressed hairs above, silky-pubescent underneath. Peduncles short, terminal axillary or extra-axillary and all apparently 1-flowered in the specimens seen, but perhaps sometimes bearing a raceme of 2 or 3. Calyx silky-villous, 3 to 4 lines long, like that of some *Crotalarias*, the 2 upper lobes falcate and united in a concave upper lip. Petals not exceeding the calyx. Pod 1 to 1½ in. long, 3 or 4 lines broad, glabrous and smooth. Seeds transverse.—*Crotalarioides fruticosa*, Soland. ms.; *Glycine lamprocarpa*, A. Cunn. Herb.

Queensland. Endeavour river, *Banks and Solander, A. Cunningham.*

34. TEPHROSIA, Pers.

Calyx-teeth or lobes nearly equal, or the 2 upper ones more united, or the lowest the longest. Petals clawed; standard nearly orbicular, usually reflexed; wings slightly adhering to the keel; keel incurved, obtuse or scarcely acute. Upper stamen free at the base, usually geniculate and at first united with the others in the middle in a tube or sheath, often quite free as the flowering advances; anthers uniform. Ovary sessile, with many or rarely 1 or 2 ovules; style in the Australian species glabrous, incurved or inflexed, more or less flattened with a terminal stigma, often slightly penicillate. Pod linear or rarely ovate, flattened, 2-valved. Seeds often with a small strophiole.—Herbs undershrubs or, in species not Australian, shrubs. Leaves pinnate; leaflets usually opposite with a terminal odd one, sometimes reduced to a single leaflet, either sessile or articulate on the petiole, the veins in most species numerous, parallel and oblique with the midrib. Flowers red purple or white, in pairs or clusters, in terminal, leaf-opposed or rarely axillary racemes, the lower clusters occasionally or sometimes all in the axils of the leaves. Bracteoles none. Standard always and the keel sometimes pubescent or silky-villous with appressed hairs.

A large genus, widely spread over the warmer regions of the New and the Old World, and particularly numerous in species in S. Africa. The Australian species are all endemic with the exception of *T. purpurea*, and, even of that, scarcely any of the Australian varieties quite agree with the common Asiatic and African forms. The Australian species also all, with the exception perhaps of *T. flammea* and *T. crocea*, belong to the section *Reineria*, with terminal or leaf-opposed racemes or axillary clustered pedicels, and to the large subsection with subulate or small stipules, except perhaps *T. venulosa*, in which they are broad and striate, but not so much so as in the S. African *Apodynomenes*. Several species differ from all extra-Australian ones in the venation of the leaflets. One Australian species also, *T. brachycarpa*, is remarkable for having the flowers and fruit of the African *Requienias*, whilst another, *T. coriacea*, has very nearly their foliage. In general the Australian species, more even than the Asiatic ones, are extremely difficult to define; the terminal or axillary, racemose or clustered inflorescences, usually so distinct, seem to pass the one into the other or to be blended together even on the same specimen, the foliage and indumentum is more than usually diversified and variable, and when to this is added the imperfection of the specimens we possess from tropical Australia, it must be expected that further investigation may considerably modify the circumscriptions of several of the species here described.

Leaflets obovate, oval, elliptical, or oblong, the primary veins anasto-
　mosing or reticulate within the margin.
　Leaflets solitary or rarely 3, coriaceous. Flowers in axillary clus-
　　ters. Pod softly tomentose . . . . 1. *T. coriacea.*
　Leaflets mostly 5 to 11, rarely under 1 in. long. Racemes elon-
　　gated.
　　Plant softly tomentose or silky. Flowers numerous. Calyx 4
　　　to 5 lines long, softly villous, lobes longer than the tube . . 2. *T. flammea.*
　　Plant nearly glabrous. Flowers few. Calyx scarcely 2 lines
　　　long, the teeth very short . . . . 3. *T. reticulata.*
　Leaflets numerous, above ½ in. long, glabrous above, silky-pubes-
　　cent or villous underneath. Racemes long.
　　Leaflets ½ to 1 in. long, very silky underneath, the veins reticu-
　　　late. Stipules persistent. Bracts small . . . . 4. *T. crocea.*
　　Leaflets 1 to 2 in. long, silky-pubescent underneath, the primary
　　　veins parallel but anastomosing within the margin. Stipules
　　　very deciduous. Bracts linear-subulate, long . . . . . 5. *T. oblongata.*
　Leaflets numerous, not ½ in. long. Racemes long.
　　Plant loosely pubescent or villous. Stipules striate, reflexed.
　　　Leaflets 11 to 19 . . . . 6. *T. porrecta.*
　　Plant closely silky-pubescent. Stipules minute, erect. Leaflets
　　　30 to 40 or more . . . . 7. *T. polyzyga.*
Leaves all or mostly simple or 1-foliolate. Leaflets long and linear
　or cuneate-oblong, the veins mostly reaching the margin or irre-
　gular.
　Leaves sessile, long, linear-lanceolate, the veins anastomosing.
　　Flowers small . . . . 8. *T. graminifolia.*
　Leaflets shortly petiolate, long, narrow-linear, without stipellæ.
　　Flowers middle-sized . . . . 9. *T. simplicifolia.*
　Leaflets long and narrow-linear, either solitary with 2 stipellæ or 3
　　with the middle one sessile or rarely another pair lower down.
　　Flowers very small . . . . 10. *T. leptoclada.*
　Leaflets cuneate-oblong, 1 or rarely 3 or 5. Flowers large . . 19. *T. oligophylla.*
Leaves pinnate. Primary veins of the leaflets oblique, numerous, and
　parallel.
　Flowers nearly sessile, mostly axillary. Pod straight, densely and
　　softly villous. Plant softly tomentose or villous.
　　Flowers not above 2 lines long. Ovule 1. Pod ovate . . . 11. *T. brachycarpa.*
　　Flowers about 3 lines. Ovules 2 or rarely 3. Pod ½ to ¾ in.
　　　long . . . . 12. *T. Stuartii.*

Flowers about 4 lines. Ovules about 6. Pod 1 to 1½ in. long 13. *T. eriocarpa.*
Flowers in short dense terminal racemes. Leaflets narrow, silvery-
silky underneath.
Pod straight. Standard about 4 lines diameter. Leaflets usually
silky on both sides . . . . 14. *T. phæosperma.*
Pod incurved towards the end. Standard nearly 6 lines diameter.
Leaflets usually green above . . . . 15. *T. astragaloides.*
Flowers in long or slender racemes.
Leaflets small, numerous, with a long terminal one. Flowers
small. Pod long . . . . 16. *T. juncea.*
Leaflets few, or, if many, the terminal one not longer than the
others.
Pod about 1 in. long, nearly straight. Seeds orbicular.
Leaflets 7 to 15. Flowers rarely 3 lines long. Pod ob-
liquely acute, thin.
Racemes filiform, not 2 in. long, with few distant pairs
of flowers . . . . 17. *T. filipes.*
Racemes long and slender . . . . 18. *T. remotiflora.*
Leaflets solitary or rarely 3 or 5. Flowers about 6 lines
long. Pod coriaceous, almost obtuse . . . . . . 19. *T. oligophylla.*
Pod above 1 in. long, more or less incurved. Seeds transversely oblong. Racemes usually long.
Leaflets few, cuneate, glabrous, on long petioles. Calyx
large, nearly glabrous ; lobes lanceolate. Standard very
silky . . . . 20. *T. macrocarpa.*
Leaflets usually above 7. Calyx small, on a slender pedicel ; teeth subulate or very short. Pod glabrous pubescent or loosely villous . . . . 21. *T. purpurea.*
Leaflets usually few, long, and narrow. Calyx large,
densely rusty-villous ; lobes longer than the tube, incurved, acuminate. Pedicels short. Pod softly and
closely pubescent . . . . 22. *T. Bidwilli.*
Leaflets usually few, cuneate, silky on both sides. Calyx
small, softly silky. Pedicels short. Pod much curved,
scarcely flattened, densely silky-tomentose . . . . . 23. *T. rosea.*

1. **T. coriacea,** *Benth.* Branches softly pubescent. Leaflets solitary,
on a petiole of ¼ to ½ in., or rarely 3, obovate, very obtuse or retuse, 1 to 2
in. long, coriaceous, minutely pubescent, the pinnæ, veins, and reticulate veinlets prominent underneath, as in *T. flammea.* Flowers in axillary clusters,
smaller than in *T. flammea,* but imperfect in our specimens. Calyx villous,
the lobes long and acute, the lowest exceeding the others and twice as long
as the tube. Upper stamen free at the base, but not so geniculate as in
most species. Pod linear, slightly falcate, softly pubescent, above 1 in. long.
Seeds lenticular, with a small strophiole.

N. Australia. Upper Victoria and Fitzmaurice rivers, *F. Mueller.* It is possible
that in other specimens the branches may terminate in racemes, yet the flowers, as far as
known, and seeds appear to be quite different from those of *T. flammea,* the only allied species. The foliage is nearly that of the African *Requienias.*

2. **T. flammea,** *F. Muell. Herb.* An erect branching undershrub, of 3
to 4 ft., the branches clothed with a loose velvety-rusty pubescence. Leaflets
5 to 9, or 3 in the uppermost leaves, broadly elliptical-oblong or almost
ovate or obovate, very obtuse, 1 to 2 in. long, shortly and softly pubescent or
almost silky on both sides, the somewhat distant primary veins and reticulate
veinlets very prominent underneath. Racemes terminal or in the upper axils.

Flowers usually orange-red, numerous, clustered and rather large. Calyx-tube 2 lines long, the lobes lanceolate, as long as the tube. Standard fully 4 lines diameter, callous at the base above the claw; keel much curved, almost rostrate, but obtuse. Upper stamen and an adjoining portion of the staminal tube hairy. Pod long, linear, rusty-villous, but not seen perfect. Seed nearly orbicular, with a small oblong strophiole.

N. Australia. York Sound, *A. Cunningham;* Upper Victoria river, *F. Mueller.* In inflorescence this species seems to connect the sections *Brissonia* and *Reineria;* the very flat glabrous style is more that of *Reineria.* The venation of the leaflets differs from that of any extra-Australian species.

3. **T. reticulata,** *R. Br. Herb.* (*under* Galega). Rootstock perennial, with prostrate or ascending stems, of 2 ft. or more, minutely pubescent with appressed hairs. Leaflets 5 to 11 or more, petiolulate, ovate or oblong, obtuse, thinly coriaceous, the primary veins scarcely more prominent than the reticulate veinlets, glabrous or loosely pubescent underneath. Stipules sometimes lanceolate, the lower ones reflexed. Racemes long and rigid, terminal or leaf-opposed. Flowers rather small, in distant pairs, on pedicels as long as the calyx. Calyx scarcely 2 lines long, the lobes shorter than the tube. Standard very broad; keel incurved, obtuse. Pod broadly linear, nearly straight or recurved, pubescent, the upper suture thickened, the valves very flat. Seeds orbicular.

N. Australia. Islands of the Gulf of Carpentaria, *R. Brown;* Sims's Island, *A. Cunningham;* Endeavour river, *Banks and Solander* (a more glabrous form).

4. **T. crocea,** *R. Br. Herb.* (*under* Galega). Stems or branches diffuse or ascending, attaining 2 ft. or more, softly silky-villous. Leaflets usually 11 to 17, from obovate to narrow-oblong, $\frac{1}{2}$ to 1 in. long, obtuse or mucronate, nearly glabrous above, softly silky underneath, the primary veins anastomosing and reticulate. Stipules lanceolate or linear, reflexed, persistent. Racemes long, terminal, leafy at the base. Flowers (pale yellow, *R. Br.*) in distant pairs or clusters on short pedicels. Bracts small. Calyx silky-pubescent, about 3 lines long, the lobes nearly as long as the tube. Standard very silky. Pod 1$\frac{1}{2}$ to 2 in. long, incurved, softly velvety. Seeds orbicular.

N. Australia. Islands of the Gulf of Carpentaria, *R. Brown.*

5. **T. oblongata,** *R. Br. Herb.* (*under* Galega). An erect shrub or undershrub, of 5 or 6 ft., the branches angular, softly pubescent. Leaflets usually 11 to 17, oblong, obtuse, 1 to 2 in. long, nearly glabrous above, silky-pubescent underneath, the primary pinnate veins parallel and prominent underneath, but anastomosing within the margin. Stipules very deciduous. Racemes long and rigid, terminal or in the upper axils, rarely leaf-opposed. Flowers (almost orange, *R. Br.*, becoming pink when dry) in distinct clusters. Bracts linear-subulate, more conspicuous than in most species. Standard fully $\frac{1}{2}$ in. diameter, callous at the base above the claw; keel much shorter and much curved. Style flattened, glabrous. Pod not seen.

N. Australia. Islands of the Gulf of Carpentaria, *R. Brown.* A very imperfect specimen of A. Cunningham's from the N. coast, may belong to the same species, the specimens of which sometimes almost assume the aspect of a *Millettia.*

6. **T. porrecta,** *R. Br. Herb.* (*under* Galega). Rootstock thick, with elongated diffuse or ascending stems, the whole plant loosely pubescent or nearly glabrous. Leaflets usually 11 to 19, sessile, from broadly elliptical-oblong to nearly linear, obtuse or with recurved points, 3 to 8 lines long, coriaceous, the primary veins arcuate and anastomosing within the edge, and conspicuous on both sides. Stipules striate, recurved. Racemes long and slender, terminal, often leafy at the base, the floral leaves usually 3-foliolate. Flowers rather small, on pedicels longer than the calyx. Calyx pubescent, about 2 lines long, the lobes narrow-subulate, pointed, at least as long as the tube. Standard pubescent, about 4 lines diameter. Style scarcely flattened, strongly bearded. Pod about 1 to 1½ in. long, arcuate, pubescent. Seeds orbicular.

N. Australia. Islands of the Gulf of Carpentaria, *R. Brown;* Port Essington, *Armstrong.*

7. **T. polyzyga,** *F. Muell. Herb.* Stems or branches elongated, hoary or almost silky with a close tomentum. Leaflets 20 to 40 or more, on a common petiole of 3 to 5 in., oblong, obtuse, mostly 4 or 5 lines long, rigid, nearly glabrous above, silky pubescent underneath; the primary veins few and anastomosing within the margin. Flowers small, in distant clusters, in long often branched racemes. Pedicels short. Calyx silky-tomentose, 2 to 2½ lines long, the lobes rather broad, acute, shorter than the tube. Standard broad; keel obtuse. Style flat. Pod nearly straight, 1½ to 2 in. long, softly tomentose. Seeds lenticular.

N. Australia. Upper Victoria river, *F. Mueller;* islands of the Gulf of Carpentaria, *R. Brown.*

8. **T. graminifolia,** *F. Muell. Herb.* Rootstock perennial, with ascending or erect branching stems of 1 to 1½ ft., slender, angled, glabrous or loosely pubescent. Leaves few, simple, sessile, linear or linear-lanceolate, acuminate, 3 to 5 in. long, rather rigid, glabrous or nearly so, the primary veins very oblique, occasionally anastomosing. Racemes slender. Flowers small, in distant pairs or clusters. Calyx-tube not 1 line long, the teeth or lobes shorter than the tube, lanceolate. Standard broad, with a small callosity above the claw; keel much curved, obtuse. Style flattened. Pod linear, pubescent, but only seen young.

N. Australia. Providence Hill, *F. Mueller.*

9. **T. simplicifolia,** *F. Muell. Herb.* Rootstock perennial, with ascending or erect branching stems of 1 to 1½ ft., slightly angular, glabrous or loosely pubescent. Leaflets single, articulate on a petiole of from ¼ to above 1 in. long, linear or linear-lanceolate, 3 to 5 in. long, rather rigid, glabrous or pubescent, obliquely penninerved, without stipellæ. Stipules rigid. Racemes terminal, often leafy at the base. Flowers in distant pairs, larger than in *T. graminifolia.* Calyx-tube softly pubescent, 1¼ lines long; lobes lanceolate, acute, shorter than the tube. Standard almost reniform; keel curved, obtuse. Staminal sheath earlier open on the upper side than in most species, the upper filament adhering to one side in the middle. Style much flattened. Pod not seen.

N. Australia. Trap plains, Roper river, *F. Mueller.*

10. **T. leptoclada,** *Benth.* Apparently annual, erect, much branched at the base, loosely pubescent. Leaflets either single with a pair of stipellæ, or 3 digitate at the end of the petiole, or rarely 1 or 2 pairs lower down the petiole, linear, often 2, 3, or even 4 in. long when single, acutely acuminate, glabrous above, pubescent underneath. Racemes long, slender, with very small flowers in distant pairs. Pedicels slender. Calyx-teeth subulate, longer than the short tube. Standard not 3 lines diameter. Pod long, narrow, straight, pubescent. Seeds nearly orbicular.

N. Australia. Upper Victoria river, *F. Mueller.*
Queensland. Bowen river, *Bowman.*

11. **T. brachycarpa,** *F. Muell. Herb.* An erect herb, not exceeding 10 in. in any of our specimens, with spreading, softly-pubescent branches. Leaflets 3, 5, or rarely 7, oblong-cuneate obtuse or retuse, rarely above ⅓ in. closely silky-pubescent, especially underneath. Flowers very small, either in leaf-opposed clusters of 2 or 3, or 2 or 3 pairs in a raceme of ½ to 1 in., with a leaf under each pair. Calyx-lobes acute, the 2 uppermost and the lowest as long as the tube, the lateral ones shorter. Standard broad, not 2 lines long; keel slightly curved, obtuse. Ovary with a single ovule. Style flattened. Pod very villous, ovate, compressed, 3 to 4 lines long. Seed transversely oblong.

N. Australia. Desert near Hooker's Creek, *F. Mueller.* This has the flowers and fruit, although not the foliage or inflorescence, of the African *Requienias*, which it is therefore necessary to unite as a section to the genus *Tephrosia.*

12. **T. Stuartii,** *Benth.* A low erect herb, with the habit of *T. brachycarpa,* the branches softly villous. Leaflets 3 or 5, from obovate to oblong-cuneate, obtuse with a small soft point, about ½ in. long, softly silky-villous. Flowers about 3 lines long, sessile in leaf-opposed clusters of 2 or 3, or in short spikes with a leaf at the base of the cluster, so that the floral leaves often appear opposite, with the flowers in the axils of one of each pair. Calyx-lobes about as long as the tube. Standard broad ; keel incurved, obtuse. Ovary with 2 or rarely 3 ovules, style flattened. Pod oblong, straight with a hooked point, flat, softly villous, ½ to ¾ in. long. Seeds 2 or 3, orbicular.

N. Australia. In lat. 18° 35′, *M'Douall Stuart ;* Sturt's Creek, *F. Mueller :* the latter an imperfect, somewhat doubtful specimen, with the flowers in pairs on a leaf-opposed peduncle. This species connects the anomalous *T. brachycarpa* with the rest of the genus, and especially with the following species.

13. **T. eriocarpa,** *Benth.* Branches densely tomentose-villous. Leaflets 3, 5, or rarely 7, in some specimens broadly oblong-cuneate, emarginate, and mostly about 1 in. long, in others narrow-oblong or linear, 1 to 2 in. long and mucronate, nearly glabrous above, silky-villous underneath, with parallel oblique veins. Flowers in almost sessile leaf-opposed pairs, or in racemes of 2 to 3 in., leafy at the base. Calyx softly villous, the tube very short; lobes nearly 2 lines long. Standard about 4 lines diameter, but not seen perfect. Pod 1 to 1½ in. long, straight or slightly recurved, densely and softly villous, the valves convex. Seeds orbicular, usually about 6.

N. Australia. Deserts of Sturt's Creek and Victoria river, *F. Mueller.*

14. **T. phæosperma,** *F. Muell. Herb.* A shrub of about 2 ft., the branches closely but densely silky-tomentose. Leaflets 5 to 9 or rarely 11, oblong-cuneate or linear-oblong, obtuse or almost acute, mucronate, 1 to 1½ or rarely nearly 2 in. long, rather rigid, more or less silky on both sides, and strongly marked with oblique parallel veins. Flowers clustered, in dense terminal racemes of 1 to 2 in. Pedicels short. Calyx silky, the tube about 1 line long, the lobes at least as long, subulate-acuminate. Standard about 4 lines diameter ; keel much curved, almost acute. Pod straight, rarely above 1 in. long, slightly pubescent; valves broad and coriaceous. Seeds nearly orbicular.

N. Australia. N. coast, *A. Cunningham ;* Upper Victoria river, *F. Mueller.*

15. **T. astragaloides,** *R. Br. Herb.* (*under* Galega). An erect undershrub of 1 to 2 ft., the branches softly pubescent or silky. Leaflets usually 7 to 17, rather crowded, oblong-cuneate or almost linear, ½ to 1 in. long, obtuse or retuse, slightly pubescent and green above, softly silky and almost silvery underneath. Racemes short, leafy, with crowded rather large flowers, or very rarely elongated with distant clusters. Calyx 2 lines long or rather more, the lobes narrow, acute, rather shorter than the tube, the 2 upper ones united above the middle. Standard 5 or 6 lines diameter ; keel incurved, obtuse. Style less flattened than in the other Australian species, and almost terete at the end, with a penicillate stigma. Pod above 1 in. long, incurved towards the end, tomentose. Seeds orbicular.

Queensland. Shoalwater Bay, *R. Brown, A. Cunningham ;* Dunk Island, *M'Gillivray ;* Burdekin Expedition, *F. Mueller ;* Rockhampton, *Thozet, Dallachy ;* near Marlborough, *Bowman.*

Var. (?) *macrostachya.* Raceme elongated. Leaflets silky-villous on both sides. Pod of *T. astragaloides.*—Bowen river, *Bowman.*

16. **T. juncea,** *R. Br. Herb.* (*under* Galega). An annual or perennial, with erect, slender but rigid and virgate, not much branched stems, of 1½ to 2 ft., glabrous or hoary-pubescent. Leaflets above 20, on a long slender common petiole, the terminal one oblong-linear ¾ to above 1 in. long, the others very much smaller, obovate obcordate or cuneate, from under 3 to nearly 5 lines long, glabrous above, hoary or silky underneath. Flowers small, in distant pairs, in long slender terminal racemes. Calyx about 1 line long, with very short teeth. Standard nearly 3 lines diameter. Pod 1½ in. long or more, narrow, usually incurved towards the end, glabrous or slightly pubescent. Seeds more or less transversely oblong.

N. Australia. Islands of the Gulf of Carpentaria, *R. Brown.*

Queensland. Endeavour river and Bustard Bay, *Banks and Solander ;* Broad Sound, *R. Brown ;* Gould Island, *M'Gillivray ;* Wide Bay, *Bidwill ;* Rockhampton, *Thozet ;* Elliott river, *Bowman ;* also in Leichhardt's collection.

17. **T. filipes,** *Benth.* A perennial, with slender diffuse or ascending branching stems, rarely above 1 ft. long, minutely silky-hoary. Leaflets 7 to 15, narrow-oblong or linear, those of the lower leaves obtuse, of the upper leaves acute with straight or recurved points, all rather rigid, glabrous above. Racemes filiform, 1 or rarely 2 in. long, with usually only 2 distant pairs of small flowers, or the terminal ones more rigid, leafy at the base. Pedicels longer than the calyx. Calyx about 1 line long, including the short teeth.

Standard about 3 lines diameter. Style flattened. Pod about 1 in. long, straight, glabrous or nearly so. Seeds orbicular.

N. Australia. Islands of the Gulf of Carpentaria, *R. Brown, Henne.*
Queensland; Northumberland Islands, *R. Brown ;* Dawson river, *F. Mueller ;* Endeavour river and Percy Island, *A. Cunningham ;* May Day Island, *Armstrong ;* Wide Bay, *Bidwill ;* Erythrina Creek, *Leichhardt.*
Var. *latifolia.* Leaflets cuneate-oblong, silky underneath.—Endeavour river, *Banks and Solander.*

18. **T. remotiflora,** *F. Muell. Herb.* A perennial or undershrub of 1 to 2 ft. or more, slightly pubescent or silky-hoary. Leaflets usually 7 to 11, oblong-cuneate, ½ to 1 in. long, obtuse or retuse, with a minute straight or recurved point, glabrous above, hoary or silky underneath. Racemes very long, slender, with small flowers in distant clusters. Pedicels rather longer than the calyx. Calyx silky-pubescent, the tube ¾ line long, the teeth rather longer, subulate-acuminate. Standard not 3 lines diameter. Style much flattened. Pod about 1 in. long, obliquely acute or slightly falcate, glabrous or slightly pubescent ; valves thin. Seeds orbicular.

N. Australia. N. coast, *R. Brown ;* Fitzmaurice and Upper Victoria rivers, *F. Mueller ;* Albert river, *Henne.* Differs from all the varieties of *T. purpurea* in the much smaller flowers, the smaller, thinner, and straighter pod, and in the shape of the seed.

19. **T. oligophylla,** *Benth.* Rootstock woody, with ascending stems of ½ to 1 ft., glabrous or slightly pubescent. Leaflets 1, 3 or rarely 5, broadly oblong-cuneate to narrow-oblong, ¾ to 1½ in. long, obtuse or retuse with a small point, glabrous or sprinkled underneath with a few hairs, the primary veins not so close as in most species. Flowers usually numerous, rather large, the racemes not very long. Bracts subulate, persistent. Pedicels short. Calyx-tube nearly glabrous, about 1 line long, the teeth rather longer, subulate-acuminate. Standard fully 4 lines diameter ; keel much curved, obtuse. Style flattened. Pod straight or slightly recurved (not incurved), about 1 in. long, glabrous or nearly so. Seeds orbicular.

N. Australia. Cape York, *M'Gillivray ;* Albany Island, *F. Mueller.*

20. **T. macrocarpa,** *Benth.* Stems apparently loosely ascending, rather rigid, glabrous. Leaflets 3, 5 or 7, oblong-cuneate, very obtuse or emarginate, ¾ to 1½ in. long, much narrowed into a petiolule of 1 to 3 lines, glabrous, rigid, the parallel primary veins very prominent underneath. Racemes very long. Flowers rather large, in distant pairs. Pedicels at least as long as the calyx. Calyx glabrous and rather thick ; lobes lanceolate, as long as the tube, the 2 upper ones more or less united, the lowest longer and incurved. Standard silky-villous, 4 to 5 lines diameter. Style very flat. Pod glabrous, flat, falcate, 2½ to 3 in. long and nearly 3 lines broad. Seeds orbicular.

N. Australia. N.W. coast (Victoria river ?), *Bynoe ;* Sturt's Creek, *F. Mueller.*

21. **T. purpurea,** *Pers. ; W. and Arn. Prod.* 218. A perennial or undershrub of 1 to 2 ft., with spreading or decumbent branches, glabrous hoary or pubescent. Leaflets usually 7 to 11, oblong-cuneate or linear, obtuse or with a small recurved point, ½ to 1 in. long, hoary or silky underneath. Racemes terminal or leaf-opposed, the lower ones often

very short, the upper ones 6 in. long or more, with distant fascicles of 2 to 4 pinkish flowers. Calyx minutely pubescent, the tube about 1 line long, the lobes subulate-pointed, variable in length. Standard about 4 lines diameter; keel obtuse. Upper stamen slightly cohering with the others in the bud, but very soon quite free. Style much flattened. Pod glabrous or pubescent, about 1½ in. long, more or less falcate. Seeds transversely oblong.—*T. Baueri*, Benth. in A. Gray, Bot. Amer. Expl. Exped. i. 408.

N. Australia. Hills near Nichol Bay, *F. Gregory's Expedition;* Port Cooper, *Herb. F. Mueller* (very pubescent fragmentary specimens); Arnhem S. Bay, *R. Brown.*

Queensland. Bowen river, *F. Mueller;* Port Denison, *Fitzalan.*

N. S. Wales. Clarence river, *Beckler;* Hunter's River, *American Exploring Expedition.*

The species is very common in tropical Asia and E. tropical Africa, where it varies considerably as to stature, length of raceme, indumentum, etc., but where the calyx-lobes are always subulate and longer than the tube. Among the Australian specimens, those above quoted are the only ones I have seen agreeing in this respect as well as in foliage with the Asiatic ones. The following forms, which are probably varieties of the same species, do not nevertheless appear to be represented out of Australia.

Var. *brevidens.* More shrubby and erect, with the habit of the Pacific island variety usually named *T. piscatoria*, Pers.; hoary with a minute appressed pubescence. Leaflets mostly 9 to 15. Racemes usually very long. Calyx-teeth very short or the lowest nearly as long as the tube.—Various points of the N. and E. coasts, islands of the Gulf of Carpentaria, *R. Brown;* Endeavour river, *Banks and Solander;* Port Essington, *Armstrong;* Cape Upstart, *M'Gillivray;* Wide Bay, *Bidwill;* Percy Island, *A. Cunningham;* Macquarrie river, *Bowman;* the latter specimen imperfect.

Var. *rufescens.* Habit of the var. *brevidens,* but the branches densely and softly rusty-villous. Leaflets more numerous, often above 20. Racemes long and many-flowered. Calyx-teeth short as in the var. *brevidens.*—Rocky hills, Gorman Creek, Moreton Bay, *C. Stuart;* Port Bowen, *A. Cunningham;* Archer's Hill, *Leichhardt.*

Var. *longifolia.* Leaflets very narrow, obtuse acute or mucronate, often 1 to 2 in. long. Racemes long. Calyx-teeth subulate, but rather short.—Gulf of Carpentaria, *R. Brown, Landsborough;* Depôt Creek, *F. Mueller;* Albert river, *Henne.*

Var. *sericea.* Leaflets numerous, narrow, acute, silky underneath. Calyx-teeth short. —Broad Sound, *R. Brown, Bowen.*

Var. (?) *laxa.* Stems loosely decumbent. Leaflets few, broad, loosely pubescent or villous, the veins parallel above, almost reticulate underneath. Calyx-teeth subulate, but very short.—Islands of the Gulf of Carpentaria, *R. Brown, Henne.*

22. **T. Bidwilli,** *Benth.* Rootstock woody, with ascending or erect stems of 1 to 2 ft., more or less pubescent. Leaflets in the lower leaves 3 or 5, oblong or lanceolate, in the others 7 to 11, linear or linear-lanceolate, acutely acuminate, 1 to 2 or even 3 in. long, narrowed at the base, in distant pairs along a slender rigid petiole, green on both sides, but slightly pubescent underneath. Racemes usually on long peduncles, the lower pairs of flowers distant. Calyx densely rusty-pubescent, the tube about 1½ lines long, the lobes longer, incurved, narrow, but not subulate, the 2 upper ones united above the middle. Standard fully 5 lines broad; keel much incurved, almost acute. Style much flattened. Pod 1½ to 2 in. long, slightly incurved, softly pubescent. Seeds transversely oblong.

Queensland. Wide Bay, *Bidwill;* Burnett river, *F. Mueller.*

N. S. Wales. Clarence and Macleay rivers, *Beckler.*

Var. (?) *densa.* Leaflets shorter and more silky; inflorescence dense, but with the calyx of *T. Bidwilli.*

N. Australia. Hills near Nichol Bay, *F. Gregory's Expedition.*

23. **T. rosea,** *F. Muell. Herb.* A perennial or undershrub, with ascending branches, closely but rather densely silky-tomentose. Leaflets usually 5 or 7, oblong-cuneate or obovate-oblong, very obtuse or retuse, green and slightly pubescent above, silky underneath. Racemes long and rather rigid. Flowers small, in clusters of 2 or 3. Pedicels usually shorter than the calyx. Calyx silky-tomentose, the tube about 1 line long, the teeth or lobes about as long. Standard about 3 lines diameter, the claw short. Style much flattened. Pod narrow, densely silky-pubescent, much curved, the sutures scarcely thickened, the valves very convex. Seeds transversely oblong.

N. Australia. Montague Sound, N.W. coast, *A. Cunningham;* Victoria river and Depot Creek, *F. Mueller.*
Var. (?) *angustifolia.* Leaflets linear, elongated. Pod softly villous.
N. S. Wales. Between Darling river and Cooper's Creek, *Neilson.* The specimen insufficient for accurate determination.

35. MILLETTIA, W. and Arn.

Calyx broad, truncate or with short teeth or lobes, the 2 upper ones often united. Standard broad, usually reflexed; keel incurved, obtuse. Upper stamen free or cohering with the others in the middle; anthers uniform. Ovary sessile or rarely stipitate, surrounded at the base by an angular or cupshaped disk, with several ovules; style inflexed, terete, glabrous, with a small terminal stigma. Pod broadly linear-lanceolate or oblong, flat and hard, or if convex, thick and woody, opening at length in 2 valves. Seeds orbicular or reniform, not strophiolate.—Trees, tall shrubs or woody climbers. Leaves unequally pinnate; leaflets penniveined with reticulate veinlets, usually stipellate. Stipules small. Racemes terminal or paniculate at the ends of the branches. Flowers usually purple pink or white, clustered or scattered along the rhachis. Bracts and bracteoles usually very deciduous.

A large genus, ranging over the warmer regions of Asia and Africa, with one endemic Australian species. It differs from the North American and Japanese genus *Wistaria,* only in the hard, usually flat or thick pod, not opening so readily, although not absolutely indehiscent as in the *Dalbergieæ.*

1. **M. megasperma,** *F. Muell.* (*under* Wistaria). A tall evergreen woody climber, glabrous except a slight hoariness on the young shoots and panicles. Leaflets 7 to 13, obovate or obovate-oblong, shortly acuminate, 1¼ to 2 in. long, somewhat coriaceous and green on both sides. Racemes 4 to 6 in. long, several in a terminal almost leafless panicle. Flowers scattered, purple. Calyx about 2½ lines long, the lobes nearly as long as the tube, the 2 upper ones united into a very broad truncate upper lip. Standard above ½ in. broad, minutely silky-pubescent, with a slightly prominent transverse callous appendage inside above the claw. Upper stamen quite free. Ovary stipitate. Pod about 6 in. long, thick, hard, almost woody, densely velvety outside. Seeds large and thick.—*Wistaria megasperma,* F. Muell. Fragm. i. 10.

Queensland. Pine river, near Moreton Bay, *W. Hill, F. Mueller;* Nurrum-Nurrum Creek, *Leichhardt.*
N. S. Wales. Richmond river, *C. Moore.*

36. SESBANIA, Pers.

(Agati, *Desv.*)

Calyx-tube broad, truncate, or with nearly equal teeth or lobes. Standard orbicular or ovate, spreading or reflexed ; keel incurved, obtuse or acuminate, the claws much longer than those of the other petals. Upper stamen free, geniculate near the base, the others united in a sheath angled near the base ; anthers uniform or nearly so. Ovary with several ovules ; style glabrous, with a small terminal stigma. Pod long and linear (or in some species not Australian oblong), 2-valved or indehiscent, the endocarp continuous with spurious transverse partitions separating the seeds. Seeds without any strophiole.—Herbs or shrubs, sometimes arborescent, but of very few years' duration. Leaves abruptly pinnate, with numerous entire leaflets, the stipellæ minute or none. Stipules setaceous, usually very deciduous. Flowers yellow, red, variegated or white, in short loose axillary racemes ; pedicels slender. Bracts and bracteoles very rarely persistent to the time of flowering.

The genus is widely spread over the tropical regions both of the New and the Old World. Of the four Australian species, three are the commonest Asiatic ones, two of them extending also over tropical Africa, the fourth is endemic.

Flowers very large (nearly 3 in. long), the petals narrowed at the
end . . . . 1. *S. grandiflora.*
Flowers not 1 in. long. Petals broad.
Racemes pendulous. Stem shrubby . . . . :. 2. *S. ægyptiaca.*
Racemes erect. Stem herbaceous.
Bracts and bracteoles very deciduous. Calyx-teeth very short. 3. *S. aculeata.*
Bracts and bracteoles setaceous, often persistent. Calyx-teeth
subulate-pointed, nearly as long as the tube . . . . . . 4. *S. simpliciuscula.*

1. **S. grandiflora,** *Pers. Syn. Pl.* ii. 316. A tall shrub or small tree of very few years' duration, glabrous and more or less glaucous. Leaflets 10 to 30 pairs, oblong or elliptical, obtuse and often mucronate, 1 to 1½ in. long. Racemes short, with 2 to 4 very large flowers, white in our Australian specimens. Calyx-tube ¾ in. long, without the turbinate base, the teeth or lobes short and broad. Petals 2 to nearly 3 in. long ; standard ovate, rather shorter than the others ; keel much incurved, ending in an obtuse beak. Pod upwards of a foot long, nearly 3 lines broad.—*Agati grandiflora,* Desv. ; DC. Prod. ii. 266 ; W. and Arn. Prod. 215 ; *A. formosa,* F. Muell. Fragm. ii. 88.

N. Australia. Near Nichol Bay, *F. Gregory's and Ridley's Expeditions;* Glenelg river, N.W. coast, *Clarkson;* Fitzmaurice river, Arnhem's Land, *F. Mueller.*
The red-flowered variety, *S. coccinea,* Pers. l. c., or *Agati coccinea,* Desv., is not amongst the Australian species I have seen. Both varieties are frequent in India, but perhaps only about villages and other places where they have been planted ; they both appear to be really indigenous in the Archipelago. The size of the flowers with the petals narrower in proportion, has induced the separation of this species as a genus, but there is no other character to distinguish it from *Sesbania.* The Sandwich Island *S. tomentosa (Agati tomentosa,* Nutt.) is quite intermediate between the two.

2. **S. ægyptiaca,** *Pers. ; DC. Prod.* ii. 264. A shrub of 5 or 6 ft., becoming, in India at least, a tree of twice that size, but of very few years' duration, glabrous and somewhat glaucous, the branches terete or obscurely

angled. Leaflets usually under 20 and often not 10 pairs, oblong, obtuse, 4 to 8 lines long or when luxuriant nearly 1 in. Flowers rather large, yellow or with a purple vexillum, in loose pendulous racemes, shorter than the leaves. Bracts and bracteoles very deciduous. Calyx about 3 lines long, without the narrow-turbinate almost stalk-like base, the teeth very short and broad. Standard about ¾ in. broad; keel much incurved, broad, obtuse, with an acute angle at the base. Pod when perfect 8 to 10 in. long and 2 to 2½ lines broad, but often much shorter by the abortion of many of the ovules. —W. and Arn. Prod. 214; Wight, Ic. t. 32; *S. picta*, Pers.; Bot. Reg. t. 873.

N. Australia. Gulf of Carpentaria, *F. Mueller.* The species is common in tropical Asia and Africa.

3. **S. aculeata,** *Pers.; DC. Prod.* ii. 265. An erect herb, usually of 4 to 5 ft., but sometimes twice that size, glabrous or the young shoots slightly pubescent, the branches terete or slightly angular. Leaflets from 20 to nearly 50 pairs, narrow-oblong, obtuse, mucronate, 4 to 8 lines long, on a common petiole often ½ to 1 ft. long and sometimes armed with small tubercles or prickles, which are however often very minute or quite wanting. Flowers yellow, much smaller than in *S. ægyptiaca*, in loose erect racemes, shorter than the leaves. Bracts and bracteoles very deciduous. Calyx about 2½ lines long, including the short turbinate base, the teeth short, broad and acute. Standard scarcely ½ in. broad; keel very much incurved, broadly obtuse in front. Pod long, narrower than in *S. ægyptiaca*, the sutures more thickened.—W. and Arn. Prod. 214; *S. australis*, F. Muell. in Trans. Vict. Inst. i. 36.

N. Australia. Sturt's Creek, *F. Mueller;* Gulf of Carpentaria, *Landsborough*, also *M'Kinlay's Expedition.*
Queensland. In the interior, *Mitchell;* Rockhampton, *Dallachy.*
N. S. Wales. Darling river, *Herb. F. Mueller.*
S. Australia. Cooper's Creek, *Howitt's Expedition.*
Var. *sericea.* Young branches and foliage silky-pubescent.—N.W. coast, *Bynoe;* Nichol Bay, *F. Gregory's Expedition;* Sturt's Creek and Flinders river, *F. Mueller;* islands of the Gulf of Carpentaria, *R. Brown.*
Var. (?) *erubescens.* Flowers rather longer, the standard pinkish. Branches more angular.—Sturt's Creek. *F. Mueller.* This may possibly be near *S. punctata*, Pers., which however is scarcely specifically distinct from *S. aculeata.*
Var. (?) *parviflora.* Flowers very much smaller and more numerous.—Albert river, *Henne;* Newcastle Water, *M'Douall Stuart's Expedition.*

4. **S. simpliciuscula,** *F. Muell. Herb.* An erect herb, of 5 to 10 ft., the branches angular, glabrous or pubescent. Leaves long, with very numerous leaflets as in *S. aculeata*, but the leaflets usually longer, linear-oblong and more mucronate. Flowers yellow, nearly as large as in *S. ægyptiaca*, but more numerous, in short loose erect racemes. Stipules bracts and bracteoles setaceous and more persistent than in any other species. Calyx-teeth about 2½ lines long, without the turbinate base, the teeth subulate-pointed, nearly as long as the tube. Standard not spotted. Pod not seen.

N. Australia. Upper Victoria river, *F. Mueller.*

37. CLIANTHUS, Soland.

(Donia, *G. Don.*)

Calyx-teeth nearly equal. Standard acuminate, closely reflexed over the calyx; wings shorter, lanceolate; keel about as long as the standard, erect, incurved, acute. Upper stamen free, the others united in a sheath; anthers reniform. Ovary stipitate, with many ovules; style subulate, incurved, longitudinally bearded along the inside towards the end; stigma minute, terminal. Pod turgid, oblong-acuminate, 2-valved. Seeds reniform, not strophiolate.—Herbs or undershrubs. Leaves pinnate. Stipules herbaceous. Flowers large, red, in short axillary racemes.

Besides the Australian species, which is endemic, the genus comprises one other from New Zealand. The Norfolk Island climber, described as *C. carneus*, forms the very distinct genus *Streblorhiza*, Endl.

1. **C. Dampieri,** *A. Cunn. in Trans. Hort. Soc. Lond. ser.* 2, i. 522. A perennial, with stout procumbent or ascending stems, of 2 ft. or more, densely villous, with long soft hairs. Leaflets about 15 to 21, obovate elliptical or oblong, obtuse or almost acute, mostly ½ to 1 in. long, nearly glabrous above, villous underneath. Stipules broad, embracing the stem. Peduncles rarely exceeding the leaves, bearing a short dense almost umbel-like raceme of large red pendulous flowers. Bracts lanceolate. Pedicels about ½ in. long, with short linear bracteoles. Calyx hirsute, nearly ¾ in. long, the lobes lanceolate, acuminate, longer than the tube. Standard 2½ in. long, with a deep purple or black shining blotch at the base; wings 1½ in. long, acute; keel nearly as long as the standard. Pod narrow-oblong, 2 to 2½ in. long, coriaceous, the seminal suture indented, softly pubescent outside, glabrous inside. Seeds small and numerous.—R. Br. in App. Sturt, Voy. 8; Bot. Mag. t. 5051; Lindl. in Paxt. Fl. Gard. t. 10; *C. Oxleyi,* A. Cunn. in Trans. Hort. Soc. l. c.; *Donia speciosa* and *D. formosa,* G. Don, Gen. Syst. ii. 468.

N. Australia. N.W. coast, *Bynoe;* Dampier's Archipelago, *A. Cunningham;* near Nichol Bay, *F. Gregory's Expedition.*

N. S. Wales. Lachlan to Darling rivers, *A. Cunningham, Sturt, Dallachy and Goodwin.*

S. Australia. Mount Arden and Lake Torrens, *F. Mueller;* Gawler range, *Eyre;* Flinders range, *Howitt's Expedition.*

38. SWAINSONA, Salisb.

(Cyclogyne, *Benth.;* Diplolobium, *F. Muell.*)

Calyx-teeth nearly equal. Standard nearly orbicular, on a short claw; wings oblong, falcate or slightly twisted, free; keel broad, incurved, obtuse or produced into a twisted beak. Upper stamen entirely free, the others united in a sheath; anthers reniform. Ovary sessile or stipitate, with many ovules; style incurved, subulate or curled inwards at the end, more or less longitudinally bearded along the inner edge, the stigma small or inconspicuous at or near the end. Pod either ovoid membranous and inflated; or narrow and coriaceous, but turgid, the upper suture occasionally impressed, or the pod divided by a longitudinal partition. Seeds usually small, reniform, withoùt any strophiole.—Herbs or undershrubs, glabrous or clothed, especially the young shoots, with short rather rigid appressed hairs. Leaves unequally pinnate, leaflets usually numerous, small, entire, without stipellæ. Stipules herbaceous, oblique with a broad base, rarely almost subulate. Flowers violet-purple, blue, red, white or yellowish, in axillary racemes. Bracts membranous, usually small. Bracteoles sometimes close to the calyx and persistent, sometimes on the pedicel, and very small or none.

The genus is limited to Australia, with the exception of a single New Zealand species, allied to *S. lessertiifolia*. The European and Asiatic *Coluteas* are however only to be distinguished by their shrubby habit and large prominent lateral stigma, and the S. African *Lessertias* are some of them so near to *S. lessertiifolia* and its allies, as to make it very difficult to draw any but a geographical line between the two genera.

A. *Standard with prominent oblique or longitudinal plate-like calli above the claw. Pod stipitate, thin, inflated.*

Style bearded only along the inner side. Flowers large. Pod acute, 1 to 2 in. long.
    Calyx densely white-tomentose . . . .   1. *S. Greyana.*
    Calyx glabrous, or nearly so. . . . .   2. *S. galegifolia.*
Style with a tuft of small hairs behind the stigma on the back, besides the longitudinal beard. Pod under ½ in. long. Flowers small . .   3. *S. brachycarpa.*

B. *Standard with transverse or confluent callosities on the top of or close above the claw, or the top of the claw much thickened. Pod sessile or nearly so, turgid, often coriaceous.*

Keel incurved, but neither twisted nor oblique. Style slender. Ovary silky-villous.
  Leaflets usually more than 9. Calyx-lobes lanceolate.
    Plant hoary or almost mealy-pubescent. Leaflets linear or oblong. Flowers large, few, on long peduncles . . . . . .   4. *S. phacoides.*
    Plant densely villous with white woolly hairs. Leaflets obovate. Flowers rather small, numerous, extending nearly to the base of the peduncle . . . .   5. *S. Burkittii.*
  Leaflets usually under 9 (except in *S. Burkei*). Calyx-lobes subulate or very short. Plants usually low or procumbent.
    Plants softly villous. Leaflets obovate. Racemes dense, ovoid before expanding. . . . .   6. *S. Burkei.*
    Plant slightly hoary. Leaflets obovate. Flowers small, few, in short racemes . . . .   7. *S. oligophylla.*
    Plant glabrous or slightly hoary. Leaflets lanceolate or linear, acute. Flowers few, in short racemes . . . .   15. *S. oroboides.*
Keel oblique or laterally twisted. Style firm, readily twisting. Ovary glabrous or nearly so.
    Leaflets few, lanceolate, acute. Ovary quite sessile . . . . .   8. *S. campylantha.*
    Leaflets numerous, oblong, obtuse. Ovary shortly stipitate. Flowers numerous. Pod short, broad, divided into 2 closed hemicarpels . . . .   9. *S. occidentalis.*
    Leaflets numerous, small, cuneate, emarginate. Ovary shortly stipitate. Flowers few, small, distant . . . .   10. *S. gracilis.*

C. *Standard without any callosities, the claw usually short broad and thin. Pod various.*

Keel spirally twisted, without callosities. Pod sessile, oblong.
    Standard ¾ to 1 in. broad. Pod above 1 in. long, the upper suture intruded, but not completely dividing it . . . .   11. *S. procumbens.*

Standard about ¼ in. broad. Pod under 1 in. long, completely di-
vided longitudinally into 2 cells . . . . 12. *S. Drummondii.*
Keel incurved, not twisted, with a large callosity on each side at the
end. Pod nearly sessile, oblong, tomentose . . . . 13. *S. canescens.*
Keel neither twisted nor callous.
Style firm, flattened, hooked or inflexed at the end, bearded only
along the inner side. Plant hoary or mealy. Leaflets narrow.
Stipules broad . . . . 14. *S. phacifolia.*
Style slender, not hooked, bearded only along the inner side.
Racemes pedunculate, exceeding the leaves.
Pod rarely under ¾ in. long. Keel very obtuse.
Leaflets usually more than 9, obtuse. Standard with a short
broad claw. Calyx usually with black hairs . . . . 16. *S. lessertiifolia.*
Leaflets 3, 5, or rarely 7, lanceolate, acute. Standard with
a rather thick narrow claw. Calyx rarely with black
hairs . . . . 15. *S. oroboides.*
Pod less than ½ in. long. Flowers small. Keel much in-
curved. Leaflets small or narrow . . . . 17. *S. monticola.*
Racemes few-flowered, much shorter than the leaves. Flowers
very small . . . . 18. *S. luteola.*
Style slender, not hooked, with a tuft of hairs at the end on the back
behind the stigma, besides the longitudinal beard, which is
often slight.
Ovary and pod sessile.
Leaflets linear, acute, 1 in. long or more . . . . 19. *S. parviflora.*
Leaflets obcordate or cuneate-emarginate, under 4 lines long . 20. *S. microphylla.*
Ovary and pod distinctly stipitate.
Flowers purple or whitish. Pod above 1 in. long, on stipes
much longer than the calyx . . . . 21. *S. Fraseri.*
Flowers yellow. Pod about ½ in. long, on a stipes not exceed-
ing the calyx . . . . 22. *S. laxa.*

Among the specimens from Hammersley Range, collected in F. Gregory's Expedition, is a
single raceme of what may be a *Swainsona*, but with large flowers, differently shaped from
those of any of the above species. The fragment is however insufficient for accurate determination.

S. Fræbelii, Regel, Gartenfl. iii. 178, is only known to me from the diagnosis in Walp.
Ann. iv. 495, which gives no character different from those of *S. lessertiifolia,* and its
allies. I have only been able to find in our libraries the first two vols. of Regel's Garten-
flora.

1. **S. Greyana,** *Lindl. Bot. Reg.* 1846, *t.* 66. A perennial or undershrub, with erect or ascending stems of 2 to 3 ft., the young shoots and leaves
white-tomentose, becoming glabrous when full-grown. Leaflets 11 to 21,
oblong obtuse or retuse, ¾ to 1 in. or sometimes 1½ in. long. Flowers large,
pink, in long erect pedunculate racemes. Bracts ovate or lanceolate. Pedicels
shorter than the calyx. Bracteoles close to the calyx and often as long as its
tube. Calyx densely cottony-white, 3 to 4 lines long, the teeth short.
Standard ¾ in. diameter, with 2 prominent erect plate-like calli above the
claw; wings shorter; keel incurved, obtuse, not so broad as in *S. galegifolia.*
Pod inflated, membranous, attaining 1½ to 2 in., on a stipes of ¼ to ¾ in.—
Bot. Mag. t. 4416; *S. grandiflora,* R. Br. in App. Sturt, Exped. 11.

N. S. Wales. Flats on the Darling river, *Victorian Expedition, Dallachy,* etc.;
Mudgee and Dubba, *Bowman.*

Victoria. Murray river, *Mitchell, Grey.*

S. Australia. Near Adelaide, *Herb. Hooker.*

The precise form and proportions of the teeth of the calyx and bracteoles prove too variable to admit of distinguishing, even as constant varieties, the two forms described by R. Brown.

2. **S. galegifolia,** *R. Br. in Ait. Hort. Kew. ed.* 2, iii. 327. A glabrous perennial or undershrub, with erect flexuose branches, sometimes under 1 ft., sometimes ascending or even climbing to the height of several feet. Leaflets 11 to 21 or rarely more, oblong, obtuse or emarginate, mostly 4 to 8 lines long. Stipules small, reflexed. Racemes pedunculate, exceeding the leaves and sometimes twice as long. Flowers rather large, deep red in the original variety. Pedicels rarely longer than the calyx, with minute bracteoles near the top. Calyx glabrous, 2½ to 3 lines long, the lobes acute, short or nearly as long as the tube. Standard 6 to 8 lines diameter, with 2 oblique or almost longitudinal plate-like prominent callosities above the claw ; wings shorter; keel broad, obtuse. Style subulate, acute, not inflexed at the end, bearded longitudinally without any terminal tuft. Pod much inflated, membranous, 1 to 2 in. long, on a stipes varying from 2 to 6 lines.—DC. Prod. ii. 271 ; *Vicia galegifolia,* Andr. Bot. Rep. t. 319 ; *Colutea galegifolia,* Sims, Bot. Mag. t. 792 ; *S. Osbornii,* Moore, in Gard. Comp. t. 65, copied into Lemair. Jard. Fleur. t. 304.

Queensland. Shoalwater Bay, *R. Brown;* Moreton Bay, *Fraser;* Wide Bay, *Bidwill, Leichhardt;* Peak Downs, *F. Mueller;* Mantuan Downs and Balonne river, *Mitchell;* Burdekin river, *Fitzalan;* plains of the Condamine, *Leichhardt.*

N. S. Wales. Port Jackson, *Sieber, n.* 508, and others ; Hunter's River, *R. Brown, Backhouse;* northwards to New England, *C. Stuart;* Macleay, Hastings, and Clarence rivers, *Beckler;* and in the interior to the Macquarrie and Darling, *F. Mueller* and others.

S. Australia, *Herb. F. Mueller,* without the precise station.

The species varies with light purplish-pink flowers, *S. coronillæfolia,* Salisb. Parad. Lond. t. 28; DC. Prod. ii. 271 ; Bot. Reg. t.1725 ; and with white flowers, Bot. Reg. t. 994, *S. albiflora,* G. Don, Gen. Syst. ii. 245 ; Lodd. Bot. Cab. t. 1642. The differences in the length of the stipes of the pod do not, as had been supposed, coincide with the differences in the colour of the flower.

3. **S. brachycarpa,** *Benth.* A perennial, glabrous or nearly so, with the habit of *S. galegifolia,* but much smaller and more slender. Leaflets numerous, oblong, mostly narrow, 2 to 3 or rarely 4 lines long. Stipules small. Racemes on long peduncles exceeding the leaves. Flowers small. apparently purple or red. Pedicels about as long as the calyx. Calyx scarcely above 1 line long, broad with short acute teeth. Standard about 5 lines diameter, with the oblique almost longitudinal plate-like appendages of *S. galegifolia,* and the wings and keel also similarly-shaped, but the style is only.very slightly bearded longitudinally and has the dorsal tuft of hairs immediately behind the stigma of *S. microphylla, laxa,* and *Fraseri.* Pod inflated, membranous, globular or ovoid, 4 to 5 lines long, on a stipes exceeding the calyx, but perhaps not quite full grown in our specimen.

Queensland. Condamine river and Darling Downs, *Leichhardt;* Burnett river, *F. Mueller.*

N. S. Wales. New England, *C. Stuart;* Clarence river, *Beckler.*

4. **S. phacoides,** *Benth. in Mitch. Trop. Austr.* 363. A perennial, with procumbent or ascending stems of 1 to 1½ ft., hoary-pubescent as well as the leaves, the young shoots silky. Leaflets 9 to 13, narrow-oblong or linear,

obtuse or retuse, 4 lines to 1 in. long. Stipules lanceolate or subulate-pointed. Flowers rather large, yellow according to Mitchell, but apparently purple in most of our specimens, in short racemes on long peduncles. Bracts small. Pedicels very short. Calyx silky-villous, about 3 lines long, the lobes lanceolate, at least as long as the tube. Standard about 6 lines diameter, with thick almost confluent callosites almost on the claw; wing short, rather broad; keel incurved, obtuse. Style slender and much incurved, but not involute at the end. Pod sessile, oblong-linear, about 1 in long, turgid, but coriaceous, silky-pubescent, the upper suture slightly indented.

Queensland. E. coast, *R. Brown;* Mount Owen and Maranoa river, *Mitchell.*
N. S. Wales. Darling river, *Victorian Expedition.*
Victoria. Murray river, *F. Mueller.*
S. Australia. Neale's River, *M'Douall Stuart ;* south of Wells' Creek, *Howitt's Expedition.*
Var. *parviflora.* Leaflets fewer, usually narrow. Flowers smaller.—Darling and Lachlan rivers, *Neilson* and others.
Var. *grandiflora.* Pubescence whiter, almost silky or mealy. Leaflets broadly oblong. Flowers large.
N. Australia. Nichol Bay and De Grey river, *Ridley's Expedition.*
The callosities at the top of the claw in this and some of the following species, are variable in shape and consistence, but are always very different from the distinct plates of *S. galegifolia,* and never disappear entirely as in *S. phacifolia, lessertiifolia,* etc.

5. **S. Burkittii,** *F. Muell. Herb.* Stems rather rigid and flexuose, densely clothed as well as the foliage and inflorescence, with soft white woolly hairs. Leaflets 13 to 21 or more. obovate, rarely exceeding 4 lines, very obtuse. Stipules broad. Racemes longer than the leaves, but often flowering from nearly the base of the peduncle, the flowers numerous, rather small, on very short pedicels. Calyx densely and softly villous, about 3 lines long, the lobes lanceolate-acuminate, about as long as the tube. Standard about 5 lines diameter, with transverse or oblique contiguous callosities close above the claw; wings much shorter; keel exceeding the wings, incurved, obtuse or almost acute. Style slender, much inflexed at the end. Pod sessile, oblong, turgid, obtuse, rather above ½ in. long, very densely woolly-tomentose, the seminal suture slightly intruded.

N. S. Wales. Between the Lachlan and Darling rivers, *Burkitt.*
S. Australia. N.E. of Lake Gairdner, *Herb. Mueller.*

6. **S. Burkei,** *F. Muell. Herb.* Apparently procumbent, the stems foliage and inflorescence clothed with soft white spreading or almost woolly hairs. Leaflets 7 to 11 or rarely more, obovate or broadly oblong, very obtuse, ¼ to ½ in. long or rather more. Stipules rather broad. Racemes dense and ovoid before expanding, somewhat lengthened afterwards, on a peduncle exceeding the leaves. Pedicels short. Bracteoles linear or subulate, rather long. Calyx hirsute with soft white hairs, 3 to 4 lines long, the lobes subulate-acuminate, much longer than the tube. Standard 5 or 6 lines diameter, with 2 more or less prominent contiguous callosities close above the claw; wings shorter ; keel exceeding the wings, much incurved, obtuse. Style slender, inflexed at the end. Young pod sessile, very woolly.

N. Australia. Burke's Creek, Newcastle Water, etc., *M'Douall Stuart.* It appears to vary in the size of the flowers ; the standard, in the dried state, always looks purplish ;

the keel in the smaller-flowered specimens appears to be yellow. The species is nearly allied to *S. oligophylla*, but is larger, coarser, with a more dense inflorescence, and the indumentum almost of *S. Burkittii*, from which it differs in its fewer leaflets, in inflorescence and calyx.

7. **S. oligophylla,** *F. Muell. Herb.* Apparently perennial, with diffuse or ascending stems, under 1 ft. long, slightly hoary. Leaflets 5 to 9, from obovate to cuneate-oblong, very obtuse, ¼ to ½ in. long. Stipules small. Flowers small, in short racemes, on peduncles sometimes scarcely exceeding the leaves, sometimes twice as long. Calyx-tube very short, with a prominent minutely-hispid nerve descending from each lobe, the lobes narrow, almost subulate, 3 or 4 times as long as the tube. Standard about 4 lines diameter, with a transverse callosity or thickening of the top of the claw; wings shorter; keel much incurved, obtuse, slightly exceeding the wings. Style slender, much incurved. Pod sessile, broadly oblong, turgid, about ½ in. long, hoary-pubescent, the upper suture slightly indented, but not seen very perfect.

N. S. Wales. Darling river, *Victorian Expedition.*
S. Australia. Cooper's Creek, *A. C. Gregory*; towards Spencer's Gulf, *Warburton*; N.W. interior, *M'Douall Stuart.*

8. **S. campylantha,** *F. Muell. in Rep. Greg. Pl. 6.* Glabrous and somewhat glaucous, with rather rigid stems of about 1 ft. in our specimens. Leaflets usually about 5, lanceolate or linear, acute, 1 to 1½ in. long, or those of the lower leaves short and obtuse. Stipules small. Flowers (purple?) not numerous, in pedunculate racemes longer than the leaves. Bracts small. Pedicels rather short. Bracteoles subulate, close to the calyx. Calyx 2 to 2½ lines long, nearly glabrous outside; lobes acute, nearly as long as the tube, pubescent inside. Standard about 5 lines diameter, with a callosity or thickening of the top of the claw; wings twisted, nearly as long as the standard and always exceeding the keel; keel with a short obtuse oblique beak. Style thick, much inflected at the base, hooked and almost involute at the end. Young pod sessile, glabrous.

Queensland. Bowen river, *Bowman.*
S. Australia. Cooper's Creek, *A. C. Gregory*; Bagot range, *M'Douall Stuart's Expedition.*

9. **S. occidentalis,** *F. Muell. Fragm.* iii. 46. A glabrous or pubescent perennial, attaining 2 or 3 ft. but often shorter, the stems usually erect and bent in zigzag at the nodes. Leaflets 11 to 17 or in some specimens more, oblong, obtuse or acute, from 3 or 4 lines to nearly 1 in. long. Flowers purple, numerous, in long pedunculate racemes. Bracts small; bracteoles minute. Calyx sprinkled with a few white or rarely black hairs, about 2½ lines long, the lobes shorter than the tube, hirsute inside. Standard 6 lines broad, but not so long, with 2 prominent transverse callosities inside above the claw; wings obovate, almost as long; keel much incurved, almost rostrate, but obtuse. Ovary stipitate, glabrous or hairy at the base, with few ovules; style hard, flattened, inflexed and readily twisting, the slender extremity often hooked but not involute. Pod almost sessile, broadly ovate, about 5 lines long, hard and rugose, the lower suture on the outer face keeled, the upper suture indented, divided internally by a complete longitudinal partition

and separating when ripe into 2 closed hemicarpels, each ripening usually only 1 or 2 seeds.—*Diplolobium Walcottii*, F. Muell. in Trans. Bot. Soc. Edinb. vii. 489.

N. Australia. N.W. coast, Depuech Island, *Bynoe;* frequent in sterile places about Nichol Bay, *F. Gregory's and Ridley's Expeditions.*

W. Australia. Raised in our gardens from *Drummond's* seeds ; limestone hills, Murchison river, *Oldfield* (these specimens not in fruit).

10. **S. gracilis,** *Benth.* Glabrous, with slender ascending or erect stems of about 1 ft. Leaflets 9 to 15, from obcordate to linear-cuneate, emarginate, rarely 3 lines long. Racemes loose and slender, with few small purple flowers. Pedicels almost as long as the calyx, clothed with short thick black hairs. Calyx glabrous or nearly so, not 2 lines long ; the lobes shorter than the tube, slightly ciliate. Standard nearly 4 lines broad, but not so long, with 2 transverse callosities above the claw ; wings nearly as long ; keel much inflexed, slightly twisted and almost rostrate but obtuse. Ovary shortly stipitate, slightly hairy at the base ; style flattened, inflexed, and twisting readily, but not involute. Pod not seen.

W. Australia. Murchison river, *Oldfield.*

11. **S. procumbens,** *F. Muell. Fragm.* iii. 46. Glabrous or the young shoots and foliage slightly silky, or sometimes pubescent or hirsute, with procumbent ascending or erect stems of 1 to 3 ft. Leaflets 11 to 21 or more, varying from oblong or almost linear and $\frac{1}{4}$ to $\frac{1}{2}$ in. long, to lanceolate or linear-acute and above 1 in. long. Stipules herbaceous, rather large. Flowers large, fragrant, violet or blue, in a loose raceme on a peduncle often attaining 1 ft. Bracts often as long as the pedicels ; bracteoles lanceolate, shorter than the calyx-tube. Calyx about 3 lines long, the lobes at least as long as the tube, ciliate inside. Standard in the ordinary form above 1 in. broad, deeply emarginate, without callosities, the claw very short; wings shorter, narrow, slightly twisted ; keel much incurved, produced into a long obtuse spirally twisted beak. Style very long and slender, spirally twisted with the keel, the slender tip sometimes hooked but not involute. Pod sessile, above 1 in. long, acute, turgid, very coriaceous, often incurved, the seminal suture either depressed or slightly prominent.—*Cyclogyne swainsonioides*, Benth. in Mitch. Trop. Austr. 397 ; *C. procumbens*, F. Muell. in Linnæa, xxv. 393 ; *S. violacea*, Henders. Illustr. Bouq. t. 19.

Queensland. Plains of the Condamine, *Leichhardt;* near Ipswich, *Nernst.*
N. S. Wales. Liverpool and Dundas plains, *Fraser, M'Arthur, Leichhardt;* open downs on the Gwydir, *Mitchell;* Darling river, *Goodwin and Dallachy;* Castlereagh, *Moore.*
Victoria. Wimmera, *Dallachy.*
S. Australia. Towards St. Vincent's Gulf, *F. Mueller.*

Var. (?) *minor.* Leaflets shorter, broader, and more frequently hirsute. Flowers smaller, the keel less twisted. Pod shorter and more turgid.—Wimmera, *Dallachy.*

12. **S. Drummondii,** *Benth.* Slightly pubescent, Leaflets numerous, narrow-oblong, obtuse, above $\frac{1}{2}$ in. long in our specimen. Stipules broad. Racemes loose. Flowers much smaller than in *S. procumbens*, but nearly similar in shape. Calyx-teeth or lobes ciliate inside. Standard about $\frac{1}{2}$ in. broad, but much shorter, deeply emarginate, without callosities, on a short

claw; wings nearly as long; keel produced into an obtuse spirally involute beak. Pod nearly sessile, oblong, acuminate, above ½ in. long, but not quite ripe in our specimen, pubescent, the seminal suture much intruded and produced into a double dissepiment, completely dividing the pod into 2 longitudinal cells.

W. Australia, *Drummond.* The only specimen seen, raised many years since in the garden of the Horticultural Society from Drummond's seeds.

13. **S. canescens,** *F. Muell. Fragm.* iii. 46. Stock woody, with erect, rigid, but herbaceous stems of 1 to 2 ft., softly tomentose-pubescent. Leaflets 9 to 15, obovate or oblong-elliptical, obtuse or retuse, ½ to 1 in. long, nearly glabrous above, softly pubescent or silky underneath. Stipules broad, herbaceous. Racemes many-flowered, on long silky-villous peduncles. Flowers nearly sessile, blue or violet-purple, variegated with pink, and a green blotch at the base of the standard. Calyx about 2½ lines long, silky-hairy, the lobes about as long as the tube. Standard about ½ in. diameter, on a very short claw, without prominent callosities; wings short; keel much curved, obtuse, with a thick callous appendage on each side of the tip. Ovary shortly stipitate; style much curved, involute at the end. Pod almost sessile, oblong, very softly tomentose-villous, in our specimens 7 to 8 lines long, rather coriaceous, with an indented upper suture, but not quite ripe.—*Cyclogyne canescens,* Benth. in Lindl. Swan Riv. App. 16; Paxt. Mag. Bot. vii. 199, with a fig.

W. Australia. Swan River, *Drummond, 1st Coll.* The general aspect, style, and pod of this species are so different from those of the few *Swainsonas* originally known, that they appeared to warrant the establishment of a distinct genus; the species subsequently discovered have, however, connected it by so many gradations with the others, that *Cyclogyne* can no longer be maintained even as a section.

14. **S. phacifolia,** *F. Muell. in S. Austral. Reg.* 1850. A perennial, with ascending or erect stems, often exceeding 1 ft., and sometimes much branched, usually hoary or white with short hairs, giving it sometimes a silky or almost mealy appearance. Leaflets usually 7 to 11, linear or narrow-oblong, acute, rather obtuse or emarginate. Stipules broad, especially the upper ones, which are often toothed. Flowers few in the raceme, on long peduncles, larger than in *S. lessertiifolia.* Calyx hoary or rarely with black hairs, nearly 3 lines long, the lobes acute or subulate-acuminate, usually about as long as the tube. Standard thin at the base, with a broad short claw and without any callosities as in *S. lessertiifolia;* keel much incurved, but obtuse; wings as long as the keel. Ovary sessile, villous; style much more rigid than in *S. lessertiifolia,* flattened in the lower portion, distinctly hooked inflexed or almost involute at the end. Pod narrow-oblong, ½ to nearly 1 in. long, often incurved, the upper suture slightly indented.—*S. stipularis,* F. Muell. in Linnæa, xxv. 393.

N. S. Wales. Darling river, *Victorian Expedition;* Flinders range, *Howitt's Expedition.*

S. Australia. Akaba, *F. Mueller;* between Stokes range and Cooper's Creek, *Wheeler.*

This species sometimes resembles some specimens of *S. phacoides,* but has no callosities whatever on the vexillum; it is more nearly allied to *S. lessertiifolia,* but the indumentum, the large stipules, and larger flowers, give it a very different aspect. The keel is also much more curved, and the broad rigid style is peculiar.

15. **S. oroboides,** *F. Muell. Herb.* A small perennial, sometimes
appearing annual, scarcely exceeding 6 in. in any of our specimens, the young
parts silky-pubescent, at length nearly glabrous. Leaflets usually 3 or 5, lanceolate, acute, the terminal one often above 1 in. long, the lateral. ones
smaller, in the lower leaves often solitary, shorter, and more obtuse, in the
upper leaves sometimes 7, smaller and linear. Stipules subulate. Flowers
small, usually few in a very short raceme, or almost umbellate on a rigid peduncle, shortly exceeding the leaves. Calyx silky-pubescent, about 2 lines
long, the lobes rather longer than the tube, but not so fine as in *S. oligophylla.*
Standard 4 to 5 lines diameter, with a very slight callosity at the top of the
claw, sometimes scarcely perceptible; wings short; keel exceeding the wings,
broad, incurved, obtuse. Ovary villous; style slender, incurved. Pod sessile, ovoid ovoid-globular or shortly oblong, often incurved, membranous, inflated, pubescent, about ½ in. long.

Queensland. Near Warwick, *Beckler.*
N. S. Wales. In the interior, *Howitt's Expedition, C. Moore;* head of the Gwydir,
Leichhardt; New England, *C. Stuart.*
Allied on the one hand to *S. oligophylla,* on the other to *S. lessertiifolia;* it is readily
distinguished from both by the foliage.

16. **S. lessertiifolia,** *DC. Prod.* ii. 271. A perennial, with diffuse or
ascending stems of 1 to 1½ ft., or shorter in mountain stations, glabrous except the young shoots and foliage, or more or less clothed with a grey, rather
rigid, appressed pubescence. Leaflets 9 to 15 or rarely more, oblong, obtuse
mucronate or almost acute, 4 to 8 lines or rarely 1 in. long. Stipules rather
broad, obtuse or acutely acuminate. Flowers rather small, violet-purple, in
short racemes, sometimes reduced to umbels or heads, on peduncles longer than
the leaves. Bracts small. Pedicels usually short. Bracteoles minute. Calyx
more or less pubescent with appressed black hairs, 2 lines long or rather
more, the teeth acute, shorter than the tube. Standard about 5 lines broad,
without callosities; wing shorter; keel very obtuse, almost hood-shaped.
Style not involute. Pod sessile, inflated, ¾ to 1 in. long, transversely veined.
—Hook. f. Fl. Tasm. i. 100; *S. viciæfolia,* F. Muell. in Dietr. Fl. Univ.
n. ser. t. 17.

a. *normalis.* Foliage nearly glabrous. Stipules broad. Keel not much curved.
N. S. Wales. Near Nangas, *M'Arthur.*
Victoria. Common on the S. coast, *Robertson, F. Mueller, Gunn,* and others.
Tasmania. Kent's group, Bass's Straits, *R. Brown;* common near Woolnorth and in
the islands of Bass's Straits, *J. D. Hooker.*
S. Australia. Guichen Bay, St. Vincent's Gulf, etc., *F. Mueller* and others.
b. *tephrotricha.* Leaves clothed on both sides with ash-grey hairs. Stipules lanceolate
acute or subulate-acuminate. Flowers rather larger, with a more incurved but still very ob-
tuse keel.—*S. tephrotricha,* F. Muell. in Linnæa, xxv. 392.
N. S. Wales. Blue Mountains and open forest land in the interior, *A. Cunningham.
Fraser, M'Arthur,* and others; head of the Gwydir, *Leichhardt;* New England, *C. Stuart*
Darling desert, *Neilson.*
Victoria. Near Melbourne, *Adamson;* Broughton, Hutt, and Hill rivers to Port
Phillip, Glenelg and Murray rivers, *F. Mueller.*
S. Australia. Encounter Bay, *Whittaker;* near Bethanie, *Behr;* St. Vincent's Gulf,
Mount Remarkable, Burra-Burra, *F. Mueller.*
The two varieties appear to pass one into the other by small gradations, but many speci-
mens are doubtful, being very rarely in fruit.

17. **S. monticola,** *A. Cunn.; A. Gray, Bot. Amer. Expl. Exped.* i. 411.
A diffuse ascending or erect perennial, closely resembling the more glabrous
forms of *S. lessertiifolia,* but appears to be constantly distinct in the shape of
the keel and the small pod. Leaflets either small, or narrower and more acute
than in that species. Racemes usually looser and more elongated. Calyx
with scarcely any black hairs. Keel broad, much incurved, rather acute or
almost rostrate. Pod sessile and inflated as in *S. lessertiifolia,* but not attaining ½ in. in length in any of our specimens.

N. S. Wales. Blue Mountains, *A. Cunningham* and others ; Nangas, *M'Arthur ;*
Cow pastures, *R. Brown* (not in fruit) ; ridgy ground between Curacau and Canowmdra, *C.
Moore.* The species requires further investigation from good fruiting specimens.

18. **S. luteola,** *F. Muell. Fragm.* i. 75. A small species, hoary or silky-
pubescent, with branching, erect or ascending stems of 4 to 8 in. in our specimens. Leaflets 7 to 13, obovate or oblong, obtuse, not above ½ in. long.
Stipules broadly lanceolate. Flowers small, yellowish, few, in almost sessile
racemes or interrupted spikes shorter than the leaves. Bracts small ; bracteoles inconspicuous. Calyx pubescent, narrower than in the other species,
not 2 lines long, the lobes acute, about as long as the tube. Standard nearly
3 lines diameter, rather longer than broad, without callosities ; wings shorter ;
keel nearly straight, obtuse. Style short, inflexed, almost involute at the extremity. Pod sessile, oblong, acuminate, ¾ to 1½ in. long, membranous and
inflated, but narrow, with the seminal suture more or less indented.

Queensland. Basaltic plains, Peak Downs, *F. Mueller.* The small narrow flowers
and close short inflorescence give to this plant a very different aspect from that of the rest
of the genus.

19. **S. parviflora,** *Benth.* Nearly glabrous, with erect slender stems
of about 1 ft. and few leaves. Leaflets 5 to 9, linear-acute, mostly 1 in.
long or more. Racemes slender, with small distant flowers. Bracts minute.
Pedicels about as long as the calyx. Calyx-tube about 1 line long, the teeth
shorter, narrow, acute. Standard without callosities, about 3 lines broad and
not so long ; wings as long as the keel, which is much curved, but obtuse.
Style much inflexed, but not involute at the end, with a small tuft of hairs
round the stigma. Pod sessile, ovoid, membranous, much inflated, about ½
in. long.

Queensland. Wide Bay, *Bidwill.*

20. **S. microphylla,** *A. Gray, Bot. Amer. Expl. Exped.* i. 410. Much
branched at the base, with ascending or erect branches of ½ to 1 ft., or rarely
more, glabrous or minutely pubescent. Leaflets numerous, obovate obcordate
or cuneate-oblong, usually emarginate, sometimes all under 1 line, more frequently 2 to 3 and rarely 4 lines long. Flowers (purple ?) small, rather numerous, in erect racemes much longer than the leaves. Pedicels very short.
Bracteoles minute. Calyx shortly pubescent, with a few small black hairs,
scarcely above 1 line long, the teeth very short. Standard about 3 lines
diameter, without callosities, the claw very short ; keel obtuse. Style much
curved, but not involute at the end, with a small tuft of hairs at the top behind
the stigma, besides the longitudinal beard of the genus. Pod sessile, ovoid or

nearly globular, 3 to 5 lines long, much inflated, more or less incurved, the base of the style much so, the seminal suture slightly intruded.

Queensland. Rockhampton, *Dallachy ;* Connor's River, *Bowman.*
N. S. Wales. Hunter's River district, *American Exploring Expedition ;* sandy plains between Wellington and Dubbo, *C. Moore;* between the Darling and Cooper's Creek, *Neilson.*
Victoria. Murray river, *F Mueller.*
S. Australia. Along the Murray, *F. Mueller:* towards Lake Gairdner, *Babbage.*

21. **S. Fraseri,** *Benth.* A tall species, often attaining 5 or 6 ft. Leaflets 11 to 21 or more, distinctly petiolulate, ovate or oblong, very obtuse, mostly ¼ to ½ in. long, green on both sides. Flowers violet-purple or nearly white, rather small, numerous, in long loose racemes. Calyx glabrous or slightly hairy, nearly 2 lines long, the teeth rather longer than in *S. laxa,* but not exceeding the tube. Standard about 5 to 6 lines diameter, without callosities ; wings shorter than the very obtuse keel. Style incurved, with a very conspicuous tuft of hairs on the top behind the stigma, and a few short hairs on the inner side. Pod inflated, membranous, acute, above 1 in. long, on a stipes much longer than the calyx.

Queensland. Moreton Bay, *C. Stuart.*
N. S. Wales. Macquarrie river, *Fraser ;* Hastings river, *Beckler.*

22. **S. laxa,** *R. Br. in App. Sturt, Exp.* 18. Apparently a rather tall species, with terete branches, glabrous or the young shoots slightly silky. Leaflets 11 to 21, distinctly petiolulate, from broadly ovate to oblong, very obtuse, rarely above ½ in. long, and often very small. Stipules broad and falcate. Flowers yellow, rather small, in long loose racemes flowering from near the base. Bracts very small. Pedicels short, with minute bracteoles below the calyx. Calyx glabrous or slightly hairy, 1½ lines long, the teeth acute, shorter than the tube. Standard about 5 lines diameter, without callosities ; wings much shorter; keel nearly as long as the standard, very obtuse. Style not involute, bearing a prominent tuft of hairs round or behind the stigma, especially at the back, besides the longitudinal beard of the genus. Pod glabrous, inflated, acute, fully ½ in. long, on a stipes usually shorter than the calyx-tube.

N. S. Wales. On the Darling, *Victorian Expedition, Dallachy.*
S. Australia. Murray scrub, towards Moorundi, *Behr.*
Var. (?) *rigida.* Leaflets small. Peduncles very long and thick. Calyx silky-pubescent. —Near the Darling river, *Victorian Expedition, Dallachy,* etc.
This species closely connects *Swainsona* with *Lessertia,* of which it has the style.

39. GLYCYRRHIZA, Linn.

(Clidanthera, *R. Br.*)

Calyx-lobes equal or the 2 upper ones shorter and more united. Petals narrow ; standard ovate or oblong, nearly sessile; keel shorter, obtuse or almost acute, the petals scarcely cohering. Upper stamen free or slightly cohering with the others in an open sheath ; anther-cells confluent at the top, the alternate smaller anthers opening deeply in two unequal valves. Ovary sessile, with 2 or more ovules; style incurved, glabrous, with a terminal stigma. Pod ovate oblong or shortly linear, flattened or turgid, glandular muricate or rarely smooth, indehiscent or opening tardily in 2 valves.   Seeds not strophiolate.—Herbs, with the root often sweet.   Leaves unequally pinnate or rarely 3-foliolate, without stipellæ.   Stipules narrow, membranous, deciduous.   Flowers blue violet white or yellowish, sessile or very shortly pedicellate, in axillary racemes.   Bracts narrow, very deciduous.   Bracteoles none.

The majority of the species are from the E. Mediterranean region, and temperate and subtropical Asia; and one is found in extratropical S. America.   The Australian species is endemic, although nearly allied to some of the Mediterranean ones.   The exceptional anthers first observed by R. Brown, upon which he founded his genus *Clidanthera* as distinguished from *Psoralea*, are to be seen, in a greater or less degree, in all the species of *Glycyrrhiza*, which moreover differ essentially from *Psoralea* in habit, in the ovules always more than one, and in the seed, even when solitary, never adhering to the pericarp.

1. **G. psoraleoides,** *Benth.*   An erect herb or undershrub of 2 ft. or more, glabrous or nearly so, but more or less glandular-viscid.   Leaflets usually 9 or 11, from elliptical-oblong to linear, ¾ to 1 in. long or rarely more, bordered with minute glandular teeth.   Flowers small, in pedunculate racemes or interrupted spikes.   Calyx about 1½ lines long; petals about twice as long.   Ovules 2.   Pod about 3 lines long, flattened, muricate, the sutures slightly thickened, apparently indehiscent, containing 1 or 2 seeds.—*Indigofera acanthocarpa,* Lindl. in Mitch. Three Exped. ii. 17; *Clidanthera psoraleoides,* R. Br. App. Sturt, Exped. 11; *Psoralea acanthocarpa,* F. Muell. Fragm. iii. 45, and Pl. Vict. ii. t. 26.

N. S. Wales.   On the Lachlan and Darling rivers, *Mitchell; Goodwin and Dallachy.* **Victoria.**   On the Murray and Murrumbidgee rivers, *F. Mueller.*

TRIBE VI. HEDYSAREÆ.—Pod separating into 1-seeded articles, or the whole pod 1-seeded and indehiscent, that is, reduced to a single article.

The above is the artificial character by which this tribe has been universally distinguished from all other *Papilionaceæ*.   It is, however, divisible into about 6 subtribes, several of which are more naturally allied to other tribes than they are to each other.   The Australian genera belong, however, to three of the most distinct subtribes: 1, *Æschynomeneæ*, with the pinnate leaves of *Galegeæ*, but the upper stamen very rarely free, including *Ormocarpum, Æschynomene,* and *Smithia;* 2, *Stylosantheæ*, with few leaflets, persistent stipular bracts, monadelphous stamens, and dimorphous anthers, of which *Zornia* is the only Australian genus; and 3, *Desmodieæ*, with the leaves usually 3-foliolate and stipellate as in *Phaseoleæ*, but the stem rarely twining, and the stipules usually dry.   The latter comprise the Australian genera *Desmodium, Uraria, Lourea,* and *Alysicarpus,* together with some genera with the pods reduced to a single article, connecting *Hedysareæ* with *Psoralea.*   Of these only one is Australian: *Lespedeza,* which moreover has exceptionally no stipellæ.   I have also included *Pycnospora* among *Hedysareæ,* although it has not the characteristic pod of the tribe, because it is in every other respect closely connected with *Desmodium,* and has no immediate affinity with any genus in any other tribe.   Some species of *Desmodium* itself have a pod which does not always separate into articles, but opens more or less constantly in 2 valves, although the articles are distinguishable by transverse lines.

40. **ORMOCARPUM,** Beauv.

Calyx 2 upper lobes connivent or shortly connate, the lowest rather longer. Standard orbicular; keel broad, incurved, as long as the wings.   Stamens all united in a sheath open on the upper side, and often splitting also on the lower side; anthers uniform.   Ovary sessile, with several ovules; style inflexed, filiform.   Pod linear, flattened, dividing into 2 or more oblong or

elongated indehiscent articles, narrowed at each end and longitudinally furrowed, often only one coming to perfection.—Tall shrubs. Leaves pinnate
with small leaflets, or in a species not Australian, of 1 large leaflet. Stipules
striate. Flowers yellow, white, or streaked with purple, solitary or few together in axillary racemes. Bracts and bracteoles persistent.

Besides the Australian species, which has a wide range in tropical Asia, there are 2 or 3
from tropical Africa, and 2 or 3 less perfectly known from Mexico.

1. **O. sennoides,** *DC. Prod.* ii. 315, var. *lævis.* Perfectly glabrous,
without the glandular pubescence of the common E. Indian form. Leaflets
9 to 15, broadly oblong, very obtuse, $\frac{1}{2}$ to $\frac{3}{4}$ in. long. Stipules lanceolate-
acuminate, sometimes very small, sometimes broad and 2 lines long. Pedun-
cles axillary, either short and 1-flowered, or rather longer, bearing 2 or 3
flowers. Pedicels 2 to 3 lines long, with 2 small bracteoles above the middle;
Flowers yellow, about 5 lines long. Pod usually of 3 or 4 articles, but some
remaining small and imperfect, and 1 or 2 only ripening, attaining in this
variety above 1 in. in length, and about 3 lines broad in the middle.—*Æschynomene coluteoides,* A. Rich. Sert. Astrol. 87. t. 32.

Queensland. Endeavour river ?, *Banks and Solander, R. Brown.* I have seen Australian specimens only in Herb. R. Brown, and a coloured figure in Sir J. Banks's unpublished plates, neither with the precise station. This glabrous variety extends over the Archipelago to Siam and the Philippines. The common E. Indian form (Wight, Ic. t. 297)
has usually a glandular-pubescent inflorescence, and the pods with shorter articles more or
less glandular-warted ; but there appear to be intermediates preventing the maintaining the
two forms as independent species.

41. ÆSCHYNOMENE, Linn.

Calyx-lobes nearly equal or united into two lips, either entire or the
upper one 2-lobed, the lower 3-lobed. Standard orbicular; keel much curved
and almost beaked, or rarely obovate and nearly straight. Stamens usually
all united in a sheath more or less split both on the upper and lower edge;
dividing the stamens into two bundles of 5 each; anthers reniform. Ovary
stipitate, with 2 or more ovules ; style filiform or subulate. Pod stipitate,
separating into 2 or more short flat usually indehiscent reticulate articles.—
Herbs undershrubs or in non-Australian species shrubs. Leaves unequally
pinnate, with small leaflets, without stipellæ. Stipules free. Flowers yellow,
often streaked with red, in axillary or rarely terminal racemes.

A considerable tropical genus, the species numerous in America, fewer in Africa, and only
two in Asia. Of the two Australian ones, one is common in Asia and Africa, the other in
South America and South Africa.

Leaflets numerous. Stipules produced below their insertion. Calyx deeply
2-lipped . . . . 1. *Æ. indica*
Leaflets 7 to 11. Stipules striate, persistent, not produced below their insertion. Calyx-lobes nearly equal . . . . 2. *Æ. falcata.*

1. **Æ. indica,** *Linn. ; DC. Prod.* ii. 320. A diffuse or erect annual of
1 to 2 ft., or when luxuriant in very wet places 3 ft. high, usually glabrous,
but the stem occasionally bearing a few asperities. Leaflets usually 40 to 60,
linear-oblong, obtuse, 2 to 3 or rarely 4 lines long. Stipules lanceolate,
acute, produced below their insertion into a rounded appendage. Racemes
shorter than the leaves, loosely 2- to 4-flowered, and often bearing a pinnate

leaf below the flowers. Pedicels slender. Bracts like the stipules but often denticulate; bracteoles short, persistent. Calyx about 2 lines long, deeply divided into 2 lips, the upper one 2-toothed, the lower shortly 3-lobed. Petals about 4 lines long, glabrous; keel much curved, almost acute. Pod on a long stipes, the upper suture straight, the lower slightly indented between the seeds; articles about 2 lines long, glabrous, smooth or more or less warted or muricate in the centre.—W. and Arn. Prod. 219; Wight, Ic. t. 405; *Æ. cachemiriana*, Camb. in Jacquem. Voy. 40. t. 48.

N. Australia. Upper Victoria river, *F. Mueller;* in the interior, *M'Douall Stuart;* also in *R. Brown's* collection without any label.

Queensland. Burdekin river, *Bowman.* The species is widely spread over tropical Asia and Africa.

2. **Æ. falcata,** *DC. Prod.* ii.322; var. *paucijuga, Benth. in Mart. Fl. Bras. Papil.* 67, *t.* 14. Stems from a woody stock diffuse decumbent or ascending, often under 1 ft. and rarely nearly 2 ft. long, more or less pubescent. Leaflets 7 to 11, obovate-oblong or cuneate, truncate or emarginate, usually oblique at the base, and about 3 to 4 lines long. Stipules acute, striate, not produced below their insertion. Peduncles slender, as long as or longer than the leaves, mostly 2- or 3-flowered. Pedicels much longer than the calyx. Bracts and bracteoles small, striate. Calyx 1½ lines long, the lobes all equally divided, as long as the tube, the 2 upper ones rather broader. Standard 3 to 4 lines diameter; wings broad; keel much curved, almost rostrate. Pod on a slender stipes of 2 to 4 lines, sprinkled with short hairs, the upper suture nearly straight and continuous, the lower edge deeply indented between the seeds; articles 4 to 6, 1½ to 2 lines diameter, opening in 2 valves on the lower edge and scarcely separating from each other.—*Æ. micrantha*, DC. Prod. ii. 321; Harv. and Sond. Fl. Cap. ii. 226, with all the synonyms there adduced.

Queensland. Broad Sound, *R. Brown;* on the Burdekin, *F. Mueller;* Wide Bay, *Bidwill;* plains of Rockhampton, *Bowman, Dallachy;* Moreton Bay, *Bidwill, F. Mueller.* The species is common in Brazil, where it diverges into a number of varieties mentioned in Martius's above-quoted Flora. The Australian form appears to me quite identical with the variety there named *paucijuga*, which is the most common in S. Brasil and Montevideo, and which is also the one found in S.E. Africa and Madagascar.

42. SMITHIA, Ait.

Calyx deeply divided into 2 lips, the upper one entire or notched, the lower entire 3-toothed or 3-lobed. Standard nearly orbicular, narrowed into a short claw, wings and keel nearly as long as the standard. Stamens united in a sheath open on the upper side and soon splitting also on the lower side; anthers reniform. Ovary sessile or stipitate, with several ovules; style filiform, with a small terminal stigma. Pod consisting of 2 or more flattened articles, separated by very narrow contractions and folded over each other within the calyx.—Diffuse herbs or in some African species shrubs. Leaves pinnate, without stipellæ. Stipules membranous or scarious. Flowers yellow, in axillary racemes or clusters. Bracts and bracteoles scarious or striate, persistent.

The genus has a considerable number of tropical Asiatic and E. African species, the only Australian one is one of the commonest in E. India.

1. **S. conferta,** *Sm. in Rees' Cyclop.* xxxiii. A procumbent or diffuse perennial of 1 to 1½ ft. or rarely more, glabrous except a few long rigid hairs or bristles on the young branches, petioles, margins and midribs of the leaflets, and on the calyx. Leaflets 7 to 15 or more, rather crowded on a short common petiole, oblong or linear; oblique, under ½ in. long. Stipules produced below their insertion into a subulate-acuminate appendage longer than the upper part. Racemes reduced to clusters of 3 to 5 flowers, almost sessile in the upper axils. Bracteoles broad, striate, above half the length of the calyx. Calyx 3½ to 4 lines long, the lips slightly falcate, acute, mucronate, finely striated. Ovules about 6. Pod not protruding from the calyx.—*S. capitata,* Desv. Journ. Bot. i. (iii.) 121; *S. sensitiva,* var. β. W. and Arn. Prod. 220.

Queensland? E. Coast, *R. Brown.* Common in E. India and in the Archipelago.

43. ZORNIA, Gmél.

Calyx small and thin, the 2 upper lobes united, the 2 lateral ones small, the lowest narrow. Standard orbicular; wings obovate or oblong; keel incurved, almost rostrate. Stamens united in a closed tube; anthers alternately long and short. Ovary sessile with several ovules; style filiform, with a small terminal stigma. Pod with the upper suture continuous, the lower one much indented; articles several, flat, smooth muricate or bristly.—Herbs: Leaves of 2 or 4 digitate leaflets, without stipellæ. Stipules striate. Flowers in terminal and axillary loose spikes. Bracts in pairs, enclosing the flowers, striate and oblique like the stipules, but broader and larger; bracteoles none.

The genus is chiefly American, one species found also in South Africa, and another widely dispersed over the warmer regions of the New and the Old World, including Australia.

1. **Z. diphylla,** *Pers. Syn.* ii. 318. A low herb, sometimes annual, sometimes forming a thick rootstock of several years' duration, the branches decumbent, ascending or nearly erect, 6 in. to 1 or 2 ft. long. Leaflets 2 at the end of the petiole varying from ovate and only 2 or 3 lines long in the lower leaves, to lanceolate or linear from ½ to 1 in. long in the upper ones, rarely all ovate acute and rather larger, or all linear. Flowers in the Australian varieties 3 to 4 lines long, almost enclosed in the narrow or ovate bracts, which like the stipules are produced into a short auricle below their insertion, and are often, as well as the leaves, marked with a few pellucid dots. Pod longer or shorter than the bracts, of 3 to 6 articles, quite smooth and reticulate or pubescent or muricate with hooked or pubescent bristles or prickles.

N. Australia. Victoria river, *F. Mueller*; Islands of the Gulf of Carpentaria, *R. Brown*; Port Essington, *Armstrong*; Sweers Island and Albert river, *Henne.*

Queensland. Burnett, Dawson, and Brisbane rivers, *F. Mueller*; from Broad Sound to Northumberland Islands, *R. Brown*; Port Curtis, *M'Gillivray*; Rockhampton, *Thozet* and others; Dogwood Creek, *Leichhardt.*

N. S. Wales. Port Jackson, *R. Brown*; Paramatta, *Woolls*; Clarence and Hastings rivers, *Beckler*; New England, *C. Stuart.*

The species is common in most hot countries in both the New and the Old World. Of the numerous varieties enumerated in Mart. Fl. Bras. Papil. 79, the following at least occur in Australia:—

a. *vulgaris.* The common Asiatic form, with the leaflets of the lower leaves small and ovate, those of the upper ones lanceolate or linear, the bracts rather narrow and flowers small.

b. *zeylonensis.* Stems elongated and loose. Leaflets rather larger, all ovate or ovate-lanceolate. Bracts rather broad. Flowers rather larger.

c. *gracilis.* Stems more erect, glabrous or hairy as well as the leaves. Leaflets mostly lanceolate or linear or even all linear. Bracts rather narrow. Flowers small.

In all the varieties the pod may be found smooth or muricate, glabrous or pubescent, and in one of the forms of the var. *gracilis*, from Sturt's Creek, *F. Mueller* (*Z. chætophora, F. Muell.* in Trans. Phil. Inst. Vict. iii. 56); the pods are rather larger and covered with rigid setæ much longer than in any other *Zornia* I have seen.

44. DESMODIUM, Desv.

(Dendrolobium, *W. and Arn.*; Dicerma, *DC.*; Nicolsonia, *DC.*)

Calyx-tube short, the 2 upper lobes more or less united. Standard from oblong to orbicular, narrowed at the base; wings oblong, usually adhering in the middle to the keel; keel obtuse. Upper stamen free or more or less united with the others in a sheath or tube; anthers uniform. Ovary sessile or stipitate with 2 or more ovules; style incurved, subulate. Pod longer than the calyx, flat, one or both sutures indented between the seeds, separating into indehiscent 1-seeded articles, or rarely the articles opening on the lower edge in 2 valves, and then not always readily separating.—Herbs shrubs or rarely small trees. Leaves pinnately 3-foliolate or 1-foliolate, with stipellæ. Stipules usually dry, striate, membranous. Flowers purple, blue, pink or white, usually small, in terminal racemes or panicles, or rarely in axillary umbels or clusters.

A very large genus widely dispersed over the tropical regions both of the New and the Old World, extending beyond the tropics into N. America, and a very few species into extratropical S. America, S. Africa, and extratropical Australia. Of the 16 Australian species, six have a wide range in East India and the Archipelago, one is common to Australia and New Caledonia, the remaining 9 are endemic but partaking of the general character of the Asiatic species, with the exception of *D. acanthocladum*, which is singular in the genus for its spinescent branchlets.

This genus is readily divisible into from 12 to 15 tolerably well-marked sections, many of which have been proposed by myself or others as distinct genera; but as they have proved to be distinguished some by habit only without marked floral or carpological characters, others by variations in the fruit, not always constant nor easily appreciated, I found it more convenient, on a general review for the Floras of Brazil and Hongkong, to retain them all under one generic name.

* *Wings usually free from the keel. Pod glabrous or silky-hairy.*

Flowers white, in dense axillary shortly pedunculate umbels. Pod-articles rather thick (Sect. **Dendrolobium**) . . . . . . 1. *D. umbellatum.*

Flowers small, in dense umbels or heads along the branches of a leafy panicle, each umbel almost enclosed in a 2-foliolate leaf. Pod-articles 2, nearly orbicular (Sect. **Phyllodium**) . . . 2. *D. pulchellum.*

Flowers in leafless racemes. Pedicels short crowded. Pod-articles 2, nearly orbicular. Leaflets *digitate or nearly so* (Sect. **Dicerma**) . . . . 3. *D. biarticulatum.*

** *Wings adhering to slight lateral protuberances or membranous appendages of the keel.*

Branchlets *spinescent.* Flowering branches reduced to axillary spines with 1 or 2 pairs of flowers below the summit. Pod-articles of *Heteroloma*, but usually 1 or 2 only . . . . . 4. *D. acanthocladum.*

Flowers in racemes or panicles. Ovules several, rarely 2 only. Pod of several articles (unless by abortion) indehiscent, the upper

suture straight or slightly indented, the lower suture much
indented between the seeds (Sect. **Heteroloma**).
Bracts narrow, persisting at least till the flower expands. Pe-
dicels usually in pairs (*Leptostachya*).
　Leaves all 1-foliolate.　Pod-articles small, nearly glabrous　.　5. *D. gangeticum.*
　Leaves all (except sometimes the lowest) 3-foliolate.
　　Pod-articles flat, prehensile-pubescent.
　　　Stems rather rigid, erect or ascending.　Fruiting-pedicels
　　　　reflexed, not longer than the calyx　.　.　.　.　.　.　6. *D. brachypodum.*
　　　Stems slender, diffuse.　Fruiting-pedicels slender, spreading,
　　　　much longer than the calyx.　Plant slightly pubescent.
　　　　Ovules and pod-articles several　.　.　.　.　.　.　.　7. *D. varians.*
　　　　Ovules and pod-articles 2 only.　.　.　.　.　.　.　.　8. *D. flagellare.*
　　　Stems diffuse or procumbent, rusty-villous.　Leaves softly
　　　　villous, rhomboid ovate.　Pedicels slender, spreading
　　　　rather longer than the calyx　.　.　.　.　.　.　.　.　9. *D. rhytidophyllum.*
　　Pod-articles somewhat turgid, slightly pubescent.
　　　Stem trailing.　Leaflets lanceolate, 2 to 3 in. long　.　.　10. *D. campylocaulon.*
Bracts broad membranous, falling off long before the flower ex-
pands (*Strobilifera*).
　Tall and erect.　Leaflets oblong or elliptical, 1½ to 2½ in. long.
　　Fruiting-pedicels mostly in pairs, rigid, reflexed, not longer
　　than the calyx　.　.　.　.　.　.　.　.　.　.　.　.　11. *D. nemorosum.*
　Diffuse and slender.　Pedicels mostly solitary, filiform,
　　spreading, longer than the calyx.
　　Pod-articles thin, strongly reticulate.
　　　Leaflets narrow-oblong or linear　.　.　.　.　.　.　.　12. *D. neurocarpum.*
　　　Leaflets broadly obcordate　.　.　.　.　.　.　.　.　.　13. *D. trichostachyum.*
　　Pod-articles scarcely separating, very finely veined, the
　　　upper suture thickened (see below, sect. *Nicolsonia*).
Flowers in terminal racemes or panicles. Bracts of the *Strobilifera.*
　Ovules several.　Pod very flat, the upper suture straight, the
　lower slightly indented and opening more or less in 2 valves
　(Sect. **Nicolsonia**).
　Fruiting-pedicels short, erect or nearly so in pairs or clusters.
　　Racemes short, dense, in a short terminal panicle.　Hairs
　　　short, usually appressed　.　.　.　.　.　.　.　.　.　.　14. *D. polycarpum.*
　　Racemes elongated.　Hairs of the stem and rhachis long and
　　　spreading　.　.　.　.　.　.　.　.　.　.　.　.　.　.　.　15. *D. trichocaulon.*
　Fruiting-pedicels slender, spreading, solitary and distant.
　　Stem loosely diffuse.　Leaves not crowded, leaflets oblong.
　　　Hairs long and spreading　.　.　.　.　.　.　.　.　.　16. *D. Muelleri.*
　　Stems procumbent, pubescent.　Leaves crowded; leaflets small.
　　　Racemes filiform, few-flowered　.　.　.　.　.　.　.　17. *D. parvifolium.*

1. **D. umbellatum,** *DC. Prod.* ii. 325.　A bushy shrub occasionally
growing into a small tree, the young shoots silky.　Leaflets 3, ovate or oval-
oblong, obtuse or rarely almost acute, mostly 1½ to 2 in. long, glabrous or
nearly so above, pale or silky-pubescent underneath, with prominent primary
veins.　Stipules very deciduous.　Flowers white, in dense axillary umbels
on a common peduncle, rarely attaining ½ in.　Bracts very deciduous.　Pedicels as long as the calyx.　Bracteoles persistent, as long as the calyx-tube.
Calyx silky, about 2 lines long, the lobes acute, not longer than the tube.
Standard broad, twice as long as the calyx ; wings much shorter ; keel as long
as the standard, without lateral protuberances.　Pod of 3 or 4 thickish almost
fleshy articles, each 3 or 4 lines long and not so much in breadth, not reticu-

late, indehiscent.—*D. australe,* DC. Prod. ii. 326 ; *Dendrolobium umbellatum,* W. and Arn. Prod. 224 (under *Desmodium*) ; Benth. in Pl. Jungh. 16 ; *Ormocarpum oblongum,* Desv. in Ann. Soc. Linn. 1825, 307.

Queensland. Barnard Isles, *M'Gillivray ;* Port Denison and Edgecumbe Bay, *Dallachy, also in R. Brown's Collection.* The species is widely spread over East India and the Archipelago.

2. **D. pulchellum,** *Benth. Fl. Hongk.* 83. A tall branching perennial or undershrub, the branches pubescent or villous. Leaflets 3, ovate, obtuse, the margins sometimes slightly sinuate, the terminal one usually 3 to 4 in. long, the lateral ones smaller, all slightly pubescent or nearly glabrous above, softly pubescent underneath. Flowers small, in dense umbels or heads, sessile along the branches of a large terminal leafy panicle, each umbel almost enclosed in a 2-foliolate leaf-like bract at its base, each leaflet broadly ovate or orbicular, ½ to ¾ in. long and very oblique at the base. Pod usually of 2 flat nearly orbicular small articles, glabrous or nearly so except a few hairs along the edge, both edges of the pod, especially the lower one, indented between the seeds.—*Dicerma pulchellum,* DC. Prod. ii. 339 ; Wight, Ic. t. 418 ; *Phyllodium pulchellum,* Desv. ; Benth. in Pl. Jungh. 217.

N. Australia. N. coast, *R. Brown.* Widely spread over East India and the Archipelago, extending northwards to S. China.

3. **D. biarticulatum,** *F. Muell. Fragm.* ii. 121. A rigid undershrub with prostrate decumbent or almost erect branches of 1 to 2 or rarely 3 ft., the young shoots softly pubescent or silky. Leaflets 3, oblong or on the lower leaves narrow-obovate, ½ to 1 or rarely 1½ in. long, rather rigid, digitate or nearly so at the end of a stiff short petiole. Stipules brown, scarious, more or less united opposite the leaf. Flowers small, red, crowded or distant in a long narrow terminal raceme. Pedicels short, usually 2 together. Bracts narrow, acuminate, rigid and striate. Calyx about 2 lines long, the lobes rather longer than the tube, the 2 upper ones united nearly to the top. Petals twice as long ; wings scarcely adhering to the keel, which has not the lateral appendages of most *Desmodia.* Ovary with only 2 ovules. Pod sessile, flat, silky-pubescent ; articles 2 or rarely 1, nearly orbicular, not 2 lines diameter, reticulate and indehiscent.—*Dicerma biarticulatum,* DC. Prod. ii. 339 ; Wight, Ic. t. 419.

N. Australia. Brunswick Bay, N.W. coast, *A. Cunningham ;* Albert and Nicolson rivers, *F. Mueller ;* islands of the Gulf of Carpentaria, *R. Brown, Henne.* **Queensland.** Burdekin river, *F. Mueller.;* Bowen river, *Bowman.* Common in E. India. Several of the Australian specimens are more erect and taller, with longer stipules bracts and bracteoles than the Indian ones, but they do not otherwise differ, and others are precisely like the Indian form figured by Wight.

4. **D. acanthocladum,** *F. Muell. Fragm,* ii. 122. A glabrous undershrub or small shrub, with numerous slender but rigid angular branches, the smaller ones ending in a fine thorn. Leaflets 3, oblong or lanceolate, the terminal one ½ to 1 in. long, the lateral ones smaller, the common petiole short. Stipules small. Flowering branches reduced to axillary leafless spines, usually shorter than the leaves, and bearing 1 or 2 pairs or clusters of flowers near the extremity. Pedicels short. Bracts very small. Flowers about 4 lines long. Calyx-lobes about as long as the tube. Wings strongly adhering

to the lateral protuberances of the keel. Ovules usually 3 or 4. Pod rarely of more than 2 articles and often only 1, pubescent with clinging hairs, the upper suture straight, the lower deeply and broadly indented, each article 5 to 6 lines long and about 2 broad, tapering to each end, flat and indehiscent.

N. S. Wales. Woods on the Clarence river, *Beckler*. This species, different from all others of the genus in its thorny branchlets, is otherwise more nearly allied to the section *Heteroloma*, subsection *Podocarpia*, than to *Dicerma*.

5. **D. gangeticum,** *DC. Prod.* ii. 327. A decumbent or erect herb or undershrub, the large-leaved forms attaining 2 or 3 ft., the small ones slender and under 1 ft., sprinkled with a few hairs. Leaves all 1-foliolate, in the large forms ovate or ovate-lanceolate 3 or 4 in. long, in the smaller ones broadly ovate-cordate or almost orbicular ½ to 1 in. long. Racemes long and slender, terminal or in the upper axils. Flowers small, the pedicels in pairs, under 2 lines long. Bracts linear-subulate, persistent to the time of flowering, but falling off soon after. Calyx about 1 line long, the lobes longer than the tube. Petals twice as long. Pod sessile, minutely pubescent, the upper margin slightly, the lower deeply indented ; articles 4 to 6, 1 to 1½ lines long and broad, flat, thin and indehiscent.—W. and Arn. Prod. ii. 125.

N. Australia. Victoria river, *F. Mueller.*
Queensland. Endeavour river, *Banks and Solander ;* Northumberland Islands, *R. Brown ;* Moreton Bay, *C. Stuart ;* Rockhampton, *Dallachy ;* Broad Sound, *Bowman.*
The species is widely spread over E. India and the Archipelago.

6. **D. brachypodum,** *A. Gray, Bot. Amer. Expl. Exped.* 434. A rather rigid, erect or decumbent perennial, of 1 to 2 ft., slightly pubescent, the specimens often assuming a bluish-black tint when dry. Leaflets 3 or in the lowest leaves solitary, from broadly ovate almost orbicular to oval-oblong, very obtuse, mostly 1 to 2 in. long, rather stiff and strongly reticulate, the stipellæ long. Stipules rather broad, striate, acuminate. Flowers small, usually in pairs, the lower ones distant, in a long terminal rigid raceme. Pedicels very short and recurved. Bracts subulate-acuminate, persistent to the time of flowering but falling off soon after. Calyx 1½ lines long, the lobes not longer than the tube. Petals about twice as long. Pod sessile or shortly stipitate, pubescent with clinging hairs, the upper suture slightly the lower deeply indented ; articles 4 to 6, about 2 lines long and nearly as broad, thin, reticulate and indehiscent.

Queensland. Burdekin river, *F. Mueller ;* Port Curtis, *M'Gillivray ;* Percy Island, *A. Cunningham ;* Rockhampton, *Dallachy, Bowman ;* Moreton Bay, *F. Mueller, Leichhardt.*
N. S. Wales Port Jackson to the Blue Mountains, *R. Brown, Banks and Solander, A. Cunningham, Woolls,* etc. ; Hunter's River, *American Exploring Expedition ;* New England, *C. Stuart ;* head of the Gwydir, *Leichhardt.*

7. **D. varians,** *Endl. in Ann. Wien. Mus.* i. 185. Stock woody with prostrate diffuse or ascending slender stems of ½ to 1½ ft., the whole plant pubescent or nearly glabrous. Leaflets 3, in the lower leaves or sometimes all broadly obovate or almost orbicular or obcordate, ¼ to ½ in. long, the upper ones or sometimes nearly all ovate oblong or almost linear, ¾ to 1 in. long. Stipules small, acute. Flowers very small, in distant pairs, in slender terminal racemes. Pedicels filiform, short when in flower, spreading and nearly

½ in. long when in fruit. Bracts small, persistent. Calyx-lobes acuminate, as long as the tube. Petals rarely 3 lines long. Pod sessile, the upper suture very slightly, the lower deeply indented; articles 3 to 6, obliquely ovate, about 2 lines long and not so broad, flat, indehiscent, clothed with short clinging hairs.—*Hedysarum varians*, Labill. Sert. Austr. Caled. 71, t. 71.

Queensland. Broad Sound and Keppel Bay, *R. Brown;* Moreton Bay, *F. Mueller, C. Stuart.*

N. S. Wales. Port Jackson, *Banks, R. Brown,* and others; Hastings river, *Beckler;* Liverpool Plains, *Woolls;* head of the Gwydir, *Leichhardt.*

Victoria. Macalister-river mountains, Ovens and Delatite rivers, Bacchus Marsh, etc., *F. Mueller.*

Tasmania. N. coast, *J. D. Hooker.*

Var. *angustifolia.* Leaves mostly linear.—*D. spartioides,* DC. Prod. ii. 337 (from a specimen formerly seen in Herb. DC.).

Var. *Gunnii.* Leaves all broadly obovate orbicular or obcordate.—*D. Gunnii,* Hook. f. Fl. Tasm. i. 101. The differences I had formerly observed in the fruit disappear in the fully ripe pod. This variety is the only one in Tasmania and occurs also in Victoria and N. S. Wales, but these pass frequently into the narrow-leaved form.

8. **D. flagellare,** *Benth.* Stems slender, trailing, hirsute when young as well as the racemes with soft loose hairs. Leaflets 3, mostly broadly obovate, but varying from almost fan-shaped to narrow-obovate, broadly truncate or emarginate, ¾ to 1½ in. long, glabrous above, sprinkled with rather long appressed hairs underneath. Stipules subulate-acuminate. Racemes long, slender, flexuose. Bracts lanceolate-subulate, persistent long after the flower opens. Pedicels mostly solitary, spreading, longer than the calyx. Calyx ciliate-hirsute, about 1 line long, the lobes thin and rather broad. Petals not seen perfect. Ovary sessile, hirsute, with 2 ovules. Pod of 1 or 2 articles, each about 2 lines long and broad, the upper suture nearly straight, the lower arcuate, thin, flat, densely pubescent with clinging hairs, indehiscent.

N. Australia. Beagle Valley, *F. Mueller.*

The specimens are too far advanced for a satisfactory description, but the species appears to be allied to *D. varians*, notwithstanding that the ovules and pod-articles are 2 only.

9. **D. rhytidophyllum,** *F. Muell. Herb.* A perennial with long procumbent almost trailing branches, softly rusty-tomentose or velvety-villous. Leaflets 3, ovate-rhomboid or the upper ones rather narrow, obtuse, mostly 1 to 2 in. long, rather thick and softly villous on both sides. Stipules lanceolate, striate, often reflexed. Racemes long. Flowers rather small, in distant pairs. Pedicels slender, rather longer than the calyx. Bracts subulate-acuminate, persistent. Calyx about 1½ lines long, the lobes longer than the tube. Petals about twice as long. Pod almost sessile, the upper suture slightly, the lower more deeply indented; articles 3 to 6, about 1½ lines long and nearly as broad, flat, indehiscent, clothed with short clinging hairs.

Queensland. Granite rocks between Dawson and Burnett rivers, *F. Mueller;* near Rockhampton, *Dallachy.*

N. S. Wales. Port Jackson, *R. Brown, Backhouse;* Paramatta, *Woolls;* Hastings, Macleay and Clarence rivers, *Beckler.*

The species is allied to *D. varians*, but much larger and coarser, with a different foliage and indumentum.

10. **D. campylocaulon,** *F. Muell. Herb.* Stem diffuse or trailing,

elongated, rather stout, slightly pubescent. Leaflets 3, lanceolate, obtuse or acute, 2 to 3 in. long, glabrous or nearly so, strongly veined underneath, the stipellæ very conspicuous. Stipules striate, thin. Racemes mostly leaf-opposed, pedunculate. Flowers numerous. Pedicels solitary or in pairs, slender but short. Bracts narrow, usually persistent. Calyx nearly 1½ lines long, the lobes longer than the tube. Petals twice as long. Pod sessile, pubescent when young with short clinging hairs, the upper suture continuous, the lower indented; articles 3 to 6, about 1½ lines long and broad, membranous turgid or almost inflated when ripe, slightly reticulate, indehiscent.

N. Australia. Fertile plains, Sturt's Creek, *F. Mueller.*

11. **D. nemorosum,** *F. Muell. Herb.* Stems apparently tall, erect, woody at the base, clothed as well as the under side of the leaves with soft silky appressed hairs. Leaflets 3 or solitary in the lowest leaves, oblong-elliptical, very obtuse, 1½ to 2½ in. long or the lateral ones smaller, glabrous above. Stipules rather long, striate. Racemes terminal. Bracts broad, membranous, acuminate, falling off long before the flowering. Flowers solitary or in pairs; pedicels very short, rigid, recurved after flowering and not exceeding the calyx when in fruit. Calyx nearly 2 lines long, the lobes rather broad, acute. Petals nearly twice as long, the lateral appendages of the keel very prominent. Pod sessile, the upper suture continuous, the lower rather deeply and broadly indented; articles few, flat, 3 to 4 lines long and about half as broad, indehiscent, pubescent with short clinging hairs.

Queensland. Brisbane river, *F. Mueller;* Pine river, *Fitzalan,* also in *Leichhardt's* collection. The foliage and habit are nearly those of the E. Indian *D. concinnum,* but the pod and flowers are very different.

12. **D. neurocarpum,** *Benth.* Stems slender, diffuse, loosely villous with spreading hairs. Leaflets 3 or the lower ones solitary, narrow-oblong, very obtuse, sometimes 1½ to 2 in. long, but often under ½ in. Stipules lanceolate, subulate, quite free. Flowers small, few, in long filiform racemes; pedicels solitary, filiform, distant. Bracts small, lanceolate, falling off long before the flowers expand. Calyx about 1 line long; lobes acute, rather longer than the tube. Pod sessile, the upper suture very slightly, the lower one more deeply sinuate; articles 2 to 4, as broad as long, flat, indehiscent, strongly reticulate, sprinkled with a few hairs.

N. Australia. Upper Victoria river, *F. Mueller.* The inflorescence and habit are those of *D. Muelleri,* with a very different pod.

Var. *gracile.* Slender and apparently annual. Leaflets linear. Flowers and pod small. —N. coast, *R. Brown.*

13. **D. trichostachyum,** *Benth.* Stems prostrate, filiform, nearly glabrous. Leaflets 1 or 3, very broadly obcordate, 2 to 4 lines or rarely ⅓ in. long, and sometimes broader than long. Stipules subulate-acuminate. Flowers very small, distant, in filiform terminal simple or branched racemes; pedicels all solitary and filiform. Bracts membranous, lanceolate, falling off long before the flowers open. Calyx about ½ line long, divided nearly to the base into narrow acute lobes. Pod sessile, the upper suture straight, the lower rather deeply indented; articles 3 or 4, small, as broad as long, thin, glabrous, strongly reticulate.

N. Australia. Islands of the Gulf of Carpentaria, *R. Brown ;* Arnhem's Land, *F. Mueller ;* Port Essington, *Armstrong.*
Queensland. Endeavour river, *Banks and Solander ;* E. coast, *A. Cunningham.*

14. **D. polycarpum,** *DC. Prod.* ii. 334. An erect decumbent or ascending perennial or undershrub, 1 to 2 or 3 ft. high or rarely more, more or less pubescent with short appressed or scarcely spreading hairs. Leaflets 3, the terminal one obovate or elliptical, 1½ to 2 in. long, the lateral ones usually smaller. Stipules striate, acuminate. Racemes terminal, dense, 1 to near 3 in. long, often several together forming a short panicle. Bracts broad lanceolate, imbricate at first, but falling off before the flowers expand. Flowers purple, crowded, 3 to 4 lines long. Pods crowded, erect, hairy or glabrous, about ¼ to ¾ in. long, the upper suture continuous, the lower indented ; articles about 4 to 6, flat, usually opening at the lower edge when ripe.—W. and Arn. Prod. 227 ; Wight, Ic. t. 406.

Queensland. Sandy Cape, Broad Sound, and Northumberland Island, *R. Brown ;* Providence Hill, *F. Mueller :* Rockhampton, *Thozet.* Extends over the whole of E. India, the Archipelago, and the Pacific Islands. To the numerous synonyms adduced by Wight and Arnott must probably be added *Hedysarum tuberculosum,* Labill. Sert. Austr. Caled. t. 72.

15. **D. trichocaulon,** *DC. Prod.* ii. 335. Very nearly allied to *D. polycarpum,* with a similar foliage and the erect pods the same, but the stems more generally decumbent, more slender, and clothed as in *D. Muelleri* with long soft spreading hairs, and the racemes much looser and slender.

Queensland. Brisbane river, Moreton Bay, *F. Mueller, Leichhardt, C. Stuart.* Not uncommon in E. India, where the above-mentioned differences appear to be constant, although it may possibly prove to be a variety only of *D. polycarpum.*

16. **D. Muelleri,** *Benth.* Stems branching at the base, apparently ascending or erect, clothed as well as the racemes with long soft spreading hairs, the young shoots almost silky. Leaflets 3, oblong, obtuse, ½ to 1½ in. long, glabrous or loosely pubescent. Stipules lanceolate, acuminate, softly hairy. Racemes terminal, slender ; pedicels distant, solitary, filiform, spreading. Bracts broad, lanceolate, acuminate, imbricate at first, but falling off long before the flower expands. Calyx nearly 1 line long, the subulate-acuminate lobes longer than the tube. Pod sessile, rather broad, the upper suture straight and slightly thickened, the lower very slightly indented between the seeds ; articles 4 to 6, as broad as long, truncate at both ends, thin and flat, with fine transverse veins, separating but apparently opening sometimes at the lower suture when ripe.

N. Australia. Upper Victoria river, *F. Mueller.*
Queensland. Bowen river, *Bowman.*

17. **D. parvifolium,** *DC. Prod.* ii. 334. A very much-branched diffuse or prostrate slender annual or perennial, sprinkled with a few silky hairs. Leaves usually small and crowded ; leaflets 3 or rarely solitary, obovate or oblong, ¼ to nearly ½ in. long or rarely more, on a short filiform common petiole. Stipules acuminate, brown and scarious. Flowers small, in short filiform racemes, usually terminating short lateral branches ; pedicels solitary, filiform, distant. Bracts membranous, acuminate, falling off long before the flower expands and seldom seen. Calyx about 1½ lines long, the lobes acu-

minate, much longer than the tube, the 2 upper ones only shortly united.
Petals scarcely exceeding the calyx. Pod sessile, glabrous or minutely pubescent, the upper suture straight or slightly indented, and often more or less
dilated, the lower more deeply indented; articles 2 to 4, thin, flat, with very
fine transverse reticulations, scarcely separating from each other and sometimes perhaps opening on the lower edge.

N. Australia. Arnhem N. Bay, *R. Brown.*
Queensland. Moreton Bay, *F. Mueller* ; Archer's Station, *Leichhardt.*
The species is common in India, extending over the Archipelago and into S. China. This
and *D Muelleri* seem to connect the section *Sagotia,* founded on the common tropical *D.
triflorum,* DC., with *D. trichostachyon* and *D. neurocarpum,* which I have referred doubt-
fully to *Heteroloma,* although they have the solitary pedicels of *Sagotia.* They all come
very near in habit to some of the looser-flowered species of the section *Nicolsonia,* but the
pod is much less disposed to open on the lower edge.

45. PYCNOSPORA, R. Br.

Calyx 2 upper lobes united into one. Standard nearly orbicular, narrowed
at the base; wings adhering to the keel; keel obtuse, with small lateral
appendages. Upper stamen free or at first united with the others, anthers
uniform. Ovary sessile, with several ovules. Style subulate, with a terminal
stigma. Pod oblong, turgid, 2-valved, transversely veined. Seeds several,
not strophiolate.—An undershrub, with the habit of *Desmodium.* Leaves
pinnately 3-foliolate or 1-foliolate, with stipellæ. Flowers small, in terminal
racemes or panicles.

The genus consists of a single species, extending over the Indian Archipelago to S. China.
It is very nearly allied to *Desmodium,* except in the pod (nearly that of *Crotalaria*), which
would technically remove it from *Hedysareæ,* but it has no immediate affinities in any
other tribe.

1. **P. hedysaroides,** *R. Br. in W. and Arn. Prod.* 197. Stock perennial, with several decumbent or ascending branched stems, 1 to 2 ft. long,
pubescent or hairy. Leaves nearly those of *Desmodium concinnum;* leaflets
obovate or obovate-oblong, the terminal one in some specimens scarcely ½ in.,
in others above 1 in. long, the lateral one usually smaller or sometimes
wanting. Stipules striate, subulate-acuminate, frequently deciduous. Flowers
about 2 lines long, purplish, in terminal slender racemes of 2 to 3 in., or occasionally longer and branching into panicles; pedicels short, in pairs.
Bracts rather broad, acuminate, membranous, striate, falling off long before
the flower expands. Pod 3 to 4 lines long, very turgid, slightly pubescent,
the valves thin, with very fine transverse reticulations. Seeds 6 to 8, small,
reniform.—*P. nervosa,* W. and Arn. Prod. 197.

N. Australia. Gulf of Carpentaria, *R. Brown;* Copeland Island, Arnhem's Land,
A. Cunningham ; Port Essington, *Armstrong,*
   Queensland. Broad Sound, *R. Brown, Bowman;* Dunk Island, *M'Gillivray;*
Rockhampton, *Thozet, Dallachy.*

46. URARIA, Desv.

Calyx-lobes subulate-acuminate, spreading, the 2 upper ones (lowest by
the resupination of the flower) shorter. Standard orbicular or obovate, narrowed into the claw; wings adhering to the obtuse keel. Upper stamen

free, the others united ; anthers uniform. Ovary sessile or nearly so, with 2 or more ovules ; style filiform with a capitate terminal stigma. Pod nearly sessile, contracted between the seeds ; articles ovate, folded back upon each other within the calyx.—Herbs or undershrubs. Leaves pinnate of 3, rarely 5 or 7 leaflets, or sometimes of a single terminal leaflet, usually prominently reticulate, with stipellæ. Stipules free, acuminate, striate. Flowers purplish or yellowish, in terminal racemes either slender and elongated or dense and spike-like, the pedicels in pairs, inflexed at the top so as to reverse the flowers. Bracts usually broad, acuminate ; bracteoles none.

An Asiatic and African tropical genus, with one or two species naturalized in some parts of tropical America. Of the Australian species, two are common Asiatic ones, the third appears to be endemic.

Upper leaves of 3 or 5 long narrow leaflets. Raceme long and slender.
Pod of 3 to 6 articles . . . . 1. *U. picta.*
Leaves mostly of 3 oblong leaflets. Raceme cylindrical dense and
spike-like. Pod of 2 articles. Bracts persistent . . . . . . 2. *U. cylindracea.*
Leaves mostly of 1 very broad leaflet. Raceme oblong dense and spike-
like. Pod of 2 articles. Bracts deciduous . . . . 3. *U. lagopoides.*

1. **U. picta,** *Desv. ; DC. Prod.* ii. 324. An undershrub with ascending or erect stems of 1 to 3 ft., loosely pubescent or villous. Lower leaves occasionally of 1 ovate leaflet, the others of 3, 5, or rarely 7 leaflets, from ovate-lanceolate to narrow oblong-lanceolate, 2 to 4 or even 5 in. long, obtuse or almost acute, glabrous or scabrous-pubescent, the Asiatic specimens often variegated with white along the midrib. Racemes long and slender, often attaining 6 to 8 in. in fruit. Bracts ovate, falling off long before the flower expands. Pedicels short, hispid-villous. Calyx-lobes setaceous, plumose, rather above 1 line long, the upper ones rather shorter. Petals more than twice as long. Pod of 3 to 6 small glabrous articles.—Wight, Ic. t. 411.

Queensland. Broad Sound, *R. Brown ;* Rockhampton, *Thozet, Dallachy ;* Bowen river, *Bowman.* Widely spread over tropical Asia and Africa and introduced into the West Indies. The Australian specimens have the leaflets usually all green, and often 3 or 1 only, but in some the leaves are nearly all 5-foliolate, as in the Asiatic ones.

2. **U. cylindracea,** *Benth.* An undershrub with decumbent or ascending stems, loosely pubescent or rusty-villous. Leaflets 3 or very rarely 1, ovate-oblong, obtuse, the terminal one usually 1½ to 3 in. long, the lateral ones smaller, slightly scabrous above, softly pubescent underneath. Racemes dense, but more elongated than in *U. lagopoides,* often attaining 3 in. when in fruit. Bracts broadly ovate, softly villous, persistent. Pedicels rather longer than the calyx. Calyx-lobes subulate-plumose as in *U. lagopoides,* but the upper ones much shorter. Pod of 2 articles, the pericarp thin, but strongly reticulate.

N. Australia. Islands of the N. coast, *R. Brown ;* Upper Victoria river, *F. Mueller ;* Port Essington, *Armstrong ;* Sweers Island, *Henne.*
Queensland, *Bowman ;* Port Denison, *Fitzalan.*
With the inflorescence and habit of *U. lagopus,* DC., which has not yet been found in Australia, this species has the 2-ovulate ovary and the pod of *U. lagopoides.*

3. **U. lagopoides,** *DC. Prod.* ii. 324. Stock short and woody or shortly creeping, with procumbent or ascending stems of ½ to 1½ ft., pubescent or loosely villous. Leaflets solitary or 3, the single or terminal one from orbi-

cular-reniform to broadly cordate-ovate, always very obtuse, 1 to 2 in. long, slightly scabrous or loosely pubescent, the lateral ones, when present, smaller. Stipules subulate-acuminate. Racemes contracted into a very dense oblong obtuse hirsute spike, of 1 to 2 in., nearly sessile above the last leaves. Bracts broadly ovate, acuminate, usually very deciduous, except sometimes at the base of the spike. Pedicels shorter than the calyx. Calyx lower-lobes (turned upwards by the inflexion of the pedicel) subulate-plumose, 2 to 3 lines long, the upper ones much shorter with a broad base. Petals not much longer than the calyx, on slender claws. Ovules 2. Pod of 2 ovate, somewhat turgid, reticulate articles, each about 1½ lines long, glabrous or rarely pubescent.—Wight, Ic. t. 289 ; *U. cercifolia,* Desv. ; DC. Prod. ii. 325.

Queensland. Broad Sound, *R. Brown ;* Brisbane river, *F. Mueller ;* Rockhampton, *Thozet.* Widely dispersed over E. India and the Archipelago.

47. LOUREA, Neck.

Calyx broadly campanulate, enlarged after flowering, the lobes broad and equal. Standard obovate or obcordate, narrowed into the claw ; wings adhering to the obtuse keel. Upper stamen free, the others united ; anthers reniform. Ovary with 2 or more ovules ; style subulate, with a capitate stigma. Pod stipitate or nearly sessile, contracted between the seeds ; articles ovate, folded back upon each other within the calyx.—Herbs. Leaflets 1 or 3, often broader than long, with stipellæ ; stipules free. Flowers in terminal racemes, the pedicels usually in pairs. Bracts very deciduous.

A genus of 3 or 4 species, natives of tropical Asia, one of which extends into Australia.

1. **L. obcordata,** *Desv. ; DC. Prod.* ii. 324. Stems slender, prostrate, usually shortly hairy, 1 to 2 ft. long. Leaflets usually 3, the terminal one broadly obovate orbicular or reniform, ½ to 1 in. broad, rather rigid and strongly reticulate in the Australian specimens, less so in most Asiatic ones, sprinkled with a few small hairs, the lateral ones smaller, ovate or obovate. Racemes slender, either simple and 2 to 6 in. long, or shorter and paniculate. Flowers small, shortly pedicellate. Calyx at first not above 1 line long and hairy, but after flowering attaining 3 lines and completely enclosing the fruit. Pod usually of 2 articles, each of about 1½ lines long, much reticulate.— *L. reniformis,* DC. Prod. ii. 324.

N. Australia. Upper Victoria river, *F. Mueller.* The species extends over the Indian Archipelago to S. China.

48. ALYSICARPUS, Neck.

Calyx deeply cleft, the lobes stiff and dry, the two uppermost often united into one. Standard obovate or orbicular, narrowed into the claw ; wings adhering to the obtuse keel. Upper stamen free, the others united ; anthers reniform. Ovary sessile or nearly so, with several ovules ; style filiform, with a capitate stigma. Pod erect, straight, nearly terete, or somewhat flattened but thick, narrowed between the seeds or equal ; articles ovate, globular, or truncate at both ends, indehiscent.—Herbs either glabrous or loosely hairy. Leaves of a single leaflet (or very rarely in species not Australian, 3-foliolate), with stipellæ. Stipules dry, striate, acuminate, free, or united opposite the leaf. Flowers small, in slender terminal or rarely axillary racemes, the pedicels usually in pairs. Bracts scarious.

The genus is generally spread over tropical Asia and Africa, one species having also established itself in some parts of America. The three Australian species are all common Indian ones.

Calyx small, with very narrow lobes. Pod several times longer, not contracted, but with slightly raised transverse lines between the seeds . . . . 1. *A. vaginalis*.
Calyx with narrow-lanceolate striate lobes, not overlapping. Pod about twice as long, scarcely contracted between the seeds, the articles slightly and irregularly wrinkled . . . . 2. *A. congifolius*.
Calyx with lanceolate, rigid, not striate lobes overlapping each other. Pod shortly exserted, much contracted between the seeds, articles deeply marked with transverse wrinkles . . . . 3. *A. rugosus*.

The common Indian *A. monilifer*, DC., with smooth globular bead-like articles to the pod, may very likely be found also in tropical Australia.

1. **A. vaginalis,** *DC. Prod.* ii. 353. A perennial, tufted or much branched at the base, the stems decumbent or ascending, from a few inches to above a foot long, glabrous or slightly pubescent. Leaves on short slender petioles, the lower ones cordate-orbicular or oval, not ½ in. long, the upper ones from oval-oblong to lanceolate-linear, and often 1 in. long or more, all obtuse. Racemes slender, terminal or at length leaf-opposed ; pedicels short, in rather distant pairs. Flowers very small. Calyx about 2 lines long, the lobes very narrow, ending in a subulate almost hair-like point, the 2 upper ones less united than in most species. Petals scarcely exceeding the calyx. Pod often ¾ in. long or rather more, slightly compressed, obscurely wrinkled, the separation of the articles marked by transverse raised lines, without any or rarely with a slight contraction.—W. and Arn. Prod. 233 ; *A. nummulariæfolius*, DC. Prod. ii. 353 ; W. and Arn. Prod. 232.

Queensland. Burdekin river and Broad Sound, *Bowman.* Common in E. India and the Archipelago, and introduced into other parts of the world.

2. **A. longifolius,** *W. and Arn. Prod.* 233. Nearly glabrous, except the raceme, which is more or less clothed with soft hairs. Stems erect, rather slender, but rigid, attaining 2 ft. or more. Leaves linear-lanceolate, acute, 2 to 4 in. long, or the lower ones shorter broader and more obtuse. Stipules usually longer than the petioles. Racemes slender, terminal ; pedicels in pairs, shorter than the calyx. Calyx about 3 lines long, the lobes lanceolate-subulate, striate, scarcely overlapping each other, the 2 upper ones almost completely united into one rather broader than the others. Pod about ½ in. long, compressed, not at all contracted between the seeds, but without raised transverse lines, and slightly and irregularly wrinkled.—Wight, Ic. t. 251.

N. Australia. Arnhem N. Bay, *R. Brown.* Also in the E. Indian peninsula.

3. **A. rugosus,** *DC. Prod.* ii. 353. An annual or biennial, with procumbent ascending or erect stems, attaining 1 to 2 ft., but sometimes low and short, pubescent or loosely hairy. Leaves articulate on a short petiole, the lowest ovate, obtuse, ½ to 1 in. long, the upper ones lanceolate or linear, 1 to 3 in. Racemes in the Australian form rather long, softly hairy. Bracts

ovate-lanceolate, striate, but falling off as in the other species long before the flower expands. Pedicels in distant pairs, much shorter than the calyx. Calyx about 3½ lines long, divided almost to the base into 4 lanceolate acute dry stiff lobes, overlapping each other on the edges and not striate, the upper one (formed of 2 united) slightly notched. Petals scarcely exceeding the calyx or rather shorter. Pod scarcely exceeding the calyx, contracted between the seeds, into 3, 4 or rarely 5 articles, as broad as or broader than long, strongly marked with transverse wrinkles.—*Hedysarum rugosum,* Willd. Sp. Pl. iii. 1172 ; *A. cylindricus,* Desv. in Ann. Linn. Soc. Par. 1825, 301, as quoted by him in Ann. Sc. Nat. ix. 417.

N. Australia. Upper Victoria river and Sturt's Creek, *F. Mueller.*
Queensland. Bowen river, *Bowman.*

The species is common in tropical Asia and Africa, where it varies much, sometimes low and diffuse, with almost all the leaves obovate or broadly oblong, sometimes tall and almost erect, with nearly all the leaves lanceolate or linear. It includes *A. styracifolia,* W. and Arn. Prod. 234, with short dense very hairy racemes and *A. Wallichii,* W. and Arn. l. c., with elongated nearly glabrous racemes. The Australian specimens have the habit of the latter with the hairs nearly of the former. De Candolle's specimen of *A. styracifolia* is nearer to *A. Wallichii,* W. and Arn. The original *Hedysarum styracifolium,* Linn., is very properly referred by W. and Arn. to a very different plant, *Desmodium retroflexum,* DC., which is surely a true *Desmodium* (sect. *Nicolsonia*), not au *Uraria. Alysicarpus Heyneanus,* W. and Arn. l. c., must probably be considered as another form of *A. rugosus.*

49. LESPEDEZA, Mich.

Calyx-lobes or teeth nearly equal or the 2 upper ones shortly united. Standard orbicular, obovate or oblong, narrowed into a claw, or rarely obtuse at the base; wings free; keel obtuse or rostrate. Upper stamen free or rarely united with the others ; anthers reniform. Ovary sessile or stipitate, with a single ovule ; style filiform, with a small terminal stigma. Pod ovate or orbicular, flat, reticulate, indehiscent.—Herbs undershrubs or shrubs. Leaves pinnately 3-foliolate or rarely 1-foliolate; leaflets entire, without stipellæ. Stipules free, usually small or very deciduous. Flowers purple pink or white, in axillary clusters or, in species not Australian, in axillary racemes or terminal panicles.

The genus is spread over North America, temperate, especially eastern Asia, and the mountains of E. India and the Archipelago. Of the two Australian species, one is Asiatic, the other endemic.

Hoary-pubescent or nearly glabrous. Calyx 1½ lines long. Pod small, orbicular . . . . 1. *L. cuneata.*
Densely and softly velvety-tomentose or woolly. Calyx 4 lines long. Pod
rather longer, obliquely semi-ovate . . . . 2. *L. lanata.*

1. **L. cuneata,** *G. Don, Gen. Syst.* ii. 307. Rootstock thick and woody; stems several, decumbent ascending or erect, stiff and but little branched, usually 1 to 2, but sometimes 3 or 4 ft. long, hoary-pubescent or at length glabrous. Leaves usually crowded, the leaflets linear-cuneate, mostly under ½ in., but occasionally ¾ or even 1 in. long, hoary or silky underneath, the common petiole 1 to 3 or rarely 4 lines long. Stipules small, subulate. Flowers pink-purple, in dense axillary clusters ; those in the upper axils nearly all complete, about 3 lines long, those of the lower

clusters mostly apetalous, with imperfect stamens. Calyx 1½ lines long, the lobes rigid, very acute, longer than the tube, the 2 upper ones united to the middle. Bracteoles ovate-lanceolate, acute. Keel curved, obtuse. Pod sessile, nearly orbicular, slightly acute, 1 to 1½ lines diameter.—*L. juncea*, DC. Prod. ii. 348, in part; Miq. Fl. Ind. Bat. i. part 1, 230, but not the Siberian *L. juncea*, Pers.

N. S. Wales, *Bynoe;* New England, *C. Stuart;* Macleay river, *Beckler;* Argyle county, *Lhotsky.*
Victoria. Ovens, King, and Broken rivers, *F. Mueller.*

2. **L. lanata,** *Benth.* Apparently shrubby, the branches leaves and calyxes densely clothed with a soft white or brownish velvety cotton. Leaves crowded; leaflets 3, cuneate-oblong, acute or rather obtuse, softly mucronate, the terminal one about 1 in. long, the lateral ones shorter, with a short common petiole. Stipules and bracts softly linear-subulate. Flowers crowded, in sessile axillary clusters. Calyx about 4 lines long, the tube very short, the lobes linear-subulate, soft, shortly plumose. Petals shorter than the calyx, tomentose outside; standard orbicular, obtuse at the base; keel slightly incurved, obtuse. Upper stamen united with the others to the middle. Pod sessile, flat, obliquely semiovate, about 5 lines long, narrowed at the end, softly tomentose.

S. Australia. Mount Strzelecki, *M'Douall Stuart.*

TRIBE * VICIEÆ.—Herbs. Leaves abruptly pinnate, the common petiole usually ending in a tendril or in a fine point. Flowers and fruit of *Phaseoleæ.* Peduncles or racemes axillary.

There are no true *Vicieæ* indigenous in Australia, unless the anomalous genus *Abrus*, which I have inserted below at the end of *Phaseoleæ*, be referred to *Vicieæ*, with which it is in many respects connected. I only give here the tribal characters because some European species of *Vicia* have established themselves in waste and cultivated places in some of the settled colonies. The genus is characterized chiefly by the style which is not bearded longitudinally on the inner side as in *Lathyrus*, but has usually a tuft of hairs at the top on the outside or all round under the stigma. Two species have been sent as naturalized, especially about Adelaide in South Australia, *V. sativa*, Linn., var. *segetalis*, Ser. in DC. Prod. ii. 361, an annual or biennial from a few in. to 1 or 2 ft. high with about 4 to 7 pairs of leaflets, and sessile usually solitary purple flowers; and *V. hirsuta*, Koch (*Ervum hirsutum*, Linn.), a very slender hairy annual with very small pale-blue flowers, 2 or 3 together on slender axillary peduncles, the pod short and 2-seeded.

TRIBE VII. PHASEOLEÆ.—Herbs usually twining or prostrate, rarely erect or shrubby at the base, very rarely trees. Leaves pinnately 3-foliolate or 1-foliolate, rarely 5- or 7-foliolate, with stipellæ, very rarely digitate or without stipellæ. Upper stamen usually free, at least at the base, or all but the base. Anthers uniform or nearly so, or very rarely alternately smaller or wanting. Pod not articulate, 2-valved. Cotyledons usually thick and fleshy.

The tribe is a natural one but very difficult to define, as it passes gradually through a few exceptional genera or species into several others. The great majority of species are at once known by their twining stems with pinnately 3-foliolate stipellate leaves, and among Australian species there are very few exceptions. Where there are 5 or 7 leaflets the stipellæ and twining habit readily distinguish them from *Galegeæ*. *Erythrinas* are arborescent but with the foliage of *Phaseoleæ*. Some species of *Atylosia*, *Rhynchosia*, and *Flemingia* are

242   XL. LEGUMINOSÆ.

erect or shrubby, and the latter genus has digitate leaflets, but these have all the habit of *Genisteæ* rather than of *Galegeæ*, and are distinguished from the former by their upper stamen free. *Pycnospora*, placed in *Hedysareæ*, has the technical characters almost of *Phaseoleæ*, but is too evidently allied to *Desmodium* to be removed far from it. *Abrus*, with its numerous leaflets, is a still further departure from the normal characters of *Phaseoleæ*, but is placed at the end of the tribe as closing the series of more or less herbaceous Papilionaceæ.

50. CLITORIA, Linn.

(Neurocarpum, *Desv.*)

Calyx tubular, the 2 upper lobes slightly connate, the lowest narrow. Standard large, erect, open, narrowed at the base without auricles; wings shorter, spreading, adhering to the keel in the middle; keel shorter, incurved, acute. Upper stamen free or more or less united with the others; anthers uniform. Ovary stipitate, with several ovules, style elongated incurved, more or less dilated upwards and bearded longitudinally on the inner side. Pod linear, flattened, the upper or both sutures thickened, the sides flat or convex, occasionally bearing a raised longitudinal rib. Seeds globose or flattened, without any strophiole.—Herbs or shrubs, short and erect or with long twining branches. Leaves pinnate with 3 or several leaflets, or occasionally only 1, usually stipellate. Stipules persistent, striate. Flowers large, solitary or clustered in the axils, or in pairs crowded in short racemes. Bracts stipule-like, persistent, the lower ones in pairs, the upper ones united into one. Bracteoles like the bracts or larger, persistent.

A considerable American genus, with a few African and Asiatic tropical species. The Australian one is endemic, its nearest affinities being S. American. The *C. Ternatea*, with pinnate leaves, the most common Asiatic and African one, very generally cultivated for ornament, has not yet been found in Australia. The genus is readily distinguished by its large tubular calyx.

1. **C. australis,** *Benth.* Stems herbaceous but hard, erect, flexuose, 1 to 2 ft. high, scarcely branched, pubescent with appressed silky hairs. Leaflets 1 or 3, ovate, obtuse, rarely shortly acuminate, 1½ to 2½ in. long, glabrous above, silky-pubescent underneath, the lateral ones when present smaller and at a distance from the terminal one. Stipules broadly lanceolate. Peduncles axillary, very short, bearing a cluster of 2 or 3 pairs of white flowers nearly 1½ in. long. Bracts narrow, acuminate. Calyx about ¾ in. long, the lobes acuminate and acute, about as long as the tube. Standard nearly 1½ in. long, wings and keel scarcely exceeding the calyx. Pod not seen.

N. Australia. Arnhem S. Bay, *R. Brown* (*Herb. R. Br.*). I have been unable to adopt Brown's MS. name which is now preoccupied in the genus.

51. GLYCINE, Linn.

(Leptolobium, *altered to* Leptocyamus, *Benth.*)

Calyx 2 upper lobes united in a 2-toothed or 2-lobed upper lip. Standard nearly orbicular, without inflexed auricles at the base; wings narrow, slightly adhering to the keel; keel obtuse, shorter than the wings. Upper stamen at first united with the others in a closed tube, often becoming at length free; anthers uniform. Ovary nearly sessile, with several ovules; style short, incurved, with a terminal stigma. Pod linear or falcate, 2-valved, with a pithy substance between the seeds, the base of the style forming a very short straight or rarely hooked point. Seeds not strophiolate.—Twining or prostrate herbs, with a perennial often thick or woody rootstock and usually pubescent or villous. Leaflets 3, or rarely 5 or 7, in opposite pairs, entire or rarely sinuately 3-lobed, stipitate. Flowers usually very small, in axillary racemes, singly scattered along the rhachis, the lower ones often solitary or clustered in the axils without a common peduncle, and sometimes without or almost without petals. Bracts small, setaceous; bracteoles narrow or minute, usually persistent.

The genus as now limited, comprising *Soja*, DC., and *Johnia*, Wight and Arn., neither of them Australian, extends over tropical and subtropical Africa and Asia. The Australian species belong to a section *Leptocyamus*, differing from the others only in the flowers being distinct from each other, not clustered along the rhachis of the raceme. Two of these species are also found in the Indian Archipelago, the remaining four are endemic, some of them perhaps reducible to varieties of *G. clandestina* or *G. tabacina*. They have by some been included in *Kennedya*, and supposed to have the strophiole of that genus. I have, however, never found any real strophiole, although the funicle, as in many other *Phaseoleæ*, expands into a thin white membrane covering the hilum, fragments of which may remain attached to the seed after its separation from the funicle.

Terminal leaflets sessile between the 2 others or the 3 very shortly and
　　equally petiolulate.
　Stems short. Leaflets 1 to 2 in. long, very hirsute. Pods falcate,
　　broad. Upper calyx-lobes free from the middle . . . . . . 1. *G. falcata.*
　Stems slender, twining. Pod linear, nearly straight. Upper calyx-
　　lobes free from the middle . . . . 2. *G. clandestina.*
　Stems short. Leaflets orbicular or obovate. Pod linear, nearly
　　straight. Upper calyx-lobes united nearly to the top . . . . 3. *G. Latrobeana.*
Terminal leaflet inserted at some distance from the lateral ones.
　Pubescent hirsute or nearly glabrous. Leaflets of the lowest leaves
　　short and broad, of the upper ones ovate-lancolate, lanceolate or
　　almost linear . . . . 4. *G. tabacina.*
　Silky with closely appressed pubescence. Leaflets linear acute . . 5. *G. sericea.*
　Softly tomentose or villous. Leaflets ovate or oblong, all obtuse . 6. *G. tomentosa.*

1. **G. falcata,** *Benth.* Stems in our specimens short, decumbent or perhaps erect, hirsute with reflexed hairs. Leaflets 3, the central one sessile between the others, all lanceolate or oblong, 1 to 2 in. long, villous, on a long hirsute common petiole. Stipules striate, larger than in the other species. Flowers all racemose, on long hirsute erect peduncles. Pedicels very short. Calyx silky-villous, 2 to 2½ lines long, the lobes nearly as long as the tube, the 2 upper ones united to about the middle. Standard rather narrower than in the other species. Pods reflexed, very hirsute, falcate, ½ to ¾ in. long and fully 2 lines broad, but not ripe in the specimens seen.

N. Australia. Sturt's Creek, *F. Mueller.*
Queensland. Sutton river, *D'Orsay.*
S. Australia, Cooper's Creek, *Bowman, Neilson.*

2. **G. clandestina,** *Wendl. Bot. Beob.* 54. Stems slender, twining, more or less hirsute with reflexed hairs. Leaflets 3, the terminal one inserted close between the 2 lateral ones or very rarely here and there slightly raised above them, those of the lower leaves often broadly obovate, about ½ in. long, those of the upper ones narrow-lanceolate or linear, ½ to 1½ in. long or more, acute, either nearly glabrous above and pubescent with appressed hairs underneath,

or silky-villous on both sides. Stipules minute. Racemes in the upper axils usually exceeding the leaves, the flowers about 4 lines long, scattered along the upper half of the peduncle, the pedicels either very short or nearly as long as the calyx ; in the lower part of the plant the flowers are smaller, often without any or with imperfect petals, and solitary or clustered in the axils, without a common peduncle. Calyx about 2 lines long, the 2 upper lobes united to the middle or nearly distinct. Pod linear, straight, ½ to 1 in. long, with a minute terminal straight or hooked point. Seeds nearly orbicular or transversely oblong, smooth or rough with raised dots, often different in the racemose and in the axillary pods.—DC. Prod. ii. 241 ; *Leptolobium clandestinum*, Benth. in Ann. Wien. Mus. ii. 125 ; *Leptocyamus clandestinus*, Benth. in Hook. f. Fl. Tasm. i. 102 ; *Teramnus clandestinus*, Spr. Syst. Veg. iii. 235 ; *Leptolobium microphyllum*, Benth. in Ann. Wien. Mus. ii. 125 ; *Glycine minima*, Willd. Enum. 756, from the diagnosis copied in DC. Prod. ii. 241.

Queensland. Keppel Bay, Broad Sound, *R. Brown ;* near Moreton Bay, *Leichhardt.*

N. S. Wales. Port Jackson to the Blue Mountains, *R. Brown,* and others; northward to New England, *C. Stuart ;* Clarence river, *Beckler ;* and southward to Twofold Bay, *F. Mueller.*

Victoria. Port Phillip, *R. Brown ;* common from the neighbourhood of Melbourne to Gipps' Land, *Adamson, F. Mueller,* and others.

Tasmania. Port Dalrymple, *R. Brown ;* common in all the northern parts of the island, ascending to 3000 ft., *J. D. Hooker.*

S. Australia. Gawler town, *Behr. ;* Buffalo range, Mount Remarkable, Spencer's Gulf, etc., *F. Mueller.*

W. Australia. Gairdner river, *Maxwell ;* and probably also from the S. coast towards the east, *Drummond, 4th Coll. n.* 39, *5th Coll. n.* 92.

Var. *sericea.* Silky-pubescent or villous. Calyx very rusty-villous. Pedicels very short. To this form belong all the W. Australian, most of the S. Australian specimens, and a few only of those from the other colonies.

3. **G. Latrobeana,** *Benth.* Considered by F. Mueller as a form of *G. clandestina,* and much resembles the undeveloped states of that species. Stems short, prostrate or scarcely twining at the ends. Leaflets all obovate or orbicular, or very few of the upper ones narrow. Stipules larger than in *G. clandestina.* Flowers rather larger, the racemose ones more crowded towards the end of long peduncles. Upper lobes of the calyx united to near the top. Standard broader than in *G. clandestina.* Pod the same as in that species. —*Zichya Latrobeana,* Meissn. in Pl. Preiss. i. 94 ; *Leptocyamus Tasmanicus,* Benth. in Hook. f. Fl. Tasm. i. 102. t. 17.

Victoria. Sandy pastures and meadows, Port Phillip to the Grampians, *Latrobe, Adamson, F. Mueller,* and others.

Tasmania. Pastures in the northern parts of the island, *J. D. Hooker.*

S. Australia. Mount Gambier, Rivoli Bay, *F. Mueller.*

4. **G. tabacina,** *Benth.* Slender, twining, pubescent or villous. Leaflets 3, the lateral ones always at a distance from the terminal one, those of the lower leaves orbicular obovate or oblong and usually obtuse, in the upper leaves ovate-lanceolate, lanceolate or almost linear and usually acute, mostly ¾ to 1 in. rarely 1½ to 2 in. long. Stipules small. Racemes slender, elongated, the flowers distant, usually about 4 lines long, on very short pedicels; in the lower part of the plant the flowers often axillary and solitary or 2 or 3 together as in other species. Calyx-lobes subulate-acuminate, shorter than

the tube, the 2 upper ones united to about the middle.   Pod straight, glabrous
or villous, ¾ to above 1 in. long, the terminal point very short, or rarely
rather longer and hooked.   Seeds smooth or tuberculate.—*Kennedya tabacina*,
Labill. Sert. Austr. Caled. 70. t. 70 ;   *Leptolobium tabacinum*, and *L. elonga-
tum*, Benth. in Ann. Wien. Mus. ii. 125 ; *Desmodium Novo-Hollandicum*, F.
Muell. in Linnæa, xxv. 394.

Queensland.   Bustard Bay, *Banks and Solander ;* Keppel Bay, Thirsty Sound, *R.
Brown ;* Moreton Bay, *F. Mueller, C. Stuart ;* Rockhampton, *Dallachy ;* in the interior
on the Maranoa, *Mitchell ;* and Condamine river, *Leichhardt.*
N. S. Wales.   Port Jackson to the Blue Mountains, *R. Brown, A. Cunningham,*
and others, northward to New England, *C. Stuart ;* and Clarence, Hastings, and Macleay
rivers, *Beckler ;* to the southward, *A. Cunningham.*
Victoria.   Rocky pastures near Melbourne, *Robertson,* also *F. Mueller.*
S. Australia.   Crystal Brook and Rocky River, *F. Mueller.*
W. Australia.   Port Gregory, *Oldfield.*
The species is also in New Caledonia, the Feejee and other islands of the South Pacific.
The most slender and glabrescent forms approach *G. clandestina,* but may be always known
by the terminal leaflet distinctly raised above the others ; the more common hirsute variety
differs from *G. tomentosa* in the upper leaflets almost always narrow and acute, the less
deeply divided calyx, etc.   Among the numerous varieties the two following are the most
prominent :—
    Var. *uncinata.*   Very hirsute.   Pod hooked at the end, although not so much so as in
Teramnus.—Rockhampton, *Thozet.*
    Var. *latifolia.*   Leaflets more obtuse and villous, almost connecting the species with *G.
tomentosa.*—*Leptocyamus latifolius,* Benth. in Mitch. Trop. Austr. 361.—To this belong
several Queensland specimens.   Where the pod is present it appears to be always longer and
more slender than in *G. tomentosa.*

5. **G. sericea,** *Benth.* Stems trailing or twining, the whole plant hoary
or white with a close appressed silky pubescence, occasionally mixed on the
calyx only with rust-coloured hairs.   Leaflets 3, linear or linear-lanceolate,
acute, mostly 1 to 2 in. long, the terminal one at a distance from the others.
Flowers rather larger than in *G. tabacina,* but otherwise like them.   Pod
usually above 1 in. long, densely silky-pubescent with appressed hairs.   Seeds
smooth.—*Leptocyamus sericeus,* F. Muell. in Hook. Kew Journ. viii. 45, and
in Trans. Phil. Inst. Vict. i. 40.

N. S. Wales.   On the Darling, *Dallachy ;* between Stokes Range and Cooper's Creek,
Wheeler.
Victoria.   Sand-hills on the Murray, *F. Mueller.*

6. **G. tomentosa,** *Benth.* Twining or prostrate, resembling the coarser
varieties of *G. tabacina,* but always more tomentose-villous, and often densely
and softly velvety-tomentose.   Leaflets 3, ovate or oblong, very obtuse, 1 to
2 in. long, not passing into the lanceolate acute form of the upper leaves of
G. tabacina.   Flowers very shortly pedicellate in the raceme as in that species, and of the same size, but usually more approximate at the end of the
peduncle.   Calyx very villous, with the lobes longer than the tube.   Pod ½ to
¾ in. long or rarely more.   Flowers in the lower axils solitary or clustered as
in all other Australian species except *G. falcata.*—*Leptolobium tomentosum,*
Benth. in Ann. Wien. Mus. ii. 125.

N. Australia.   Upper Victoria river, *F. Mueller ;* Arnhem N. Bay, *R. Brown.*
Queensland.   Endeavour river, *Banks and Solander ;* Broad Sound, *R. Brown ;* Port
Curtis, *M'Gillivray ;* Peak Downs, *F. Mueller.*
N. S. Wales.   Between the Darling and Cooper's Creek, *Neilson.*
We have the same species from the Philippine Islands and from S. China.

52. HARDENBERGIA, Benth.

Calyx-teeth short, the 2 upper ones more or less united. Standard broadly orbicular, emarginate, without inflexed auricles; wings obovate-falcate, adhering to the keel; keel obtuse, shorter and usually very much shorter than the wings. Upper stamen quite free, the others united in a sheath; anthers reniform. Ovary sessile, with several ovules; style short thick, incurved, attenuate at the end, with a terminal stigma. Pod linear, compressed or turgid. Seeds ovoid or oblong, laterally attached to a short funicle, strophiolate. —Glabrous twining herbs or undershrubs. Leaves of 1, 3 or 5 entire stipellate leaflets. Stipules small, striate. Flowers small, violet white or pinkish, with a yellowish or greenish spot on the standard, in axillary racemes, the pedicels usually in pairs or small clusters. Bracts small, deciduous, or rarely persistent. Bracteoles none.

A small genus limited to Australia and distinguished from *Kennedya* by the short calyx-teeth and (except the doubtful *H. retusa*) by the small keel, and still more by the habit and numerous small flowers of a very different colour, giving it more the aspect of a *Glycine* than of a true *Kennedya*.

Leaflets cordate ovate lanceolate or linear. Keel much shorter than the wings.
Leaflets solitary. Pod flat, with dry pulp inside . . . . . 1. *H. monophylla*.
Leaflets 3 or 5. Pod turgid, clear inside . . . . 2. *H. Comptoniana*.
Leaflets obovate truncate obcordate or broadly 2-lobed. Keel rather shorter than the wings . . . . 3. *H. retusa*.

1. **H. monophylla,** *Benth. in Hueg. Enum.* 41. Leaflets always solitary, usually ovate or lanceolate, 2 to 3 or even 4 in. long, obtuse or rather acute, often coriaceous and strongly reticulate, but varying from broadly cordate-ovate to narrow-lanceolate, more or less cordate or rounded at the base, articulate on a petiole of ½ to 1 in. Flowers usually numerous, about 5 lines long, on pedicels rather longer than the calyx, in pairs or rarely 3 together, the upper racemes often forming a terminal panicle. Calyx about 1½ lines long. Pod sessile, flat, attaining about 1½ in., coriaceous, more or less filled between the seeds with a pithy pulp. Seeds very oblique, almost transverse. —Maund, Botanist, t. 84; *Glycine bimaculata*, Curt. Bot. Mag. t. 263; *Kennedya monophylla*, Vent. Jard. Malm. t. 106; DC. Prod. ii. 384; Bot. Reg. t. 1336; Lodd. Bot. Cab. t. 758; *K. longiracemosa*, Lodd. Bot. Cab. t. 1940; *K. ovata*, Sims, Bot. Mag. t. 2169; DC. Prod. ii. 384; *K. cordata*, Lindl. Bot. Reg. t. 944; *Hardenbergia monophylla*, and *H. ovata*, Benth. in Ann. Wien. Mus. ii. 124; Hook. f. Fl. Tasm. ii. 361; *H. cordata*, Benth. in Ann. Wien. Mus. ii. 124.

Queensland. Moreton Island, *F. Mueller.*

N. S. Wales. Port Jackson to the Blue Mountains, *R. Brown, Sieber, n.* 378, *and Fl. Mixt. n.* 552, and others; northward to Hastings river, *Beckler;* and southward to Twofold Bay, *F. Mueller.*

Victoria. Dandenong ranges, etc., *F. Mueller;* Bendigo diggings, where the thick root is used for Sarsaparilla, *Adamson.*

Tasmania. Rocky hills near Frogmore, Richmond, *Oldfield.*

S. Australia. Near Bethanie, *Oswald;* towards Spencer's Gulf, *F. Mueller.*

In some specimens from near Bathurst, *Herb. F. Mueller,* with broadly cordate leaves, the bracts are numerous and persistent; in the ordinary forms, including those with the same broad cordate leaves, the bracts are all fallen off long before the flowers expand.

2. **H. Comptoniana,** *Benth. in Hueg. Enum.* 41. Leaflets 3 or 5, and in the latter case the lateral ones in 2 opposite pairs, not opposite in distant pairs as in other 5-foliolate *Phaseoleæ,* varying from ovate to linear-lanceolate, rather obtuse, usually 1½ to 3, sometimes 4 to 5 in. long, rounded or truncate at the base. Flowers of the same size, colour and structure as those of *H. monophylla,* in pairs or clusters of 3 or 4 along the racemes. Pod cylindrical, very turgid, coriaceous, attaining 1½ in. in length, quite free from any pithy pulp inside. Seeds almost sessile and longitudinal.—Meissn. in Pl. Preiss. i. 94; *Glycine Comptoniana,* Andr. Bot. Rep. t. 602; Bot. Reg. t. 298; *Kennedya Comptoniana,* Link, Enum. ii. 235; DC. Prod. ii. 383; *K. macrophylla,* Lindl. Bot. Reg. t. 1862; *Hardenbergia Huegelii,* Benth. in Hueg. Enum. 41; *H. digitata,* Lindl. Bot. Reg. 1840, t. 60; *H. Lindleyi,* Meissn. in Pl. Preiss. i. 94; *H. Makoyana,* Ch. Lem. Illustr. Hortic. v. t. 179.

W. Australia. King George's Sound, *R. Brown ;* and thence to Swan River, *Drummond, 1st Coll. and n.* 271, *Huegel, Preiss, n.* 1093, 1094, and others.

3. **H. (?) retusa,** *Benth.* A tall twiner, the young shoots and inflorescence silky-pubescent, the foliage at length glabrous, the branches usually angular. Leaflets 3, broadly obovate-truncate, obcordate or broadly and obtusely 2-lobed, the midrib usually produced into a short point, 1½ to 3 in. long, somewhat coriaceous, rather shining above, pale underneath. Stipules ovate or lanceolate, striate, reflexed. Flowers like those of the other species, or rather larger and more numerous, usually several together in each cluster, the rhachis of the cluster sometimes slightly developed, the racemes axillary or in terminal panicles as in the other species. Calyx about 2 lines long, hoary-pubescent, the teeth very short and obtuse. Standard nearly 5 lines diameter, broad and emarginate; wings nearly as long, falcate; keel rather shorter, much incurved, obtuse. Ovary nearly sessile, with about 10 ovules. Style rather thickened and inflexed at the base, then straight and slender, with a small terminal stigma. Pod broadly linear, flattened, silky-villous, about 2¼ in. long, without any pithy partitions inside. Seeds strophiolate.—*Dolichos obcordatus,* A. Cunn. Herb. ; *Glycine retusa,* Soland. mss.

Queensland. Endeavour river, *Banks and Solander,* A. *Cunningham ;* Dunk Island, *M'Gillivray ;* Albany Island, *F. Mueller ;* Cape York, *W. Hill.* The keel is rather larger than in the other species, but the other characters and habit are quite those of *Hardenbergia.*

53. KENNEDYA, Vent.

(Physolobium, *Hueg.* ; Zichya, *Hueg.* ; Amphodus, *Lindl.*)

Calyx-lobes about as long as the tube, the 2 upper ones united in an emarginate or 2-toothed upper lip. Standard obovate or orbicular, narrowed into a short claw, with minute inflexed auricles ; wings falcate, adhering to the keel ; keel incurved, obtuse or rather acute. Upper stamen free, the others united ; anthers uniform. Ovary nearly sessile or shortly stipitate, with several ovules ; style filiform, rarely toothed at the top, stigma terminal. Pod linear, flattened cylindrical or turgid, 2-valved, more or less divided by a pithy substance between the seeds. Seeds ovoid or oblong, laterally attached, with a very prominent strophiole.—Perennials, with prostrate trailing or twining stems, usually pubescent or villous. Leaves pinnately 3-foliolate or very rarely with an additional pair or reduced to 1 ; leaflets entire or obscurely 3-lobed, with stipellæ. Stipules persistent, striate or veined. Flowers red or rarely almost black, in racemes, umbels, pairs, or solitary on axillary peduncles. Bracts either stipule-like and persistent, or small and very deciduous. Bracteoles none. Disk round the ovary obscurely annular or none at all.

The genus is entirely Australian, and, with *Hardenbergia*, distinguished amongst *Phaseoleæ* by the prominently strophiolate seeds. I had formerly divided it into 3 genera, which, as far as the materials we then had could show, appeared quite distinct, viz. *Kennedya*, with a narrow standard and flat pods ; *Physolobium*, with an orbicular standard, flowers in pairs, and a turgid pod ; and *Zichya*, with an orbicular standard, flowers umbellate, and a flat pod ; but the numerous additional specimens and some new species since examined, show that the above characters are too variously combined to be relied on even for good sections, and that the forms which I had considered as constituting several species of each genus were in many instances not even varieties. Taking the standard and the pod as the primary character, the species might be arranged as follows :—

Standard narrow-obovate. Pod flat (**Kennedya**).—1. *K. nigricans* ; 2. *K. rubicunda.*
Standard broadly-obovate. Pod cylindrical.—4. *K. prostrata.*
Standard orbicular. Pod flat (**Zichya**).—5. *K. eximia* ; 6. *K. coccinea* ; 7. *K. microphylla.*
Standard orbicular. Pod turgid (**Physolobium**).—8. *K. parviflora* ; 9. *K. Stirlingii* ; 10. *K. glabrata* ; 11. *K. macrophylla.*
But these groups are not natural, and where the fruit is unknown, as in the case of *K. procurrens*, the place of a species cannot be fixed.

Standard narrow-obovate. Keel almost acute. Pod compressed.
  Flowers above 1 in. long, racemose.
  Flowers nearly black, in a one-sided raceme. Wings spreading at
    the top . . . . 1. *K. nigricans.*
  Flowers red. Wings erect . . . . 2. *K. rubicunda.*
Standard broadly-obovate, almost orbicular. Keel obtuse. Flowers
    not above ¾ in. long.
  Flowers racemose. Pedicels very short. Bracts very small. (Pod
    unknown) . . . . 3. *K. procurrens.*
  Peduncles 1- or 2- flowered. Pedicels long. Bracts stipule-like, per-
    sistent. Pod cylindrical, coriaceous . . . . 4. *K. prostrata.*
  Flowers umbellate, usually few. Bracts small, very deciduous.
    Pod narrow, flat, with a thickened upper suture . . . . . 5. *K. eximia.*
Standard orbicular or broader than long. Flowers rarely above ½ in.
    long.
  Flowers umbellate or shortly racemose. Bracts small, very de-
    ciduous.
    Stipules broad, leafy. Keel about as long as the wings.
      Pod narrow, flat with a thickened upper suture. Flowers um-
        bellate, usually few. Keel obtuse . . . . 5. *K. eximia.*
      Pod very turgid.
        Flowers umbellate or nearly so. Leaflets cuneate. Keel
          almost acuminate . . . . 10. *K. glabrata.*
        Flowers distinctly racemose. Leaflets obovate or orbicular.
          Keel obtuse . . . . 11. *K. macrophylla.*
    Stipules very small. Keel considerably shorter than the wings.
      Flowers umbellate . . . . 6. *K. coccinea.*
  Peduncles 1- or 2-flowered, or with 2 rarely 3 pairs of flowers.
    Bracts small, very deciduous. Flowers small, solitary. Pod flat. 7. *K. microphylla.*
    Bracts broad, stipule-like, persistent. Pod very turgid.
      Peduncles usually 1-flowered. Pod under 1 in. long . . . 8. *K. parviflora.*
      Peduncles mostly with 1, 2, or 3 pairs of flowers. Pod above
        1 in. long . . . . 9. *K. Stirlingii.*

K. Baumanni and *K. splendens,* Meissn. in Pl. Preiss. i. 89, described from garden plants supposed erroneously to have been of Australian origin, are S. American species of *Camptosema.*

Steudel's 'Nomenclator' contains several other names of *Kennedyas,* extracted from garden catalogues, which, being unaccompanied by descriptions, it is impossible to identify.

1. **K. nigricans,** *Lindl. Bot. Reg. t.* 1715. A large twining species, usually pubescent, but less so than *K. rubicunda,* which it nearly resembles. Leaflets broadly ovate or rhomboidal, obtuse or emarginate, 2 to 3 in. long, and very often only one to each leaf. Stipules small, striate, reflexed. Flowers above 1 in. long, in racemes shorter than the leaves, but narrower and less pendulous than in *K. rubicunda,* all turned to one side, and of a deep violet-purple almost black colour, with a large greenish-yellow blotch on the standard. Standard narrow-obovate, reflexed; wings narrow, about as long as the keel, with the tips spreading; keel rather acute. Pod as in *K. rubicunda,* flattened, glabrous or slightly pubescent. Seeds almost transverse, laterally attached to a funicule protruding far into the cavity.—Bot. Mag. t. 3652; Meissn. in Pl. Preiss. i. 89.

W. Australia. King George's Sound to Cape Riche, *Drummond, 4th Coll. n.* 38, *Preiss, n.* 1091, *Oldfield,* and others,

2. **K. rubicunda,** *Vent. Jard. Malm. t.* 104. A large twining species, pubescent or villous. Leaflets 3, usually ovate and 3 to 4 in. long, but varying from broadly rhomboid obovate or almost orbicular, to ovate-lanceolate or lanceolate and acute, the smaller ones often under 2 in. and when very luxuriant attaining 5 or 6 in., rarely nearly glabrous, sometimes softly silky on both sides. Stipules small, striate, reflexed. Flowers of a dull or dark red, 1¼ to 1½ in. long, in pedunculate racemes rarely exceeding the leaves. Pedicels usually in pairs, rather longer than the calyx. Bracts small, very deciduous. Calyx silky-villous, 5 to 6 lines long. Standard narrow-obovate, abruptly reflexed from about the middle; wings narrow, erect, adhering to the keel above the middle; keel narrow, as long as the wings, rather acute. Pod flat or the valves slightly convex, 2 to 4 in. long, usually villous. Seeds oblong, almost transverse, laterally attached to a funicle protruding far into the cavity.—DC. Prod. ii. 383; Lodd. Bot. Cab. t. 954; *Glycine rubicunda,* Curt. Bot. Mag. t. 268; *Amphodus ovatus,* Lindl. Bot. Reg. t. 1101; *Kennedya phaseolifolia,* Hoffm. from the descr. in Linnæa, xvi. Litt. Ber. 281.

Queensland. Moreton Bay, *Fraser, F. Mueller,* and others; Wide Bay, *Bidwill;* Ipswich, *Nernst.*

N. S. Wales. Port Jackson to the Blue Mountains, *R. Brown, Sieber, n.* 381, and others; northward to Clarence, Hastings, and Macleay rivers, *Beckler,* and southward to Twofold Bay, *F. Mueller.*

Victoria. Gipps' Land, from Lake King to the limits of the colony, *F. Mueller.*

3. **K. procurrens,** *Benth. in Mitch. Trop. Austr.* 365. Stems prostrate, pubescent or villous. Leaflets 3, ovate or elliptical, obtuse or mucronate, 1 to 2 in. long, rugose, slightly pubescent above, more so underneath. Stipules broadly lanceolate, reflexed. Flowers rather above ½ in. long, in a short raceme at the end of a rather long peduncle. Bracts small. Pedicels short. Calyx about 2½ lines long, the lobes shorter than the tube. Standard broadly obovate, emarginate; keel rather narrow, obtuse; wings narrower and scarcely so long. Pod not seen.

Queensland. Mount Kennedy, on the Maranoa, *Mitchell.* Only a single specimen seen, unless one without flowers from Keppel Bay, *R. Brown,* belongs to the same species.

4. **K. prostrata,** *R. Br. in Ait. Hort. Kew. ed.* 2, iv. 299. Stems prostrate or sometimes twining in the large variety, pubescent or hirsute. Leaflets 3, broadly obovate or orbicular, under 1 in. long in the ordinary variety, often undulate, pubescent or hirsute. Stipules leafy, broadly cordate, acute or acuminate, free or united. Peduncles 1- or 2-flowered, rarely with 2 pairs of flowers. Pedicels usually longer than the calyx, with stipule-like bracts at the base. Flowers scarlet, nearly ¾ in. long. Calyx pubescent, about 4 lines long. Standard obovate; keel incurved, obtuse; wings much narrower and rather shorter, adhering only near the base. Pod nearly cylindrical, very coriaceous, pubescent, 1½ to 2 in. long. Seeds attached by a very short funicle.—Hook. f. Fl. Tasm. i. 101; *Glycine coccinea,* Curt. Bot. Mag. t. 270.

N. S. Wales. Port Jackson. *R. Brown, Sieber, n.* 377, and others.
Victoria. Very common throughout the colony, *F. Mueller.*
Tasmania. Port Dalrymple, *R. Brown;* abundant throughout the island in dry and stony places, etc., *J. D. Hooker.*
S. Australia. Lofty Ranges and Spencer's Gulf, *F. Mueller.*
W. Australia. S. coast, E. of King George's Sound, *Maxwell.*
Var. *major,* DC. Larger and usually more hirsute, sometimes twining. Leaves ovate or rhomboidal, much undulate, often attaining 2 in. Stipules often very large, attaining even 1 in. diameter. Pod above 2 in. long.—*K. bracteata,* Gaud. in Freyc. Voy. 286, t. 113; Meissn. in Pl. Preiss. i. 90; *K. stipularis,* Desv. in Ann. Sc. Nat. ix. 421; *K. Marryattæ,* Lindl. Bot. Reg. t. 1790; Maund, Botanist, t. 83; Meissn. in Pl. Preiss. i. 90; *K. arenaria,* Hueg. Enum. 38.—King George's Sound, *R. Brown,* and thence to Vasse, Swan, and Murchison rivers, *Drummond, Preiss, n.* 1095 *and* 1098, and others; Geographe Bay, *Herb. Mus. Par.*

5. **K. eximia,** *Lindl. in Paxt. Mag.* xvi. 35, *with a fig.* Prostrate or twining, silky-villous or at length nearly glabrous. Leaflets 3, ovate, or obovate and all under 1 in. long, or, in luxuriant specimens, broader, very obtuse, and fully 1½ in. long. Stipules broad, leafy, acute, veined. Flowers scarlet, not above 7 or 8 lines long, 2, 3 or more together in an umbel or very short raceme. Bracts very deciduous. Calyx usually silky-villous, about 3 lines long, with lobes shorter than the tube, but liable to become enlarged and leafy. Standard very broadly obovate or almost orbicular, but not so broad as in the several following species; keel much curved, obtuse; wings fully as long, but not so broad. Pod glabrous or slightly pubescent, usually curved, 1½ to 2 in. long, narrow and much flattened, with the upper suture thickened. Seeds small.

W. Australia, *Drummond, 5th Coll. n.* 91, *and Suppl. n.* 45; moist places, Gales Brook, and Phillips ranges, *Maxwell.* This species has the stipules of *Physolobium,* the pod of *Zichya,* and in the shape of the flower is intermediate between that and *K. prostrata.*

6. **K. coccinea,** *Vent. Jard. Malm. t.* 105. Twining or trailing, always pubescent rusty or silky-villous, and often densely so. Leaflets 3 or very rarely 5, the additional pair lower down, usually ovate or oblong, very obtuse and under 2 in. long, but varying from broadly ovate to narrow-oblong, entire or slightly sinuately 3-lobed. Stipules very small. Flowers scarlet, about ½ in. long, several together in an umbel or very short umbel-like raceme on long axillary peduncles, with 1 or 2 small deciduous bracts at their base.

Pedicels rarely longer than the calyx. Calyx rusty-villous, 3 to 3½ lines long. Standard broadly orbicular; wings shorter; keel still shorter, very obtuse. Style sometimes filiform at the end as in the other species, sometimes dilated at the back immediately behind the stigma into a tooth-like appendage. Pod flattened, the upper suture thickened, 1½ to 2 in. long. Seeds on a very short funicle.—DC. Prod. ii. 383 ; Sweet, Fl. Austr. t. 23 ; Bot. Mag. t. 2664 ; Lodd. Bot. Cab. t. 1126 ; Paxt. Mag. ii. 99, with a fig.; *Zichya coccinea,* Hueg.; Benth. in Hueg. Enum. 40, and in Ann. Wien. Mus. ii. 123 ; Maund, Botanist, t. 120 ; *K. inophylla,* Lindl. in Bot. Reg. t. 1421; *Zichya inophylla,* Hueg.; Benth. l. c. (a broad-leaved very silky form); *K. dilatata,* A. Cunn.; Lindl. in Bot. Reg. t. 1526; *K. heterophylla,* Mackay, in Loud. Hort. Brit. 314 ; *Zichya Molly,* Hueg. Bot. Arch. t. 1; Benth. l. c.; Meissn. in Pl. Preiss. i. 93 ; *Z. sericea,* Hueg.; Benth. l. c.; Meissn. l. c.; *Z. tricolor,* Lindl. in Bot. Reg. 1839, t. 52 ; Meissn. l. c.; *Z. angustifolia,* Lindl. in Bot. Reg. 1839, under t. 52 ; Meissn. l. c. ; *Z. villosa,* Lindl. in Bot. Reg. 1842, t. 68 ; Meissn. l. c. 92 ; *Z. glabrata,* Meissn. in Pl. Preiss. i. 92, but not of Benth. ; *Z. pannosa,* Paxt. Mag. viii. 147, with a fig.

W. Australia. King George's Sound and neighbourhood to Esperance Bay, *R. Brown, Drummond, Preiss, n.* 1086, 1088, 1090, 1096, 1097, *Oldfield,* and others.

The long list of synonyms is owing to great differences in the shape and size of the leaflets, in the indumentum, and in the style with more or less of an appendage or entirely without, which was very striking when we had only single specimens of various forms ; but we now find that the leaves are very variable even on the same plant, there is every intermediate between the slightly pubescent and the very silky-villous specimens, and I find styles with and without the appendage in combination with each of the most prominent varieties as to foliage.

7. **K. microphylla,** *Meissn. in Pl. Preiss.* i. 91. A small prostrate species, sprinkled with a few short stiff hairs. Leaflets broadly obcordate, in some specimens all under 2 lines long, in others nearly attaining ½ in. Stipules broad, herbaceous, striate. Flowers the smallest of the genus, rarely attaining 5 lines, solitary or rarely 2 together on axillary peduncles; pedicels short; bracts small and deciduous. Calyx nearly 2 lines long. Standard orbicular ; keel obtuse, as long as the wings. Pod linear, flat or with convex valves, the upper suture slightly thickened. Seeds small.

W. Australia, *Drummond, 5th Coll. n.* 94; borders of Lake Matilda, *Preiss, n.* 1092; Kalgan river, *Maxwell.*

8. **K. parviflora,** *Meissn. in Pl. Preiss.* i. 91, ii. 222. Prostrate or trailing, glabrous or hirsute with spreading hairs. Leaflets 3, obovate or broadly obcordate-truncate, sometimes as small as in *K. microphylla,* but often above 1 in. long. Stipules leafy, acute, veined, 3 to 6 lines broad. Flowers smaller than in any species except *K. microphylla,* and usually solitary, the pedicel longer than the peduncle, with small persistent leafy bracts at the articulation. Calyx 2 to 2½ lines long. Standard orbicular, nearly ¼ in. diameter; keel much curved, obtuse, nearly as long as the wings. Pod usually about ¾ in. long, very turgid, coriaceous, glabrous or pubescent. Funicle of the seeds very short.—*Physolobium carinatum,* Benth. in Hueg. Enum. 39, and in Ann. Wien. Mus. ii. 124 ; Meissn. in Pl. Preiss. i. 91.

W. Australia. King George's Sound and neighbourhood to Cape Riche, *Huegel, Baxter, Drummond, 2nd Coll. n.* 120, *Preiss, n.* 1084, 1099 ; Vasse river, *Oldfield* (with

pubescent pods). The smaller specimens of this species have a distinct aspect, with the peduncles almost all 1-flowered; the larger ones seem almost to pass into *K. Stirlingii.*
K. physolobioides, Walp. Ann. i. 251, from the diagnosis given, is probably this species.

9. **K. Stirlingii,** *Lindl. Bot. Reg. t.* 1845. Trailing or twining to a considerable length, hairy with soft spreading or silky hairs. Leaflets 3, ovate or orbicular, very obtuse, usually above 1 in. long. Stipules broadly cordate, usually large and much veined. Flowers scarlet, usually in 1, 2, or 3 pairs, on axillary peduncles, the pedicels long and slender, with persistent stipular leafy bracts at their base. Calyx 3 to 3½ lines long, hirsute. Standard orbicular, fully ½ in. diameter; keel broad, much incurved, very obtuse, nearly as long as the wings. Pod 1 to 2 in. long, very turgid, coriaceous. Funicle of the seeds very short.—*Physolobium elatum,* Hueg. Bot. Arch. t. 2 ; Benth. in Hueg. Enum. 39, and in Ann. Wien. Mus. ii. 124; Meissn. in Pl. Preiss. i. 92 ; *P. Stirlingii,* Benth. l. c.

W. Australia. Swan River, *Huegel, Drummond, n.* 270 ; *Preiss, n.* 1087, 1089.

10. **K. glabrata,** *Lindl. in Bot. Reg. t.* 1838.· A slender twining species, glabrous or with a few spreading hairs. Leaflets 3, cuneate or obovate-truncate, mucronate, ½ to 1 in. long. Stipules broad, veined. Flowers scarlet, several together in a small umbel on axillary peduncles. Bracts none or very deciduous. Calyx scarcely 3 lines long, the lobes short, the upper ones forming an obtuse emarginate upper lip. Standard orbicular, about 5 lines diameter; wings much falcate; keel much incurved, almost acute or shortly acuminate. Pod glabrous, very turgid, under 1 in. long.—*Zichya glabrata,* Benth. in Hueg. Enum. 40, and in Ann. Wien. Mus. ii. 123 ; Bot. Mag. t. 3956.

W. Australia. King George's Sound, *Baxter.* This species has the habit and inflorescence of *Zichya* with the pod of *Physolobium,* and may be generally known when in flower by the foliage and keel.

11. **K. macrophylla,** *Benth.* A tall, coarse, twining species, loosely hirsute with spreading hairs, silky on the young shoots. Leaflets 3, obovate or orbicular, very obtuse, often above 2 in. long. Stipules very broad, often united and attaining 1 in. in diameter; stipellæ lanceolate. Flowers red, distinctly racemose on axillary peduncles. Bracts deciduous. Pedicels rather short. Calyx not 3 lines long, with short lobes, as in *K. glabrata.* Standard orbicular, nearly ½ in. diameter; keel much curved, obtuse, nearly as long as the wings. Pod glabrous, very turgid, about 1½ in. long, acuminate with the persistent style.—*Physolobium macrophyllum,* Meissn. in Pl. Preiss. ii. 222 ; *Kennedya lateritia,* F. Muell. Fragm. iv. 78.

W. Australia, *Drummond, 2nd Coll. n.* 118. With the pod and aspect of *Physolobium,* this species has the inflorescence of the true *Kennedyas.*

54. ERYTHRINA, Linn.

Calyx campanulate or cylindrical, obliquely truncate or slit on the upper side, entire or toothed. Standard broad or long, erect or recurved, narrowed at the base, without appendages ; wings short, often minute or none ; keel short, the petals united or free. Stamens all united at the base, the upper one often free from the middle ; anthers reniform. Ovary stipitate, with several ovules; style subulate, oblique at the end, with a small stigma. Pod stipitate, linear-falcate, acuminate, narrowed at the base, more or less contracted between the seeds, 2-valved, usually pithy between the seeds. Seeds distant, ovoid or oblong, with a lateral oblong hilum, not strophiolate. —Erect trees or shrubs, rarely tall herbs, the trunk, branches, and often the petioles armed with conical prickles. Leaflets 3, usually broad, entire or 3-lobed, the stipellæ usually gland-like. Stipules small. Racemes axillary, or, if terminal, leafy at the base. Flowers large, usually red, in clusters of 2 or 3 on lateral nodes along the peduncle. Bracts small or none.

The genus is widely dispersed over tropical America, Africa, and Asia, extending into N. America and S. Africa. Of the 2 Australian species, one is a common Asiatic one, the other is endemic. The genus is a very natural one and well characterized, some botanists have, however, proposed to break it up into three or four, founded on diversities in the form of the calyx and proportions of the petals, which appear to vary so much from species to species as scarcely to serve even as sectional characters. Both the Australian ones have a spathe-like calyx, the wings and keel-petals all short and nearly free.

Leaves broadly 2- or 3-lobed. Calyx about ½ in. long. Standard
scarcely clawed . . . . 1. *E. vespertilio.*
Leaves entire. Calyx about ¾ in. long. Standard narrowed into a short
claw . . . . 2. *E. indica.*

1. **E. vespertilio,** *Benth. in Mitch. Trop. Austr.* 218. Glabrous, the branches prickly, but not the leaves. Leaflets broadly cuneate at the base, spreading to 3 or 4 in. in breadth, often but not always broader than long, usually 3-lobed, the lateral lobes spreading or recurved, obtuse, sometimes broader than long, sometimes much longer than broad, the middle one triangular or lanceolate, usually acute, broad or narrow, either longer than the lateral ones or more frequently much smaller or disappearing altogether, in which case the leaf is divided into 2 long narrow diverging or divaricate lobes. Flowers numerous, pendulous, in showy erect racemes. Calyx about ½ in. long, broad, entire or obscurely toothed, obliquely truncate and slit on the upper side. Standard ovate, erect at the base, recurved upwards, nearly 1½ in. long, narrowed but scarcely clawed at the base; wings obliquely oblong, about 4 lines long; keel-petals like the wings, but about 6 lines long, free. Style hooked at the end. Pod elongated, torulose, with few large red seeds.

N. Australia. Gulf of Carpentaria, *R. Brown;* Upper Victoria river, *F. Mueller;* frequent towards central Australia, *M'Douall Stuart's Expedition.*
Queensland. Endeavour river, *Banks and Solander;* Bay of Inlets, *R. Brown;* Brisbane river and Moreton Bay, *Fraser, F. Mueller,* etc.; Cape York, *M'Gillivray;* Port Denison, *Fitzalan;* in the interior, on the Maranoa, etc., *Mitchell;* found during the whole of *Leichhardt's* expedition, *Herb. Mus. Par.*
The Brisbane river specimens have usually large leaflets with broad short lobes, those from the north-west (*E. biloba, F.* Muell. in Hook. Kew Journ. ix. 21), have 2 narrow lobes with or without a small intermediate one, the others show every gradation from the one form to the other.

2. **E. indica,** *Lam.; DC. Prod.* ii. 412. Glabrous, the branches but not the leaves armed with prickles usually black. Leaflets entire, very broadly ovate, often 6 to 8 in. long, the terminal one rhomboidal, the lateral ones rather oblique. Flowers scarlet, nearly 2 in. long, in dense racemes. Calyx broad, fully ¾ in. long, entire or slightly toothed, obliquely truncate and slit on the upper side. Standard· ovate, scarcely recurved, narrowed into a dis-

tinct stipes. Wings and keel nearly similar, all free obovate and about $\frac{1}{2}$ in. long. Pod much contracted between the seeds, often attaining 9 in. to 1 ft. Seeds few, large, red, distant.—Wight, Ic. t. 58.

N. Australia. Islands of the N. coast, *R. Brown.*
Queensland. Port Denison, *Fitzalan.* Common in East India and the Archipelago, and often planted for shade or ornament.

55. MUCUNA, Adans.

Calyx broadly campanulate, 4-toothed, the upper one (consisting of 2 combined) broader, the lowest longer. Standard shorter than the wings; keel as long as or longer than the wings, incurved at the end, with a hardened point or beak. Upper stamen free, the others united; filaments usually dilated upwards; anthers alternately longer and erect, and shorter versatile and often bearded. Ovary sessile, with several ovules; style filiform, with a terminal stigma. Pod thick, clothed with stinging often very deciduous hairs, 2-valved. Seeds roundish with a long linear hilum, or oblong with a shorter lateral hilum.—Large twiners. Leaflets 3, stipellate. Stipules small or none. Flowers usually large, purple yellow or nearly white, in axillary racemes, the pedicels clustered along the rhachis on lateral nodes, or on short peduncles, when the raceme is converted into a corymb or dense panicle.

The genus is widely spread over the tropical regions of the New and the Old World; the only Australian species is a common Asiatic one. With considerable diversity in the pod and seeds, the species are all distinguished by the keel and stamens. The pungent hairs of the pod are rarely wanting, and all become black in drying.

1. **M. gigantea,** *DC. Prod.* ii. 405. Glabrous or slightly hairy when young. Leaflets from broadly ovate to ovate-lanceolate, acuminate, 3 to 4 in. long, the lateral ones very oblique. Flowers of a pale greenish-yellow, nearly $1\frac{1}{2}$ in. long, in short loose corymbs, on pendulous peduncles of from 2 or 3 in. to nearly 1 ft. Calyx about $\frac{1}{2}$ in. broad, hirsute with deciduous hairs. Standard ovate, concave, reflexed; wings rather longer, the edges pubescent below the middle; keel still longer, with a short indurated inflexed beak. Shorter anthers bearded. Pod rather thick but flat, above 1 in. broad, with a narrow longitudinal wing on each side of each suture, the pungent hairs usually falling off before maturity. Seeds 2 to about 6, large, orbicular, half encircled by the hilum.—W. and Arn. Prod. 254; Hook. Bot. Misc. iii. t. Suppl. 14 (wrong as to colour?).

N. Australia. Islands of the N. coast, *R. Brown.*
Queensland. Brisbane river and Moreton Bay, *Fraser, F. Mueller,* and others; Rockhampton, *Thozet;* Edgecumbe Bay, *Dallachy.*
N. S. Wales. Clarence river, *Herb. F. Mueller.*
The species is widely distributed over E. India, the Archipelago, and islands of the S. Pacific.

56. GALACTIA, R. Br.

Calyx-lobes 4 (the upper one consisting of 2 combined) acuminate, the 2 lateral shorter. Standard ovate, narrowed at the base, the margins slightly inflexed; wings narrow, slightly adhering to the keel; keel about as long. Upper stamen free, the others united; anthers uniform. Ovary nearly sessile; style filiform, with a small terminal stigma. Pod linear, straight or curved, usually flat, 2-valved, with a pithy pulp between the seeds. Seeds not strophiolate.

—Prostrate or twining herbs. Leaflets 3 or rarely 1 or 5, stipellate. Flowers in axillary racemes, clustered along the common peduncle. Bracts small, setaceous, deciduous. Bracteoles very small.

The species are chiefly American, one of them widely spread over the warmer regions of both the New and the Old World, two or three others are African or Asiatic. Of the two Australian species, one is the common cosmopolitan one, the other is endemic. The genus is readily distinguished by the acuminate calyx with the upper lobe always quite entire.

Glabrous or pubescent with spreading hairs. Flowers few in the raceme, under ½ in. long . . . . 1. *G. tenuiflora.*
Silky-pubescent or villous. Flowers numerous, 7 or 8 lines long . . . 2. *G. Muelleri.*

1. **G. tenuiflora,** *Willd. ; Wight and Arn. Prod.* 206. Stems from a thick rhizome, usually slender, twining, attaining several feet, glabrous or pubescent with spreading or reflexed hairs. Leaflets 3, ovate or oblong, usually 1 to 2 in. long and obtuse, but variable in shape and size, glabrous or pubescent especially underneath. Peduncles rarely exceeding 6 in., with few distant clusters of 2 or 3 flowers each, on a small gland-like node. Pedicels very short. Flowers pale reddish-purple or nearly white, varying from 4 to 6 lines. Calyx-lobes narrow, longer than the tube. Pod 1 to 2 in. long, linear, flat, coriaceous, with thickened margins, glabrous or pubescent. Seeds obliquely attached, smooth.

N. Australia. Upper Victoria river, *F. Mueller;* islands of the Gulf of Carpentaria and adjoining coast, *R. Brown, Henne, Landsborough;* Strangways river, *M'Douall Stuart's Expedition.*

Queensland. Endeavour river, *Banks and Solander;* Keppel Bay, *R. Brown;* Port Curtis and Keppel Island, *M'Gillivray;* Brisbane river, *F. Mueller;* Broad Sound, *Bowman.*

N. S. Wales. Clarence river, *Moore, Beckler;* between the Darling and Cooper's Creek, *Neilson.*

The species is common in tropical Asia, Africa, and America, and varies much in the breadth of the leaflets from almost orbicular to linear, in the indumentum, and in the size of the flowers. This has given rise to very numerous synonyms, of which nineteen are quoted in Mart. Fl. Bras. Papil. 143. The Australian specimens have usually larger flowers and the pod straighter than in most of those from other countries, but some are precisely like the Indian ones.

2. **G. Muelleri,** *Benth.* A larger and much stouter plant than *G. tenuiflora,* the branches softly tomentose. Leaflets oval-oblong and very obtuse or elliptical, 1½ to 2½ in. long, firm, softly silky-pubescent on both sides. Peduncles long, rigid, bearing numerous flowers, considerably larger than in *G. tenuiflora,* in distinct or distant clusters. Calyx silky, nearly 5 lines long. Standard 7 to 8 lines long. Pod silky, about 2 in. long, straight.

N. Australia. Fitzmaurice river, *F. Mueller.*

57. CANAVALIA, DC.

Calyx 2 upper lobes united into a large obtuse entire or 2-lobed upper lip, 3 lower ones into a much smaller entire or 3-lobed lower lip. Standard broad, reflexed, with 2 callosities inside above the claw; wings oblong or linear, falcate or twisted, free; keel incurved and sometimes rostrate. Stamens all united in a tube, open at the very base, where the upper one is free; anthers uniform. Ovary shortly stipitate, with several ovules; style filiform or slightly thickened, with a terminal stigma. Pod oblong or linear, broad, 2-valved, with a prominent longitudinal rib or wing on each side of the upper suture. Seeds rounded or oblong, with a linear hilum, varying in length.—Large herbs with twining or trailing stems. Leaflets 3, usually large, stipellate. Stipules minute, often gland-like or none. Flowers rather large, purplish pink or white, in axillary racemes ; pedicels very short, clustered on lateral nodes along the upper portion of the rhachis. Bracts minute. Bracteoles small, orbicular, very deciduous.

The species are widely distributed over the tropical regions of the New and the Old World, some of them cultivated in India for food, although others are very poisonous. The Australian one is a common maritime plant. The genus is readily known by the calyx, different from all except some species of *Phaseolus*, which have a very different keel, stigma, and pod.

1. **C. obtusifolia,** *DC*. *Prod*. ii. 404. Glabrous or the young shoots silky-pubescent, the stems more frequently prostrate or trailing than twining. Leaflets broadly obovate or orbicular, very obtuse or retuse, 2 to 3 in. long and rather thick, rarely thinner, attaining 4 or 5 in. and obscurely acuminate. Flowers pink or nearly white, along the upper portion of stout erect peduncles, varying from 6 in. to 1 ft. in length. Calyx nearly ½ in. long, the upper lip nearly as long as the tube, with 2 very broad rounded lobes, the lower nearly half as long with 3 small lobes. Standard orbicular, ¾ in. diameter ; keel much curved, but obtuse. Pod ¾ to 1 in. broad, the longitudinal wings very narrow. Seeds 2 to 8, the hilum oblong or shortly linear, not half the length of the seed.—Benth. in Mart. Fl. Bras. Pap. 178, t. 48, where the numerous synonyms are quoted.

N. Australia. Nichol Bay, N.W. coast, *F. Gregory's Expedition ;* Upper Victoria river, *F. Mueller ;* islands of the Gulf of Carpentaria, *R. Brown*.

Queensland. Endeavour river, Bustard Bay, Bay of Inlets, etc., *Banks and Solander, R. Brown ;* Moreton Bay, *F. Mueller ;* Ipswich, *Nernst ;* Broad Sound and Fitzroy river, *Bowman*.

N. S. Wales. Manly Beach, Port Jackson, *Woolls*.

The species is common on the sea-coasts of S. America, Africa, and tropical Asia.

F. Mueller's herbarium contains a specimen of *C. Bonariensis*, Lindl., with narrower acuminate leaflets and the lower lip of the calyx very small and entire, marked " Moreton Bay " on the label, but probably from a garden. It is a S. American and S. African, but not an Asiatic species.

58. PHASEOLUS, Linn.

Calyx 2 upper lobes or teeth united in a short entire or 2-lobed upper lip, the lowest one longer and narrow. Standard broad, recurved, often oblique or twisted ; wings obliquely obovate or oblong, adhering to the keel ; keel produced into a long linear beak, more or less spirally twisted. Upper stamen free, the others united ; anthers uniform. Ovary nearly sessile, with several ovules ; style thickened within the beak of the keel and twisted with it, more or less bearded upwards along the inner side ; stigma oblique or on the inner side of the style. Pod linear or falcate, flat or terete, 2-valved. Seeds with a small or shortly linear hilum, not strophiolate.—Herbs either annual or perennial and woody at the base, short and erect or elongated and twining in the same species. Leaves pinnately 3-foliolate or very rarely 1-foliolate, stipellate, the leaflets entire or lobed. Stipules usually persistent, striate, sometimes produced below their insertion. Flowers white yellowish purple or red, in clusters of 2, 3 or more on lateral nodes in the upper portion of long axillary peduncles. Bracts and bracteoles usually very deciduous.

A considerable genus, dispersed over the warmer regions of the New and the Old World. Several species long cultivated in various countries have given rise to many forms published as separate species, although frequently undistinguishable except by their origin. To these generally cultivated species belong the only three hitherto found in Australia.

Stipules not produced below their insertion. Pod flattened.
Bracteoles broad, striate, persistent, as long as the calyx. Flowers
    small, pale yellowish-white . . . . 1. *P. vulgaris.*
Bracteoles small and very deciduous. Flowers large, pink or whitish
    with the wings purple . . . . 2. *P. Truxillensis.*
Stipules oblong, produced below their insertion. Pod at length nearly
    cylindrical . . . . 3. *P. Mungo.*

1. **P. vulgaris,** *Linn.; DC. Prod.* ii. 392. Glabrous or silky-pubescent when young, dwarf and erect or tall and twining. Leaflets broad, acuminate, 2 to 3 in. long or more, the stipellæ small, obtuse. Stipules small, not produced below their insertion, Peduncles short, with few rather small pale yellow-green or whitish flowers. Bracteoles ovate, striate, as long as the calyx, and persistent. Calyx upper lobe very short, broad, obtuse and entire. Pod broadly linear, flattened, straight or slightly falcate.—Benth. in Mart. Fl. Bras. Papil. 182, with the synonyms quoted.

N. Australia. Nichol Bay and De Grey river, *Ridley's Expedition.* The species is the most generally cultivated in all warm and temperate countries. Its origin is uncertain, probably Asiatic. No note accompanies the Australian specimens giving any clue as to the circumstances under which they were gathered.

2. **P. Truxillensis,** *H. B. and K.; DC. Prod.* ii. 391. Twining or trailing at the base, glabrous or more or less hairy, the hairs reflexed on the stem, appressed or silky on the leaves. Leaflets usually broadly ovate, obtuse or shortly acuminate, 2 to 4 in. long, the terminal one somewhat rhomboid, the lateral ones very oblique, those of the upper leaves narrower, the stipellæ small, oblong. Stipules small, not produced below their insertion. Peduncles usually long, with few flowering nodes at or near the end. Pedicels very short. Bracteoles shorter than the calyx and very deciduous. Flowers rather large, mixed purplish or pink and white or the standard yellowish. Calyx upper lobes short, very broad and obtuse, the lowest narrow acute and as long as or longer than the tube. Standard and wings nearly 1 in. long; keel forming 2 or 3 coils. Pod flattened, straight or falcate, attaining 3 or 4 in. in length and nearly ½ in. in breadth.—*P. rostratus,* Wall. Pl. As. Rar. i. 56, t. 63; Wight, Ic. t. 34.

N. Australia. Between Fitzmaurice river and Macadam Range, *F. Mueller;* islands of the Gulf of Carpentaria, *R. Brown.*
Queensland. Bustard Bay, Endeavour river, etc., *Banks and Solander;* Port Curtis and Barnard Island, *M'Gillivray.*
The species is frequently sent as indigenous in collections from almost all tropical countries, but it is also frequently cultivated. The numerous synonyms, as far as ascertained, are quoted in Mart. Fl. Bras. Papil. 186.

3. **P. Mungo,** *Linn,; DC. Prod.* ii. 395. More or less hirsute, the hairs reflexed on the branches, appressed on the leaves, the Australian speci-

mens elongated and twining, but, like *P. vulgaris*, it is more generally cultivated as dwarf and erect. Leaflets mostly ovate, acute or acuminate, entire
or minutely 3-lobed, the lateral ones very oblique, 2 to 3 in. long. Stipules
rather large, oblong, produced below their insertion so as to appear peltate.
Flowers rather small, pale yellow, in 2 or 3 clusters at the end of the peduncle or rarely more numerous in a somewhat elongated raceme. Bracteoles
small and very deciduous. Calyx 2 upper lobes very short and broad, the
lowest nearly as long as the tube and acute. Keel spirally twisted to one
side, with a spur near the base of the lamina on the same side only. Pods
spreading or reflexed, cylindrical when ripe, 2 to 3 in. long, scarcely 3 lines
broad, hirsute or rarely glabrous.

N. Australia. Upper Victoria river, *F. Mueller*; islands of the N. coast, *R. Brown*.
Queensland. Endeavour river, *Banks and Solander*: Gould Island, *M'Gillivray*;
limestone hills, *Leichhardt*; Bowen river, *Bowman*; Rockhampton, *Dallachy*.
The Australian specimens belong chiefly to the slender twining form, which includes *P.
radiatus*, Roxb., or *P. Roxburghii*, W. and Arn. Prod. 246, and *P. trinervius*, Heyne,
W. and Arn. Prod. 245. This, like the erect form, including *P. Max*, Linn., as well as *P.
Mungo*, appears to be very abundant, wild or cultivated, in tropical Asia, but not in
America.

59. VIGNA, Savi.

(Scytalis *and* Strophostyles, *E. Mey.*; Plectrotropis, *Schum.*)

Calyx 2 upper lobes united into one, or more or less distinct. Standard
broad, spreading; wings obliquely obovate, adhering to the keel; keel incurved, either obtuse or with an obliquely incurved beak, not forming a ·com-
plete spire. Upper stamen free, the others united; anthers uniform.
Ovary nearly sessile, with several ovules; style thickened upwards and longitudinally bearded on the inner side; stigma very oblique or on the inner
side of the style. Pod linear, nearly terete when ripe, 2-valved. Seeds with
a small or shortly linear hilum, not strophiolate.—Herbs either prostrate and
trailing or twining, or short and erect in the same species. Leaves pinnately
trifoliolate, stipellate, the leaflets entire or 3-lobed. Stipules usually persistent,
rarely produced below their insertion. Flowers greenish yellow or purple, on
very short pedicels, in clusters of 2 or 3 on lateral nodes in the upper portion
of long axillary peduncles, or more frequently only 1 or 2 such clusters at
the end of the peduncle. Bracts and bracteoles usually very deciduous.

The genus is as widely distributed as *Phaseolus* over the warmer regions of the New and
the Old World, and comprises several extensively cultivated species. It only differs indeed
from *Phaseolus* in the keel not forming complete spires. Of the four Australian species,
three have a very wide range in the New and the Old World, the fourth is endemic.

Calyx-lobes acute, as long as the tube, the 2 upper united at the base
    only. Keel with a long obliquely incurved beak . . . . 1. *V. vexillata.*
Calyx-lobes short, the 2 upper ones united into one broad entire one.
    Keel rather acute, but not beaked.
    Leaflets obovate, very obtuse . . . . 2. *V. lutea.*
    Leaflets mostly ovate, acute or acuminate, or narrow and lanceolate or
        linear.
        Flowers 7 or 8 lines long. Leaflets mostly ovate . . . . . . 3. *V. luteola.*
        Flowers about 5 lines long. Leaflets mostly lanceolate or linear . 4. *V. lanceolata.*

1. **V. vexillata,** *Benth. in Mart. Fl. Bras. Papil.* 193, *t.* 50, *f.* 1.

Stems, from a tuberous rootstock, twining and hirsute as well as the leaves, the hairs reflexed on the branches, more appressed and scattered on the leaves, rarely at length glabrous. Leaflets usually ovate-lanceolate in the lower leaves, narrow-lanceolate in the upper ones, 2 to 4 in. long and entire, but varying in breadth and size. Stipules cordate-lanceolate, sometimes shortly auriculate. Flowers greenish-yellow, more or less tinged with purple, larger than in most species, 2 to 4 together at the summit of the peduncle. Calyx 4 to 5 lines long, the lobes lanceolate, acuminate, as long as the tube, the 2 upper ones shortly united at the base. Standard 10 or 11 lines diameter, reflexed ; wings rather shorter, one with a long auricle at the base, the other with scarcely any ; keel with a long incurved almost involute oblique beak, not however forming a complete spire, and with a lateral spur on one side below the beak. Stigma thick, on the inner side of the style above the beard. Pod nearly cylindrical, straight or slightly curved, 3 to 4 in. long, about 2 lines broad.—*Phaseolus vexillatus*, Linn. ; Jacq. Hort. Vind. t. 102 ; *Vigna hirta*, Hook. Ic. Pl. t. 637 ; *V. tuberosa*, A. Rich. Tent. Fl. Abyss. i. 217, t. 42.

N. Australia. Sea Range and Point Pearce, *F. Mueller ;* Albert river, *Henne.*

Queensland. Bustard Bay and Endeavour river, *Banks and Solander ;* Broad Sound, etc., *R. Brown ;* Moreton Bay, *Bidwill ;* Pine river, *Fitzalan ;* Burdekin and Burnett rivers, *F. Mueller ;* Port Curtis, *M'Gillivray ;* Bowen river, *Bowman ;* Rockhampton, *Thozet, Dallachy.*

N. S. Wales. Blue Mountains, *Miss Atkinson, Woolls ;* Hunter's River, *Sieber ;* New England, *C. Stuart ;* Macleay and Clarence rivers, *Beckler.*

The species is widely spread over tropical Asia, Africa, and America. By the obliquity of the flower and length of the beak it is intermediate in some respects between *Vigna* and *Phaseolus*, and has been placed alternately by botanists in either of these genera or in *Dolichos*, or has been proposed as a distinct genus under the name of *Plectrotropis* by Schumacher and of *Strophostyles* by E. Meyer, as appears by the extensive synonymy collected in the ' Flora Brasiliensis ' above quoted.

2. **V. lutea,** *A. Gray, Bot. Amer. Expl. Exped.* i. 454.　　Rather coarse, prostrate, trailing or shortly twining, nearly glabrous or the young shoots hoary or silky, with centrally fixed appressed hairs. Leaflets orbicular, obovate or ovate-rhomboid, usually very obtuse, 1½ to 3 in. long. Stipules short and broad ; stipellæ obtuse. Flowers yellow, like those of *V. luteola*, in few clusters crowded at the end of the peduncle. Calyx 1½ lines long, the lobes or teeth shorter than the tube, the 2 upper ones united into one short and broad one. Standard above ½ in. diameter or rather more ; keel broad, much incurved, rather acute, but not beaked. Stigma oblong, very oblique or quite on the inner side of the style above the dense beard. Pod glabrous, straight or curved, about 3 in. long, ¼ in. wide.—*Dolichos luteus*, Swartz, Fl. Ind. Occid. iii. 1246 ; *V. retusa*, Walp. Rep. i. 778 ; Harv. and Sond. Fl. Cap. ii. 242 (*Scytalis*, E. Mey.); *V. anomala*, Walp. Rep. i. 779 (*Scytalis*, Vog.).

N. Australia. Victoria river, *F. Mueller.*

Queensland. Harvey's Bay, *R. Brown ;* islands along the coast down to Moreton Bay, *M'Gillivray, F. Mueller, Thozet*, and others.

N. S. Wales. Sandy shores, Kingstown and Newcastle, *R. Brown ;* Clarence river, *Wilcox ;* Hastings river, *Beckler ;* Botany Bay, *Banks and Solander.*

The species appears to be not unfrequent in maritime sands in tropical Asia and islands of the Pacific, and in Southern Africa, more rare in the W. Indies. I have not seen Swartz's specimens, but his description leaves no doubt that A. Gray is right in referring his plant to the present one.

3. **V. luteola,** *Benth. in Mart. Fl. Bras. Papil.* 194, *t.* 50, *f.* 2. Prostrate, trailing or twining, often hirsute, especially when young, sometimes nearly glabrous, very rarely with a few of the centrally affixed hairs of *V. lutea.* Leaflets very variable, usually ovate or ovate-lanceolate, 1 to 2 in. long, entire or slightly sinuately 3-lobed, the upper ones sometimes longer and narrower, almost always acute or acuminate. Stipules small. Flowers pale-yellow, in few clusters at the end of the peduncle. Calyx scarcely 2 lines long, the 2 upper lobes united into one very short and broad one, the lowest about as long as the tube. Standard 7 or 8 lines diameter; keel as long as the wings, broad, much curved and rather acute, but not beaked. Stigma oblong, very oblique or quite on the inner side of the style above the dense beard. Pod hirsute or rarely at length glabrous, nearly terete, often torulose, 2 to 3 in. long, scarcely ¼ in. broad.—*Dolichos luteolus,* Jacq. Hort. Vind. t. 90 ; *Vigna glabra* and *V. villosa,* Savi ; DC. Prod. ii. 401.

Queensland. Endeavour river, *Banks and Solander ;* Burdekin river, *F. Mueller.*
N. S. Wales. Clarence river, *Wilcox.*
The species is common in tropical America and temperate N. America, chiefly near the sea ; it is also in tropical Africa. I have not seen any Asiatic specimens, which leads me to doubt in some measure the identification of the Australian ones. I can however detect no difference between them and some of the common American forms.

4. **V. lanceolata,** *Benth. in Mitch. Trop. Austr.* 350. Glabrous or slightly pubescent, short and erect or elongated and twining, but always more slender than *V. luteola.* Leaflets usually lanceolate, obtuse or rather acute, 1½ to 2 or rarely 3 in. long, entire or the terminal one hastately lobed on each side at the base, the lateral ones on one side only, those of the lower leaves sometimes short and ovate, in the upper ones long and linear. Flowers much smaller than in *V. luteola,* otherwise like them, in few clusters at the end of the peduncle. Calyx about 1 line long, the 2 upper lobes united into one very short and broad one, the lowest not so long as the tube. Standard about 5 lines diameter, with the 2 callosities and inflexed auricles of the allied species ; keel broad, much incurved, rather acute, but not beaked. Pod glabrous or pubescent, nearly terete, 1 to 2 in. long.—*V. suberecta,* Benth. in Mitch. Trop. Austr. 388.

N. Australia. Upper Victoria river, *F. Mueller ;* to the S. of Wills' Creek, *Howitt's Expedition ;* Arnhem S. Bay, *R. Brown ;* Albert river, *Henne.*
Queensland. Endeavour river, *Banks and Solander ;* Broad Sound, *R. Brown ;* Mount Owen and Mount Faraday, *Mitchell ;* Archer's Creek, *Leichhardt ;* Bowen river, *Bowman.*
N. S. Wales. Between Darling river and Cooper's Creek, *Neilson.*
Var. *filiformis.* Stems long slender and twining. Leaflets all linear or linear-lanceolate. Flowers smaller.—Victoria river, *F. Mueller ;* N. coast, *R. Brown ;* Port Essington, *Armstrong.*
The pod in some of R. Brown's specimens, with lanceolate leaves, is much flatter than in the other specimens seen in fruit, but I can detect no specific differences.

60. DOLICHOS, Linn.

Calyx-lobes short, the 2 upper ones united into one broad entire or emarginate one. Standard orbicular, recurved or spreading, with 2 inflexed auricles at the base and 2 callosities inside; wings obovate, falcate, adhering to the keel, nearly as long as the standard; keel much incurved, often beaked, but not spiral. Upper stamen free, the others united; anthers uniform. Ovary nearly sessile, with several ovules; style thickened upwards, either bearded longitudinally on the inner side or hairy all round, at least round the stigma, which is small and terminal. Pod flattened, usually falcate and acute, 2-valved. Seeds with a small or shortly linear hilum, not strophiolate.— Herbs often woody at the base, twining trailing or short and suberect. Leaves pinnately trifoliolate, stipellate. Stipules small. Flowers sometimes few together, on very short axillary peduncles or even solitary, with striate persistent bracts and bracteoles, almost as in *Clitoria*, more frequently in axillary racemes, clustered on lateral nodes along the peduncle, with very small and deciduous bracts and bracteoles, as in *Vigna* and *Phaseolus*.

The genus is chiefly S. African, with a few tropical Asiatic and S. American species. The only Australian one is widely spread over E. India and S.E. Africa. But, besides that, *D. lignosus*, Linn.; DC. Prod. ii. 397; Bot. Mag. t. 382 (*D. gibbosus*, Thunb., Harv. and Sond. Fl. Cap. ii. 244), a tall twiner with elegant racemes of pink and white flowers, a native of S. Africa, but cultivated for ornament in hot countries, has been sent as wild, from the neighbourhood of King George's Sound, by *Maxwell* and *Oldfield*, but is probably introduced. *D. rhynchosioides*. Schlecht. Linnæa, xxvii. 531, raised from S. Australian seeds, appears from his description to be the same introduced *D. lignosus*.

1. **D. biflorus,** *Linn.; DC. Prod.* ii. 398. Either dwarf and nearly erect or elongated and twining, softly pubescent in all its parts, or at length glabrous. Leaflets ovate, acuminate, 1 to 2 in. long, the lateral ones very oblique. Stipules ovate or lanceolate, striate, persistent. Flowers usually 1 or 2, rarely 3 or 4, clustered in the axils on a very short common peduncle, of a yellowish colour, not ½ in. long. Bracts and bracteoles narrow, almost subulate or the lower ones broader and striate, Calyx-lobes subulate, longer than the tube, the lowest longer than the others. Standard broadly obovate; wings narrow; keel much incurved, with a short obtuse beak. Style filiform, glabrous, except a small tuft of hairs round the terminal stigma. Pod falcate, usually 1½ to 2 in. long and about 4 lines broad.—*D. uniflorus*, Lam.; DC. Prod. ii. 398; *D. axillaris*, E. Mey.; Harv. and Sond. Fl. Cap. ii. 245.

N. Australia. Gilbert river, *F. Mueller.*
The species is dispersed over tropical Asia and S.E. Africa. In inflorescence, and in the style not longitudinally bearded, it differs from most species of the genus, and approaches in some respects *Clitoria.* The lowest flowers are apparently sometimes small and apetalous.

61. DUNBARIA, W. and Arn.

Calyx-lobes acuminate, the 2 upper ones united into 1 entire or slightly toothed one. Standard obovate or orbicular, erect or reflexed, with inflexed auricles at the base and 2 callosities inside; keel incurved, obtuse. Upper stamen free, the others united; anthers uniform. Ovary sessile, with several ovules; style filiform, incurved and rather thickened about the middle; stigma terminal, small. Pod linear, flat, often falcate, acuminate; valves rather thin, not indented between the seeds. Seeds nearly orbicular, with a short or oblong hilum, and a thin or small strophiole, scarcely fleshy.—Trailing or twining herbs, usually tomentose, often viscid. Leaves pinnately trifoliolate without stipellæ, leaflets usually sprinkled with resinous dots underneath.

Stipules striate or small or none. Flowers usually yellow, in axillary racemes, singly scattered along the rhachis or rarely solitary in the axils. Bracts usually broad and membranous, but very deciduous. Bracteoles none.

The genus extends, like *Atylosia*, over E. India and the Archipelago, the Australian species being one of the most widely dispersed. Nearly allied to *Atylosia*, it has the pod almost of a *Dolichos*, differing from *Rhynchosia* only in the more numerous ovules.

1. **D. conspersa,** *Benth. in Pl. Jungh.* i. 241. A slender twiner, hoary all over with a minute tomentum, scarcely becoming glabrous when old, and more or less sprinkled with resinous dots. Leaflets broadly rhomboidal, rarely 1 in. long, entire or the terminal one broadly sinuate-3-lobed. Flowers rather small, yellow, 2 together or rarely solitary in the axils of the leaves, on short pedicels. Calyx-lobes lanceolate-falcate, the upper and lower ones longer than the tube. Pod nearly straight or falcate, obliquely acuminate, 1 to 1¼ in. long, glabrous or slightly tomentose. Seeds 6 to 8, with an oblong hilum and a rather thicker strophiole than in most species.—*Dolichos (?) rhynchosioides*, Miq. Fl. Ind. Bat. i. part 1, 185.

Queensland. Dunk Island, *M'Gillivray*. The species ranges over the eastern provinces of India and the Archipelago up to S. China. It has very much the aspect of some of the common trailing *Rhynchosias*, but the pod is quite different.

62. ATYLOSIA, W. and Arn.

(Cantharospermum, *W. and Arn.*)

Calyx-lobes acuminate, the 2 upper ones united into 1 entire or slightly toothed one. Standard orbicular, reflexed, with 2 inflexed auricles at the base and often with 2 slight callosities inside; keel incurved, obtuse. Upper stamen free, the others united; anthers uniform. Ovary sessile, with several ovules; style filiform, incurved and rather thickened above the middle; stigma terminal, small. Pod oblong or broadly linear, straight, scarcely acuminate, 2-valved, with partitions between the seeds, the valves coriaceous or rarely thin, transversely indented between the seeds. Seeds ovate or orbicular, with an oblong hilum and a thick fleshy strophiole.—Trailing or twining herbs or erect shrubs, usually tomentose or softly villous. Leaves pinnately trifoliolate, without stipellæ; leaflets usually glandular-dotted underneath. Stipules small or none. Flowers yellow, solitary or clustered in the axils or at the end of axillary peduncles. Bracts usually broad and membranous, but so deciduous as to be rarely seen. Bracteoles none.

The genus extends over E. India and the Archipelago and westward to the Mauritius. Of the Australian species, one is common over the whole range of the genus, the others are endemic. F. Mueller proposes to reunite it with *Cajanus* to which it is closely allied, but the pod is differently shaped, and the strophiole appears to be constant. It only differs from some sections of *Rhynchosia* in the ovules always more than 2.

Stems trailing or twining.
  Pod broad, thin, transversely reticulate. Leaflets broadly obovate
    or orbicular . . . . 1. *A. marmorata.*
  Pod coriaceous with deeply depressed transverse lines between the
    seeds. Leaflets rhomboid ovate or elliptical.
      Pedicels axillary solitary or clustered with scarcely any common
        peduncle . . . . 2. *A. scarabæoides.*
      Peduncles axillary, often exceeding the leaves . . . . . . 3. *A. reticulata.*

Stems erect, shrubby at the base. Pod coriaceous.
  Leaflets very rugose, thick and soft. Pod villous, with long hairs.
    Terminal leaflets at some distance from the lateral one . . . **4. *A. grandifolia.***
    Terminal leaflets close between the 2 lateral ones . . . . . **5. *A. pluriflora.***
  Leaflets scarcely rugose, silky-hoary or silvery-tomentose. Pod
    tomentose, without long hairs . . . . **6. *A. cinerea.***

1. **A. marmorata,** *Benth.* Stems rather slender, trailing or twining, pubescent or hirsute. Leaflets very broadly obovate orbicular or broader than long, very obtuse, 1 to 1½ in. long, softly pubescent when young, rather rigid and nearly glabrous when old. Peduncles solitary or 2 together, often slightly exceeding the leaves, either 1-flowered or bearing 1 or 2 pairs of flowers. Calyx slightly tomentose, about 4 lines long, the lobes rather longer than the tube. Petals 5 or 6 lines long. Ovary with about 4 ovules. Pod very flat, about 1 in. long, and nearly ½ in. broad, mottled with purple and thinly hairy, the valves very thin and marked with fine transverse reticulations, the transverse depressed lines between the seeds very faint.—*Glycine marmorata,* R. Br. Herb.

N. Australia. Upper Victoria river, *F. Mueller ;* islands of the Gulf of Carpentaria, *R. Brown, Henne.*
Queensland. Nebo Creek and Bowen river, *Bowman ;* Port Denison, *Fitzalan.*
The species is allied to the E. Indian *A. platycarpa,* Benth., but the leaflets are differently shaped, the pod straighter, etc. These two species belong to the section *Rhynchosioides,* which only differs from the section *Nomismia* of *Rhynchosia,* in the number of ovules.

2. **A. scarabæoides,** *Benth. in Pl. Jurgh.* i. 242. Trailing or twining, slender, but often extending to several feet, softly tomentose in all its parts. Leaflets obovate or elliptical, obtuse or the upper ones narrower and almost acute, 1 to 1½ in. long, rugose and soft. Peduncles very short or almost none, bearing 1 to 5 shortly pedicellate flowers, about 5 lines long. Calyx-lobes falcate, acute, rather longer than the tube and nearly as long as the petals. Ovary with 6 to 8 ovules. Pod about ¾ to 1 in. long and 3 to 4 lines broad, softly tomentose and hairy, the valves coriaceous, marked between the seeds by deep transverse lines and furrows.—*Rhynchosia (?) scarabæoides,* DC. Prod. ii. 387 ; *Cantharospermum pauciflorum,* W. and Arn. Prod. 255.

Queensland. Port Denison, Burdekin river, etc., *Bowman, Dallachy.* The species is common in E. India and the Archipelago, extending from the Mauritius to S. China. The Australian specimens have the leaflets rather narrow, but do not otherwise differ from the common form. This and all the following species belong to the section *Cantharospermum,* corresponding to the section *Ptychocentrum* of *Rhynchosia.*

3. **A. reticulata,** *Benth.* Stems elongated, trailing or rarely twining, rather coarse, rusty-hirsute or velvety. Leaflets ovate or rhomboidal, obtuse, 1½ to 3 in. long, rugose and softly velvety-tomentose. Peduncles usually 2 or 3 in each axil, of unequal lengths, bearing at the end a cluster or short raceme of about 3 to 9 flowers, the pedicels at first very short, but often at length as long as the calyx. Calyx rusty-villous, fully 6 lines long, the lobes linear acuminate, the upper or lower ones about 3 times as long as the tube, the lateral ones rather shorter. Petals scarcely exceeding the calyx. Ovary with about 6 ovules. Pod about 1 in. long and 4 lines wide, very villous, with long hairs, the valves coriaceous, marked with deep rather oblique trans-

verse furrows between the seeds.—*Dolichos reticulatus,* Ait. Hort. Kew. ed. 1,
iii. 33; DC. Prod. ii. 400.

N. Australia. Islands of the Gulf of Carpentaria, *R. Brown.*
Queensland. Endeavour river, *Banks and Solander, A. Cunningham;* Shoalwater
Bay, *R. Brown;* Rockhampton, *Dallachy.*

4. **A. grandifolia,** *F. Muell. Herb.* An erect branching, rusty-tomentose or softly villous shrub, closely resembling in other respects *A. reticulata.*
Leaflets broadly ovate or rhomboidal, obtuse or rather acute, 2 to 3 in. long,
very rugose and soft. Flowers in irregular clusters or short racemes, pedunculate in the upper axils, the upper peduncles short, forming an irregular terminal leafy panicle. Pedicels at first very short, or rarely attaining the length
of the calyx. Calyx rusty villous, about ½ in. long, the lobes about twice as
long as the tube. Petals scarcely exceeding the calyx. Pod as in *A. reticulata,* about 1 in. long and 4 lines wide, very villous with long hairs, the
valves coriaceous, marked with deep transverse furrows between the seeds.

N. Australia. Upper Victoria river, *F. Mueller;* islands of the Gulf of Carpentaria,
R. Brown.
Queensland. Burnett ranges, *F. Mueller;* Burdekin Expedition, *Fitalan;* Fitzroy
river, *Bowman;* Port Denison, *Dallachy.*
Var. *calycina.* Leaflets very thick and rugose. Calyx-lobes longer and broader than in
the ordinary form, the tube very short. Pod very broad.—Victoria river, *F. Mueller.*

5. **A. pluriflora,** *F. Muell. Herb.* Erect and shrubby, clothed with a
soft woolly or silky tomentum, usually very white on the branches. Leaflets
from obovate, very obtuse, and about 1 in. long, to elliptical-oblong, obtuse
or almost acute and near 2 in. long, more or less rhomboid, very rugose and
soft, the terminal one inserted close between the lateral ones instead of being
raised at some distance above them as in all the other Australian species.
Peduncles in the upper axils bearing an irregular head or umbel-like cluster
of flowers, the pedicels at first very short, at length nearly as long as the
calyx. Calyx softly villous, 4 to 5 lines long, the lobes not much longer
than the tube. Petals exceeding the calyx, attaining about 7 lines. Ovary
with 4 to 6 ovules. Pod about 1 in. long, very villous with long hairs, the
valves coriaceous, marked with deep transverse furrows between the seeds.

Queensland. Broad Sound, *R. Brown;* Burdekin Expedition, *Fitzalan;* Rockhampton, *Thozet;* near Princhester, *Bowman;* Thozet's River, *Dallachy.*

6. **A. cinerea,** *F. Muell. Herb.* A shrub of 2 or 3 ft., more slender
than the last two species, hoary or silvery with a soft but close and short tomentum. Leaflets ovate obtuse or rather acute, 1 to 1½ in. long, strongly
reticulate underneath, and sometimes rugose above, but much less so than in
A. grandifolia. Peduncles axillary, bearing usually a short irregular raceme
of rather large flowers. Pedicels rather thick, often as long as the calyx.
Calyx tomentose or silky, 4 to 5 lines long, the lobes lanceolate, rather longer
than the tube. Standard and keel 8 or 9 lines long. Ovary with 4 to 6
ovules. Pod 1 in. long or rather more, nearly 4 lines broad, softly tomentose, but without long hairs, the valves coriaceous, with transverse furrows
between the seeds.

N. Australia. Nichol Bay, *F. Gregory's Expedition;* Victoria river, *F. Mueller*
Some of these specimens may at first sight be confounded with the more tomentose forms of
Rhynchosia acutifolia, which however has smaller flowers and only 2 ovules in all the

ovaries I have examined.—Islands of the Gulf of Carpentaria, *R. Brown* (specimens rather doubtful, scarcely canescent, and possibly a var. of *A. grandifolia*).

63. RHYNCHOSIA, Lour.

(Nomismia, *W. and Arn.*; Copisma, *E. Mey.*)

Calyx 2 upper lobes more or less united. Standard obovate or orbicular, usually with inflexed auricles at the base, rarely callous inside; keel incurved, obtuse or rarely beaked; wings narrower or rarely obovate, and often shorter. Upper stamen free, the others united; anthers uniform. Ovary sessile or nearly so, with 2 or very rarely 1 ovule; style curved above the middle and often thickened; stigma terminal. Pod flattened, oblong or falcate, 2-valved, rarely divided inside. Seeds ovoid, rounded or almost reniform, with a lateral short or oblong hilum, the funicle centrally attached, with or without a strophiole.—Trailing or twining herbs or erect undershrubs or low shrubs, often tomentose and usually sprinkled with resinous dots. Leaves pinnately 3-foliolate, rarely in species not Australian 1- or 5-foliolate, without any or with small setaceous stipellæ. Stipules broad or linear or very small or none. Peduncles axillary, bearing a raceme or rarely single flowers, usually yellow, the standard often streaked with purple. Bracts very deciduous; bracteoles none.

A considerable genus, dispersed over the warmer regions of the globe, especially S. America and Africa, with several Asiatic species, and a few in N. America. Of the six Australian species, one is very common both in the New and the Old World, the others are all endemic, although one is closely allied to a common S. American one.

Stems erect and shrubby.
　Leaflets minutely tomentose, slightly rugose. Peduncles slender,
　　1- or few-flowered. Keel obtuse. Seeds strophiolate . . . 　2. *R. acutifolia.*
　Leaflets softly tomentose, very rugose. Pedicels short, axillary.
　　Keel beaked . . . . 　3. *R. rostrata.*
Stems trailing or twining. Flowers racemose.
　Pod nearly as broad as long, membranous, transversely reticulate.
　　Seeds strophiolate. Calyx-lobes much longer than the tube . 　1. *R. rhomboidea.*
　Pod falcate, much longer than broad. Seeds not strophiolate.
　　Pod tomentose, depressed between the seeds. Leaflets mostly
　　　above 2 in. long. Calyx-lobes shorter than the tube . . . 　4. *R. Cunninghamii.*
　　Pod nearly glabrous or hairy. Leaflets not much above 1 in.
　　　long. Calyx-lobes rather longer than the tube.
　　　　Flowers scarcely above 3 lines long . . . . 　5. *R. minima.*
　　　　Flowers nearly 5 lines long . . . . 　6. *R. australis.*

1. **R. rhomboidea,** *F. Muell. Herb.* Diffuse, trailing or slightly twining, pubescent and apparently somewhat viscid. Leaflets broadly ovate, rhomboid or almost orbicular, obtuse or scarcely acute, mostly under 1 in. long. Peduncles slender, not exceeding the leaves, bearing 2 to 4 small distant flowers. Bracts striate, deciduous. Pedicels very short. Calyx hirsute, 2½ lines long, the lobes narrow, much longer than the tube, the upper ones united to the middle. Standard about 3 lines long; keel obtuse. Ovules 2; style filiform. Pod flat, 6 or 7 lines long, 4 to 5 lines broad, not acuminate, the valves thin, with fine transverse reticulations. Seeds with a thick fleshy strophiole.

N. Australia. Victoria river, *F. Mueller.* Nearly allied to *R. nummularia*, DC., in which, however, the flowers are large and the ovules solitary. These species belong to the section *Nomismia*, proposed as a genus by Wight and Arnott, differing from the section *Rhynchosioides* of *Atylosia*, in having never more than 2 ovules.

2. **R. acutifolia,** *F. Muell. Herb.* (*under* Atylosia). Erect and shrubby but slender, minutely and closely tomentose or rarely softly pubescent. Leaflets ovate or ovate-lanceolate, ¾ to 1½ in. long, almost acute or shortly acuminate, sometimes slightly rugose. Peduncles axillary, slender, bearing 2, 3 or few flowers in an irregular raceme. Pedicels at first short, often at length as long as the calyx and filiform. Flowers yellow, 5 or 6 lines long. Calyx-lobes rather longer than the tube. Standard with 2 callosities inside; keel obtuse, but very much incurved. Ovules 2. Pod coriaceous not acuminate, minutely tomentose, slightly depressed between the seeds, 6 to 8 lines long, fully 3 lines broad. Seeds with a thick fleshy strophiole.

N. Australia. Victoria river, *F. Mueller, Bynoe ;* Gilbert river, *F. Mueller* (a fragment with loose fruits, one having had 3 ovules); N.W. Coast, *A. Cunningham ;* Nichol Bay, *F. Gregory's Expedition* (more tomentose, with the aspect of *Atylosia cinerea,* but with the small flowers and 2 ovules of *R. acutifolia*). The species belongs to the section *Ptychocentrum,* W. and Arn., differing from the shrubby *Atylosias* of the section *Scarabæoides* in the 2 ovules. From the above-mentioned pod with the Gilbert River specimen it appears that there may be occasionally a third ovule as occurs also in *R. minima ;* but I found 2 only in all the ovaries I examined.

3. **R. (?) rostrata,** *Benth.* Erect and shrubby, densely woolly-tomentose or velvety. Leaflets ovate or rhomboidal, thick, soft, and very rugose, rather acute, not 1½ in. long in our specimens, but probably larger in more perfect ones. Flowers apparently clustered in the axils on a very short peduncle. Calyx densely clothed with white wool, the tube about 2 lines, the lobes narrow, the upper and lower ones about 4 lines long, the lateral ones shorter. Standard not exceeding the calyx, slightly tomentose outside, with 2 prominent callosities inside ; wings obovate, much falcate ; keel much curved, with a narrow obtuse beak longer than broad. Ovary very hirsute, with 2 ovules in the only flower examined. Pod unknown.

N. Australia. York Sound, N.W. coast, *A. Cunningham.* The true affinities of this species cannot be determined until the fruit and seeds are known, but it appears to belong to the section *Pseudocajan* of *Rhynchosia.*

4. **R. Cunninghamii,** *Benth.* Twining to a considerable length, shortly and softly pubescent or velvety. Leaflets broadly ovate-rhomboidal, acuminate, 2 to 4 in. long, rather thin. Stipules subulate-acuminate, and stipellæ often present. Racemes axillary, not exceeding the leaves, the pedicels 2 to 4 lines long, often 2 together or the upper ones almost clustered. Calyx tomentose, 2 to 2½ lines long, the lobes obtuse, shorter than the tube. Standard fully twice as long, slightly callous inside ; keel obtuse. Pod slightly falcate and shortly acuminate, slightly contracted between the seeds, nearly ¾ in. long, ¼ in. broad, densely tomentose and often also hairy, valves coriaceous. Seeds ovoid, bluish-black, the hilum short, lateral, without any strophiole.

Queensland. Endeavour river, *Banks and Solander ;* N.E. coast, *A. Cunningham;* Moreton Bay, *C. Stuart ;* Mount Elliott and Thozet's Creek, *Fitzalan, Dallachy.* The species is very closely allied to the common S. American *R. phaseoloides,* DC., but besides some slight differences in the size of the flowers, the seeds in that species have usually, if not always, a large scarlet spot round the hilum.

5. **R. minima,** *DC. Prod.* ii. 385.　Slender, trailing or twining, minutely tomentose or nearly glabrous.　Leaflets broadly ovate-rhomboidal, mostly about 1 in. long when full-grown but often much smaller, with minute or without any stipellæ.　Stipules also very small or none.　Racemes mostly longer than the leaves, bearing in their upper portion scattered pendulous yellow flowers rarely exceeding 3 lines in length, on very short pedicels; in the lower axils the peduncles are sometimes very short and few-flowered. Calyx about 2 lines long, the lobes rather longer than the tube, the 2 upper ones united to the base or to the middle.　Standard obovate, without callosities; keel obtuse.　Style slightly thickened upwards.　Pod falcate, shortly acuminate or acute, narrowed at the base, ½ to nearly ¾ in. long, shortly tomentose-pubescent.　Seeds without any strophiole.—Mart. Fl. Bras. Papil. t. 54. f. 2; *R. punctata,* DC. Mem. Leg. t. 56, and Prod. ii. 385; *R. nuda, R. ervoidea, R. medicaginea,* and *R. rhombifolia,* DC. Prod. ii. 385, 386; *R. laxiflora,* Camb. in Jacquem. Voy. t. 54; and numerous other synonyms cited in Mart. Fl. Bras. Papil. 204.

N. Australia. Nichol Bay, *F. Gregory's and Ridley's Expeditions;* Victoria river, *F. Mueller;* Goulburn Islands, *A. Cunningham;* New Year's Island, *R. Brown.*
Queensland. Broad Sound, Keppel Bay, *R. Brown;* Bowen and Burdekin rivers, *Bowman, Dallachy;* Moreton Bay, *C. Stuart.*
N. S. Wales. Hastings and Macleay rivers, *Beckler;* between the Darling and Cooper's Creek, *Neilson;* Mount Goningberi, *Beckler.*
The species appears to be abundant in almost all tropical or subtropical countries.

6. **R. australis,** *Benth.*　Slender with trailing or twining stems as in *R. minima,* with the same foliage and inflorescence, but the flowers are much larger, usually nearly 5 lines long, bright yellow, without streaks on the standard.　Calyx-lobes all narrow, subulate-acuminate, rather longer than the tube, the 2 upper lobes more united than in *R. minima.*　Pod falcate, acuminate, narrowed at the base, above ¾ in. long, hirsute with long hairs besides the minute tomentum of *R. minima.*　Seeds not strophiolate.

N. Australia. Port Essington, *Armstrong.*
Queensland. Moreton Bay, *Bidwill;* Rockhampton, *Thozet, Dallachy, Bowman.*
N. S. Wales. Clarence river, *Beckler.*
If a variety of *R. minima,* this is yet different from any of the numerous forms assumed in other countries by that ubiquitous species, approaching in some respects the S. African *R. gibba.*

64. ERIOSEMA, DC.

Calyx 2 upper lobes usually free.　Standard obovate or oblong, with inflexed auricles at the base, rarely callous inside; keel slightly incurved, obtuse; wings usually longer, narrow.　Upper stamen free, the others united; anthers uniform.　Ovary sessile, with 2 ovules; style filiform, incurved above the middle and sometimes slightly thickened.　Pod flattened, obliquely orbicular or broadly oblong, 2-valved, not divided inside.　Seeds oblong, oblique, not strophiolate, the funicle attached to one end of a long linear hilum.— Herbs or undershrubs, erect prostrate or rarely twining, tomentose or glabrous, the resinous dots less conspicuous than in *Rhynchosia.*　Leaves pinnately 3-foliolate or 1-foliolate, usually without stipellæ.　Stipules lanceolate, free or united opposite the leaf.　Flowers usually yellow, in axillary racemes, or clusters, the standard often silky-villous. Bracts very deciduous; bracteoles none.

The genus is numerous in S. American and African species, with a single tropical Asiatic one which is the same as the only Australian one. Formerly considered as a section of *Rhynchosia*, it differs in the position of the seeds and generally in habit.

1. **E. chinense,** *Vog. in Pl. Meyen.* 31. Rhizome a perennial oblong tuber. Stems erect, ½ to 1 ft. high, simple or branching at the base only, more or less hirsute with long rust-coloured hairs, intermixed with a short pubescence. Leaflets solitary, nearly sessile, from oblong-lanceolate to linear, 1 to 2 in. long, sprinkled with a few long hairs on the upper surface and on the veins of the lower one, otherwise glabrous above, hairy or glaucous underneath. Peduncles axillary, exceedingly short, with 1, 2 or rarely 3 flowers, about 4 lines long. Bracts small, narrow. Calyx pubescent or villous, the lobes either shorter than the tube, or with long subulate points. Pod 4 to 6 lines long, 3 to 4 lines broad, covered with long rusty hairs.—*Pyrrhotrichia tuberosa*, W. and Arn. Prod. i. 238; *Rhynchosia virgata*, Hamilt, in Wall. Cat. n. 5503.

N. Australia. Arnhem N. Bay, *R. Brown;* Port Essington, *Armstrong*.
Queensland. Endeavour river, *Banks and Solander, A. Cunningham.*
The species is frequent in the hilly districts of N. India, also in Ceylon, Burmah, the Philippines, and S. China.

65. FLEMINGIA, Roxb.

Calyx-lobes nearly equal or the lowest longer. Standard oval obovate or orbicular, with inflexed auricles at the base, often callous inside; keel incurved, obtuse or acute; wings usually rather shorter. Vexillary stamen free, the others united; anthers uniform. Ovary short, sessile or nearly so, with 2 ovules; style filiform, incurved above the middle and often slightly thickened; stigma terminal. Pod very oblique, short, turgid, 2-valved. Seeds with a short hilum without any strophiole.—Herbs undershrubs or shrubs, rarely twining, usually tomentose or pubescent and sprinkled with resinous dots. Leaves digitately 3-foliolate or 1-foliolate, without stipellæ, the veins of the leaflets very prominent underneath. Stipules usually dry, striate, deciduous. Flowers purple-reddish, or mixed with yellow, in panicles or spike-like racemes. Bracts either like the stipules or (in species not Australian) large, leafy, concave, enclosing the flowers. Bracteoles none.

The genus is most numerous in tropical Asia, with one or two African species. Of the Australian species two are common to E. India and the Archipelago, the other two are endemic. The species with large leafy bracts forming the section *Ostryodium*, DC., common in the Archipelago, have not yet been found in Australia. The habit and foliage of the genus are almost those of some *Genisteæ*, from which tribe, however, it is readily distinguished by the free upper stamen. Several species also resemble some.*Psoraleas*, but the ovary and pod are quite different.

Flowers in small loose panicles . . . . 1. *F. lineata.*
Flowers 2 or 3 on a short axillary peduncle . . . . 2. *F. pauciflora.*
Flowers in axillary oblong spike-like racemes . . . . 3. *F. parviflora.*
Flowers in dense globular sessile heads . . . . 4. *F. involucrata.*

1. **F. lineata,** *Roxb. ; DC. Prod.* ii. 351. An erect undershrub or low shrub with slender branches, the young parts and inflorescence minutely rusty

tomentose, the foliage at length nearly glabrous. Leaflets 3, from obovate-cuneate to elliptical-oblong or broadly lanceolate, obtuse or acute, $1\frac{1}{2}$ to 3 in. long. Stipules and bracts small, usually persistent. Flowers small, secund and loosely racemose along the branches of small irregular axillary or terminal panicles. Calyx minutely tomentose, about 3 lines long, the lobes much falcate, longer than the tube. Standard broad, scarcely exceeding the calyx; keel at least as long, much curved, acute; wings rather shorter. Pod very oblique, about 4 to 6 lines long and 3 lines broad.—Wight, Ic. t. 327.

N. Australia. Victoria river, Treachery Bay, Gulf of Carpentaria, *F. Mueller;* Port Essington, *Armstrong.*

Queensland. Rockhampton, *Thozet ;* Burdekin river and Denison Creek, *Bowman ;* Port Denison and Edgecombe Bay, *Dallachy.*

The species is widely spread over E. India and the Archipelago.

2. **F. pauciflora,** *Benth.* A low perennial or undershrub, with the habit of *F. parviflora*, but softly silky-pubescent or villous all over. Leaflets 3, from obovate to elliptical-oblong, obtuse or softly mucronate, 1 to 2 in. long. Stipules narrow, acuminate, often persistent. Flowers small, 2 or 3 together, almost sessile, on short axillary peduncles. Bracts narrow, silky, persistent. Calyx silky, the tube very short, the lobes narrow, acuminate, often 3 lines long. Petals shorter than the calyx-lobes. Standard rather broad; keel obtuse. Pod very oblique, as broad as long, attaining nearly $\frac{1}{2}$ in.

N. Australia. Victoria river, *F. Mueller;* Gulf of Carpentaria, *Landsborough.*

3. **F. parviflora,** *Benth.* A low perennial or undershrub, with rather slender prostrate or ascending stems, rarely exceeding 1 ft., loosely pubescent, almost silky when young. Leaflets 3, from obovate-cuneate and 1 to $1\frac{1}{2}$ in. long, to ovate-lanceolate and 3 to 4 in. long, usually very rugose. Stipules very deciduous. Flowers small, pink, in short dense spike-like racemes sessile in the axils. Bracts lanceolate, silky-hairy, very deciduous. Calyx silky-pubescent, about 2 lines long, the upper lobe rather longer than the tube, the lowest still longer. Petals shortly exceeding the calyx; standard ovate, obtuse; keel nearly straight, obtuse. Pod 4 to 5 lines long, not 3 lines broad.

Queensland. Shoal Bay Passage, *R. Brown ;* Brisbane river, Moreton Bay, *F Mueller ;* Burdekin river, *Bowman ;* Port Denison, *Dallachy ;* Lynedoch valley, *Leichhardt.* Nearly allied to the E. Indian *F. prostrata*, Roxb., but in the latter species the flowers are considerably larger, although the petals are shorter than the calyx.

4. **F. involucrata,** *Benth. in Pl. Jungh.* i. 246. An erect stout perennial or undershrub of 2 to 4 ft., more or less villous with soft spreading hairs. Leaflets 3, ovate, rather acute, 2 to 3 in. long. Stipules lanceolate, very deciduous. Flowers in dense globose heads of 1 in. diameter or rather more, sessile or very shortly pedunculate in the upper axils and at the ends of the branches. Outer bracts ovate-lanceolate, striate, pubescent, forming an involucre round the head ; inner ones much narrower. Calyx covered with long soft hairs, the upper lobes about 4 lines, the lowest nearly 6 lines long and much broader than the others. Standard rather shorter than the calyx-lobes, obovate ; keel scarcely shorter, almost acute. Pod enclosed in the calyx, scarcely 3 lines long, usually 1-seeded by abortion.—*F. capitata*, Zoll. ; Miq. Fl. Ind. Bat. i. part i. 166.

Queensland. Endeavour river, *R. Brown.* Also in N. E. India and in Java.

66. ABRUS, Linn.

Calyx campanulate, truncate or shortly and broadly toothed. Standard ovate, the short claw adhering to the base of the staminal tube; keel much curved, the petals united from the base, often longer than the wings. Stamens 9 united in a sheath open on the upper side, the upper one deficient. Ovary sessile, with several ovules; style short, incurved; stigma terminal. Pod oblong or linear, flat, 2-valved, with cellular partitions between the seeds, Seeds not strophiolate.—Stems usually twining or trailing, woody at the base. Leaves abruptly pinnate, with several pairs of small leaflets, without stipellæ, the common petiole ending in a short point. Racemes terminal or axillary, the flowers in clusters on lateral thickened nodes. Bracts minute or none; bracteoles none.

A small genus dispersed over the tropical regions both of the New and the Old World, the only Australian species extending over the whole range. It is in some measure inter-mediate between the tribes *Vicieæ*, *Phaseoleæ*, and *Dalbergieæ*.

1. **A. precatorius,** *Linn. ; DC. Prod.* ii. 381. Glabrous or slightly pubescent. Leaflets in 7 to 10 pairs, oblong-elliptical or rarely obovate, usually about ½ in. long. Racemes with 1 or 2 leaves, or at least with a leaf-less pair of stipules below the flowers, the flowering part 1 in. or rather more in length, the nodes rather crowded. Flowers pink, or rarely white or pur-ple, 5 to 6 lines long, the keel narrow, longer than the wings. Pod sessile, 1 to 1½ in. long, 6 to 7 lines broad, almost squared at both ends and at-tached by the inner angle, glabrous or scaly outside. Seeds usually black with a large scarlet spot, sometimes brown with a darker spot, or white and unspotted.—Lam. Illustr. t. 608. f. 1 ; W. and Arn. Prod. i. 236 ; *A. pauci-florus*, Desv. in Ann. Sc. Nat. ix. 418 ; *A. squamulosus*, E. Mey. Comm. Pl. Afr. Austr. 126.

N. Australia. Islands of the Gulf of Carpentaria, *R. Brown, Henne.*
Queensland. Burdekin Expedition and Port Denison, *Fitzalan.* Very common in India and the Archipelago, extending into tropical and southern Africa, and frequent also, but perhaps naturalized, in several parts of S. America.

TRIBE VIII. DALBERGIEÆ.—Trees or woody climbers. Leaves pinnate, with 5 or more leaflets or sometimes one large leaflet, very rarely 3. Stipellæ none, or small and subulate. Stamens all united in a sheath or tube or into two parcels of 5, very rarely the upper one free. Pod indehiscent.

This tribe is closely connected on the one hand with the arborescent *Galegeæ*, from which it is distinguished by the indehiscent pod, and on the other hand with *Sophoreæ*, differing in the united stamens. The genera are all tropical or subtropical, American, Asiatic or African.

67. DALBERGIA, Linn.

Calyx-teeth short, the lowest rather longer. Standard obovate or orbicular; keel obtuse. Stamens all united in a sheath open on the upper side, or in 2 equal bundles, or reduced to 9, the upper one deficient; anthers small, erect, didymous, opening at the top. Ovary stipitate, with 1 or few ovules; style incurved, with a terminal stigma. Pod thin and flat, oblong, linear or rarely falcate, indehiscent, the margins neither thickened nor winged. Seeds single or few and distant along the centre of the pod, very flat and reniform. —Trees or woody climbers. Leaves pinnate, without stipellæ, the leaflets usually alternate. Flowers small, usually numerous in axillary or terminal dichotomous cymes or irregular panicles. Bracts and bracteoles usually minute.

A large genus dispersed over the tropical regions of the New and the Old World. The only Australian species has also been found in New Guinea.

1. **D. densa,** *Benth. in Hook. Lond. Journ.* ii. 217. A small tree, with the branches sometimes weak or climbing. Leaflets 7 to 15, broadly oblong, or oval-elliptical, very obtuse, ¾ to 1½ in. long, glabrous above, minutely pubescent underneath. Panicles or clusters of racemes, under 2 in. long, not much branched. Flowers scarcely above 2 lines long. Calyx nearly glabrous, about 1 line long, the teeth very short and broad. Claws of the petals as long as the calyx. Ovary glabrous; style rather slender. Pod very thin, obtuse, 1½ to nearly 3 in. long, about ½ in. wide, slightly reticulate along the centre, on a stipes of about 2 lines.

Queensland. Prince of Wales Islands, *R. Brown;* Albany Island, *W. Hill.* Also in New Guinea. The Australian specimens have rather more leaflets than the New Guinea ones, but do not otherwise differ. The species is allied to the common *D. tamarindifolia,* Roxb., from E. India and the Archipelago, and has the same pod, but much larger broader and fewer leaflets and smaller flowers.

68. LONCHOCARPUS, H. B. and K.

Calyx truncate or very broadly and shortly toothed. Standard orbicular obovate or rarely oblong; wings usually slightly adhering to the keel; keel nearly straight or incurved, obtuse. Upper stamen free at the base, connate with the others in the middle; anthers uniform. Ovary with 2 or more ovules; style incurved, stigma small, terminal. Pod very flat, oblong or broadly linear, thin or rather thick and coriaceous, indehiscent, the upper or both margins sometimes thickened or bordered by a prominent nerve, but not winged. Seeds usually 1 or 2, rarely more, flat, reniform or orbicular.— Trees or woody climbers. Leaves pinnate; leaflets usually few, opposite, with a terminal odd one; stipellæ none or small and setaceous. Stipules small or none. Racemes or raceme-like panicles axillary or forming large terminal panicles. Flowers violet purple or white, usually in pairs or in clusters along the rhachis. Bracts small, deciduous; bracteoles also small, but often more persistent.

A numerous genus in S. America and tropical Africa, but as yet unknown in Asia. The only Australian species is endemic, coming, however, very near to some African paniculate species, generically distinguished by Fenzl under the name of *Philenoptera*, and by Klotzsch under that of *Capassa.* In flower, *Lonchocarpus* cannot always be distinguished from *Millettia,* but the pod is always thinner and indehiscent, the pod alone distinguishing the genus from *Derris* and *Pongamia.*

1. **L. Blackii,** *Benth.* A tall woody climber, the young branches and foliage rusty-pubescent, at length glabrous. Leaflets 7 to 11, ovate, shortly and obtusely acuminate, 1 to 1½ or rarely above 2 in. long, on rather long petiolules, with small setaceous stipellæ. Flowers dark purple, in long loose

racemes, forming large terminal panicles, the pedicels usually 2 together on a short common peduncle. Calyx about 2 lines long, slightly silky-pubescent, with short broad teeth. Standard about 4 lines broad, narrowed into a long claw; keel nearly as long, obliquely ovate; wings very small. Ovary very villous, with numerous ovules; style short, inflexed. Pod very thin, 2 to 5 in. long, ½ to ¾ in. broad, with 2 to 4 or 5 broad flat reniform seeds.—*Millettia Blackii*, F. Muell. Fragm. ii. 123.

Queensland. Brisbane river, Moreton Bay, *F. Mueller, W. Hill, Leichhardt;* Ipswich, *Nernst;* Broad Sound and Bowen river, *Bowman;* Rockhampton, *Dallachy.*

N. S. Wales. Clarence and Hastings rivers, *Beckler.*

69. DERRIS, Lour.

(Brachypterum, *W. and Arn.*)

Calyx truncate or very shortly and broadly toothed. Standard obovate or orbicular; keel slightly incurved. Upper stamen usually free at the base, united with the others in the middle; anthers uniform. Ovary sessile or shortly stipitate, with several ovules; style incurved, with a small terminal stigma. Pod flat, oblong or linear, straight or slightly incurved, thin or coriaceous, indehiscent, the upper or both sutures bordered by a narrow wing. Seeds 1, 2, or rarely 3, very flat, orbicular or reniform. —Tall woody climbers or rarely trees. Leaves pinnate; leaflets opposite, with a terminal odd one; stipellæ small and setaceous or none. Stipules small or none. Flowers white yellowish or rarely violet, usually clustered on lateral nodes along the rhachis of axillary racemes. Bracts and bracteoles small and deciduous.

A tropical genus, comprising a considerable number of Asiatic species with three S. American ones, one only of the Asiatic species extending into Africa. The Australian species are both common Indian ones. The genus differs from *Lonchocarpus* and *Pongamia* chiefly in the winged fruit.

Leaflets 9 to 13, usually obtuse. Racemes long and slender. Pod lanceolate, narrowed at both ends . . . . 1. *D. scandens.*
Leaflets 3 to 7, usually acuminate. Racemes rather short and crowded.
Pod short and broad, obliquely rounded at both ends . . . . . . 2. *D. uliginosa.*

1. **D. scandens,** *Benth. Syn. Dalb.* 103. A tall woody climber, sometimes rusty-pubescent or almost villous at first, nearly glabrous when full-grown. Leaflets 9 to 13, broadly oblong, obtuse, retuse or shortly and obtusely acuminate, 1 to 2 in. long. Racemes rather slender, from 4 or 5 in. to nearly 1 ft. long. Flowers about 5 lines long, in clusters of 3 to 6, the pedicels filiform. Pod either 1-seeded and about 1½ in. long, or when several-seeded attaining 3 in. or more, 5 to 6 lines broad, glabrous or minutely pubescent, acute at both ends, very thin, with a narrow wing along the upper suture.—*Dalbergia scandens*, Roxb. Pl. Corom. t. 192; Wight, Ic. t. 275.

Queensland. Wide Bay, *Bidwill;* Brisbane river, Moreton Bay, *F. Mueller, C Stuart;* Ipswich, *Nernst.*

N. S. Wales. Clarence river, *Beckler.*

Common in E. India and the Archipelago.

2. **D. uliginosa,** *Benth. in Pl. Jungh.* i. 252, *and Syn. Dalb.* 107. A tall woody climber, glabrous in all its parts. Leaflets, in the common variety,

5 or 7, ovate or oval-oblong, $1\frac{1}{2}$ to 3 in. long, shortly and obtusely acuminate, somewhat coriaceous and shining. Axillary racemes 1 to 3 in. long, the terminal one longer. Flowers 4 to 5 lines long, on short pedicels, the clusters rather crowded. Standard broadly ovate; wings and keel scarcely shorter, but narrow-oblong. Ovules usually 4 to 6, all in the lower part of the ovary. Pod very flat and thin, 1 to $1\frac{1}{2}$ in. long, very obtuse at both ends so as to become nearer square than round, but very oblique, sometimes as broad as long, but in some varieties narrower. Seeds 1 or 2.—*Pongamia uliginosa*, DC. Prod. ii. 416.

N. Australia. Islands of the Gulf of Carpentaria, *R. Brown;* Albert river, *Henne;* Fitzmaurice river, *F. Mueller.*
Queensland. Cape York, *W. Hill.*
Common in E. India and the Archipelago, extending from S.E. Africa to S. China.

70. PONGAMIA, Vent.

Calyx truncate. Standard orbicular, with inflexed auricles at the base; keel slightly incurved, obtuse. Upper stamen free at the base, connate with the others in a tube in the middle; anthers uniform. Ovary nearly sessile, with 2 ovules; style incurved, stigma small, terminal. Pod broadly and obliquely oblong or slightly falcate, thick but flat, 1-seeded, indehiscent, the sutures obtuse, without wings. Seed reniform.—Tree. Leaves pinnate, without stipellæ. Flowers in axillary racemes. Bracts very deciduous; bracteoles minute or none.

The genus is limited to a single species widely diffused over tropical Asia.

1. **P. glabra,** *Vent. Jard. Malm. t. 28.* Glabrous except a very slight pubescence on the inflorescence. Leaflets 5 or 7, ovate, shortly and obtusely acuminate, usually broad, about 3 in. long, on a rather long petiole, but variable in size. Racemes loose, about 3 to 5 in. long. Flowers in pairs, the pedicels 2 to 4 lines long. Standard about $\frac{1}{2}$ in. diameter, lower petals shorter. Pod usually $1\frac{1}{2}$ to 2 in. long and about 1 in. broad, sessile or nearly so, often somewhat falcate or with a very short incurved point.—Benth. Syn. Dalb. 117.

N. Australia. Fitzmaurice river, *F. Mueller;* Raffles Bay, *A. Cunningham.*
Queensland. Cape York and Fitzroy Island, *M'Gillivray;* Port Denison, *Fitzalan;* Edgecombe Bay, *Dallachy.*
Var. *minor.* Leaflets small and narrow.—Gulf of Carpentaria, *Leichhardt;* Cape Grafton, *A. Cunningham;* Port Denison, *Fitzalan.*
The species occurs throughout Southern India on the coast and plains to the foot of the hills, and is abundant in the Archipelago.

TRIBE IX. SOPHOREÆ.—Trees woody climbers or rarely tall shrubs, or, in one or two species not Australian, almost herbaceous. Leaves pinnate, with several leaflets, usually without stipellæ, or reduced to one large leaflet. Stamens all free or scarcely united at the base.

This tribe is very near *Dalbergieæ*, differing only in the free stamens; from *Podalyrieæ* it is chiefly distinguished by the habit and foliage. It also forms the passage from *Papilionaceæ* to *Cæsalpinieæ*, presenting the greatest differences in the radicle, from short and quite straight to long and accumbent even in the same genus. *Barklya* has at first sight the almost regular flowers of some *Cæsalpinieæ*, but the upper petal is outside, not inside, a constant distinction between the two suborders.

71. SOPHORA, Linn.

Calyx-teeth very short. Standard broad, erect or reflexed; wings oblong, erect, free; keel-petals like the wings or rather larger, overlapping each other at the back but scarcely united. Stamens 10, free, or 9 of them slightly connected in a ring at the base; anthers uniform. Ovary shortly stipitate, with several ovules; style incurved, with a minute terminal stigma. Pod moniliform, fleshy, coriaceous or woody, indehiscent or opening at length in 2 valves, each seed enclosed in a separate cell. Seeds globular, oblong or flattened; cotyledons fleshy; radicle very short and straight or more or less elongated and inflexed.—Trees shrubs or rarely undershrubs. Leaves unequally pinnate, without stipellæ or with very small setaceous ones. Stipules small. Flowers white yellow or rarely violet-blue, in racemes either simple and terminal or forming large terminal panicles. Bracts small, deciduous; bracteoles usually none.

The genus is dispersed over the warmer regions of the New and the Old World, extending also into New Zealand and S. Chili, where it assumes the form distinguished by some as a genus under the name of *Edwardsia*, with a shorter standard and exserted stamens. The two Australian species belong to the true *Sophoras*, with a larger standard and the stamens enclosed in the keel. One is a common tropical seacoast tree or shrub, the other is endemic.

Hoary. Leaflets under 18, broadly ovate or orbicular, rather thick . . 1. *S. tomentosa.*
Softly pubescent. Leaflets above 20, oval-oblong, thin . . . . . . 2. *S. Fraseri.*

1. **S. tomentosa,** *Linn.; DC. Prod.* ii. 95. A tall shrub or small tree, hoary all over with a minute close tomentum. Leaflets 11 to 17, broadly ovate or orbicular, very obtuse or retuse, about 1 in. long or rather more, rather thick and sometimes almost silky, rarely becoming glabrous. Flowers pale yellow, in loose simple terminal racemes; pedicels as long as the calyx. Calyx very broad, about 3 lines long, truncate with scarcely prominent teeth. Standard broad, 9 to 10 lines diameter, spreading or reflexed above the middle; wings and keel rather shorter, covering the stamens. Pod indehiscent, much contracted between the seeds, appearing to consist of 5 to 10 nearly globular articles, each enclosing a globular seed with a hard shining testa; radicle scarcely prominent and straight.—Benth. in Mart. Fl. Bras. Pap. 314, t. 124, with the synonymy there cited.

Queensland. Keppel Bay, Broad Sound, etc., *R. Brown;* on the seacoast and adjoining islands, from the Brisbane to the Burdekin, *F. Mueller, M'Gillivray, Fitzalan, Henne,* and others.

N. S. Wales. Hastings river, *Beckler.*

2. **S. Fraseri,** *Benth.* An erect shrub of 4 to 6 ft., the branches softly pubescent or tomentose, more slender than in *S. tomentosa.* Leaflets 21 to 31, oblong or rarely oval, obtuse or retuse, from under ½ in. to about ¾ in. long, rather thin, pubescent. Flowers rather smaller than in *S. tomentosa,* in similar loose terminal simple racemes. Calyx broad, 2 to 2½ lines long; the teeth prominent though very short and broad. Petals and stamens of *S. tomentosa,* except that 9 of the stamens appear to be very shortly connected in a ring at the base. Pod tomentose, much less contracted between the seeds than in *S. tomentosa,* the articles more oblong. Seeds ovoid-oblong, shining; radicle prominent and slightly incurved.

Queensland. Moreton Bay, *Fraser ;* Murrum-Murrum Creek, *Leichhardt ;* Pine river, *Fitzalan ;* Ipswich, *Nernst.*
N. S. Wales. Clarence river, *Beckler.*

72. CASTANOSPERMUM, A. Cunn.

Calyx-teeth very short and broad. Standard obovate-orbicular, recurved, narrowed into a claw; wings and keel-petals shorter than the standard, all free and nearly similar, erect, oblong. Stamens 10, all free ; anthers linear, versatile. Ovary on a long stipes, with several ovules, tapering into an incurved style ; stigma small, terminal. Pod large, coriaceous, almost woody, turgid, 2-valved, spongy inside. Seeds large, nearly globular ; cotyledons thick ; radicle scarcely prominent, straight.—Tree. Leaves large, unequally pinnate. Flowers large, yellow, in loose axillary or lateral racemes. Bracts small ; bracteoles none.

The genus is limited to a single species, endemic in Australia,

1. **C. australe,** *A. Cunn. in Hook.· Bot. Misc.* i. 241, *t.* 51, 52. A tall glabrous tree. Leaves 1 to 1½ ft. long ; leaflets 11 to 15, ovate-elliptical or broadly oblong, shortly acuminate, 3 to 5 in. long, shortly petiolulate. Racemes under 6 in. long, either in the axils of the older leaves or on the leafless older wood ; pedicels nearly 1 in. long. Calyx about 8 lines long, including the turbinate base. Standard above 1 in. diameter. Pod 8 or 9 in. long, about 2 in. broad, slightly falcate, almost terete, the valves hard and thick, the spongy substance inside dividing it into 3 to 5 cells, each containing a large chestnut-like seed,

Queensland. Endeavour river, *Banks and Solander ;* Brisbane river, Moreton Bay, *A. Cunningham, F. Mueller ;* Pine river, *Fitzalan.*
N. S. Wales. Clarence river, *Beckler.*
The seeds are eaten roasted, like chestnuts.

73. BARKLYA, F. Muell.

Calyx-teeth very short and obtuse. Petals all free, obovate, erect, similar and nearly equal, on long claws, the upper outer one or standard rather broader than the others. Stamens 10, all free, longer than the petals ; anthers sagittate. Ovary stipitate, with several ovules, tapering into a short style with a minute terminal stigma. Pod stipitate, flat, the valves thin and scarcely separating. Seeds flat, albuminous ; cotyledons obovate, flat ; radicle inflexed.—Tree. Leaves simple (unifoliolate), petiolate. Flowers small, yellow, in dense racemes. Bracts very small ; bracteoles none.

The genus is limited to a single species, endemic in Australia, approaching in habit and in the small regular flowers to some *Cæsalpinieæ* allied to *Bauhinia,* but with the floral æstivation and the embryo of *Papilionaceæ.*

1. **B. syringifolia,** *F. Muell. in Journ. Linn. Soc.* iii. 158, *and Fragm.* i. 109, *t.* 3. A handsome tree, attaining from 20 to 60 ft., glabrous or the young shoots and inflorescence rusty-tomentose. Leaves very broadly cordate, shortly acuminate, 2, 3, or even 4 in. long and often as broad as long, entire, 5- to 7-nerved, on a petiole of 1 to 2 in., slightly thickened at the base and at the top. Stipules small, ovate, deciduous. Flowers of a bright golden yellow, in dense racemes of 6 to 9 in., forming handsome terminal

panicles. Pedicels short. Calyx about 2 lines long. Petal-claws rather longer than the calyx, the lamina about as long. Ovary glabrous, with 3 or 4 ovules. Pod oblong-lanceolate, oblique or slightly falcate, $1\frac{1}{2}$ to 2 in. long and about $\frac{1}{2}$ in. broad, narrowed at the base, with 1 or 2 seeds.

Queensland. Woods near Brisbane, *W. Hill, Fitzalan;* Rockhampton, *Dallachy;* Wide Bay, *Leichhardt.*

SUBORDER II. CÆSALPINIEÆ.

Flowers usually 5-merous, very rarely 4-merous or 3-merous, the sepals united at the base into a short tube, lined by the disk, bearing at its margin the petals and stamens, rarely forming a campanulate or tubular calyx with the stamens near the base, as in *Papilionaceæ,* the free part of the sepals or lobes of the calyx, imbricate or rarely valvate. Corolla irregular or nearly regular, either with the 5 (or 4 or 3) petals variously imbricate in the bud, but the upper one never outside and usually quite inside, or, in genera not Australian, some or all of the four lower petals wanting. Stamens 10 or fewer, or, in genera not Australian, indefinite, free or rarely more or less united, all perfect or several of them reduced to staminodia. Ovules anatropous or nearly so. Radicle of the embryo short and straight.

The tropical genera of this suborder are numerous, and have been distributed into several tribes, but these are not sufficiently marked to render it necessary to apply them to the few genera found in Australia. *Barklya* amongst *Sophoreæ* has the regular corolla of some *Cæsalpinieæ,* but with the æstivation of *Papilionaceæ,* and *Erythrophlœum* amongst *Mimoseæ* has the imbricate æstivation of *Cæsalpinieæ,* but in a very slight degree, with the inflorescence characteristic of *Mimoseæ.*

74. GUILANDINA, Linn.

Sepals 5, shortly united at the base, much imbricate, nearly equal or the lowest rather larger and more concave. Petals 5, nearly equal, sessile, spreading. Stamens 10, free; anthers uniform. Ovary sessile, with 2 ovules; style subulate, with a small stigma. Pod ovate, compressed, covered with prickles, 2-valved, thickly coriaceous. Seeds ovoid or globular, with a very thick hard fleshy testa and no albumen.—Woody climbers, armed with prickles. Leaves twice pinnate. Flowers yellow, in simple or paniculate racemes. Bracts very deciduous.

A genus of 2 species, both of them dispersed over the tropical regions of the New and the Old World, one only hitherto found in Australia.

1. **G. Bonducella,** *Linn. Spec.* 545. A shrub, with loose spreading or climbing branches, pubescent or villous in all its parts, armed with numerous scattered hooked prickles, especially on the petioles. Leaves with a common petiole of 1 to $1\frac{1}{2}$ ft.; pinnæ in 4 to 6 distant pairs, each 4 to 6 in. long; leaflets 8 to 12 pairs, oblong, $\frac{3}{4}$ to 1 in. long or rarely nearly twice that size. Stipules lobed and leafy, deciduous. Racemes 4 to 6 in. long, simple or branched in the upper axils. Flowers shortly pedicellate and crowded in the upper part. Bracts with a long recurved point. Calyx about 4 lines long. Petals scarcely longer. Pod 2 to 3 in. long, about $1\frac{1}{2}$ in. broad. Seeds large, of a bluish-grey or lead colour.—Lam. Illustr. t. 336.

Queensland. Cumberland Islands, *R. Brown;* Barnard Island, *M Gillivray;* Edgecombe Bay, *Dallachy ;* Low Island, *Henne* (the latter specimen a leaf only, and therefore doubtful). Widely spread and often very common, especially near the sea, in tropical Asia, Africa, and America. It is usually confounded with *G. Bonduc,* Linn., which is a much rarer plant, although equally found, indigenous or introduced, in East India, in the Archipelago, and in the West Indies. It is nearly glabrous, has usually larger leaflets, no stipules ; the bracts are erect, not recurved, and the seeds are said to be yellow, not grey. It remains to be ascertained how far these differences are constant.

75. CÆSALPINIA, Linn.

Sepals 5, shortly united at the base, much imbricated, the lowest one larger and concave. Petals 5, spreading, rather unequal, the upper inner one the smallest, the 2 lowest outer ones the largest. Stamens 10, free ; anthers uniform, ovate. Ovary with 2 or more ovules ; style subulate, with a small stigma. Pod flattened, obliquely ovate, oblong or broadly linear-falcate, without prickles, 2-valved. Seeds thick or flattened, with a very small hilum ; albumen none ; radicle short, straight.—Shrubs trees or woody climbers, often armed with scattered recurved prickles. Leaves abruptly bipinnate. Flowers yellow, in racemes, either single in the upper axils or forming terminal panicles. Filaments more or less hairy at the base.

A considerable genus, spread over the tropical regions of the New and the Old World. The Australian species are both of them common Asiatic ones.

Glabrous. Pinnæ 2 to 4 pairs. Leaflets 2 or 3 pairs, ovate, rather acute.
  Pod ovate, 1-seeded . . . . 1. *C. nuga.*
Pubescent or tomentose. Pinnæ 6 to 8 pairs. Leaflets 8 to 12 pairs,
  oblong, very obtuse. Pod oblong, 6 to 8-seeded . . . . . . 2. *C. sepiaria.*

1. **C. nuga,** *Ait. ;- DC. Prod.* ii. 481. A woody climber, glabrous in all its parts, armed with a few scattered recurved prickles, especially on the petioles. Pinnæ 2, 3, or 4 pairs ; leaflets 2 or 3 pairs to each pinna, ovate, 1½ to 2 in. long, usually rather acute, but occasionally obtuse, coriaceous and shining. Racemes 4 to 8 in. long, forming large terminal panicles. Pedicels slender. Lowest sepal about 5 lines long, the others shorter. Petals scarcely exceeding the lowest sepal. Ovary with 2 ovules. Pod obliquely oval, acuminate, flat, with coriaceous valves. Seed usually solitary, flat, broad.— *C. paniculata,* Desf. ; DC. Prod. ii. 481 ; Wight, Ic. t. 36.

Queensland. Barnard Islands, *McGillivray.* Generally distributed over E. India and the Archipelago, extending to S. China.

2. **C. sepiaria,** *Roxb. ; W. and Arn. Prod.* 282. A woody climber, the branches petioles and racemes more or less tomentose or pubescent and armed with numerous scattered recurved prickles. Pinnæ 6 to 10 pairs ; leaflets 8 to 12 pairs, oblong, very obtuse, rarely exceeding ½ in. in length, pubescent or villous when young, often glabrous when old. Stipules broad and semi-sagittate, but very deciduous, or sometimes none. Flowers numerous, yellow, in axillary and terminal racemes of 5 or 6 in. Pedicels longer than the calyx. Bracts ovate, acute, very deciduous. Lowest sepal about 5 lines long. Petals about 6 lines. Ovary with several ovules. Pod oblong-linear, 2 to 3 in. long and nearly 1 in. broad, rounded at the end, with a long narrow point, glabrous. Seeds 6 to 8, ovoid, thick, the hilum very small at one end. —Wight, Ic. t. 37.

Queensland. Near Brisbane, *Henne.* An E. Indian species, much planted for hedges, and now naturalized in the W. Indies and some other tropical countries. It may therefore possibly be an introduced plant also in Australia.

76. MEZONEURUM, Desf.

Sepals 5, shortly united at the base, much imbricated, the lowest one larger and concave. Petals 5, spreading, rather unequal, the upper inner one the smallest, the 2 outer lower ones the largest. Stamens free; anthers uniform, ovate or oblong. Ovary with 2 or more ovules; style subulate, with a very small terminal stigma. Pod quite flat, very thin or coriaceous, indehiscent or opening tardily in 2 valves, the upper suture bordered by a wing. Seeds very flat reniform or orbicular, with a small lateral hilum; albumen none.—Woody climbers (or rarely erect?), sparingly armed with small prickles, usually only at the base of the pinnæ of the leaves. Leaves abruptly bipinnate. Flowers (yellow?) in racemes, either simple in the upper axils or forming large terminal panicles. Filaments glabrous or slightly hairy.

The genus is dispersed over tropical Asia and Africa, with one Australian species which appears to be endemic.

1. **M. brachycarpum,** *Benth.* Apparently climbing, the branches and petioles tomentose-pubescent, unarmed except a minute prickle under each raceme and a rather larger recurved one under each pinna of the leaf. Leaves often above 1 ft. long; pinnæ 3 to 8 pairs, each 3 to 4 in. long; leaflets 4 to 10 pairs, ovate-oblong, very obtuse or retuse, ½ to ¾ in. long, very oblique at the base, glabrous above, pubescent underneath. Racemes paniculate, about ½ ft. long, clothed with a golden-yellow pubescence; pedicels very short. Flowers much smaller than in the rest of the genus, the calyx-lobes not above 2 lines long and the petals scarcely exceeding them. Filaments rather longer, slightly bearded at the base. Style filiform, with a slightly dilated stigma. Ovules 2 (or sometimes 3?). Pod obliquely and broadly obovate or almost orbicular, nearly 2 in. long and 1½ in. broad, more coriaceous than in most species, and perhaps dehiscent, the wing of the upper suture about 2 lines broad. Seed large, solitary, very flat, reniform, with a very small hilum.

Queensland. Wide Bay, *Bidwill* (in fruit and leaves); Araucaria Range, Moreton Bay, *F. Mueller* (leaves only).
N. S. Wales. Richmond river, *C. Moore* (flowers and fragments of leaves).

77. PTEROLOBIUM, R. Br.

Sepals 5, united in a cup at the base, much imbricate, the lowest longer and concave. Petals 5, spreading, the 2 lowest rather larger than the others. Stamens 10, free; anthers ovate, uniform. Ovary sessile, with a single ovule. Style filiform or slightly clavate, with a truncate stigma. Pod sessile, samaroid, the lower seed-bearing part obliquely ovate or lanceolate, indehiscent, ending in an oblique oblong or falcate membranous wing. Seed attached near the apex of the cell, flat, without albumen.—Trees or woody climbers, armed with scattered hooked prickles, especially at the base of the pinnæ of the leaves. Leaves abruptly bipinnate. Flowers in racemes, either simple or forming terminal panicles. Filaments usually bearded.

The genus contains very few species, dispersed over tropical Asia and Africa. The Australian one is endemic, but not quite certain as to its genus until the fruit has been seen.

1. **P. nitens,** *F. Muell. Herb.* A handsome woody climber, the branchlets and rhachis of the leaves minutely rusty-pubescent. Prickles very small, except under the pinnæ of the leaves. Pinnæ 3 to 5 pairs; leaflets 3 to 5 pairs, obliquely obovate or almost rhomboid, very obtuse, rarely exceeding ½ in., shining above, glabrous or ciliate on the edge. Racemes rusty-pubescent, apparently paniculate; pedicels much shorter than in the other species, scarcely exceeding 1 line. Flowers rather small. Calyx lower lobe considerably longer than the others. Filaments bearded at the base. Style slightly clavate. Fruit not seen.

Queensland. Mount Müeller, near Edgecombe Bay, *Dallachy.* Although I have not seen the fruit, the 1-ovulate ovary, bearded stamens and style, leave little doubt that this belongs to *Pterolobium.*

78. PELTOPHORUM, Vog.

(Cæsalpinia, *sect.* Brasilettia, *DC.*)

Sepals 5, united in a cup at the base, much imbricate, nearly equal or the lowest rather larger and more concave. Petals 5, spreading, undulate, the 2 lower outer ones rather larger than the others. Stamens 10, free; anthers uniform, oblong-linear. Ovary sessile, with 2 or more ovules; style filiform, with a broad peltate stigma. Pod oblong-lanceolate, thin and flat, indehiscent, tapering at the base, the margins thin and faintly marked within them by a longitudinal nerve, but not distinctly winged. Seeds 1, 2 or rarely more, very flat, without albumen.—Tall hard-wooded trees, without prickles. Leaves twice pinnate, with numerous leaflets. Flowers yellow, in racemes forming terminal panicles.

The genus consists of two or three tropical American species, one in S.E. Africa, and one or perhaps two in the Indian Archipelago, one of which is the Australian one.

1. **P. ferrugineum,** *Benth.* A large tree, the young branches petioles and inflorescence densely rusty-tomentose. Pinnæ of the leaves 8 to 10 pairs; leaflets 10 to 20 pairs or fewer on the lowest pinnæ, oblong, very obtuse or retuse, oblique at the base, ½ to ¾ in. long, shining above, minutely rusty-tomentose underneath. Racemes 5 or 6 in. long or more, forming a large terminal panicle. Pedicels very short. Bracts small, lanceolate, deciduous. Calyx 4 to 5 lines long, globular before opening. Petals rather longer, obovate, undulate, villous at the base. Pod about 3 to 4 in. long, ¾ to nearly 1 in. broad, shortly acuminate, narrowed at the base, glabrous or nearly so, with 1, 2, or 3 seeds.—*Cæsalpinia ferruginea,* Dcne. Herb. Tim. Descr. 134; Miq. Fl. Ind. Bat. 1, part i. 111 and 1081; *C. arborea,* Zoll. in Miq. l. c. 112.

N. Australia, *Bynoe;* islands of the N. coast, *R. Brown* (in fruit); Earl De Grey's Island, *Armstrong.* In the Archipelago from Timor to the Philippines. Some Malacca specimens, perhaps belonging to a different although closely allied species, have the pods twice as long, with 3 or 4 seeds.

79. CASSIA, Linn.

(Cathartocarpus, *D. Don.*)

Sepals 5, somewhat unequal, much imbricate, the outer ones the smallest, scarcely connected at the base. Petals 5, spreading, nearly equal or the lower outer ones rather larger. Stamens usually 10, free, either all nearly equal and perfect or 2 or 3 lower ones larger or on longer filaments, and 3 or 4 upper ones reduced to small staminodia ; anthers when perfect opening at the end in pores or in short lateral slits. Ovary with several ovules, incurved, tapering into a short style. Pod cylindrical or flat, indehiscent or 2-valved. Seeds oblong or obovate, transverse, with fleshy albumen ; coty-ledons flat or rarely folded, usually cordate ; radicle short, straight.—Trees shrubs or herbs. Leaves abruptly pinnate, the leaflets opposite. Flowers yellow or very rarely reddish-purple or white, in axillary or terminal racemes or solitary. Bracts usually deciduous. Bracteoles none.

A large genus, widely distributed within the tropical and subtropical regions of both the New and the Old World, but particularly numerous in America. Of the 27 Australian species, 5 are widely spread over tropical Asia and Africa, 1 is American also found in Africa, the remaining 21 are all endemic. The genus is divided into several sections, founded chiefly upon the fruit ; but as some are only represented in Australia by single species, and the per-fect pod rarely accompanies the specimens, the chief divisions in the following synopsis are, for convenience, selected also from other characters.

A. *Flowers in pedunculate racemes or umbels, either axillary or forming a terminal panicle or compound raceme. (The Australian species all shrubs or trees.)*

Stamens 7 perfect, of which 2 or 3 lower ones larger or on longer filaments ; 3 small and imperfect staminodia.
Lower stamens with long filaments and short ovate anthers, the other perfect ones with oblong-linear anthers. Pod very long and thick, with horizontal seeds (**Cathartocarpus**) . . . 1. *C. Brewsteri.*
Perfect anthers all oblong-linear, the lower ones longer.
Racemes short, almost corymbose, axillary or in a narrow ter-minal panicle. Pod thick or turgid. Seeds mostly hori-zontal. (**Chamæfistula.**)
Leaflets 3 or 4 pairs, with glands between those of each pair, but none on the petiole below . . . . 2. *C. lævigata.*
Leaflets 4 to 10 pairs, with a gland at the base of the petiole, but none between the leaflets . . . . 3. *C. Sophera.*
Racemes short, loose, on long peduncles, forming a large ter-minal panicle. Leaflets 10 to 20 pairs, pubescent. Pod very flat (**Chamæsenna**) . . . . 4. *C. laxiflora.*
Racemes elongated, on long axillary peduncles. Bracts large, deciduous. Pod very flat. (**Chamæsenna.**)
Glabrous. Leaflets 4 to 8 pairs, large, broad, very obtuse, reticulate. Stipules broad, obtuse . . . 5. *C. magnifolia.*
Pubescent. Leaflets 9 to 15 pairs, oblong or ovate, obtuse, mucronate. Stipules ovate-cordate, acuminate, rigid. Bracts broad, obtuse . . . . 6. *C. venusta.*
Pubescent. Leaflets 9 to 15 pairs, ovate-lanceolate, acute, mucronate. Stipules narrow. Bracts acuminate . . . 7. *C. notabilis.*
Glabrous. Leaflets 4 or 5 pairs, oblong-linear. Stipules small, subulate. Bracts broad, obtuse . . . . . . 8. *C. pleurocarpa.*
Stamens 10, all with oblong-linear perfect anthers, all equal or the lower ones rather longer. (**Psilorhegma.**)
Glands between the leaflets (at least of the lowest pair), oblong, subulate or stipitate, very rarely wanting.
Bracts acuminate.
Glabrous or minutely pubescent. Leaflets 6 to 10 pairs, obovate. Bracts lanceolate, often rather broad . . . . 9. *C. suffruticosa.*
Softly pubescent. Leaflets 4 to 6 pairs, obovate or cuneate, mostly emarginate. Bracts very narrow . . . . . . 10. *C. retusa.*

Bracts small, broad, obtuse. Leaflets oblong-lanceolate or
    linear.
Leaflets usually 6 to 10 pairs (Eastern species) . . . . 11. *C. australis.*
Leaflets usually 3 to 5 pairs (Western species) . . . . 12. *C. Chatelainiana.*
Glands between the leaflets sessile, flat, obscure or none (ovoid in
    C. leptoclada).
Very glutinous, otherwise glabrous. Leaflets usually 4 or 5
    pairs . . . . 13. *C. glutinosa.*
Glabrous or glaucous. Stipules leafy, semicordate. Leaflets
    usually 3 to 5 pairs . . . . 14. *C. pruinosa.*
Glabrous, glaucous, hoary, or white-tomentose. Stipules small
    subulate or none. Flowers in very short corymbose ra-
    cemes.
Leaves all simple, phyllodineous. Glands none or on the
    upper edge about the middle.
Leaves usually slender and green. Peduncles 1- or 2-
    flowered. Pod very much curved or annular . . . . 15. *C. circinata.*
Leaves usually thick, hoary or white, vertically compressed.
    Peduncles several-flowered. Pod straight or slightly
    curved . . . . 16. *C. phyllodinea.*
Leaflets 1 or more pairs, rarely none in the lower leaves and
    then the phyllodineous petiole has a gland at the end.
Leaflets mostly 1 or 2 pairs, terete or linear . . . . 17. *C. eremophila.*
Leaflets mostly 3 to 6 pairs, linear-terete, channelled above 18. *C. artemisioides.*
Leaflets mostly 3 to 6 pairs, linear-lanceolate, cuneate,
    elliptical or almost obovate . . . . 19. *C. Sturtii.*
Leaflets 1, 2, or rarely 3 pairs, ovate, obovate, or broadly
    oblong. Pod not above ½ in. broad . . . . . . 20. *C. desolata.*
Leaflets 2 or rarely 1 pair, broadly obovate. Pod nearly
    ¾ in. broad, very obtuse . . . . 21. *C. oligophylla.*
Softly pubescent. Leaflets 2 to 4 pairs, elliptical-oblong. Sti-
    pules small, setaceous. Flowers in an umbel of 4 to 6 . . 22. *C. oligoclada.*
Glabrous and glaucous. Stems slender. Leaflets 2 pairs, obo-
    vate or oblong. Glands ovoid. Stipules minute. Pedun-
    cles 2-flowered . . . . 23. *C. leptoclada.*

D. *Flowers in simple racemes, either terminal or becoming lateral by the elongation of
the branch. Stamens 5 to 10, all perfect. Pod flat.* (**Absus.**)
Herbaceous. Leaflets 2 pairs, obovate . . . . 24. *C. Absus.*

C. *Peduncles 1-flowered, solitary or 2 or 3 together in or just above the axils. Stamens
5 to 10, all perfect. Pod flat.* (**Chamæcrista.**)
Leaflets usually under 12 pairs. Gland stipitate below the lowest
    pair. Sepals rather obtuse. Anthers 5.
Petals scarcely longer than the calyx. Stigma peltate . . . . 25. *C. pumila.*
Petals twice as long as the calyx. Stigma small . . . . . . 26. *C. concinna.*
Leaflets above 20 pairs. Sepals very acute. Anthers 6 to 10 . . 27. *C. mimosoides.*

Besides the above there are in Dallachy's Queensland collection fragments of a tree, apparently of some species of *Cassia* of the sections *Cathartocarpus* or *Chamæfistula*, but insufficient for identification or description.

C. tomentosa, Linn. f., a shrubby species allied to *C. Sophera* but more tomentose, with
more oblong leaflets and a flatter pod, occurs in some Australian collections as introduced or
cultivated.

SECTION I. CATHARTOCARPUS.—Sepals obtuse. Stamens 10, 3 or 4 upper
ones small and imperfect; 5 or 4 equal, perfect, with short filaments and oblong or linear anthers, opening in terminal pores and sometimes also in short
slits; 2 or 3 lower ones with long filaments and short ovate anthers opening

on the inner face in short slits. Pod long, hard, thick or terete, usually indehiscent. Seeds more or less flattened and lying horizontally in the pod
(the flat sides at right angles to the valves), separated by complete partitions.
—Usually trees. Flowers in axillary pedunculate racemes.

1. **C. Brewsteri,** *F. Muell.* 4*th Ann. Rep.* 17. A tree, attaining 30 to
40 ft., usually glabrous in all its parts. Leaflets 2 to 4 pairs, from narrow-
ovate or obovate and about ¾ in. long to narrow-oblong or oblong-lanceolate
and 2 in. long, obtuse or emarginate, narrowed at the base, the common petiole without glands. Racemes 3 to 6 in. long. Bracts minute or none.
Pedicels slender. Sepals about 3 lines long. Petals stipitate, narrow-ovate,
rather obtuse, about 4 lines long. Filaments of the 3 long lower stamens
longer than the petals, swollen into a globular appendage about the middle,
with ovate anthers, the other stamens shorter than the petals. Pod (only one
seen) nearly 1 ft. long, about 8 lines wide, thick but slightly compressed, the
edges persistent after the inside has fallen away. Seeds thick, ovoid, the
testa pulpy when soaked; albumen copious.—*Cathartocarpus Brewsteri,* F.
Muell. Fragm. i. 110.

Queensland. Hilly pastures and river-banks on the Burdekin, *F. Mueller :* Rockhampton, *Thozet ;* Port Denison and Fitzroy river, *Bowman.*

Var. *tomentella.* Branches, under side of the leaflets and inflorescence minutely hoary-
tomentose. Leaflets short and broad. Flowers rather small.—Castle Creek, *Bowman.*

The seeds of this and some other species of *Fistula* and *Chamæfistula* appear to be flattened at right angles to the embryo, which, as in the other sections of *Cassia,* lies thus parallel to the valves. In others, such as the African *C. goratensis,* I have seen the cotyledons
so folded as to have no particular relative position, but I have as yet been able to examine
but very few perfect seeds in either of these sections.

Section II. CHAMÆFISTULA.—Sepals obtuse. Stamens 10, 3 upper ones
small and imperfect, 7 perfect, the 2 or 3 lower ones often larger or on longer
filaments than the others ; anthers oblong-linear, the cells opening in terminal
pores. Pod terete or turgid, or if compressed thick, woody coriaceous. or
membranous, indehiscent or 2-valved. Seeds some or all more or less flattened and lying horizontally in the pod (at right angles to the valves), separated by complete or incomplete partitions or pulp.—Shrubs, or in species not
Australian, tall herbs. Flowers in axillary pedunculate racemes or terminal
panicles.

2. **C. lævigata,** *Willd. ; Vog. Syn. Cass.* 19. An erect glabrous shrub
of several feet. Leaflets 3 or 4 or rarely 2 pairs, ovate to lanceolate, usually
acuminate, 1½ to 3 in. long, with an oblong or slender gland between those
of each pair. Racemes axillary, pedunculate, short and almost corymbose,
the upper ones forming a short terminal panicle. Sepals unequal, the inner
ones 4 or 5 lines long. Petals broad, very obtuse, varying from ½ to ¾ in.
Perfect anthers 4, almost sessile, 1 on a short and 2 on much longer filaments.
Pod 2 to 3 in. long, membranous or slightly coriaceous, cylindrical or more
or less inflated when ripe, 2 to 3 in. long, opening at length in 2 valves. Seeds
crowded and horizontal or the upper ones less crowded and almost vertical.—
F. Muell. Fragm. iv. 14.

Queensland. Near Brisbane, *Herb. F. Mueller.*
N. S. Wales. Hastings and Clarence rivers, *Beckler.*

A common species in tropical America, occurring also in tropical Africa, but probably introduced there and perhaps not really indigenous in Australia.

3. **C. Sophera,** *Linn.; Vog. Syn. Cass.* 20, var. *schinifolia.* An erect shrub or undershrub of several feet, usually glabrous. Leaflets 4 to 10 pairs, lanceolate, mostly acute, 1 to 2 in. long, with an obovate or ovate acute gland on the petiole near the base. Racemes short and few-flowered, on short peduncles in the upper axils, and forming a narrow terminal almost raceme-like panicle. Sepals 3 to 4 lines long. Petals broad, obtuse. Perfect anthers 2 larger than the others, all on short filaments. Pod 2 to 4 in. long, at first flat but thick, when ripe terete or turgid, 2-valved. Seeds crowded and mostly or all horizontal.—*C. schinifolia,* A. DC. 7th Not. Pl. Rar. Hort. Gen. 35 ; *C. Barclayana,* Sweet, Fl. Austral. t. 32 ; Vog. Syn. Cass. 45 ; F. Muell. Fragm. iv. 14.

Queensland. Burdekin river, *F. Mueller;* near Fort Cooper, *Thozet ;* Moreton Bay, *Dallachy, C. Stuart ;* Ipswich, *Nernst.*

N. S. Wales. Hunter's River, *R. Brown ;* New England, *C. Stuart ;* Hastings river, *Tozer.*

Var. *pubescens.* Branches petioles and inflorescence more or less pubescent. Petiolar gland occasionally disappearing.—Broad Sound, *R. Brown, Bowman;* Ottley's Station, *Leichhardt ;* Paramatta, *Woolls.*

The species, in its glabrous form, is common in E. India and the Archipelago and in tropical Africa. It is there often confounded with *C. occidentalis,* of which I had formerly considered it a variety, and under which it is included in Hooker and Thomson's Indian distributions. The latter species is, however, annual, with the few leaflets of *C. lævigata,* but with the petiolar gland of *C. Sophera,* and the pod remains flat, although thick when ripe.

SECTION III. CHAMÆSENNA. Sepals obtuse. Stamens of *Chamæfistula.* Pod very flat and thin. Seeds flattened parallel to the embryo, and lying vertically in the pod (parallel to the valves), separated by more or less complete partitions or thin pulp. Shrubs. Flowers in axillary pedunculate racemes or terminal panicles.

4. **C. laxiflora,** *Benth.* A tall erect shrub; softly tomentose-pubescent. Leaflets 10 to 20 pairs, elliptical-oblong, ¾ to 1 in. long, without any glands on the common petiole. Stipules inconspicuous. Flowers in loose racemes pedunculate in the upper axils, the upper ones forming a loose pyramidal terminal panicle. Bracts inconspicuous: Sepals obtuse, the inner ones fully 3 lines long. Petals not twice as long. Perfect anthers 7, nearly equal, on very short filaments, opening by terminal pores, 3 very small and empty. Style slightly thickened at the end with a truncate stigma. Pod stipitate glabrous, thin and flat, acuminate, about 3 in. long and 4 lines wide.

N. Australia. Arnhem N. Bay, *R. Brown.*

5. **C. magnifolia,** *F. Muell. Fragm.* i. 166. Glabrous. Leaflets 4 to 8 pairs, broadly ovate, very obtuse and emarginate, broad and oblique at the base, 2 to 3 in. long, coriaceous and strongly veined on both sides, the common petiole ½ to 1 ft. long; glands between the leaflets obscure or none. Stipules persistent, ovate, the margins recurved at the base. Peduncles attaining 1 ft. in length, rigid, bearing a raceme in the upper part. Flowers not seen. Fruiting pedicels 1 in. long. Pod very flat, 3 to 4 in. long, ¾ in. broad, glabrous, with thin valves. Seeds flat, on slender funicles; albumen scanty.

Queensland. Rocky granite ridges, Upper Gilbert river, *F. Mueller.*

6. **C. venusta,** *F. Muell. Fragm.* i. 165. A tall shrub or small tree, the young parts softly silky-pubescent or villous, becoming at length nearly glabrous. Leaflets 10 to 15 pairs, or in smaller specimens 7 to 10 pairs, oblong or ovate-oblong, obtuse and finely mucronate, 1 to 2 in. long, very obliquely rounded at the base, rather coriaceous; glands very small between the leaflets of most pairs. Stipules ovate-cordate, acuminate, rigid and persistent, especially at the base of the peduncles, the margins usually revolute at the base. Peduncles in the upper axils ½ to 1 ft. long, rigid, bearing in their upper portion a raceme of flowers on short pedicels. Bracts membranous, orbicular, imbricate before flowering, but soon falling off. Sepals 5 to nearly 6 lines long. Petals rather longer. Perfect anthers 7 on short filaments, 2 of them nearly twice as large as the others, 3 small imperfect stamens. Ovary villous. Pod very flat, about 3 in. long and ½ in. broad. Seeds flat, rather distant.

N. Australia. Cambridge Gulf and Dampier's Archipelago, *A. Cunningham;* granite hills, Nichol Bay, and Hammersley Range, *F. Gregory's Expedition;* sandstone table-land, Arnhem's Land, *F. Mueller;* Attack Creek, *M'Douall Stuart's Expedition;* raised also in the Melbourne garden from seeds gathered in *M'Kinlay's Expedition.*

7. **C. notabilis,** *F. Muell. Fragm.* iii. 28. Villous with long soft hairs. Leaflets 9 to 15 pairs, ovate-lanceolate or oval-oblong, acute or the lower ones obtuse and mucronate, sessile, very obliquely rounded or truncate at the base, 1 to 1½ in. long; glands very small between the leaflets of most of the pairs. Stipules narrow and deciduous. Racemes on elongated peduncles in the upper axils. Bracts lanceolate, acuminate, very deciduous. Pedicels short. Sepals villous, about 3 lines long. Petals shortly exceeding the calyx. Perfect anthers 7 on very short filaments, of which 2 larger than the others; 3 small imperfect stamens. Ovary glabrous. Pod not seen.

N. Australia. Between Bonney river and Mount Morphett, *M'Douall Stuart's Expedition,* and probably the same species, leaves only, from Strangways River.

8. **C. pleurocarpa,** *F. Muell. Fragm.* i. 223 *and* ii. 182. A tall erect glabrous shrub. Leaflets usually 4 or 5 rather distant pairs, oblong-linear, 1½ to 2 in. long, rather thick; glands none. Stipules small, subulate, deciduous. Flowers loosely racemose in the upper portion of axillary peduncles. Bracts membranous, broad, obtuse, very deciduous. Sepals thin, broad, 3 to 3½ lines long. Petals unequal, rather longer than the calyx. Perfect anthers 7, on short filaments, 2 of them nearly twice as large as the others and incurved; 3 small imperfect stamens. Ovary glabrous. Pod stipitate, flat, very obtuse, about 2 in. long and ½ in. wide, the valves thin with a raised longitudinal line along the centre, interrupted between each seed. Seeds thick, cuneate-oblong, truncate, with a raised line across each near the end, corresponding to that on the pod; albumen copious.

N. S. Wales. Mount Goningberi, Barrier Range, *Victorian Exploring Expedition.*
W. Australia. Murchison river, *Oldfield.*

SECTION IV. PSILORHEGMA. Sepals obtuse. Stamens 10, all perfect and similar or the lower ones rather larger; anthers oblong-linear, opening in slits either short and terminal or extending more or less down the sides. Pod very flat and thin. Seeds flattened parallel to the embryo and lying vertically in the pod (parallel to the valves) separated by more or less complete partitions or thin pulp. Shrubs. Flowers in very short corymbose racemes or umbels pedunculate in the axils, rarely reduced to 2 flowers.

9. **C. suffruticosa,** *Kœn.; Vog. Syn. Cass.* 30. A tall weak shrub, quite glabrous or the young branches inflorescence and under side of the leaves pubescent. Leaflets 6 to 10 pairs, obovate or broadly oblong, obtuse, mostly 1 to 1½ in. long; glands oblong or slender, usually stipitate, between those of the 1, 2 or 3 lowest pairs. Stipules linear or subulate. Flowers in short umbel-like racemes in the upper axils. Bracts lanceolate, acuminate, rather persistent. Sepals very obtuse, the inner ones 3 to 4 lines long. Petals broad, ½ in. long or more, 2 or 3 lower ones rather larger than the others. Anthers all on short filaments, 2 or 3 rather larger than the others. Pod 3 to 4 in. long, 4 to 5 lines broad.—W. and Arn. Prod. 289; *C. acclinis,* F. Muell. Fragm. iv. 13.

N. Australia. Islands of the N. Coast, *R. Brown.*
Queensland. Percy islands, *A. Cunningham;* Rockhampton, *Dallachy;* Edgecombe Bay and Port Denison, *Fitzalan;* Ipswich, *Nernst.*
N. S. Wales. Hastings river, *Beckler.*
Although more variable in aspect than the Indian specimens, some of the Australian specimens, especially some of Brown's, cannot be distinguished from Asiatic ones. The species is said to be in cultivation only in E. India, but indigenous in the Archipelago. It is very closely allied to, and perhaps a variety of *C. glauca.*

10. **C. retusa,** *Soland.; Vog. in Linnæa,* xv. 72. Shrubby and softly pubescent, especially the young parts. Leaflets 4 to 6 pairs, obovate to oblong-cuneate, very obtuse or emarginate, ½ to above 1 in. long; glands slender or stipitate between those of the 1, 2 or 3 lowest pairs. Stipules linear, acuminate, deciduous. Flowers crowded in short almost umbellate racemes, on axillary peduncles shorter than the leaves. Bracts narrow, acuminate. Sepals broad, very obtuse. Petals not twice as long. Anthers all nearly equal. Pod stipitate, 2 to 4 in. long, 4 to 5 lines broad.

Queensland. Bustard Bay, *Banks and Solander;* Shoalwater Bay, Broad Sound and Thirsty Sound, *R. Brown.* The species is closely allied on the one hand to *C. suffruticosa,* on the other to some forms of *C. australis.*

11. **C. australis,** *Sims, Bot. Mag. t.* 2676. A tall erect shrub, either quite glabrous or loosely pubescent, the young branches more or less angular. Leaflets usually 8 to 10 pairs, in some specimens reduced to 6 or 7, in others increased to 11 or 12 pairs, oblong lanceolate or almost linear, obtuse or acute, ½ to ¾ or rarely 1 in. long, the margins usually recurved and sometimes revolute; glands slender or stipitate between the leaflets of most or only of the lower pairs, or rarely almost none. Stipules subulate, deciduous. Flowers 2 to 6 in a loose umbel on peduncles usually shorter than the leaves, but sometimes longer. Bracts small, broad, obtuse. Sepals very obtuse, 2 to 3 lines long. Petals broad, ½ in. long or rather more. Anthers 2 or 3 often rather larger than the others. Pod shortly stipitate, glabrous, 3 to 4 in. long, 3 to 4 lines broad, straight or curved into a half-circle. Seeds shining black. —Bot. Reg. t. 1322; *C. umbellata,* Reichb. Icon. Exot. t. 206; *C. Schultesii,* Colla, Hort. Ripul. App. ii. 344, and iii. t. 10; *C. Barrenfieldii* (afterwards corrected to *C. Fieldii*), Colla, Hort. Ripul. App. iv. 23, t. 11; *C. coronilloides,* A. Cunn.; Benth. in Mitch. Trop. Austr. 384.

Queensland. Broad Sound, *R. Brown, A. Cunningham;* Burdekin river and Peak Downs, *F. Mueller;* Rockhampton, *Thozet;* Comet and Condamine rivers, *Leichhardt.*

N. S. Wales. Paramatta, *Woolls;* Blue Mountains and Hunter's River, *R. Brown, A. Cunningham,* and others; New England, *C. Stuart;* Hastings, Macleay and Clarence rivers, *Beckler;* Boyd river, *Leichhardt.*
Victoria. Gipps' Land, *F. Mueller.*
Var. *revoluta.* Leaflets narrow-linear and acute, the margins much revolute, glabrous or pubescent.—*C. revoluta,* F. Muell. in Trans. Vict. Inst. 1852, 120; *C. aciphylla,* Benth. in A. Gray, Bot. Amer. Expl. Exped. i. 465. To this belong most of the southern and several of the N. S. Wales specimens from the interior.
Var. *pedunculata.* Peduncles much longer than the leaves.—St. George's River, *R. Brown;* Blue Mountains, *A. Cunningham,* and others.
Var. (?) *glaucescens.* Slender and glaucous. Leaflets few.
N. Australia. Hooker's Creek, *F. Mueller.*

12. **C. Chatelainiana,** *Gaud. in Freyc. Voy.* 485, *t.* 111. An erect glabrous shrub of several ft. Leaflets 3, 4 or 5 rather distant pairs, linear, obtuse, about ¾ to 1 in. long, rather thick, flat; gland usually long and subulate between those of the lowest pair only, but sometimes also of the next pair, or rarely wanting. Flowers rather large, in umbels at the end of short axillary peduncles with sometimes 1 or 2 pedicels below the umbel. Bracts ovate or oblong, very obtuse. Sepals nearly 3 lines long. Petals broad, ½ to ¾ in. long. Lower stamens rather larger than the others. Pod straight, often ½ in. broad.—Vog. Syn. Cass. 47.

W. Australia. Sharks Bay, *Gaudichaud;* Murchison River, *Oldfield;* also *Drummond.* Very nearly allied on the one hand to *C. australis,* in which, however, the leaflets when linear are usually revolute, and on the other to *C. eremicola,* which has fewer leaflets without the subulate gland.

13. **C. glutinosa,** *DC. Prod.* ii. 495. An erect glabrous shrub, the specimens very glutinous and brittle. Leaflets usually 4 or 5 pairs, oblong-linear, rather obtuse, ½ to 1 in. long or rather more, flat and rather thick; gland flat and broad between the leaflets of the lowest 1 or 2 pairs. Flowers umbellate on axillary peduncles usually shorter than the leaves. Sepals coloured, obtuse, 3 to 3½ lines long. Petals twice as long. Anthers 2 or 3 lower ones upon rather longer filaments than the others. Pod straight, about 4 lines broad, as glutinous as the rest of the plant.—Vog. Syn. Cass. 47.

N. Australia. Attack Creek, *M'Douall Stuart's Expedition.* Described by De Candolle probably from specimens gathered on the N. coast in Baudin's Expedition, but I have not seen them.

14. **C. pruinosa,** *F. Muell. Fragm.* iii. 48. A tall erect shrub, glabrous but often more or less glaucous. Leaflets 3 to 5 pairs, oblong-elliptical, rather obtuse, ½ to ¾ in. long, rather thick and flat; gland small and flat, usually between the leaflets of the lowest 1 or 2 pairs. Stipules broad, leafy, semi-cordate. Flowers rather large, 2 to 5 together in umbels on axillary peduncles usually shorter than the leaves. Sepals coloured, obtuse, the inner ones fully 3 lines long. Petals twice as long. Anthers 3 or 4 rather longer than the others. Pod stipitate, straight, apparently about 4 lines broad, but not seen perfect.

N. Australia. N.W. coast, *Bynoe;* Rocky Hills, Nichol Bay, *F. Gregory's Expedition.*
N. S. Wales. Between Stokes Range and Cooper's Creek, *Wheeler.*

15. **C. circinata,** *Benth. in Mitch. Trop. Austr.* 384. An erect bushy shrub of several ft., glabrous or hoary with a minute silky tomentum. Leaves

all phyllodineous without leaflets, linear-terete, often almost filiform or very slightly vertically flattened, 1 to 1½ in. long, often clustered on the nodes of the previous year's wood, and then sometimes not half so long; gland none, or very obscure at or above the middle of the phyllodium. Peduncles short, axillary, bearing 1 or 2 flowers on slender pedicels. Bracts minute. Sepals obtuse, 2 to 2½ lines long. Petals twice as long. Anthers 2 or 3 rather larger than the others and on longer filaments. Pod fully 5 lines broad, very flat and thin as in the rest of the section, but usually curved into a complete circle.—R. Br. in App. Sturt, Exped. 15.

Queensland. Balonne river, *Mitchell;* Burdekin river, *F. Mueller;* Suttor river, *D'Orsay;* Edgecombe Bay, *Dallachy.*

N. S. Wales. Mount Flinders, *A. Cunningham;* Darling river to Barrier Range, *Victorian Exploring Expedition.*

16. **C. phyllodinea,** *R. Br. in App. Sturt, Exped.* 15. An erect rigid bushy shrub, hoary or white with a close silky tomentum. Leaves all phyllodineous, linear, vertically compressed but thick, obliquely obtuse truncate or even shortly 2-lobed at the end, 1 to 1½ in. long, narrowed at the base; gland none, or a faint one on the upper edge. Peduncles short, axillary, bearing a very short raceme of several flowers on slender pedicels. Bracts very small. Sepals obtuse, 2 to 2½ lines long. Petals twice as long. Anthers 2 or 3 rather larger than the others. Pod stipitate, straight or slightly curved, 5 to 6 lines broad, obtuse.

N. S. Wales. Lachlan and Darling desert to Barrier Range and Cooper's Creek, *Victorian Exploring Expedition, Dallachy and Goodwin,* and others.

S. Australia. S. coast, *R. Brown;* Flinders range, *F. Mueller.*

F. Mueller is disposed to consider this and the preceding phyllodineous species, together with the five following ones as forms of one species, and it is true that we occasionally meet with specimens apparently connecting them, but so it is with the whole of the section from *C. glauca* to *C. circinata,* which we certainly should not be justified in uniting. Those specimens of *C. eremophila,* var. *platypoda,* in which the lower leaves are phyllodineous without leaflets, can generally if not always be distinguished from *C. phyllodinea* by the glands at the end of the phyllodia where the leaflets have aborted.

17. **C. eremophila** (by a clerical error **nemophila**), *A. Cunn. in Vog. Syn. Cass.* 47. An erect bushy shrub, glabrous or slightly hoary but never so white as some of the allied species. Leaflets 1 or 2 pairs, very narrow-linear, thick, terete and channelled above or slightly flattened out, sometimes very short, usually about 1 in. long, and often more, the petiole terete or vertically flattened; gland depressed between the lowest or the only pair; the lower leaves sometimes reduced to a flattened phyllodium with the gland at the end where the leaflets have aborted. Peduncles short, or rarely as long as the leaves, bearing a short almost corymbose raceme of several flowers on slender pedicels. Bracts very small. Sepals obtuse, rarely 2 lines long. Petals usually more than twice as long. Anthers 2 or 3 lower ones rather larger or on longer filaments than the others. Pod straight or slightly curved, 3 to 4 lines broad or rarely more.—R. Br. in App. Sturt. Exp. 14; *C. canaliculata,* R. Br. l. c.; *C. heteroloba,* Lindl. in Mitch. Three Exped. ii. 122.

Queensland. On the Maranoa, *Mitchell;* desert of the Suttor and Burdekin, *F. Mueller.*

N. S. Wales. Near Port Jackson, *Herb. F. Mueller;* New England, *C. Stuart;*

common in the desert interior from the Lachlan and Darling to the Barrier Range, *A. Cunningham, Victorian Exploring. Expedition,* etc.

Victoria. Murray desert and Wimmera, *F. Mueller, Mitchell, Dallachy,* and others.

S. Australia. S. coast, *R. Brown ;* from the Murray to Flinders' Range and Spencer's Gulf, *F. Mueller ;* Venus Bay and Mount Serle, *Warburton.*

W. Australia, *Drummond, Roe ;* Stirling Range, Phillips Range, etc., *Maxwell.*

A very variable species, of which specimens occur occasionally with here and there an additional pair of leaflets, showing an approach towards *C. artemisioides,* and some of the western ones with the gland rather more prominent are at first sight like reduced forms of *C. Chatelainiana.* The two following varieties which have been distinguished as species, are very inconstant; they both occur mixed with the common form.

Var. *platypoda.* Petioles vertically compressed, the lower ones often without leaflets.— *C. platypoda,* R. Br. in App. Sturt Exped. 15.

Var. *zygophylla.* Leaflets 1 or 2 pairs, linear, flat, often 1 to 2 lines broad.—*C. zygophylla,* Benth. in Mitch. Trop. Austr. 288.

18. **C. artemisioides,** *Gaud. in DC. Prod.* ii. 495. An erect bushy shrub, hoary or white with a minute silky tomentum. Leaflets 3 to 6 pairs, linear-terete and more or less channelled above, slender but rigid, usually $\frac{3}{4}$ to 1 in. long, but sometimes longer or shorter ; glands small and flat between those of the lower 1 or 2 pairs. Flowers in a short dense raceme on peduncles much shorter than the leaves. Bracts small, ovate. Sepals obtuse, 2 to $2\frac{1}{2}$ lines long. Petals about twice as long. Anthers 2 or 3 longer than the others on longer filaments. Pod straight, 2 to 3 in. long, about 4 lines broad. —*C. teretifolia,* Lindl. in Mitch. Three Exped. i. 289 ; *C. teretiuscula,* F. Muell. in Linnæa, xxv. 389.

Queensland. Dawson river, *F. Mueller.*

N. S. Wales. In the interior, *Fraser ;* near Mount Flinders, *A. Cunningham ;* in the Darling desert, *Mitchell* and others, and thence to the Barrier Range, *Victorian Exploring Expedition* and others.

S. Australia. Near Cudnaka and towards Lake Torrens, *F. Mueller ;* Mount Serle, *Warburton.*

Nearly allied to *C. eremophila* and *C. Sturtii,* this differs from the former chiefly in the more numerous leaflets, from the latter in their shape and in the narrower pod. If the three were united, it is the name of *C. artemisioides* that has the priority.

19. **C. Sturtii,** *R. Br. in App. Sturt Exped.* 14. A bushy shrub, glabrous or more frequently glaucous hoary or white with a close tomentum. Leaflets usually 3 to 5 pairs, linear, lanceolate, cuneate, elliptical or almost obovate, $\frac{1}{2}$ to 1 in. long, thick, flat or concave, sometimes all small and almost ovate, the lower leaves rarely with only 2 pairs ; glands small between the leaflets of the lowest 1 or 2 pairs. Stipules small and deciduous as in all the allied species. Flowers in short axillary dense racemes as in *C. eremophila,* but usually more numerous on a longer peduncle. Sepals obtuse, 2 to 3 lines long, frequently tomentose. Petals twice as long. Pod when perfect fully $\frac{1}{2}$ in. broad, straight or slightly curved and very obtuse.

Queensland, *Bowen ;* Suttor river, *F. Mueller.*

N. S. Wales. Darling river to the Barrier Range and Cooper's Creek, *Victorian Exploring Expedition, Howitt's Expedition,* etc.

Victoria. Murray scrub and Wimmera, *Dallachy.*

S. Australia. Near Cudnaka, *F. Mueller.*

W. Australia. *Drummond.*

Var. (?) *coriacea.* Leaflets usually 4 or 5 pairs, small oblong or obovate, very obtuse, thick and green or glaucous.—S. coast, *R. Brown ;* Mount Flinders, *A. Cunningham ;* Darling and Murray desert and S. Australia.

Var. (?) *tomentosa.* Leaflets oblong or narrow-obovate, very white. Sepals much longer and tomentose.— Mount Murchison and Barrier Range. This may be a variety of *C. desolata.* The specimens of these forms, although numerous, are often fragmentary, and rarely have good fruit, or show at once the foliage of the barren and of the flowering branches.

20. **C. desolata,** *F. Muell. in Linnæa,* xxv. 389. Shrubby, the young parts hoary or white, becoming glabrous with age. Leaflets 1, 2 or very rarely 3 pairs, ovate obovate or broadly oblong, $\frac{1}{2}$ to 1 in. long or more, coriaceous ; gland depressed between those of the lowest or of both pairs, rarely wanting. Flowers in a very short raceme, on short axillary peduncles. Bracts ovate, concave. Sepals about 3 lines long, usually pubescent or tomentose. Petals twice as long. Anthers nearly equal or 2 or 3 lower ones scarcely longer. Pod not seen perfect but apparently more like that of *C. Sturtii* than of *C. oligophylla.*

N. Australia. Depuech Island, N.W. coast, *Bynoe ;* Victoria river, *F. Mueller ;* Central Mount Stuart, *M'Douall Stuart.*

N. S. Wales. Barrier Range, *Victorian Exploring Expedition.*

S. Australia. Spencer's Gulf, *Warburton.*

Some of the specimens are very doubtful and may belong to *C. oligophylla,* which this species closely resembles in foliage and flowers ; and it would require more perfect materials than I have seen to establish satisfactorily the distinction between this species and *C. Sturtii* on the one hand and *C. oligophylla* on the other.

21. **C. oligophylla,** *F. Muell. Fragm.* iii. 49. A tall shrub, glabrous or minutely pubescent. Leaflets 2 or rarely only 1 pair, broadly obovate, very obtuse, $\frac{3}{4}$ to 1 in. long, coriaceous ; glands depressed and rather large between those of each pair. Flowers in short dense racemes, on axillary peduncles, rather more numerous than in the preceding species, with rather larger oblong or lanceolate bracts. Sepals obtuse, pubescent, about 3 lines long. Petals not twice as long. Anthers 2 or 3 lower ones rather longer than the others. Pod 2 to $2\frac{1}{2}$ in. long, nearly $\frac{3}{4}$ in. broad, very obtuse.

N. Australia. Sandy plains, Nichol Bay, *F. Gregory's Expedition.* From the few specimens gathered, this appears to be closely allied to *C. desolata,* differing chiefly in the pod twice as broad as in *C. artemisioides ;* half as broad again as in *C. Sturtii,* and probably as in *C. desolata ;* but better specimens are required to confirm the species.

22. **C. oligoclada,** *F. Muell. Fragm.* iii. 49. A shrub of 1 to 3 ft., softly pubescent in all its parts. Leaflets 3 or 4 pairs, or in slender starved specimens only 1 or 2 pairs, elliptical-oblong, obtuse or almost acute, shortly mucronate, $\frac{3}{4}$ to 1 in. long ; glands none. Stipules small, setaceous. Flowers 4 to 6, umbellate, on slender axillary peduncles, about as long as the leaves ; pedicels almost filiform. Bracts minute, deciduous. Sepals obtuse, the largest about 2 lines long. Petals twice as long or the upper inner ones rather less. Anthers 3 a little larger than the others. Pod stipitate, falcate, pubescent, 1 to $1\frac{1}{2}$ in. long and about 4 lines broad. Seeds 4 to 6, on very short funicles.

N. Australia. Cambridge Gulf, N.W. coast, *A. Cunningham ;* Upper Victoria river, *F. Mueller ;* Gulf of Carpentaria, *R. Brown.*

Var. (?) *gracilis.* Very slender and quite glabrous. Leaflets 1 or 2 pairs. Flowers smaller.—Attack Creek, *M'Douall Stuart.*

The short broad falcate pod of this species reminds one at first sight of that of the *Sennas.*

23. C. leptoclada, *Benth.* A shrub of 3 or 4 ft., glabrous and very glaucous, with very slender often purplish branches. Leaflets 2 pairs, obovate to oblong-elliptical, very obtuse and sometimes emarginate, those of the upper pair ½ to 1 in. long, of the lower pair smaller or wanting in the lower leaves; glands small, ovoid, between those of each pair. Stipules very minute. Peduncles in the upper axils very short, bearing 2 flowers on filiform pedicels, or sometimes the peduncle adnate to the branch, the pedicels then proceeding from a little above the axil. Sepals obtuse, not 2 lines long. Petals deep yellow, above 4 lines long. Anthers 2 or 3 rather larger than the others. Pod stipitate, flat and glabrous, either nearly orbicular and 1-seeded, or 2-seeded and about ½ in. long and 4 lines broad.

N. Australia. Islands of the Gulf of Carpentaria, *R. Brown.* (*Herb. R. Br.*)

SECTION V. ABSUS.—Sepals usually obtuse. Stamens 5 to 10, all perfect and similar, the anthers opening in slits either short and terminal or extending down the sides of the cells. Pod flat, oblique and obliquely acute. Seeds lying vertically in the pod. Herbs or (in species not Australian) shrubs, often glandular-pubescent. Flowers in simple terminal racemes, becoming sometimes lateral by the elongation of the branch.

24. C. Absus, *Linn.; Vog. Syn. Cass.* 50. A viscidly pubescent much-branched annual or biennial, rarely exceeding 1 ft. Leaflets 2 pairs, obliquely and broadly obovate, obtuse, ½ to 1 in. long or rarely more, the common petiole rather long and slender; glands small between the leaflets of the lowest or of both pairs. Stipules narrow. Flowers small, in short terminal or at length lateral racemes. Bracts small, reflexed. Sepals narrow, obtuse, pubescent, about 3 lines long. Petals scarcely longer. Stamens usually 5. Style dilated at the end, with a rather broad fringed stigma. Pod 1 to 1¼ in. long and about ¼ in. broad. Seeds with very little albumen.—F. Muell. Fragm. iii. 50.

N. Australia. Upper Victoria river, *F. Mueller;* islands of the Gulf of Carpentaria, *R. Brown.*

Queensland. Bustard Bay, *Banks and Solander;* Port Denison, *Fitzalan;* Bowen river, *Bowman.*

The species is common in tropical Asia and Africa.

SECTION VI. CHAMÆCRISTA.—Sepals obtuse or acute. Stamens 5 to 10, all perfect; anthers opening in slits either short and terminal or extending more or less down the sides of the cells. Pod linear, flat, straight or falcate. Seeds lying vertically in the pod. Herbs or undershrubs. Peduncles axillary, 1-flowered, solitary or 2 or 3 together.

25. C. pumila, *Lam.; Vog. Syn. Cass.* 65. A diffuse, more or less pubescent perennial of short duration, with a hard almost woody base, the stems rarely exceeding 1 ft. Leaflets much fewer than in *C. mimosoides,* in the Australian specimens usually 8 to 12 pairs, linear-falcate, 2 to 3 lines long; gland stipitate on the petiole below the lowest pair. Pedicels axillary, solitary, shorter than the leaves, with minute bracteoles above the middle. Sepals rather obtuse, about 2 lines long. Petals scarcely exceeding the sepals. Stamens 5, nearly equal, obtuse. Style short, slightly thickened at the end, with a broadly peltate stigma. Pod narrow, 1 to 1½ in. long, oblique or slightly curved.

Queensland. Port Curtis, *M'Gillivray;* Bowen river, *Bowman.* Pedicels longer than in the Indian specimens, but the style and other essential characters are quite the same.

26. **C. concinna,** *Benth.* A diffuse perennial, woody at the base, more or less pubescent, with the habit of *C. pumila,* but at once known by the much larger flowers and by the style. Leaflets 8 to 10 or rarely 12 to 15 pairs, rather crowded, linear-falcate, mucronate, 2 to 3 lines long ; gland stipitate below the lowest pair. Peduncles 1-flowered, solitary, usually longer than the leaves, with minute bracteoles at or above the middle. Sepals 3 lines long, obtuse or minutely mucronate. Petals nearly twice as long. Stamens 5 ; anthers nearly equal. Style incurved, not thickened, with a small terminal stigma. Pod rarely above 1 in. long, about 2 lines wide.—*C. pumila,* F. Muell. Fragm. iii. 47, not of Lam.

N. Australia. Upper Victoria river, *F. Mueller.*
Queensland. Keppel Islands, *M'Gillivray;* Wide Bay, *Bidwill;* Broad Sound, *Bowman;* Rockhampton, *Thozet, Dallachy;* Moreton Bay, *C. Stuart.*
N. S. Wales. Mount Flinders, *Leichhardt.*

27. **C. mimosoides,** *Linn.; Vog. Syn. Cass.* 68. An annual or perennial of short duration, with a hard almost woody base, and numerous diffuse or ascending wiry stems, of 1 to 2 ft. or rarely more, usually pubescent. Leaves 1½ to 2 in. long ; leaflets numerous (20 to 50 pairs), linear-falcate, mucronate, seldom above 2 lines long ; gland depressed, below the lowest pair. Pedicels axillary, solitary or 2 or 3 together, unequal, but rarely above ½ in. long. Sepals very acute, above 3 lines long. Petals 3 to 4 or rarely 5 lines long. Stamens 7 to 10 ; anthers all similar, but rather unequal in size. Style slightly dilated at the end with a truncate stigma. Pod 1½ to 2 in. long, scarcely 2 lines broad, oblique or slightly curved.—F. Muell. Fragm. iii. 48.

N. Australia. Victoria river, *F. Mueller;* Port Essington, *Armstrong;* islands of the Gulf of Carpentaria, *R. Brown.*
Queensland. Broad Sound and Northumberland islands, *R. Brown;* common in the colony in moist pastures, *A. Cunningham, F. Mueller,* and others.
N. S. Wales. Clarence river, *Beckler.*

80. PETALOSTYLES, R. Br.

Sepals 5, much imbricate, somewhat unequal, scarcely connected at the base. Petals 5, spreading, nearly equal. Stamens 3 perfect ; filaments very short ; anthers linear, the cells opening inwardly in longitudinal slits ; 2 small staminodia, with acuminate imperfect anthers. Ovary nearly sessile, with several ovules ; style large and petal-like, saccate immediately above the ovary, with 3 erect lobes, 2 short ones in front, the other much longer, concave, the midrib prominent inside and terminating at the top in a small stigma. Pod flat, oblong-linear, oblique, 2-valved. Seeds ovate-oblong, compressed ; testa shining ; funicle expanded into a fleshy appendage distinct from the seed ; albumen copious ; cotyledons flat.—Shrubs. Leaves simply pinnate. Flowers yellow, on axillary peduncles.

The genus is limited to a single species, endemic in Australia, very nearly allied to *Cassia* (sect. *Chamæcrista*) and to *Labichea*, but distinguished especially by the very singular style.

1. **P. labicheoides,** *R. Br. in App. Sturt Exped.* 17. An erect, bushy, nearly glabrous, somewhat glaucous shrub of several feet, the young shoots minutely silky. Leaflets from about 11 to above 30, mostly alternate along the rhachis with an odd terminal one, narrow-oblong, mucronate, $\frac{1}{2}$ to $\frac{3}{4}$ in. long, narrowed at the base, but not oblique, thick, somewhat concave, the midrib only conspicuous underneath. Stipules narrow and very deciduous. Peduncles axillary, 1-flowered, with 2 small very deciduous bracteoles. Sepals acute, $\frac{1}{2}$ in. long, green and glabrous. Petals obovate, nearly $\frac{3}{4}$ in. long. Ovules 4 to 6. Style deep-yellow, like the petals, and not much shorter. Pod 1 to $1\frac{1}{2}$ in. long.

N. Australia. Dampier's Archipelago, *Bynoe ;* Nichol Bay, *F. Gregory's Expedition.*
Queensland. Suttor Range, rare, *F. Mueller.*
N. S. Wales. Barrier Range, *Victorian Exploring Expedition.*
S. Australia. Akava river, *F. Mueller ;* Mount Serle and towards Spencer's Gulf, *Warburton.*
Var. *cassioides.* Leaflets smaller, numerous, obovate or oblong, obtuse or retuse.—Sturt's Creek and Gulf of Carpentaria, *F. Mueller.*

81. LABICHEA, Gaudich.

Sepals 4 or 5, much imbricate, somewhat unequal, scarcely connected at the base. Petals as many as sepals, spreading, nearly equal. Stamens 2 ; filaments very short ; anthers oblong-linear, opening in terminal pores, either both alike or one of them produced into a tube exceeding the other and filled with pollen at the base only. Ovary sessile or shortly stipitate, with 2 or rarely 3 ovules, tapering into a short style, with a small terminal stigma. Pod oblong or lanceolate, oblique, flat, 2-valved. Seeds obovate or oblong, with a hard shining testa; funicle (in the species examined) expanded below the top into a globular fleshy appendage ; albumen copious ; cotyledons flat. —Shrubs or undershrubs. Leaves unequally pinnate, or from the common petiole not being developed, consisting of 3 or 5 digitate leaflets or reduced to the terminal leaflet. Stipules small, deciduous. Flowers yellow, few together, in short loose axillary racemes. Bracts small and deciduous. Bracteoles none.

The genus is limited to Australia, and is very nearly allied to *Cassia.*
Sepals 5. Petals 5. Anthers unequal.
  Leaves pinnate, with an elongated common petiole . . . . . . 1. *L. cassioides.*
  Leaflets digitate, without any common petiole . . . . 2. *L. nitida.*
Sepals 4. Petals 4.
  Bushy shrub. Leaves simple or digitate. Anthers unequal . . . 3. *L. lanceolata.*
  Bushy shrub. Leaves simple or digitate. Anthers equal . . . . 4. *L. rupestris.*
  Undershrub with nearly simple stems. Leaves all simple. Anthers equal . . . . 5. *L. punctata.*

1. **L. cassioides,** *Gaud. in DC. Prod.* ii. 507. A bushy glabrous shrub. Leaves pinnate, with the common petiole $\frac{1}{2}$ to 1 in. long; leaflets 5 to 15, or reduced to 3 in the upper leaves, narrow-oblong or linear, $\frac{1}{4}$ to 1

in. long, coriaceous, with a pungent point. Racemes loose, often as long as the leaves. Sepals 5, the outer ones 5 lines long, the innermost rather shorter and more petal-like. Petals 5, rather longer than the calyx, nearly orbicular. Anthers incurved, one oblong-linear, the other much longer and polliniferous at the base only. Ovary silky-pubescent, with 2 ovules. Pod not seen, but represented in Gaudichaud's figure like that of the other species, except that there is no swelling of the funicle.—Gaud. in Freyc. Voy. 485, t. 112 ; *L. tephrosiæfolia*, Meissn. in Bot. Zeit. 1855, 12.

W. Australia. Sharks' Bay, *Gaudichaud;* between Moore and Murchison rivers, *Drummond, 6th Coll. n.* 7.

2. **L. nitida,** *Benth.* A rigid shrub, with divaricate slightly pubescent branches. Leaflets usually 5, digitate, without a common petiole, from obo-vate-oblong to elliptical, obtuse with a pungent point, coriaceous and shining above, the central one ½ to 1 in. long, the lateral ones smaller, all shortly pe-tiolulate. Racemes short and loose. Flowers much larger than in *L. rupes-tris.* Sepals 5, about 4 lines long. Petals 5, the lower ones fully ½ in. long, the upper ones rather smaller. One anther at least half as long again as the other. Ovary very villous, with 3 ovules in both the flowers examined. Pod not seen.

N. Australia. Victoria river, *Bynoe.*
Queensland, *Burdekin Expedition.*

3. **L. lanceolata,** *Benth. in Hueg. Enum.* 41. An erect glabrous shrub, of 6 ft. or more. Leaflets sometimes all or mostly solitary, from narrow-ob-long to lanceolate or linear, obtuse or acute, with a pungent point, 1½ to 4 in. long, coriaceous and shining, on a petiolule of 1 to 2 lines, at the base of which are occasionally 1 or 2 very small leaflets ; in other specimens the cen-tral leaflet is much smaller, and there are 1 or 2 less disproportionate ones on each side. Racemes loose, nearly as long as the leaves. Sepals 4, about 4 lines long. Petals 4, longer than the calyx. One anther nearly twice as long as the other, without pollen above the middle. Ovules 2. Pod about 1 in. long. Seeds with a testa marked with longitudinal rows of glandular dots, and becoming swollen and pulpy when soaked.—*L. diversifolia*, Meissn. in Pl. Preiss. i. 23 ; Lindl. and Paxt. Fl. Gard. t. 52 ; *L. bipunctata*, Paxt. Mag. x. 150, with a fig.

W. Australia. King George's Sound to Swan River, *A. Cunningham, Fraser, Huegel, Drummond, 1st Coll. n.* 161, *Preiss, n.* 1027, and others; Murchison river, *Oldfield.*

4. **L. rupestris,** *Benth. in Mitch. Trop. Austr.* 342. A small hard bushy shrub, the branches pubescent or at length glabrous. Leaflets in some specimens mostly 3-foliolate; the terminal one linear-oblong, 1 to 2 in. long, coriaceous, with a pungent point, the lateral ones much smaller, in other spe-cimens most or all digitate, with 3 or 5 less unequal leaflets, without any com-mon petiole. Racemes short dense and few-flowered. Sepals 4, about 2¼ lines long. Petals about the same length. Anthers both nearly of the same size, scarcely shorter than the petals. Ovules 2. Pod short, acuminate, fre-quently 1-seeded only, but not seen ripe.—*L. digitata*, Benth. in Mitch. Trop. Austr. 273.

Queensland. Sandstone rocks and ravines about Mount Pluto, *Mitchell;* Newcastle

Range, *F. Mueller* (in leaf only). The two forms I had distinguished, with 1-foliolate or very unequally 3-foliolate leaves, and with nearly equal 5-foliolate leaves, may be found on different branches of the same specimen.

5. **L. punctata,** *Benth. in Lindl. Swan Riv. App.* 15. An undershrub, with ascending or erect almost simple stems, of 1 to 1½ ft., more or less flattened and usually glabrous. Leaves all simple, the lower ones sometimes ovate, 1 to 2 in. long, the upper ones lanceolate or linear, attaining often 4 or 5 in., coriaceous, reticulate and scabrous above, with minute dots, which however are often visible in some other species, glabrous or slightly pubescent underneath. Racemes very short and several-flowered. Sepals 4, about 4 lines long, the outer ones concave and rather acute, the inner more petal-like. Petals 4, rather longer than the calyx. Anthers both of the same size. Ovary slightly pubescent, with 2 ovules. Pod not seen.—Meissn. in Pl. Preiss. i. 24.

W. Australia. Swan River and Darling Range, *Collie, Drummond,* 1st *Coll. and* 2nd *Coll. n.* 279, *Preiss, n.* 1025, 1026 ; Harvey River, *Oldfield.*

82. TAMARINDUS, Linn.

Sepals 4, united at the base into a turbinate tube, the free portion or segments much imbricate. Petals 3, the lateral ones ovate, the upper inner one narrower, concave. Stamens incurved, united in a sheath to the middle, 3 or rarely 2 only perfect, with ovate anthers, 4 or 5 others reduced to short teeth at the top of the sheath. Ovary stipitate, with several ovules ; style inflexed, rather thick, with a truncate stigma. Pod linear or oblong-linear, curved, thick, but slightly compressed, the epicarp crustaceous and fragile, the mesocarp pulpy, the endocarp thick and fleshy, forming complete partitions between the seeds. Seeds broadly obovate, flattened ; testa rather thick ; albumen none ; embryo straight, with a short radicle.—Tree. Leaves abruptly pinnate. Flowers in terminal racemes.

The genus contains only one species, widely spread over tropical regions.

1. **T. indicus,** *Linn. ; DC. Prod.* ii. 488. A hard-wooded tree, with a spreading head and a pale or glaucous glabrous foliage. Leaflets 10 to 20 pairs, oblong-linear, obtuse, ½ to ¾ in. long. Stipules small, deciduous. Flowers yellow, the racemes short and loose, really terminal, but usually on very short branchlets so as to appear lateral and shorter than the leaves. Bracts very deciduous. Calyx-segments about 4 lines long. Petals rather longer. Pod about 1 in. broad, varying in length according to the number of seeds ripened, usually 2 or 3.

N. Australia. On the cliffs at the entrance to Victoria river, *F. Mueller* (fruits only) ; ·Port Essington, *Leichhardt* (fragments of a raceme). Common, wild or cultivated, in tropical Asia and Africa, and introduced into the West Indies. This tree supplies the well-known *Tamarinds,* used as conserves and in medicine. The Australian specimens being fragmentary only, I have described it from Indian ones.

83. BAUHINIA, Linn.

Sepals united at the base into a short or long disk-bearing tube, the free part separating into 5 or fewer valvate or induplicate lobes. Petals 5, inserted at the summit of the tube, usually clawed, more or less unequal. Stamens 10, free, either all perfect or some reduced to small staminodia. Ovary stipitate, the stalk adnate to one side of the calyx-tube, with several ovules; style usually filiform, with a capitate, broad or oblique, terminal stigma. Pod linear or oblong, compressed, 2-valved. Seeds compressed; albumen usually thin; radicle short and straight.—Trees or woody climbers. Leaflets either 2 distinct from the base, or (in the majority of species not Australian) united into an entire or 2-lobed leaf, with 5 to 11 digitate nerves. Racemes terminal.

A large genus, distributed over the tropical regions of the New and the Old World. It has been divided into several genera, which I have in other works been disposed to adopt; but it may be more convenient to follow De Candolle and others in considering them as subgenera. The following Australian species, all endemic, with one or two nearly allied Asiatic ones, form a small group, with the 2 leaflets quite distinct, and in their flowers and fruit agreeing with *Phanera*, except that the 10 stamens are all perfect.

Calyx disk-bearing base very short, free part campanulate, deeply lobed. Outer petals 5 to 6 lines long. Pod 2 in. broad . . . 1. *B. Cunninghamii.*
Calyx disk-bearing base turbinate, free part as long, shortly lobed. Petals 6 or 7 lines long. Pod not 1½ in. broad . . . . . . 2. *B. Carronii.*
Calyx disk-bearing base cylindrical, free part as long, divided to the base. Outer petals 1½ in. long. Pod 1 to 1¼ in. broad . . . 3. *B. Hookeri.*

1. **B. Cunninghamii,** *Benth.* A tree of 20 ft. or more, the young branches slender, rarely short and spinescent, the young shoots and leaves pubescent, at length glabrous. Leaflets quite distinct, broadly falcate-ovate, very obtuse, ¾ to 1½ in. long, and more than half as broad, finely 5- to 7-nerved. Flowers 2 or 3 together on a very short common peduncle, the pedicels shorter than the calyx. Calyx about 5 lines long, velvety-tomentose, the disk-bearing base very short and broad, the free part broadly campanulate, thick, divided below the middle into 5 ovate equal lobes. Petals silky-tomentose, ovate, the 2 outer lower ones exceeding the calyx by above 3 lines, the lateral ones by about 2 lines, and the uppermost inner one scarcely at all. Stamens 10, longer than the petals, the lowest the longest. Ovary on a long stipes, with 8 to 10 ovules; stigma large, capitate. Pod very flat, thinly coriaceous, about 2 in. broad, 6 in. long or shorter, according to the number of seeds ripened.—*Phanera Cunninghamii,* Benth. in Pl. Jungh. i. 264; *Bauhinia Leichhardtii,* F. Muell. in Trans. Vict. Inst. iii. 50, partly.

N. Australia. Careening Bay and Vansittart Bay, *A. Cunningham;* N.W. coast, *Bynoe;* Oakover river, Nichol Bay, *F. Gregory's Expedition;* Arnhem's Land, *F. Mueller.*

2. **B. Carronii,** *F. Muell. in Trans. Vict. Inst.* iii. 49. A tree, with the foliage and inflorescence of *B. Cunninghamii,* or with rather narrower leaflets, but undistinguishable without the flowers or fruit, both of which are narrower. Calyx slightly tomentose, about ½ in. long, the disk-bearing base narrow-turbinate, the free portion about the same length, very shortly 5-lobed. Petals obovate, silky outside, the lower ones exceeding the calyx by about 5 lines, the others rather shorter. Stamens and pistil of *B. Cunninghamii.* Pod coriaceous, not quite 1½ in. broad, the valves coriaceous, slightly convex when ripe.

Queensland. On the Burdekin, *F. Mueller;* in the interior, *Mitchell.*
N. S. Wales. Towards Cooper's Creek, *Howitt's Expedition.*

Some specimens of Leichhardt's, in leaf only and therefore not determinable, referred by F. Mueller to the preceding species, appear to me rather to belong to the present one.

3. **B. Hookeri,** *F. Muell. in Trans. Vict. Inst.* iii. 51. A large tree, with a spreading head, usually quite glabrous. Leaflets quite distinct, very obliquely and broadly ovate or obovate, very obtuse, ¾ to 1½ in. long, finely 5- to 7-nerved, with a small thick point terminating the petiole between them. Flowers white, edged with crimson, few in short terminal racemes, the pedicels very short. Calyx glabrous or nearly so, 1 in. long or even more, the disk-bearing base narrow-cylindrical, the free part about as long, dividing nearly to the base into 5 narrow lobes. Petals clawed, ovate, nearly equal, the lamina nearly 1½ in. long, slightly villous outside near the base. Stamens 10, rather longer than the petals. Ovary on a long stipes; stigma large. Pod stipitate, flat, 1 to 1¼ in. broad.

N. Australia. Arnhem N. Bay, *R. Brown*; Port Essington, *A. Cunningham.*
Queensland. Broad Sound, *R. Brown*; Gilbert river, *F. Mueller*; Suttor river, *D'Orsay*; Rockhampton, *Dallachy*; islands of Torres Straits, *Henne.*
Var. *puberula.* Young shoots slightly pubescent. Calyx tomentose, the free part shorter than the disk-bearing base. Pod large and broad.—Burdekin river, *F. Mueller.*
The latter specimens come very near to *B. (Phanera) Blancoi*, Benth. in Pl. Jungh. i. 264, which we have from Siam and from the Philippine Islands, and of which *B. Hookeri* may prove to be a variety only.

84. CYNOMETRA, Linn.

Sepals very shortly united at the base, the free part separating into 4 imbricate segments, the upper one rather broader (consisting of 2 sepals?). Petals 5, oblong-lanceolate, nearly equal, the upper one innermost. Stamens 10 or more, free; filaments filiform; anthers small. Ovary nearly sessile, with 2 ovules; style subulate, with a small terminal stigma. Pod obliquely and broadly semiorbicular, thick, fleshy, and turgid, 2-valved. Seed usually solitary, thick; radicle short, straight.—Trees or shrubs. Leaves abruptly pinnate, with 1, 2, or rarely more pairs of leaflets. Flowers small, usually reddish, in axillary or lateral clusters or short racemes.

The genus is distributed over the tropical regions of the New and the Old World. The only Australian species is a common Asiatic one.

1. **C. ramiflora,** *Linn.; DC. Prod.* ii. 509, var. *bijuga.* A tree, either glabrous or the young shoots and inflorescence rusty-pubescent. Leaflets in the Australian specimens 2 pairs or rarely 1 pair only, obliquely obovate-oblong, very obtuse or shortly and obtusely acuminate, coriaceous, penninerved, the terminal ones 2 to 3 in. long, the lower ones smaller. Flowers in very short axillary racemes or clusters. Bracts dry, concave, at first imbricate, but very deciduous. Pedicels 2 to 3 lines long. Calyx and petals not 2 lines long. Stamens 10, rather longer. Ovary very villous. Fruit as broad as long, very thick and fleshy, ½ to ¾ in. long and very rugose in the dried specimens, probably larger and smoother when fresh.—W. and Arn. Prod. i. 293 ; *C. bijuga*, Spanoghe, in Miq. Fl. Ind. Bat. i. part i. 78.

Queensland. Endeavour river, *Banks and Solander*, *A. Cunningham.*—Widely spread over E. India and the Archipelago, most frequently with 2 pairs of leaves in Ceylon and the Archipelago, with one pair only on the continent of India, but the two can scarcely be distinguished, even as varieties.

SUBORDER III. MIMOSEÆ.

Flowers 5-merous, 4-merous, or rarely 3-merous or 6-merous, small, regular, sessile in spikes or heads or very rarely shortly pedicellate, often polygamous. Sepals valvate, often united. Petals valvate, except in *Erythrophlœum*, often united. Stamens equal to or double the number of petals or indefinite. Seeds usually flattened, with a hard shining testa. Albumen none or very scanty. Radicle of the embryo short and straight. Leaves bipinnate, except in the American genus *Inga*.

The genera of this well-marked suborder are not numerous in proportion to the species, and are technically distributed into four tribes :—*Parkieæ*, with definite stamens and slightly imbricate petals, including *Erythrophlœum* ; *Eumimoseæ*, with definite stamens, including *Entada, Adenanthera, Dichrostachys*, and *Neptunia* ; *Acacieæ*, with indefinite free stamens, consisting chiefly of the vast genus *Acacia* ; and *Ingeæ*, with indefinite monadelphous stamens, including *Albizzia* and *Pithecolobium*.

85. **ERYTHROPHLŒUM**, Afzel.

(Fillæa, *Guillem. and Perr.* ; Laboucheria, *F. Muell.*)

Calyx-teeth 5, valvate in the bud. Petals 5, small, very slightly imbricate. Stamens 10, inserted with the petals on a perigynous disk, free, longer than the petals, all equal and perfect ; anthers ovate, without glands. Ovary stipitate, with several ovules ; style short, with a terminal stigma. Pod oblong, linear, flat, coriaceous, 2-valved. Seeds ovate, compressed, transverse ; funicle filiform ; testa pulpy outside ; albumen thin ; radicle short, straight. Trees. Leaves bipinnate. Flowers small, almost sessile in long cylindrical spikes, forming a terminal panicle. Bracts inconspicuous.

A small genus, containing, besides the Australian species, which is endemic, two or three from tropical Africa. In the slight imbrication of the petals it approaches *Cæsalpinieæ*, and especially *Mora*, and may be considered, with *Parkia*, as intermediate between that suborder and true *Mimoseæ*.

1. **E. Laboucherii,** *F. Muell. Herb.* A hard-wooded tree, the branches and foliage glabrous. Pinnæ opposite, in 2 or 3 pairs ; leaflets 4 to 9, alternate, obliquely obovate or orbicular, very obtuse or retuse, mostly 1½ to 2 in. long. Spikes rather dense, nearly sessile, 1 to 3 in. long. Flowers 2 to 2½ lines long. Calyx sprinkled and ciliate with a few hairs. Petals rather longer than the calyx, with woolly edges. Stamens more than twice as long as the petals, distinctly inserted in 2 rows. Ovary shortly stipitate, hairy, with about 10 ovules. Pod 4 to 6 in. long, 1 to 1½ in. broad, flat, with thinly-coriaceous valves. Seeds nearly orbicular.—*Laboucheria chlorostachys,* F. Muell. in Journ. Linn. Soc. iii. 159.

N. Australia. Careening Bay and Vansittart Bay, N.W. coast, *A. Cunningham ;* Victoria river, *Bynoe, F. Mueller ;* islands of the Gulf of Carpentaria, *R. Brown ;* Strangways river, *M'Douall Stuart.*

Queensland. Endeavour river, *Banks and Solander, A. Cunningham ;* Burdekin and Gilbert rivers, *F. Mueller ;* also in Leichhardt's collection, and said to be his " Leguminous Iron-bark Tree."

86. **ENTADA,** Adams.

Calyx very shortly 5-toothed. Petals 5, valvate, more or less united or free. Stamens 10, shortly exserted, free ; anthers tipped by a gland. Ovary nearly sessile, with several ovules ; style filiform, with a truncate stigma. Pod large and long, flat, coriaceous or woody, the sutures thick and forming a persistent replum, the valves falling away separately and divided transversely into 1-seeded articles, the endocarp separating from the epicarp and persisting round the large orbicular flat seeds.—Tall woody climbers, unarmed. Leaves abruptly bipinnate, the pinnæ of the upper pair often converted into tendrils without leaflets. Flowers small, sessile in long slender spikes, either solitary in the upper axils or forming a terminal simple panicle. Bracts very small.

The genus is common to the New and the Old World within the tropics. The only Australian species is the same as the most generally diffused Asiatic one.

1. **E. scandens,** *Benth. in. Hook. Journ. Bot.* iv. 332. A woody climber, stretching over the largest trees, the young parts and inflorescence slightly pubescent, at length glabrous. Leaves usually consisting of a common petiole of 2 to 6 in., terminating in 2 simple tendrils, which are not however always developed, and bearing below them 1 or 2 pairs of pinnæ ; leaflets on each pinna 2 or 3 rarely 4 or even 5 pairs, obovate-oblong obtuse or emarginate, often very oblique, 2 to 5 in. long when few, smaller when more numerous. Spikes varying in length from 1 or 2 in. to nearly 1 ft. Flowers about 1½ in. long. Calyx very small, truncate or minutely toothed. Petals lanceolate, rigid, becoming at length quite separate. Gland of the anthers very deciduous. Pod woody, attaining 2 to 4 ft. in length and 3 to 4 in. in breadth. Seeds 10 to 30, nearly 2 in. diameter.—*Mimosa scandens,* Linn. Sp. Pl. 1501 ; *Entada Pursætha,* DC. Prod. ii. 425 ; Miq. Fl. Ind. Bat. i. part i. 45.

Queensland. Cape York, *M'Gillivray,* small specimens in flower and leaf only, the description completed from Indian ones. The species is widely diffused over tropical Asia and Africa and the West Indies, the seeds being carried very far by ocean currents without losing their power of germination. The opinion now generally adopted that the East and West Indian species are the same is, however, not universally admitted ; and our herbarium specimens, numerous as they are, are mostly too imperfect, the foliage, flowers, and pods too rarely matched to determine the question with any approach to certainty.

87. ADENANTHERA, Linn.

Calyx 5-toothed. Petals 5, valvate in the bud, cohering at first, at lengt free. Stamens 10, free ; anthers ovate, tipped by a deciduous gland. Ovary sessile, with several ovules ; style filiform, with a small terminal stigma. Pod linear, compressed, falcate, 2-valved, the endocarp often forming more or less complete partitions between the seeds ; valves somewhat convex. Seeds thick, with a hard, shining, red or red-and-black testa, surrounded usually by a thin pulp ; funicle slightly thickened ; albumen scanty ; radicle short, straight.—Unarmed trees. Leaves abruptly bipinnate, with several pairs of pinnæ and of leaflets. Flowers small, white or yellowish, always pedicellate in long spike-like racemes, either solitary in the axils or forming a simple terminal panicle.

A genus of few species, natives of the tropical regions of the Old World, one of them introduced and now naturalized in the West Indies. It is closely allied to *Prosopis* and several others separated from *Mimosa* and *Acacia,* differing chiefly in the pedicellate flowers and in the seeds resembling those of *Ormosia,* and externally those of *Abrus,* and the original *A. pavonina,* Linn., has moreover a remarkably long twisting pod.

Fragments of a plant evidently belonging to this genus, but insufficient for determination, are in F. Mueller's collection from Gilbert river. One imperfect leaf is like that of *A. pavonina*, but with it are two loose pods which I can only compare with those of a Borneo plant in the Kew herbaria. They are short, slightly falcate, and the seeds more like those of the Cingalese *A. bicolor*, which, however, has very different leaves.

88. DICHROSTACHYS, W. and Arn.

(Caillea, *Guillem. and Pers.*)

Calyx 5-toothed. Petals 5, valvate, usually cohering. Stamens in the perfect flowers 10, free; anthers ovate, tipped with a deciduous gland. Ovary nearly sessile, with several ovules; style short or filiform, with a small terminal stigma. Lower flowers of the spike neuter, with long, linear, white or coloured staminodia, and a small rudimentary ovary. Pod linear, compressed, variously twisted, indehiscent or the valves irregularly separating from the persistent sutures.—Rigid shrubs, the branchlets occasionally spinescent. Leaves abruptly bipinnate, with a stipitate gland between the pinnæ of the lowest or of all the pairs; leaflets small. Stipules subulate or acuminate, often imbricate on the short flowering branches. Flowers sessile, in dense cylindrical spikes, either terminal or apparently axillary by the shortness of the branchlet, the upper flowers of the spike hermaphrodite and yellow, the lower ones neuter and white pink or purple.

The genus extends over tropical Africa and Asia. Of the Australian species one is the common Indian one, the other is endemic. F. Mueller proposes to reunite it with *Neptunia*, but the difference in the pod is accompanied by too marked a difference in habit to be neglected, unless the whole of the *Mimoseæ* with definite stamens and gland-tipped anthers be united into one genus.

Pinnæ 8 to 10 pairs. Leaflets 12 to 20 pairs . . . . 1. *D. cinerea.*
Pinnæ 1 or 2 pairs. Leaflets 4 to 6 pairs . . . . 2. *D. Muelleri.*

1. **D. cinerea,** *W. and Arn. Prod.* 271. Glabrous or the branches and petioles pubescent. Pinnæ 8 to 10 pairs, ½ to 1 in. long; leaflets 12 to 20 pairs, crowded, oblong-linear, slightly falcate, 1 to 1⅓ or rarely 2 lines long. Spikes shortly pedunculate, usually nodding, about 1 in. long. Hermaphrodite flowers about 1 line long without the stamens, which are twice or thrice as long. Staminodia of the neuter flowers still longer and white. Pod 2 to 3 in. long, 3 to 4 lines wide, irregularly twisted, viscid-pubescent or glabrous.—Wight, Ic. t. 357.

N. Australia. Port Essington, *Armstrong.* Extends over E. India and the Archipelago.

2. **D. Muelleri,** *Benth.* Glabrous or slightly pubescent. Pinnæ 1 or 2 pairs, rather distant; leaflets 4 to 6 pairs, oblong-linear, 3 to 5 lines long, rather coriaceous and pale-coloured. Spikes pedunculate, about 1¼ in. long. Hermaphrodite flowers about 1 line long without the stamens. Neuter flowers in the lower part of the spike not so crowded as the others, with long pale whitish-purple staminodia.—*Neptunia spicata*, F. Muell. Fragm. iii. 151.

N. Australia. Arnhem's Land, *Waterhouse in M'Douall Stuart's Expedition.*

89. NEPTUNIA, Lour.

Calyx 5-toothed. Petals 5, valvate in the bud, cohering or free. Stamens in the perfect flowers 10 or (in the Australian species) 5, free; anthers ovate, tipped with a deciduous gland, very rarely wanting; pollen granular. Ovary stipitate, with 2 or several ovules; style filiform, with a truncate or concave stigma. Lower flowers of the head male or more frequently neuter, with long linear staminodia. Pod short and broad, flat, turned downwards, 2-valved, without pulp inside. Seeds transverse, flattened, ovate or orbicular, the funicle not dilated.—Procumbent or floating perennials or undershrubs. Leaves abruptly bipinnate, with small leaflets, either without glands or with a depressed gland below the pinnæ. Stipules membranous, obliquely cordate. Peduncles axillary, usually with 2 distant stipule-like deciduous bracteoles, and bearing a single globular or ovoid flower-head. Flowers small, sessile, mostly hermaphrodite, but a few of the lower ones either like the others but male by the abortion of the ovary, or neuter with a smaller calyx and corolla, and long linear almost petal-like staminodia.

A small genus, widely diffused over the tropical regions of the New and the Old World, extending also into N. America. The two Australian species are endemic, and differ from all others in their stamens always 5 only instead of 10.

Peduncles slender, 1 to 3 in. long. Ovules several. Pod oblong, with
    several seeds . . . . 1. *N. gracilis.*
Peduncles very short. Ovules 2. Pod orbicular, 1-seeded . . . . 2. *N. monosperma.*

1. **N. gracilis,** *Benth. in Hook. Journ. Bot.* iv. 355. Stock perennial, often woody, with procumbent or ascending stems of about 1 ft. or rarely twice as long, glabrous or slightly pubescent. Pinnæ usually 1 or 2 pairs; leaflets 6 to 20 pairs, oblong-linear, falcate, 2 to 3 or rarely 4 lines long; glands none in the ordinary form, but small setaceous stipellæ under the pinnæ. Stipules leafy, obliquely cordate, acuminate. Peduncles 1 to 3 in. long, with 2 broad cordate bracteoles, one about the middle, the other much lower down, both very deciduous. Flower-head small, globular, with very few of the male or neuter flowers at the base and sometimes none at all. Hermaphrodite flowers smaller than in the extra-Australian species, always with 5 stamens only and few or sometimes none of the neuter ones at the base of the head. Gland of the anthers small, sometimes perhaps quite wanting. Ovules 6 to 8. Pod, when perfect, ¾ in. long, about 4 lines wide, with 4 to 6 seeds.

N. Australia. Islands of the N. coast, *R. Brown.*

Queensland. Shoalwater Bay and Broad Sound, *R. Brown;* Moreton Bay, *C. Stuart;* near Warwick, *Beckler;* in the interior, on the Maranoa, etc., *Mitchell, Leichhardt,* and others.

N. S. Wales. Between the Darling and Cooper's Creek, *Neilson.*

Var. *major.* Larger and more erect. Leaves with a depressed gland below the lowest pair of pinnæ, but less conspicuous than in *N. monosperma.* Flowers rather large.—Bay of Inlets, *Banks and Solander;* Rockhampton and Burdekin river, *Herb. F. Mueller.*

Var. *villosula.* More or less pubescent. Pinnæ 3 or 4 pairs, without glands. Flower-heads rather large.—Sturt's Creek, *F. Mueller;* Gulf of Carpentaria, *Landsborough;* also *Neilson's* N. S. Wales specimens.

2. **N. monosperma,** *F. Muell. Herb.* Stems apparently herbaceous, but stouter and taller than in *N. gracilis* (except perhaps in the var. *major*).

Pinnæ 2 or 3 pairs; leaflets 20 to 30 pairs, mostly 3 to 5 lines long; gland large, depressed below the lowest pair of pinnæ. Stipules and bracteoles very much smaller than in *N. gracilis* and very deciduous. Peduncles very short, rarely exceeding ½ in. Flowers small, all with 5 stamens only, with very few or scarcely any of the neuter ones at the base of the head. Ovules always 2 only. Pod nearly orbicular, 3, 4, or rarely 5 lines diameter, with a single seed.

N. Australia. Upper Victoria river, *F. Mueller;* Gulf of Carpentaria, *Landsborough, Henne.*

Queensland, *Bowman.*

90. ACACIA, Willd.

(Vachellia, *W. and Arn.;* Tetracheilos, *Lehm.;* Chithonanthus, *Lehm.*)

Sepals 5, 4, or 3, free or united (wanting in *A. Huegelii* and *A. squamata*). Petals as many, free or united (wanting in *A. squamata*). Stamens indefinite, usually very numerous, free or slightly connected at the very base. Pod linear or oblong, flat or nearly cylindrical, straight, falcate or variously twisted, opening in 2 valves or indehiscent. Seeds more or less flattened, usually marked in the centre of each face with an oval or horseshoe-shaped depression or opaque spot or ring, sometimes very obscure. Funicle usually thickened into a fleshy aril under or round the seed.—Trees, shrubs, climbers, or rarely undershrubs, with or without prickles or stipular spines. Leaves twice pinnate or reduced to a simple phyllodium or dilated petiole. Flowers usually yellow or white, in globular heads or cylindrical spikes, often polygamous.

A very large genus, dispersed over the warmer regions of the globe, and in Australia the most numerous in species of all Phænogamous genera. Of the Australian species, one only, *A. Farnesiana,* is common to the warmer regions of the New and the Old World, the remainder are all endemic. Of these by far the greater number belong to the phyllodineous series, which is entirely Australian, with the exception of a very few from New Caledonia, the Indian Archipelago, and the Pacific Islands, none of which can be specifically identified with any Australian ones, although very near some of the tropical species. Acacias are also very generally distributed over every part of Australia, but are entirely absent from New Zealand.

Taken as a whole, the genus is the most marked of those which have been dismembered from the Linnæan *Mimosa,* being at once distinguished from *Inga* and its allies by the free stamens, and from the true *Mimoseæ* by their indefinite number; but, for its subdivision, notwithstanding considerable differences in the flowers and more striking ones in the fruit, it has been found impossible to establish upon these differences any definite sections, even among those species where both flowers and fruit are well known, and in the majority of specimens gathered, the pod is neglected by collectors. Species with the most discrepant pods are sometimes almost identical in foliage, and, on the other hand, pods apparently identical, sometimes belong to species widely different in foliage and even in flower. I have therefore on each of the three occasions when I have gone through the genus in detail, with a large number of specimens before me, in vain sought for any better mode of distributing the species than in *Series,* founded chiefly upon foliage and inflorescence. There are only one or two species in which the cylindrical spike appears to pass into the globular head, and the venation of the phyllodia is nearly, though not quite, as constant. The glands on the upper edge of the phyllodia and on the common petiole in the compound leaf seldom afford even a specific distinction, and the bracts in the flower-head still less so, and I have therefore in the descriptions seldom mentioned them. The bracts are almost always narrow, closely packed in with the flowers, and more or less dilated at the end, sometimes quite peltate; generally the outer ones of each head are flatter, the inner ones more slender and proportionally more dilated at the end, where they are usually ciliate and sometimes acuminate. The characters derived from the united or free sepals must be used with caution, for the sepals, at first united, often separate as the flowering advances. The forms assumed by the pod are tolerably constant in species, although rarely available for classing them in groups; so it is also with the seeds, transverse or longitudinal, and with the infinite variety of forms assumed by the funicle. This funicle on the ripe seed rarely remains short and filiform, it almost always forms two or three folds under the seed, the end of the last fold or the whole of the last and more or less of the lower folds being thickened into a variously-shaped small fleshy aril, usually described as a strophiole, but always a part of the funicle and continuous with the lower filiform part, or forming the whole funicle ; occasionally the thickened part is much elongated extending round one side of the seed, returning on the same side and forming another double fold on the other side, or completely encircling the seed in a double fold returning on the same side, or extending twice round without a return, or even encircling it in a triple fold. All these and other modifications appear to be constant in each species, but only rarely available for specific diagnosis, for in many species the funicle is as yet.unknown ; it is often unsafe to rely on it unless the seed is quite ripe, and then the thin part of the funicle is so brittle that it is often destroyed merely by the elastic opening of the pod.

Leaves all or mostly reduced to flat terete or subulate phyllodia or minute scales without leaflets. (**Phyllodineæ.**)

Flowers in globular heads.
　Phyllodia none or reduced to minute scales.
　　Branches spinescent . . . . 11. *A. spinescens.*
　　Branches rush-like, not spinescent . . . . IV. CALAMIFORMES.
　Phyllodia (either small and tooth-like or vertically flattened or
　　elongated) decurrent on or continuous with the branches.
　　Branches flat or 2-winged by the decurrent phyllodia . . . I. ALATÆ.
　　Branches 3-winged or the phyllodia (usually pungent) very
　　　shortly or scarcely decurrent, but not articulate . . . . II. CONTINUÆ.
　Phyllodia articulate on the stems, at least when old.
　　Phyllodia rigid, tapering into pungent straight points, usually
　　　narrow or short, not whorled (except in 43, *A. verticillata*) III. PUNGENTES.
　　Phyllodia linear-subulate, terete or tetragonous, rarely slightly
　　　flattened, obtuse or with incurved or innocuous points, not
　　　whorled . . . . IV. CALAMIFORMES.
　　Phyllodia terete or slightly flattened, usually short, all whorled
　　　or crowded and irregularly whorled or clustered . . . . V. BRUNIOIDEÆ.
　　Phyllodia vertically flattened, broader than thick, obtuse acute
　　　or with incurved or innocuous points.
　　　Phyllodia 1-nerved, the veins pinnate, reticulate or rarely 1
　　　　or 2 secondary small nerves from the base on one side of
　　　　the midrib . . . . VI. UNINERVES.
　　　Phyllodia with 2, 3 or more parallel nerves . . . . . VII. PLURINERVES.
Flowers in cylindrical or oblong spikes.
　Phyllodia rigid, tapering into pungent points.
　　Phyllodia several-nerved, decurrent on the stem . . . . 10. *A. triptera.*
　　'Phyllodia 1- or 3-nerved, articulate on the stem . . . . . III. PUNGENTES.
　Phyllodia obtuse or with a callous, innocuous or hooked point . VIII. JULIFLORÆ.

Leaves all bipinnate. Flowers in globular heads or rarely in spikes. (**Bipinnatæ.**)

Stipules none or brown and scarious. Spines none or axillary.
　Flower-heads several, in axillary or paniculate racemes . . . IX. BOTRYOCEPHALÆ.
　Flower-heads or spikes single, on axillary solitary or clustered
　　peduncles . . . . X. PULCHELLÆ.
Stipules all or some of them spinescent. Flower-heads single on
　axillary peduncles . . . . XI. GUMMIFERÆ.

Div. I. *PHYLLODINEÆ.*—Leaves mostly phyllodineous without leaflets.

Series I. **Alatæ.**—*Phyllodia bifariously decurrent, forming 2 opposite wings to the stem, the free part short, broad, and acute, or rigid and pungent-pointed. Flowers in globular heads, on axillary simple peduncles, rarely appearing racemose by the reduction of the flowering branches.*

Wings broad, continuous with the next below, removing the axillary
   bud to a distance from the stem . . . . 1. *A. bossiæoides.*
Wings interrupted at each node, the axillary bud sessile on the stem.
  Wings broad, free part of the phyllodia usually short and broad,
    with an almost central nerve.
    Phyllodia without glands. Flowers numerous in a small com-
     pact head . . . . 2. *A. glaucoptera.*
    Phyllodia with a glandular angle on the upper edge. Heads
     6 to 12-flowered . . . . 3. *A. alata.*
  Wings narrow, free part narrow or with the nerve almost marginal.
    Phyllodia erect, incurved. Flowers numerous in the head,
     5-merous. Petals smooth . . . . 4. *A. diptera.*
    Phyllodia rigid, recurved. Flowers few in the head, 4-merous.
     Petals strongly striate . . . . 5. *A. stenoptera.*

Series II. **Continuæ.**—*Phyllodia narrow, rigid, tapering into a pungent point, continuous with the stem and shortly and trifariously or irregularly decurrent. Flowers in heads or spikes on axillary simple peduncles.*

Flowers in globular heads.
  Phyllodia linear or linear-lanceolate, flattened, 1-nerved.
    Heads nearly sessile, with 4 to 8 flowers . . . . 6. *A. incurva.*
    Heads pedunculate, compact, with above 40 flowers . . . 7. *A. trigonophylla.*
  Phyllodia terete, very shortly decurrent.
    Lower phyllodia 1 to 2 in. long, upper ones short and distant.
     Pod curved or twisted, about 2 lines broad . . . . . 8. *A. continua.*
    Phyllodia crowded on the branchlets, slender and rigid, 2 to
     4 in. long. Pod nearly 1½ in. broad . . . . 9. *A. Peuce.*
Flowers in cylindrical spikes. Phyllodia lanceolate, thick, several-
  nerved . . . . 10. *A. triptera.*

(106, *A. anceps,* has some of the phyllodia only half articulate on the angles of the stem, and 59, *A. extensa,* has the branches often narrowly winged, with phyllodia scarcely to be distinguished from them.)

Series III. **Pungentes.**—*Rigid shrubs, branches in some species spinescent. Phyllodia articulate on the stem, rigid, tapering into pungent points, subulate, linear or lanceolate, or rarely none. Flowers in heads or spikes, on axillary simple peduncles.*

(Besides the following species, a few of the short-leaved *Calamiformes* and of the small rigid-leaved *Plurinerves* might almost be classed among the *Pungentes.*)

A. **Aphyllæ.**—*Phyllodia none. Branches spinescent. Heads globular, sessile* . . . . 11. *A. spinescens.*

B. **Plurinerves.**—*Phyllodia 2- or more nerved, or terete and nerveless. Heads globular.*
Phyllodia falcate, not exceeding ½ in. (See Series VII. *Plurinerves.*)
Phyllodia linear-lanceolate, 3- or more nerved. Petals smooth or
  with prominent midribs.
  Sepals distinct, narrow, spathulate (Western species except *A.*
   Oswaldi).
    Phyllodia attached by a broad base, very rigid and pungent . 12. *A. latipes.*
    Phyllodia slightly contracted at the base, very rigid and pungent 13. *A. cochlearis.*
    Phyllodia linear-lanceolate, with 3 prominent nerves, scarcely
     pungent. Petals without prominent midribs . . . . 177. *A. heteroclita.*

Phyllodia rigid, but scarcely pungent, several-nerved. Flower-
　heads almost sessile (Eastern species) . . . . . . 184. *A. Oswaldi.*
Sepals united (Eastern or tropical species).
Seeds longitudinal (extratropical species).
Peduncles short. Pod 4 to 5 lines broad. Branches usually
　woolly . . . . 14. *A. lanigera.*
Peduncles slender. Pod 1 to 2 lines broad . . . . . 16. *A. trinervata.*
Seeds obliquely transverse. Branches glabrous, somewhat
　viscid (tropical species) . . . . 15. *A. phlebocarpa.*
Phyllodia nearly terete, striate with several nerves or nerveless.
Petals smooth.
Stipules lanceolate, small and very deciduous.
Phyllodia divaricate, straight or slightly incurved. Peduncles
　very short. Petals narrow, with raised midribs . . . . 17. *A. colletioides.*
Phyllodia straight, short, with not more than 2 nerves. Pe-
　duncles very short. Petals smooth . . . . 34. *A. sphacelata.*
Phyllodia divaricate, straight or recurved. Peduncles as long
　as or longer than the heads. Petals with raised midribs . 18. *A. striatula.*
Phyllodia crowded, incurved, mostly under 1 in. Petals with-
　out raised midribs . . . . 21. *A. sulcata.*
Stipules setaceous or spinescent.
Stipules slender. Phyllodia divaricate, recurved or flexuose.
　Pod flat, with raised margins . . . . 19. *A. campylophylla.*
Stipules rigidly spinescent. Phyllodia erect, almost obtuse,
　with short pungent points. Petals concave and thickened
　at the top . . . . 20. *A. teretifolia.*
Phyllodia linear or lanceolate, recurved, with 2 thick marginal or
　nearly marginal nerves. Stipules setaceous or spinescent.
Petals striate.
Phyllodia mostly ¼ in. long, rather broad . . . . . . 22. *A. costata.*
Phyllodia ½ to 1 in. long, narrow . . . . 23. *A. barbinervis.*
Phyllodia 1 to 2 in. long, very narrow, with hooked points . 24. *A. ataxiphylla.*
Petals smooth. Phyllodia ½ to 1 in., rather broad . . . . 25. *A. Baxteri.*
Phyllodia subulate, terete, finely striate. Heads sessile, ovoid.
Flowers 5-merous . . . . 228. *A. aciphylla.*

C. Uninerves.—*Phyllodia* 1-*nerved. Heads globular.*
Petals prominently striate. Pod thick, with broad smooth margins.
　Flowers 5-merous.
Phyllodia crowded, under ¼ in. long, obliquely oblong-lanceolate,
　often falcate . . . . 26 *A. auronitens.*
Phyllodia scattered, ¾ to 1½ in. long, linear-tetragonous . . . 27. *A. quadrisulcata.*
Petals membranous, smooth or with a prominent midrib.
　Flowers mostly 5-*merous.*
Phyllodia oblong-lanceolate, under ¼ in. long. Branches
　spinescent, woolly . . . . 28. *A. erioclada.*
Phyllodia linear or lanceolate, broader than thick.
Pod 2½ to 3 lines broad, not contracted between the seeds
　when perfect (E. and S. species) . . . . 29. *A. siculiformis.*
Pod 1½ lines broad, regularly contracted between every seed
　(N.W. species) . . . . 30. *A. patens.*
Phyllodia linear-subulate, acicular or 4-gonous.
Phyllodia mostly recurved. Branches tomentose. Pedun-
　cles slender. Heads small. Pod-valves very convex, striate 31. *A. laricina.*
Phyllodia straight, clustered at the old nodes. Peduncles
　slender. Funicle encircling the seed . . . . . . 32. *A. tetragonophylla.*
Phyllodia straight, often numerous, but not clustered. Pe-
　duncles very short . . . . 34. *A. sphacelata.*

Phyllodia scattered, erect or slightly spreading, often 1 in.
long. Peduncles slender, mostly clustered . . . . . 33. *A. genistoides.*
Phyllodia scattered, divaricate, rather broader or with a
small glandular angle at the base.
Petals without any prominent nerve. Peduncles slender 35. *A. ingrata.*
Petals with a prominent midrib. Peduncles slender.
Funicle filiform . . . . 36. *A. juniperina.*
Petals without any prominent nerve. Heads almost sessile 37. *A. asparagoides.*
(See also 93, *A. vomeriformis,* with the glandular angle
more prominent.)
Flowers mostly 4-merous or 3-merous.
Sepals free, narrow. Funicle shortly thickened . . . . . 38. *A. tenuifolia.*
Calyx toothed or lobed.
Peduncles longer than the flower-head.
Phyllodia linear. Funicle folded and thickened under
the seed . . . . 39. *A. diffusa.*
Phyllodia linear-lanceolate, broad at the base. Funicle
dilated and clasping the base of the seed . . . . 40. *A. rupicola.*
Flower-heads almost sessile, few-flowered.
Flowers 6 to 10, mostly 4-merous . . . . 41. *A. rhigiophylla.*
Flowers 1 to 4, mostly 3-merous . . . . 45. *A. axillaris.*

D. **Spicatæ.**—*Phyllodia* 1- *to* 3-*nerved. Flowers* 4- *or* 3-*merous in cylindrical oblong
or ovoid spikes.*

Spikes or heads few-flowered, almost sessile.
Flowers 6 to 10, mostly 4-merous . . . . 41. *A. rhigiophylla.*
Flowers 1 to 4, mostly 3-merous . . . . 45. *A. axillaris.*
Spikes pedunculate, many-flowered.
Spikes cylindrical or ovoid, dense. Flowers mostly 4-merous.
Phyllodia linear-lanceolate, very rigid, 3- or 4-nerved and striate 42. *A. oxycedrus.*
Phyllodia subulate, linear, or oblong, mostly 1-nerved and ver-
ticillate . . . . 43. *A. verticillata.*
Spikes interrupted, slender. Flowers mostly 3-merous . . . 44. *A. Rioeana.*

SERIES IV. **Calamiformes.**—*Phyllodia rarely none, more frequently narrow-linear
or subulate, terete tetragonous or very slightly flattened, articulate on the stem, obtuse or
with short innocuous or recurved points,* 1- *or several-nerved. Flowers in globular heads
on simple axillary peduncles, or rarely several heads in a short raceme, or irregularly
racemose by the abortion of the floral phyllodia.*

(Some of Ser. VIII., *Julifloræ,* have similar phyllodia, but cylindrical or ovoid flower-
spikes. 21, *A. sulcata,* and a few allied *Pungentes,* have varieties with the phyllodia almost
obtuse, bringing them near to the short-leaved *Calamiformes.*)

A. **Subaphyllæ.**—*Phyllodia none or very few and slender, resembling the branches.*
Flowers 1 to 4 in the head, 4-merous. Petals striate. Pod 4-angled 46. *A. tetragonocarpa.*
Flowers numerous, 5-merous. Petals smooth. Pod not angled.
Flower-heads racemose . . . . 47. *A. restiacea.*

B. **Plurinerves.**—*Phyllodia striate, with* 2, 3, *or more nerves on each side.*
No calyx or corolla. Flower-heads in a raceme, enclosed when
young in imbricate deciduous scales . . . . 48. *A. squamata.*
Calyx and corolla present. Peduncles 1-headed.
Phyllodia numerous, under ½ in. long, obtuse or with a short
straight point. Peduncles 3 to 4 lines long . . . . . . 49. *A. brachyphylla.*
Phyllodia mostly nearly 1 in. long, with a hooked or recurved
point. Peduncles 3 to 4 lines long.
Petals narrow, membranous (tropical species) . . . . . 50. *A. Bynoeana.*
Petals concave and thickened at the end (Western species) . 20. *A. teretifolia.*
Phyllodia mostly 1½ to 3 in. or longer. Peduncles very short.

Nerves of the phyllodia few and prominent, usually 3.
　Sepals narrow-spathulate, free (Western species) . . . . 51. *A. triptycha.*
　Sepals united (Eastern species) . . . . 178. *A. elongata.*
Nerves of the phyllodia numerous, very fine, often only visible
　under a lens.
　Phyllodia slender, terete, scarcely hooked. Sepals narrow-
　　spathulate, nearly free. Pod unknown (Western species) 52. *A. leptoneura.*
　Phyllodia thick, slightly flattened, hooked. Flowers un-
　　known. Pod broad, flat, membranous (Southern species) 54. *A. papyrocarpa.*
　Phyllodia terete, rigid, scarcely hooked. Sepals spathulate,
　　united to the middle. Pod very narrow, twisted, convex,
　　coriaceous (Eastern species) . . . . 53. *A. rigens.*
　(185, *A. lineolata,* has sometimes nearly the phyllodia of *A. leptoneura,* but decidedly
flattened and often much broader.)

C. **Uninerves.**—*Phyllodia 1-nerved or nerveless.*

Phyllodia mostly elongated, above 1½ in. long.
　Peduncles 1-headed. Flowers 1 to 4 in the head, 4-merous.
　　Petals striate . . . . 46. *A. tetragonocarpa.*
　Peduncles 1-headed. Sepals spathulate (Eastern species).
　　Phyllodia 1½ to 2 or rarely 3 in. long; nerve prominent.
　　　Petals with a prominent midrib . . . . 55. *A. pugioniformis.*
　　Phyllodia 3 to 6 in. long, obscurely nerved. Petals without
　　　any prominent nerve . . . . 56. *A. juncifolia.*
　Peduncles mostly bearing a raceme of 2 or 3 heads. Calyx
　　shortly toothed or lobed (Eastern species) . . . . . 57. *A. calamifolia.*
　Peduncles 1-headed or rarely irregularly racemose. Calyx trun-
　　cate (Western species).
　　Phyllodia terete, obscurely nerved . . . . 58. *A. scirpifolia.*
　　Phyllodia prominently nerved. Branches acutely angular,
　　　almost winged . . . . 59. *A. extensa.*
Phyllodia short, rarely exceeding 1 in.
　Phyllodia very rigid, tetragonous, with very prominent nerves, 1
　　in. long or rather more . . . . 60. *A. gonophylla.*
　Phyllodia mostly under 1 in., terete or slightly flattened hori-
　　zontally, usually nerveless, and often petiolate.
　　Branches not thorny.
　　　Phyllodia mostly ½ to ¾ in., narrowed but not petiolate.
　　　　Peduncles slender . . . . 61. *A. ericifolia.*
　　　Phyllodia mostly 1 in. long, distinctly petiolate. Peduncles
　　　　short . . . . 62. *A. uncinella.*
　　Branches spinescent . . . . 63. *A. oxyclada.*
　(147, *A. subulata,* has sometimes the long slender phyllodia scarcely broader than in *A.
juncifolia,* but is readily distinguished by the small flower-heads several in a raceme.)

SERIES V. **Brunioideæ.**—*Phyllodia numerous, small, linear-subulate (except* A. con-
ferta), *verticillate, clustered or crowded, obtuse or with innocuous or rarely rigid points.
Flowers in globular heads or simple axillary peduncles, usually exceeding the phyllodia.*
Phyllodia all regularly verticillate.
　Phyllodia many in the whorl, rigid, almost pungent, ¼ to 1 in.
　　long. Petals smooth . . . . 64. *A. cedroides.*
　Phyllodia 8 to 10 or more in the whorl, slender, under ½ in. long.
　　Petals striate.
　　Phyllodia recurved at the end, sulcate. Pod sessile . . . 65. *A. lycopodifolia.*
　　Phyllodia recurved at the end, not sulcate. Pod on a stipes
　　　of 2 to 3 lines . . . . 67. *A. galioides.*
　　Phyllodia straight at the end, sulcate. Pod sessile . . . 66. *A. hippuroides.*

Phyllodia 5 to 7 in the whorl, recurved at the end, not sulcate.
  Petals not striate, with a prominent midrib . . . . . . 68. *A. Baueri*
Phyllodia in clusters of about 3. Pod almost woody, narrowed
  into a long stipes . . . . 69. *A. subternata.*
Phyllodia crowded, but scattered or irregularly verticillate, under
  ½ in. long.
  Phyllodia linear-subulate . . . . 70. *A. bruniades.*
  Phyllodia flattened . . . . 71. *A. conferta.*

SERIES VI. **Uninerves.**—*Phyllodia vertically flattened, either narrow and obtuse or with a short oblique point, or broad and obtuse, acute or rarely pungent-pointed, with 1 central or nearly marginal nerve, or very rarely 2-nerved. Flowers in globular heads, either on simple axillary peduncles, solitary, in pairs or clusters, or several in axillary racemes.*

A. **Spinescentes.**—*Rigid shrubs, with spinescent branches. Phyllodia small, usually narrow, not pungent, without marginal glands. Stipules minute or none (except sometimes in A. scabra). Peduncles 1-headed.*

Phyllodia narrow-linear.
  Scabrous-pubescent. Phyllodia scattered. Stipules setaceous . 72. *A. scabra.*
  Nearly glabrous. Phyllodia clustered. Sepals free . . . . 73. *A. nodiflora*
  Glabrous. Phyllodia scattered.
    Flowers not above 10, in a very small head. Calyx very short,
      truncate. Funicle filiform . . . . 74. *A. spinosissimà.*
    Flowers about 20 in the head. Calyx lobed, half as long as
      the corolla. Funicle club-shaped . . . . 75. *A. ulicina.*
Phyllodia obovate oblong or lanceolate, 2 to 4 lines long. Calyx
  very short, truncate . . . . 76. *A. erinacea.*

B. **Armatæ.**—*Shrubs or undershrubs, not spinescent (except sometimes A. congesta and A. idiomorpha). Phyllodia from obovate to lanceolate, rarely above 1½ in. long, more or less undulate, with a central nerve and usually nerve-like margins; marginal glands none or very small and obscure. Stipules generally persistent, either spinescent or setaceous or acuminate or phyllodia-like. Peduncles 1-headed.*

Phyllodia short, broadly recurved-falcate, pungent pointed. Calyx
  none. Petals 5, hirsute . . . . 77. *A. Huegelii.*
Phyllodia obovate to incurved-falcate. Calyx small. Petals 4,
  glabrous.
  Glabrous. Buds obtuse. Stipules spinescent . . . . . 78. *A. nervosa.*
  Scabrous-pubescent. Buds acuminate. Stipules small . . . 79. *A. obovata.*
Phyllodia various. Flowers 5-merous.
  Petals glabrous.
    Glabrous, rigid and spinescent. Phyllodia obliquely ovate-lan-
      ceolate rigid, almost pungent. Stipules small . . . . 80. *A. congesta.*
    Glabrous, rigid, unarmed except the stipules. Phyllodia ob-
      liquely oblong or cuneate, very thick or rigid, undulate . 81. *A. dermatophylla.*
    Glandular-pubescent and resinous. Phyllodia obliquely ob-
      long-linear or narrow-falcate . . . . 82. *A. aspera.*
    Hirsute or glabrous. Phyllodia obliquely semiovate or lan-
      ceolate. Stipules spinescent (wanting in some garden va-
      rieties) . . . . 83. *A. armata.*
    Densely pubescent, rigid and spinescent. Phyllodia broadly
      ovate, pungent-pointed . . . . 84. *A. idiomorpha.*
  Petals hirsute.
    Shrubby. Phyllodia obovate-orbicular, not narrowed at the
      base . . . . 85. *A. Shuttleworthii.*
    Undershrub. Phyllodia obliquely obovate or oblong, narrowed
      at the base. Stipules setaceous . . . . 87. *A. pilosa.*

Habit and phyllodia of *A. pilosa.* Bracts more acuminate.
Stipules with a broad base . . . .   86. *A. Gregorii.*
Undershrub. Phyllodia narrow, incurved and oblique, scarcely
  narrowed at the base. Stipules setaceous . . . . . .   88. *A. crispula.*
Phyllodia of *A. crispula.* Stipules like the phyllodia and about
  half their length . . . .   89. *A. crassistipula.*

Some garden specimens of *A. armata* have some of the phyllodia 2-nerved; the other 2-
or more nerved species, formerly included in *Armatæ,* are now transferred to the *Plurinerves.*

107, *A. hispidula,* and 108, *A. undulifolia,* among *Brevifoliæ,* have the undulate phyllodia, but not the stipules of *Armatæ,* and 120, *A. sentis,* and 121, *A. dentifera,* among
Angustifoliæ, have occasionally spinescent stipules.

C. **Triangulares.**—*Shrubs usually rigid and occasionally spinescent. Phyllodia
small, rigid, the nerve either near the lower margin or rarely central, the upper side more
or less dilated, the margin rounded or angular, with usually a gland at the angle. Stipules spinescent or setaceous or minute. Peduncles 1-headed.*

Phyllodia tapering to a pungent point, the glandular angle below
  the middle or diverging from the base.
Phyllodia hastate or lanceolate, the nerve nearly central. Flowers
  4-merous, 3 to 5 (usually 4) in the head.
Petals glabrous . . . .   90. *A. hastulata.*
Petals densely villous . . . .   91. *A. horridula.*
Phyllodia semilanceolate or divaricately 2-lobed, the nerve near
  the lower margin.
Flowers 4-merous, 8 to 12 in the head . . . .   92. *A. divergens.*
Flowers 5-merous, 30 or more in the head . . . . . .   93. *A. vomeriformis.*
Phyllodia truncate-triangular or trapezoid, the nerve often excen-
  trical, produced into a small pungent point, the upper angle
  above the middle.
Flowers 4-merous, 2, rarely 3 or 4 in the head . . . . .   94. *A. biflora.*
Flowers 4-merous, 6 to 10 or more in the head.
  Phyllodia triangular or rhomboidal, not much longer than
    broad at the top . . . .   95 *A. decipiens.*
  Phyllodia cuneate, much longer than broad . . . . . .   96. *A. cuneata.*
Flowers mostly 5-merous, 8 to 20 or more in the head. Glan-
  dular angles of the phyllodia broad and rounded, the pointed
  angle at the end of the lower side scarcely pungent.
Calyx lobed.
  Phyllodia very coriaceous. Petals hirsute . . . . .   97. *A. dilatata.*
  Phyllodia usually small, slightly coriaceous. Petals gla-
    brous.
    Phyllodia obovate or broadly cuneate-oblong . . . .   98. *A. bidentata.*
    Phyllodia narrow-cuneate oblong . . . .   99. *A. acanthoclada.*
Sepals distinct, spathulate. Phyllodia scarcely triangular.
  (See the first 3 species of the following subseries.)

D. **Brevifoliæ.**—*Shrubs, never spinescent. Phyllodia either broad ovate or falcate,
or narrow, oblong or linear, short (under 1 in. or larger in A. anceps), obtuse or with a
small recurved innocuous point, occasionally undulate and more pointed. Stipules minute,
or none. Peduncles 1-headed.*

Phyllodia from obovate-orbicular to linear, obtuse but with the
  nerve ending in a small recurved point. Sepals free, linear-
  spathulate.
Phyllodia rarely exceeding ½ in. Branches scarcely angular.
  Flowers 10 to 15 or rarely 20 in the head (Eastern species).

Phyllodia broadly obovate or orbicular . . . . 100. *A. obliqua.*
Phyllodia oblong, with a minute gland at the end . . . . 101. *A. acinacea.*
Phyllodia linear . . . . 102. *A. lineata.*
Phyllodia mostly ½ to 1 in. Branches acutely angular. Flowers
　numerous (30 or more) in the head (Western species).
Phyllodia linear. Pod narrow. Funicle short . . . . . 103. *A. triquetra.*
Phyllodia lanceolate . . . . 104. *A. ligustrina.*
Phyllodia obovate-oblong or obliquely cuneate. Pod broad.
　Funicle long, much-folded . . . . 105. *A. Meissneri.*
Phyllodia crowded, linear, obtuse, under ½ in. long. Sepals linear-
　spathulate . . . . 71. *A. conferta.*
Phyllodia from orbicular to cuneate or linear, very obtuse or mu-
　cronate. Sepals united in a toothed calyx (except *A. dura*).
Branches with 3, rarely 2 acute angles. Phyllodia very thick,
　rather large, broad or narrow. Flower-heads dense, on thick
　peduncles . . . . 106. *A. anceps.*
Branches scarcely angular.
Phyllodia oblong falcate, often undulate, the nerve-like margin
　scabrous denticulate . . . . 107. *A. hispidula.*
Phyllodia broadly ovate or orbicular, mucronate, much undu-
　late, entire. Peduncles slender . . . . 108. *A. undulifolia.*
Phyllodia linear or linear-oblong, very obtuse, with a thick
　nerve within the lower edge, and a very thick nerve-like
　upper edge. Flower-heads small, few-flowered.
Calyx shortly lobed or toothed. Branches usually hoary
　(Eastern species) . . . . 109. *A. flexifolia.*
Sepals free, spathulate. Branches glabrous. Phyllodia very
　rigid (Western species) . . . . 110. *A. dura.*
Phyllodia cuneate-oblong, very obtuse, thick and fleshy,
　almost nerveless . . . . 111. *A. spathulata.*
Phyllodia oblong-linear, obtuse, equally but not prominently
　2-nerved . . . . 113. *A. montana.*

(155, *A. brachybotrya* has occasionally the peduncles 1-headed, and is then near *A. undulifolia.*)

E. **Angustifoliæ.**—*Shrubs or trees, not spinescent. Phyllodia oblong-lanceolate or
linear, occasionally falcate, with 1 central nerve or rarely 2-nerved, mostly above 1 in.
long. Stipules minute or none, or rarely spinescent. Peduncles 1-headed. Some of the
species very resinous.*

Peduncles very short, rarely above ¼ in. long. Usually resinous.
Phyllodia oblong-linear or lanceolate, faintly 1-nerved. Sepals
　free, narrow . . . . 112. *A. microcarpa.*
Phyllodia 2-nerved. Calyx toothed or lobed.
Phyllodia very obtuse, mostly under 1½ in. . . . . . . 113. *A. montana.*
Phyllodia mostly acute, above 1½ in. . . . . 114. *A. verniciflua.*
Phyllodia 1-nerved. Calyx toothed or lobed.
Phyllodia mostly acute or mucronate; lateral veins fine and
　not very conspicuous . . . . 115. *A. leprosa.*
Phyllodia mostly obtuse ; lateral anastomosing veins very nu-
　merous and conspicuous . . . . 116. *A. stricta.*
Peduncles usually nearly or above ½ in. long. Phyllodia always
　1-nerved.
Very resinous. Lateral veins few, rather prominent, anastomo-
　sing. Calyx toothed or lobed.
Phyllodia lanceolate or linear-lanceolate . . . . 117. *A. dodonæifolia.*
Phyllodia narrow-linear, numerous . . . . 118. *A. Gnidium.*
Not resinous. Lateral veins scarcely conspicuous.

Phyllodia narrow-linear, thick, the midrib obscure. Funicle
linear, thick, scarcely folded. (Flowers unknown) . . . 119. *A. ramosissima.*
Phyllodia with a prominent midrib.
Calyx short, truncate . . . . 138. *A. salicina.*
Sepals narrow, free.
Pedicels solitary or in pairs, the upper ones often race-
mose from the abortion of the phyllodia. Stipules
often spinescent.
Phyllodia rarely above 2 in. long . . . . 120. *A. sentis.*
Phyllodia very narrow, 3 to 8 in. long . . . . . 121. *A. dentifera.*
Pedicels slender, clustered in the axils. Stipules not
spinescent. Phyllodia lanceolate-falcate, 4 to 6 in.
long . . . . 122. *A. fasciculifera.*

F. **Racemosæ.**—*Not spinescent. Phyllodia not pungent (except in a few broad-
leaved species), with one central nerve or very rarely a second shorter or fainter one, the
veinlets when visible diverging from the midrib or reticulate. Flower-heads globular, few
or all or nearly all in axillary racemes, very rarely a few solitary in the lower axils of a
young branch. Flowers usually small and 5-merous in all except* A. myrtifolia. *Petals
not striate.*

Phyllodia mostly long, falcate-lanceolate or almost linear, narrowed
at the ends, more or less distinctly penniveined. Calyx about
half as long as the corolla.
Eastern species. Flower-heads small.
Phyllodia mostly long lanceolate-falcate.
Sepals free, narrow. Gland of the phyllodia at the base or
none. Funicle surrounding the seed . . . . . 123. *A. falcata.*
Sepals united. Gland of the phyllodia at the base or none.
Veinlets transverse; funicle short . . . . . . 124. *A. macradenia.*
Sepals united. Gland of the phyllodia distant from the
base, a secondary nerve often leading to it. Veinlets ob-
tuse. Funicle surrounding the seed . . . . . . 125. *A. penninervis.*
Phyllodia mostly long linear-falcate.
Sepals united. Gland distant from the base or none.
Funicle surrounding the seed . . . . 126. *A. retinodes.*
Sepals free or nearly so. Gland at the base of the phyl-
lodia minute or none. Funicle short, the last fold
clavate . . . . 127. *A. neriifolia.*
Western species (the first near *A. sentis,* the second near *A. fal-
cata,* the others near *A. pycnantha*).
Phyllodia 1-nerved.
Sepals narrow, quite distinct . . . . 121. *A. dentifera.*
Sepals united, at least at first.
Flower-heads small, on short slender peduncles. Funi-
cle long, folded . . . . 128. *A. microbotrya.*
Flower-heads rather large, on rather stout peduncles.
Funicle clavate, as long as the seed, scarcely folded
at the base.
Phyllodia more or less prominently penniveined . . 130. *A. cyanophylla.*
Phyllodia smooth, the veins scarcely conspicuous . . 129. *A. saligna.*
Phyllodia 2-nerved . . . . 174. *A. bivenosa.*
Phyllodia long falcate-lanceolate or almost linear or thick and nearly
straight. Flower-heads dense, rather large. Calyx at least
long as the corolla.
Phyllodia penniveined, much falcate. Seeds longitudinal, funicle
clavate, scarcely folded at the base . . . . 131. *A. pycnantha.*
Phyllodia thick, straight or falcate, obtuse. Seeds transverse, the
funicle almost encircling them . . . . 132. *A. notabilis.*

Phyllodia straight or falcate, thick, oblong-linear and obtuse or
rarely lanceolate acute, the lateral veins inconspicuous or very
obscure. Calyx about half as long as the corolla, readily separating into distinct sepals.
　Phyllodia 3 to 6 in. long, obtuse, incurved. Sepals narrow, quite
　　distinct. Two folds of the funicle encircling the seed . . . 133. *A. gladiiformis.*
　Phyllodia 1½ to 3 in. long, obtuse, nearly straight. Racemes
　　very short. Sepals at first cohering . . . . 134. *A. obtusata.*
　Phyllodia about 3 in. long, rather acute, usually with 1 gland.
　　Racemes many-headed. Sepals at first cohering . . . . 135. *A. rubida.*
　Phyllodia (on flowering branches) 1½ to 2½ in. long, usually with
　　2 or 3 distant glands. Racemes many-headed. Sepals at first
　　cohering. Three folds of the funicle encircling the seed . . 136. *A. amœna.*
　Phyllodia 2 to 5 in. long, scarcely falcate, thick, obscurely reti-
　　culate. Racemes few-headed. Sepals at first cohering. Fu-
　　nicle clavate, scarcely folded at the base . . . . 137. *A. hakeoïdes.*
　(See also the longest-leaved forms of *A. crassiuscula* and *A. decora.*)
Phyllodia linear, thick, rarely falcate-lanceolate, usually without
glands, the lateral veins reticulate and obscure. Seeds longi-
tudinal.
　Calyx short, thin, truncate.
　　Phyllodia mostly oblong-linear, obtuse. Pod thick . . . 138. *A. salicina.*
　　Phyllodia lanceolate-linear with an oblique or recurved callous
　　　point (western species) . . . . 139. *A. rostellifera.*
　Calyx lobed, thin, half as long as the corolla (Western species).
　　Phyllodia all narrow-linear, erect. Pod narrow, flat, thinly
　　　coriaceous. Funicle folded into a small fleshy aril . . . 140. *A. pycnophylla.*
　　Phyllodia all narrow-linear. Flower-heads small. Pod ½ in.
　　　broad, very flat, coriaceous. Funicle long, folded . . . 141 *A. Harveyi.*
Phyllodia linear or lanceolate, usually thick. Sepals very thin or
narrow, distinct. Pod very flat, obtuse. Seeds transverse.
Young racemes often enclosed in imbricate scales.
　Young branches acutely triquetrous. Pod broad coriaceous and
　　glaucous. Sepals very thin.
　　Phyllodia mostly above 3 in. long (Eastern species) . . . 142. *A. suaveolens.*
　　Phyllodia mostly under 3 in. long (Western species) . . . 143. *A. subcærulea.*
　　Phyllodia long, lanceolate, penniveined (Western species) . . 144. *A. Lindleyi.*
　Young branches scarcely angular. Pod membranous. Sepals
　　very narrow.
　　Phyllodia under 2 in. long, oblong-linear or lanceolate (Wes-
　　　tern species) . . . . 145. *A. leptopetala.*
　　Phyllodia several inches long, narrow linear (Eastern species). 146. *A. Murrayana.*
　Phyllodia very narrow-linear, rather thin, 3 to 6 in. long. Flowers
　　of *A. linifolia* . . . . 147. *A. subulata.*
Phyllodia rarely exceeding 1½ in., linear lanceolate or obliquely
oblong, not very thick, without thick margins.
　Pod flat 4 to 6 lines broad, not contracted between the seeds.
　　Seeds along the centre.
　　Glabrous or pubescent. Flowers under 15 in the head.
　　　Phyllodia linear, straight . . . . 148. *A. linifolia.*
　　　Phyllodia lanceolate-falcate or slightly oblong . . . . 149. *A. prominens.*
　　Hispid with long hairs. Flowers above 20 in the head . . 150. *A. Leichhardtii.*
　Pod rather thick, coriaceous, 2 to 2⅔ lines broad, contracted be-
　　tween the seeds. Flowers about 20 in the head . . . . 151. *A. crassiuscula.*
　Pod flat, about 3 lines broad, often contracted between the seeds.
　　Seeds close to the edge.
　　Phyllodia linear or lanceolate, 1 to 1½ in. Flowers 15 to 20
　　　in the head . . . . 152. *A. decora.*

Phyllodia obliquely oblong or broadly lanceolate, rarely above
   1 in. long.   Flowers 8 to 15 in the head . . . . . . 153. *A. buxifolia.*
Phyllodia falcate oblong or obliquely ovate, rarely above 1 in.
   long.   Flowers 4 to 10 in the head . . . . 154. *A. lunata.*
Phyllodia ovate obovate or broadly oblong, usually under 1 in., or
   in Nos. 157 and 158 often 1½ in. long.
Racemes short with few heads.   Flowers numerous.
   Sepals free or separating.   Funicle short, clavate . . . . 155. *A. brachybotrya.*
   Calyx short, broad, shortly toothed.   Funicle long, folded
     round the seed . . . . 156. *A. Wattsiana.*
Racemes much longer than the phyllodia.
   Glaucous and scarcely pubescent.   Phyllodia ovate obovate or
     broadly oblong, 1 to 1½ in long.   Flowers numerous   157. *A. podalyriæfolia.*
   Softly pubescent.   Phyllodia elliptical falcate or broadly and
     obliquely oblong, 1½ in. or rather longer.   Flowers numerous 158. *A. uncifera.*
   Hirsute.   Phyllodia ovate-elliptical recurved obliquely aris-
     tate mostly about ½ in.   Flowers 10 to 20 . . . . . 159. *A. vestita.*
   Glaucous and glabrous.   Phyllodia falcate-ovate or curved-
     oblong, under 1 in. long.   Flowers 10 to 20 . . . . . 160. *A. cultriformis.*
   Glabrous.   Phyllodia falcate-obovate or trapezoid, usually 2-
     nerved and not exceeding ½ in.   Flowers 8 to 12 . . . 161. *A. pravissima.*
Phyllodia ovate, spinescent, 1 to 3 in. long.   Stipules spinescent.
   Racemes exceeding the leaves.   Flowers numerous . . . . 162. *A. pyrifolia.*
Phyllodia incurved-falcate, ovate lanceolate or linear, coriaceous
   with thick margins.   Flowers 4-merous, 2 to 4 or rarely 6 in
   the head.   Racemes rarely exceeding the leaves . . . . . 163. *A. myrtifolia.*

SERIES VII. **Plurinerves.**—*Phyllodia vertically flattened, obtuse or with an innocuous or recurved point (rarely pungent when the phyllodium is broad), with 2 or more
longitudinal nerves.   Flowers in globular heads on axillary peduncles, either solitary or
clustered or shortly racemose.*

A. **Armatæ.**—*Stipules spinescent.   Phyllodia falcate, subulate-acuminate or almost
pungent.*
Phyllodia ¾ to 1½ in. long, 2-nerved, the upper margin entire,
   nerve-like, with a prominent angle . . . . 164. *A. scalpelliformis.*
Phyllodia 1½ to 3 in. or longer, 2- to 4-nerved, the upper margin
   much-curved, undulate, crenulate . . . . : . . 165. *A. urophylla.*

B. **Triangulares.**—*Stipules setaceous or minute, not spinescent.   Phyllodia small
(under ½ in. long), broadly falcate-ovate or triangular with small points often pungent.*
Phyllodia triangular.   Pod narrow, curved or twisted, glabrous.
   Branches usually spinescent.   Phyllodia rather distant.   Flowers
     20 to 30 in the head . . . . 166. *A. sublanata.*
   Branches elongated, rather rigid.   Phyllodia numerous.   Flowers
     10 to 15 in the head . . . . 167. *A. amblygona.*
Phyllodia ovate-falcate.   Pod flat, glandular-hispid.
   Flowers not ¾ line long.   Calyx much shorter than the corolla . 168. *A. deltoidea.*
   Flowers above 1 line long.   Calyx nearly as long as the corolla. 169. *A. stipulosa.*

(95, *A. decipiens* and some allied 1-nerved species, have occasionally 1 or 2 secondary
nerves, but fainter and shorter than the principal one.)

C. **Brevifoliæ.**—*Phyllodia under 1 in. long, obovate ovate or broadly oblong, very
obtuse, often undulate.   Stipules inconspicuous.*
Phyllodia under ¼ in. or very rarely ½ in. long.   Petals smooth.
   Phyllodia faintly 2-nerved, very oblique . . . . 170. *A. loxophylla.*
   Phyllodia faintly several-nerved, obtuse but with an incurved
     hair-like point . . . . 171. *A. setulifera.*

Phyllodia ½ to 1 iu. long. Petals strongly striate.
Phyllodia faintly nerved. Calyx shortly toothed. Pod thick
and hard, hooked at the end, tapering into a long stipes . . 172. *A. translucens.*
Phyllodia several-nerved. Sepals spathulate, free or separating.
Pod thiuly coriaceous, flat with thickened parallel margins,
glutinous and villous . . . . 173. *A. impressa.*

D. **Oligoneuræ.**—*Phyllodia above ½ in. and mostly above 1 in.
late or linear, straight or scarcely falcate, with 2 or 3 nerves, faintly or not at all veined
between them (except* A. *Simsii), and not glutinous (except* A. *subporosa).*

Flower-heads racemose. Phyllodia coriaceous, obtuse.
Phyllodia 2-nerved. Flower-heads loosely and irregularly race-
mose. Calyx short, toothed . . . . 174. *A. bivenosa.*
Phyllodia 3-nerved. Flower-heads very shortly racemose. Se-
pals distinct . . . . 175. *A. trineura.*
Peduncles solitary or clustered.
Phyllodia rigid, prominently 3-nerved, rarely 2-nerved.
Sepals free, spathulate (Western species).
Phyllodia ½ to 1½ in. long, linear-cuneate . . . . . 176. *A. nitidula.*
Phyllodia 1½ to 3 in. loug, linear-lanceolate or linear . . 177. *A. heteroclita.*
Calyx turbinate, lobed. Phyllodia long, linear (Eastern species) 178. *A. elongata.*
Phyllodia less rigid, nerves 3 rarely 2, less prominent, and often
veins between them.
Glutinous. Phyllodia marked with numerous glandular dots . 179. *A. subporosa.*
Not glutinous nor dotted . . . . 180. *A. Simsii.*

(13, *A. cochlearis* and some allied species amoug *Pungentes*, have occasionally scarcely
pungent phyllodia, bringing them near to *A. heteroclita.*)
(113, *A. montana*, and 114, *A. verniciflua*, have usually 2 prominent uerves, but are
closely connected with and pass into some 1-nerved species; both are glutinous.)
(191, *A. Whanii*, 194, *A. iriophylla*, and some others of *F, Nervosæ*, have sometimes only
2 or 3 nerves, but usually either a greater number or they anastomose with each other.)

E. **Microneura.**—*Glabrous or glaucous and not glutinous. Phyllodia thick, veinless
or with very fine, scarcely prominent parallel veins, narrow or rarely short and obovate.*

Phyllodia cuneate-oblong or almost obovate, very obtuse, about ½
in. long . . . . 181. *A. leptospermoides.*
Phyllodia linear or linear-cuneate, obtuse, ¾ to 1½ in. long . . . 190. *A. farinosa.*
Phyllodia linear-lanceolate, 1 to 3 in. long.
Flower-heads on short peduncles.
Pod 2 to 3 lines broad, coriaceous, longitndinally striate . . 182. *A. homalophylla.*
Pod 5 lines broad, thin, flat, transversely reticulate . . . . 183. *A. pendula.*
Flower-heads sessile . . . . 184. *A. Oswaldi.*
Phyllodia linear, very narrow and rigid, 1½ to 2 in. long . . . 185. *A. lineolata.*
Phyllodia linear, ½ ft. long or more.
Veins only visible under a lens (Western species) . . . . . 186. *A. coriacea.*
Veins very fine but prominent (Eastern species) . . . . . 187. *A. stenophylla.*
Phyllodia linear-lanceolate falcate, ½ ft. loug or more . . . . 200. *A. harpophylla.*

F. **Nervosæ.**—*Often viscid, occasionally glaucous, rarely hoary or pubescent. Phyllodia straight or sometimes falcate, coriaceous or thin, with several prominent nerves
and, when broad, reticulate between them, the nerves rarely reduced to 3 when the phyllodium is narrow.*

Glaucous. Phyllodia coriaceous, oblong-cuneate, obtuse; nerves
3 to 5, slightly prominent, veins reticulate . . . . . . 188. *A. hemignosta.*
Glabrous, not viscid. Phyllodia linear, coriaceous with a prominent
midrib and several fine parallel nerves . . . . 192. *A. heteroneura.*

Glabrous, often viscid. Phyllodia rigid, narrow, obtuse, with few
prominent, more or less anastomosing nerves.
Phyllodia ¾ to 1 in. long. Veins scarcely anastomosing. Se-
pals free.
    Nerves prominent. Peduncles glabrous . . . . 189. *A. sclerophylla.*
    Nerves faint. Peduncles tomentose . . . . 190. *A. farinosa.*
    Phyllodia 1½ to 2 in. long. Veins more anastomosing. Sepals
      very thin, united . . . . 191. *A. Whanii.*
Very viscid. Nerves or veins numerous.
Nerves parallel scarcely anastomosing.
    Phyllodia narrow-linear . . . . 193. *A. viscidula.*
    Phyllodia linear-lanceolate or oblong . . . . 194. *A. ixiophylla.*
Reticulate veins very prominent, anastomosing with the nerves . 195. *A. dictyophleba.*
Not viscid. Nerves or veins usually numerous.
Phyllodia nearly straight, coriaceous, many-nerved, strongly re-
ticulate.
    Phyllodia oblong-lanceolate. Funicle folded and thickened
      under the seed (Eastern species) . . . . 196. *A. venulosa.*
    Phyllodia oblong, obtuse. Funicle encircling the seed in a
      double fold (Western species) . . . . 197. *A. cyclopis.*
    Phyllodia falcate-oblong or lanceolate, obtuse. Veins nume-
      rous. Funicle encircling the seed (Eastern species) . . 198. *A. melanoxylon.*
Phyllodia long and falcate, coriaceous.
Reticulate veins numerous.
    Pod broad. Funicle encircling the seed in a double fold . 198. *A. melanoxylon.*
    Pod narrow, twisted. Funicle folded under the seed . . 199. *A. implexa.*
Reticulate veins few, scarcely conspicuous, nerves several, fine 200. *A. harpophylla.*
Phyllodia rather thin, straight, with several nerves and few fine
intermediate veins.
    Branches terete . . . . 201. *A. excelsa.*
    Branches flattened, 2-edged or 2-winged . . . . . . 202. *A. complanata.*

G. **Dimidiatæ.**—*Phyllodia, usually broad, and often long, falcate or very oblique,
with 2, 3, or 4 prominent distant nerves, and reticulately penniveined between them.*
Glabrous. Phyllodia rather thin (3 to 4 in.). Flower-heads in
axillary racemes growing out into leafy branches. Pod 6 to 8
lines broad . . . . 203. *A. binervata.*
Glabrous. Phyllodia coriaceous, long-falcate (6 to 8 in.). Flower-
heads in axillary clusters or very short racemes . . . . 204. *A. latescens.*
Hoary or glaucous. Phyllodia large, broad (3 to 4 in.). Pod
above 1 in. broad . . . . 205. *A. sericata.*
Young shoots hoary or yellowish-tomentose. Phyllodia large broad
(4 to 8 in.). Flower-heads small, in a terminal panicle. Pod ¾
in. broad . . . . 206. *A. flavescens.*
Woolly or velvety-tomentose. Phyllodia obovate or orbicular (1½
to 2 in.). Flower-heads in a terminal raceme . . . . . 207. *A. retivenia.*

Series VIII. **Julifloræ.**—*Phyllodia vertically flattened or in a few species terete,
several-nerved or rarely 1-nerved, obtuse acute or pointed, rarely slightly pungent. Flowers
in cylindrical dense or interrupted spikes, rarely, when sessile, shortly oblong.*

A. **Rigidulæ.**—*Phyllodia flat, often short, straight oblique or shortly falcate. Spikes
dense (except A. megalantha). Flowers 5-merous. Species all tropical except the last 2,
which are eastern.*
Phyllodia small, rarely above ½ in. long, undulate.
    Phyllodia mostly clustered, narrow, 2 to 4 lines long. Spikes
      sessile. Sepals free . . . . 208. *A. amentifera.*
    Phyllodia broad, about ½ in. long. Spikes pedunculate. Calyx
      sinuate-toothed. Seeds oblique . . . . 209. *A. Wickhami.*

Phyllodia narrow, obtuse with a short point, ½ to 1¼ in. long.
Phyllodia obscurely 3- to 5-nerved. Pod 3 to 6 lines broad.
Seeds oblique . . . . 210. *A. lysiphlœa.*
Phyllodia obscurely 1-nerved. Pod 1½ lines broad. Seeds
longitudinal . . . . 211. *A. linarioides.*
Phyllodia rather broad, coriaceous, mostly 1 to 3 in. rarely 4 in.
long, obtuse or with a glandular callous point, straight oblique
or shortly falcate.
Tomentose or pubescent. Stipules conspicuous. Phyllodia 1 to
2 in. long, 2- to 4-nerved, with anastomosing veins. Pod
narrow . . . . 212. *A. stipuligera.*
Resinous. Phyllodia 1 to 1½ in. long, 5 to 9-nerved, without
intermediate veins . . . . 213. *A. ptychophylla.*
Glabrous, except the young shoots. Phyllodia obliquely narrowed at both ends, somewhat undulate, with a terminal
gland often large.
Phyllodia finely 3- to 5-nerved. Spikes pedunculate. Buds
very striate. Calyx sinuate-toothed . . . . 214. *A. stigmatophylla.*
Phyllodia very coriaceous, 5- to 9-nerved. Spikes sessile.
Pod terete, turgid. Seeds oblique . . . . 215. *A. umbellata.*
Phyllodia finely 3- to 5-nerved. Spikes pedunculate. Buds
smooth. Calyx deeply lobed . . . . 216. *A. leptophleba.*
Phyllodia broadly oblong-falcate, obtuse or with a hooked callous
point, very coriaceous.
Spikes dense. Flowers not 1 line long. Pod flat. Seeds ob-
lique.
Pod sessile, but narrowed at the base . . . . 217. *A. limbata.*
Pod stipitate . . . . 218. *A. brevifolia.*
Spikes interrupted. Flowers 2 lines long . . . . . . 219. *A. megalantha.*
Phyllodia nearly straight, coriaceous, obtuse, 3 to 4 in. long. Pod
narrow. Seeds longitudinal.
Branches very angular.
Spikes pedunculate . . . . 220. *A. gonoclada.*
Spikes sessile . . . . 221. *A. pycnostachya.*
Branches terete or nearly so . . . . 222. *A. subtilinervis.*

B. **Tetramerœ.**—*Phyllodia flat, coriaceous or thin, straight or falcate, several-nerved
or 1-nerved when very narrow. Spikes often loose. Flowers 4-merous. Seeds longitudinal.*

Spikes very dense and sessile.
Phyllodia linear-lanceolate, very rigid, 5- to 7-nerved. Pod spi-
rally twisted (Western species) . . . . 223. *A. cochliocarpa.*
Phyllodia long, broadly falcate, 2- to 5-nerved and much reticulate 224. *A. Dallachiana.*
Spikes usually slender or interrupted.
Phyllodia broadly and obliquely obovate, ¾ to 1½ in. long . . 225. *A. alpina.*
Phyllodia rarely falcate, 2 to 6 in. long or more, 3- to 5-nerved,
broadly oblong and much reticulate or narrow and scarcely veined 226. *A. longifolia.*
Phyllodia long, narrow-linear, mostly 1-nerved . . . . . . 227. *A. linearis.*

(44, *A. Riceana,* and 45, *A. axillaris,* differ from the short-leaved forms of *A. linearis*
in having the phyllodia more rigid and pungent.)
(A few species of the following *Stenophyllœ* have 4-merous flowers, but with linear-terete
finely-striate phyllodia.)

C. **Stenophyllæ.**—*Phyllodia linear-subulate or narrow-linear, straight or slightly
curved, terete or flat but thick, rarely under 2 in. long or above 1½ lines broad. Spikes
dense, short or slender, with small 5-merous, or in terete-leaved species often 4-merous
flowers.*

Spikes sessile. Flowers 5-merous or 4-merous. Pod narrow, with

longitudinal seeds in *A. aciphylla* and *A. ephedroides*, broad
with oblique seeds in *A. brachystachya*, unknown in the
others.

Phyllodia linear-subulate, terete or nearly so.
  Phyllodia pungent-pointed, minutely striate under a lens.
    Spikes ovoid or oblong. Flowers 5-merous . . . . . 228. *A. aciphylla*.
  Phyllodia few-nerved. Spikes ovoid or oblong. Flowers 4-
    merous . . . . 229. *A. ephedroides*.
  Phyllodia minutely striate under a lens. Spikes 3 to 4 lines
    long. Flowers 4-merous . . . . 230. *A. Burkittii*.
  Phyllodia very long, minutely striate under a lens. Spikes ½
    in. long. Flowers 5-merous . . . . 232. *A. cyperophylla*.
  Phyllodia with few prominent nerves. Spikes above ½ in.
    Flowers 4-merous . . . . 233. *A. multispicata*.
Phyllodia ½ to 1 line broad, flat but thick. Spikes under 4 lines
  long.
  Phyllodia finely striate with the central nerve prominent.
    Flowers 4-merous. Sepals united . . . . 231. *A. microneura*.
  Phyllodia minutely striate under a lens. Flowers 5-merous.
    Sepals free . . . . 241. *A. brachystachya*.
Spikes pedunculate. Flowers usually 5-merous. Pod narrow, with
  longitudinal seeds in *A. pityoides* and *A. oncinophylla*, hard,
  with valves rolling back elastically and oblique seeds, in *A.*
  xylocarpa, A. gonocarpa, A. drepanocarpa, and *A. arida ;*
  flat, broad, thin, with oblique or transverse seeds, in *A. aneura*
  and *A. brachystachya*.
  Phyllodia terete or scarcely flattened, very finely striate under a
    lens.
    Calyx thin and deeply divided, the lobes not spathulate . . 234. *A. pityoides*.
    Sepals free, narrow-spathulate . . . . 240. *A. aneura*.
  Phyllodia terete, nerveless or obscurely 1-nerved. Calyx lobed. 235. *A. xylocarpa*.
  Phyllodia very narrow but flat, 1-nerved. Sepals free . . . 236. *A. gonocarpa*.
  Phyllodia flat but thick, prominently 3- or 5-nerved.
    Phyllodia with a hooked or rarely straight point (Western
      species) . . . . 237. *A. oncinophylla*.
    Phyllodia obtuse (tropical species) . . . . 238. *A. drepanocarpa*.
  Phyllodia flat, minutely striate under a lens.
    Calyx shortly lobed . . . . 239. *A. arida*.
    Sepals free, linear-spathulate.
      Spikes cylindrical . . . . 240. *A. aneura*.
      Spikes ovoid or oblong, 2 to 3 lines long . . . . . . 241. *A. brachystachya*.

D. **Falcatæ.**—*Phyllodia usually long or large, more or less falcate, narrowed at each
end, with numerous parallel nerves or veins either all equal or the central one or several
more prominent than the others, the smaller ones occasionally anastomosing. Spikes
slender, dense or rarely interrupted. Flowers mostly 5-merous.*

(Several species of this group cannot be distinguished without the fruit.)
Pod (where known and probably in all the species) narrow, with
  longitudinal seeds, or rarely broader with the longitudinal
  seeds along the centre.
  Phyllodia narrow-lanceolate or linear-lanceolate, rather thick,
    slightly falcate, with very fine parallel nerves, the midrib
    usually more prominent.
    Loosely pubescent. Stipules conspicuous. Phyllodia not
      above 3 in. long . . . . 242. *A. conspersa*.
    Glabrous or the young shoots silky-pubescent. Stipules in-
      conspicuous. Phyllodia mostly above 3 in. long.
      Eastern species. Pod unknown . . . . 243. *A. doratoxylon*.

Western species.
Young shoots usually silky-pubescent. Midrib of the
phyllodia prominent. Pod narrow, convex over the
oblong seeds , ... 244. *A. acuminata.*
Quite glabrous. Nerves of the thick phyllodia nearly
equal. Pod unknown . . . . 245. *A. stereophylla.*
Glabrous and glaucous. Pod rather broad, with thickened
margins. Seeds ovate . . . . 246. *A. signata.*
Phyllodia narrow-lanceolate, not so thick and rather more falcate
than in the preceding species, usually with about 3 nerves
more prominent than the rest.
Phyllodia sprinkled with a few hairs. Pod flat, slightly
convex over the seeds . . . . 247. *A. delibrata.*
Phyllodia very glabrous, the smaller veins between the 3 principal nerves scarcely conspicuous. Pod unknown . . . 248. *A. oligoneura.*
Phyllodia glabrous. Nerves numerous. Pod very convex
over the seeds and moniliform . . . . 249. *A. torulosa.*
Phyllodia glabrous. Pod spirally twisted into numerous coils 250. *A. julifera.*
Phyllodia glabrous. Spikes interrupted, 2 to 3 in. long (slender but dense in the preceding species) . . . . . . 251. *A. Solandri.*
Phyllodia more falcate than in the preceding species, often
broader or longer, with more nerves. Pod narrow or flat,
straight or twisted.
Branches scarcely angular. Phyllodia coriaceous, often hoary
with numerous very fine nerves, all free from the base.
Calyx pubescent . . . . 253. *A. glaucescens.*
Branches very angular. Phyllodia with 1 or 2 of the principal
nerves confluent with the lower margin of the base. Pod
twisted . . . . 254. *A. Cunninghamii.*
Branches terete or nearly so. Flowers glabrous.
Pod very narrow and straight . . . . 255. *A. leptocarpa.*
Pod broad, very flexuose or twisted, not spiral. Seeds along
the centre . . . . 256. *A. polystachya.*
Branches terete or nearly so. Flowers pubescent. Pod long
and slender, longitudinally striate and furrowed . . . 257. *A. holcocarpa.*
Pod (where known and probably in all the species) rather broad,
coriaceous woody or rarely rather thin. Seeds very oblique or
transverse. Parallel veins of the phyllodia usually numerous
and closely packed.
Phyllodia long, narrow, and slightly falcate (as in *A. julifera*).
Pod flat, with straight margins and undulate valves. Seeds
orbicular.
Branches acutely angular . . . . 258. *A. plectocarpa.*
Branches scarcely angular . . . . 259. *A. pachycarpa.*
Phyllodia broad, falcate or very oblique. Pod nearly terete and
turgid . . . . 260. *A. tumida.*
Phyllodia long, as in *A. julifera*, but usually more falcate. Pod
hard, flat, contracted to the base.
Phyllodia 3 to 5 lines broad. Pod obtuse, not hooked . . 261. *A. loxocarpa.*
Phyllodia 6 to 8 lines broad. Pod obtusely hooked at the end 262. *A. oncinocarpa.*
Phyllodia broad, falcate or very oblique. Pod broad, hard, and
woody, obliquely veined.
Branchlets 3-angled. Pod obtusely recurved or hooked at the
end, much narrowed at the base . . . . 264. *A. aulacocarpa.*
Branchlets very flat. Pod not hooked, narrowed at the base . 265. *A. calyculata.*
Branchlets nearly terete or slightly angular. Pod broad, very
hard, obliquely truncate at the base.
Pod flat or scarcely twisted, the outer margin entire . . 266. *A. crassicarpa.*

Pod much twisted, the outer margin deeply sinuate . . . 267. *A. auriculiformis.*
Pod quite uncertain.
　Phyllodia narrowed at both ends, with numerous parallel veins
　　or nerves, and usually falcate, as in the *Falcatæ*, but under
　　3 in. long . . . . 252. *A. leptostachya.*
　Phyllodia of *A. julifera* and others of the *Falcatæ*, but the small
　　lateral parallel veins appearing, under a lens, very much ana-
　　stomosed into a fine network . . . . 263. *A. retinervis.*

E. **Dimidiatæ.**—*Phyllodia large, broad, very oblique or falcate, with 3 or more dis-
tant prominent nerves, more or less confluent at or near the lower margin at the base,
pinnately net-veined between them.*
Branchlets very acutely angular or almost winged.
　Glabrous and glaucous. Spikes pedunculate. Flowers mostly
　　4-merous . . . . 268. *A. latifolia.*
　Hoary or silky-pubescent. Spikes sessile. Flowers mostly 5-
　　merous . . . . 269. *A. holosericea.*
Branchlets terete or scarcely angular.
　Spikes 1 to 2 in. long. Flowers glabrous. (Erect shrub) . . 270. *A. dimidiata.*
　Spikes scarcely ½ in. long, dense. Flowers densely pubescent.
　　Stem prostrate or diffuse . . . . 271. *A. humifusa.*

Div. II. *BIPINNATÆ.*—*Leaves all bipinnate.*

Series IX. **Botryocephalæ.**—*Leaves bipinnate. Stipules small or none. Flower-
heads globular, in axillary racemes or terminal clusters.*
Pinnæ 2 to 4 pairs, rarely 5 or 6 pairs. Leaflets above ¼ in. long.
　Young shoots golden-pubescent. Leaflets 8 to 12 pairs, lan-
　　ceolate, acute, 1 to 2 in. long . . . . 272. *A. elata.*
　Glabrous and glaucous. Leaflets 12 to 20 pairs, oblong or linear,
　　scarcely acute, about ½ in. long. Gland distant from the
　　lowest pinnæ . . . . 273. *A. pruinosa.*
　Glabrous and glaucous. Leaflets 4 to 8 pairs, obovate-oblong,
　　very obtuse, 4 to 6 lines long. Gland close to the lowest pinnæ 274. *A. spectabilis.*
　Foliage pubescent. Leaflets 6 to 10 pairs or more, obtuse, 3 to
　　4 lines long. Gland at the base of the petiole . . . . . 275. *A. polybotrya.*
　Foliage glabrous. Leaflets 10 to 15 pairs, obtuse or acute, firm,
　　pale underneath, 3 to 4 lines long . . . . 276. *A. discolor.*
Pinnæ 8 to 15 pairs. Leaflets very numerous, narrow-linear and
　very small, or 3 to 4 lines long and subulate.
　Glabrous, or, if tomentose, pubescent, the young shoots of a yel-
　　lowish or golden tinge. Pod (always ?) under 4 lines broad,
　　often contracted between the seeds . . . . 277. *A. decurrens.*
　Silvery-tomentose or very glaucous. Pod (always ?) above 4
　　lines broad, not contracted between the seeds . . . . . 278. *A. dealbata.*
Pinnæ 12 to 15 pairs. Leaflets 6 to 10 pairs, ovate-cordate,
　under 1 line long. Branches pubescent . . . . 279. *A. cardiophylla.*
Pinnæ 3 to 6 pairs. Leaflets 6 to 10 pairs, oblong, under 2 lines
　long. Branches glabrous or hispid. Flowers numerous, in
　dense heads . . . . 280. *A. leptoclada.*
Pinnæ 3 to 10 pairs. Leaflets 6 to 20 pairs, linear, under 2 lines
　long. Branches hirsute. Flowers few in the heads . . . . 281. *A. pubescens.*

Series X. **Pulchellæ.**—*Leaves bipinnate. Stipules none or smaller, setaceous, not
spinescent. Flower-heads globular or cylindrical, on simple axillary solitary or clustered
peduncles.*
Seeds longitudinal. Flower-heads globular. Petals not striate.
　Spines axillary, rarely entirely wanting. Pinnæ 1 pair. Leaf-
　　lets 4 to 7 pairs . . . . 282. *A. pulchella.*

No spines.
  Branches pubescent. Pinnæ 2 or 3 pairs. Leaflets 3 to 6
    pairs, oblong, 1 to 2 lines long. Sepals free or nearly so . 283. *A. Mitchelli.*
  Glabrous. Pinnæ 2 to 5 pairs. Leaflets 20 to 30 pairs,
    with a broad oblique base, 1 to 3 lines long. Calyx
    shortly toothed . . . . 284. *A. pentadenia.*
  Glabrous. Pinnæ 1 pair. Leaflets 4 to 6 pairs, above 3 lines
    long. Flowers large, 4-merous, 3 to 8 in the head . . . 285. *A. Gilberti.*
Seeds transverse. No spines. Flower-heads globular. Petals
    usually striate. Pinnæ usually 1 or 2 pairs.
  Glabrous. Leaflets 5 to 10 pairs, 3 to 5 lines long. Flowers
    20 or more in the head . . . . 286. *A. nigricans.*
  Branches pubescent or hirsute. Flowers 12 to 15 in the head.
    Leaflets 5 to 10 pairs, 2 to 3 or rarely 4 lines long . . . 287. *A. obscura.*
    Leaflets 1 to 4 pairs, 1 to 2 lines long . . . . 288. *A. strigosa.*
Seeds transverse. No spines. Flower-spikes cylindrical or rarely
    ovoid. Petals smooth. Pinnæ 2 pairs. Leaflets 2 to 6 pairs . 289. *A. Drummondii.*

(Branches of 292, *A. Bidwilli* and of 293, *A. pallida*, are sometimes almost or quite
without stipular spines.)

SERIES XI. **Gummiferæ.**—*Leaves bipinnate. Stipules of some or all the leaves per-
sistent and spinescent. Flower-heads globular, on solitary or clustered simple peduncles.*
Bracts small, close under the flower-heads.
  Pinnæ 4 to 6 pairs. Leaflets small, 10 to 20 pairs. Pod thick,
    cylindrical or spindle-shaped, indehiscent, pithy between the
    seeds . . . . 290. *A. Farnesiana.*
  Pinnæ 1 or 2 pairs. Leaflets small, 8 to 12 pairs. Pod narrow-
    linear, 2-valved . . . . 291. *A. suberosa.*
Bracts forming a little 4-lobed ring round the middle of the pe-
    duncle.
  Pinnæ 15 to 25 pairs. Leaflets scarcely 1 line long. Flowers
    4-merous. Pod coriaceous; valves slightly convex, striate
    lengthwise . . . . 292. *A. Bidwilli.*
  Pinnæ 3 to 10 pairs. Leaflets 3 to 4 lines long. Flowers 5-
    merous . . . . 293. *A. pallida.*

Mimosa obliqua, Pers. Syn. ii. 261, not of Wendl. nor of Lam., a phyllodineous Australian *Acacia*, is insufficiently described to determine even to which series it belongs.
  A. visneoides, Colla, Hort. Ripul. App. ii. 339, is described from a garden specimen in
leaf only and is quite unrecognizable.
  Numerous manuscript names of A. Cunningham's, F. Mueller's, and others, unaccompanied by descriptions, but quoted in G. Don's ' General System,' in my own papers in
Hooker's ' London Journal of Botany,' i. 318, and in the ' Linnæa,' xxvi. 603, in F. Mueller's
paper in the ' Journal of the Linnean Society,' iii. 114, in Seemann's work, ' Die in Europa
eingeführten Acacien,' 1852, in Steudel's ' Nomenclator,' or in garden catalogues, are here
omitted as unpublished.

DIVISION I. *PHYLLODINEÆ.*

Leaves all (except on young seedlings and occasionally one or two on young
branches) reduced to *phyllodia*, that is to the petiole either terete or angular
or more or less vertically dilated so as to assume the appearance of a rigid
simple leaf, with an upper and a lower edge or margin, and two lateral similar
surfaces, and either sessile or contracted at the base into a short petiole, the
upper edge often bearing 1, 2, or rarely 3 or more shield-shaped or tubercular or depressed glands.

SERIES I. ALATÆ.—Phyllodia bifariously decurrent, forming 2 opposite wings to the stem, the free part either short broad and acute, or rigid and pungent-pointed. Flowers in globular heads, on axillary simple peduncles, the heads rarely appearing racemose by the reduction of the flowering branches.

This series differs from the *Continuæ* by the phyllodia always distichous, much more decurrent, and often reduced apparently to lobes or teeth of the wings. A few other species, such as *A. complanata*, have flattened stems, but the phyllodia are distinctly articulate on the node.

1. **A. (?) bossiæoides,** *A. Cunn. Herb. ; Benth. in Hook. Lond. Journ.* i. 323 ; *not of Seem.* A tall shrub, glabrous and glaucous, apparently allied to *A. alata*, but somewhat doubtful, the flowers and fruits being unknown. Phyllodia short, triangular, bifarious and decurrent along the stem, each one continued beyond the next below, forming continuous opposite wings, the axillary buds thus removed to 2 or 3 lines from the stem in the centre, as in some species of *Brachysema* and *Bossiæa,* but the venation is that of the winged *Acacias*. Stipules small, lanceolate, oblique and almost semisagittate at the base.

N. Australia. Liverpool river, N.W. coast, *A. Cunningham.* A very remarkable species, originally published as an *Acacia*, on the authority of Cunningham, which I now regret, although I know of no other genus to which it is likely to belong.

2. **A. glaucoptera,** *Benth. in Linnæa,* xxvi. 604. A much-branched, glabrous and glaucous shrub. Phyllodia erect, bifarious, oblong-falcate, decurrent along the stem, each one continued to the next below, but not beyond, forming opposite wings, notched at each node, where the axillary bud is sessile on the stem ; the free part of the phyllodium from ¾ to 1½ in. long, and 3 to 4 lines broad at the base, with a central prominent nerve. Stipules small, rigid, but not spinescent. Peduncles solitary or 2 together, under ½ in. long, bearing each a globular compact head of very numerous (above 30) small flowers, mostly 5-merous. Sepals distinct, linear-spathulate, scarcely half as long as the corolla. Petals distinct, rather narrow, smooth. Pod not seen.—*A. bossiæoides*, Seem. Eingef. Acac. t. 1, not of A. Cunn.

W. Australia. Towards Cape Riche, *Drummond, 5th Coll. n.* 1, and in leaf only, *4th Coll. n.* 1 ; Clay flats, Fitzgerald, Gardner and Phillips ranges, *Maxwell.*

3. **A. alata,** *R. Br. in Ait. Hort. Kew. ed.* 3, v. 464. A tall shrub, attaining 5 or 6 ft. or more, but flowering when only 1 or 2 ft. high, glabrous or more or less hirsute. Phyllodia falcate-ovate, erect or spreading, bifariously decurrent along the stem, each one continued to the next below, but not beyond, forming opposite wings, notched at each node, varying in breadth from 1 or 2 lines to ½ in., the free part, in the ordinary form, about ½ to ¾ in. long, usually with a gland-bearing angle on the upper edge, the central nerve terminating in a pungent point. Stipules spinescent. Peduncles solitary or 2 together, bearing each a globular head of about 6 to 12 comparatively large flowers, mostly 5-merous. Bracts ovate, sessile. Calyx sometimes minute, sometimes nearly one-third as long as the corolla, broad, more or less lobed. Petals 5, smooth, united to the middle. Pod stipitate, oblong-falcate or lanceolate, acuminate, incurved, with thickened margins, glabrous or hirsute ; valves convex. Seeds ovate, transverse, the funicle with few short folds, the last thickened into a small aril.—Wendl. Comm.

Acac. t. 1; Bot. Reg. t. 396; DC. Prod. ii. 448; Colla, Hort. Ripul. t. 17; Reichb. Ic. et Descr. Pl. t. 88, f. 1; Meissn. in Pl. Preiss. i. 4.

W. Australia. King George's Sound and adjoining districts, *R. Brown, Fraser,* and others; Swan and Canning rivers, *Oldfield, Preiss, n.* 997, *Drummond, 1st Coll. n.* 306 *and* 307.

Var. *platyptera,* Meissn. Phyllodia larger narrower and more distant, recurved at the end and scarcely pungent.—*A. platyptera,* Lindl. Bot. Reg. 1841, Misc. 3; Bot. Mag. t. 3933; *A. uniglandulosa,* Seem. and Schmidt, in Flora, 1844, 495.—Raised from Drummond's seeds.

Var. *biglandulosa.* Phyllodia short, broad, coriaceous, mostly with 2 or sometimes 3 gland-bearing angles or teeth on the upper edge.—Amongst Oldfield's and other specimens.

4. **A. diptera,** *Lindl. Swan Riv. App.* 15. A glabrous and glaucous or pubescent shrub, with virgate branches, more slender and elongated than in *A. alata.* Phyllodia few, distant, bifariously decurrent, forming long opposite wings, sometimes very narrow, sometimes 2 to 4 lines broad, the free part falcate, varying from a minute tooth to ¾ in. in length, the nerve close to the upper edge, which has no gland. Stipules minute or none. Peduncles slender, solitary or 2 together in the axils, or appearing racemose from the flowering branch being reduced to a slender leafless rhachis, bearing each a globular head of about 20 flowers, mostly 5-merous. Calyx very small, shortly lobed. Petals united at the base, slightly striate. Pod stipitate, slightly falcate, flat with nerve-like margins, about 2 in. long and 4 lines broad, but not quite ripe in our specimens. Seeds transverse.—Hook. Ic. Pl. t. 369; Meissn. in Pl. Preiss. i. 4; *A. Willdenowiana,* Wendl. Verz. K. Berggart. 1845, 5, according to Seem. Eingef. Acac. 9.

W. Australia. Swan River, *Drummond, 1st Coll., Preiss, n.* 995, 996, and others; Vasse river, *Mrs. Molloy, Oldfield;* Gordon plains, *Oldfield;* northward to Moore and Murchison rivers, *Drummond, 6th Coll. n.* 3, and southward to Mount Wuljenup, *Preiss, n.* 393, and Cheynes Beach, *Maxwell.*

Var. *erioptera,* Grah. in Bot. Mag. t. 3939. Pubescent. Flowers rather large.—Swan River, *Drummond.*

5. **A. stenoptera,** *Benth. in Hook. Lond. Journ.* i. 325. A rigid undershrub or low bushy shrub, quite glabrous or slightly scabrous. Phyllodia bifariously decurrent, forming long opposite wings, 1½ to 2 or even 3 lines broad, the free part lanceolate or linear-falcate, rigid, tapering to a pungent point, 1-nerved, without marginal glands. Stipules small, not pungent. Peduncles under ½ in. long, bearing each a head of 6 to 10 flowers, larger than in *A. alata* and mostly 4-merous. Calyx less than half as long as the corolla, shortly lobed. Petals rather rigid and strongly striate. Pod only seen unripe, very much falcate, acuminate, ½ in. broad in the middle, narrowed into a stipes of nearly 1 in., and apparently very convex, with a longitudinal wing on each side of each suture.—Meissn. in Pl. Preiss. i. 5, ii. 199.

W. Australia. Swan River to King George's Sound, *Fraser, Drummond, 2nd Coll. n.* 145 *and* 146, *Preiss, n.* 991 *and* 992, and others; Vasse river, *Mrs. Molloy.* The flowers and fruit are nearly those of *A. tetragonocarpa,* but the habit and phyllodia are very different.

SERIES II. CONTINUÆ.—Phyllodia narrow, rigid, tapering into a pungent point, continuous with the stem and shortly and trifariously or irregularly decurrent. Flowers in heads or spikes, on axillary simple peduncles.

This small series connects the *Alatæ* with several others, and I had much doubt whether it would be most convenient to unite the species with *Alatæ* as in my former papers, or with *Pungentes*, with which they more generally agree, or distribute them into several of the series, placing *A. incurva* and *A. trigonophylla* in the *Alatæ*, *A. continua* in the *Pungentes*, *A. Peuce* in the *Calamiformes*, and *A. triptera* in the *Juliflorœ*, but on the whole it appeared to me that these large series remain better defined, if the *Continuæ* are collected into a small intermediate group.

6. **A. incurva,** *Benth. in Hook. Lond. Journ.* i. 325. A rigid shrub of 1½ to 2 ft., glabrous or slightly scabrous. Phyllodia continuous with the stem and shortly and trifariously decurrent, linear or linear-lanceolate, incurved or recurved, mostly 1½ to 2 in. long, rigid and tapering into a pungent point, 1-nerved, without marginal glands. Stipules minute or shortly setaceous. Flower-heads globular, mostly solitary and almost sessile, containing 4 to 8 flowers, mostly 4-merous. Calyx very short. Petals membranous, not striate, the buds very angular. Pod unknown.—Meissn. in Pl. Preiss. i. 5.

W. Australia. Vasse river, *Mrs. Molloy*; sandy plains, near Erwin, *Preiss, n.* 990. Var. *brachyptera.* Phyllodia divaricate, under ½ in. long.—*A. brachyptera*, Benth. in Hook. Lond. Journ. i. 325.—King George's Sound, *A. Cunningham;* near Mount Desmond, *Maxwell*, also *Oldfield.*

7. **A. trigonophylla,** *Meissn. in Pl. Preiss.* ii. 199. A rigid glabrous shrub, resembling at first sight the coarse specimens of *A. incurva*, but with very different flower-heads. Phyllodia continuous and shortly decurrent, linear-lanceolate, spreading or recurved, rigid and tapering into a pungent point, often 1 to 1½ in. long and 1 to 2 lines broad, with one nearly central nerve. Peduncles solitary, ¼ to ½ in. long, bearing each a globular compact head of 40 to 50 or more flowers, mostly 5-merous. Calyx nearly half as long as the corolla, splitting into linear-cuneate sepals. Petals rather thick, with prominent midribs. Pod straight, flat with thickened margins, 2 or 3 in. long, 2½ lines broad, much contracted between the seeds. Seeds not seen, but evidently longitudinal.—*A. pteroelada*, F. Muell. Fragm. iv. 3.

W. Australia, *Drummond. 2nd Coll. n.* 144; dense thickets in gravelly soil, Champion Bay, *Walcott.*

8. **A. continua,** *Benth.* A rigid shrub of 1 to 2 ft., glabrous in all its parts, the young branches angular-striate. Phyllodia continuous with the stem and usually shortly decurrent, nearly terete, rigid, tapering into pungent points, with several raised nerves disappearing with age, the lower ones often 1 to 2 in. long and erect, the upper ones much shorter, spreading or recurved, or even reduced to short spines. Peduncles usually very short, bearing each a globular head of above 30 flowers, mostly 5-merous. Outer bracts broadly cuneate, inner ones small. Sepals spathulate, distinct, about half as long as the corolla. Petals smooth, at length free. Pod flat, much curved or twisted, about 2 lines broad.—*A. colletioides*, F. Muell. Pl. Vict. ii. 5, not of A. Cunn.

Victoria. Ridges of the Murray desert, *F. Mueller.*
S. Australia. Ranges and scrubs, from the Murray to St. Vincent's and Spencer's Gulfs and Mount Remarkable, *F. Mueller;* Lake Gilles, *Burkitt.*

I had, in the 'Linnæa,' xxvi. 609, followed F. Mueller in referring this to *A. colletioides*, of which it has the pod, but I find, on further examination, that the character of the phyllodia, continuous not articulate on the stem, now pointed out by F. Mueller, does not exist in Cunningham's plant, which has moreover a different venation.

9 (?). **A. Peuce,** *F. Muell. Fragm.* iii. 151.   A tree of 15 to 30 ft., with the aspect of a Pine, quite glabrous and somewhat glaucous.   Phyllodia rather crowded, not articulate, but very shortly decurrent on the stem, subulate-terete, rigid, tapering into pungent points, mostly 2 to 3 in. long or rather more, straight and erect, with few slightly prominent nerves.   Stipules very minute and deciduous.   Flowers . . .   Pod (from a single imperfect one) very flat, glaucous, several in. long and nearly 1½ in. broad.   Seeds broadly ovate, flat, the funicle filiform to the end.

N. Australia. N. of Will's Creek, *Howitt's Expedition.* The pod is not attached to the specimen, and closely resembles that of *A. sericata*, and some doubt therefore attaches itself to the species.

10. **A. triptera,** *Benth. in Hook. Lond. Journ.* i. 325.   A dense rigid shrub of 3 or 4 ft., quite glabrous.   Phyllodia numerous, lanceolate, recurved-falcate, decurrent on the stem, rigid and tapering into a pungent point, ½ to 1 in. long, striate with several prominent nerves, without any marginal gland.   Stipules scarcely any.   Peduncles short, solitary or 2 together, bearing each a cylindrical spike of ½ to ¾ in.   Flowers not crowded, very small and globular in the bud, mostly 4-merous.   Calyx short, broad, lobed.   Petals membranous and smooth.   Pod unknown.

Queensland. Sandstone ridges, Mantuan Downs, *Mitchell ;* on the lower Macquarrie, *Bowman,* also in *Leichhardt's Collection.*

N. S. Wales. Barren lands, N. of Arbuthnot Range, *Fraser.*

SERIES III. PUNGENTES.—Rigid shrubs, the branches in some species spinescent. Phyllodia articulate on the stem, rigid, tapering into pungent points, usually subulate linear or lanceolate, or rarely none. Flowers in heads or spikes, on axillary simple peduncles.

The pungent phyllodia give these species in general so peculiar a habit, that it appeared more convenient to unite them in a separate group than to distribute them in the great series founded on venation and inflorescence, at the same time it must be admitted that, in a few species, the pungent character is variable. On the other hand, the adoption of the two series *Pungentes* and *Calamiformes* disposes of those terete-leaved species where the venation is uncertain, and of *A. verticillata*, where the spicate passes into the globular inflorescence.

A. APHYLLÆ.—Phyllodia none except minute scales. Branches spinescent.

The general aspect of this single species is so much that of the *Pungentes*, that it is placed here, although no phyllodia appear on any of the specimens seen. The two other species formerly described as aphyllous, occasionally produce a few branch-like phyllodia, which indicate their place among *Calamiformes.*

11. **A. spinescens,** *Benth. in Hook. Lond. Journ.* i. 323.   A glabrous, rigid, bushy, diffuse shrub, rarely exceeding 2 ft., the branches terete, striate and mostly spinescent.   Leaves none, replaced on the young shoots by small brown decidous scales.   Flower-heads globular, sessile along the branches, containing 2 to 6 flowers, mostly 5-merous.   Calyx truncate, shortly and

broadly toothed, about half as long as the corolla, Petals free, striate. Pod shortly stipitate, linear, curved, flat with slightly thickened margins, about 1½ lines broad ; valves convex over the seed and contracted between them. Seeds longitudinal, the funicle expanded into a broadly clavate lateral aril, and scarcely folded below it.—*F.* Muell, Pl. Vict. ii. 3.

N. S. Wales. In the S.W. interior, *Fraser.*
Victoria. Desert tracts of the N.W. of the colony, *F. Mueller.*
S. Australia. Memory Cove, *R. Brown ;* Encounter Bay, *Whittaker ;* Murray desert, Onkaparinga river, Barossa ranges and towards St. Vincent's Gulf, *F. Mueller.*

B. PLURINERVES.—Phyllodia 2- or more nerved, or terete and nerveless. Heads globular.

Where the phyllodia are terete and 2-nerved, the total number of apparent nerves is 6, 2 on each side, and the 2 nerve-like edges of the phyllodium nearly equally prominent.

12. **A. latipes,** *Benth. in Hook. Lond. Journ.* i. 334. A glabrous rigid shrub, very nearly allied to *A. cochlearis,* and perhaps a variety, differing chiefly in the thick rigid phyllodia more divaricate, and attached by a broad base, rarely 1 in. long, about 1 line broad, with 3 or 4 prominent nerves as in that species. Peduncles about 1 to 2 lines long, bearing a globular head of 30 to 40 or more flowers, mostly 5-merous. Sepals narrow-linear, with spathulate concave tips. Petals rather firm, smooth, united above the middle. Pod not seen.

W. Australia. Swan River, *Drummond ;* Darling range, *Oldfield.*

13. **A. cochlearis,** *Wendl. Comm. Acac.* 15. A rigid shrub of several ft., glabrous and often somewhat viscid, or the young shoots woolly-pubescent. Phyllodia linear or linear-lanceolate, rigid, tapering into a pungent point, 1 to 1½ or rarely 2 in. long, rarely above 2 lines broad, more or less narrowed at the base, with 3 or 4 not very prominent nerves on each side. Peduncles solitary or 2 together, 2 to 3 lines long or rarely more, bearing a globular head of numerous (above 30) flowers, mostly 5-merous. Sepals narrow, spathulate, free. Petals smooth, free. Pod about 1 to 2 in long, 2 lines wide, flat with thickened margins. Seeds longitudinal, the funicle thickened at the end, but not seen perfect.—DC. Prod. ii. 451 ; Meissn. in Pl. Preiss. i. 11 ; *Mimosa cochlearis,* Labill. Pl. Nov. Holl. ii. 85, t. 234 ; *A. eglandulosa,* DC. Mem. Leg. 445 ; Prod. ii. 450 (from the diagnoses given).

W. Australia. King George's Sound to Vasse and Swan rivers, *Labillardière, Wakefield, Drummond,* 1st *Coll. and* 2nd *Coll. n.* 139, *Preiss, n.* 933 *and* 949, and others.
A. Benthamii, Meissn. in Pl. Preiss. i. 11, ii. 202, is a slight variety, quite glabrous, with narrower phyllodia ; *A. heteroclita,* Meissn., is nearly allied, but has much less pungent phyllodia, with the points usually recurved, and 2 or 3 prominent nerves, and is therefore placed amongst *Plurinerves oligoneuræ.*

14. **A. lanigera,** *A. Cunn. in Field, N. S. Wales,* 345. A rigid shrub of several feet, the young shoots usually woolly-pubescent. Phyllodia linear or lanceolate, rigid, tapering into a pungent point, 1 to 1½ or rarely 2 in. long, mostly 2 to 3 lines broad, with several nerves, occasionally anastomosing or all parallel. Peduncles exceedingly short, solitary, 2 together or almost clustered, bearing a globular head of about 30 flowers, mostly 5-merous. Calyx campanulate, with broad obtuse lobes, not half as long as the corolla. Petals smooth, united to the middle. Pod attaining 6 to 8 in.

in length, and 4 or even 5 lines in breadth at the seeds, much contracted between them ; the valves slightly convex. Seeds longitudinal, last short fold of the funicle and the end of the next much thickened, the remainder of the latter and the third fold filiform and extending some way round the seed, but not seen perfect.—Bot. Mag. t. 2922.

N. S. Wales. Frequent in rocky barren ranges of the interior, from the Blue Mountains to Liverpool plains, Lachlan river and to the southward, *A. Cunningham, Fraser, Huegel, Mitchell,* and others.

Victoria. Forest Creek, *F. Mueller.*

Var. *gracilipes.* Peduncles rather longer, although shorter than in *A. trinervata.* The species has much the aspect of *A. cochlearis,* but has a very different calyx and pod.

A. multinervia, DC. Mem. Leg. 445, Prod. ii. 450, answers, in the short character given, to *A. lanigera,* with the exception of the gland on the upper edge of the phyllodium which I do not find in that or any other allied species known to me.

15. **A. phlebocarpa,** *F. Muell. in Journ. Linn. Soc.* ii. 119. A glabrous somewhat viscid shrub of 2 to 5 ft. ; branches nearly terete. Phyllodia narrow-lanceolate, rigid, tapering into a pungent point, 1 to 1½ in. long, mostly 2 to 3 lincs broad, narrowed at the base, with several parallel nerves occasionally anastomosing, 1 to 3 more prominent. Stipules small. Peduncles solitary, attaining ¼ in. when in fruit. Flowers 5-merous, but only seen withered. Calyx turbinate, lobed, half as long as the corolla. Petals apparently striate. Pod curved, hard, rather flat, with much thickened margins and obliquely veined between them, depressed between the seeds. Seeds rounded, compressed, oblique ; funicle with the last 2 or 3 folds much dilated into a cup-shaped apparently 2-lobed aril, enclosing the base of the seed.

N. Australia. Rocky places at the sources of Seven Emu river, Gulf of Carpentaria, *F. Mueller.* The species appears to be closely allied to *A. lanigera.*

16. **A. trinervata,** *Sieb. in DC. Prod.* ii. 451. A tall shrub, glabrous or the young shoots slightly pubescent, branches angular. Phyllodia linear, spreading, thick, rigid, tapering into pungent points, 1 to 1½ in. long or rarely all under 1 in., 2- or 3 nerved, slightly contracted at the base. Stipules minute. Peduncles solitary, about ½ in. long, slender, bearing each a small globular head of numerous small flowers, mostly 5-merous. Calyx about half as long as the corolla, very thin, irregularly toothed or lobed. Petals smooth, united to the middle. Pod straight or scarcely curved, flat with scarcely thickened margins ; attaining 3 to 5 in. in length. 1 to 1½ or rarely 2-lines broad. Seeds oblong, longitudinal, funicle dilated from the base into an almost membranous aril consisting of 3 or 4 folds—*A. taxifolia,* A. Cunn. in Field, N. S. Wales, 344, not of Willd.; *A. Cunninghamii,* G. Don, Gen. Syst. ii. 404.

N. S. Wales. Grose river, *R. Brown;* Blue Mountains, *Sieber, n.* 445, *A. Cunningham,* and others.

Var. *brevifolia.* Phyllodia ½ to ¾ in. long.—*A. genistifolia,* Link, Enum. Hort. Berol. ii. 442 (?) ; Benth. in Hook. Lond. Journ. i. 355. The character given by Link applies nearly as well to some forms of *A. juniperina.*

Var. *angustifolia.* Phyllodia very narrow, with the 2 or 3 nerves scarcely prominent, but shorter than in *A. elongata,* and always pungent.—Blue Mountains, *Miss Atkinson.*

17. **A. colletioides,** *A. Cunn.; Benth. in Hook. Lond. Journ.* i. 336. A very rigid shrub, the branches terete or nearly so and glabrous. Phyllodia

very spreading, linear-terete or slightly compressed, rigid, tapering into pungent points, striate with several nerves, distinctly articulate at the base as in the allied species, ¾ to 1½ in. long. Stipules small, membranous or none. Peduncles solitary, not above 2 lines long, bearing each a small globular head of very small 5-merous flowers. Sepals linear-spathulate, distinct, at least half as long as the corolla. Petals distinct, smooth, without prominent midribs. Pod linear, flat, the margins not thickened, curved or twisted, about 2 to 2½ lines broad and slightly contracted between the seeds. Seeds longitudinal, the 2 or 3 last folds of the funicle dilated into a yellow cupular arillus, membranous at the edge and half enclosing the seed.

N. S. Wales. Harrington plains in the interior, *A. Cunningham*, specimens in fruit only.

Var. *nyssophylla.* Nerves of the phyllodia scarcely prominent but not otherwise differing from A. Cunningham's specimens.—*A. nyssophylla*, F. Muell. Fragm. iv. 4, and Pl. Vict. ii. 9.

Victoria. In the N.W. desert, *F. Mueller.*

S. Australia. Near Lake Gairduer, *Babbage.*

I had formerly confounded specimens of this variety with some of those of *A. continua*, but F. Mueller has well pointed out the character of the continuous decurrent phyllodia distinguishing the latter.

18. **A. striatula,** *Benth. in Hook. Lond. Journ.* i. 336. A rigid shrub, allied to *A. colletioides* and *A. campylophylla.* Branches slightly angular and minutely pubescent. Phyllodia spreading or recurved, linear-terete or slightly compressed, rigid and tapering into pungent points, in our specimens mostly about ½ in. long, but probably often longer, striate with several nerves. Stipules small, lanceolate, membranous, very deciduous. Peduncles short, bearing each a small globular head of rather numerous flowers, mostly 5-merous. Sepals free, very thin, linear-spathulate. Petals smooth, but with prominent midribs as in *A. juniperina.* Pod unknown.—Meissn. in Pl. Preiss. ii. 201.

W. Australia, *Drummond;* South Hutt river, *Oldfield.*

19. **A. campylophylla,** *Benth. in Linnæa,* xxvi. 605. A rigid bushy shrub, apparently low, and quite glabrous, branches angular. Phyllodia numerous, linear-terete, rigid but recurved or flexuose, tapering into pungent points, mostly ¾ to 1 in. long, striate with several nerves. Stipules small, setaceous, almost spinescent. Flowers not seen perfect, but from their remains they appear to have been many in the head, with narrow smooth petals. Fruiting peduncles ½ to ¾ in. long. Pod stipitate, linear, narrow, flat with nerve-like margins and not contracted between the seeds, but not seen full-grown.

W. Australia, *Drummond, 4th Coll. n.* 134, *and n.* 41, of some other set. Evidently allied to *A. colletioides*, but differing in the stipules, and apparently in the pod.

20. **A. teretifolia,** *Benth. in Hook. Lond. Journ.* i. 326. A glabrous shrub of 1 or 2 ft., with virgate angular-striate branches, the smaller branchlets rigidly flexuose. Phyllodia linear-terete or scarcely flattened, rigid but less pungent than the allied species, ¾ to 1 or rarely 1¼ in. long, very obscurely or irregularly nerved. Stipules shortly subulate-spinescent, but soon wearing off. Peduncles solitary, shorter than the phyllodia, bearing each a globular head of flowers, mostly 5-merous. Sepals small, narrow-linear, united at the

base only. Petals linear-spathulate, concave and thick at the top, giving the
buds a peculiar turbinate truncate shape. Pod (only seen in Herb. Sonder.)
sessile, linear, incurved, terete or compressed at right angles to the valves,
acuminate and narrowed at the base, 2 to 4 in. long, hard and quite smooth,
neither contracted between the seeds nor showing prominent sutures. Seeds
not seen.—Meissn. in Pl. Preiss. i. 6, ii. 200.

W. Australia, *Drummond, 2nd Coll. n.* 140 ; Swan and Canning rivers, *Preiss, n.* 975
and 998 ; Gordon river, *Oldfield.*—Resembles sometimes *A. genistoides* and *A. ericifolia,*
but the pod is totally different, the phyllodia, although often sulcate, have no prominent
nerve, and the stipules are usually more or less spinescent.

21. **A. sulcata,** *R. Br. in Ait. Hort. Kew. ed.* 3, v. 460. A dense
bushy shrub, attaining sometimes several ft., quite glabrous, the branches
slightly angular. Phyllodia usually crowded, linear-terete, incurved, with a
short pungent point, ½ to 1 in. long, deeply sulcate-striate. Stipules minute.
Peduncles usually 2 together, rather shorter than the phyllodia, bearing each
a globular head of about 10 to 15 flowers, mostly 5-merous. Sepals distinct,
linear-spathulate, not half as long as the corolla. Petals smooth, without
prominent midribs, completely separating. Pod flat, but very flexuous, much
curved or twisted, rarely above 1 in. long, 1½ to 2 lines broad. Seeds nearly
orbicular; funicle very shortly filiform and folded at the base, dilated into a
cup-shaped fleshy arillus under the seed.—DC. Prod. ii. 450; Bot. Reg. t.
928 ; Wendl. Comm. Acac. t. 10 ; Meissn. in Pl. Preiss. i. 11.

W. Australia. King George's Sound and adjoining districts, *R. Brown, Drummond,*
5th Coll. n. 3, *Preiss, n.* 978, and others, and eastward to E. Mount Barren, *Maxwell.*
Some specimens appear almost to pass into *A. ericifolia* and its allies amongst *Calami-*
formes.

22. **A. costata,** *Benth. in Hook. Lond. Journ.* i. 339. A low shrub,
with thick rigid striate often spinescent branches, pubescent or glabrous.
Phyllodia narrow-lanceolate, 2 to 4 lines long, very rigid and thick, tapering
into pungent points, the upper edge convex but without any gland, both margins thick and nerve-like. Stipules minute, setaceous. Peduncles rather
short, bearing each a globular head of about 10 to 15, 5-merous flowers.
Calyx small, deeply lobed or at length separating into distinct ciliate acute
sepals. Petals rigid, striate, united at the base. Pod not seen.

W. Australia. Swan River, *Drummond, 1st Coll. ;* Darling range, *Oldfield.*
Some specimens of Drummond's have much resemblance to this, with the same spinescent
branches, but the phyllodia are occasionally slightly angular, and the flowers are nearer those
of *A. divergens.*

23. **A. barbinervis,** *Benth. in Hook. Lond. Journ.* i. 326. A low shrub,
often under 1 ft., usually loosely pubescent with short spreading hairs, the
branches angular. Phyllodia linear-falcate, recurved, rigid, tapering into a
pungent point, ¾ to 1 in. long or the lower ones shorter and broader, with
very prominent nerve-like margins. Stipules very small, setaceous, sometimes spinescent. Peduncles shorter than the phyllodia, bearing each a globular head with usually 8 to 10, rarely 12 to 15, mostly 5-merous flowers.
Calyx broad, shortly toothed, ciliate, not half so long as the corolla. Petals
thick and prominently striate. Pod not seen.—Meissn. in Pl. Preiss. i. 7.

W. Australia. Swan River, *Drummond, 1st Coll.; Preiss, n.* 988.

24. **A. ataxiphylla,** *Benth. in Linnæa,* xxvi. 605. A low rigid under-shrub or shrub, the branches prominently and acutely angled, the young shoots hoary-pubescent but soon glabrous. Phyllodia linear or subulate, rigid but flexuose, tapering into a hooked pungent point, 1½ to 2 in. long, with a very prominent nerve on each side. Stipules small, setaceous or pungent. Peduncles scarcely ⅓ in. long, bearing each a globular head of 20 to 30 or more flowers, mostly 5-merous. Calyx deeply and acutely lobed, less than half as long as the corolla. Petals thick and more or less prominently striate or veined. Pod not seen.

W. Australia, *Drummond, 4th Coll. n.* 6.

25. **A. Baxteri,** *Benth. in Hook. Lond. Journ.* i. 327 (misspelt *Bagsteri*). A glabrous rigid shrub, branches prominently and acutely angled or striate. Phyllodia linear-oblong or lanceolate, somewhat recurved-falcate, very rigid and tapering into a pungent point, ⅓ to 1 in. long, with very thick nerve-like margins, the lower one sometimes slightly within the edge. Stipules seta-ceous, at length deciduous. Peduncles short, solitary or 2 or 3 together, bearing each a globular head of numerous (above 40) flowers, mostly 5-merous. Sepals distinct, linear-spathulate, thickened at the end, half as long as the co-rolla. Petals quite smooth. Pod not seen.

W. Australia. S. coast, *Baxter;* Salt river, *Maxwell.* The aspect is nearly that of *A. auronitens* or of some varieties of *A. siculiformis.* It also resembles *A. crassistipula,* but without the leafy stipules of that species.

C. UNINERVES.—Phyllodia with 1 nerve on each side, central or nearly so. Flowers in globular heads.

When the phyllodia are very narrow they are usually 4-gonous, the four angles formed by the upper and lower nerve-like margins and by one nerve on each side.

26. **A. auronitens,** *Lindl. Swan Riv. App.* 15. A rigid bushy shrub, the branches nearly terete and usually shortly hirsute. Phyllodia numerous, rigid, glabrous, linear-oblong, oblique or curved, tapering into a pungent point and narrowed at the base, mostly about ⅓ in. long, with a nearly central nerve, and scarcely thickened nerve-like margins. Stipules short, setaceous or pungent. Peduncles nearly as long as the leaves, bearing each a globular head of about 15 to 20 flowers, mostly 5-merous. Sepals distinct, linear-spathulate, very thin. Petals, soon separating, striate but not thick. Pod about 1 in. long, and 3 lines broad, straight, very thick, not contracted be-tween the seeds, with very broad smooth margins, and obscurely veined be-tween. Seeds thick, obovoid; funicle short, not folded, expanding into a short broad concave aril at the end of the seed.

W. Australia. Swan River, *Drummond, 1st Coll.;* also 4th *Coll. n.* 7.

27. **A. quadrisulcata,** *F. Muell. Fragm.* iii. 127. A rigid straggling shrub, the branches terete, slightly viscid-pubescent, at length glabrous. Phyllodia narrow-linear, almost tetragonous, the single nerve on each side very prominent, very rigid and tapering into a pungent point, ¾ to 1½ in. long. Stipules minute. Peduncles solitary, shorter than the phyllodia, bearing each a globular head of 15 to 20 flowers, mostly 5-merous, but occasionally 6-merous. Sepals distinct, very small and narrow, linear-spathulate. Petals striate, united above the middle. Pod straight or slightly curved, 1 to 1½

in. long, nearly 3 lines broad, very thick, with broad smooth margins and obscurely veined between them. Seeds thick, obovoid-globose, often mottled; funicle short, not folded, expanding into a short broad concave arillus at the end of the seed.

W. Australia. Murchison river, *Oldfield.* Differs from *A. auronitens* chiefly in the form of the phyllodia.

28. **A. erioclada,** *Benth. in Linnæa,* xxvi. 606. A rigid spreading shrub, the branches numerous, not very stout, mostly spinescent and densely woolly, becoming glabrous with age. Phyllodia narrow-oblong, oblique or curved, rigid and tapering into a pungent point, under ½ in. long, narrowed at the base, the midrib and nerve-like margins prominent, and sometimes with a faint nerve between. Stipules setaceous, sometimes spinescent, deciduous. Peduncles 2 to 3 lines long, bearing each a globular head of above 30 flowers, much smaller than in the last two species, mostly 5-merous. Sepals spathulate with a dark thickened top, but often united. Petals quite smooth. Pod not seen.

W. Australia, *Drummond, 4th Coll. n.* 7; Vasse river, *Oldfield.*

29. **A. siculiformis,** *A. Cunn.; Benth. in Hook. Lond. Journ.* i. 337. An erect or diffuse rigid glabrous shrub, attaining several ft., the branches nearly terete. Phyllodia linear or linear-lanceolate, oblique or slightly curved, rigid and tapering into a pungent point, ½ to 1½ in. long, 1 to 2 lines broad, with a prominent central nerve and the margins sometimes slightly thickened, narrowed at the base. Stipules minute, membranous. Peduncles solitary, nearly ¼ in. long in the original form, bearing a globular head of numerous flowers, mostly 5-merous. Sepals free, narrow-spathulate, ciliate, half as long as the corolla. Petals smooth, shortly united at the base. Pod stipitate, oblong-linear, very flat, not contracted between the seeds when perfect, rarely above 1 in. long, 2½ to 3 lines broad. Seeds transverse or oblique, the funicle filiform to the end.—F. Muell. Pl. Vict. ii. 6.

N. S. Wales. Rocky Hills S.W. of Lake George, *A. Cunningham;* Mount Mitchell, Clarence river, *Beckler.*

Var. *bossiæoides,* Benth. in Hook. Lond. Journ. i. 337. More diffuse, peduncles scarcely ¼ in. long or shorter.—*A. Stuartiana,* F. Muell.; Benth. in Linnæa, xxvi. 609; Hook. f. Fl. Tasm. i. 104. t. 19; Dietr. Fl. Univers. N. Ser. t. 82.

Victoria. Along subalpine streamlets in many parts of the Australian Alps, ascending to 5000 ft., *F. Mueller.* One specimen in a young state from the Tummut Valley has the long peduncles of the original form.

Tasmania. Derwent river, *R. Brown;* Western Mountains, and S. Esk and Derwent rivers at an elevation of 3000 to 4000 ft., *J. D. Hooker.*

30. **A. patens,** *F. Muell. in Journ. Linn. Soc.* iii. 120 (under *A. siculi-formis,* to which I had erroneously referred it). A tall glabrous shrub, the branchlets somewhat glutinous. Phyllodia numerous, linear-lanceolate, rather rigid and tapering into a pungent point, narrowed at the base and often somewhat falcate, ½ to ¾ in. long, 1 to 1½ lines wide, 1-nerved or obscurely penniveined, usually with a small gland near the middle of the upper edge. Stipules small, lanceolate. Peduncles often slightly exceeding the leaves, bearing each a dense globular head of flowers, mostly 5-merous, but only seen in bud. Sepals distinct, linear-spathulate, ciliate. Petals narrow, almost free.

Pod linear, straight, flat, 1 in. long or rather more, 1½ lines broad, and contracted between every seed. Seeds ovate, almost longitudinal; funicle filiform
to the end.—*A. Maitlandi*, F. Muell. Fragm. iii. 46.

N. Australia. Stony places, Hammersley Range, Nichol Bay, *F. Gregory's Expedition;* Hooker and Sturt's Creeks, *F. Mueller.* The foliage and young flowers cannot be distinguished from those of some specimens of *A. siculiformis*, but the pod is certainly different.

31. **A. laricina,** *Meissn. in Pl. Preiss.* i. 6. A bushy shrub of 1 to 2
ft., the branches nearly terete, rather slender, hoary-tomentose. Phyllodia
numerous, recurved-spreading, narrow-linear, almost tetragonous from the
very prominent nerve on each side, ½ to ¾ or rarely 1 in. long, with a short
pungent point. Stipules setaceous, almost spinescent. Peduncles slender,
shorter than the phyllodia, bearing each a small globular head of 15 to 20
small flowers, mostly 5-merous. Calyx turbinate, shortly toothed, about half
as long as the corolla. Petals not striate, but with prominent midribs, usually
cohering to the middle. Young pod narrow, glabrous or pubescent, incurved,
acuminate and contracted at the base, not torulose, the sides very convex and
marked with longitudinal prominent nerves.

W. Australia. In the interior, *Preiss, n.* 973, *Drummond, 3rd Coll. n.* 101, *4th
Coll. n.* 5; E. Mount Barren, *Maxwell.*

32. **A. tetragonophylla,** *F. Muell. in Journ. Linn. Soc.* iii. 121 (under *A. sphacelata*), *and Fragm.* iv. 3. A tall spreading shrub or small tree,
glabrous; branches terete. Phyllodia usually clustered on the old nodes,
linear-subulate, rigid, pungent-pointed, ½ to 1 in. long or rarely more, with 1
or 2 nerves on each side. Stipules small, deciduous. Peduncles solitary or
2 together, nearly as long as the phyllodia, bearing a globular head of numerous (often above 50) 5-merous flowers. Sepals linear-spathulate, half as
long as the corolla. Petals smooth, usually cohering to the middle. Pod
much curved or twisted, flat with thickened margins, nearly 3 lines broad.
Seeds longitudinal; funicle yellow, shortly flexuose and much thickened at
the base, then completely encircling the seed in a single fold more or less dilated the whole length.

N. S. Wales. From the Darling to the Barrier Range, *Victorian and other Expeditions.*

S. Australia. Dry pastures on the Cudnaka, Flinders and Elders Ranges, *F. Mueller;*
towards Spencer's Gulf, *Warburton.*

33. **A. genistoides,** *A. Cunn. Herb.* A tall glabrous shrub, very rigid
and sometimes spinescent; branches terete or slightly angular. Phyllodia
erect or scarcely spreading, linear-subulate, with a very prominent nerve on
each side, shortly pungent-pointed, mostly rather above 1 in. long. Stipules
small, fine-pointed, deciduous. Peduncles solitary or 2 together, slender,
usually above ½ in. long, bearing each a very compact globular head of numerous small flowers, mostly 5-merous. Sepals free, narrow-linear, spathulate, more than half as long as the corolla. Petals smooth, with a prominent
midrib. Pod somewhat incurved, 1 to 3 in. long, 2½ to 3 lines broad, the
valves convex, not striate, slightly contracted between the seeds. Seeds ovate,
longitudinal; funicle yellow, thickened almost from the base, nearly twice
encircling the seed, one fold within the other without any return.

W. Australia. Dirk Hartog's Island, *A. Cunningham;* Sharks Bay, *Milne;* Murchisou and S. Hutt rivers, *Oldfield.* Very nearly allied to *A. tetragonophylla*, and some-times resembles *A. teretifolia*, but the stipules are not spinescent, the phyllodia are promi-nently 1-nerved, and the pod is quite different.

34. **A. sphacelata,** *Benth. in Hook. Lond. Journ.* i. 338. A rigid shrub, the branches not very stout, nearly terete, glabrous or pubescent. Phyllodia scattered, linear-subulate, erect or spreading, rigid and tapering into a pungent point, mostly ½ to ¾ in. long, with 1 or rarely 2 prominent nerves on each side. Stipules minute. Peduncles mostly solitary, short, bearing each a small globular head of 15 to 20 or more flowers, mostly 5-merous. Sepals distinct, linear-spathulate, with dark tips, half as long as the corolla. Petals free, smooth. Pod not seen.—Meissn. in Pl. Preiss. i. 12.

W. Australia, *Drummond, 1st Coll. n.* 299; Mount Currie, *Preiss, n.* 985 (which I have not seen).

Var. *sessilis.* Branches woolly. Flower-heads almost sessile. Pod, according to Meiss-ner, linear, straight, flat, 1½ in. long, 1½ lines broad, with thickened margins, woolly when young, at length glabrous.—*A. sessilis*, Benth. in Hook. Lond. Journ. i. 336; Meissn. in Pl. Preiss. i. 11, ii. 202.—Swan River, *Drummond, 1st Coll., Preiss, n.* 979, 980, 982; Murchison river, *Oldfield.*

Var. *retrorsa.* Phyllodia reflexed.—*A. retrorsa*, Meissn. in Bot. Zeit. 1855, 10.—Be-tween Moore and Murchison rivers, *Drummond, 6th Coll. n.* 4.

Some specimens have almost the aspect of *A. striatula*, but differ in their 1- or rarely 2-nerved phyllodia, and very short peduncles.

35. **A. ingrata,** *Benth.* Branches rigid, glabrous or minutely pubes-cent. Phyllodia divaricate or reflexed, very rigidly linear-subulate, with a rather broad base and tapering into a pungent point, rarely ½ in. long, with a prominent nerve on each side. Peduncles short, solitary or 2 together, bear-ing each a small globular head of about 5 or 6 flowers, mostly 5-merous. Calyx very short and broad, thin, slightly sinuate toothed. Petals smooth, without raised midribs, cohering at the base. Pod unknown.

W. Australia. F. Mount Barren, *Maxwell (Herb. F. Mueller).* The specimens are insufficient for a satisfactory description, they have some resemblance to some forms of *A. juniperina*, but the flowers are different.

Some small specimens of Drummond's, alluded to above under *A. costata*, look much like *A. ingrata* also, and have the same flowers, only rather more numerous in the head, but the branches are spinescent and the phyllodia have the very thick margins of *A. costata*, and occasionally the upper one has an angle as in *A. horridula*, but the flowers differ from those of either of the latter species. A small specimen from Murchison river, *Oldfield*, without flowers, has the foliage nearly of *A. ingrata* and a single pod, very thick and hard as in *A. auronitens*, but longer and rather narrower, without the distinct broad margin of that species.

36. **A. juniperina,** *Willd. Spec. Pl.* iv. 1049. A rigid bushy divari-cate shrub, attaining several feet, the branches pubescent or in some varieties glabrous. Phyllodia scattered, often numerous, divaricate, linear-subulate, rigid and tapering into a pungent point, rarely above ½ in. long, with a pro-minent nerve on each side and a rather broad base. Peduncles often exceed-ing the leaves, bearing each a dense globular head of numerous (20 to 50) flowers, mostly 5-merous. Bracts more or less acuminate. Sepals narrow-spathulate, at first united but readily separating. Petals also separating, smooth but with prominent midribs. Pod more or less falcate, flat, 1 to 2

in. long, about 2 lines broad, usually contracted between the seeds. Seeds longitudinal, the funicle but little folded and filiform to the end.—*Mimosa juniperina*, Vent. Jard. Malm. t. 64; *M. ulicina*, Wendl. Coll. ii. 25, t. 6; *M. ulicifolia*, Salisb. Prod. 324?; *A. juniperina*, Lodd. Bot. Cab. t. 398; DC. Prod. ii. 449; Hook. f. Fl. Tasm. i. 105; F. Muell. Pl. Vict. ii. 7; *A. verticillata*, Sieb. Pl. Exs. not of Willd.; *A. echinula*, DC. Prod. ii. 449; *A. pungens*, Spreng. Syst. iii. 134.

Queensland. Moreton Island and Brisbane river, *F. Mueller.*
N. S. Wales. Port Jackson to the Blue Mountains, *R. Brown, Sieber, n.* 447, 449, and *Fl. Mixt. n.* 602, and others; northward to New England, *C. Stuart*, and Clarence river, *Beckler;* southward to Twofold Bay, *F. Mueller.*
Victoria. Rocky especially granitic hills, Mount Hunter, Corner Inlet, Mount Liger, Genoa river, Muddy Creek, Grampians, etc., *F. Mueller.*
Tasmania. Port Dalrymple, *R. Brown;* light sandy soil near George Town and West head, *J. D. Hooker.*
Var. *Brownei.* Branches glabrous; peduncles slender.—*A. acicularis*, R. Br. in Ait. Hort. Kew. ed. 3, v. 460, not of Willd.; altered to *A. pugioniformis* by Wendl. in Flora, 1819, 139, but not *A. pugioniformis*, Wendl. Comm.; *A. Brownii*, Steud.; DC. Prod. ii. 449; Lodd. Bot. Cab. t. 1333; *A. arceuthos*, Spreng. Syst. iii. 134.—Port Jackson, *Sieber, n.* 463 and others.
A. genistifolia, Link, Enum. Hort. Berol. ii. 442; DC. Prod. ii. 449, above referred with doubt to *A. trinervata*, may, from the very incomplete description given, be almost equally referable to some forms of *A. juniperina.*

37. **A. asparagoides,** *A. Cunn. in Field, N. S. Wales,* 343. A glabrous rigid shrub; branches nearly terete. Phyllodia spreading, linear-subulate, thick and rigid, tapering into a pungent point, $\frac{1}{2}$ to $\frac{3}{4}$ in. long, with a prominent nerve on each side, and a scarcely prominent glandular angle at the base on the upper edge. Stipules minute. Flower-heads almost sessile, solitary, globular, with about 20 to 30 small flowers, mostly 5-merous. Sepals linear, slightly spathulate, cohering at first, but readily separating. Petals distinct, narrowed into a claw at the base, smooth, without any prominent nerve. Pod unknown.

N. S. Wales Rare on the rocky verge of Regent's Glen, Blue Mountains, *A. Cunningham.*

38. **A. tenuifolia,** *F. Muell. in Trans. Phil. Soc. Vict.* i. 37; *Pl. Vict.* ii. 8. A diffuse or procumbent shrub; branches terete, glabrous or pubescent. Phyllodia linear-subulate, rigid but slender, tapering into a pungent point, with a strong raised nerve on each side, rarely above $\frac{1}{2}$ in. long. Stipules small, deciduous. Peduncles slender, shorter than the leaves, bearing each a globular head of about 20 small flowers, mostly 4-merous. Sepals free or united at the base only, small and narrow. Petals soon separating, smooth, the midrib scarcely prominent. Pod straight or curved, scarcely contracted between the seeds, 1 to 2 in. long and nearly 2 lines broad, the valves convex. Seeds ovate, longitudinal, the last short fold of the funicle and sometimes part of the next thickened into a small aril.

Victoria. Dry Stringy-bark and Iron-bark Ranges towards the Upper Yarra, Goulburn, Broken and Ovens rivers, and near Ballarat, *F. Mueller.* Readily distinguished from *A. juniperina*, which it resembles, by the 4-merous flowers, the petals not ribbed, and the thickened funicles.

39. **A. diffusa,** *Lindl. Bot. Reg. t.* 634. A glabrous divaricately-branched

or diffuse shrub, either quite low or rising to 5 or 6 ft. ; branchlets angular. Phyllodia linear, thick and rigid, tapering into a pungent point, prominently 1-nerved, mostly ¾ to 1 in. long, 1 to 1½ lines broad, slightly narrowed at the base. Peduncles usually 2 or 3 together, under ½ in. long, bearing each a globular head of about 20 flowers, mostly 4-merous, rarely 5-merous or 3-merous. Calyx with short broad somewhat thickened lobes, not half as long as the corolla. Petals membranous, without the prominent nerve of *A. juniperina*, cohering or at length separating. Pod stipitate, often attaining 3 or 4 in., about 2 lines broad, flat or the valves at length convex, Seeds longitudinal ; funicle much folded, thickened either from the middle upwards or nearly from the base.—DC. Prod. ii. 450 ; Bot. Mag. t. 2417 ; Hook. f. Fl. Tasm. i. 105 ; F. Muell. Pl. Vict. ii. 6 ; *A. prostrata,* Lodd. Bot. Cab. t. 631.

Victoria. Heathy ground, Stringy-bark forests and other barren localities throughout the greater part of the colony, *F. Mueller.*

Tasmania. Port Dalrymple and Derwent river, *R. Brown ;* abundant in dry places especially by roadsides throughout the island, *J. D. Hooker.*

Var. *cuspidata.* Phyllodia more slender, often not broader than thick, sometimes all under ¾ in., sometimes on barren shoots attaining 2 in.—*A. cuspidata,* A. Cunn. ; Benth. in Hook. Lond. Journ. i. 337.

N. S. Wales. Brushy hills of the southern districts, *A. Cunningham, Fraser, Huegel.*

40. **A. rupicola,** *F. Muell. ; Benth. in Linnæa,* xxvi. 610. A glabrous shrub, attaining 5 ft. ; branches slightly angular. Phyllodia linear or linear-lanceolate, rigid and tapering into a pungent point, prominently 1-nerved, rarely above ½ in. long, broader at the base than in *A. diffusa,* which this species closely resembles. Stipules minute or none. Peduncles about ½ in. long, bearing each a small globular head of numerous flowers, mostly 4-merous. Calyx shortly lobed, half as long as the corolla. Petals membranous, without any prominent midrib. Pod linear, flattened, straight or curved, 1 to 2 in. long, about 2 lines broad, not contracted between the seeds. Seeds longitudinal ; funicle much folded and thickened nearly from the base, the upper folds forming a broad aril clasping the base of the seed.—F. Muell. Pl. Vict. ii. 8 ; Dietr. Fl. Univers. N. Ser. t. 8.

Victoria. Rocky mountains near the Wimmera, *Dallachy.*

S. Australia. Mount Lofty Ranges, *F. Mueller ;* towards Bremer river, Mount Barker Creek, *L. Fischer.*

D. SPICATÆ.—Phyllodia 1- 2- or 3-nerved. Flowers 4-merous or 3-merous, in cylindrical or oblong rarely ovoid spikes, on axillary peduncles or almost sessile.

Some species of this group, especially *A. rhigiophylla* and *A. verticillata* var. *ovoidea,* seem in their inflorescence to connect the spicate with the capitate species, which in almost all species of *Mimoseæ,* Australian or extra-Australian, are so constantly distinct.

41. **A. rhigiophylla,** *F. Muell. ; Benth. in Linnæa,* xxvi. 611. A compact rigid shrub of 4 or 5 ft., with numerous spreading terete branches, slightly glutinous-pubescent when young, at length glabrous. Phyllodia linear or linear-lanceolate, thick and very rigid, tapering to a pungent point, ½ to 1 in. long, with 2 or 3 raised nerves on each side. Flower-heads small and nearly sessile, less compact than in most species and often oblong, with

6 to 10 flowers, mostly 4-merous. Calyx campanulate, with broad short lobes, not half as long as the corolla. Petals quite smooth and readily separating. Pod unknown.—F. Muell. Pl. Vict. ii. 9.

S. Australia. Desert on the lower Murray, towards Mount Barker Range, *F. Mueller.*

42. **A. oxycedrus,** *Sieb. in DC. Prod.* ii. 453. A tall rigid spreading shrub, the branches usually pubescent and nearly terete. Phyllodia scattered or rarely irregularly verticillate, rigid, tapering into a pungent point, rather broad at the base, ½ to ¾ or rarely above 1 in. long, with 3 or rarely 4 prominent nerves on each side. Stipules small, often spinescent. Peduncles short, bearing each a dense cylindrical spike, often above 1 in. long. Flowers mostly 4-merous. Calyx short, the lobes obtuse. Petals smooth, readily separating. Pod incurved, acuminate and narrowed at the base, about 2 lines wide and often above 3 in. long, pubescent or at length glabrous, the valves very convex, striate, thicker and harder than in *A. verticillata.* Seeds oblong, longitudinal; funicle much folded, and usually more or less thickened from the base.—Sweet, Fl. Austr. t. 6; Bot. Mag. t. 2928; Reichb. Icon. Exot. t. 120; Paxt. Mag. Bot. vii. 151, with a fig.; F. Muell. Pl. Vict. ii. 10.

N. S. Wales. Blue Mountains, *Sieber, n.* 427, *A. Cunningham, Fraser.*

Victoria. Not unfrequent in heathy tracts throughout the southern part of the colony, ascending also into the mountains, *F. Mueller.*

Tasmania, *Fitzalan (Herb. F. Mueller,* but doubtful).

S. Australia. Between Mount Gambier and Rivoli Bay, *F. Mueller.*

43. **A. verticillata,** *Willd. Sp. Pl.* iv. 1049. A shrub, rather low and spreading or erect and bushy, sometimes growing into a small tree; branches angular-striate, pubescent or rarely glabrous. Phyllodia scattered or more often verticillate, linear-subulate, lanceolate or oblong, rigid and tapering into a pungent point, about ½ in. long or shorter, rarely ¾ in. long, with a prominent central nerve and rarely 1 or 2 slender lateral ones. Stipules minute. Peduncles short or slender, bearing each, in the common variety, a dense cylindrical spike of ½ to 1 in. long, or rarely longer and loose. Flowers 4-merous, globular in the bud. Calyx short, broad, irregularly lobed. Petals smooth, united at the base. Pod flat, with slightly thickened margins, straight or curved, acute at each end, 1½ to 2 in. long when perfect, 2 lines wide and scarcely contracted between the seeds. Seeds oblong, longitudinal; funicle much folded, thickened sometimes from the base, but always much more so at the end or about the middle.—*Mimosa verticillata,* L'Hér. Sert. Angl. 30; Vent. Jard. Malm. t. 63; Bot. Mag. t. 110; *A. verticillata,* DC. Prod. ii. 453; Lodd. Bot. Cab. t. 535; Hook. f. Fl. Tasm. i. 106; F. Muell. Pl. Vict. ii. 10; *A. semiverticillata,* Knowl. and Westc. Fl. Cab. ii. 27 (from the char. in Walp. Rep. i. 922).

Victoria. Port Phillip, *R. Brown;* extremely abundant in all the more humid parts of the colony, *F. Mueller.*

Tasmania. Derwent river, Port Dalrymple, and islands of Bass's Straits, *R. Brown;* abundant in moist situations throughout the island, *J. D. Hooker.*

Var. *latifolia.* Phyllodia lanceolate or oblong.—*A. ruscifolia,* A. Cunn. in G. Don, Gen. Syst. ii. 407; Bot. Mag. t. 3195; *A. mœsta,* Lindl. Bot. Reg. 1846, t. 67. With the common variety, especially in Tasmania.

Var. *ovoidea.* Phyllodia slender. Spikes very short, sometimes reduced to small ovoid-globular heads. Flowers small. Calyx-lobes narrow.—*A. ovoidea,* Benth. in Hook. Lond. Journ. i. 339 ; Hook. f. Fl. Tasm. i. 105, t. 20 ; Dietr. Fl. Univers. N. Ser. t. 8.

Victoria. Grampians, Gipps' Land, and other parts of the colony, but less frequent than the cylindrical-spiked variety, *F. Mueller.*

Tasmania. Amongst grass, etc., in dry places, Woolnorth, Circular Head, etc., *J. D. Hooker.*

S. Australia. Not unfrequent in the colony, extending northward to Mount Remarkable, *F. Mueller.*

Distinct as are the Tasmanian specimens from the common variety, the Continental ones appear to pass into the more slender forms of the true *A. verticillata.*

44. **A. Riceana,** *Henslow, in Maund, Botanist,* iii. t. 135. A handsome dark-green, tall shrub or small tree, glabrous with angular branchlets. Phyllodia scattered or almost whorled, linear or subulate, sometimes all rather broad and ½ to ¾ in. long, sometimes very narrow, and 1 to 1½ in. long, tapering into pungent points, 1-nerved. Stipules minute. Spikes slender and loose, often above 1 in. long, the flowers often distant, 3-merous or rarely 4-merous, ovoid and obtuse or acute in the bud. Calyx short, with broad obtuse ciliate lobes. Petals smooth, readily separating. Pod usually curved, acuminate, often 2 to 3 in. long and scarcely 1½ lines broad, slightly pubescent when young, but soon glabrous ; valves very convex, coriaceous, contracted between the seeds. Seeds oblong, longitudinal ; funicle much folded and thickened nearly from the base.—Hook. f. Fl. Tasm. i. 106 ; *A. setigera,* Hook. Ic. Pl. t. 316 ; *A. erythropus,* Ten. Cat. Hort. Neap. Annot. 77 (from the description given) ; *A. taxifolia,* Lodd. Bot. Cab. t. 1225 ? (from the figure).

Tasmania. Derwent river, *R. Brown ;* moist shady places in the southern parts of the island, *J. D. Hooker.*

45. **A. axillaris,** *Benth. in Hook. Lond. Journ.* i. 341. Very near the long-leaved forms of *A. Riceana,* and perhaps a variety as suggested by F. Mueller, but the phyllodia are often still longer, showing faint lateral nerves, and the inflorescence is different ; the spikes are sessile or nearly so and contain but very few flowers, rarely more than 4, otherwise like those of *A. Riceana.* I have not as yet seen any intermediate specimens.—Hook. f. Fl. Tasm. 1. 107.

Tasmania, *Gunn ;* Brook's Head, St. Paul's River, *C. Stuart.*

SERIES IV. CALAMIFORMES.—Phyllodia rarely none, more frequently narrow-linear or subulate, terete, tetragonous or very slightly flattened, articulate on the stem, obtuse or with short innocuous or recurved points, 1- or several-nerved. Flowers in globular heads, on simple axillary peduncles, or rarely several heads in a short raceme, or irregularly racemose by the abortion of the floral phyllodia.

Although the long slender phyllodia, not broader than thick, give a peculiar character to the majority of the species included in this series, a few, by their more rigid phyllodia with straighter points, pass into the *Pungentes,* the shorter-leaved ones connect them with the narrowest among the *Uninerves brevifoliæ,* some almost pass into the *Plurinerves microneuræ,* and a few of the *Julifloræ* only differ in their flower-heads oblong or cylindrical not globular.

A. SUBAPHYLLÆ.—Phyllodia none or very few and slender, resembling the branches, chiefly subtending the inflorescences.

46. **A. tetragonocarpa,** *Meissn. in Pl. Preiss.* i. 4. A glabrous undershrub, with erect slender rush-like stems of 1 to $2\frac{1}{2}$ ft., slightly branched and nearly leafless. Phyllodia mostly reduced to small striate scales at the base of the branches, 1 or 2 often filiform, terete, resembling the branches, the lower ones of the young plant occasionally bipinnate as in other phyllodineous species. Peduncles short, often 2 or 3 together, bearing a head of 3 or 4 flowers, occasionally reduced to a single one, 4-merous and oblong in the bud. Calyx short. Petals rigid, striate, nearly $1\frac{1}{2}$ lines long. Pod stipitate, linear-falcate, about 1 in. long, with a very prominent longitudinal angle on each side of each suture. Seeds longitudinal, the last fold of the funicle dilated into a small turbinate almost cup-shaped aril under the seed, with a filiform fold below it.—*Tetracheilos Meissneri*, Lehm. in Pl. Preiss. ii. 368.

W. Australia. King George's Sound, *R. Brown*; Swan River, *Fraser;* boggy ground near Spencer, Plantagenet district, *Preiss, n.* 866 ; Israeliti Bay, *Maxwell*.

Var. *scabra.* Phyllodia long, rigid, prominently 1-nerved, glabrous. Pod not seen. Flowers as in the ordinary form.—A single specimen from *Oldfield*.

The longitudinally 4-angled pod, upon which Lehmann founded the genus *Tetracheilos*, occurs also in *A. stenoptera*, a species widely different in several other respects, and to a certain degree also in some of the *Julifloræ*.

4⁷. **A. restiacea,** *Benth. in Hook. Lond. Journ.* i. 3. A glabrous undershrub or shrub, with rush-like branches somewhat angular or striate, leafless or with a few slender terete phyllodia, scarcely to be distinguished from the branches. Flower-heads globular, several in a raceme, covered when very young by imbricate scales very early deciduous. Flowers mostly 5-merous, usually 20 to 30 in the head. Calyx small, thin, irregularly lobed. Petals smooth. Pod (according to Lehmann) stipitate, linear, moniliform, with reticulately veined coriaceous valves.—Meissn. in Pl. Preiss. i. 3, ii. 199 ; *Chithonanthus restiaceus*, Lehm. in Pl. Preiss. ii. 368.

W. Australia. Swan River and adjoining districts, *Drummond, 2nd Coll. n.* 143, *Preiss, n.* 971, 972 ; Preston and Murchison rivers, *Oldfield*.

A. squamata, Morren, in Ann. Soc. Hort. Gand, iii. 209, t. 134, a work which I do not find in our libraries, appears to be, from the characters copied into Walp. Ann. i. 264, the same as *A. restiacea*, and certainly not *A. squamata*, Lindl.

The imbricate scales covering the very young racemes, which induced Lehmann to propose his genus *Chithonanthus*, occur also in *A. squamata*, in *A. suaveolens*, and in some other species widely differing in other characters.

B. PLURINERVES.—Phyllodia either striate with 2, 3, or more prominent nerves, or with numerous very fine parallel nerves, scarcely visible without a lens.

48. **A. squamata,** *Lindl. Swan Riv. App.* 15. Glabrous with erect, terete, rush-like branches. Phyllodia few, distant, linear-terete, slightly striate with fine nerves, resembling the branchlets, but not continuous with them, with short recurved points, attaining about 2 in. Flower-heads in short axillary racemes, with imbricate striate scales, falling off as the flowers open. Calyx and corolla none or reduced to a minute membranous ring. Stamens very densely packed, above 100 in each flower, and 6 to 10 flowers in the head. Pod unknown.—Hook. Ic. Pl. t. 367.

W. Australia. Swan River, *Drummond, 1st Coll. ;* between Perth and King George's Sound, *Harvey ;* hills near Coogenup and Gordon Plains, *Oldfield*.

49. A. brachyphylla, *Benth. in Linnæa,* xxvi. 615. Apparently a diffuse shrub, the branches woolly-pubescent or at length glabrous. Phyllodia numerous, linear-terete, incurved, obtuse or with short innocuous points, under $\frac{1}{2}$ in. long, striate with several nerves. Stipules small, deciduous. Peduncles shorter than the phyllodia, bearing each a small globular head of apparently 5-merous flowers, but not seen perfect. Pod flexuose, 1 to 2 in. long, scarcely $1\frac{1}{2}$ lines broad, flat, with slightly thickened nerve-like margins. Seeds small, obovoid, oblique or longitudinal, the last fold of the funicle thickened into a short fleshy aril.

W. Australia, *Drummond.*

50. A. Bynoeana, *Benth. in Linnæa,* xxvi. 614. Shrubby, with numerous branches, loosely pubescent and sometimes glutinous. Phyllodia numerous, linear-terete, striate with several nerves, usually recurved at the point, rarely above 1 in. long. Stipules small, deciduous. Peduncles 3 to 4 lines long, bearing each a small globular head of about 20 flowers, mostly 5-merous. Calyx with narrow ciliate lobes. Petals narrow, smooth, not much longer than the calyx and quite distinct. Pod much curved, flat with thickened margins, scarcely above 1 line broad. Seeds oblong, longitudinal, the last fold of the funicle, and sometimes part of the next also, thickened into a fleshy aril.—*A. leptophylla,* F. Muell. Fragm. iv. 9.

N. Australia. N.W. coast, *Bynoe;* Gulf of Carpentaria, *F. Mueller.* The latter are the specimens alluded to by F. Mueller, Pl. Vict. ii. 12, as nearly resembling *A. Wilhelmsiana.* The corresponding ones, both in Herb. Hooker and in Herb. Sonder, were, by some mistake, labelled as *A. Wilhelmsiana* from the Murray scrub, and were mentioned by me in Linnæa, xxvi. 613, as a var. of *A. nematophylla,* F. Muell. The latter is, however, a short-leaved form of *A. calamifolia,* which has never more than 1 nerve on each side of the phyllodium.

51. A. triptycha, *F. Muell. Herb.* A spreading shrub, very nearly allied to *A. leptoneura,* with linear-subulate, spreading or flexuose phyllodia of $1\frac{1}{2}$ to 3 in., but with only 2 or 3 prominent nerves on each side. Peduncles short, bearing a small head of about 20 flowers, mostly 5-merous. Sepals free, narrow-spathulate. Petals smooth. Pod unknown.

W. Australia, *Drummond (4th Coll. ?), n. 132, and 5th Coll. n.* 5 ; Kalgau river, *Oldfield ;* Termination rock, *Maxwell.*

52. A. leptoneura, *Benth. in Hook. Lond. Journ.* i. 341. Shrubby. Branches slightly angular, glabrous or hoary-pubescent when young. Phyllodia linear-subulate, rather rigid, nearly terete, straight or flexuose, finely striate with numerous scarcely prominent nerves, almost obtuse or with short recurved points, mostly about 2 in. long. Peduncles short, bearing each a globular head of 20 to 30 flowers, mostly 5-merous. Sepals narrow-spathulate, distinct or very shortly united in a cup at the base. Petals smooth, readily separating. Pod not seen.—Bot. Mag. t. 4350; Meissn. in Pl. Preiss. i. 12 (except the var. β).

W. Australia, *Drummond ;* Fitzgerald ranges, *Maxwell.*

53. A. rigens, *A. Cunn. in G. Don, Gen. Syst.* ii. 403. A tall shrub, either quite glabrous or pale or hoary with a minute pubescence ; branchlets somewhat angular. Phyllodia linear-subulate rather rigid, nearly terete, straight or incurved, usually 2 to 3 in. long and very finely striate with 3 to

5 scarcely prominent nerves, with a short, innocuous, oblique or recurved point, but in some specimens 3 nerves on each side are prominent, at least at the base. Peduncles very short, bearing each a globular head of about 20 flowers, mostly 5-merous. Sepals spathulate, united to about the middle. Petals smooth. Pod linear, straight or curved, flat, about 1½ lines broad, much contracted between the seeds, the valves coriaceous and convex at the seeds. Seeds ovate, longitudinal; funicle with several folds, the last dilated into a turbinate almost cup-shaped aril.—*A. chordophylla,* F. Muell. in Linnæa, xxvi. 612, and Pl. Vict. ii. 11.

N. S. Wales. Low flat country on the Lachlan river, *A. Cunningham ;* Darling river and thence to the Barrier Range, *Victorian Expedition.*
Victoria. Desert country along the Murray and Wimmera, *F. Mueller.*
S. Australia. Port Lincoln, *Wilhelmi ;* towards Lake Alexandrina, *F. Mueller.*
Distinguished from *A. leptoneura* only in the sepals rather more united, from the narrow-leaved forms of *A. elongata* in the phyllodia still narrower and less flattened.
Var. *longifolia.* Phyllodia slender, often 6 in. long. Heads almost sessile, with nume-rous flowers.—In *Leichhardt's* collection.

54. **A. papyrocarpa,** *Benth.* A small tree, of about 25 ft., the branches nearly terete, glabrous or minutely hoary-pubescent. Phyllodia linear-subulate, rigid, thick but slightly flattened, 2 to 3 in. long, tapering into a recurved but not pungent point, narrowed at the base, striate with numerous fine parallel nerves only visible under a lens, slightly hoary-tomen-tose especially along the centre, without any midrib. Flowers not seen, but from the scars they must have been in globular heads on very short peduncles. Pod flat, falcate or flexuose, 3 to 4 in. long, 4 to 5 lines broad ; valves thin, almost membranous. Seeds ovate, longitudinal; funicle with very short folds, gradually and not much thickened from the base.

S. Australia. S. coast, *R. Brown* (*Herb. R. Br.*). The phyllodia at first sight re-semble those of *A. rigens,* or of some forms of *A. calamifolia,* but the veins are finer and more numerous and the pod is very different, more like that of *A. Murrayana.*

C. UNINERVES.—Phyllodia 1-nerved or nerveless.

Where the nerves (one on each side) are prominent, the phyllodium becomes tetra-gonous.

55. **A. pugioniformis,** *Wendl. Comm. Acac.* 38, *t. 9, but not the syn. given in Flora,* 1819, 139. A tall glabrous shrub, with slender slightly an-gular branches. Phyllodia rather numerous, straight or slightly curved, mostly erect, linear-subulate, 1¼ to 2 rarely 3 in. long, abruptly terminating in a short straight point, nearly tetragonous by a prominent nerve on each side. Stipules minute. Peduncles solitary or 2 together, 2 to 3 lines or rarely ½ in. long, bearing each a globular head of numerous flowers, mostly 5-merous. Sepals linear-spathulate, ciliate, at length free, about half as long as the co-rolla. Petals smooth, with a prominent midrib. Pod unknown.—DC. Prod. ii. 450 ; *A. quadrilateralis,* DC. Prod. ii. 451.

Queensland. Brisbane and Logan rivers, *A. Cunningham ;* also in *Leichhardt's* col-lection.
N. S. Wales. Port Jackson to the Blue Mountains, *R. Brown, R. Cunningham, Sieber, n.* 442.
The *A. quadrilateralis* inserted by Decaisne in the Herb. Tim. Descr. 132 as a Timor plant, which I have not seen, is more likely to be the following species.

56. A. juncifolia, *Benth. in. Hook. Lond. Journ.* i. 341. A tall glabrous shrub, with slender branches, quite terete. Phyllodia linear-subulate, erect or spreading, slightly flattened with a scarcely prominent nerve on each side, 3 to 6 in. long or even more, with a very short erect or curved point or obtuse. Stipules minute. Peduncles solitary or 2 together, rarely ½ in. long, bearing each a small globular head of numerous flowers, mostly 5-merous. Sepals spathulate, at length free, half as long as the corolla. Petals smooth, without the prominent midrib of *A. pugioniformis.* Pod straight, flat or flexuous, often 3 or 4 in. long, 1½ to 2 lines wide. Seeds obovate-oblong, longitudinal; funicle not folded, slightly thickened towards the end.—*A. pinifolia,* Benth. in Mitch. Trop. Austr. 342.

N. Australia. Islands of the Gulf of Carpentaria, *R. Brown;* barren stony places on the Macarthur, Gulf of Carpentaria, *F. Mueller.*
Queensland. E. coast, *R. Brown;* Port Bowen, *A. Cunningham;* near Mount Pluto, *Mitchell;* also in *Leichhardt's* collection.
N. S. Wales. Barren lands, N.W. interior, *Mitchell, A. Cunningham, Fraser.*
Var. *planifolia.* Phyllodia flatter, nearly a line broad, with a more prominent midrib, almost like those of *A. subulata,* but the peduncles all simple.—In *Mitchell's* collection.

57. A. calamifolia, *Sweet, in Lindl. Bot. Reg.* t. 839. A tall shrub, glabrous and often glaucous or slightly mealy, the branches rather slender and terete. Phyllodia linear-subulate, in the northern specimens very slender and mostly 3 to 4 in. long, in the more southern ones usually about 2 in. and from that to 3 in., rarely shorter, and then often slightly flattened and nearly 1 line broad but thick, sometimes slender as in the long ones, always tapering into a fine recurved point which only wears away with age, nerveless or with one fine nerve on each side. Flower-heads globular, smaller than in *A. pugioniformis,* and usually 3 or 4 in a short raceme, more rarely solitary. Flowers numerous, mostly 5-merous. Calyx thin and transparent, with short, broad, ciliate lobes, often splitting into spathulate sepals. Petals smooth, distinct. Pod often 5 or 6 in. long, usually curved, 2 to 2½ lines broad but much contracted between the distant seeds, the valves hard and convex over the seeds. Seeds oblong, longitudinal; funicle long, often almost encircling the seed, then bent back and returning within the previous fold, thickened at the end into a long clavate or shortly turbinate fleshy aril. —Lodd. Bot. Cab. t. 909 ; F. Muell. Pl. Vict. ii. 12 ; *A. pulverulenta,* A. Cunn.; Benth. in Hook. Lond. Journ. i. 342 (the shorter-leaved southern specimens).

N. S. Wales. In the interior, on the Macquarrie, Lachlan, etc., *A. Cunningham, Fraser, Mitchell;* Liverpool plains, *C. Moore;* also in *Leichhardt's* collection.
Victoria. Not unfrequent in the N.W. desert, *F. Mueller.*
S. Australia. Desert land from the Murray to Spencer's Gulf and Kangaroo Island, extending northward to Lake Torrens, *F. Mueller* and others.
Var. *Wilhelmsiana.* Phyllodia shorter, peduncles longer.—*A. Wilhelmsiana,* F. Muell. in Trans. Phil. Inst. Vict. i. 37.—S. Australia, *R. Brown, F. Mueller.*
Under the name of *A. nematophylla,* F. Muell., I had, in Linnæa, xxvi. 612 (owing partly to a wrong label originally sent with F. Mueller's specimens), confounded this variety with the northern *A. Bynoeana,* which is at once known by the venation of the phyllodia.

58. A. scirpifolia, *Meissn. in Bot. Zeit.* 1855, 10. Quite glabrous or the young shoots very sparingly pubescent; branchlets angular. Phyllodia

linear-subulate, obtuse or with a minute reourved innocuous point, mostly 3
to 5 in. long, terete or slightly flattened, obscurely 1-nerved. Stipules very
deciduous. Peduncles solitary, bearing a small globular head of about 20
flowers, mostly 5-merous. Calyx turbinate, truncate, obtusely sinuate-
toothed, about half as long as the corolla. Petals smooth. Pod long, straight
or nearly so, about 2 lines broad and much contracted between the seeds,
valves hard, very convex over the seeds. Seeds oblong, longitudinal; funicle
with short folds, 2 at least of the last much thickened or dilated.

W. Australia. Between Moore and Murchison rivers, *Drummond, 6th Coll. n.* 5,
Oldfield. The calyx and pod are nearly those of *A. salicina* and its allies, but the phyllodia and seeds are more those of some of the *Calamiformes*.

59. **A. extensa,** *Lindl. Swan Riv. App.* 15. An erect glabrous shrub
of several ft., with elongated branches, always angular and sometimes almost
winged. Phyllodia linear-subulate, rigid but slender, often scarcely to be
distinguished from branchlets, erect spreading or recurved, sometimes few
and short, more frequently from 3 or 4 in. to twice that length, with a prominent nerve on each side. Peduncles solitary or shortly and irregularly racemose, bearing each a globular head of 20 or more flowers, mostly 5-merous.
Calyx turbinate, thin, truncate or shortly and broadly lobed. Petals smooth,
united at the base. Pod long, linear, straight or curved, about 1½ lines broad
and contracted between the seeds; valves thinly coriaceous. Seeds oblong,
longitudinal; funicle with short folds, the last 2 usually much thickened.—
Meissn. in Pl. Preiss. i. 6 ; *A. graminea,* Lehm. Del. Sem. Hort. Hamb.
1842 ; Meissn. in Pl. Preiss. i. 5 ; *A. pentaedra,* Regel, Gartenfl. i. 228,
t. 24.

W. Australia. Swan River, *Drummond, 1st Coll.;* Vasse river, *Mrs. Molloy;*
Princess Royal Harbour, *Preiss, n.* 983 ; Gordon ranges, *Maxwell;* Preston and King
rivers, *Oldfield.*

60. **A. gonophylla,** *Benth. in Linnæa,* xxvi. 613. A rigid glabrous
shrub, allied in some respects to *A. sulcata;* branches angular. Phyllodia
not crowded, linear, incurved, obtuse with a short point, rigid and acutely tetragonous by the very prominent nerves, narrowed at the base, 1 to 1½ in.
long. Stipules minute. Peduncles usually in pairs, 2 to 4 lines long, bearing each a globular head of about 12 to 20 flowers, mostly 5-merous. Calyx
very thin, often separating into distinct sepals, half as long as the corolla.
Petals smooth. Pod linear, flat, slightly contracted between the seeds, about
2 lines broad. Seeds oblong, longitudinal, the last fold of the funicle very
short and turbinate, the next long and considerably dilated.

W. Australia. Towards Cape Riche, *Drummond, 5th Coll. n.* 4 ; Phillips Range and
Bremer Bay, *Maxwell.*
Var. *crassifolia.* Lower phyllodia sometimes dilated, linear-cuneate or almost oblong,
2 in. long, very thick and much narrowed at the base, upper ones as in the common form.
Heads rather larger, with more flowers.—Towards the Great Bight, *Maxwell.*
A fragment from Kojonerup range has the stipules almost spinescent; another, from
Phillips range, has the phyllodia not ½ in. long ; neither are sufficient for accurate determi-
nation.

61. **A. ericifolia,** *Benth. in Hook. Lond. Journ.* i. 345. A low bushy
shrub, attaining sometimes 3 or 4 ft., glabrous or loosely hirsute. Phyllodia
numerous, shortly linear-terete or slightly flattened horizontally, obtuse,

mostly ½ to ¾ very rarely 1 in. long, narrowed at the base, nerveless or obscurely 1-nerved. Stipules deciduous. Peduncles short, solitary or in pairs, bearing each a head of about 15 to 20 or rather more flowers, mostly 5-merous. Sepals free, narrow-spathulate. Petals smooth, united to the middle. Young pod hard and terete, but not seen fully formed.—*A. Hookeri,* Meissn. in Pl. Preiss. ii. 202.

W. Australia. Swan River, *Drummond, 1st Coll.* (*n.* 300, according to Meissner), *2nd Coll. n.* 141, *Preiss, n.* 981; Murchison river, *Oldfield.* Narrow-leaved specimens of *A. leptospermoides* might at first be mistaken for this species, but may be known by at least 3 nerves visible under a lens, at least at the base of the leaf.

62. **A. uncinella,** *Benth. in Linnæa,* xxvi. 613. A low bushy shrub, glabrous or the young shoots minutely pubescent; branches terete or slightly angular. Phyllodia shortly linear-subulate, terete or nearly so, mostly about 1 in. long, faintly 1- or 3-nerved on one side, narrowed at the base and as it were petiolate. Stipules minute, deciduous. Peduncles 2 to 4 lines long, rather slender, solitary or in pairs, bearing each a small globular head of about 20 flowers, mostly 5-merous. Sepals distinct, narrow-spathulate. Petals smooth. Pod (from the remains on one specimen) not above 1¼ lines broad, not contracted between the seeds.

W. Australia. In the interior, *J. S. Roe ;* W. tributary to the Oldfield river, *Maxwell.*

63. **A. oxyclada,** *F. Muell. Herb.* A bushy, heath-like, glabrous shrub of 3 or 4 ft., the very numerous slender branches mostly ending in fine thorns. Phyllodia shortly linear-terete or slightly flattened horizontally, mostly 3 to 4 lines long, nerveless on one side, faintly 1- or 3-nerved on the other, almost petiolate. Flowers not seen. Fruiting peduncles 3 to 4 lines long, bearing the scars of a head of several flowers. Pod about 1 in. long, 1½ lines broad, flexuose. Seeds orbicular; funicle short, thickened at the end, but not seen perfect.

W. Australia. Murchison river, *Oldfield.* The species has the habit of the *Uninerves spinescentes,* but the phyllodia are almost terete.

Series V. BRUNIOIDEÆ.—Phyllodia numerous, small, linear-subulate (except in *A. conferta*), verticillate, clustered or crowded, obtuse or with innocuous or rarely rigid points. Flowers in globular heads, on simple axillary peduncles usually exceeding the phyllodia.

Distinct as are the majority of the species of this group, they pass into the *Uninerves brevifoliæ* through *A. conferta.* Verticillate phyllodia occur also in *A. verticillata,* but they are there rigid and pungent.

64. **A. cedroides,** *Benth. in Linnæa,* xxvi. 615. A much-branched shrub of 3 or 4 ft., the specimens having the aspect of a short-leaved Fir or of an Equisetum; branches shortly villous. Phyllodia verticillate, linear-subulate, 4-gonous, rigid, almost pungent-pointed, ½ to 1 in. long. Stipules setaceous. Peduncles 2 to 4 lines long, bearing each a small head of 20 to 30 flowers, mostly 5-merous, the receptacle densely ciliate. Calyx broad, obtusely lobed, not half so long as the corolla. Petals narrow, smooth, usually cohering to the middle. Pod sessile, incurved, acuminate, 1½ to 2 in. long, about 2 lines broad, the valves hard, longitudinally striate, with broad

smooth margins. Seeds oblong, longitudinal; funicle folded and thickened at the end into a turbinate aril, but not seen perfect.

W. Australia. Southern districts, *Drummond, 4th Coll. n.* 4; among rocks, Fitz-gerald Range, Mount Bland, etc., *Maxwell.*

65. **A. lycopodifolia,** *A. Cunn. in Hook. Ic. Pl. t.* 172. A much-branched diffuse or divaricate shrub, clothed with very short spreading hairs and more or less viscid. Phyllodia verticillate, about 8 to 10 in the whorl, subulate, rarely above 3 lines long and often only 1 to 2 lines, sulcate with a prominent vein on each side, erect at the base, recurved at the end with a fine glabrous viscid point, sometimes very short, sometimes nearly as long as the phyllodium. Stipules setaceous. Peduncles longer than the phyllodia, bearing each a globular head of numerous flowers, mostly 5-merous, the bracts protruding when young. Calyx very short, with small acute teeth. Petals several times as long, striate, pubescent, united above the middle. Pod sessile or very shortly contracted at the base, quite flat, straight or slightly curved, 1 to 1½ in. long, 2½ to 3 lines broad. Seeds nearly orbicular, oblique, the last fold of the funicle thickened into a fleshy aril.

N. Australia. Cambridge Gulf, N.W. coast, *A. Cunningham;* Hammersley Range, Nichol Bay, *F. Gregory's Expedition;* Victoria river, *Bynoe, F. Mueller;* Arnhem S. Bay, *R. Brown;* also in *Leichhardt's* collection.

Var. *glabrescens.* Pubescence much shorter or disappearing. Phyllodia rather longer, but sulcate, with recurved points, as in the ordinary form. Calyx rather more prominent. Pod sessile.—*A. asperulacea,* F. Muell. in Journ. Linn. Soc. iii. 123.—Victoria river, *F. Mueller.*

66. **A. hippuroides,** *Heward; Benth. in Hook. Lond. Journ.* i. 344. Pubescent, with verticillate, subulate, sulcate phyllodia, as in *A. lycopodifolia,* of which it may possibly prove a variety, but the phyllodia are much longer, attaining 5 or 6 lines, straight not recurved at the end, and the calyx, thinly membranous, is at least one-third as long as the corolla. Petals striate and pod sessile, as in *A. lycopodifolia.*

N. Australia. Usborne's Harbour, N.W. coast, *Voyage of the Beagle;* Attack Creek, *M'Douall Stuart's Expedition.*

A fragment from M'Douall Stuart's Expedition differs in the phyllodia more rigid and short, but with the same straight points, the stems more viscid, awnless, pubescent, and especially in the extreme tenuity of the calyx, which it is rather difficult to find, and in the petals membranous and smooth, as in *A. cedroides,* but broader and glabrous.

67. **A. galioides,** *Benth. in Hook. Lond. Journ.* i. 344. Pubescent or tomentose, with verticillate finely subulate phyllodia, as in *A. lycopodifolia,* but the phyllodia are more slender, slightly striate only, not sulcate with prominent nerves, 2 to 5 lines long, recurved at the end but apparently without viscid points. Flowers mostly 5-merous, in globular heads on peduncles exceeding the leaves, as in that species, but the calyx is at least one-third as long as the corolla, and the pod is always borne on a stipes of 2 to 3 lines.

N. Australia. Victoria river, *F. Mueller;* islands of the Gulf of Carpentaria, *R. Brown;* and a variety with rather stouter phyllodia, Sweers Island, *Henne.*

68. **A. Baueri,** *Benth. in Hook. Lond. Journ.* i. 344. Apparently an undershrub, with erect or ascending stems, under 1 ft. high, the terete branchlets minutely pubescent, otherwise glabrous. Phyllodia in whorls of 5 to 7,

linear-subulate, terete, without prominent nerves, but often a slight furrow underneath, recurved at the end and obtuse or with a minute point, about 4 to 6 lines long. Stipules minute or none. Peduncles rather longer than the phyllodia, bearing a very small head of 10 to 20 flowers, mostly 5-merous and scarcely ¾ line long. Calyx fully half as long as the corolla, with acuminate teeth. Petals with a prominent midrib, but not striate. Pod falcate, narrowed at each end, 1½ to 2 lines broad, hard, longitudinally striate. Seeds not seen.

N. Australia. N. coast, *R. Brown*, without the precise station. The aspect of the plant is that of *A. subternata*, but the phyllodia are verticillate as in the preceding species, although fewer in the whorl.

69. **A. subternata,** *F. Muell. in Journ. Linn. Soc.* iii. 124. A glabrous shrub of 3 to 5 ft.; branchlets angular, sulcate, slightly viscid when young. Phyllodia mostly in clusters of 2, 3 or 4, linear-terete or very slightly compressed, with short recurved or hooked points, 3 to 6 lines long, without prominent nerves and scarcely furrowed. Stipules minute or none. Peduncles scarcely longer than the phyllodia, bearing each a globular head of numerous flowers, mostly 5-merous. Sepals rather rigid, linear-spathulate, fully half as long as the corolla, united in a 5-nerved cup at the base. Petals slightly striate, united to the middle. Pod flat, rigidly coriaceous, much narrowed into a long stipes, about 2 lines broad in the upper part, somewhat viscid, very obliquely striate, with thickened margins and oblique partitions between the seeds. Seeds oblong, obliquely transverse; funicle straight, gradually thickened from the base to the end.

N. Australia. Table land, Upper Victoria river, *F. Mueller.* Some young specimens of *Leichhardt's* appear to be referable to the same species.

70. **A. bruniades,** *A. Cunn. in G. Don, Gen. Syst.* ii. 404. A heathlike shrub, glabrous or the terete branches minutely pubescent. Phyllodia crowded, but scattered or irregularly verticillate, linear-terete, 2 to 4 lines long, with short straight points, without prominent nerves or furrows. Stipules minute or none. Peduncles longer than the phyllodia, bearing each a globular head of rather small flowers, mostly 5-merous. Calyx turbinate, angular, half as long as the corolla, with short obtuse minutely ciliate lobes. Petals free, smooth with rather prominent midribs. Pod not seen.

Queensland. Brisbane river, *A. Cunningham;* Mounts Hooker and Lindsay, *Fraser.*

71. **A. conferta,** *A. Cunn.; Benth. in Hook. Lond. Journ.* i. 345. A tall heath-like shrub, with terete slightly pubescent branches. Phyllodia crowded, scattered or irregularly verticillate, linear, compressed, 3 to 4 or rarely 5 lines long, rigid, mostly obtuse, without nerves or with slightly thickened nerve-like margins, ½ to 1 line broad. Stipules minute or none. Peduncles longer than the phyllodia, bearing each a globular head of numerous small flowers, mostly 5-merous. Sepals linear-spathulate, distinct or slightly united at the base. Petals distinct, smooth, but with slightly prominent midribs. Pod very flat, stipitate, obtuse, 1 to 1½ in. long, about 5 lines broad, glaucous. Seeds nearly transverse, ovate, the last fold of the funicle dilated into a cup-shaped aril.

Queensland. Shoalwater Bay, *R. Brown;* Dawson river, *F. Mueller;* tributaries of the Macquarrie, *A. Cunningham;* on the upper Maranoa, *Mitchell;* also in *Leichhardt's Collection.*

SERIES VI. UNINERVES.—Phyllodia vertically flattened, either narrow and obtuse or with a short oblique or innocuous point, or broad and obtuse acute or rarely pungent-pointed, with 1 central or nearly marginal nerve, or very rarely 2-nerved. Flowers in globular heads, either on simple axillary peduncles, solitary or in pairs or clusters, or several in axillary racemes.

This series comprises all the species with 1-nerved phyllodia and globular flower-heads, where the phyllodia are not continuous with the stem as in the *Alatæ* and *Continuæ,* nor narrow and pungent as in the *Pungentes,* nor terete or tetragonous as in the *Calamiformes,* nor whorled as in the *Brunioideæ.* The nerve is usually central or nearly so, with the small veins, when conspicuous, pinnate or reticulate ; where the nerve is excentrical or near the lower margin, one or two of the principal veins will sometimes arise from near the base on the upper side, but diverge from the midrib and are not continued to the end of the phyllodium as in the *Plurinerves. A. verniciflua,* and a very few others, have most or some of their phyllodia 2-nerved as in the 2-nerved *Plurinerves,* but are placed among the *Uninerves,* either because the second nerve is very inconstant or from their close affinity to species where it does not exist.

A. SPINESCENTES.—Rigid shrubs with spinescent branches. Phyllodia small, usually narrow, not pungent, without marginal glands. Stipules minute or none (except sometimes in *A. scabra*). Peduncles 1-headed.

Spinescent branches occur also in a very few of the *Armatæ* and of the *Triangulares* of the present series, in some of the *Pungentes,* and in *A. amblygona* among *Plurinerves.*

72. **A. scabra,** *Benth.* in *Linnæa,* xxvi. 605. A shrub with rigid divaricate spinescent branches, more or less scabrous-pubescent. Phyllodia linear-oblong, very oblique or falcate, obtuse or with a small recurved point, about ½ in. long, 1 to 2 lines broad, rather thick and 1-nerved, scabrous-pubescent or at length glabrous. Stipules small, setaceous or almost spinescent. Peduncles about as long as the leaves, bearing each a globular head of above 30 flowers, mostly 5-merous. Calyx thin, the lobes obtuse and ciliate, readily separating into spathulate sepals, fully half as long as the corolla. Petals thin, with slightly prominent midribs. Pod (not seen ripe) narrow-linear, flat, with thickened nerve-like margins.

W. Australia, *Drummond, 2nd Coll. n.* 162. Allied on the one hand to the following species, on the other to some of the narrow-leaved *Armatæ.*

73. **A. nodiflora,** *Benth.* in *Linnæa,* xxvi. 621. A shrub with long rigid virgate branches and divaricate branchlets, often spinescent and nearly glabrous. Phyllodia clustered on the old nodes, linear, oblique or falcate, obtuse or with a small oblique point, rarely above ½ in. long, ½ to 1 line broad, 1-nerved. Stipules minute or none. Peduncles slender, about as long as the phyllodia, bearing each a small globular head of numerous flowers, mostly 5-merous. Sepals narrow-linear, spathulate, free. Petals smooth, readily separating. Pod not seen.

W. Australia, *Drummond, 4th Coll. n.* 8.

74. **A. spinosissima,** *Benth.* in *Linnæa,* xxvi. 621. A glabrous shrub, with numerous slender striate spinescent branches. Phyllodia linear-falcate, obtuse or mucronulate, mostly 2 to 3 lines long, and under 1 line broad, obscurely 1-nerved. Peduncles slender, about as long as the phyllodia, bearing

each a very small head of 6 to. 10 small flowers, mostly 5-merous. Calyx very short, truncate or sinuately toothed. Petals smooth. Pod linear, flat, with nerve-like margins, 1½ or scarcely 2 lines broad, on a rather long stipes. Seeds nearly globular, the funicle filiform to the end, but not seen quite ripe.

W. Australia, *Drummond, 5th Coll. n.* 51.

75. **A. ulicina,** *Meissn. in Pl. Preiss.* ii. 202. A rigid spreading glabrous shrub of 2 to 4 ft., the branches sulcate-striate, the smaller short ones spinescent or reduced to small thorns. Phyllodia linear, obtuse or obliquely mucronate, often 1 in. long on the main branches, ¼ to ½ in. on the side ones, obscurely 1-nerved. Peduncles short, bearing each a globular head of about 20 flowers, mostly 5-merous. Calyx obtusely lobed, half as long as the corolla. Petals smooth, but with prominent midribs. Pod linear, very flexuose, 1 to 1½ lines broad, much contracted between the seeds. Seeds obovate-oblong, thick; the funicle with a first short filiform fold, the next thickened into a club-shaped, almost hood-shaped aril, almost as long as the seed.

W. Australia, *Drummond, 2nd Coll. n.* 147; Bowes river and S. Hutt river, *Oldfield.*

76. **A. erinacea,** *Benth. in Hook. Lond. Journ.* i. 360. A rigid spreading glabrous shrub, with striate spinescent branches. Phyllodia obliquely obovate-oblong or lanceolate, obtuse or mucronulate, 3 to 5 lines long and 1 to 2 broad, thick, rigid and obscurely 1-nerved. Peduncles 2 to 4 lines long, bearing each a head of about 20 or fewer flowers mostly 5-merous. Calyx very short, truncate or minute-toothed. Petals smooth, without prominent midribs. Pod not seen.

W. Australia, *Drummond (2nd or 3rd Coll.), n.* 163.

B. **ARMATÆ.**—Shrubs or undershrubs, not spinescent (except sometimes in *A. congesta* and *A. idiomorpha*). Phyllodia from obovate to lanceolate, rarely above 1½ in. long, more or less undulate, with a central nerve and usually nerve-like margins, the marginal gland very small and obscure or none at all. Stipules generally persistent, either spinescent or subulate, or acuminate or phyllodia-like. Peduncles 1-headed.

Persistent rigid stipules and short undulate phyllodia are the characteristics of this group, but there are some species in which the stipules occasionally disappear, and spinescent stipules exist also in several of the *Triangulares* and in *A. sentis* and *A. dentifera* among *Angustifoliæ;* the undulate phyllodia occur also in *A. hispidula* and *A. undulifolia* among *Brevifoliæ.* I formerly considered *Armatæ* as a primary series, but finding the characters so vague, I have now thought it more convenient to divide them into subseries of *Uninerves* and *Plurinerves.*

77. **A. Huegelii,** *Benth. in Hueg. Enum.* 42. An erect bushy shrub, with nearly terete shortly hirsute branches. Phyllodia semiovate, recurved-falcate, ½ to ¾ or rarely 1 in. long, 2 to 5 lines broad, tapering into a pungent point, narrowed at the base, often undulate, pubescent, with a curved central nerve. Stipules setaceous, almost spinescent. Peduncles short, bearing each a small head of 20 to 30 flowers, mostly 5-merous. Calyx none. Petals united above the middle, hirsute outside. Pod (according to Meissner) sessile, linear, falcate, 1 in. long, 2 lines wide, flat with nerve-like margins, pubescent.—Hueg. Bot. Archiv, t. 10; Meissn. in Pl. Preiss. i. 7.

W. Australia. Swan River, *Huegel, Fraser;* Swan River and Port Leschenault, *Preiss, n.* 953, 961; Tone and Vasse rivers, *Oldfield.*

78. **A. nervosa,** *DC. Mém. Leg.* 444; *Prod.* ii. 449. A glabrous erect shrub or undershrub of 1 to 2 or 3 ft., branches prominently and acutely angular and striate. Phyllodia broadly lanceolate, incurved-falcate, rather rigid, but usually undulate, tapering into a short point, 1 to 1½ or rarely 2 in. long, with a nearly central nerve and nerve-like margins. Stipules spinescent. Peduncles short, usually 2 or 3 together, bearing each a head of 8 to 12 rather large flowers, mostly 4-merous. Calyx very short and broad, lobed, or sometimes scarcely any. Petals several times longer, smooth, spreading, united at the base only. Pod not seen.—Field and Gardn. Sert. Pl. t. 4; Meissn. in Pl. Preiss. i. 7, ii. 200.

W. Australia. Geographe Bay, *Baudin's Expedition;* Swan River, *Drummond,* 1st *Coll. n.* 290 or 298, *Fraser,* and others; Gordon river, *Oldfield;* summit of Green Mountain, *Preiss, n.* 936. The flowers resemble those of *A. myrtifolia,* but the stipules and inflorescence are quite different.

79. **A. obovata,** *Benth. in Hook. Lond. Journ.* i. 329. An undershrub or shrub of 1 to 1½ ft., more or less scabrous-pubescent, with angular-striate branches. Lower phyllodia obovate or oblong, obtuse, 1 to 1½ in. long, the upper ones broadly incurved-falcate, acute and pungent-pointed, all much undulate, with thickened nerve-like margins, a prominent midrib and pinnate veins. Stipules setaceous. Peduncles solitary or 2 or 3 together, about as long as the phyllodia, bearing each a head of about 6 to 8 flowers, mostly 4-merous, with a very short calyx as in *A. nervosa,* but more acuminate in the bud. Pod not seen.—Meissn. in Pl. Preiss. i. 8.

W. Australia. Cape Leewin, *Collie;* gravelly places, Green Mountain, *Preiss, n.* 931.

80. **A. congesta,** *Benth. in Hook. Lond. Journ.* i. 327. A rigid shrub, with divaricate terete pubescent branches, often spinescent. Phyllodia obliquely oval-oblong or falcate-lanceolate, obtuse or acute and shortly mucronate, ⅓ to ¾ or rarely nearly 1 in. long, 2 to 4 lines broad, usually undulate, quite glabrous, not thick, with a prominent midrib and obscurely pinnate veins. Stipules very short, usually spinescent. Peduncles solitary or clustered, bearing each a globular head of numerous (40 to 50) small flowers, mostly 5-merous. Sepals free, linear-spathulate, fully half as long as the corolla. Petals smooth, glabrous. Pod linear, compressed but thick, much curved and contracted between the seeds. Seeds longitudinal, the last 2 folds of the funicle expanding into an oblique cup-shaped aril at the end of the seed.—Meissn. in Pl. Preiss. ii. 200; *A. Baxteri,* Meissn. in Pl. Preiss. i. 7, not of Benth.

W. Australia, *Drummond,* 2nd *Coll. n.* 293; near York, *Preiss, n.* 977. Very much resembles some varieties of *A. armata,* but the stipules are smaller and the calyx and pod different.

81. **A. dermatophylla,** *Benth.* Quite glabrous. Branchlets angular. Phyllodia oblong-cuneate, somewhat falcate, undulate, with a very short rigid or callous point, ¾ to 1½ in. long, very thick and rigid, with a central nerve and thick nerve-like margins, scarcely prominent. Stipules rigid, subulate, persistent. Peduncles shorter than the phyllodia, bearing each a dense glo-

bular head of about 20 flowers, mostly 5-merous. Sepals narrow, linear-spathulate, free or shortly united at the base. Petals smooth. Pod unknown.

W. Australia. Murchison river, *Oldfield.* The specimens are small, but the smooth thick phyllodia distinguish them from all the *Armatæ*, except some varieties of *A. nervosa*, from which it is easily known by the broader phyllodia, less pungent stipules, and much more numerous 5-merous flowers.

82. **A. aspera,** *Lindl. in Mitch. Three Exped.* ii. 139. A spreading shrub, attaining several ft., very resinous and rough with a glandular viscid pubescence ; branches rigid, nearly terete, striate. Phyllodia oblong-linear, oblique and more or less undulate, obtuse or with an incurved innocuous point, ½ to 1 in. or rarely 1½ in. long, with a central nerve and slightly thickened nerve-like margins. Stipules small, setaceous. Peduncles solitary or in pairs, not above ¼ in. long, rather thick, bearing each a dense globular head of 30 to 50 or more flowers, mostly 5-merous. Bracts usually acuminate and protruding beyond the buds. Sepals spathulate, glandular-pubescent at the end, more or less united to the middle. Petals smooth, glabrous, united to the middle. Pod linear, curved, glandular-hispid, 1 to 2 in. long, about 2 lines broad, contracted between the seeds. Seeds longitudinal, oval-oblong ; funicle with short folds, the last 2 thickened into an almost cup-shaped aril at the base of the seed.—F. Muell. Pl. Vict. ii. 21 ; *A. erythrocephala*, A. Cunn. ; Benth. in Hook. Lond. Journ. i. 362 (a narrow-leaved form).

N. S. Wales. Lachlan river, *A. Cunningham.*
Victoria. Junction of the Loddon and Murray, *Mitchell ;* open barren forest-lands and scrubby ridges, Serra and Victoria ranges, Grampians, Forest Creek, Black Forest, Goulburn and Broken rivers, *F. Mueller.*
Var. *densifolia.* Phyllodia smaller. Flower-heads almost sessile. Bracts much less acuminate.—*A. strigosa*, Lindl. in Mitch. Three Exped. ii. 185, not of Link ; *A. densifolia*, Benth. in Hook. Lond. Journ. i. 360.—Near Mount Zero, *Mitchell.*
This species has been sometimes confounded with *A. hispidula*, which is however readily known by the tuberculate almost denticulate margins of the phyllodia, the smaller heads with fewer flowers and especially by its short broad thick straight 1- or 2-seeded pod.

83. **A. armata,** *R. Br. in Ait. Hort. Kew. ed.* 3, v. 463. A tall bushy shrub, attaining sometimes 10 ft. or more ; branches angular-striate, hirsute-pubescent or rarely glabrous. Phyllodia semiovate, obliquely oblong or incurved-lanceolate, undulate, obtuse or with a very short oblique point, with a nearly central midrib and pinnate veins, varying from 4 lines to above 1 in. in length, and in breadth from one-fifth to nearly half their length. Stipules straight, divaricate and spinescent, often 4 to 5 lines long. Peduncles usually about as long as the phyllodia, bearing a globular head of rather numerous 5-merous flowers. Calyx thin, lobed, but not usually separating into sepals, about half as long as the corolla. Petals narrow, glabrous, smooth. Pod straight or curved, 1½ to 2 in. long, 2 to 3 lines broad, not contracted between the seeds, softly villous or rarely glabrous or hispid. Seeds oblong, longitudinal, the funicle slightly dilated nearly from the base, forming 3 or 4 folds, scarcely more thickened under the seed.—Bonpl. Jard. Malm. t. 55 ; DC. Prod. ii. 449 ; Bot. Mag. t. 1653 ; Lodd. Bot. Cab. t. 49 ; F. Muell. Pl. Vict. ii. 3 ; *A. furcifera*, Lindl. in Mitch. Three Exped. ii. 267.

N. S. Wales. Blue Mountains and in the interior to Peel's Range, *A. Cunningham ;* northward to New England, *C. Stuart.*

Victoria. Barren ridges from near Port Phillip to Goulburn and Broken rivers, near Forest Creek and towards the mouth of the Glenelg, *F. Mueller.*

S. Australia. From the Murray to Spencer's and St. Vincent's Gulfs and Mount Remarkable, *F. Mueller;* Kangaroo Island, *R. Brown, Waterhouse.*

W. Australia. King George's Sound (or to the E.?), *Baxter;* Murchison river, *Oldfield.*

Var. *angustifolia.* Phyllodia narrower.—*A. paradoxa,* DC. Prod. ii. 449 ; *A. undulata,* Wild. Enum. Hort. Berol. Suppl. 68 ; Wendl. Comm. Acac. t. 3 ; Bot. Reg. t. 843 ; Lodd. Bot. Cab. t. 753 ; Reichb. Ic. et Descr. Pl. t. 89.

This species is now an old inmate of our gardens, where it varies much, and is said to have been frequently hybridized. Some of these forms have a second nerve to some or all the phyllodia, or have the stipules very small or none. These garden forms include *A. ornithophora,* Sweet, Fl. Austral. t. 24; Lodd. Bot. Cab. t. 1469 ; *A. hybrida,* Lodd. Bot. Cab. t. 1342 ; *A. micracantha,* Dietr. in Allgem. Gart. Zeit. i. 83 ; *A. tristis,* Grah. in Bot. Mag. t. 3420. It is possible that a few of the Western or out-of-the-way stations given for the species, may have been erroneously founded on cultivated specimens sent as wild.

84. **A. idiomorpha,** *A. Cunn.; Benth. in Hook. Lond. Journ.* i. 329. A shrub with very rigid divaricate often spinescent branches, softly villous as well as the phyllodia. Phyllodia broadly and obliquely ovate, undulate, pungent-pointed, about ½ in. long, with a strong central nerve and nerve-like margins. Stipules spinescent, recurved. Flowers (only seen in a single globular head sent with the specimen but not attached to it) numerous, 5-merous, glabrous or nearly so.

W. Australia. Dirk Hartog's Island, *A. Cunningham.*

85. **A. Shuttleworthii,** *Meissn. in Pl. Preiss.* i. 7. Apparently shrubby, hirsute or hispid with short hairs ; branches terete or slightly angular. Phyllodia broadly ovate or orbicular, very oblique, mucronate, thick rigid and much undulate, ¼ to ½ in. long, obliquely obtuse at the base, sessile, 1-nerved ; the gland, if any, at or below the middle. Stipules minute, setaceous, sometimes spinescent. Peduncles very short, bearing each a head of about 8 to 12 flowers, mostly 5-merous. Calyx lobed, about half as long as the corolla. Petals free, hispid. Pod flat, but thick and coriaceous, about 3 lines broad, either 1-seeded and nearly orbicular, or 2-seeded and twice as long, very obtuse, scabrous-pubescent, perhaps sometimes longer as there are several ovules.

W. Australia, *Drummond, n.* 294.

86. **A. Gregorii,** *F. Muell. Fragm.* iii. 47. A diffuse or procumbent shrub or undershrub of 1 to 1½ ft., softly pubescent. Phyllodia obovate or oblong with a small usually recurved point, narrowed at the base, mostly ½ to ¾ in. long, undulate, somewhat coriaceous, 1-nerved and penniveined. Stipules small, lanceolate, acuminate, dry and persistent. Peduncles solitary, nearly as long as the phyllodia, bearing a globular head of 30 to 40 or more flowers, mostly 5-merous. Bracts acuminate. Calyx more than half as long as the corolla, with narrow thin ciliate lobes. Petals slightly pubescent, smooth with prominent midribs, cohering to the middle. Pod unknown.

N. Australia. Nichol Bay, N.W. coast, *F. Gregory's Expedition.* A second label describes it as a very rigid prickly shrub of 6 to 8 ft., but has evidently been misplaced.

87. **A. pilosa,** *Benth. in Linnæa,* xxvi. 607. A decumbent branching undershrub or shrub, often under 1 ft., but probably sometimes taller, more

or less hirsute with rather long spreading hairs. Phyllodia oblique, obovate or broadly oblong, shortly acute or mucronate, much narrowed at the base, much undulate, 1-nerved, mostly ½ to 1 in. long. Stipules setaceous, almost spinescent. Peduncles about as long as the phyllodia, slender, bearing each a small globular head of about 12 to 20 flowers, mostly 5-merous. Calyx-lobes ciliate, fully half as long as the corolla. Petals hirsute, cohering to the middle. Pod oblong, flat, very coriaceous, obtuse, hispid, few-seeded, but not seen ripe.

W. Australia, *Drummond, Suppl. to 3rd Coll. n.* 35, *5th Coll. n.* 12 ; Gordon and Kalgan rivers, *Oldfield ;* Phillips Ranges, *Maxwell.*

88. **A. crispula,** *Benth. in Linnæa,* xxvi. 607. Very near *A. pilosa* in habit and essential characters, but the phyllodia are much narrower, falcate-oblong, rarely exceeding ½ in. in length, 1 to 1½ lines broad, obtuse, shortly acute or mucronate, narrowed at the base, undulate, rather thick and 1-nerved. Stipules setaceous or spinescent. Peduncles slender, bearing a head smaller than in *A. pilosa,* though often with more flowers, 5-merous and hirsute as in that species. Pod unknown.

W. Australia, *Drummond, n.* 78. Possibly a narrow-leaved variety of *A. pilosa.*

89. **A. crassistipula,** *Benth. in Hook. Lond. Journ.* i. 326. Apparently a small shrub, more or less hirsute with soft spreading hairs ; branches angular. Phyllodia oblong-linear, falcate, undulate, with a short incurved or hooked point, ½ to ¾ in. long, 1-nerved with nerve-like margins. Stipules like the phyllodia and about half their size, persistent. Peduncles rather shorter than the phyllodia, bearing each a small globular head of about 30 to 40 flowers, mostly 5-merous. Calyx turbinate, shortly lobed, more than half as long as the corolla. Petals pubescent or hirsute, smooth, but with a prominent midrib, cohering to the middle. Pod unknown.

W. Australia, *Drummond, n.* 295. Allied to the last 3 species, but distinguished from the whole genus by the phyllodineous stipules.

C. TRIANGULARES.—Shrubs usually rigid and occasionally spinescent. Phyllodia small, rigid, the nerve either near the lower margin or rarely central, the upper side more or less dilated, the margin rounded or angular, with usually a gland at the angle. Stipules spinescent or setaceous or minute. Peduncles 1-headed.

These are generally characterized by the small phyllodia with the lower margin nearly straight, the upper one forming a very prominent obtuse or acute angle. They pass into the small-leaved *Pungentes* through some varieties of *A. vomeriformis,* where the glandular angle almost disappears, and into the subseries *Brevifoliæ,* through some forms of *A. bidentata,* which come very near to *A. obliqua.* As in the case of the *Armatæ,* I had originally established the *Triangulares* as a distinct series, but it now appears more natural to separate the many-nerved from the one-nerved species, and consider them as subseries only of the *Plurinerves* and *Uninerves.*

90. **A. hastulata,** *Sm. in Rees Cycl.* xxxix. *Suppl.* A shrub of 2 or 3 ft. ; branches rather slender, elongated, virgate or divaricate, terete, pubescent or at length glabrous. Phyllodia numerous, hastate-lanceolate or almost cordate, tapering into pungent points, 2 to 3 lines long, with 1 central nerve, the lower margin rounded near the base, the upper one more angular and usually bearing a gland. Stipules setaceous, persistent. Peduncles very

short and slender, bearing each a head of 3 to 5, usually 4, flowers, mostly 4-merous. Calyx very short, with broad obtuse lobes. Petals smooth, glabrous. Pod 1 to 2 in. long, scarcely above 1 line broad, curved, acuminate, nearly terete, longitudinally striate, coriaceous, glabrous or sprinkled with a few hairs. Seeds oblong, longitudinal; the last fold of the funicle thickened into a short aril under the seed.—DC. Prod. ii. 449; Bot. Mag. t. 3341; Meissn. in Pl. Preiss. i. 10.

W. Australia. Sandy and rocky places, King George's Sound and adjoining districts, *R. Brown, Menzies, Preiss, n.* 959, and others.

91. **A. horridula,** *Meissn. in Pl. Preiss.* i. 9. A shrub of 2 or 3 ft.; branches virgate, rigid, terete or slightly sulcate, pubescent. Phyllodia numerous, obliquely ovate-lanceolate or lanceolate, tapering into a pungent point, 3 to 4 or rarely 5 lines long, the upper angle near the base slightly prominent, with or without a small gland, rigid, with a nearly central nerve, glabrous or pubescent. Stipules setaceous, rigid. Peduncles shorter than the phyllodia, bearing each a head of 3 to 5, usually 4, flowers, mostly 4-merous. Calyx very short, with broad obtuse lobes. Petals acute, densely villous. Pod pubescent when young, not seen fully formed.

W. Australia, *Drummond, Preiss, n.* 1151; Canning river, *Preiss, n.* 965; Harvey river, *Oldfield.* Although very near *A. hastulata,* this species appears to be distinguishable by its more rigid, more oblique, less acuminate phyllodia, and narrower villous petals. Some specimens of Harvey's from King George's Sound appear, however, to be almost intermediate.

92. **A. divergens,** *Benth. in Hook. Lond. Journ.* i. 331. Glabrous or pubescent; branches angular, divaricate, rather slender. Phyllodia numerous, triangular or 2-lobed, tapering to a pungent point, 2 to 4 lines long, the upper angle or lobe diverging at right angles from the base of the phyllodium and usually bearing a gland, the midrib adjoining the lower straight margin. Stipules small, setaceous, almost spinescent, diverging from the base of the midrib. Pedicels filiform, often rather longer than the phyllodia, bearing each a small globular head of 8 to 12 flowers, mostly 4-merous. Calyx short, open, broadly lobed. Petals smooth, glabrous. Pod not seen.

W. Australia, *Drummond, 2nd Coll. n.* 159; Vasse river, *Mrs. Molloy.* Resembles in foliage some forms of *A. vomeriformis,* but the flowers are quite different.

93. **A. vomeriformis,** *A. Cunn.; Benth. in Hook. Lond. Journ.* i. 332. A diffuse or procumbent rigid shrub; branches terete, pubescent or glabrous. Phyllodia numerous, from obliquely lanceolate to broadly triangular or 2-lobed, rigid and pungent-pointed, 2 to 4 lines long, the upper angle below the middle, either short and rounded or very prominent, rarely bearing a gland, the midrib usually adjoining the lower straight margin. Peduncles slender, about as long as the phyllodia or longer, bearing each a globular head of above 30 flowers, mostly 5-merous. Sepals spathulate, united above the middle or at length free, thick and dark at the top, half as long as the corolla. Petals glabrous, smooth, readily separating. Pod linear, often elongated, flat, coriaceous, glabrous, about 2 lines broad, contracted between the seeds. Seeds nearly orbicular; funicle short, filiform to the end.—F. Muell. Pl. Vict. ii. 7; Dietr. Fl. Univers. N. Ser. t. 82; *A. Gunnii,* Benth. in Hook. Lond. Journ. i. 332; Hook. f. Fl. Tasm. i. 104. t. 18.

N. S. Wales. Blue Mountains and westward towards the Macquarrie, *A. Cunningham, Fraser, Huegel, Miss Atkinson ;* New England, *C. Stuart.*

Victoria. Barren mountains, forests and scrubs,.Forest Creek, Grampians, Mitta-Mitta and Macalister rivers, sources of the Genoa river, etc., ascending to 4000 ft., *F. Mueller.*

Tasmania. Port de l'Espérance, *R. Brown ;* S. Esk river, near Hobarton, near Campbeltown, etc., *J. D. Hooker.*

S. Australia. Lofty Range, *Whittaker, F. Mueller;* Tattiara country, *J. E. Woods.*

94. **A. biflora,** *R. Br. in Ait. Hort. Kew. ed.* 3, v. 463. A diffuse or bushy shrub, branches scarcely angular, pubescent or rarely glabrous. Phyllodia more or less triangular, 3 or 4 lines or rarely ½ in. long, with a small pungent point, the principal nerve near the lower straight margin, the upper margin forming a prominent angle above the middle, usually bearing a gland, and occasionally a secondary nerve tending towards it. Stipules setaceous or spinescent. Peduncles short, bearing each 2 or very rarely more 4-merous flowers, acuminate in the bud. Calyx short, ciliate, with broad lobes. Petals rather rigid but not striate. Pod flat with thickened nerve-like margins, often ½ in. long, 1½ to 2 lines broad, acuminate and narrowed at the base, valves coriaceous. Seeds longitudinal, the last fold of the funicle thickened into a small aril under the seed.—Wendl. Comm. Acac. t. 2 ; DC. Prod. ii. 449 ; Reichb. Ic. et Descr. Pl. t. 12 ; *A. triangularis,* Benth. in Hueg. Enum. 42 ; Meissn. in Pl. Preiss. i. 10.

W. Australia. King George's Sound and eastward to Cape Riche, *R. Brown, Baxter, Drummond, 4th Coll. n. 3, Preiss, n. 963 and 966,* and others; Vasse river, *Mrs. Molloy.* It varies in the buds nearly obtuse or much acuminate ; and very rarely most of the flower-heads have 3, 4 or even 5 flowers.

95. **A. decipiens,** *R. Br. in Ait. Hort. Kew. ed.* 3. v. 463. A bushy shrub, sometimes low and diffuse, but frequently attaining 10 to 12 ft., glabrous or rarely sparingly hirsute. Phyllodia triangular or irregularly trapeziform, rigid, usually 4 to 8 lines long and nearly as broad at the top, the principal nerve near the lower straight margin and ending in a small point, the upper margin forming 1 or rarely 2 very prominent angles, tipped with a gland, and occasionally there are 1 or 2 faint secondary nerves. Stipules here and there spinescent. Peduncles short, bearing each a head of 6 to 10 flowers mostly 4-merous, obtuse in the bud. Calyx short, broadly lobed. Petals free, glabrous and smooth. Pod thick and hard, glabrous, much incurved, 1 to 2 in. long, 1 to 1½ lines broad, acuminate and narrowed at the base, scarcely contracted between the seeds. Seeds oblong, longitudinal ; funicle with the last folds dilated into a thick obliquely turbinate aril.—*Mimosa decipiens,* Kœn. in Sm. Ann. Bot. i. 366. t. 8 ; *A. decipiens,* DC. Prod. ii. 449 ; Bot. Mag. t. 1745, 3244 ; Meissn. in Pl. Preiss. i. 8 ; Reichb. Ic. et Descr. Pl. t. 12, and 88 ; *A. dolabriformis,* Colla, Hort. Ripul. 1, not of Wendl. ; *A. incrassata,* Hook. Ic. Pl. t. 370 ; *A. biflora,* Paxt. Mag. ix. 221, with a fig., not of R. Br.

W. Australia. King George's Sound and adjoining districts, *R. Brown, A. Cunningham, Drummond, n. 296, Preiss, n. 955, 962, 967, 968,* and others; Vasse river, *Mrs. Molloy ;* Gordon river, *Oldfield.* It varies much in the comparative length and breadth of the phyllodia, the broadest part always above the middle, occasionally the upper angle is broadly rounded, projecting much beyond the mucronate angle, approaching the prevailing form in *A. bidentata,* the phyllodium then usually much longer, sometimes nearly 1 in. long.

96. **A. cuneata,** *Benth. in Hueg. Enum.* 42. A tall shrub, glabrous or

with loosely hirsute angular branches. Phyllodia cuneate-oblong, truncate at the end, ½ to 1 in. long, the nerve much more central than in *A. decipiens,* curved and ending in a small point at the lower angle, the upper one usually also acute and sometimes longer, tipped with a gland. Stipules setaceous. Peduncles nearly as long as the phyllodia, bearing each a globular head of 8 to 15 flowers, mostly 4-merous, smaller than in *A. decipiens.* Calyx broadly turbinate, half as long as the corolla, with broad obtuse lobes. Petals smooth. Pod much curved, hirsute, coriaceous, with thickened margins, 2 to 3 in. long, about 2 lines broad.—Meissn. in Pl. Preiss. i. 9.

W. Australia. Swan River and Rottenest island, *Fraser, Huegel, Drummond, n.* 257 (or 297 ?), *Preiss, n.* 954, 956, 957, and others. Nearly allied to *A. decipiens,* but the differences both in phyllodia and flowers appear to be constant.

97. **A. dilatata,** *Benth. in Linnæa,* xxvi. 608. A rigid shrub, softly pubescent. Phyllodia broadly triangular-cuneate, rigidly coriaceous, ½ to ¾ in. long and almost as broad at the top, the principal nerve near the lower straight margin, with usually 1 or 2 other nerves diverging from the base, the upper angle obtuse, usually without any gland. Stipules setaceous or spinescent. Peduncles rarely ½ in. long, bearing each a globular head of about 20 flowers, mostly 5-merous. Calyx half as long as the corolla, with narrow ciliate lobes. Petals strongly striate, hispid with a few rigid hairs. Pod not seen.

W. Australia, *Drummond.* The shape of the phyllodia is not unlike that of some of the thick and broad-leaved forms of *A. decipiens,* but the flowers are very different.

98. **A. bidentata,** *Benth. in Hook. Lond. Journ.* i. 333. A very rigid divaricate shrub; branches terete, pubescent, or rarely nearly glabrous, occasionally spinescent. Phyllodia obovate or cuneate-oblong, 2 to 4 lines or rarely ½ in. long, thick and rigid, the principal nerve near the lower straight margin terminating in a small point, with a secondary nerve usually diverging from the base, the upper margin forming a broad rounded or rarely acute lobe or angle, often longer than the point. Stipules minute. Peduncles often exceeding the phyllodia, bearing each a small globular head of about 8 to 15 flowers, either 4-merous or 5-merous. Calyx very small and thin. Petals smooth. Pod (only seen loose) much curved or twisted, about 1¼ lines broad, with flat thinly coriaceous valves. Seeds longitudinal.—Meissn. in Pl. Preiss. i. 10.

W. Australia, *Drummond, Preiss, n.* 958, 969; Kojonerup and Gardner ranges, *Maxwell.*

99. **A. acanthoclada,** *F. Muell. Fragm.* iii. 127, *and Pl. Vict.* ii. 4. A rigid divaricate spinescent shrub, with the habit, characters, small rigid phyllodia and flowers of *A. bidentata,* and very probably a variety only of that species, differing only in the phyllodia all narrow-cuneate slightly notched at the end, one angle or lobe acute or mucronate, the other obtuse. Neither species (or variety) has been seen in good fruit.

Victoria. Sand ridges of the Murray desert, especially about Kielkoyne, *Howitt's Expedition.*

W. Australia, *Drummond.* Specimens apparently precisely similar to those from the Murray.

D. **BREVIFOLIÆ.**　Shrubs, never spinescent.　Phyllodia either broad, ovate or falcate, or narrow-oblong or linear, short, mostly under 1 in. long, obtuse or with a small recurved innocuous point, or sometimes undulate and more pointed.　Stipules minute or none.　Peduncles 1-headed.

These species pass into the *Triangulares* through *A. obliqua*, into the short-leaved *Calamiformes* through *A. lineata*, and some forms of *A. dura ;* and the shorter-leaved forms of *A. montana* and some others of the *Angustifoliæ*, might almost be included in the present subseries.　*A. anceps* is very variable in the form of its phyllodia, and they are usually larger than in other *Brevifoliæ*, but on the whole, it appears to be better placed here than in any other subseries of *Uninerves.*

100.　**A. obliqua,** *A. Cunn. ; Benth. in Hook. Lond. Journ.* i. 334.　A much-branched shrub, attaining sometimes several ft., glabrous or slightly pubescent.　Phyllodia obliquely obovate or orbicular, from ¼ to nearly ½ in. long, the principal nerve scarcely prominent, near the lower margin, and terminating in a minute recurved point, with often 1 or 2 fainter nerves diverging from the base.　Stipules minute.　Peduncles slender, solitary or in pairs, usually exceeding the phyllodia, bearing each a globular head of about 8 to 15 flowers, mostly 5-merous.　Sepals distinct, linear-spathulate, fully half as long as the corolla.　Petals smooth.　Pod linear, twisted, 1½ to 2 lines broad, with coriaceous valves.　Seeds ovate, longitudinal; funicle not folded, scarcely so long as the seed, gradually thickened almost from the base into a clavate fleshy aril.—*A. cyclophylla*, Schlecht. in Linnæa, xx. 663 ; *A. rotundifolia*, Hook. Bot. Mag. t. 4041 ; Paxt. Mag. xv. 123, with a fig.

N. S. Wales.　Near Bathurst, *Fraser ;* between the Lachlan and Macquarrie rivers, *A. Cunningham ;* Nangas, *M'Arthur.*
Victoria.　At and near the Ovens ranges, *F. Mueller.*
S. Australia.　Scrub of the Murray, *Behr ;* Encounter Bay, *Whittaker ;* Lofty, Barossa, and Flinders ranges, *F. Mueller.*
Some specimens resemble some of the broader-leaved forms of *A. bidentata*, but are less rigid, never spinescent, and the calyx is different.　The flowers and fruit are those of *A. acinacea*, with which F. Mueller proposes to unite it as a broad-leaved variety.

101.　**A. acinacea,** *Lindl. in Mitch. Three Exped.* ii. 267.　A much-branched shrub of several feet, usually glabrous, with the habit and essential characters of *A. obliqua* and *A. lineata*, and scarcely differing from either except in the shape of the phyllodia, which are obliquely oblong or somewhat falcate, rarely above ½ in. long, obtuse at the end, with a small recurved point usually under a small gland, the nerve somewhat excentrical, the marginal gland below the middle or none.　Peduncles slender, often exceeding the phyllodia, bearing each a globular head of about 10 to 20 flowers, mostly 5-merous.　Sepals distinct, linear-spathulate.　Petals smooth.　Pod linear, curved or twisted, coriaceous, 1½ to 2 lines broad.　Seeds longitudinal, the funicle thickened into a clavate fleshy aril, scarcely so long as the seed.—F. Muell Pl. Vict. ii. 5 ; *A. Latrobei*, Meissn. in Pl. Preiss. i. 10.

Victoria.　Near Mount William, *Mitchell ;* grassy and somewhat scrubby ridges throughout the greater part of the colony, except the eastern districts, *F. Mueller.*
S. Australia.　Memory Cove, *R. Brown ;* near Spencer's Gulf and Flinders Range, *F. Mueller.*

102.　**A. lineata,** *A. Cunn. in G. Don, Gen. Syst.* ii. 403.　A bushy shrub of several feet ; branches nearly terete, usually pubescent or villous and sometimes slightly resinous.　Phyllodia linear, with a small hooked point, about ½ in. long or rarely ¾ in. or rather more, the nerve very near the lower margin and usually without any gland. Stipules minute. Peduncles slender, rarely exceeding the phyllodia, bearing each a small globular head of 10 to 15 or rarely more flowers, mostly 5-merous. Sepals distinct, linear-spathulate. Petals smooth. Pod linear, curved or twisted, coriaceous, 1½ to 2 lines broad. Seeds longitudinal, the funicle thickened into a clavate fleshy aril scarcely so long as the seed.—Bot. Mag. t. 3346 ; *A. runciformis,* A. Cunn. in G. Don, Gen. Syst. ii. 404; F. Muell. Pl. Vict. ii. 21; *A. dasyphylla,* A. Cunn.; Benth. in Hook. Lond. Journ. i. 359 (a more pubescent form) ; *A. imbricata,* F. Muell. Fragm. i. 5, ii. 177.

N. S. Wales. Liverpool Plains, Wellington Valley, etc., *A. Cunningham.*
Victoria. N.W. desert, *Lockhart Morton.*
S. Australia. Port Lincoln and Dombey Bay, *F. Mueller.*
The habit, inflorescence, flowers and fruit are those of *A. obliqua* and *A. acinacea,* from which this species only differs in the narrow phyllodia. I have not followed F. Mueller in taking up the name of *runciformis,* because that of *lineata,* of the same date, is universally adopted by gardeners as well as botanists, and does not appear to me to be in itself objectionable.

103. **A. triquetra,** *Benth. in Hook. Lond. Journ.* i. 358. A glabrous erect shrub ; branches acutely angled. Phyllodia linear, obtuse or with a minute recurved point, mostly under 1 in. long, narrowed at the base, the nerve nearly central, the marginal gland minute or none. Stipules minute. Peduncles rarely exceeding the phyllodia, bearing each a small globular head of numerous very small flowers, mostly 5-merous. Sepals narrow-linear, spathulate, ciliate. Petals smooth, narrow, distinct, almost clawed. Pod much curved, flat with nerve-like margins, 2 to 3 in. long, 1¼ lines broad. Seeds oblong, longitudinal, the funicle not folded, thickened into a short club-shaped fleshy aril.—*A. Meissneri,* var. *angustifolia,* Meissn. in Pl. Preiss. i. 13.

W. Australia, *Baxter, Drummond, n.* 109 *and* 292. Differs from *A. lineata* in its glabrous very angular branches, from *A. Meissneri* in its narrow phyllodia and in the pod.

104. **A. ligustrina,** *Meissn. in Pl. Preiss.* ii. 203. A glabrous shrub with acutely angled branches, very closely allied to *A. triquetra,* but apparently differing in the more coriaceous obliquely lanceolate phyllodia, often above 1 in. long, narrowed at both ends, the lateral veins quite inconspicuous, and usually with 2 or 3 small distant glands on the upper margin. Flowers numerous, mostly 5-merous, in small globular heads ; sepals linear-spathulate and petals smooth as in *A. triquetra.* Pod unknown.

W. Australia, *Drummond,* 2nd *Coll. n.* 150.

105. **A. Meissneri,** *Lehm. Del. Sem. Hort. Hamb.* 1842. A tall shrub, quite glabrous and often glaucous, the young shoots very angular. Phyllodia obliquely obovate-oblong or almost cuneate, obtuse or with a small hooked point, ½ to 1 in. long, 2 to 4 lines broad, much narrowed at the base, coriaceous but thinner than in *A. anceps,* without nerve-like margins, 1-nerved and slightly penniveined. Peduncles shorter than the phyllodia, bearing each a small globular head of numerous (about 30) small flowers, mostly 5-merous. Sepals free, narrow, linear-spathulate, slightly ciliate at the top. Petals very narrow, smooth, quite distinct. Pod almost sessile, elongated, flat, 3 to 4 lines broad, the margins not thickened. Seeds oblong, longitudinal; funicle

very long and much folded, the last fold almost encircling the seed and returning, but thickened only at the end.—Meissn. in Pl. Preiss. i. 13.

W. Australia. Muddy sandy places near York, *Preiss, n.* 930 ; Swan River, *Mylne.*

106. **A. anceps,** *DC. Mem. Leg.* 446, *and 'Prod.* ii. 451. An erect shrub of several feet, quite glabrous, and often of a glaucous or purplish hue, the branches rigid, very acutely angled when young. Phyllodia from broadly ovate to oblong, attached by a broad base and only partially articulate, the lower edge continuous and shortly decurrent, 1½ to 2 in. long and ½ to 1½ in. broad, or even larger on barren branches, very rigid, often undulate, 1-nerved, penniveined, with thickened margins. Peduncles from under ½ to nearly 1 in. long, thick, bearing each a rather large globular head of numerous flowers, mostly 5-merous. Calyx more than half as long as the corolla, turbinate, broadly and obtusely toothed. Petals smooth, readily separating. Pod stipitate, straight, flat, rigidly coriaceous, very obtuse, about 1½ in. long and ½ in. broad. Seeds longitudinal ; funicle long, filiform, much folded, shortly thickened at the end, but not seen perfect.—*A. Muelleri,* Benth. in Linnæa, xxvi. 603.

S. Australia. St. Peter's Island, Nuyts' Archipelago, *Baudin's Expedition;* Port Lincoln, *R. Brown, F. Mueller;* towards Spencer's Gulf, *Warburton.*

Var. (?) *angustifolia.* Branches rather less angular. Phyllodia from narrow-obovate to linear-oblong, 1 to 2 in. long, occasionally with a prominent gland above the middle. Peduncles under ½ in. long.—S. coast, . *Brown ;* towards Spencer's Gulf, *Warburton.* This variety has sometimes almost the phyllodia of *A. notabilis,* but the peduncles are always simple.

107. **A. hispidula,** *Willd. Spec. Pl.* iv. 1054. A rigid spreading shrub, scabrous all over with very short stiff hairs or tubercles. Phyllodia numerous, broadly falcate, with a minute point, cuneate at the base, mostly ¼ to ¾ in. long, 2 to 3 lines broad, with a central nerve and thickened nerve-like margins more or less tuberculate or almost denticulate. Peduncles short, bearing each a small globular head of 12 to 20 flowers, mostly 5-merous. Calyx lobed, about half as long as the corolla. Petals smooth, connate to the middle. Pod ovate and 1-seeded or oblong and 2-seeded, very obtuse, about 4 lines broad, flat but thickly coriaceous, without prominent margins. Seeds oval-oblong, longitudinal ; funicle with the last fold much thickened and nearly as long as the seed, and shortly folded below it.—*Mimosa hispidula,* Sm. Bot. Nov. Holl. 59, t. 16 ; *A. hispidula,* DC. Prod. ii. 450 ; Lodd. Bot. Cab. t. 823 ; Hook. Ic. Pl. t. 161.

N. S. Wales. Port Jackson, *R. Brown* and others. This species has been confounded with *A. aspera,* which has scabrous-pubescent but not tuberculate phyllodia, denser flower-heads and a very different pod, besides the stipules usually persistent.

108. **A. undulifolia,** *A. Cunn. in G. Don, Gen. Syst.* ii. 404. A shrub sometimes low and bushy, but often attaining a great size, and very handsome from its long pendulous garland-like flowering branches ; branchlets slightly angular but soon terete, pubescent, hirsute or rarely glabrous. Phyllodia numerous, ovate or almost orbicular, very obliquely truncate or narrowed at the base and often petiolate, usually about ½ in. but varying from ¼ to nearly 1 in. long, coriaceous, undulate, 1-nerved and penniveined, the margins thickened, terminating in a short or fine point. Peduncles slender, often

2 A 2

exceeding the phyllodia, bearing each a globular head of 20 to 30 or more flowers, mostly 5-merous. Calyx very short, toothed. Petals smooth, united above the middle. Pod shortly stipitate, 7 to 9 lines broad, very flat, with nerve-like margins. Seeds flat, ovate, oblique; funicle with the last fold thickened and not half so long as the seed, and short folds below it.—Bot. Mag. t. 3394; Lodd. Bot. Cab. t. 1544; Lemaire, Jard. Fleur. t. 282; *A. uncinata*, Lodd.; Lindl. Bot. Reg. t. 1332; *A. piligera*, A. Cunn. in Bot. Mag. under n. 3394; Hook. Ic. Pl. t. 166 (*A. setigera*, A. Cunn.).

Queensland. Near Brisbane, *F. Mueller, Leichhardt.*

N. S. Wales. Blue Mountains, *R. Brown;* abundant in thickets, barren woods, etc., in the N.W. interior, beyond Bathurst, Cox's River, Liverpool Plains, etc., *A. Cunningham, Fraser, Mitchell,* and others.

Var. *sertiformis.* More glabrous, with larger phyllodia, not contracted at the base.—*A. sertiformis*, A. Cunn. in Bot. Mag. under n. 3394; Hook. Ic. Pl. t. 159.—Liverpool Plains.

Var. *dysophylla.* Softly villous. Phyllodia large.—*A. dysophylla*, Benth. in Hook. Lond. Journ. i. 346.—Pine ridge near Croker's Range, *A. Cunningham.*

Var. *humilis.* Diffuse and low, glabrous. Phyllodia not above 3 or 4 lines long, very oblique and often recurved, nearly as broad as long.—N.W. interior of N. S. Wales, *Fraser;* also the Brisbane specimens from *F. Mueller and Leichhardt.*

A. plagiophylla, F. Muell. in Journ. Linn. Soc. iii. 131, not of Sieber, belongs probably to one of the varieties of *A. undulifolia.*

109. **A. flexifolia,** *A. Cunn.; Benth. in Hook. Lond. Journ.* i. 359. Shrubby, the branches terete or nearly so, hoary with a minute tomentum. Phyllodia numerous, linear, obtuse, scarcely mucronate, $\frac{1}{2}$ to $\frac{3}{4}$ in. long, rigidly coriaceous and glabrous, narrowed at the base, with an impressed gland and often bent in at a little distance from the base, with a prominent nerve very near the lower margin, the upper margin thickened and nerve-like. Stipules minute. Peduncles very short, solitary or in pairs, tomentose, bearing each a small globular head of 6 to 10 very small flowers, mostly 5-merous. Calyx thin, shortly lobed, not half as long as the corolla. Petals smooth, united to the middle.

N. S. Wales. Cugeegong river and between Lachlan and Macquarrie rivers, *A. Cunningham.*

110. **A. dura,** *Benth. in Linnæa*, xxvi. 622. A low rigid glabrous shrub; branches slightly angular when young. Phyllodia linear or linear-cuneate, very obtuse, $\frac{3}{4}$ to rather more than 1 in. long, thick and very rigid, narrowed at the base and often bent in with an impressed gland below the middle, with a very prominent nerve, and the upper margin thick and nerve-like or forming a slightly intramarginal nerve, the smaller phyllodia straight, without any gland. Peduncles not 2 lines long, mostly solitary, bearing each a small globular head of 6 to 10 flowers, mostly 5-merous. Sepals very thin, spathulate, distinct. Petals narrow, smooth, distinct.

W. Australia, *Drummond.* Allied on the one hand to *A. flexifolia,* on the other to *A. nitidula,* differing from the former in the more rigid phyllodia and in the calyx, from the latter in the marginal nerve, small flower-heads, etc.

111. **A. spathulata,** *F. Muell. Herb.* A diffuse or bushy shrub of 3 or 4 ft.; branches usually crowded, nearly terete, glabrous or slightly pubescent. Phyllodia linear-cuneate or oblong-spathulate, very obtuse, 4 to 8 lines

long, thick and almost fleshy, faintly 1-nerved. Peduncles solitary or in pairs, bearing each a globular head of numerous flowers, only seen in young bud. Pod oblong-linear, straight, flat but thick and hard, very obtuse, with broad margins, 1 to 1½ in. long, 2 to 3 lines broad. Seeds oval-oblong, longitudinal; funicle short, thickened into a club-shaped oblique aril, with the edges more or less dilated over the seed.

N. Australia. Bay of Rest, N.W. coast, *A. Cunningham.*
W. Australia. Dirk Hartog's Island and Sharks Bay, *Milne*; Murchison river, *Oldfield* (phyllodia small, narrow; pod narrow) ; Murray river, *Oldfield* (phyllodia larger ; pod broader). Some of the smaller specimens have some resemblance to *A. leptospermoides*, but the venation (often only visible under a lens at the base of the phyllodium) is different.

E. ANGUSTIFOLIÆ.—Shrubs or trees, not spinescent. Phyllodia oblong-lanceolate or linear, occasionally falcate, with 1 central nerve or rarely 2-nerved, mostly above 1 in. long. Stipules minute or none, or spinescent in *A. sentis* and *A. dentifera.* Peduncles 1-headed. Some species very resinous.

A few of this section have often, and *A. verniciflua* almost constantly 2 nerves to the phyllodia, thus connecting the *Uninerves* with the *Plurinerves*, to which however I have transferred most of the 2-nerved species I had formerly included in the *Uninerves.*

112. **A. microcarpa,** *F. Muell. Fragm.* i. 6, *and Pl. Vict.* ii. 17. A shrub of 2 to 5 ft., glabrous or the young shoots minutely pubescent with silky or golden hairs. Phyllodia oblong-linear or linear-lanceolate, oblique or falcate, obtuse or with a small recurved point, mostly 1 to 1½ rarely 2 in. long, coriaceous, rather thick, quite smooth, besides a scarcely prominent central nerve. Peduncles mostly in pairs, rarely above ¼ in. long and often much shorter, bearing each a globular head of rather numerous flowers, mostly 5-merous. Sepals free, narrow-linear, spathulate, ciliate. Petals smooth, distinct. Pod linear, curved or twisted, acuminate, often 2 or 3 in. long, but not above 1½ lines broad at the seeds and much contracted between them ; valves rather coriaceous. Seeds longitudinal; funicle thickened into a small turbinate oblique aril.

N. S. Wales. Plains of the Darling, *Victorian Expedition.*
Victoria. Salt-bush country on the Murray, Avoca, and Wimmera rivers, *F. Mueller.*
S. Australia. Near Port Lincoln, *Wilhelmi.*
Some specimens from the Melbourne Botanic Garden have much smaller phyllodia, under 1 in. long, small flower-heads, and the pods scarcely above 1 line broad, thus approaching *A. acinacea* in aspect.

113. **A. montana,** *Benth. in Hook. Lond. Journ.* i. 360. A resinous viscid shrub of 4 to 6 ft.; branches slightly angular and pubescent. Phyllodia oblong, oblong-lanceolate or broadly linear, very obtuse, narrowed at the base, coriaceous, more or less distinctly 2-nerved, in the N. S. Wales specimens 1 to 1½ in. long, 2 to 3 lines broad, in the Victorian ones often much smaller, narrower, approaching towards those of *A. lineata.* Peduncles usually 2 to 3 lines long, solitary or in pairs, bearing each a small globular head of numerous flowers, mostly 5-merous. Calyx thin, lobed and ciliate, sometimes separating into distinct sepals. Petals smooth, distinct. Pod flat, densely tomentose, 1 to 2 in. long and about 2 lines broad, not contracted between the seeds. Seeds obovate-oblong, longitudinal; funicle with the 2 or 3 last folds thickened into an oblique almost hood-shaped aril, at least half

as long as the seed.—F. Muell. Pl. Vict. ii. 22; *A. clavata,* Schlecht. Linnæa, xx. 662.

N. S. Wales. High lands near Liverpool Plains, *Fraser.*
Victoria. Low stony and scrubby ridges and barren plains, Mount Korong, Avoca, Murray, and Wimmera rivers, *F. Mueller.*
S. Australia. East declivity of the scrub on the Murray, *Behr;* Lake Alexandrina and St. Vincent's Gulf, *F. Mueller.*

114. **A. verniciflua,** *A. Cunn. in Field, N. S. Wales,* 344. A tall resinous viscid shrub, branches slightly angular, glabrous. Phyllodia from oblong to linear-lanceolate, narrowed at each end and mostly acute, usually falcate, 2 to 4 in. long, with 2 or very rarely only 1 prominent nerve. Peduncles short, in pairs or clusters, bearing each a globular head of numerous mostly 5-merous flowers. Calyx turbinate, shortly toothed or lobed, about half as long as the corolla. Petals smooth, united. Pod linear, usually straight, flat, 2 to 3 lines broad, glabrous or viscid, pubescent. Seeds rather small, longitudinal; funicle short, the last folds forming a very small aril under the seed.—Bot. Mag. t. 3266; Hook. f. Fl. Tasm. i. 108; F. Muell. Pl. Vict. ii. 22; *A. graveolens,* A. Cunn. in G. Don, Gen. Syst. ii. 404; Bot. Mag. t. 3279; Lodd. Bot. Cab. t. 1460; *A. virgata,* Lodd. Bot. Cab. t. 1246.

N. S. Wales. Rocky hills of the interior, *A. Cunningham, Fraser, Huegel,* and others; head of the Gwydir river, *Leichhardt.*
Victoria. Widely distributed over the colony in mountain and forest regions, rocky hills, etc., *F. Mueller.*
Tasmania. Derwent river, *R. Brown;* common in many parts of the island, especially about Hobarton, also St. Patrick's River, Launceston, and near Yorktown, *J. D. Hooker.*
S. Australia. Upper valleys of the Torrens and Onkaparinga rivers, *F. Mueller.*
The species is nearly allied to *A. leprosa,* differing chiefly in the broader phyllodia almost always 2-nerved.
Var. *latifolia.* Phyllodia shorter and broader.—*A. exudans,* Lindl. in Mitch. Three Exped. ii. 214; Dietr. Fl. Univers. N. Ser. t. 83.—Plains of the Glenelg, *Mitchell.*

115. **A. leprosa,** *Sieb. in DC. Prod.* ii. 450. A tall shrub or small tree, with pendulous branchlets, more or less glutinous, otherwise glabrous or the young shoots minutely pubescent. Phyllodia narrow, linear-lanceolate, acute or obtuse with a small callous point, narrowed at the base, 1½ to 3 in. long, 1-nerved, with anastomosing veins, those of the barren shoots broader, often ½ in. long in the middle, and thinner with fine veins oblique on the midrib and connected in an intramarginal almost continuous vein. Peduncles mostly in pairs or clusters, hoary-pubescent, rarely above ¼ in. long, bearing each a globular head of numerous flowers, mostly 5-merous. Calyx half as long as the corolla, with short obtuse ciliate lobes. Petals united to the middle, with rather thick smooth tips. Pod falcate or rarely straight, flat, 2 to 2½ lines broad. Seeds oval-oblong, longitudinal; funicle with the last fold thickened into an irregularly turbinate or cup-shaped aril under the seed.—Bot. Reg. t. 1441 (rather doubtful); F. Muell. Pl. Vict. ii. 23; *A. reclinata,* F. Muell. First Gen. Rep. 12.

N. S. Wales. Port Jackson, *Sieber, n.* 455, *M'Arthur.*
Victoria. Dandenong Ranges, *F. Mueller.*
Var. *tenuifolia.* Branches erect. Phyllodia very narrow, linear-falcate, with recurved points.—Between the Goulburn and Broken rivers, Victoria, *F. Mueller.*

116. **A. stricta,** *Willd. Spec. Pl.* iv. 1052. A shrub of 2 or 3 ft. with

creeping roots in Tasmania, attaining 5 or 6 ft. further north, glabrous and often slightly viscous; branchlets erect, angular. Phyllodia linear, obtuse, rarely with a short oblique point, 2 to 4 or even 5 in. long, from 2 to 4 or 5 lines broad, coriaceous, 1-nerved, with numerous fine anastomosing pinnate veins. Peduncles in pairs or clusters, rarely 2 lines long, bearing each a globular head of about 20 to 30 small flowers, mostly 5-merous. Calyx turbinate, half as long as the corolla, shortly and obtusely lobed. Petals smooth, united to the middle. Pod elongated, flat, obtuse or acuminate, 1½ to 2½ lines broad, not contracted between the seeds. Seeds oval-oblong, longitudinal; funicle with the last fold thickened into an irregularly turbinate or cup-shaped aril under the seed.—DC. Prod. ii. 450; Reichb. Ic. et Descr. Pl. t. 90 (the venation not represented); Lodd. Bot. Cab. t. 99 (a doubtful narrow-leaved form); *Mimosa stricta,* Andr. Bot. Rep. t. 53; Bot. Mag. t. 1121; *Acacia emarginata,* Wendl. Comm. Acac. 27; DC. Prod. ii. 450.

N. S. Wales. Port Jackson to the Blue Mountains, *R. Brown, Sieber, n.* 456, *Fl. Mixt. n.* 594, and others; Newcastle, *Leichhardt;* New England, *C. Stuart;* and southward to Gabo Island, *Maplestone.*

Victoria. Widely distributed over the southern and eastern parts of the colony, from the sandy or rocky coasts to stony mountains, wet valleys, or heath or forest ground, *F. Mueller.*

Tasmania. Common throughout the island in dry soil, *J. D. Hooker.*

117. **A. dodonæifolia,** *Willd. Enum. Suppl.* 68. A tall shrub, glabrous but very resinous, the branches soon terete. Phyllodia oblong-linear or lanceolate, obtuse or with a small recurved point, mostly 2 to 4 in. long and 2 to 4 lines broad, much narrowed towards the base, 1-nerved, with the lateral anastomosing veins much fewer and more prominent than in *A. stricta,* 1 or 2 glands on the upper margin sometimes very prominent but often wanting. Peduncles solitary or in pairs, often above ½ in. long, bearing a globular head of numerous flowers, usually 5-merous. Calyx more than half as long as the corolla, with short thick lobes sometimes separating into distinct sepals. Petals smooth, rather thickened at the tips, usually united to the middle. Pod elongated, nearly flat, straight or falcate, obtuse, when perfect about 2½ lines broad. Seeds oblong, longitudinal, the last 2 or 3 folds of the funicle much dilated into an irregularly cup-shaped aril.—*Mimosa dodonæifolia,* Pers. Syn. Pl. ii. 261; *A. viscosa,* Schrad. in Wendl. Diss. Acac. 30, t. 7; *A. dodonæifolia,* DC. Prod. ii. 450; Reichb. Ic. et Descr. Pl. t. 91; Colla, Hort. Ripul. t. 27; *A. visciflua,* F. Muell. Pl. Vict. ii. 24.

S. Australia. Kangaroo Island, *Baudin's Expedition;* Memory Cove, *R. Brown;* Port Lincoln, *Wilhelmi.* The species has long been cultivated on the Continent, originally raised from seeds collected in Baudin's Expedition, but *A. stricta* now frequently represents it in botanic gardens.

118. **A. Gnidium,** *Benth.* A small tree, glabrous with the young shoots glutinous; branches erect, virgate, soon becoming terete. Phyllodia numerous, erect, narrow-linear, obtuse with a small callous hooked point, 1 to 2 in. long and not above 1 line broad, rigid with a scarcely prominent midrib, the lateral veins obscure, anastomosing, and sometimes almost parallel. Peduncles slender, nearly ½ in. long, bearing each a globular head of 15 to 20 flowers, mostly 5-merous, but often also 4-merous. Calyx shortly and broadly lobed, half as ong as the corolla. Petals smooth. Pod unknown

360 XL. LEGUMINOSÆ. [*Acacia*.

Queensland. Under sandstone hills near Mount Pluto, *Mitchell.* On a hasty survey, I had formerly put this aside as a variety of *A. viscidula*, from which however on examination I find it to differ essentially in the venation of the phyllodia as well as in the flowers. The nearest affinity appears to be with *A. dodonæifolia*, from which our specimens chiefly differ in the very narrow phyllodia.

119. **A. ramosissima,** *Benth. in Hook. Lond. Journ.* i. 356 (*partly*). Apparently shrubby, with slender, pubescent, slightly angular branches. Phyllodia numerous, narrow-linear, obtuse with a minute hooked point, 1 to 1½ in. long, about 1 line broad, narrowed at the base, rather thick, obscurely 1-nerved. Stipules minute but often persistent. Flowers not seen. Fruiting peduncles about ½ in. long, with the scars of a globular head. Pod linear, straight, flat but thickly coriaceous, scarcely contracted but transversely depressed between the seeds, without thickened margins, 1 to 1½ in. long, 2 lines broad. Seeds nearly orbicular ; funicle thickened from near the base into a scarcely fleshy linear aril, and scarcely folded below it.

W. Australia, *Drummond,* 3rd (*or* 4th *?) Coll. n.* 79. Under the name of *A. ramosissima,* I had confounded several species, closely resembling each other in foliage, but which prove to have different flowers and fruits. The one for which I now retain the name differs from all the thick narrow 1-nerved western ones in the pod and in the pubescent branchlets, but our specimens are very indifferent.

120. **A. sentis,** *F. Muell. in Journ. Linn. Soc.* iii. 128, *and Pl. Vict.* ii. 18. A divaricately-branched rigid shrub or small tree, branchlets nearly terete, glabrous or pubescent when young. Phyllodia lanceolate-oblong or linear, mostly oblique falcate or curved, 1-nerved and more or less penniveined, in some specimens ¾ in. long and 2 or 3 lines broad, in others more than 2 in. long and about 1 line broad, usually glabrous, the marginal gland near the base or none. Stipules either subulate-spinescent or very small or none. Peduncles rather slender, solitary or in pairs, axillary or by the abortion of the phyllodia in terminal racemes, bearing each a small globular head of 20 to 30 flowers, mostly 5-merous. Sepals linear-spathulate, free. Petals smooth. Pod thin, flat, ½ to ¾ in. broad. Seeds broadly ovate, longitudinal, along the centre of the pod ; funicle transverse, gradually thickened from the base upwards, straight or shortly folded under the seed.—*A. Victoriæ,* Benth. in Mitch. Trop. Austr. 333.

N. Australia. Victoria river and Plains of Promise, Gulf of Carpentaria, *F. Mueller.*
Queensland. Bargoo river, *Mitchell.*
N. S. Wales. From the Darling river to the Barrier range, *Victorian and other Expeditions.*
Victoria. Low sandhills and arid salt-bush plains towards the junction of the Murray and Darling, *F. Mueller.*
S. Australia. Base of Flinders range, towards Spencer's Gulf and in the interior, *F. Mueller.*

121. **A. dentifera,** *Benth. in Maund, Botanist,* iv. *t.* 179. A tall glabrous shrub, branchlets striate or slightly angular. Phyllodia narrow-linear, acute or obtuse with a small recurved point, 3 to 8 in. long, and 1 to 2 lines broad, with a prominent midrib and obscurely veined. Stipules small and tooth-like or none. Peduncles slender, mostly above ½ in. long, solitary or in pairs, the upper ones often forming a raceme by the abortion of the phyllodia, each bearing a globular or ovoid head of densely packed flowers, mostly 5-merous. Sepals very narrow, free. Petals smooth, united above the

middle. Pod straight or curved, flat, 2 to 3 in. long, about 3 lines broad, not contracted between the seeds. Seeds oblong, longitudinal; funicle with the last 2 or 3 folds dilated into an orange-coloured aril under the seed.—Meissn. in Pl. Preiss. 1. 17.

W. Australia. Swan River to King George's Sound, *Drummond, 1st Coll., 2nd Coll. n.* 298, *5th Coll. n.* 7; Canning river, Darling range and Murray river, *Preiss, n.* 932, 944, 946; Tweed river, *Oldfield;* Stirling ranges, *Maxwell.*

A. longifolia, Paxt. Mag. xii. 269, with a fig. (*A. dentifera,* Bot. Mag. t. 4032) is a garden variety, with more falcate phyllodia, and the flowering branches when young forming long leafless racemes, often exceeding the leaves and producing numerous flower-heads from the base.

122. **A. fasciculifera,** *F. Muell. Herb.* A tree, glabrous in all its parts; branchlets slightly angular. Phyllodia lanceolate-falcate, acuminate, with a callous point, narrowed at the base, mostly 4 to 6 in. long, coriaceous, with a prominent midrib and nerve-like margins, the pinnate veins scarcely prominent. Peduncles filiform, ½ to 1 in. long or even more, clustered in the axils, bearing each a globular head of 20 to 30 or more flowers, mostly 5-merous. Sepals narrow, linear-spathulate, free. Petals smooth, soon separating. Pod not seen.

Queensland. Rockhampton and Moreton Bay, *Dallachy.* The phyllodia are nearly those of *A. falcata,* but rather more coriaceous, the inflorescence is very different, showing an affinity to *A. harpophylla* and *A. complanata.*

F. RACEMOSÆ.—Shrubs or trees, not spinescent. Phyllodia not pungent except in a very few broad-leaved species, with 1 central nerve or very rarely a second shorter or fainter one, the veinlets when visible diverging from the midrib or reticulate. Flower-heads globular, all or nearly all in axillary racemes, very rarely a few solitary in the lower axils of a young branch. Flowers usually small and 5-merous in all except *A. myrtifolia.* Petals not striate.

This subseries is distinguished from the other *Uninerves* by the inflorescence. The phyllodia are, in the species commencing the subseries, long and falcate as in the falcate *Juliflorae,* and in the larger *Plurinerves,* in the latter species of the subseries they generally correspond in shape to those of the *Brevifoliæ.*

123. **A. falcata,** *Willd. Spec. Pl.* iv. 1053. A tall shrub or tree, glabrous, with angular branchlets. Phyllodia lanceolate-falcate, acuminate, much narrowed towards the base, 3 to 6 in. long or even more, 1-nerved, obliquely penniveined, the margins slightly thickened without any gland or with an obscure one at the base. Racemes much shorter than the phyllodia, usually with 10 to 20 small globular heads of about 20 small flowers, mostly 5-merous; peduncles short and slender. Sepals free, narrow-spathulate, ciliate. Petals smooth, soon separating. Pod flat, with slightly thickened margins, 2 to 3 in. long; about 3 lines broad. Seeds ovate, longitudinal, close to the margin; funicle slightly dilated and coloured from the base, very long, extending round the seed and bent back on the same side, encircling it in a double fold and thickened at the end into a short fleshy aril.—*Mimosa obliqua,* Wendl. Bot. Beob. 57; *A. falcata,* DC. Prod. ii. 451; Wendl. Comm. Acac. 20. t. 14; Lodd. Bot. Cab. t. 1115; *A. plagiophylla,* Spreng. Syst. iii. 135.

Queensland. Brisbane river, Moreton Bay, *A. Cunningham, Leichhardt,* and others.

N. S. Wales. Port Jackson to the Blue Mountains, *R. Brown, Sieber, n.* 450, and others.

124. **A. macradenia,** *Benth. in Mitch. Trop. Austr.* 360. A shrub of 10 to 12 ft., glabrous, with angular branchlets. Phyllodia lanceolate-falcate, rather acute, much narrowed towards the base, coriaceous, 1-nerved, with thickened margins and often a gland at the base, like those of *A. falcata,* but usually longer, attaining from 6 in. to 1 ft., and the fine veins more numerous, prominent, and transverse. Racemes short with several small globular heads of flowers mostly 5-merous. Calyx turbinate, more than half as long as the corolla, shortly and broadly toothed, ciliate. Petals smooth, pubescent. Pod long, flat, 5 to 6 lines broad. Seeds orbicular; funicle slightly thickened from the base, but not enlarged under the seed, not folded, about half as long as the seed.

Queensland. Beds of rivers near Mount Pluto, *Mitchell* (in flower); Rockhampton, *Thozet* (in fruit).

125. **A. penninervis,** *Sieb. in DC. Prod.* ii. 452. A tree attaining sometimes 40 ft. but usually smaller, glabrous in all its parts in the common variety, with angular branchlets. Phyllodia from oblong to lanceolate-falcate, more or less acuminate, usually 3 to 4 in. long, but sometimes twice that length, much narrowed towards the base, 1-nerved and more or less prominently and finely penniveined, the margins usually nerve-like, and often but not always a short secondary nerve terminating in a marginal gland much below the middle. Racemes rather short but loose, with several small globular heads of about 20 flowers, mostly 5-merous. Calyx truncate or shortly toothed, not half so long as the corolla. Petals smooth. Pod flat, straight or curved, with slightly thickened margins, often 4 or 5 in. long, nearly ½ in. broad. Seeds ovate, longitudinal; funicle long, dilated and coloured nearly from the base, extending round the seed and bent back on the same side, so as to encircle it in a double fold.—Bot. Mag. t. 2754; F. Muell. Pl. Vict. ii. 14; *A. impressa,* Lindl. Bot. Reg. t. 1115; Lodd. Bot. Cab. t. 1319.

Queensland. Brisbane river, Moreton Bay, *Fraser, F. Mueller;* Sandstone ridges near Mount Pluto, *Mitchell.*

N. S. Wales. Blue Mountains, *Sieber, n.* 458, and others; and inland to the Macquarrie, *A. Cunningham, Fraser;* northward to Hastings river, *Beckler;* and southward to Twofold Bay, *F. Mueller.*

Victoria. Granitic ranges and mountains on the Broken, Ovens, and Snowy rivers, *F. Mueller.*

Tasmania. Brown's Road, Mount Wellington, *Oldfield.*

Var. *falciformis.* Phyllodia usually longer and more falcate, young shoots and inflorescence minutely hoary or golden-pubescent. Pod nearly ¾ in. broad.—*A. falciformis,* DC. Prod. ii. 452; *A. astringens,* A. Cunn. in G. Don, Gen. Syst. ii. 405.—From Twofold Bay, *F. Mueller,* to Moreton Bay, *Leichhardt,* and New England, *C. Stuart;* and from the Blue Mountains, *Sieber, n.* 616, and others, to the Lachlan river, *A. Cunningham, Fraser.* Called *Blackwood* by the western colonists, and the bark used for tanning.

126. **A. retinodes,** *Schlecht. in Linnæa,* xx. 664. A moderate-sized tree usually glabrous, the branchlets at first very angular. Phyllodia linear-lanceolate, more or less falcate, with a small recurved point or obtuse, much narrowed towards the base, mostly 3 to 5 in. long, 1-nerved, finely penniveined, the marginal gland above the base rarely wanting, those of barren shoots sometimes short and broad. Racemes much shorter than the phyllodia, almost always branched, with several, often 10 to 20, small globular heads of 12 to 20 flowers, mostly 5-merous. Calyx thin, not half so long as

the corolla, with very short broad ciliate lobes. Petals smooth, united to the middle, but readily separating. Pod shortly stipitate, usually straight, nearly flat, 3 to 8 in. long, 3 to 4 lines broad. Seeds oblong, longitudinal, the funicle dilated and coloured nearly from the base, extending round the seed and bent back on the same side, encircling it in a double fold.—F. Muell. Pl. Vict. ii. 13.

Victoria. Port Phillip, *R. Brown;* grassy ridges and open valleys throughout the greater part of the colony, *F. Mueller.*

S. Australia. Memory Cove, *R. Brown;* very frequent in rich soils near water in the valleys, *Behr;* Kangaroo island, *Waterhouse;* and northward to Flinders range, *F.·Mueller.*

· The species differs from *A. penninervis* chiefly in the narrow phyllodia, from *A. neriifolia* in the fewer smaller flowers, the sepals more united, and especially in the narrower pod and in the funicle, and usually from both species in the more compact branched racemes; but some flowering specimens are uncertain in this respect.

127. **A. neriifolia,** *A. Cunn.; Benth. in Hook. Lond. Journ.* i. 357. A tall and handsome shrub or small tree; branchlets slender, slightly angular, glaucous or mealy-tomentose when young, but soon glabrous. Phyllodia linear-lanceolate, more or less falcate, with a small callous point often recurved, much narrowed towards the base, mostly 3 to 5 in. long and 2 to 4 lines broad, 1-nerved, obscurely penniveined, with 1 or sometimes 2 or 3 distant marginal glands rarely all wanting. Racemes always simple, rather slender, much shorter than the phyllodia, the rhachis and peduncles usually tomentose. Flower-heads globular, small, with 30 to 40 flowers, mostly 5-merous. Sepals spathulate, more than half as long as the corolla, ciliate, free or slightly adnate below the middle. Petals smooth, usually free. Pod flat, straight or nearly so, several inches long, about 4 lines broad, often slightly contracted between the seeds. Seeds oval-oblong, longitudinal; funicle with the last fold appressed and thickened from the middle upwards into a club-shaped aril, the lower folds short and filiform.

Queensland. Open forests on the Balonne river, *Mitchell.*

N. S. Wales. Detached whinstone hills, Liverpool plains, *A. Cunningham, Fraser;* very common about Tenterfield, New England, *C. Stuart;* Head of the Gwydir river, *Leichhardt.*

S. Australia. Some specimens in flower in the Hookerian herbarium appear to belong to this species rather than to *A. retinodes,* and *A. iteaphylla,* F. Muell., Benth. in Linnæa, xxvi. 617, in fruit only, from Arkaba, has the pod and seeds of *A. neriifolia;* but neither can be identified with certainty until the flowers and fruit shall have been properly matched.

128. **A. microbotrya,** *Benth. in Hook. Lond. Journ.* i. 353. A tall shrub, quite glabrous except the inflorescence, branches slightly angular, soon becoming terete. Phyllodia lanceolate-falcate, acuminate, acute, obtuse or with a short incurved point, much narrowed towards the base, mostly 3 to 5 in. long,but very variable in size, and when small sometimes scarcely falcate, 1-nerved, penniveined, the nerve-like margin fine or scarcely prominent, with 1 or 2 small marginal glands often wanting. Racemes ½ to 1½ in. long, with from 3 or 4 to above 20 small globular heads of numerous small flowers, mostly 5-merous, the rhachis and peduncles when young minutely silky or mealy-tomentose. Calyx very thin, lobes very short or slightly spathulate, often readily separating into distinct sepals. Petals glabrous or minutely pubescent, the midribs prominent. Pod doubtful.—Meissn. in Pl. Preiss. i. 15.; *A. myriobotrya* and *A. leiophylla,* var. *microcephala,* Meissn. l. c.; *A.*

subfalcata and *A. daphnifolia*, Meissn. in Bot. Zeit. 1855, 11; *A. rostellifera*, Seem. Eingef. Acac. t. 2, not of Benth., and therefore probably also *A. pterigoidea*, Seem. in Verhandl. K. K. Gartenb. 1846, 11, quoted by Seem. l. c. 33.

W. Australia. Swan River, *Drummond, 1st Coll. and n.* 286, *Preiss, n.* 923; between Moore and Murchison rivers, *Drummond, 6th Coll. n.* 1 and 2; Murchison river, *Oldfield;* and probably Kalgan river, *Oldfield.* This species appears to represent *A. penninervis* in the west, from which it differs in the usually narrower phyllodia, the flowers more numerous although in very small heads, and probably in the narrow pod and more filiform funicles. It is also very near the following *A. saligna*, but the flower-heads are not half the size, the peduncles much more slender, the calyx longer, etc., and the pod probably different. Of the numerous specimens I have seen there are only two in fruit, both apparently agreeing with the flowering ones in branches foliage and inflorescence, in one the pods, not yet ripe, are straight, flat, about 2 in. long and 1½ lines wide, with longitudinal seeds and a filiform funicle forming several long folds, in the other the pods are old, 5 to 6 in. long, 6 to 8 lines broad, coriaceous, convex over the seeds, often narrowed between them; the seeds all shed. Flowering specimens of some forms of this species, of *A. neriifolia* and of *A. retinodes*, are often very difficult to distinguish.

129. **A. saligna,** *Wendl. Comm. Acac.* 26. A tall shrub or tree, quite glabrous, with angular branchlets. Phyllodia falcate-lanceolate, rather obtuse, much narrowed towards the base, many inches long, rather thick, 1-nerved, obscurely or rarely more distinctly penniveined, with nerve-like margins, the marginal gland distinct from the base, often wanting. Racemes short, with few globular heads, larger than in the preceding six species, the rhachis rather stout and flexuose. Flowers about 25 to 30 in the head, often slightly pubescent, mostly 5-merous. Calyx not half so long as the corolla, truncate or sinuate-toothed. Petals smooth. Pod flat, often 5 or 6 in. long, usually straight, scarcely 3 lines broad, not at all or slightly contracted between the seeds. Seeds oval-oblong, longitudinal; funicle short, gradually thickened almost from the base into a club-shaped fleshy aril.—DC. Prod. ii. 450; *Mimosa saligna*, Labill. Pl. Nov. Holl. ii. 86. t. 235; *A. leiophylla*, Benth. in Hook. Lond. Journ. i. 351.

W. Australia. King George's Sound, rare, *R. Brown;* to the eastward, *Baxter;* towards the Great Bight, *Maxwell.* I have not seen authentic specimens of Labillardière's plant, but have every reason to believe that it was this species, gathered on the same coast, and not a Tasmanian plant that he figured and described. The species is nearly allied on the one hand to *A. cyanophylla*, on the other to *A. microbotrya.*

A. Blomei, Ohlend. in Neue Allgem. Gartenzeit. 1845, 369, described from a specimen in leaf only, is referred by Seemann, Eingef. Acac. 30, to *A. leiophylla*, Benth.

130. **A. cyanophylla,** *Lindl. Bot. Reg.* 1835, *Misc.* 49. A tall handsome shrub, glabrous and often more or less glaucous, emitting, at least in cultivation, suckers from the roots, branches scarcely angular. Phyllodia from linear-oblong to lanceolate-falcate, the lower ones sometimes above a foot long, the upper ones ½ ft. or less and narrower, much narrowed towards the base, 1-nerved, penniveined, with nerve-like margins, the marginal gland usually obscure or none. Racemes short, with 3 to 5 heads of numerous (above 40) flowers, mostly 5-merous, the common rhachis rather stout and flexuose, the peduncles 3 to 6 lines long. Sepals cohering to the middle in a turbinate lobed calyx. Petals smooth, but with prominent midribs. Pod several inches long, flat, 2 to 2½ or rarely 3 lines broad, contracted between the seeds. Seeds oblong, longitudinal; funicle as long as the seed, the last

fold slightly thickened into a somewhat clavate aril, with very short folds below it.—Meissn. in Pl. Preiss. i. 15.

W. Australia. Swan River, *Mangles, Huegel, Preiss, n.* 925, *Drummond,* 1*st Coll. and n.* 284, and others. Closely allied on the one hand to *A. saligna,* to which I had referred it in Hueg. Enum. 42, and on the other to *A. pycnantha.*

131. **A. pycnantha,** *Benth. in Hook. Lond. Journ.* i. 351. A small or middle-sized tree, quite glabrous; branches terete or nearly so. Phyllodia lanceolate-falcate, obtuse or rather acute, much narrowed towards the base, 3 to 6 in. long, the larger ones often 1 in. broad in the middle, coriaceous, 1-nerved, penniveined with nerve-like margins, the marginal gland rather large near the base. Racemes short, with a few dense globular heads of 50 to 100 flowers, mostly 5-merous, the rhachis and peduncles rather stout. Calyx shortly lobed, ciliate, usually about ⅔ as long as the corolla. Petals smooth, glabrous, distinct or readily separating. Pod straight or slightly curved, several inches long, about 3 lines broad, flat and rather thin. Seeds oval-oblong, longitudinal; funicle not so long as the seed, thickened upwards, either not folded, or with 1 or 2 very short folds at the base.—Schlecht. Linnæa, xx. 664; F. Muell. Pl. Vict. ii. 15; Dietr. Fl. Univers. N. Ser. t. 86; *A. petiolaris,* Lehm. Novit. Hort. Hamb. in Linnæa, xxv. 306; *A. falcinella,* Meissn. in Bot. Zeit. 1855, 11.

Victoria. Frequent throughout the greater part of the colony in open forest country or scrubs, *F. Mueller.*

S. Australia. Common especially on undulating hills, exuding abundance of gum and furnishing bark for tanning, *Behr, F. Mueller.*

Var. (?) *angustifolia.* Branchlets angular, phyllodia narrower, flower-heads fewer and smaller.—Memory Cove, *R. Brown;* Spencer's Gulf, *F. Mueller,* referred here on the authority of F. Mueller, but from the inspection of the specimens it appears somewhat distinct. Pod unknown.

132. **A. notabilis,** *F. Muell. Fragm.* i. 6. A tall handsome shrub, glabrous and often glaucous, the branchlets terete or nearly so. Phyllodia from lanceolate-falcate to almost linear, narrowed at the base, usually oblique, 4 to 8 in. long, or the lower ones shorter and broader, thickly coriaceous, 1-nerved with thick nerve-like margins, obscurely veined, the marginal gland near the base not very conspicuous. Racemes short, with dense globular heads of above 50 flowers as in *A. pycnantha,* and the calyx as in that species ⅔ as long as the corolla. Petals silky-pubescent, united to the middle. Pod straight, flat, glaucous, 1½ to 3 in. long, 4 to 5 lines wide, somewhat coriaceous with nerve-like margins. Seeds transverse; funicle long and filiform, encircling the seed in a double fold, only very shortly thickened at the end into a small fleshy aril.

N. S. Wales. Towards the Barrier range, *Victorian Expedition.*

S. Australia. S. coast, *R. Brown;* Port Lincoln, *Wilhelmi;* Flinders range, *F. Mueller.*

133. **A. gladiiformis,** *A. Cunn.; Benth. in Hook. Lond. Journ.* i. 354. A tall shrub, quite glabrous; branchlets angular. Phyllodia linear-lanceolate or almost spathulate, curved, very obtuse or with a small hooked point, 3 to 6 in. long, much narrowed towards the base, thickly coriaceous, 1-nerved, smooth and shining, the margins thickened and usually 2 or more marginal glands. Racemes short, the rhachis rigid and flexuose, with several dense

globular heads of above 30 flowers, mostly 5-merous. Sepals narrow-linear, spathulate, with dark concave tips more than half the length of the corolla. Petals smooth. Pod elongated, flat but flexuose, about 3 lines broad, coriaceous. Seeds longitudinal; funicle long, slightly dilated, encircling the seed in a double fold, but not returning the third time as in *A. amœna.*

N. S. Wales. Blue Mountains and rocky hills to the westward, *A. and R. Cunningham.*

134. **A. obtusata,** *Sieb. in DC. Prod.* ii. 453. A tall shrub, quite glabrous; branchlets angular. Phyllodia oblong-linear or almost spathulate, usually straight, very obtuse, 1½ to 3 in. long, rigidly coriaceous, 1-nerved, with thickened nerve-like margins, the veinlets inconspicuous, with or without marginal glands. Racemes short, with few densely packed heads of above 30 flowers, mostly 5-merous. Sepals thick, spathulate, half as long as the corolla, at first united but readily separating when fully out. Pod unknown.

N. S. Wales. Blue Mountains, *Sieber, n.* 441; *A. Cunningham, Fraser.* Allied on the one hand to *A. amœna,* on the other to *A. gladiiformis,* but apparently distinct from both.

135. **A. rubida,** *A. Cunn. in Field, N. S. Wales,* 344. A tall shrub, quite glabrous, allied to *A. amœna,* and perhaps a variety; branchlets angular. Phyllodia lanceolate, often falcate, rather acute, much narrowed towards the base, mostly about 3 in. long, rather thick, 1-nerved, with nerve-like margins, the veinlets inconspicuous and never more than 1 marginal gland. Racemes shorter than the phyllodia, with several, often 10 to 12, rather small heads of 10 to 15 flowers, mostly 5-merous. Sepals half as long as the petals, usually coherent. Petals smooth. Pod unknown.—*A. amœna,* Sieb. Pl. Exs., not of Wendl.

N. S. Wales. Port Jackson to the Blue Mountains, *Sieber, n.* 452; head of the Gwydir, *Leichhardt;* Clarence river, *Beckler?* (specimens not in flower).

136. **A. amœna,** *Wendl. Comm. Acac.* 10, *t.* 4. A tall shrub, quite glabrous, young branches pubescent. Phyllodia obliquely lanceolate or oblanceolate, straight or falcate, obtuse or with a small recurved point, much narrowed towards the base, not very thick, 1-nerved with nerve-like margins and more or less distinctly veined, with 1, 2, or 3 often prominent distant marginal glands, 1½ to 2½ in. long on the flowering shoots, longer on the barren branches. Racemes usually shorter than the phyllodia, with several small globular heads of about 8 to 12 flowers, mostly 5-merous. Sepals short, broad, usually separating when the flower is fully out. Petals 5, distinct, smooth with prominent midribs. Pod flat, straight or curved, with nerve-like margins, several in. long, 3 to 4 lines broad, not contracted between the seeds. Seeds ovate, longitudinal; funicle dilated and reticulate from near the base, very long, extending round the seed, returning on the same side and bent back a third time, encircling the seed in a triple fold, and thickened at the end into a fleshy aril, two-thirds the length of the seed. —DC. Prod. ii. 452; Fl. Muell. Pl. Vict. ii. 17.

N. S. Wales. Blue Mountains, *R. Brown;* Illawarra and banks of the Lachlan and Macquarrie, *A. Cunningham, Fraser.*

Victoria. Rocky mountains along Macalister river and its tributaries, granitic banks of Snowy River and adjoining mountains, at an elevation of 2000 to 4000 ft., *F. Mueller.*

The funicle completely encircling the seed a third time does not occur in any other species which I have been able to observe, and is in all the seeds I have examined of *A. amœna ;* it remains, however, to be ascertained whether it is really so constant a character as it appears to be.

137. **A. hakeoides,** *A. Cunn.; Benth. in Hook. Lond. Journ.* i. 354. A tall shrub, glabrous or nearly so, the branches scarcely angular. Phyllodia linear-spathulate or narrow oblong-lanceolate, obtuse, much narrowed at the base, 2 to 5 in. long, rather thick, 1-nerved, obscurely marked with longitudinal reticulations, the margins scarcely prominent, usually with a gland towards the middle. Racemes shorter than the phyllodia, with a few globular heads of about 20 flowers, mostly 5-merous. Sepals spathulate, cohering at first but readily separating, half as long as the corolla. Petals smooth, glabrous or minutely pubescent. Pod flat, usually curved, 2 to 3 lines broad, much contracted between the seeds. Seeds oblong, longitudinal ; funicle half as long as the seed, the last fold thickened into a clavate, keeled, fleshy aril almost from the base, with 2 or 3 very minute folds below it.—F. Muell. Pl. Vict. ii. 16.

N. S. Wales. Lachlan and Dumaresq rivers, *A. Cunningham.*
Victoria. Murray desert, *Prince Paul William of Wirtemberg. Dallachy.*
Allied in flowers to *A. obtusata* and *A. crassiuscula,* and in foliage to *A. salicina,* but differing in several points from each of these species.

138. **A. salicina,** *Lindl. in Mitch. Three Exped.* ii. 20. A tall shrub or small tree, with branches often pendulous, the foliage of a pale or glaucous hue and quite glabrous ; branchlets scarcely angular. Phyllodia mostly straight or nearly so, oblong-linear or lanceolate, obtuse or slightly acuminate, much narrowed at the base, 2 to 5 in. long and not above ½ in. broad, but in some varieties occasionally broader or falcate, always rather thick, the midrib scarcely prominent, the lateral veins obscurely reticulate, the margins scarcely thickened, the gland very rare. Racemes short, irregularly bearing 2 or 3 dense globular heads or reduced to a single head. Flowers 20 or more, mostly 5-merous. Calyx short, truncate, entire or minutely toothed. Petals quite smooth. Pod straight, 1 to 3 in. long, in the ordinary form not above 3 lines broad ; valves somewhat convex, hard and thick. Seeds orbicular, longitudinal; funicle thickened and usually scarlet almost from the base, forming several folds under the seed.—F. Muell. Pl. Vict. ii. 12 ; Dietr. Fl. Univers. N. Ser. t. 83 ; *A. ligulata,* A. Cunn. ; Benth. in Hook. Lond. Journ. i. 362.

N. Australia. Banks of creeks, Arnhem's Land, *F. Mueller* ; Curtis Island, *Henne.*
Queensland. Open forest lands on the Balonne, *Mitchell ;* Suttor river, *F. Mueller.*
N. S. Wales. On the Lachlan and thence to the Barrier Range, *A. Cunningham, Mitchell, Victorian Expedition.* etc.; Liverpool plains, *Leichhardt.*
Victoria. In the N.W. desert, *F. Mueller.*
S. Australia. From the Murray to St. Vincent's and Spencer's Gulfs, and northward to the desert interior, *F. Mueller;* Memory Cove, *R. Brown.*
W. Australia. Dirk Hartog's Island, *A. Cunningham ;* Sharks Bay, *Milne ;* Murchison river, *Oldfield ;* also a specimen from *Baudin's Expedition,* in Herb. R. Brown.

Var. *varians.* Branches more spreading. Phyllodia more veined, the lower ones often much broader and almost penninerved, as in *A. penninervis,* but without the thickened margin or gland. Pod about 4 lines broad, the seeds often oblique and the folds of the funicle extending up one side.—*A. varians,* Benth. in Mitch. Trop. Austr. 132. To this belong all the tropical and subtropical specimens. It is generally a very distinct form, and it is with some hesitation that I have followed F. Mueller in considering it a variety only of *A. salicina*.

139. **A. rostellifera,** *Benth. in Hook. Lond. Journ.* i. 356. A tall shrub or small tree, nearly allied to *A. salicina*, with which it is united by F. Mueller, but the aspect is different, the nerve of the phyllodia is much more prominent, and the pod is unknown. Branchlets angular, flexuous, quite glabrous. Phyllodia linear-lanceolate, 2 to 5 in. long, 3 to 4 lines broad, straight or falcate, with an oblique or recurved usually callous point, rather thick, 1-nerved, very obscurely veined. Flower-heads few, in short racemes, as in *A. salicina*, with numerous 5-merous flowers. Calyx short, truncate. Petals quite smooth, without the prominent midribs of *A. cyanophylla*, which the long-leaved specimens sometimes resemble.—*A. subbinervia*, Meissn. in Pl. Preiss. i. 16.

W. Australia. Swan River, *Drummond,* 1*st Coll. n.* 285, 2*nd Coll. n.* 103; Murchison river, *Oldfield;* Rottenest Island, *Preiss, n.* 924. The second nerve of the phyllodia, from whence Meissner derived his name, very seldom occurs, and was therefore in some measure exceptional in the specimen described by him. The species requires further investigation.

140. **A. pycnophylla,** *Benth.* A glabrous, erect shrub of several ft., the branches slightly angular. Phyllodia usually numerous, erect, linear, obtuse or with a small straight point, 1½ to 3 in. long, thick and nerveless besides the midrib, narrowed at the base, without marginal glands. Flower-heads generally 2 or 3, in short axillary racemes on short thick peduncles, globular, with 10 to 15 flowers, mostly 5-merous. Calyx half as long as the corolla, shortly and broadly lobed. Petals smooth, but angular in the bud. Pod linear, flat with thickened margins, not contracted between the seeds, 1 to 1½ lines broad; valves thinly coriaceous. Seeds longitudinal, the last 2 or 3 short folds of the funicle thickened into a small fleshy aril under the seed. —*A. crassiuscula*, Meissn. in Pl. Preiss. i. 16, not of Sieber.

W. Australia. King George's Sound and adjoining districts, *Drummond,* 3*rd Coll. n.* 98, *Preiss, n.* 929, *Oldfield*.

Var. *angustifolia*. Phyllodia narrower; peduncles rather longer; petals thinner. Pod the same.—Cape Paisly and Cape Legrand, *Maxwell*.

The foliage is nearly that of *A. ramosissima* and *A. Harveyi*, but the pod is very different. The lobed calyx as well as the pod readily distinguish it from the narrowest-leaved forms of *A. salicina*.

141. **A. Harveyi,** *Benth.* Quite glabrous, with erect slender branches, slightly angular when young. Phyllodia narrow-linear, obtuse or with a short hooked point, 2 to 3 in. long, 1 to 1½ lines broad, narrowed at the base, rather thick, the midrib scarcely prominent and the veinlets very obscure. Flower-heads small, globular, several in short slender racemes, the partial peduncles 1 to 2 lines long. Flowers small, 20 to 30 in the head, mostly 5-merous. Calyx thin, shortly lobed, more than half as long as the corolla. Petals smooth. Pod very flat, with scarcely thickened margins, not contracted between the seeds, 3 in. long or more, above ½ in. broad, coriaceous and transversely reticulate. Young seeds along the centre of the pod, with a long funicle folded and thickened under the seed, but not seen ripe.—*A. ramosissima*, Benth.; Meissn. in Pl. Preiss. i. 16, partly.

W. Australia. Between King George's Sound and Cape Riche, *Harvey* (in flower), *Drummond,* 4*th Coll. n.* 130 (in fruit). Allied in foliage to *A. pycnophylla* and *A. ramo-*

sissima, but very different in fruit. Of Preiss's specimens I have only seen n. 941, which appears to me to belong to the present species, but cannot be absolutely determined without the fruit.

142. **A. suaveolens,** *Willd. Spec. Pl.* iv. 1050. A shrub, attaining about 3 to 6 ft., quite glabrous, often glaucous, with acutely angled branchlets. Phyllodia linear or almost lanceolate, mostly 3 to 4 in. but sometimes 6 in. long, 2 to 4 lines broad, obtuse or mucronulate, narrowed towards the base, rather thick, 1-nerved with nerve-like margins, obscurely veined. Flower-heads small, in axillary racemes, at first enclosed in imbricate scaly bracts, which fall off very early. Rhachis and peduncles slender. Flowers 6 to 10 in the head, mostly 5-merous. Sepals thin, narrow linear-spathulate, quite distinct. Petals thin, quite smooth. Pod oblong, flat, coriaceous, glaucous, very obtuse, 1 to 1½ in. long, 6 to 8 lines broad. Seeds oblong, transverse; funicle filiform nearly till maturity, when it is contracted into short folds more or less thickened under the seed into a small fleshy aril.— *Mimosa suaveolens,* Sm. in Trans. Linn. Soc. i. 253; Labill. Pl. Nov. Holl. ii. 87, t. 236; *M. obliqua,* Lam. in Journ. Hist. Nat. i. 89, t. 5, according to Wendl. Comm. Acac. 33, but not of Pers.; *M. angustifolia,* Jacq. Hort. Schœnbr. iii. 74, t. 391; *A. suaveolens,* DC. Prod. ii. 453; Lodd. Bot. Cab. t. 730; Reichb. Ic. et Descr. Pl. t. 46; Hook. f. Fl. Tasm. i. 107; F. Muell. Pl. Vict. ii. 14; *A. angustifolia,* Wendl. Comm. Acac. 34; DC. Prod. ii. 453.

Queensland. Moreton Bay, *Fitzalan.*
N. S. Wales. Port Jackson to the Blue Mountains, *R. Brown, Sieber, n.* 462, *and Fl. Mixt. n.* 595; northward to Hastings river, *Beckler;* southward to Twofold Bay, *F. Mueller.*
Victoria. Bushy sand ridges, barren scrubby plains, and rocky coast declivities in the southern and eastern parts of the colony, *F. Mueller.*
Tasmania. Dry soils, N. coast, and islands of Bass's Straits, also on the Derwent, *J. D. Hooker.*

143. **A. subcærulea,** *Lindl. Bot. Reg. t.* 1075. An erect or spreading shrub, quite glabrous and often very glaucous; branchlets acutely angular. Phyllodia oblong-linear, lanceolate or narrow-linear, obtuse or mucronate, 1½ to 3 in. long, 2 to 4 lines broad, straight, thick and coriaceous, 1-nerved and veinless, or, when broad, rather thinner, slightly falcate, and faintly penninerved, the nerve-like margin much less prominent than in *A. obtusata* and its allies, and sometimes altogether disappearing. Racemes often as long as the phyllodia, enclosed when young in very deciduous bracts or scales, the rhachis and peduncles slender. Flower-heads globular, with above 30 flowers, mostly 5-merous. Sepals very thin, narrow and distinct or sometimes broader and slightly cohering, about half as long as the corolla or shorter. Petals smooth, at length separating. Pod 1 to 1½ or rarely 2 in. long, ½ to near ¾ in. broad, very obtuse, flat, coriaceous and glaucous, as in *A. suaveolens.* Seeds oblong, transverse, the funicle thickened into a few short closely packed folds under the seed.—*A. hemiteles,* Benth. in Linnæa, xxvi. 619; *A. apiculata,* Meissn. in Pl. Preiss. i. 17.

W. Australia. Sandy and stony places, King George's Sound to Cape Riche, *Drummond, 4th Coll. n.* 10, *Preiss, n.* 919, *Oldfield,* and eastward towards the Great Bight, *Maxwell;* Point Possession, *Collie.*

Var. *parvifolia.* Phyllodia shorter, pod rather longer.—Near Cape Riche, *Harvey;* "Norabup," *Oldfield.*

144. **A. Lindleyi,** *Meissn. in Pl. Preiss.* i. 14. A glabrous, glaucous shrub, evidently allied to *A. subcærulea,* but somewhat uncertain until the fruit has been seen. Branches much stouter, very acutely 3-angled when young. Lower phyllodia 6 to 10 in. long, 1½ to 2 in. broad, lanceolate, much narrowed at the base, 1-nerved, with fine but prominent pinnate veins; upper ones smaller and narrower, more like those of *A. subcærulea.* Racemes short, probably enclosed when young in imbricate scales, but all fallen off in the specimens seen; heads few, globular, containing numerous 5-merous flowers. Sepals narrow.

W. Australia. In the interior, *Preiss, n. 947, Drummond (4th Coll.), n. 25.* The specimens I have seen are not satisfactory.

145. **A. leptopetala,** *Benth. in Linnæa,* xxvi. 619. A bushy shrub, attaining 8 or 10 ft., glabrous and often somewhat glaucous; branches slightly angular. Phyllodia oblong linear or lanceolate, obtuse, narrowed at the base, 1 to 2 in. long, coriaceous, with a scarcely prominent nerve and very obscurely reticulate veins. Peduncles slender, in short axillary racemes or rarely solitary, bearing each a small globular head of numerous (above 30) flowers, mostly 5-merous. Sepals very narrow, distinct, spathulate at the end. Petals rather narrow, free, smooth. Pod flat, straight, 2 to 3 in. long, about ½ in. wide, obtuse; valves membranous. Seeds transverse, ovate; funicle thickened from the base, and forming an aril of about 3 folds under the seed.

W. Australia, *Drummond, Suppl. to 5th Coll. n. 52;* Murchison river, *Oldfield.* The pod seen in the latter specimens only, which otherwise agree with Drummond's.

146. **A. Murrayana,** *F. Muell. Herb.* Glabrous; branchlets slightly angular. Phyllodia linear, straight or nearly so, obtuse or .with a callous hooked point, 5 to 6 in. long, 1 to 1½ lines wide, thick, with a slightly prominent nerve and very obscurely veined. Peduncles about ½ in. long, 2 or 3 together in a short raceme, bearing each a globular head of numerous (above 50) flowers, mostly 5-merous. Sepals very thin, narrow, linear-spathulate, fully half as long as the corolla. Petals free or nearly so, narrow, very thin. Pod linear, 2 to 3 in. long, about 4 lines broad, flat, obtuse; valves membranous. Seeds transverse, ovate; funicle filiform or slightly thickened from the base, forming 2 or 3 short folds under the seed.

Queensland. Open Forest, St. George's Bridge on the Balonne, *Mitchell.*
S. Australia. Cooper's Creek, *Murray, in Howitt's Expedition.*
Murray's specimen, a single one, is in flower only; Mitchell's, of which we have several, are in fruit only; it is possible, therefore, that they may not have been correctly matched, but they appear all to belong to one species, differing from *A. leptopetala* in little besides the long narrow phyllodia.

147. **A. subulata,** *Bonpl. Jard. Malm.* 110, *t.* 45. A tall glabrous shrub, attaining 10 ft. or more, with erect, slender, slightly angular branches. Phyllodia narrow-linear, mucronulate, narrowed at the base, 3 to 6 in. long, scarcely 1 line broad, straight or nearly so, rather thin, 1-nerved. Flower-heads several, globular, small, in slender axillary racemes, the peduncles almost filiform. Flowers about 12 to 20, very small, mostly 5-merous.

Calyx thin, turbinate, usually toothed, fully half as long as the corolla. Petals smooth. Pod not seen.—DC. Prod. ii. 453.

N. S. Wales. Forests of the W. branches of Hunter's River and plains of Daley, *A. Cunningham;* also in *Leichhardt's* collection. The phyllodia are sometimes scarcely broader than in the flat-leaved variety of *A. juncifolia*, but the inflorescence is different.

148. **A. linifolia,** *Willd. Sp. Pl.* iv. 1051. A tall shrub, glabrous or minutely pubescent, sometimes glaucous when young; branchlets angular. Phyllodia linear or linear-lanceolate, narrowed at each end, 1 to 1½ in. long, 1½ to 2 or rarely 2½ lines broad, rather thin, 1-nerved, slightly veined, the slender nerve-like margins and midrib often minutely ciliate, with a small gland above the base. Racemes scarcely exceeding the phyllodia or shorter, comprising several small globular heads of about 8 to 12 flowers, mostly 5-merous. Calyx short, broadly lobed. Petals smooth, separating nearly to the base. Pod linear, very flat, 2 to 4 in. long, 4 to 6 lines broad and not contracted between the seeds; valves thinly coriaceous, with nerve-like margins. Seeds longitudinal, along the centre of the pod, the last fold of the funicle thickened into a club-shaped lateral aril, the other folds minute.—*Mimosa linifolia*, Vent. Jard. Cels. t. 2 ; Andr. Bot. Rep. t. 394 ; *M. linearis*, Wendl. Bot. Beob. 56, and Hort. Herrenh. 8, t. 18, not of Sims; *A. ab etina*, Willd. Sp. Pl. iv. 1051 ; DC. Prod. ii. 453 ; *A. linifolia*, Bonpl. Jard. Malm. 56, t. 16 ; DC. Prod. ii. 453 ; Bot. Mag. t. 2168 ; Lodd. Bot. Cab. t. 383 ? (this fig. looks more like *A. subulata*).

Queensland. Brisbane river, Moreton Bay, *A. Cunningham, Fraser, F. Mueller*, and others ; Wide Bay, *Leichhardt ;* Broad Sound, *Bowman.*

N. S. Wales. Port Jackson to the Blue Mountains, *Sieber, n.* 465, *and Fl. Mixt. n.* 597, and others ; Liverpool plains, *A. Cunningham.*

The broad-leaved forms of this species, with the margins less ciliated or not perceptibly so, connect it with the following, *A. prominens.*

149. **A. prominens,** *A. Cunn. in G. Don, Gen. Syst.* ii. 406. A tall shrub, glabrous and usually glaucous, with angular branchlets. Phyllodia from linear-lanceolate to oblong-falcate, when narrow nearly those of *A. linifolia*, but not so decidedly ciliate, more acute and the marginal gland further from the base, and passing from that to nearly those of *A. lunata*, but always much thinner than the latter, with the pinnate veins as well as the gland more conspicuous, mostly 1 to 1½ in. long, from 2 lines broad in the narrow form to 3, 4, or even 5 in the broad ones. Racemes about as long as the phyllodia, with very small globular heads of about 8 to 10 or rarely 12 to 15 small flowers, mostly 5-merous. Calyx very short, broadly lobed. Petals smooth or nearly so. Pod very flat, 2 to 3 in. long when perfect, 3, 4, or rarely 5 lines broad. Seeds longitudinal along the centre, the last fold of the funicle thickened into a fleshy clavate lateral aril, the other folds very small. —Bot. Mag. t. 3502 ; Dietr. Fl. Univers. N. Ser. t. 83 ? ; *A. fimbriata*, A. Cunn. in G. Don, Gen. Syst. ii. 406.

N. S. Wales. Blue Mountains, *Caley, A. Cunningham*, and others.

S. Australia? Between Rocky Creek and Mount Remarkable, *F. Mueller.* Specimens very young and doubtful.

This species may prove to be a broad-leaved variety of *A. linifolia.* F. Mueller refers it to *A. lunata*, describing the pod very accurately, but that is not the fruit of the true *A. lunata*, which has always the seed lying close to the upper suture, not in the centre of the pod.

150. **A. Leichhardtii,** *Benth.* Branches slender, terete, hispid with long spreading hairs. Phyllodia linear-lanceolate, falcate, mucronate, rarely above 1 in. long, the midrib and nerve-like margins prominent and ciliate, otherwise veinless, the marginal gland about the middle, often wanting. Racemes much longer than the phyllodia, slender, with numerous small heads of 20 or more flowers, mostly 5-merous. Calyx turbinate, shortly lobed, half as long as the corolla. Petals smooth. Pod very flat, 2 to 4 in. long, about 4 lines broad, hispid with long hairs, but not seen ripe.

Queensland. Expedition range, *Leichhardt,*

151. **A. crassiuscula,** *Wendl. Comm. Acac.* 31, *t.* 8. A shrub of several feet, glabrous and often rather glaucous when young ; branches usually acutely angled. Phyllodia numerous, linear, often falcate, with a small oblique point or the lower ones obtuse and almost lanceolate, rather thick, 1-nerved and veinless, the nerve-like margins often but not always ciliate, 1½ to 2 in. long in some specimens, above 3 in. in others, 1½ to 2 or in larger ones 3 lines broad, the marginal gland below the middle. Racemes shorter than the phyllodia, with several small dense globular heads of 20 or more flowers. Calyx turbinate, fully half as long as the corolla, ciliate and readily separating into spathulate sepals. Petals often separating, with prominent midribs. Pod linear, rather thick, about 2½ lines broad, contracted between the seeds. Seeds in the centre of the pods, but not seen perfect.—DC. Prod. ii. 453 ; Hook. f. Fl. Tasm. i. 108 ; *A. adunca,* A. Cunn. in G. Don, Gen. Syst. ii. 406 ; *A. Sieberiana.* Tausch. in Flora, 1836, 420, not of DC.

Queensland. Moreton Bay, *Fitzalan,*
N. S. Wales. Port Jackson to Blue Mountains, *R. Brown, Sieber, n.* 464 ; rocky barren brushy hills of the Blue Mountains, *A. Cunningham, Fraser.*
Tasmania. Flinders Island, Bass's Straits, *J. D. Hooker.*
Although evidently distinct from all other species I have examined, it is impossible to define this one satisfactorily until the fruit shall be more certainly known. It is certainly very distinct from the Western plant referred to in ' Plantæ Preissianæ.'

152. **A. decora,** *Reichb. Icon. Exot. t.* 199. A shrub of several feet, glabrous or slightly glaucous-tomentose ; branchlets angular. Phyllodia lanceolate or linear, narrowed at the base, straight or slightly falcate, 1 to 2 in. long, thicker than in *A. linifolia,* 1-nerved, slightly penniveined, with nerve-like margins and usually with a gland below the middle. Racemes usually numerous, longer than the phyllodia, with several globular heads of about 20 flowers, mostly 5-merous. Calyx shortly lobed, not half so long as the corolla. Pod straight or curved, flat, about 3 lines broad. Seeds longitudinal, close to the upper suture ; last fold of the funicle thickened into a lateral club-shaped aril, the lower folds very small.

Queensland. Keppel Bay, *R. Brown* ; Dawson river, *F. Mueller ;* near Mount Pluto, *Mitchell.*
N. S. Wales. Liverpool plains, *A. Cunningham ;* New England, *C. Stuart.*
Perhaps a variety of *A. buxifolia,* differing only in the longer phyllodia and more numerous flowers in the heads. Some specimens from the Melbourne Botanic Garden, apparently of this species, have the phyllodia still longer and narrower.

153. **A. buxifolia,** *A. Cunn. in Field, N. S. Wales,* 344. A glabrous shrub with angular branchlets. Phyllodia obliquely oblong-lanceolate, somewhat falcate, narrowed at each end, usually under 1 in., rarely 1½ in. long,

rather thick, with a scarcely prominent nerve and obscure veins, the marginal gland small or none. Racemes scarcely exceeding the phyllodia, with several small globular heads of 8 to 12 or rarely more flowers, mostly 5-merous. Calyx short, broad. Petals smooth. Pod straight or curved, flat, 3 or 4 lines broad. Seeds longitudinal, close to the upper suture ; last fold of the funicle thickened into a lateral club-shaped aril, the lower ones very small.—Hook. Ic. Pl. t. 164.

N. S. Wales. Hunter's and Macquarrie rivers, *A. Cunningham, Fraser ;* New England, *C. Stuart ;* Clarence river, *Beckler.*

Nearly allied to *A. lunata,* and perhaps a variety with narrower straighter phyllodia, and some specimens appear almost to pass into *A. decora.*

154. **A. lunata,** *Sieb. in DC. Prod.* ii. 452. A glabrous shrub of several feet, with angular branchlets, often glaucous. Phyllodia oblong-falcate or almost ovate, but very oblique, obtuse or with a minute oblique or recurved point, rarely 1 in. long, 3 to 6 lines broad, coriaceous, 1-nerved, obscurely veined, the margins scarcely thickened, the gland minute or none. Racemes longer than the phyllodia, with several small heads of 4 to 10 comparatively large flowers, mostly 5-merous. Calyx short and broad. Petals smooth. Pod flat, glaucous, straight or curved, 3 to 4 lines broad. Seeds longitudinal, close to the upper suture, the last fold of the funicle thickened into a lateral club-shaped aril, the lower folds very small.—Bot. Reg. t. 1352 ; Lodd. Bot. Cab. t. 384 ; Sweet, Fl. Austr. t. 42 ; F. Muell. Pl. Vict. ii. 17 (partly) ; *A. falcinella,* Tausch. in Flora, 1836, 419 ; *A. brevifolia,* Lodd. Bot. Cab. t. 1235 ? ; *A. oleæfolia,* A. Cunn, in G. Don, Gen. Syst. ii. 405 ; *A. dealbata,* A. Cunn. in Field, N. S. Wales, 345, not of Lindl. ; *A. furfuracea,* G. Don, Gen. Syst. ii. 405.

Queensland. Moreton Bay, *A. Cunningham.*

N. S. Wales. Port Jackson to the Blue Mountains, *R. Brown, Sieber, n.* 461, *and Fl. Mixt. n.* 600, and others; extending to the Macquarrie river, and Argyle County, *A. Cunningham, Backhouse.*

Victoria. Barren scrubby ridges between Mayday Hills and Ovens river, *F. Mueller* (specimens not in fruit).

Without the fruit this species may readily be confounded with *A. prominens,* the phyllodia are however more coriaceous, with the veins less conspicuous and the flowers in the heads usually rather fewer and larger.

155. **A. brachybotrya,** *Benth. in Hook. Lond. Journ.* i. 347. A handsome shrub of several feet, glabrous glaucous or silvery-white with a close silky-pubescence ; branchlets slightly angular, soon terete. Phyllodia obliquely obovate or oblong, obtuse or rarely mucronulate, ½ to 1 in. or in very luxuriant specimens twice as long, narrowed at the base, coriaceous, 1-nerved, penniveined, the marginal gland near the middle, small or often wanting. Peduncles rather short, solitary, or more frequently 2 to 5 on a very short common peduncle, often growing out into a leafy branch, each bearing a globular head of numerous (20 to 50) flowers, mostly 5-merous. Sepals linear-spathulate, free or connected by a thin membrane. Petals smooth, distinct or readily separating. Pod linear, straight or undulate, 3 to 5 lines broad, flat but the valves often alternately convex and concave over the seeds and sometimes much warted. Seeds longitudinal ; funicle thickened at the end into a club-shaped lateral aril and once folded below it.—*A. argyrophylla,* Hook. ; F. Muell. Pl. Vict. ii. 18.

N. S. Wales. Peel's Range, *A. Cunningham;* from the Darling to the Barrier range, *Victorian Expedition.*

Victoria. Not uncommon in the N.W. desert, *F. Mueller.*

S. Australia. Towards Spencer's Gulf, *F. Mueller ;* Mount Hall, *Warburton.*

The following forms appear at first sight very distinct, but they pass too gradually into each other to be separable as species.

a. *argyrophylla.* Silvery-silky, turning sometimes to a golden-yellow. Phyllodia mostly ¾ to 1½ in. long. Flower-heads often solitary, usually with more than 30 flowers.—*A. argyrophylla,* Hook. Bot. Mag. t. 4384 ; *A. bombycina,* Benth. in Lindl. and Paxt. Fl. Gard. ii. 101, f. 186.—Chiefly in S. Australia.

b. *glaucophylla.* Glaucous and more or less pubescent. Phyllodia mostly ½ to ¾ in. long. Flower-heads mostly 2 to 5, shortly racemose, with about 20 to 30 flowers.—*A. brachybotrya,* Benth., as above ; *A. dictyocarpa,* Benth. in Linnæa, xxvi. 616.—N. S. Wales and Victoria.

c. *glabra.* Quite glabrous. Phyllodia small and narrow. Flower-heads small.—Murray desert.

156. **A. Wattsiana,** *F. Muell. Herb.* A dense bushy glabrous shrub, of 4 to 5 ft. ; branchlets angular. Phyllodia obovate-oblong, very obtuse, narrowed at the base, ½ to 1 in. long, coriaceous, 1-nerved, obscurely penni-veined, the marginal gland below the middle sometimes wanting. Racemes as long as the phyllodia, with usually few globular heads of about 15 to 20 flowers, mostly 5-merous. Calyx short, broad, toothed. Petals smooth. Pod falcate, coriaceous, 2 to 4 in. long, 3 to 4 lines broad, not contracted between the seeds. Seeds longitudinal along the centre ; funicle long and much folded round the seed, but not seen quite ripe.

S. Australia. Between Broughton and Rocky Creeks, *F. Mueller.* The foliage and habit are those of *A. brachybotrya,* it has also much resemblance in habit and calyx to the shorter leaved forms of *A. lunata* (*A. oleæfolia,* A. Cunn.), to which I had referred it in the 'Linnæa,' xxvi. 616, but differs essentially from them both in the pod and especially in the elongated funicle.

157. **A. podalyriæfolia,** *A. Cunn: in G. Don, Gen. Syst.* ii. 405. A tall shrub, more or less mealy-glaucous, and minutely pubescent, rarely quite glabrous. Phyllodia obovate ovate or oblong, more or less oblique, obtuse or narrowed at one or both ends, mostly 1 to 1½ in. long, 1-nerved, with 1 or 2 marginal glands. Racemes much longer than the phyllodia, with several, often 10 to 20, small globular heads of numerous small mostly 5-merous flowers. Calyx turbinate, sinuate-toothed, not half so long as the corolla. Petals free or very slightly cohering, hirsute, with prominent midribs. Pod very flat, nearly ¾ in. broad, 1 to several in. long, glabrous or pubescent. Seeds ovate, longitudinal ; funicle rather long, in short folds under the seed, the last fold slightly thickened.—*A. Fraseri,* Hook. Ic. Pl. t. 171 ; *A. Caleyi,* A. Cunn. ; Benth. in Hook. Lond. Journ. i. 317 (a stunted specimen).

Queensland. Brisbane river, Moreton Bay, *A. Cunningham, Fraser,* and others ; in the interior about Lake Salvator, etc., *Mitchell ;* between the Suttor and Dawson rivers, *F. Mueller ;* Wide Bay, *C. Moore.*

158. **A. uncifera,** *Benth. in Mitch. Trop. Austr.* 341. A shrub of about 5 ft., softly velvety-pubescent ; branchlets nearly terete. Phyllodia obliquely oblong or elliptical-falcate, narrowed at both ends, with a hooked point, 1½ to 2 in. long, ½ to ¾ in. broad, 1-nerved, with 1 to 3 minute marginal glands. Racemes rather longer than the phyllodia, with several glo-

bular heads, smaller than in *A. podalyriæfolia,* on shorter peduncles. Flowers also smaller, otherwise like those of that species, of which this may prove to be a variety. Pod unknown.

Queensland. Foot of sandstone rocks near Mount Pluto, *Mitchell.*

159. **A. vestita,** *Ker, in Bot. Reg. t.* 698. A tall bushy shrub, softly pubescent or villous, usually 8 to 10 ft. high, but attaining twice that size. Phyllodia numerous, obliquely ovate-elliptical, more or less recurved-falcate, undulate, mostly about ½ in. long, with a fine but not pungent point, cuneate at the base, 1-nerved. Racemes much longer than the phyllodia, forming a terminal leafy panicle, each with several small globular heads of 10 to 20 or sometimes more flowers, mostly 5-merous. Calyx turbinate, shortly and obtusely lobed, nearly half as long as the corolla, rarely separating into distinct sepals. Petals smooth, glabrous, usually free. Pod very flat, straight, glabrous and glaucous, 1½ to 3 in. long, 4 to 5 lines broad. Seeds oval-oblong, longitudinal along the centre of the pod; last fold of the funicle thickened into an oblong-clavate lateral aril nearly as long as the seed, with 2 or 3 short filiform folds below it, and transverse at the base.—DC. Prod. ii. 452.

N. S. Wales. Blue Mountains, *Sieber, n.* 444, *A. Cunningham,* and others.

160. **A. cultriformis,** *A. Cunn. in G. Don, Gen. Syst.* ii. 406. A tall bushy shrub, glabrous and often mealy-glaucous when young; branchlets angular. Phyllodia numerous, obliquely obovate-lanceolate, recurved-falcate or almost triangular, mucronulate, narrowed at the base, ½ to ¾ or rarely 1 in. long, coriaceous, 1-nerved, with thickened margins and usually 1 marginal gland, sometimes on a prominent angle as in the *Triangulares.* Racemes numerous, much longer than the phyllodia, consisting often of 10 to 20 globular heads, forming a terminal leafy panicle. Flowers 10 to 20 in the head, mostly 5-merous. Calyx broad, lobed, scarcely half as long as the corolla. Petals smooth. Pod very flat, glabrous, glaucous, 2 or 3 in. long when perfect, 3 or rarely 4 lines broad. Seeds longitudinal, near the suture; last fold of the funicle thickened into a lateral club-shaped aril, with very small folds below it.—Hook Ic. Pl. t. 170; Paxt. Mag. xi. 113, with a fig.; *A. scapuliformis,* A. Cunn. in G. Don, Gen. Syst. ii. 405 (specimens with rather broader and shorter phyllodia).

N. S. Wales. Rocky ridges and brushy forest ground, Hunter's and Dumaresq rivers, *A. Cunningham;* sources of the M'Intyre, on the borders of Queensland, *Herb. F. Mueller.*

A. glaucifolia, Baum. Cat. Hort. Bollv., from Meissner's description, Pl. Preiss. i. 14, must be the short-leaved form of *A. cultriformis.* In Pl. Preiss. ii. 202, a specimen in Herb. Lehmann, is mentioned as believed to have been gathered in West Australia, but this must be a mistake, at any rate I know of no Western species at all like it.

161. **A. pravissima,** *F. Muell. Fragm.* i. 5, *and Pl. Vict.* ii. 5, *t.* 24. A tall shrub or small tree, quite glabrous; branchlets angular, slender. Phyllodia numerous, broadly and very obliquely falcate obovate or almost trapezoid, recurved, 3 to 5 lines or rarely ½ in. long, the lower terminal angle acute or mucronate, the upper one rounded as in the *Triangulares,* but the gland not placed at the angle, but much below it, rather thick and usually 2-nerved. Racemes much longer than the leaves, with 10 to 20 small globular flower-heads, the whole forming a handsome terminal leafy panicle. Flowers about 8 to 12 in the head, mostly 5-merous. Calyx small, obtusely lobed. Petals smooth. Pod flat, glabrous. Seeds longitudinal along the centre of the pod;

funicle very shortly thickened into a small fleshy aril under the seed, with several small folds below it.

Victoria. Valleys of the Australian Alps and banks of streams descending from them, foot of Buffalo ranges, Snowy and Macalister rivers, etc., *F. Mueller.* The affinity of this species is evidently with *A. cultriformis,* although the phyllodia are nearly those of the *Triangulares.*

162. **A. pyrifolia,** *DC. Mem. Leg.* 447, *and Prod.* ii. 452. A glabrous shrub, often glaucous, with slightly flattened branches. Phyllodia broadly ovate, with a pungent point, 2 to 3 in. long and almost as wide in perfect specimens, but often not half that size, thinly coriaceous, 1-nerved, penniveined. Stipules spinescent. Racemes numerous, often more than twice as long as the phyllodia, with 10 to 12 or more rather small but dense globular heads of very numerous flowers, often above 100, mostly 5-merous. Sepals distinct, linear-clavate. Pod flat, more or less curved or circinate, 2 to 3 in. long when perfect, 4 to 6 lines broad, often slightly contracted between the seeds, with rather thick margins; valves convex over the seeds. Seeds ovate, rather thick; funicle thickened and much folded under them.— F. Muell. Fragm. iii. 17.

N. Australia. Dampier's Archipelago, *A. Cunningham;* Nichol Bay, *F. Gregory's Expedition.*

W. Australia. Sharks Bay, *Baudin's Expedition.*

The large phyllodia and spinescent stipules distinguish this from all others with long racemes of globular heads.

163. **A. myrtifolia,** *Willd. Spec. Pl.* iv. 1054. A tall glabrous shrub, slightly glaucous in some varieties; branches acutely angular. Phyllodia oblique or falcate, obovate, ovate-lanceolate, lanceolate or linear, usually acute or mucronate and narrowed at the base, 1 to 2 in. long or much longer when narrow, coriaceous, 1-nerved, with thickened nerve-like margins, the pinnate veins rarely conspicuous, the marginal gland below the middle. Racemes rarely exceeding the phyllodia, with several almost sessile flower-heads, consisting of only 2, 3 or 4 rather large flowers, almost always 4-merous. Calyx very short, broadly lobed. Petals smooth, separating nearly to the base. Pod linear, curved, flattened, but thick, with very thick margins, usually 1 to 2 in. long, about 2 lines broad; valves hard and almost woody. Seeds oblong, longitudinal; funicle very short, scarcely folded, thickened nearly from the base into an almost cup-shaped fleshy aril.—*Mimosa myrtifolia,* Sm. in Trans Linn. Soc. i. 252, and Bot. Nov. Holl. 51, t. 15; *A. myrtifolia,* DC. Prod. ii. 452; Sw. Fl. Austr. t. 49; Lodd. Bot. Cab. t. 772; Hook. f. Fl. Tasm. i. 107; Meissn. in Pl. Preiss. i. 14; F. Muell. Pl. Vict. ii. 19.

N. S. Wales. Port Jackson to the Blue Mountains, *R. Brown, Sieber, n.* 437, *and Fl. Mixt. n.* 602, and others. and southward to Twofold Bay, *F. Mueller.*

Victoria. Frequent in barren places, as well in the lowlands as in the mountains throughout the colony, *F. Mueller.*

Tasmania. Port Dalrymple, *R. Brown;* abundant in dry soil throughout the colony, *J. D. Hooker.*

S. Australia. Rocky soils in the hill land, *Behr;* Encounter Bay, *Whittaker;* Mount Torrens, *F. Mueller.*

W. Australia. King George's Sound and to the eastward, *R. Brown, Preiss, n.* 927, and others.

The three following forms, distinguished by the breadth of the phyllodia, are usually considered as species, but they all pass into each other in W. Australia.

a. *celastrifolia.* Phyllodia mostly 1¼ to 2 in. long and often 1 in. broad.—*A. celastrifolia,* Benth. in Hook. Lond. Journ. i. 349 ; Bot. Mag. t. 4306 ; Meissn. in Pl. Preiss. i. 14 ; *A. Pawlikowskyana,* Ohlend. in Neue Allgem. Gartenzeit, 1845, 369, and *A. Ludwigii,* Ohlend. Verz. 1844, 74, according to Seem. Eingef. Acac. 28.—Swan River, *Drummond, 1st Coll. n.* 281 ; Grantham district, W. Australia, *Preiss, n.* 915, 916.

b. *normalis.* Phyllodia mostly 1 to 2 in. long and about ½ in. broad. The common form in the Eastern and Southern colonies.

c. *angustifolia.* Phyllodia mostly 2 to 4 in. long, 2 to 4 lines broad.—*A. marginata,* R. Br. in Ait. Hort. Kew. ed. 3, v. 462 ; Wendl. Comm. Acac. 19, t. 5 A ; DC. Prod. ii. 452 ; Meissn. in Pl. Preiss. i. 14 ; *A. trigona,* A. DC. Not. 8. Pl. Rar. Jard. Gen. 20.— To this belong the great majority, but not all of the King George's Sound specimens, occurring very rarely in the other colonies.

SERIES VII. PLURINERVES.—Phyllodia vertically flattened, obtuse or with an innocuous or recurved point (rarely pungent when the phyllodium is broad), with 2 or or more longitudinal nerves. Flowers in globular heads on axillary peduncles, either solitary or clustered or shortly racemose.

A. ARMATÆ.—Stipules spinescent. Phyllodia falcate, subulate-acuminate or almost pungent.

These correspond to the subseries *Armatæ* of *Uninerves,* except that the phyllodia are larger.

164. **A. scalpelliformis,** *Meissn. in Pl. Preiss.* ii. 200. A tall shrub, glabrous or the striate branches pubescent. Phyllodia obliquely triangular-lanceolate, acute and pungent-pointed, ¾ to 1½ in. long, 3 to 5 lines broad, tapering at the base, 2-nerved, penniveined, with nerve-like margins, with a gland on the prominent angle of the upper edge. Stipules setaceous-spinescent, spreading. Peduncles solitary, bearing each a globular flower-head. Flowers not seen perfect, but the withered remains have a broad short lobed calyx and 5 petals separating almost to the base. Pod very long, scarcely 1¼ lines broad, readily twisting, with broad very thick obtusely dilated sutures. Seeds oval-oblong, longitudinal, the last 2 or 3 folds of the funicle thickened into a small aril under the seed.

W. Australia, *Drummond, 2nd Coll. n.* 161. Differing from *A. urophylla* chiefly in the phyllodia.

165. **A. urophylla,** *Benth. in Bot. Reg.* 1841, *Misc.* 24, *and in Hook. Lond. Journ.* i. 329. A shrub of several ft., glabrous or slightly hirsute ; branches angular striate. Phyllodia semiovate or broadly and very obliquely ovate-lanceolate, subulate, acuminate, undulate, 1½ to 3 in. long and ¾ to 1 in. broad below the middle, in luxuriant cultivated specimens often twice that size, 2- to 4-nerved, the lower margin nearly straight, the upper one much curved and crenulate, the marginal gland near the base. Stipules setaceous-spinescent. Peduncles 2 to 4 lines long, solitary clustered or forming very short racemes, bearing each a small globular head of 8 to 12 flowers, either 5-merous or 4-merous. Calyx short, truncate. Petals smooth. Pod often 5 or 6 in. long, scarcely 1½ lines broad, readily twisting, with very thick obtusely dilated sutures. Seeds oval-oblong, longitudinal, the last 2 or 3 folds of the funicle thickened into a small aril under the seed.—Bot. Mag. t. 4573 ; Meissn. in Pl. Preiss. i. 8 ; *A. smilacifolia,* Field. and Gardn. Sert. Pl. t. 3.

W. Australia. King George's Sound, *Menzies ;* Swan River and Darling Range, *Drummond, 2nd Coll. n.* 282, *Preiss, n.* 913, 918 ; Harvey and Gordon rivers, *Oldfield.*

B. TRIANGULARES.—Stipules setaceous or minute, not spinescent. Phyllodia small (under ½ in. long), broadly falcate-ovate or triangular, with small points often pungent.

These correspond to the subseries *Triangulares* of *Uninerves*, with which I had formerly united them as an independent series.

166. **A. sublanata,** *Benth. in Hueg. Enum.* 42. A rigid shrub, woollypubescent when young, at length glabrous ; branches striate. Phyllodia from broadly triangular-falcate and mucronate to lanceolate-falcate and tapering into a pungent point, rarely ½ in. long, occasionally approaching in form those of *A. decipiens,* but the upper angle much more obtuse, without any or with a very minute gland, and usually much narrower and always with 3 to 5 or even more nerves. Stipules small, deciduous. Peduncles longer than the phyllodia, bearing each a globular head of 20 to 30 flowers, mostly 5-merous, glabrous or slightly hirsute, the bract-points projecting beyond the young buds. Calyx shortly toothed, half as long as the corolla. Pod linear, much twisted, about 2 lines broad.—*A. pravifolia,* F, Muell. Fragm. i. 4, and in Journ. Linn. Soc. iii. 117.

S. Australia. S. coast, *R. Brown ;* Crystal Brook, Flinders and Elders Ranges, *F. Mueller* (good specimens only in *Herb. R. Br.*). Some young specimens of Oldfield's from Port Gregory in W. Australia are also like this species, but insufficient for determination.

167. **A. amblygona,** *A. Cunn. ; Benth. in Hook. Lond. Journ.* i. 332. Branches elongated, diffuse, rather rigid, terete, pubescent. Phyllodia falcate-lanceolate or almost triangular, 3 to 4 lines long, 1½ to 3 lines broad, several-nerved, the lower nerve produced into a sharp point, the upper margin much curved but without any glandular angle. Peduncles rarely exceeding the phyllodia, bearing each a globular head of about 10 to 15 flowers, mostly 5-merous. Sepals broadly cuneate, cohering or at length separating, fully half as long as the corolla. Petals smooth, readily separating. Pod linear, usually curved, 1½ to 2 lines broad, contracted between the seeds. Seeds ovate, longitudinal; funicle thickened into a lateral oblong or club-shaped aril, with a short fold below it.—*A. Nernstii,* F. Muell. Fragm. iv. 3.

Queensland. Brisbane river, Moreton Bay, *A. Cunningham, C. Stuart, Nernst ;* towards Mount Pluto, *Mitchell ;* between Suttor river and Peak Range, *F. Mueller.*

N. S. Wales. Highlands west of Macquarrie river, *Fraser ;* rocky hills on Wellington and Lachlan rivers, *A. Cunningham.*

168. **A. deltoidea,** *A. Cunn. in G. Don, Gen. Syst.* ii. 401. A muchbranched shrub, glabrous or pubescent when young, with terete branches. Phyllodia numerous, obliquely triangular-ovate, acute, mucronate, 2 to 3 lines long and almost as broad, thick, several-nerved, the upper margin much curved but without any gland-bearing angle. Stipules setaceous, persistent. Peduncles slightly exceeding the phyllodia, bearing each a small globular head of numerous (30 to 50) very small flowers, mostly 5-merous. Calyx deeply divided into narrow thin lobes. Petals slightly thickened and pubescent at the tips, cohering above the middle. Stamens few. Pod elongated, slightly curved, flat, coriaceous, glandular-hispid, nearly 4 lines broad. Seeds not seen.

N. Australia. Greville Island, Montague Sound, and Barren Islands, Regent's Inlet, N.W. coast, *A. Cunningham.*

169. **A. stipulosa,** *F. Muell. in Journ. Linn. Soc.* iii. 119. Glandular-pubescent or hirsute; branches terete. Phyllodia obliquely ovate-falcate or almost triangular, tapering into a pungent point, 2 to 3 or rarely 4 lines long and almost as broad, thick and faintly several-nerved, the upper angle very obtuse, with a small gland about the middle or often none. Stipules setaceous, spreading. Peduncles usually exceeding the phyllodia, pubescent, bearing each a dense head of numerous (above 30) flowers, mostly 5-merous but sometimes 6- to 8-merous. Calyx thin, striate, pubescent, as long as the corolla, at length separating into distinct sepals. Petals narrow, cohering. Stamens very numerous. Pod linear, curved, flat, glandular-hispid, 2 to 3 lines broad, scarcely contracted between the seeds. Seeds very oblique; funicle with the last 3 folds dilated into an almost membranous aril at the base of the seed.

N. Australia. Upper Victoria river and Sturt's Creek, *F. Mueller.* Very near *A. deltoidea,* differing chiefly in the flowers twice as large, and in the proportion of the calyx and corolla.

C. BREVIFOLIÆ.—Phyllodia under 1 in. long, obovate ovate or broadly oblong, very obtuse, often undulate. Stipules inconspicuous.

170. **A. loxophylla,** *Benth. in Linnæa,* xxvi. 622. A diffuse shrub, very resinous, but otherwise glabrous. Phyllodia numerous, very obliquely obovate or oblong, incurved, very obtuse, but often minutely mucronulate, in some specimens not exceeding $\frac{1}{4}$ in., in others narrower and $\frac{1}{4}$ to $\frac{1}{2}$ in. long, rather thick, with 2 faint nerves, otherwise veinless. Stipules obsolete. Peduncles 1 to 2 lines long, bearing each a globular head of above 20 flowers, mostly 5-merous. Calyx turbinate, rather thick, above half as long as the corolla, shortly and broadly lobed and ciliate. Petals smooth, narrow, free or readily separating. Pod curved, linear, about 2 lines broad; valves convex over the seeds, flat between them. Seeds ovate, longitudinal; funicle dilated almost from the base into a cup-shaped almost membranous aril of 2 or 3 folds.

W. Australia. Towards Cape Riche, *Drummond, 5th Coll. n.* 14 (with short broad phyllodia), *Maxwell* (with longer narrower phyllodia).

171. **A. setulifera,** *Benth. in Linnæa,* xxvi. 625. Apparently diffuse, with the aspect almost of *A. Wickhami,* quite glabrous. Phyllodia numerous, obliquely ovate, undulate, obtuse with a very oblique bristle-like point, rarely above 2 lines long and broad, rigid, with several very faint nerves, of which 2 or 3 are rather more prominent. Stipules deciduous. Peduncles not exceeding the phyllodia, rigid, bearing each a small globular dense head of about 12 to 20 flowers, mostly 5-merous. Calyx thin, toothed, readily separating into distinct sepals. Petals rather thick. Pod not seen.

N. Australia. N.W. coast, *Bynoe.*

172. **A. translucens,** *A. Cunn. in Hook. Ic. Pl. t.* 160. A bush shrub or small tree, glabrous or the young shoots pubescent; branchlets terete or slightly angular. Phyllodia from obliquely obovate to narrow-oblong or almost linear, incurved, usually much undulate, mostly about $\frac{1}{2}$ in. but the lower ones sometimes 1 in. long, obtuse with an oblique or recurved terminal gland, coriaceous, obscurely several-nerved. Peduncles rigid, $\frac{1}{2}$ to 1 in. long, bearing each a dense globular head of numerous flowers, mostly 5-merous.

Calyx broad, eup-shaped, shortly toothed, scarcely half as long as the corolla. Petals striate, but smaller and less rigid than in *A. impressa.* Pod 1 to 1½ in. long, flat but thick and almost woody, obtusely hooked at the end, about 2 lines broad above the middle and gradually narrowed into a long stipes, obliquely veined, partitioned inside between the seeds. Seeds oblong, oblique; funicle slightly folded and gradually dilated into a cup-shaped aril at the base of the seed.

N. Australia. Montague Sound and Bay of Rest, N.W. coast, *A. Cunningham;* Sturt's Creek, *F. Mueller;* Islands of the Gulf of Carpentaria, *R. Brown, Henne.* The pod is that of some *Juliflora,* but has only been seen in the narrow-leaved specimens.

173. **A. impressa,** *F. Muell. in Journ. Linn. Soc.* iii. 133. A tall shrub, the branches slightly angular, more or less pubescent as well as the foliage. Phyllodia obovate or obovate-oblong, very oblique, undulate, obtuse or with a small glandular point, ½ to ¾ or rarely 1 in. long, with 3 to 5 nerves more prominent than in *A. translucens,* and anastomosing veins. Peduncles about as long as the phyllodia, bearing each a globular head of about 12 to 20 flowers, mostly 5-merous. Sepals distinct or slightly coherent, spathulate. Petals rigid and striate, united at the base. Pod straight or nearly so, thinly coriaceous, flat with thickened margins, about ½ in. broad, very glutinous and villous. Seeds oval-oblong, transverse, the central area much depressed; funicle forming several folds, the last 2 or 3 dilated into a rather small aril at the base of the seed.

N. Australia. Sturt's Creek, Victoria and Van Alphen rivers, *F. Mueller;* Short's Range, *M'Douall Stuart.* Very near *A. translucens* in foliage and flowers, but with a very different fruit.

D. OLIGONEURÆ.—Phyllodia above ½ in. and mostly above 1 in. long, oblong lanceolate or linear, straight or slightly falcate, with 2 or 3 nerves faintly or not at all veined between them and usually not glutinous.

Although this subseries is generally distinct from the next two by the phyllodia with only 2 or 3 prominent nerves, yet the *A. Simsii* is very variable in this respect, connecting the *Oligoneuræ* with the *Nervosæ,* and *A. subporosa,* a very faintly nerved species, has the resinous foliage of several of the *Nervosæ.*

174. **A. bivénosa,** *DC. Prod.* ii. 452. A large bushy shrub, glabrous and often glaucous, or the upper leaves and inflorescence of a golden-yellow. Phyllodia from obovate to oblong-lanceolate, obtuse and mostly with a callous recurved point, 1½ to 3 in. long in the commonest form, 1 to 1½ in the short-leaved varieties, narrowed at the base, coriaceous, with 2 more or less prominent nerves, penniveined and with nerve-like margins. Racemes of few globular heads with a flexuose rhachis and short peduncles, or reduced to a single peduncle with a single head, of 20 to 30 flowers, not very small and mostly 5-merous. Calyx sinuate-toothed and petals smooth as in *A. salicina.* Pod elongated, nearly straight, flat, coriaceous, about 3 lines broad. Seed not seen perfect.—*A. binervosa,* DC. Mem. Leg. 448; *A. xanthina,* Benth. in Hook. Lond. Journ. i. 355; Meissn. in Pl. Preiss. i. 16; *A. elliptica,* A. Cunn.; Benth. in Hook. Lond. Journ. i. 347.

N. Australia. N.W. coast, Admiralty Bay, *Baudin's Expedition;* Bay of Rest and Dampier's Archipelago, *A. Cunningham;* Depuech Island, *Bynoe;* Hearson Island, Nichol Bay, *F. Gregory's Expedition.*

W. Australia. Swan River, *Baudin's Expedition, Drummond,* 1*st Coll. n.* 283, *Preiss, n.* 928; Sharks Bay, *Milne;* Dirk Hartog's Island, *A. Cunningham.* There are two forms, one (*A. elliptica,* A. Cunn.) more glaucous with short obovate oblong phyllodia faintly 2-nerved, the other (*A. xanthina,* Benth.) with longer phyllodia more prominently nerved and often of a golden colour. Both are in the Paris herbarium, from Baudin's Collection, and we have both from the N.W. coast, as well as from the west, the yellow one chiefly from.Swan River; and, different as they look, some specimens are quite intermediate or combine the two.

175. **A. trineura,** *F. Muell. Pl. Vict.* ii. 25, *and Fragm.* iv. 5. A strongly scented bushy shrub, glabrous and glaucous; branchlets slightly angular. Phyllodia narrow, cuneate-oblong, straight or slightly curved, very obtuse, mostly 1½ to 2 in. long, 3 to 4 lines broad, narrowed at the base, with 3 prominent nerves and a few oblique veins. Flower-heads small, globular, in very short racemes of 3 to 6. Flowers above 20 in the head, very closely packed, glutinous and mostly 5-merous. Sepals linear-spathulate. Petals smooth, usually free. Pod unknown.

Victoria. Sandy banks of the Wimmera, *Dallachy;* N.W. desert, *Lockhart Morton.*

176. **A. nitidula,** *Benth.* A low and diffuse or erect and bushy glabrous shrub, occasionally slightly glutinous; branchlets terete or nearly so. Phyllodia linear-cuneate, obtuse or with a minute callous point, mostly ½ to 1½ in. long, rather thick, rigid and prominently 2- or 3-nerved. Peduncles slender, mostly in pairs, 2 to 4 lines long, bearing each a small globular head of 12 to 20 flowers, mostly 5-merous. Sepals free, narrow, linear-spathulate, ciliate. Petals smooth, distinct from the base or nearly so. Pod not seen.

W. Australia, *Drummond,* 3*rd Coll. n.* 128 (phyllodia 1 to 1½ in.); along granite boulders, Goose Island Bay, *R. Brown;* Cape Arid, *Maxwell* (phyllodia ½ to ¾ in.). The species approaches in some respects *A. dura,* but differs in the nearly straight phyllodia with at least 2 nerves, both distinct from the margin, more numerous flowers, etc.

177. **A. heteroclita,** *Meissn. in Pl. Preiss.* i. 318. An erect shrub of 3 or 4 ft., glabrous or the young shoots minutely silky-pubescent; branchlets more or less angular. Phyllodia from linear-lanceolate, often oblique and 1½ to 2 in. long, to linear and exceeding 3 in., rather rigid, tapering into a straight or recurved point, but scarcely pungent, narrowed at the base, with 3 fine but prominent nearly equal nerves, and veinless or nearly so between them. Peduncles solitary or in pairs, bearing each a globular head of above 20 flowers, mostly 5-merous. Sepals narrow, linear-spathulate. Petals smooth, cohering to the middle. Pod narrow-linear, flat, with thickened nerve-like margins, usually straight, 2 to 3 in. long, 1½ lines broad. Seeds longitudinal; funicle with the last fold dilated into a short fleshy oblique lateral aril and short filiform folds below it.—*A. trissoneura,* F. Muell. Fragm. iv. 6.

W. Australia. Towards Cape Riche, *Preiss, n.* 938, *Drummond,* 5*th Coll. n.* 8 (with narrow phyllodia), *n.* 11 (with broad phyllodia); Fitzgerald Range, *Maxwell;* Swan River, *Oldfield* (phyllodia short and broad), *Drummond* (1*st Coll. ?*), *n.* 288. The latter specimens, intermediate between the narrow and broad forms, have been mixed up in some sets with specimens of *A. cochlearis,* from which this species differs in the phyllodia less rigid, with more prominent nerves and more abruptly contracted into a less pungent point.

178. **A. elongata,** *Sieb. in DC. Prod.* ii. 451. A tall shrub, glabrous or the young shoots silky-pubescent. Phyllodia narrow-linear, obtuse or

with a small oblique or hooked point, 2 to 3 or even 4 in. long, 1 to 1½ or rarely above 2 lines broad, with 3 prominent nerves, and, when broad, a few oblique veins between them. Peduncles solitary or in pairs, not exceeding ½ in. and usually much shorter, slender and pubescent, bearing a globular head of numerous (30 or more) flowers, mostly 5-merous. Calyx obtusely lobed, fully half as long as the corolla. Petals smooth, with the midrib slightly prominent. Pod linear, straight, flat, 1½ to 2 lines broad. Seeds longitudinal, the last folds of the short funicle dilated into a small aril at the base of the seed.—Bot. Mag. t. 3337 ; F. Muell. Pl. Vict. ii. 24 ; *A. hebecephala,* A. Cunn. in Lond. Hort. Brit. 406.

N. S. Wales. Port Jackson to the Blue Mountains, *R. Brown, Sieber, n.* 443, *and Fl. Mixt. n.* 598, *A. Cunningham,* and others ; southward to Illawarra, *A. Cunningham,* and Twofold Bay, *Huegel.*
Victoria. Granite ridges near the Genoa river, *F. Mueller.*
This species differs from *A. trinervata* chiefly in the long narrow phyllodia, not pungent, when very narrow they are almost like those of the *Calamiformes.* In some garden specimens the nerves almost disappear as represented in the above quoted figure ; and then it is not very easy to distinguish them from those specimens of *A. viscidula* in which the nerves are very faint.

179. **A. subporosa,** *F. Muell. Pl. Vict.* ii. 24, *and Fragm.* iv. 5. A tree attaining the height of 40 ft. ; branchlets slender, viscid when young, scarcely angular. Phyllodia linear-lanceolate or linear, often slightly falcate, acute with the point usually incurved, 1½ to 3 or rarely 4 in. long, from 1 line broad in the narrow, to 3 or 4 lines in the broad variety, with 2 or 3 slightly prominent nerves and when broad a few faint ones between them, and often marked with semitransparent glandular dots. Peduncles solitary or in pairs, slender, ½ in. long or shorter, bearing a globular head of 20 to 30 flowers, mostly 5-merous. Calyx lobed, not half so long as the corolla. Petals smooth. Pod unknown.

N. S. Wales. Near Mount Imlay, Twofold Bay, *F. Mueller.*
Victoria. Forest gullies on the barren range at the eastern boundary of the colony, *F. Mueller.*
Var. *linearis.* Tall shrub. Phyllodia narrow-linear. Flower-heads much smaller, with 12 to 20 smaller flowers.—With the broad-leaved variety, *F. Mueller.*

180. **A. Simsii,** *A. Cunn.; Benth. in Hook. Lond. Journ.* i. 368. A tall glabrous shrub with slender branches scarcely angular. Phyllodia linear or lanceolate, usually falcate, obtuse or mucronate, much narrowed towards the base, 2 to 5 in. long, 1½ to 2 or rarely 3 lines broad, in the ordinary form with 3 or sometimes only 2 prominent nerves, smooth and finely veined between them. Peduncles solitary or in pairs, slender, rarely above 3 lines long, bearing each a globular head of 20 to 30 flowers, mostly 5-merous. Calyx nearly half as long as the corolla, lobed and readily separating into distinct sepals. Petals smooth, distinct. Pod straight, flat or undulate, acuminate, 2 to 3 lines broad. Seeds compressed-globular, longitudinal ; funicle with the last fold shortly thickened into a clavate aril about half as long as the seed, with a short filiform fold below it.

N. Australia. Islands of the Gulf of Carpentaria, *R. Brown, Henne.*
Queensland. Bay of Inlets, *Banks and Solander ;* rocky hills, Cleveland Bay, *A. Cunningham ;* Cape Upstart, *Burdekin Expedition ;* Port Denison, *Fitzalan ;* Edgecombe Bay, *Dallachy ;* bed of the Belyando, *Mitchell,* also in *Leichhardt's* Collection.

Var. *multisiliqua.* Phyllodia shorter, rather broader, and nearly straight, with 3 prominent nerves and scarcely veined between them.   Pod narrow.—On first seeing R. Brown's
specimens from the Carpentaria Islands, they looked so different from the ordinary form that
I was inclined to adopt them as a distinct species under Brown's name of *A. multisiliqua,*
but I have since found that the two forms pass one into the other there as in other localities,
and both phyllodia may be found even on the same specimen on different branches, depending perhaps sometimes on differences in comparative luxuriance.

E. MICRONEURÆ.—Glabrous or glaucous, not glutinous. Phyllodia thick,
apparently veinless or with very fine scarcely prominent parallel veins or
nerves, often scarcely visible without a lens, narrow or rarely short and
obovate.

181. **A. leptospermoides,** *Benth. in Linnæa,* xxvi. 626.   A low
much-branched glabrous shrub, with slender terete branches.   Phyllodia numerous, cuneate-oblong, obtuse, mostly ½ in. long or rather more, narrowed at
the base on a short petiole, thick, apparently veinless, the few very fine nerves
very rarely conspicuous except towards the base and under a lens.   Peduncles
solitary, about 2 lines long, bearing each a small globular head of above 20
flowers, mostly 5-merous.   Sepals free, narrow-spathulate, ciliate.   Petals
smooth, free, rather narrow.   Pod unknown.

W. Australia, *Drummond, 4th Coll. n.* 11 ; East Mount Barren, *Maxwell (Herb.
Oldfield).*   The narrow-leaved specimens bear some resemblance to *A. ericifolia* and the
broader ones to *A. spathulata,* but have a different venation.

182. **A. homalophylla,** *A. Cunn.; Benth. in Hook. Lond. Journ.* i.
365 (there spelt *omalophylla*).   A small tree, glabrous or the foliage minutely
hoary or pale ; branchlets at first slightly angular.   Phyllodia lanceolate-falcate, narrow-oblong or linear, obtuse with a small oblique point, narrowed at
the base, 1 to 3 in. long, 1 to 4 lines broad, thick, very finely striate with
parallel nerves only to be seen under a lens.   Peduncles in pairs or clustered
on a very short common peduncle, bearing dense globular heads of numerous
flowers, mostly 5-merous.   Sepals cuneate or spathulate, free or slightly connate, more than half as long as the corolla.   Petals smooth, free.   Pod linear,
usually glaucous, slightly curved, 2 to 3 lines broad, longitudinally veined ;
valves coriaceous, convex over the seeds, contracted between them.   Seeds
oval-oblong, longitudinal ; funicle short, much folded and dilated almost from
the base into a short oblique aril.—F. Muell. Pl. Vict. ii. 28.

N. S. Wales.   Abundant on the barren heaths of the interior, from the Lachlan to
the Barrier Range ; one of the spear-woods of the natives, *A. Cunningham, Victorian Expedition,* and others.

Victoria.   Salt-bush flats on the Murray, yielding the hard dark and fragrant ' Myall-
wood,' *F. Mueller.*

S. Australia ?   Spencer's Gulf, *Wilhelmi.*   A single specimen in leaf only, and therefore doubtful.

183. **A. pendula,** *A. Cunn. in G. Don, Gen. Syst.* ii. 404.   A handsome
tree, the foliage pale or ash-coloured, with a minute pubescence ; branchlets
usually pendulous, slightly angular, soon terete.   Phyllodia linear-lanceolate,
falcate, acuminate, narrowed towards the base, 2 to 3 in. long, rigidly coriaceous, very finely striate, with numerous parallel nerves, only to be seen under
a lens.   Peduncles usually clustered on a very short common peduncle, rarely
above 2 lines long, bearing each a small globular head of about 12 to 20 flowers,

mostly 5-merous, much smaller than in *A. homalophylla.* Calyx turbinate and lobed, but readily separating into distinct sepals. Petals smooth. Pod linear, but very flat, and fully 5 lines broad, thinly coriaceous, transversely reticulate, the sutures bordered by a very narrow wing. Seeds nearly orbicular ; funicle thickened into a narrow clavate aril, and scarcely folded below it, but not seen perfect.—*A. leucophylla,* Lindl. in Mitch. Three Exped. ii. 13.

Queensland. On the Maranoa, *Mitchell.*

N. S. Wales. Morasses of the Lachlan, *A. Cunningham ;* the only timber-tree in these immense morasses, *Fraser.*

Without the fruit the specimens are very difficult to distinguish from those of *A. homalophylla.* In both species, but especially in this one, 3 of the nerves of the phyllodia are sometimes slightly prominent.

184. **A. Oswaldi,** *F. Muell. Pl. Vict.* ii. 27, *and Fragm.* iv. 5. A rigid bushy shrub, attaining 8 to 10 ft., glabrous or the young shoots hoary or silky-pubescent ; branchlets slightly angular. Phyllodia falcate-lanceolate, varying to linear or oblong-lanceolate, mostly 1½ to 2 in. long, rigid, with a short usually incurved innocuous or scarcely pungent point, much narrowed at the base, with numerous slightly prominent nerves, parallel or anastomosing when the phyllodium is broad. Flower-heads small, globular, sessile or nearly so, solitary or in pairs or clusters, containing about 10 to 15 flowers, mostly 5-merous. Sepals linear-cuneate or spathulate, free. Petals smooth, usually pubescent. Pod long and much curved or twisted, 3 to 4 lines broad, hard or almost woody ; valves convex over the seeds, slightly contracted between them. Seeds large, ovate, longitudinal ; last fold of the funicle dilated into a broad, obliquely cup-shaped, fleshy aril, the lower folds short and filiform or slightly dilated.

Queensland? Towards Broad Sound, a small specimen in *Herb. F. Mueller,* without collector's name.

N. S. Wales. Interior desert, from the Lachlan and Darling to the Barrier range and Cooper's Creek, *Victorian Expedition,* and others.

Victoria. Desert on the Murray, from the Murrumbidgee to the W. frontier, *F. Mueller.*

S. Australia. S. coast, *R. Brown ;* from the Murray desert to St. Vincent's Gulf, *F. Mueller.*

Var. *abbreviata.* Phyllodia rigid, about 1 in. long, almost like those of *A. lanigera.*— S. coast, *R. Brown.*

185. **A. lineolata,** *Benth. in Linnæa,* xxvi. 626. Young shoots silky-pubescent, but soon glabrous. Phyllodia numerous, narrow-linear, erect, mucronulate, mostly 1½ to 2 in. long, 1 to 2 lines broad, rigid, finely striate with numerous parallel nerves, scarcely visible without a lens, 3 occasionally rather more prominent. Peduncles usually in pairs, recurved, very short, bearing each a globular head of numerous flowers, mostly 5-merous. Sepals narrow-spathulate, free or shortly united at the base, half as long as the corolla. Petals smooth. Pod unknown.

W. Australia, *Drummond, 4th Coll. n.* 12 *and* 13. Very nearly allied to the Eastern *A. homalophylla,* as well as to *A. leptoneura* among the *Calamiformes,* and *A. microneura* among the *Julifloræ.*

A specimen from Murchison river, *Oldfield,* has broader phyllodia and fewer and larger flower-heads, with the sepals connate, but is insufficient to determine whether it be a species or variety.

Acacia.]

186. **A. coriacea,** *DC. Mem. Leg.* 446, *and Prod.* ii. 451. Ashy-grey, with the young shoots silky-hoary or almost golden; branchlets terete. Phyllodia long-linear, straight or curved, obtuse, narrowed towards the base, often ½ ft. long or more, 1 to 2½ lines wide, thickly coriaceous, with numerous fine and closely packed longitudinal nerves, only visible under a lens. Peduncles usually in pairs, ¼ to ½ in. long, bearing each a globular head of 20 to 25 flowers, mostly 5-merous, hoary-pubescent in the bud. Calyx ¾ line long, tubular, with ciliate lobes. Petals rather longer, united above the middle. Pod 6 to 9 in. long, moniliform; valves coriaceous, very convex, 4 to 5 lines broad, oblong and striate over the seeds, much contracted between then. Seeds longitudinal, distant; funicle folded and dilated under the seed, but not seen perfect.

N. Australia. Bay of Rest, N.W. coast, *A. Cunningham;* Depuech island, *Bynoe;* Nichol Bay, *F. Gregory's Expedition.*
W. Australia. Sharks Bay, *Baudin's Expedition;* Dirk Hartog's Island and Sharks Bay, *Milne.*

187. **A. stenophylla,** *A. Cunn.; Benth. in Hook. Lond. Journ.* i. 366. A very hard-wooded tree, quite glabrous, with angular branchlets. Phyllodia long-linear, acuminate or falcate, much narrowed at the base, 6 in. to 1 ft. long, about 2 to 2½ lines broad, thinly coriaceous, not at all hoary, finely striate, with numerous prominent parallel nerves. Peduncles under ½ in. long, usually in short racemes of 3 to 6, but sometimes solitary, bearing each a globular head of 20 to 30 or more flowers, mostly 5-merous. Calyx half as long as the corolla, with short broad densely ciliate lobes. Petals pubescent. Pod long, moniliform; valves coriaceous, 4 to 5 lines broad and convex over the seeds, but not striate, much narrowed between them. Seeds ovate, longitudinal; funicle in short folds, the last slightly thickened into a small aril.— F. Muell. Pl. Vict. ii. 26.

N. Australia. Hooker's and Sturt's Creeks, *F. Mueller.*
Queensland. Maranoa and Narran rivers, *Mitchell.*
N. S. Wales. Lachlan river, *A. Cunningham;* thence to the Darling river, Barrier range, and Cooper's Creek, *Victorian Expedition,* etc.
Victoria. Banks of the Murray, *F. Mueller.*
S. Australia. Murray desert, *F. Mueller.*
A. sericophylla, F. Muell. in Journ. Linn. Soc. iii. 122, is probably a narrow-leaved form of this species.

F. NERVOSÆ.—Foliage often viscid, occasionally glaucous, rarely hoary or pubescent. Phyllodia straight or sometimes falcate, coriaceous or thin, with several prominent nerves and, when broad, reticulate between them, the nerves rarely reduced to three when the phyllodium is narrow.

The nerves are always either more numerous and nearer together than in the *Oligoneuræ,* or the reticulations between them numerous and prominent.

188. **A. hemignosta,** *F. Muell. in Journ. Linn. Soc.* iii. 134. A tall shrub or small tree, more or less glaucous or pale; branchlets slender, slightly striate. Phyllodia falcate-lanceolate or oblong, obtuse, much narrowed towards the base, 2 to 4 in. long and often ½ to ¾ in. broad above the middle, thinly coriaceous, with 3 or sometimes 4 or 5 fine but slightly prominent nerves and more or less reticulate between them. Peduncles slender, 3 to 5 lines long, solitary or in terminal racemes through the abortion of the upper

phyllodia, bearing each a small globular head of about 20 to 30 flowers, mostly 5-merous. Sepals very narrow, free. Petals smooth, glabrous. Pod flat, 2 to 4 in. long, about ½ in. wide; valves thin, the sutures bordered by an acute edge. Seeds orbicular; funicle oblique, not folded, filiform or very slightly thickened.

N. Australia. Cambridge Gulf, *A. Cunningham*; Albert, Victoria, Gilbert, and Roper rivers, *F. Mueller*.

189. **A. sclerophylla,** *Lindl. in Mitch. Three Exped.* ii. 139. A glabrous bushy shrub of several ft., not glutinous, according to F. Mueller, but with the strong resinous smell of the viscid species; branchlets terete or nearly so. Phyllodia oblong, linear or more or less cuneate, thick and rigid, striate with several prominent nerves, much narrowed at the base, usually about ¾ in. long, and from that to 1 or 1½ in. in the typical form. Peduncles exceedingly short, usually in pairs, bearing each a small globular head of about 12 to 20 flowers, mostly 5-merous. Sepals very thin, narrow linear-spathulate. Petals also thin and free. Pod (only seen in the long-leaved variety) linear, often about 2 in. long, not 1½ lines broad; valves convex, rather hard, longitudinally striate. Seeds oval-oblong, longitudinal, the last 2 folds of the funicle dilated into a small cup-shaped aril.—F. Muell. Pl. Vict. ii. 25; Dietr. Fl. Univers. N. Ser. t. 85 (incorrect as to nervation).

Victoria. In the Murray desert, *Mitchell, F. Mueller*.
S. Australia. From the Murray desert to Lake Alexandrina and St. Vincent's Gulf, also Spencer's Gulf, near Tumbay Bay, *F. Mueller*.
Var. *longifolia*. Phyllodia 1½ to 2 in. long. With the typical form.

190. **A. farinosa,** *Lindl. in Mitch. Three Exped.* ii. 146. A much-branched diffuse or bushy shrub, attaining several feet, glabrous or mealy-glaucous when young; branchlets terete or slightly angular. Phyllodia linear-cuneate, obtuse, much narrowed at the base, thick and rigid, ¾ to 1½ in. long, much like those of *A. sclerophylla*, but usually narrower, with much finer nerves. Peduncles very short, mealy-tomentose, usually in pairs, bearing each a small globular head of 12 to 20 or more flowers, mostly 5-merous. Sepals free, linear-spathulate. Petals smooth, free or readily separating. Pod linear, curved, about 2 lines broad or rather more; valves coriaceous and hard, not striate; margins thickened, slightly contracted between the seeds. Seeds ovoid, longitudinal; funicle short with the last folds dilated into a cup-shaped aril under the seed.

Victoria. On the Murray, *Mitchell*.
S. Australia. Murray desert to Spencer's Gulf, *F. Mueller, Wilhelmi*; Kangaroo Island, *Waterhouse*.
Very closely allied to *A. sclerophylla*, and perhaps a variety, chiefly distinguished by the nerves of the phyllodia scarcely prominent or quite inconspicuous, and thus connecting this subseries with the *Microneuræ*.

191. **A. Whanii,** *F. Muell. Herb.* Shrubby, with angular, minutely pubescent branchlets. Phyllodia oblong-linear, often incurved, very obtuse, with a minute point, 1½ to 2 in. long, much narrowed at the base, rigidly coriaceous, striate with several nerves, parallel but here and there anastomosing; upper margin slightly thickened, with usually an indented gland below the middle. Peduncles exceedingly short, in pairs or clusters, bearing

each a globular head of about 20 flowers, mostly 5-merous. Calyx rather short, thin, lobed and readily separating into distinct sepals. Petals smooth, united to near the middle. Pod unknown.

Victoria. Near Skipton, *W. Whan.* Very near *A. sclerophylla,* but the phyllodia broader, with the veins often anastomosing and the sepals and petals more united.

192. **A. heteroneura,** *Benth. in Linnæa,* xxvi. 624. A glabrous shrub, with angular branchlets. Phyllodia narrow-linear, obtuse or obliquely mucronate, much narrowed towards the base, 1½ to 3 in. long, 1 to 2 lines broad, rigid, with many parallel nerves, the central one very prominent, with 2 to 5 fine ones on each side. Peduncles slender, 2 to 3 lines long, mostly in pairs, bearing each a small globular head of from 12 to 20 flowers, 5-merous or occasionally 4-merous. Calyx short, very thin, with short broad lobes not ciliate in the bud. Petals smooth, with the midribs slightly prominent. Pod unknown.

W. Australia, *Drummond, 4th Coll. n.* 138.

193. **A. viscidula,** *A. Cunn.; Benth. in Hook. Lond. Journ.* i. 363. A shrub, more or less pubescent and resinous-viscid; branchlets terete or nearly so. Phyllodia narrow-linear, with a small usually hooked point, rather incurved, narrowed at the base, 2 to 2½ in. long, 1 to 1½ lines broad, coriaceous, several-nerved. Peduncles very short, generally in pairs, rarely clustered. Flowers numerous, in dense globular or slightly ovoid heads, mostly 5-merous, but often 4-merous. Bracts acuminate. Sepals narrow-spathulate, quite free or scarcely connected at the base. Petals pubescent. Pod linear, straight, acuminate, pubescent, about 2 lines broad; valves nearly flat, with thickened margins. Seeds oblong, longitudinal; funicle with the last 2 or 3 folds thickened into an obliquely cup-shaped or apparently 2-lobed aril at the base of the seed.

N. S. Wales. Banks of the Lachlan, *Fraser;* Blue Mountains?, *Caley in Herb. R. Br.*

Var. *angustifolia.* Phyllodia about ¾ line broad. Flower-heads smaller.

194. **A. ixiophylla,** *Benth. in Hook. Lond. Journ.* i. 364. A glabrous or pubescent glutinous shrub of several feet. Phyllodia oblong, lanceolate or broadly linear, usually oblique or falcate, obtuse or with a small callous recurved point or gland, ¾ to 1½ or rarely nearly 2 in. long, 2 to 3 or rarely 4 lines broad, coriaceous, striate, with numerous fine but prominent nerves, anastomosing when the phyllodium is broad. Peduncles in pairs on short racemes of 3 or 4, bearing each a small globular head of 15 to 20 or rarely more flowers, mostly 5-merous. Sepals narrow-spathulate, quite free. Petals free or readily separating. Pod (only known in the Western specimens) very flexuose, hispid or glabrous, 2 to 3 lines broad. Seeds oblong, longitudinal; funicle dilated into an obliquely oblong or club-shaped aril, not one-third as long as the seed, and very shortly filiform and folded below it.—*A. glutinosa,* F. Muell. Fragm. iv. 6 (the Western specimens).

Queensland. In the interior towards Mount Pluto, *Mitchell;* between Severn and Condamine rivers, *Leichhardt.*

N. S. Wales. N. of Liverpool plains, *A Cunningham.*

W. Australia, *Drummond (4th Coll. ?), n.* 129, and 5*th Coll. n.* 13, *Maxwell.* I am still unable to detect any difference between this Western plant and the Eastern *A. iriophylla*. Both the narrow and the broad-leaved forms occur both in the East and in the West.

195. **A. dictyophleba,** *F. Muell. Fragm.* iii. 128. Glabrous but very resinous; branchlets nearly terete. Phyllodia cuneate-oblong to lanceolate-falcate, very obtuse, with a small callous point, much narrowed at the base, 2 to 3 in. long, 2 to 5 lines broad, very coriaceous, with several nerves and intermediate reticulations, all much raised, and scabrous with a resinous exudation. Peduncles solitary, 6 to 8 lines long, bearing each a very dense globular head of 5-merous flowers. Calyx more than half as long as the corolla, very thin, with thickened resinous lobes or teeth. Pod unknown.

N. Australia. Mount Humphries, *M'Douall Stuart.*

196. **A. venulosa,** *Benth. in Hook. Lond. Journ.* i. 366. A tall shrub, softly pubescent or glabrous and sometimes slightly viscid ; branchlets angular. Phyllodia falcate-oblong or lanceolate, narrowed at each end, mostly 2 to 3 in. long, ¼ to ½ in. broad, very rigid, many-nerved and strongly veined, with about 3 nerves more prominent than the rest. Peduncles in pairs or clusters or on a very short common peduncle, mostly 2 to 4 lines long, rather thick, tomentose, bearing each a globular head of above 20 flowers, mostly 5-merous. Sepals narrow-spathulate, usually united below the middle. Petals smooth. Pod linear, straight or curved, pubescent when young, 1 to 2 in. long, about 3 lines broad ; valves convex over the seeds, depressed, but not contracted between them. Seeds ovate, longitudinal, the last 2 or 3 folds of the funicle thickened into a concave or 2-lobed aril under the seed.

N. S. Wales. Liverpool plains, *A. Cunningham ;* head of the Gwydir river, *Leichhardt ;* New England, *C. Stuart.*

197. **A. cyclopis,** *A. Cunn. in G. Don, Gen. Syst.* ii. 404. A shrub of 6 to 10 ft., usually glabrous, with angular branchlets. Phyllodia narrow-oblong, nearly straight, obtuse, narrowed at the base, 1½ to 3, rarely 4 in. long, rigidly coriaceous, with 3 to 5 nerves, and anastomosing almost longitudinal veins. Peduncles solitary or 2 or 3 in a short raceme, bearing each a dense globular head of numerous flowers, mostly 5-merous. Calyx turbinate, shortly lobed or toothed, more than half as long as the corolla. Petals smooth, free. Pod flat and 4 to 6 lines broad as in *A. melanoxylon,* but more coriaceous, curved or twisted. Seeds nearly orbicular ; funicle thickened and richly coloured from the base, encircling the seed in double folds.—Meissn. in Pl. Preiss. i. 18.

W. Australia. King George's Sound, *R. Brown, A. Cunninghan, Fraser,* and others, and eastward towards the Great Bight, *Maxwell ;* Swan River, *Toward, Drummond, Preiss, n.* 926, and others ; Preston river, *Oldfield.*

198. **A. melanoxylon,** *R. Br. in Ait. Hort. Kew. ed.* 3, v. 462. A hard-wooded tree, attaining a very large size, but sometimes flowering when under 20 ft., glabrous or the young shoots minutely pubescent; branchlets angular. Phyllodia falcate-oblong or almost lanceolate, 3 to 4 in. long in the common varieties, ½ to 1 in. broad, obtuse or rarely almost acute, much narrowed towards the base, coriaceous, with several longitudinal nerves and numerous anastomosing veins. Peduncles 3 to 4 lines long, few together in a short raceme or sometimes solitary, bearing each a very dense globular head of 30 to 50 or more flowers, mostly 5-merous and often so closely packed in the

head that the calyxes cohere.　Calyx more than half as long as the corolla, thin and shortly toothed.　Petals connate above the middle.　Pod elongated, flat, often curved into a circle, 3 to 4 lines broad, with thickened nerve-like margins.　Seeds nearly orbicular; funicle very long, dilated and coloured from the base, very flexuose, more or less encircling the seed in double folds. —Wendl. Comm. Acac. 24, t. 6; DC. Prod. ii. 452; Bot. Mag. t. 1659; Lodd. Bot. Cab. t. 630; Hook. f. Fl. Tasm. i. 109; F. Muell. Pl. Vict. ii. 28; *A. arcuata*, Sieb. Pl. Exs. and in Spreng. Syst. iii. 135 (by mistake attributed to Labillardière).

N. S. Wales.　Port Jackson to the Blue Mountains, *A. Cunningham, Sieber, n.* 459, *Fl. Mixt.* 593, and others.

Victoria.　Rich soil in valleys or grassy ranges throughout the colony, except the desert, ascending to considerable elevations, *F. Mueller.*

Tasmania.　Port Dalrymple and Derwent river, *R. Brown;* abundant throughout the island, *J. D. Hooker.*

S. Australia.　Encounter Bay, *Whittaker;* St. Vincent's Gulf, *F. Mueller.*

The wood, known to the colonists under the name of *Blackwood,* and the less appropriate one of *Lightwood,* is celebrated for hardness and durability.

A. brevipes, A. Cunn. in Bot. Mag. t. 3358, from the single specimen preserved of the cultivated plant described, appears to be a variety of *A. melanoxylon,* with longer more falcate phyllodia, attaining 5 to 7 in.

199.　**A. implexa,** *Benth. in Hook. Lond. Journ.* i. 368.　A glabrous tree, sometimes slightly glaucous; branchlets terete or nearly so.　Phyllodia lanceolate-falcate, more acuminate, more narrowed at the base and thinner than in *A. melanoxylon,* mostly 5 or 6 in. long or more, with several slender longitudinal nerves and fine veins.　Peduncles few, in a very short raceme, more slender than in *A. melanoxylon,* bearing each a small dense head of numerous flowers, mostly 5-merous.　Calyx scarcely half as long as the corolla, turbinate.　Petals smooth, united to the middle.　Pod narrow-linear, much curved and twisted, 2 or rarely nearly 3 lines broad, contracted between the seeds.　Seeds ovate-oblong, longitudinal; funicle dilated and coloured almost from the base, much folded under the seed, but not encircling it.—F. Muell. Pl. Vict. ii. 29.

Queensland.　Moreton Bay, Dawson and Burnett rivers, *F. Mueller.*

N. S. Wales.　Port Jackson, *R. Brown, Woolls;* northward to Clarence river, *C. Moore;* Mount Lindsay, *W. Hill;* southward to Shoalhaven river and Illawarra, *A. Cunningham.*

Victoria.　Open river banks and grassy ridges scattered over the colony: Yarra river, Bacchus Marsh, Snowy River, etc., *F. Mueller.*

200.　**A. harpophylla,** *F. Muell. Herb.*　Probably a tree, glabrous or the young shoots minutely hoary; branchlets slightly angular.　Phyllodia falcate-lanceolate, mostly 6 to 8 in. long, narrowed but obtuse at the end, much narrowed at the base, coriaceous, pale or glaucous, with several not very prominent nerves and scarcely veined between them.　Peduncles slender, ½ to ¾ in. long, clustered or rarely in a very short raceme, bearing each a small globular head of about 12 to 15 mostly 5-merous flowers.　Sepals spathulate, not half as long as the corolla, free or slightly connected below the middle.　Petals smooth, free.　Pod unknown.

Queensland.　Rockhampton, *Thozet.*

201. **A. excelsa,** *Benth. in Mitch. Trop. Austr.* 225. A large forest-tree; branchlets slender, terete or nearly so, glabrous or rarely minutely pubescent. Phyllodia oblong-falcate, rather obtuse or mucronulate, narrowed at the base, 2 to 3 in. long, ½ to ¾ in. broad, thinly coriaceous, with 5 to 7 nerves, and smooth or faintly veined between them. Peduncles solitary, in pairs or clusters, sometimes not 2 lines, in other specimens nearly ¼ in. long, bearing each a globular head of numerous (20 to 30) flowers, mostly 5-merous. Sepals distinct. Petals smooth. Pod straight, flat, about 3 lines broad, thinly coriaceous, the sutures narrow-edged or almost winged, not usually dehiscent but hardening over the seeds and readily breaking off between them. Seeds ovate, longitudinal; funicle short and filiform, neither folded nor enlarged.—*A. Daintreana,* F. Muell. Fragm. iv. 6.

Queensland. Near Lake Salvator, *Mitchell;* Peak Downs, *F. Mueller;* Clarke river, *Daintree;* also in *Bowman's collection.* Very closely allied to, and perhaps a variety of, *A. laurifolia,* Willd., from New Caledonia and the Pacific islands, differing chiefly in the narrower phyllodia and pods.

202. **A. complanata,** *A. Cunn.; Benth. in Hook. Lond. Journ.* i. 369. A tree, glabrous; branchlets flattened, bordered by 2 or rarely 3 acute angles or narrow wings. Phyllodia oval or oblong, obtuse, 2 to 3 in. long, ½ to 1 in. broad, thinly coriaceous, with 5 to 9 or even more longitudinal nerves and a few fine veins between them. Peduncles slender, about ½ in. long, in axillary clusters often of 6 to 8 or more, or by the abortion of the upper phyllodia forming an irregular terminal raceme, bearing each a globular head of numerous flowers, mostly 5-merous. Sepals free, spathulate. Petals smooth, free. Pod curved, acuminate, very flat, 3 to 4 lines broad, the upper suture nerve-like or with a narrow border. Seeds oblong; funicle in the Banksian specimens short and not dilated, but not quite perfect, in F. Mueller's specimens elongated, more or less dilated from near the base, and encircling the seed in a single fold.—*A. anceps,* Hook. Ic. Pl. t. 167, not of DC.

Queensland. Endeavour river, *Banks and Solander;* Wide Bay, *Bidwill, Moore;* Dumaresq river, *A. Cunningham;* Brisbane river, *Fraser, F. Mueller,* and others.
N. S. Wales. Clarence river, *Beckler.*

G. DIMIDIATÆ.—Phyllodia usually broad and often long, falcate or very oblique, with 2 or 3 prominent distant nerves and reticulately penniveined between them. Stipules minute or none.

These correspond with the subseries *Dimidiatæ* of *Julifloræ,* differing in their capitate inflorescence. The phyllodia are much larger and more oblique than in the *Oligoneuræ,* and the stipules are not spinescent as in *A. urophylla* amongst *Armatæ,* which has sometimes large phyllodia.

203. **A. binervata,** *DC. Prod.* ii. 452. A tall shrub or a tree, attaining sometimes 30 to 40 ft., glabrous, with slightly angular branchlets, soon becoming terete. Phyllodia falcate, oblong or lanceolate, narrowed at each end, mostly 3 to 4 in. long, with 2 or 3 longitudinal nerves and pinnately veined between them, the marginal gland below the middle rather conspicuous. Peduncles rather slender, 3 to 8, at first in an axillary raceme, but after flowering the raceme often grows out into a leafy branch with the peduncles at the base, each bearing a globular head of about 20 flowers, mostly 5-merous. Calyx scarcely half as long as the corolla, sinuate-toothed. Pe-

tals smooth. Pod long, flat and very thin, about ⅓ in. broad. Seeds obovate, longitudinal along the centre of the pod ; funicle folded and dilated
under the seed but not surrounding it.—*A. umbrosa*, A. Cunn. in G. Don,
Gen. Syst. ii. 405 ; Bot. Mag. t. 3338.

N. S. Wales Port Jackson or Blue Mountains, *Sieber, n.* 504; Nepean and Hastings rivers, *Fraser;* Hastings and Clarence rivers, *Beckler;* Clyde river, *C. Moore;*
southward to Illawarra, in shady woods, *A. Cunningham, Huegel.* Allied in some respects
to *A. penninervis,* differing in the venation of the phyllodia, in the pod and seeds. etc.
A. dineura, F. Muell. in Journ. Linn. Soc. iii. 130, from the Table Land, Upper Roper
river, N. Australia, appears to me to be a form of the same species, of a pale hue, with rather less oblique and more obtuse phyllodia. The specimens are small and in fruit only,
with the racemes grown out into leafy branches.
Some specimens of Mitchell's have narrow much more coriaceous phyllodia and very small
flower-heads, but without the fruit it cannot be determined whether they are a distinct
species or not.

204. **A. latescens,** *Benth. in Hook. Lond. Journ.* i. 380. A tall shrub
or tree, glabrous, with angular branchlets. Phyllodia lanceolate-falcate, obtuse, much narrowed towards the base, 6 to 8 in. long or even more, with 2
or 3 nerves, and a few reticulate or almost parallel veins between them ; marginal glands minute or none. Peduncles slender, ¼ to ½ in. long, in clusters
or short axillary racemes, bearing each a small globular head of above 30
flowers, mostly 5-merous. Calyx thin, more than half as long as the corolla,
with very short obtuse thickened lobes or teeth. Petals united above the
middle, rather thickened at the tips. Pod unknown.

N. Australia. Mayday Island, Van Diemen's Gulf, *A. Cunningham ;* Capstan Island,
Port Essington, *Armstrong.*

205. **A. sericata,** *A. Cunn. ; Benth. in Hook. Lond. Journ.* i. 380.
Pale, with a very minute almost mealy down, or glabrous and glaucous;
branchlets terete cr nearly so. Phyllodia broadly falcate, obtuse but narrowed at both ends, mostly 3 to 4 in. long, 1 to 1½ in. broad in the middle,
or on barren shoots much longer and narrower, with 3 or 4 principal nerves,
of which 1 or 2 confluent with the lower margin of the base, transversely reticulate between them, the outer or upper margin often sinuate. Flowers
not seen. Pod very flat, glaucous, 3 to 5 in. long, 1 to 1½ in. broad, sutures
bordered with a narrow edge ; valves coriaceous, hard when ripe, with raised
reticulations.. Seeds transverse, not seen perfect.—*A. platycarpa,* F. Muell.
in Journ. Linn. Soc. iii. 145.

N. Australia. Montagu and York Sounds, N.W. coast, *A. Cunningham* (foliage with
the valves of a ripe fruit) ; Victoria river (foliage only) and Gulf of Carpentaria (specimens
with unripe fruits), *F. Mueller.*

206. **A. flavescens,** *A. Cunn. ; Benth. in Hook. Lond. Journ.* i. 381.
Young shoots clothed with a hoary or yellowish almost fleecy tomentum,
soon wearing off; branchlets angular. Phyllodia broadly falcate, acuminate,
cuneate at the base, 4 to 8 in. long, 1 to 2 or even 3 in. broad, with usually
3 nerves, the lowest carried on to a terminal gland, the 2 others ending in
small glands on the upper margin, veins transversely reticulate between them.
Flowers in small globular heads, on short peduncles in an irregular terminal
panicle, and apparently 5-merous, with narrow sepals; but very imperfect in
our specimens. Pod straight or curved, very flat, 3 to 5 in. long, ¾ in. broad,

coriaceous, reticulate, with slightly thickened margins. Seeds transverse; funicle forming short slightly thickened folds under the seed, but not seen quite ripe.

Queensland. Sandy Cape, Broad Sound, Northumberland Islands, *R. Brown; Percy Islands, A. Cunningham;* sandstone ridges of Kongili, *Leichhardt.*

207. **A. retivenia,** *F. Muell. Fragm.* iii. 128. Densely clothed with a whitish woolly almost fleecy or velvety tomentum. Phyllodia obliquely obovate or orbicular, very obtuse, 1½ to 2 in. long, thick and soft, with about 4 prominent nerves and transverse reticulations. Upper phyllodia (in the only 2 branchlets seen) very much reduced, with dense globular flower-heads in their axils, on short thick tomentose peduncles. Flowers numerous, 5-merous, the buds shorter than the acute bracts, but not seen full-grown. Calyx thin, shortly lobed and ciliate. Pod unknown.

N. Australia. Short's Range, *M'Douall Stuart.*

SERIES VIII. JULIFLORÆ.—Phyllodia vertically flattened or, in a few species, terete, articulate on the stem, several-nerved or rarely 1-nerved, obtuse acute or pointed, rarely slightly pungent. Flowers in cylindrical dense or interrupted spikes, rarely, when sessile, shortly oblong.

I have united under this series all the spicate phyllodineous Acacias except *A. tripteris,* which has decurrent phyllodia continuous with the stem, and a few short narrow rigid-leaved *Pungentes,* which appeared to have no immediate affinity with the great mass of spicate species. Many of these are particularly difficult to distinguish without the fruit, and although I have passed over as doubtful numerous specimens which I had in flower only, there are still several of which I do not feel quite certain of having correctly matched the flowers and fruits.

A. RIGIDULÆ.—Phyllodia flat, often short, straight oblique or shortly falcate. Spikes dense, except in *A. megalantha.* Flowers mostly 5-merous.

The species here collected differ generally from the following subseries in their 5-merous flowers, and from the *Falcatæ* in their shorter, more coriaceous, obtuse, usually straight or undulate phyllodia, but some species almost pass into the latter subseries. The species are all tropical, except the last two of the subseries.

208. **A. amentifera,** *F. Muell. in Journ. Linn. Soc.* iii. 141. Apparently diffuse, glabrous and perhaps resinous ; branches angular-strٖte. Phyllodia often clustered, obliquely oblong, incurved, obtuse or recurved at the end, 2 to 4 lines long, coriaceous, nerveless or very obscurely 2-nerved. Spikes sessile, oblong-cylindrical, rarely as long as the phyllodia, with numerous closely-packed flowers, mostly 5-merous. Bracts acuminate, exceeding the flowers. Sepals free, narrow-linear, more than half as long as the corolla. Petals smooth, free or readily separating. Stamens few (about 20). Pod unknown.

N. Australia. Upper Victoria river, *F. Mueller.*

209. **A. Wickhami,** *Benth. in Hook. Lond. Journ.* i. 379. A glabrous shrub, often very glaucous or resinous ; branchlets angular-striate. Phyllodia numerous, obliquely ovate or falcate-oblong, obtuse with a small oblique glandular point, rarely exceeding ½ in., coriaceous, undulate, with several nerves all very faint or 1 or 3 more prominent. Spikes pedunculate, ½ to ¾ or rarely 1 in. long, densely cylindrical. Flowers mostly 5-merous. Calyx

thin, broadly sinuate-toothed, fully half as long as the corolla. Petals united below the middle, the midribs prominent. Pod flat, but thick and woody, obliquely veined, about 2 in. long, 2 lines broad above the middle, gradually tapering to the base. Seeds oblique, oblong; funicle straight, gradually thickened from the base into a narrow-turbinate aril, scarcely folded towards the end.

N. Australia. Swan Bay, N.W. coast, *Voyage of the Beagle;* table land between Victoria river and Sturt's Creek, *F. Mueller;* islands of the Gulf of Carpentaria, *R. Brown,* and adjoining mainland, *F. Mueller.* The more or less prominent nerves, the glaucous hue or resinous exudations, vary in different parts of the same specimen.

210. **A. lysiphloea,** *F. Muell. in Journ. Linn. Soc.* iii. 137. A rigid shrub of several feet or small tree, glabrous or nearly so, often very resinous. Phyllodia rather crowded, erect, obliquely linear-oblong, linear or oblanceolate, mostly obtuse but with a short rigid straight or oblique point, narrowed at the base, ½ to 1 in. long, thick and rigid, with 3 to 5 obscure or more or less prominent nerves. Spikes pedunculate, ¾ to 1 in. long, slender but dense. Flowers mostly 5-merous. Sepals very short, thin, shortly united at the base. Petals shortly united, the midribs prominent. Pod flat, oblique or falcate, 1 to 2 in. long, ¼ to ½ in. broad, hard and almost woody, reticulate and resinous. Seeds ovate, oblique; funicle shortly folded and thickened into a small aril under the base of the seed.

N. Australia. Hooker's and Sturt's Creeks, *F. Mueller;* islands of the Gulf of Carpentaria, *R. Brown;* sandy plains and valleys of the adjoining mainland, *F. Mueller.* Some of the narrow-leaved specimens have some resemblance tö *A. linarioides,* but the fruit is very different.

211. **A. linarioides,** *Benth. in Hook. Lond. Journ.* i. 371. Glabrous or slightly pubescent and viscid, with terete branchlets. Phyllodia rather crowded, linear, obtuse with a small rigid but not pungent point, ¾ to 1½ in. long, not above 1 line broad, obscurely 1-nerved. Spikes slender but rather dense, shortly pedunculate and exceeding the phyllodia. Flowers mostly 5-merous. Sepals small, thin, free or slightly connate at the base. Petals connate to the middle, with thickened tips. Pod linear, slightly curved, 1½ lines broad at the seeds and contracted between them, the valves convex, rigid, obscurely striate, with thickened nerve-like margins. Seeds oblong, longitudinal; funicle with the last 2 or 3 folds thickened into an irregularly cup-shaped aril under the seed.

N. Australia. Cavern Island, Gulf of Carpentaria, *R. Brown.* This species has the small 5-merous flowers and nearly the phyllodia of *A. lysiphloea;* with the pod more allied to that of *A. longifolia* and its allies, but more rigid.

212. **A. stipuligera,** *F. Muell. in Journ. Linn. Soc.* iii. 144. Softly tomentose or pubescent; branchlets nearly terete. Phyllodia obliquely falcate, oblong or lanceolate, shortly narrowed at each end, with a small callous or hooked point, 1 to 2 in. long, 3 to 7 lines broad, coriaceous, with 2, 3, or 4 very prominent nerves and nerve-like margins and numerous anastomosing more or less longitudinal veins, the principal nerves often scabrous with resinous exudations. Stipules brown, small, but more conspicuous than in any other *Juliflorae* except *A. conspersa.* Spikes nearly sessile, solitary or in pairs, 1 to nearly 2 in. long, dense and tomentose. Flowers mostly 5-merous. Calyx

half as long as the corolla, thin with spathulate lobes, readily separating into distinct sepals. Petals united to the middle, tomentose. Pod (only seen one loose one) long, linear, slightly twisted, not 1½ lines broad, coriaceous, with nerve-like margins.

N. Australia. Sources of the Victoria river, *F. Mueller;* in the interior, lat. 18°, " Scrub-Wattle," *M'Douall Stuart.*

213. **A. ptychophylla,** *F. Muell. in Journ. Linn. Soc.* iii. 142. Very resinous, otherwise glabrous; branchlets slightly angular. Phyllodia obliquely oblong, very obtuse, with a small terminal gland or callous point, 1 to 1½ in. long, 2 to 3 lines broad, shortly narrowed at the base, rigidly coriaceous, with 5 to 9 prominent parallel nerves without conspicuous intervening veins. Spikes shortly pedunculate, very dense, 1 in. long or rather more. Flowers mostly 5-merous. Calyx thin, more or less deeply lobed. Petals united to the middle, with prominent midribs. Pod unknown.

N. Australia. Sturt's Creek, *F. Mueller.*

214. **A. stigmatophylla,** *A. Cunn.; Benth. in Hook. Lond. Journ.* i. 377. Glabrous or scarcely hoary-pubescent when young; branchlets rather slender, acutely angled or compressed when young, soon becoming terete. Phyllodia obliquely narrow-oblong or slightly falcate, narrowed at each end, tipped with a prominent nerve or broad callous point, 1½ to 3 in. long, 3 to 4 lines broad, undulate, coriaceous, with 3 to 5 fine but prominent nerves and very fine (sometimes obscure), longitudinal, more or less anastomosing, closely-packed veins between them. Spikes shortly pedunculate, ½ to ¾ or at length 1 in. long, slender, often clustered in the upper axils, forming a terminal leafy panicle. Flowers small, mostly 5-merous. Calyx loosely campanulate, sinuate-toothed, slightly ciliate, half as long as the corolla. Petals with prominent midribs, giving the buds a strongly striate appearance. Pod unknown.

N. Australia. Brunswick Bay, N.W. coast, *A. Cunningham;* Victoria river, *F. Mueller.*

215. **A. umbellata,** *A. Cunn.; Benth. in Hook. Lond. Journ.* i. 378. A tall shrub, young shoots slightly hoary or silky but soon glabrous; branchlets nearly terete. Phyllodia from oblong and scarcely falcate to obliquely oblong-rhomboidal or broadly falcate, obtuse with a broadly callous or glandular tip, 2 to 4 in. long, ¾ to 1½ in. broad, very coriaceous, with numerous parallel nerves or veins, 5 to 9 more prominent and some of them confluent with the lower margin at the base, the others closely packed, fine and rarely anastomosing. Spikes sessile or nearly so, often clustered, rather dense, 1 to 1½ in. long. Flowers mostly 5-merous but sometimes 4-merous. Calyx pubescent, sinuate-toothed or shortly lobed. Petals with prominent midribs. Pod falcate or nearly straight, almost terete, with convex and coriaceous valves, as in *A. tumida.* Seeds ovate, oblique; funicle short, dilated into a small aril of 2 or 3 folds under the seed.

N. Australia. Islands of the Gulf of Carpentaria, *R. Brown;* Cleveland Bay and Cape Flinders, *A. Cunningham;* Seven-Emu, and Robinson rivers, *F. Mueller,* probably also Depot Creek, *F. Mueller* (specimens in flower only). Allied in foliage to *A. stigmatophylla,* in fruit to *A. tumida.*

216. **A. leptophleba,** *F. Muell. in Journ. Linn. Soc.* iii. 143. Nearly glabrous and probably resinous ; branchlets slightly angular. Phyllodia obliquely oblong-falcate, obtuse with an oblique callous or glandular point, narrowed at the base, 2 to 2½ in. long, about ½ in. broad in our specimens, coriaceous, undulate, with 3 or more slightly prominent nerves and thickened nerve-like margins, the intermediate fine veins more or less anastomosing, almost reduced to closely packed longitudinal parallel veins. Spikes pedunculate, solitary or in pairs, 1 to 1½ in. long. Flowers dense, mostly 5-merous. Calyx fully half as long as the corolla, thin, deeply lobed or divided to the base into narrow sepals. Petals thin, the midribs much less prominent than in *A. stigmatophylla.* Pod unknown.

N. Australia. Sturt's Creek, *F. Mueller.* I had, in the above quoted paper, referred these specimens to *A. aulacocarpa,* but having now been able to match much more accurately the flowering and fruiting specimens of that species, I find I was in error. The present species is much nearer to *A. stigmatophylla,* from which, in the absence of the fruit, it is distinguished by the more coriaceous phyllodia, larger flowers, deeply lobed calyx, and smoother petals.

217. **A. limbata,** *F. Muell. in Journ. Linn. Soc.* iii. 145. Branches glabrous, very acutely 2- or 3-angled and glaucous, as in *A. latifolia,* but not nearly so broad. Phyllodia oblong, falcate, obtuse or with a hooked callous point, 2 to 3 in. long, ½ to ¾ in. broad, thickly coriaceous, with several nerves, of which the lower ones are confluent with the lower margin at the base, the smaller veins reticulate but longitudinal. Flowers not seen. Fruiting spikes pedunculate. Pod flat, hard and thin, with thickened margins and obliquely veined, as in *A. aulacocarpa* and its allies, but very glabrous and smooth, obtuse, 1½ to 2 in. long, 4 lines broad in the middle and tapering to the base, but not seen quite ripe.

N. Australia, *F. Mueller.*

218. **A. brevifolia,** *Benth.* Glabrous and somewhat glaucous, with angular branchlets. Phyllodia obliquely oblong, somewhat falcate, very obtuse with a small callous point, narrowed at the base, 1½ to 2 in. long, ½ to ¾ in. broad, very coriaceous with several prominent nerves, the intermediate veins irregularly reticulate or longitudinal. Spikes short, oblong-cylindrical, pedunculate. Flowers mostly 5-merous. Calyx short, sinuate-toothed. Petals smooth. Pod flat, thick, almost woody, with oblique veins and thickened margins, very obtuse, 1½ in. long, nearly 4 lines broad, abruptly contracted below the middle into a broad stipes. Seeds oblique, but not seen ripe.—*A. leptophleba* (referred by me to *A. aulacocarpa*), var. *brevifolia,* F. Muell. in Journ. Linn. Soc. iii. 144.

Queensland. Desert of the Suttor, *F. Mueller.* Allied to *A. leptophleba* and *A. limbata,* but differing in flower from the former and in fruit from the latter, and the venation of the phyllodia does not agree with either.

219. **A. megalantha,** *F. Muell. in Journ. Linn. Soc.* iii. 143. Glabrous, with compressed but not angular branchlets. Phyllodia broadly semiovate or lanceolate-falcate, obtuse, narrowed towards the base, 2 to 4 in. long, ¾ to 1 in. wide, rather thick, coriaceous with 3 to 7 prominent nerves, and more or less distinct parallel veins between them. Spikes pedunculate, solitary or in pairs, 1 to 1½ in. long, with a stout rhachis. Flowers distinct or distant,

larger than in any other *Acacia*, the oblong buds 2 lines long before expanding, mostly 5-merous. Calyx shortly campanulate, toothed. Petals united at the base but readily separating, smooth, with thickened hood-shaped tips and inflexed points. Pod unknown.

N. Australia. Sturt's Creek; *F. Mueller.*

220. **A. gonoclada,** *F. Muell. in Journ. Linn. Soc.* iii. 140. Glabrous and glaucous, branchlets stout, with 2 or 3 much raised acute angles. Phyllodia lanceolate-oblong, slightly falcate, obtuse with oblique glandular tips, obliquely narrowed towards the base, 3 to 4 in. long, 4 to 8 lines broad, coriaceous, with 2 or 3 more prominent nerves almost confluent with the lower edge near the base, and numerous fine parallel veins scarcely anastomosing and not very closely packed. Spikes shortly pedunculate, oblong-cylindrical, dense, about ½ in. long. Flowers mostly 5-merous. Calyx half as long as the corolla, shortly toothed. Petals distinct, smooth, without prominent midrids. Pod narrow-linear, straight, 1 to 1½ in. long, 1½ lines broad, thin and flat with nerve-like margins. Seeds longitudinal; funicle dilated and folded under the seed, but not ripe in our specimens.

N. Australia. Victoria river, *F. Mueller.* Allied in some respects to *A. Cunninghamii,* but with the straighter more coriaceous phyllodia of the preceding species.

221. **A. pycnostachya,** *F. Muell. Pl. Vict.* ii. 33. An erect shrub, glabrous and sometimes glaucous ; branchlets stout with acutely raised angles. Phyllodia oblong-lanceolate, oblique, straight or slightly falcate, obtuse or with a recurved callous tip, mostly about 3 in. rarely 4 in. long, ½ to ¾ in. broad, rigidly coriaceous, with many fine nerves, 3 or 4 rather more prominent, the smaller ones scarcely anastomosing. Spikes sessile, solitary or 2 or 3 together, thickly cylindrical, 1 to 1½ in. long, very dense. Flowers mostly 5-merous. Calyx sinuate-toothed, not half as long as the corolla. Petals united at the base, rather rigid but smooth. Pod narrow-linear, erect, flat with thickened margins and longitudinal seeds, but not seen ripe.

N. S. Wales. New England, near Tenterfield, abundant at an elevation of 1000 to 1500 ft., *C. Stuart.* The foliage is that of some forms of *A. longifolia,* var. *Sophoræ,* but the inflorescence and flowers are quite distinct.

222. **A. subtilinervis,** *F. Muell. Pl. Vict.* ii. 32, *and Fragm.* iv. 8. An erect glabrous shrub, branchlets terete or nearly so and somewhat resinous. Phyllodia lanceolate, narrowed at both ends but not falcate, with a callous often incurved point, 2 to 4 in. long, 3 to 6 lines wide, coriaceous, with a slightly prominent central nerve and several parallel veins, very fine and almost transparent, but neither closely packed nor prominent. Spikes shortly pedunculate, solitary or in pairs, rather dense, ½ in. long or rather more. Flowers mostly 5-merous. Calyx sinuate-toothed, fully half as long as the corolla. Petals with a slightly prominent midrib. Pod unknown.

Victoria. Granite hills about Mount Imlay, *F. Mueller* (a single specimen in Herb. F. Muell.).

B. TETRAMERÆ.—Phyllodia flat, coriaceous or thin, straight or falcate, several-nerved or 1-nerved when very narrow. Spikes often loose. Flowers 4-merous. Seeds longitudinal.

In foliage this subseries connects the *Rigidulæ* with the *Falcatæ,* differing from both in the flowers almost universally 4-merous.

223. **A. cochliocarpa,** *Meissn. in Bot. Zeit.* 1855, 10. Glabrous and very rigid, apparently viscid when young; branchlets slightly angular, soon terete. Phyllodia linear-lanceolate, somewhat falcate, mostly acuminate, but not pungent, narrowed at the base, 2 to 4 in. long, 2 to 4 lines broad, very rigid, with 5 to 7 very prominent nerves, the central one usually thick and the margins often, but not always, much raised. Spikes sessile, solitary or in pairs, ½ to ¾ in. long. Flowers dense, mostly 4-merous. Sepals narrow-spathulate, ciliate, distinct. Petals united at the base. Pod glabrous, spirally twisted into a dense compact cylindrical coil of about 4 lines diameter; valves coriaceous, flat, with thickened margins. Seeds ovate, last 2 or 3 folds of the funicle dilated into an aril nearly as large as the seed, but under it.

W. Australia. Between Moore and Murchison rivers, *Drummond, 6th Coll. n.* 6, also in the 3*rd Coll.*

224. **A. Dallachiana,** *F. Muell. Fragm.* i. 7, *and Pl. Vict.* ii. 32. A tree of 20 to 30 ft., glabrous, with angular usually glaucous branchlets. Phyllodia lanceolate-falcate, obtuse or with a callous point, much and obliquely narrowed towards the base, 3 to 6 in. long, often 1 in. broad in the middle, coriaceous, with 2 to 5 or 6 more or less prominent primary nerves and conspicuously reticulate between them. Spikes usually in pairs, sessile, cylindrical, 1 to 1½ in. long. Bracts peltate, remarkably conspicuous and densely imbricate with the flower buds. Flowers mostly 4-merous. Calyx obtusely toothed, more than half as long as the corolla. Petals smooth, united to the middle, with rather prominent midribs. Pod linear, straight or nearly so, 2 to 2½ lines broad; valves convex over the seeds, flat and narrower between them. Seeds ovoid, longitudinal, the last 2 or 3 folds of the funicle much dilated, the last forming a cup-shaped aril under the seed.

Victoria. Between granite blocks on the summits of the Buffalo ranges, at an elevation of 4000 to 4500 ft., *F. Mueller.*

225. **A. alpina,** *F. Muell. Fragm.* iii. 129. A low divaricate glabrous shrub, nearly allied to the var. *Sophoræ* of *A. longifolia*, and included by F. Mueller (Pl. Vict. ii. 31) amongst the varieties of the latter species, but the habit and foliage appear too different to adopt that view in the absence of intermediates. Branchlets acutely angular or flattened. Phyllodia broadly and very obliquely obovate, ¾ to 1½ in. long, very obtuse or rarely with a small oblique point, coriaceous, finely 3- or 4-nerved, with numerous fine reticulations. Spikes very short and few-flowered. Pod narrow. Flowers and fruit otherwise as in *A. longifolia.*

Victoria. Alpine summits of Mount Useful, and northern plateau of Mount Wellington, *F. Mueller.*

226. **A. longifolia,** *Willd. Spec. Pl.* iv. 1052. An erect shrub, sometimes low and bushy, but attaining often a considerable size or growing into a small tree, glabrous or slightly pubescent when young; branchlets angular. Phyllodia from broadly oblong to oblong-lanceolate or linear, very obtuse or almost acuminate, usually narrowed towards the base, with 2 to 5 more or less prominent longitudinal nerves and conspicuously or faintly reticulate between them, varying in length from 2 to 3 in. in some varieties, to 5 or 6 in others. Spikes axillary, loose and interrupted, flowers not imbricate, almost

always 4-merous. Calyx very short, toothed. Petals smooth, united at the base or sometimes quite separating. Pod linear, often several in. long, 2 to 4 lines broad or rarely more; valves coriaceous, convex over the seeds, usually contracted between them. Seeds longitudinal, often distant, funicle not much folded, thickened almost from the base into a turbinate almost cup-shaped aril at the base of the seed, and sometimes nearly as large.—F. Muell. Pl. Vict. ii. 30.

Queensland. Moreton Bay, *A. Cunningham.*

N. S. Wales. Port Jackson and Blue Mountains, *R. Brown, Sieber,* and others; northward to Clarence river, *Beckler;* and New England, *C. Stuart;* southward to Twofold Bay, *F. Mueller.*

Victoria. Chiefly in the eastern and southern parts of the colony, *F. Mueller.*

Tasmania. Common in various places throughout the island, *J. D. Hooker.*

S. Australia. Chiefly near the coast, *R. Brown* and others.

Under the name of *A. longifolia,* I have followed F. Mueller in including the following forms, which, different as they generally appear, are connected by such a gradual chain of intermediates that they cannot be separated by any positive characters, excepting perhaps the first which seems to have a much broader pod, but is as yet not sufficiently known.

a. *phlebophylla,* F. Muell. Pl. Vict. ii. 31. Phyllodia resembling those of the var. *Sophoræ,* but more coriaceous, often above 1½ in. broad, very prominently reticulate. Pod nearly ½ in. broad, much flatter and less contracted between the seeds than in the other varieties.— Fissures of granite boulders on Mount Buffalo, Victoria, at an elevation of 3000 to 4000 ft., *F. Mueller.*

b. *Sophoræ,* F. Muell. Pl. Vict. ii. 30. Phyllodia obovate-oblong, very obtuse, coriaceous, about 2 or rarely 3 in. long, ½ to 1 in. broad, smaller veins reticulate. Calyx rather larger than in the other varieties. Pod usually much curved and thick, either slender and narrow, or 3 to 4 lines broad and very thick.—*Mimosa Sophoræ,* Labill. Pl. Nov. Holl. ii. 87, t. 237; *A. Sophoræ,* R. Br. in Ait. Hort. Kew. ed. 3, v. 462; DC. Prod. ii. 454.; Lodd. Bot. Cab. t. 1351; Hook. f. Fl. Tasm. i. 110.—Often abundant chiefly on the seacoast, Moreton Bay, N. S. Wales, Victoria, Tasmania, and S. Australia.

c. *typica.* Phyllodia linear-oblong or oblong-lanceolate, mostly obtuse, 4 to 5 in., or occasionally above 6 in. long, coriaceous but often less so than in the last, the reticulate veinlets more or less elongated and parallel. Pod usually long and slender.—*Mimosa longifolia,* Andr. Bot. Rep. t. 207; Vent. Jard. Malm. t. 62; *A. longifolia,* Willd.; DC. Prod. ii. 454; Bot. Reg. t. 362; Bot. Mag. t. 1827, 2166; Lodd. Bot. Cab. t. 678; Paxt. Mag. iv. 197 (very incorrect as to foliage); Maund, Botanist, t. 77; F. Muell. Pl. Vict. ii. 30; *A. obtusifolia,* A. Cunn. in Field, N. S. Wales, 345 (with narrow phyllodia); *A. spathulata,* Tausch, in Flora, 1836, 420; *A. intertexta,* Sieb. in DC. Prod. ii. 454 (with broad phyllodia).—Port Jackson to the Blue Mountains, *R. Brown, Sieber, n.* 438, 439, 453, *Fl. Mixt. n.* 592, and others; Twofold Bay and E. Gipps' Land, *F. Mueller.*

d. *mucronata,* F. Muell. Pl. Vict. ii. 31. Phyllodia linear-oblong or linear-spathulate, often very narrow, but coriaceous and obtuse, scarcely veined besides the 3 to 5 rather prominent parallel nerves, mostly 1½ to 2½ in. long, 1 to 3 lines broad.—*A. mucronata,* Willd. Enum. Hort. Berol. Suppl. 68 (name only); Wendl. Comm. Acac. 46. t. 12; DC. Prod. ii. 454; Bot. Mag. t. 2747; Hook. f. Fl. Tasm. i. 110; F. Muell. Pl. Vict. ii. 31; *A. dependens,* A. Cunn.; Benth. in Hook. Lond. Journ. i. 372.—Tasmania, *R. Brown;* common throughout the island, *J. D. Hooker;* also in Victoria on sandy heath-ridges near Latrobe river, Corner Inlet, etc., *F. Mueller.*

e. *floribunda,* F. Muell. Pl. Vict. ii. 31. Phyllodia linear or linear-lanceolate, usually narrowed at the end or acute, 3 to 4 or even 5 in. long, less coriaceous than in the preceding forms, the smaller veins less anastomosing and passing into long parallel veins scarcely finer than the principal nerve.—*Mimosa floribunda,* Vent. Choix, t. 13; *A. floribunda,* Willd. Spec. Pl. iv. 1051; DC. Prod. ii. 454; *A. angustifolia,* Lodd. Bot. Cab. t. 763, not of Jacq.; *A. intermedia,* A. Cunn. in Bot. Mag. t. 3203 (with broader phyllodia); *A. decussata,* Ten. Cat. Hort. Neap. 77.—Port Jackson to the Blue Mountains, *R. Brown, Sieber, n.* 440, and others, northward to New England, *C. Stuart;* Clarence river, *Beckler;* and southward to Tambo river, Victoria, *F. Mueller.*

f. *dissitiflora*: Phyllodia often very long and narrow as in *A. linearis*, but rather more coriaceous, with 1 or 2 nerves parallel to the principal one, and continued nearly the whole length of the leaf, connecting in some measure the var. *mucronata* with *A. linearis.*—*A. dissitiflora*, Benth. in Hook. Lond. Journ. i. 371.—N. coast of Tasmania.

227. **A. linearis,** *Sims, Bot. Mag. t.* 2156. An erect shrub of several ft., glabrous or slightly pubescent when young; branchlets angular. Phyllodia narrow-linear, from 4, 5 or 6 in. long to twice that length, scarcely above 1 line broad, with a prominent longitudinal nerve, and occasionally an additional faint one on each side. Spikes loose and interrupted, slender, 1 to 2 in. long, quite glabrous. Flowers mostly 4-merous. Calyx very short, toothed. Petals smooth, united at the base. Pod linear, nearly straight, several in. long, usually about 2 lines broad. Seeds longitudinal, but not seen perfect.—DC. Prod. ii. 454 ; Lodd. Bot. Cab. t. 595 ; Hook. f. Fl. Tasm. i. 109 ; F. Muell. Pl. Vict. ii. 31 ; *A. longissima*, Wendl. Comm. Acac. 45. t. 11 ; Bot. Reg. t. 680.

N. S. Wales. Port Jackson, *R. Brown, Sieber, n.* 451, 456, *and Fl. Mixt.* 596, and others ; northward to Hastings river, *Beckler ;* Richmond river, *C. Moore.*

Victoria. Wet forest glens and periodically flooded river-banks in the E. part of Gipps' Land, *F. Mueller.*

Tasmania. Circular Head, *Gunn.*

Enumerated by F. Mueller amongst the varieties of *A. longifolia*, and certainly very near the extreme forms of the var. *dissitiflora*, differing chiefly in the long narrow phyllodia, either str.ctly 1-nerved or with only a faint accessory nerve on each side.

C. STENOPHYLLÆ.—Phyllodia linear-subulate or narrow-linear, straight or slightly curved, terete or flat but thick, rarely under 2 in. long or above 1½ lines broad. Spikes dense or slender, with small 5-merous or, in terete-leaved species, often 4-merous flowers.

Some species of this subseries, with very short spikes, almost pass into the series *Calamiformes*, and one or two of the last species of the subseries, with rather broader phyllodia, are closely connected with the narrower-leaved species of *Falcatæ*.

228. **A. aciphylla,** *Benth. in Linnæa,* xxvi. 627. A glabrous shrub with scarcely angular branchlets, very soon terete. Phyllodia linear-subulate, rigid with a short pungent point, 2 to 4 in. long, terete and minutely striate with very fine parallel nerves scarcely visible without a lens. Spikes sessile, dense, ovoid or oblong, not above 3 lines long. Flowers mostly 5-merous. Calyx turbinate, very shortly lobed, fully half as long as the corolla. Petals readily separating. Pod only seen in a single loose valve about 1 in. long, 1¼ lines broad, coriaceous, rather hard, slightly contracted between the seeds, which were evidently longitudinal.—*A. leptoneura*, var. (?) *pungens*, Meissn. in Pl. Preiss. i. 12 ?

W. Australia, *Drummond, 4th Coll. n.* 14 ; in the interior, *Preiss, n.* 976 (I have not seen Preiss's specimens).

229. **A. ephedroides,** *Benth. in Hook. Lond. Journ.* i. 370. A shrub, with slender divaricate or flexuose branchlets, glabrous or the young shoots hoary or silky-pubescent. Phyllodia linear-filiform, terete, striate with few nerves, or obtusely 4-gonous, straight or flexuose, with a small straight or hooked point, 2 to 6 in. long. Spikes sessile, dense, ovoid-oblong or shortly cylindrical, scarcely exceeding 3 lines, or rarely 4 lines long. Flowers crowded, mostly 4-merous. Calyx shortly lobed, not half so long as the co-

rolla. Petals smooth except a very prominent midrib, readily separating. Pod very narrow, but not seen ripe.—Meissn. in Pl. Preiss. i. 18 ; *A. filifolia*, Benth. in Hook. Lond. Journ. i. 369.

W. Australia. Cape Porteraz, *Fraser ;* Swan River ? and towards Cape Riche, *Drummond, n.* 302, *2nd Coll. n.* 149 *and* 156 ; *5th Coll. n.* 2 ; Darling Range, *Preiss, n.* 974.

230. **A. Burkittii,** *F. Muell. Herb.* A glabrous shrub, branchlets slender, nearly terete. Phyllodia linear-subulate, terete or slightly compressed, with a fine usually recurved but not pungent point, 2 to 3 in. long, striate with very fine parallel nerves, the central one scarcely more prominent. Spikes oblong, sessile, solitary or in pairs, 3 to 4 lines long. Flowers mostly 4-merous. Sepals spathulate, above half as long as the corolla, united in a short cup at the base. Petals smooth, rather thick at the tips, but without prominent midribs. Pod unknown.

S. Australia. Lake Gilles in the interior, *Burkitt.* Very near *A. microneura*, but the phyllodia scarcely broader than thick and the calyx different.

231. **A. microneura,** *Meissn. in Pl. Preiss.* i. 19. Young shoots minutely ashy-pubescent, becoming glabrous, branchlets slightly angular or nearly terete. Phyllodia narrow-linear, almost subulate, with a recurved point, 2 to 3 in. long, ½ to 1 line broad, rigid, with numerous parallel nerves, the central one usually more prominent. Spikes solitary or in pairs, sessile, 3 to 4 lines long. Flowers mostly 4-merous. Calyx shortly toothed, not half so long as the corolla. Petals smooth. Pod unknown.

W. Australia. S. interior, *Preiss, n.* 942 ; *Drummond, 5th Coll. n.* 10. Differs from *A. ephedroides* chiefly in the more flattened phyllodia.

232. **A. cyperophylla,** *F. Muell. Herb.* Tall, with curly bark and dark wood, branchlets terete. Phyllodia linear-subulate with a fine usually curved point, 6 to 10 in. long, terete or very slightly compressed, striate with numerous exceedingly fine parallel nerves only visible under a lens, hoary with a very minute loose pubescence. Spikes sessile or nearly so, oblong, not ½ in. long. Flowers mostly 5-merous or 6-merous. Calyx turbinate, about half as long as the corolla, at first shortly toothed but often dividing nearly to the base. Petals smooth, glabrous. Pod unknown.

Queensland ? *Leichhardt.*
S. Australia. Stony ground, Cooper's Creek, *A. C. Gregory.*

233. **A. multispicata,** *Benth.* Young shoots minutely silky-pubescent, otherwise glabrous ; branchlets slender, nearly terete, whitish. Phyllodia linear-subulate, with fine usually recurved but not pungent points, 1½ to 2¼ in. long, rather rigid, terete, with few prominent parallel nerves and furrows. Spikes sessile, solitary or in pairs, cylindrical, not very dense, ½ to ¾ in. long. Flowers mostly 4-merous. Calyx broad, short, thin, pubescent, shortly lobed. Petals smooth, at length separating. Ovary densely tomentose. Pod unknown.

W. Australia, *J. S. Roe, Drummond ;* Hill river, *Oldfield.* Differs from *A. ephedroides* in the prominently few-nerved phyllodia, longer spikes, unribbed petals, etc.

234. **A. pityoides,** *F. Muell. in Journ. Linn. Soc.* iii. 135. Quite glabrous ; branchlets slender, terete. Phyllodia linear-subulate, rather rigid but

not pungent, 2 to 4 in. long, slender, terete and almost nerveless, or slightly flattened and striate with very fine nerves, scarcely visible without a lens. Spikes usually in pairs, pedunculate, about ½ in. long, slender but dense. Flowers small, mostly 5-merous. Calyx very thin and transparent, deeply lobed or the sepals quite free but not spathulate, fully two-thirds as long as the corolla. Petals thin, connate to the middle, without prominent midribs. Pod elongated, nearly flat, curved or twisted, 1 to 1½ lines broad, slightly contracted between the seeds; valves thinly coriaceous. Seeds obovate, longitudinal; funicle not seen perfect.

N. Australia. Sturt's Creek and Gilbert river, *F. Mueller.*
Queensland. Ridges of the Suttor, *F. Mueller.*

235. **A. xylocarpa,** *A. Cunn.; Benth. in Hook. Lond. Journ.* i. 370. A shrub of 2 to 4 ft., glabrous and slightly viscid; branchlets terete. Phyllodia linear-subulate, not pointed, 2 to 4 in. long, rather rigid, terete or rarely flattened to nearly 1 line in breadth, obscurely 1-nerved. Spikes mostly in pairs, shortly pedunculate, slender but closely packed, ½ to ¾ in. long when fully out. Flowers mostly 5-merous. Calyx shortly lobed, about half as long as the corolla. Petals united to the middle, with prominent midribs. Pod nearly terete or slightly flattened, 1½ to 3 in. long, shortly acuminate, 3 to 4 lines broad and thick near the end, gradually tapering to the base; valves hard, almost woody, striate lengthwise, opening elastically from the end downwards. Seeds oblique; funicle straight, gradually and slightly thickened from the base upwards.—*A. orthocarpa,* F. Muell. in Journ. Linn. Soc. iii. 136.

N. Australia. Dampier's Archipelago and Water Island, N.W. coast, *A. Cunningham;* Nichol Bay, *F. Gregory's Expedition;* Upper Macarthur river, Gulf of Carpentaria, *F. Mueller.*

Var. (?) *tenuissima.* Phyllodia longer and more slender. Spikes short. Pod unknown. —*A. tenuissima,* F. Muell. in Journ. Linn. Soc. iii. 135.—Sturt's Creek, *F. Mueller.*

236. **A. gonocarpa,** *F. Muell. in Journ. Linn. Soc.* iii. 136. A shrub of 4 or 5 ft., young shoots viscid, with slender flattened or angular branchlets, at length terete. Phyllodia very narrow-linear, but flat, with a small callous or hooked point, mostly 1½ to 2 or rarely 3 in. long, prominently 1-nerved. Spikes shortly pedunculate, solitary or in pairs, ¼ to ½ in. long, very slender, but with numerous closely packed very small flowers, mostly 5-merous. Sepals very narrow, linear, thin and distinct. Petals thin, cohering to the middle. Pod hard and woody, 1½ to 2 in. long, about 3 lines broad; valves opening elastically from the end downwards as in *A. xylocarpa,* but with raised acute longitudinal angles as in *A. tetragonocarpa.* Seeds not seen, the pod obliquely partitioned for their reception as in *A. xylocarpa.*

N. Australia. Arnhem N. Bay, *R. Brown;* rocky shores of the Gulf of Carpentaria, *F. Mueller.*

237. **A. oncinophylla,** *Lindl. Swan Riv. App.* 15. Glabrous or slightly viscid-pubescent; branchlets somewhat angular. Phyllodia linear-subulate, straight or slightly curved, with a straight oblique or hooked point, 3 to 5 in. long, about 1 line broad, flat but thick and rigid, with 3 prominent nerves. Spikes shortly pedunculate, mostly in pairs, very dense, rarely above ½ in. long or in cultivated specimens rather looser and ¾ in. long. **Flowers**

mostly 5-merous, but occasionally 4 merous or 3-merous. Calyx lobed, sometimes separating into distinct sepals, fuliy half as long as the corolla. Petals smooth, connate to the middle. Pod (according to Meissner) linear, slightly tortuose, 1 to 1½ in. long, 2 lines broad, flat, shortly villous. Seeds ovate, lenticular, half embraced by the folds of the funicle.—Meissn. in Pl. Preiss. i. 19; Bot. Mag. t. 4353 (the calyx overlooked).

W. Australia. Swan River, *Drummond, 1st Coll. and n.* 137; Darling Range, *Preiss, n.* 914.

238. **A. drepanocarpa,** *F. Muell. in Journ. Linn. Soc.* iii. 137. A glabrous shrub, the young shoots resinous, branchlets slender, slightly angular. Phyllodia narrow-linear, straight or slightly curved, obtuse, narrowed towards the base, 2 to 4 in. long, 1 to 2 lines broad, with a slightly prominent central nerve and 1 or 2 finer veins on each side. Spikes slender, not very dense, ½ to ¾ in. long, shortly pedunculate. Flowers mostly 5-merous. Calyx thin, with narrow lobes, half as long as the corolla. Petals connate to the middle, with prominent midribs as in *A. xylocarpa.* Pod erect, linear, 1½ to 3 in. long, 1¼ to 2 lines broad, flat but thick with much raised margins and obliquely veined between them, the almost woody valves rolling back elastically as in *A. gonocarpa.* Seeds oblique; funicle straight, gradually thickened from the base, narrow-turbinate and cup-shaped under the seed.

N. Australia. Rocks of the S.W. shore of the Gulf of Carpentaria, *F. Mueller;* Whitsunday and Palm Islands, *Henne.* Differs from *A. gonocarpa,* as *A. arida* does from *A. xylocarpa,* in its broader and flatter phyllodia.

239. **A. arida,** *Benth. in Hook. Lond. Journ.* i. 370. Glabrous or minutely hoary; branchlets slender, terete or slightly compressed. Phyllodia narrow-linear, obtuse or with a callous point, narrowed at the base, 2 to 4 in. long, 1 to 2 lines broad, flat but thick, obscurely 3-nerved. Spikes shortly pedunculate, solitary or in pairs, slender and compact, ¼ to ½ in. long. Flowers very small, mostly 5-merous. Calyx shortly lobed. Petals smooth, without prominent midribs. Pod thick but flat, nearly 4 in. long, 3 lines broad, narrowed to the base; valves hard, almost woody, rolling back elastically. Seeds not seen, but evidently oblique.

N. Australia. Parched desert shores of Cambridge Gulf, N.W. coast, *A. Cunningham.* Differs from *A. xylocarpa* chiefly in the broader flat phyllodia.

240. **A. aneura,** *F. Muell. in Linnæa,* xxvi. 627, *and Fragm.* iv. 8. Shrubby, often hoary with a very minute pubescence; branchlets terete or nearly so. Phyllodia narrow-linear, obtuse or with a recurved or oblique callous point, usually flat but thick, 1½ to 3 in. long, 1 to 1½ lines broad, but varying from short and narrow-oblong to very narrow and almost terete, without conspicuous nerves, but finely and obscurely striate under a lens. Spikes shortly pedunculate, ½ to ¾ in. long. Flowers mostly 5-merous. Sepals very narrow, linear-spathulate. Petals smooth. Pod thin, flat, obliquely oblong, very obtuse, narrowed at the base, 1 to 1½ in. long, about 4 lines broad, the sutures edged with a narrow wing. Seeds ovate, oblique or transverse; funicle with 2 or 3 short folds, expanded into a small membranous aril under the seed.

N. S. Wales. From the Darling to the west frontier, *Victorian Expedition.*

S. Australia. Flinders Range, *F. Mueller ;* Lake Gairdner, *Babbage* ; in the desert interior, forming the chief ingredient of the Mulga scrub, *M‘Douall Stuart.*
Var. (?) *stenocarpa.* Pod narrow, turgid. Seeds longitudinal, with the funicle much more dilated and folded.—Barrier Range, *Victorian Expedition.*

241. **A. brachystachya,** *Benth.* Very near *A. aneura* and perhaps a short-spiked variety, slightly glaucous or hoary, but without visible pubescence. Phyllodia linear-subulate, slightly compressed, rigid but not pungent, very finely striate, with numerous nerves scarcely visible without a lens. Spikes sessile or very shortly pedunculate, ovoid or oblong, 2 to 3 lines long. Flowers mostly 5-merous. Sepals very narrow, linear-spathulate. Petals smooth, often minutely pubescent. Pod unknown.

N. S. Wales. Mutanie Ranges, *Victorian Expedition.*

D. FALCATÆ.—Phyllodia usually long or large, more or less falcate, narrowed at each end, with numerous parallel nerves or veins, either all equal or more frequently 1, 3, or more, prominent and undivided, the smaller ones often more or less anastomosing. Spikes slender, dense or rarely interrupted. Flowers mostly 5-merous.

Some of the shorter-leaved species or specimens of *A. conspersa, A. leptostachya, A. tumida,* etc., connect this group with the *Rigidulæ,* but the phyllodia are usually less coriaceous, longer or more falcate, with more numerous nerves. Several species with very different pods appear almost identical in foliage and flowers.

242. **A. conspersa,** *F. Muell. in Journ. Linn. Soc.* iii. 140. A shrub of 5 to 10 ft., with loosely pubescent branches and conspicuous though small brown stipules as in *A. stipuligera.* Phyllodia narrow-lanceolate, mostly falcate, narrowed at each end, obtuse or with a small rigid or glandular point, 2 to 3 in. long, 2 to 4 lines broad, coriaceous, often minutely mealy or slightly pubescent, with a prominent central nerve and often 2 less prominent lateral ones, and numerous very fine parallel veins between. Flowers not seen. Pod linear, straight or slightly falcate, about 1½ lines broad, thick but flat until ripe and then the valves slightly convex and hard. Seeds oblong, longitudinal ; funicle short, the last 2 folds expanded into an aril under the seed.

N. Australia. Islands of the Gulf of Carpentaria, *R. Brown ;* and rocky sandy districts of the adjoining mainland, Upper Roper and Limmen-Blight rivers, *F. Mueller ;* Sterculia Creek, *Leichhardt.*

243. **A. doratoxylon,** *A. Cunn. in Field, N. S. Wales,* 345. A tall shrub or small tree, glabrous with an ashy hue ; branchlets at first acutely angular, but soon terete. Phyllodia elongated, slightly falcate, shortly acuminate, and often with oblique or recurved points, 4 to 8 in. long, 2, 3, or rarely 4 lines broad, narrowed towards the base, rather thick, with numerous fine parallel nerves, the central one more prominent. Spikes shortly pedunculate, solitary or clustered, rarely 1 in. long, rather dense. Flowers mostly 5-merous. Calyx sinuate-toothed, not half as long as the corolla. Petals with slightly prominent midribs. Pod unknown.

N. Australia ? Daly Waters, *M‘Douall Stuart.* In leaf only and therefore doubtful.
Queensland. On the Upper Maranoa, *Mitchell ;* Moreton Bay, *C. Moore.*
N. S. Wales. Lachlan and Macquarrie rivers, the Spear-wood of certain tribes of the interior, *A. Cunningham, Fraser.*

Victoria. Ovens Ranges, *F. Mueller,* specimens in very young bud. F. Mueller, Pl. Vict. ii. 33, describes the flowers as 4-merous or 3-merous; I found them all 5-merous in the spike I soaked, but in that very young state the petals cohere so firmly that it is very difficult to ascertain their number.

244. **A. acuminata,** *Benth. in Hook. Lond. Journ.* i. 373. A tree of 30 to 40 ft., glabrous or the young shoots silky with an almost golden pubescence; branchlets terete or nearly so. Phyllodia long-linear, somewhat falcate, narrowed at each end, from 3 in long and 2 or 3 lines wide to 10 in. or more long and then rarely above 2 lines wide, striate as in *A. doratoxylon* with numerous fine parallel nerves, the central one more prominent. ·Spikes nearly sessile, not above 1 in. long. Flowers mostly 5-merous, larger than in *A. doratoxylon.* Calyx longer in proportion and readily separating into distinct sepals. Pod linear, straight or nearly so, 2 to 2½ lines broad, flat or convex over the seeds, somewhat contracted between them. Seeds oblong, longitudinal, the last 2 or 3 folds of the funicle dilated into an aril under the seed.—*A. doratoxylon* and *A. acuminata,* Meissn. in Pl. Preiss. i. 19 ; *A. Oldfieldii,* F. Muell. Fragm. iv. 7.

W. Australia. From King George's Sound to Swan River and northward to Murchison river, *Baxter, Drummond,* 3rd *Coll. n.* 99, 4th *Coll. n.* 135, 5th *Coll. n.* 6, *J. S. Roe, Oldfield, Preiss, n.* 934, 935, and probably 945. Very near *A. doratoxylon* and perhaps a variety, said to be similarly used for making arms, and the wood to smell strongly of raspberry jam.

Var. *latifolia.* Phyllodia sometimes 4 lines wide.—Middle Island, on the S. coast, *Maxwell.*

245. **A. stereophylla,** *Meissn. in Pl. Preiss.* ii. 203. Quite glabrous, even the young shoots, with terete branchlets. Phyllodia linear, often somewhat falcate, with glandular or callous tips, 3 to 5 in. long, 1½ to 3 lines broad, narrowed towards the base, striate with fine parallel nerves as in *A. acuminata,* but thicker and more rigid and the nerves all equal or 1 or 3 rather more conspicuous. Spikes very dense, sessile or shortly pedunculate, ½ to ¾ in. long. Flowers mostly 5-merous. Sepals linear-spathulate, ciliate, free or slightly cohering. Petals readily separating, smooth, with slightly thickened tips. Pod unknown.

W. Australia, *Drummond,* 3rd *Coll. n.* 100. This may prove to be a marked variety of *A. acuminata,* if it has the same pod.

246. **A. signata,** *F. Muell. Fragm.* iv. 7. A glabrous straggling shrub of 6 to 8 ft.; branchlets compressed or terete, glaucous. Phyllodia linear-lanceolate, obtuse or with a minute callous point, 2 to 5 in. long in the specimens seen, much narrowed towards the base, rather rigid, striate with numerous fine nerves, 1 or 3 rather more prominent as in the preceding species. Flowers not seen. Fruiting-spike pedunculate. Pod shortly stipitate, linear-falcate, flat, coriaceous, with thickened margins, about 3 in. long and nearly 3 lines broad, not contracted between the seeds. Seeds ovate, longitudinal or somewhat oblique; funicle dilated from the base, forming an aril of 2 or 3 folds under the seed.

W. Australia. Murchison river, *Oldfield.* Foliage of *A. acuminata,* but the fruit different.

247. **A. delibrata,** *A. Cunn.; Benth. in Hook. Lond. Journ.* i. 374.

Branchlets slender, slightly angular, silky-pubescent when young. Phyllodia linear-lanceolate, falcate, narrowed at both ends, 4 to 6 in. long, 2 to 5 lines broad in the middle, rather thin, sprinkled with loose silky hairs, with about 3 fine but prominent nerves, and finer less conspicuous and not very numerous longitudinal veins between them occasionally anastomosing. Flowers not seen. Fruiting-spikes with a rhachis of 1 to 1½ in. Pod elongated, straight, flat with thickened margins, about 3 lines broad, the coriaceous valves rather convex over the seeds, narrowed between them. Seeds not seen, but probably longitudinal.

N. Australia. York Sound and Port Warrender, N.W. coast, *A. Cunningham.* The specimens are very imperfect, but do not match with any other we have. The fruit is nearly that of *A. torulosa,* but very much flatter and the phyllodia much thinner, etc. The bark of the old branches appears to peel off in shreds.

248. **A. oligoneura,** *F. Muell. in Journ. Linn. Soc.* iii. 139. Glabrous with angular branchlets. Phyllodia narrow-lanceolate, falcate, narrowed at both ends, 4 to 6 in. long, ½ to ¾ in. broad, with about 3 prominent nerves, a very few longitudinal less conspicuous ones, and the smaller veins more or less reticulate. Spikes pedunculate, in pairs clusters or short racemes, slender but rather dense, about ½ in. long. Flowers very small, mostly 5-merous. Calyx short, sinuate-toothed. Pod not seen.

N. Australia. Victoria river and Macadam Range, *F. Mueller.* Possibly the same as *A. delibrata.*

249. **A. torulosa,** *Benth. in Journ. Linn. Soc.* iii. 139. A tall shrub or small tree, glabrous, with angular branchlets. Phyllodia linear-lanceolate, falcate, with an oblique glandular point, narrowed towards the base, 4 to 8 in. long, 3 to 4 lines wide, coriaceous, with 3 to 5 prominent nerves and numerous very fine parallel ones between them. Spikes solitary or in pairs or threes, very shortly pedunculate, ½ to ¾ in. long, slender but rather dense. Flowers small, mostly 5-merous. Sepals narrow-linear, spathulate, ciliate, free or slightly united at the base. Petals united to the middle. Pod long, remarkably moniliferous, the valves thickly coriaceous, convex, oblong, and about 3 lines broad over the seeds, much contracted between them. Seeds oblong, longitudinal; funicle short, the last fold expanded into a small obliquely cup shaped aril under the seed.

N. Australia. Sandy banks of Roper and Nicholson rivers, Gulf of Carpentaria, *F. Mueller.*
Queensland. Dayman's Island, Endeavour Straits, *W. Hill.*
Scarcely to be distinguished from *A. julifera, A. plectocarpa,* and some others, except by the fruit.

250. **A. julifera,** *Benth. in Hook. Lond. Journ.* i. 374. A tall shrub or tree, usually glabrous except the inflorescence; branchlets slender, angular when young but soon terete. Phyllodia narrow-lanceolate, falcate, narrowed at both ends, 4 to 6 in. long, ¼ to ½ in. broad, coriaceous, with 1 to 3 fine nerves and the nerve-like margins rather more prominent than the numerous fine veins between them. Spikes dense, shortly pedunculate, 1 to 1½ in. long, solitary or 2 or 3 together on a short common peduncle. Flowers mostly 5-merous. Calyx short, more or less lobed, pubescent, woolly or rarely almost glabrous. Pod long, 1½ to 2 lines broad, spirally twisted into numerous

coils either loose and irregular or closely packed into a short cylinder; valves flat or slightly convex. Seeds longitudinal; funicle slightly thickened from the base, at first straight, forming 2 or 3 more dilated folds under the seed.

Queensland. Cumberland Islands, *R. Brown*; Rodd's Bay, *A. Cunningham*; Rockingham Bay, *W. Hill*; Edgecombe Bay, *Dallachy*. Very difficult, without the pod, to distinguish from *A. doratoxylon* and *A. plectocarpa.* Phyllodia more falcate than in the former. Branchlets much less angular than in the latter.

251. **A. Solandri,** *Benth.* A tall shrub or tree, glabrous or the young shoots slightly silky; branchlets nearly terete. Phyllodia as in *A. julifera,* narrow-lanceolate, falcate, 4 to 6 in. long, 3 to 4 lines broad, with 1 to 3 slightly prominent and numerous very fine parallel nerves. Spikes 2 to 3 in. long, slender, interrupted and glabrous or nearly so. Flowers distant as in *A. linearis,* but much smaller and all or mostly 5-merous. Calyx short and truncate. Petals smooth. Pod unknown.

Queensland. Bay of Inlets, *Banks and Solander* (*Herb. R. Br.*), and possibly a form with woolly calyxes from the head of Boyd river, *Leichhardt,* the specimens imperfect. I am unable to adopt for this species Solander's ms. name of *salicifolia,* as there already exist an *A. saligna* and an *A. salicina.*

252 ? **A. leptostachya,** *Benth.* Hoary or silvery white with a very minute pubescence or nearly glabrous; branchlets slender, slightly angular. Phyllodia linear or lanceolate mostly falcate, narrowed at each end but obtuse, 1 to 2 or rarely 3 in. long, 1 to 5 lines broad, straight or slightly oblique at the base, coriaceous and finely striate with numerous nerves all equal or 2 or 3 rather more prominent. Spikes mostly in pairs, very shortly pedunculate, slender, ¾ to above 1 in. long, glabrous or nearly so. Flowers usually distant, mostly 5-merous. Calyx short, truncate. Petals smooth, united at the base only. Pod not seen.

Queensland. Newcastle Range, *F. Mueller*; Edgecombe Heights, Port Denison, *Dallachy*; Port Denison, *Fitzalan*; Broad Sound, *Herb. F. Mueller.* Until the fruit is known the affinities of this species must remain uncertain. F. Mueller considers it as a form of *A. glaucescens,* but the phyllodia are quite different, and the specimens have more the aspect of some of the species with transverse seeds.

253. **A. glaucescens,** *Willd. Spec. Pl.* iv. 1052, *and Hort. Berol. t.* 101. A tree attaining 50 ft. or more, the foliage generally ashy or hoary with a very minute close pubescence or the young shoots yellowish; branchlets more slender and much less angular than in *A. Cunninghamii.* Phyllodia oblong-falcate or lanceolate, narrowed at both ends, mostly 4 to 6 in. long, ½ to near 1 in. broad in the middle, coriaceous, striate with numerous very fine nerves, 3 to 5 rather more prominent, the smaller ones occasionally anastomosing, and all free from the lower margin from the base. Spikes nearly sessile or shortly pedunculate, often clustered in the upper axils, 1 to 2 in. long. Flowers distinct or distant, mostly 5-merous but occasionally 4-merous. Calyx short, truncate or sinuate-toothed, pubescent or woolly. Pod (if correctly matched) linear, much twisted or irregularly coiled; valves hard, convex, about 2 lines broad. Seeds longitudinal.—DC. Prod. ii. 454; *Mimosa binervis,* Wendl. Bot. Beob. 56, quoted in Comm. Acac. 53; *A. homomalla,* Wendl. Comm. Acac. 49, t. 13 (from the figure and description); DC. Prod. ii. 454; *A. cinerascens,* Sieb. in DC. Prod. ii. 454; Bot. Mag. t. 3174; *A. leucadendron,* A. Cunn.; Benth. in Hook. Lond. Journ. i. 374.

Queensland. Brisbane river, Moreton Bay, *A. Cunningham, F. Mueller,* and others ; between the Severn and Condamine rivers, *Leichhardt.*
N. S. Wales. Frequent on Hunter's River, *R. Brown* and others; Nepean river, *A. Cunningham ;* Clarence river, *Beckler, C. Moore ;* Richmond river, *C. Moore ;* Blue Mountains, *Miss Atkinson,* also *Sieber, n.* 448 ; Bent's Basin, *Woolls.*

254. **A. Cunninghamii,** *Hook. Ic. Pl. t.* 165, *not of Don.* A shrub or small tree of 10 to 20 ft., glabrous or hoary-pubescent ; branchlets acutely 3-angled. Phyllodia falcate-oblong or lanceolate, narrowed at both ends, mostly 5 to 6 in. long and 1 to 1½ in. broad, or larger on barren shoots, with numerous parallel veins, 3 to 5 more prominent than the others, and 1 or 2 confluent with the lower margin near the very oblique base. Spikes 1½ to 3 in. long. Flowers mostly 5-merous, often distinct or distant. Calyx short, truncate or sinuate-toothed, usually glabrous. Petals smooth. Pod long, linear, very flexuose or twisted, 1 to 2 lines broad ; valves coriaceous, convex. Seeds longitudinal, but not seen ripe.

Queensland. Brisbane river, Moreton Bay, *A. Cunningham, F. Mueller,* and others ; sandy forests near Mount Owen, *Mitchell.*
N. S. Wales. Hunter's and Hastings rivers, *A. Cunningham.*
Distinguished from *A. glaucescens* by the very angular branches, the larger phyllodia and their venation.
Var. *longispicata.* Branches stout and still more angular. Phyllodia 6 to 8 in. long, 1 to 2 in. broad. Spikes 3 to 4 in. long.—*A. longispicata,* Benth. in Mitch. Trop. Austr. 298.—Near Mount Pluto and Lake Salvator, *Mitchell.*

255. **A. leptocarpa,** *A. Cunn. ; Benth. in Hook. Lond. Journ.* i. 376. Usually glabrous ; branchlets at first slightly angular, but soon terete. Phyllodia falcate-lanceolate, narrowed at each end, 4 to 6 in. long, 4 to 8 lines broad, with 3 or more fine slightly prominent nerves and very fine parallel ones between them, rarely anastomosing, and not nearly so close as in several allied species, the interval between each several times the breadth of the vein. Spikes 1½ to 2 in. long, solitary or in pairs. Flowers mostly 5-merous, usually glabrous, not very close. Calyx short, sinuate-toothed. Petals smooth, united at the base. Pod linear, straight or nearly so, several in. long, 1½ to 2 lines broad ; valves coriaceous, convex over the seeds, contracted between them. Seeds longitudinal ; funicle with the last 2 or 3 folds dilated into an oblong cup-shaped aril, nearly as long as the seed, but embracing its base only.

Queensland. Cape York, *W. Hill ;* Endeavour river and Cape Flinders, *A. Cunning ham ;* Shoalwater Bay, *R. Brown ;* Port Denison, *Fitzalan.*

256. **A. polystachya,** *A. Cunn. ; Benth. in Hook. Lond. Journ.* i. 376. Glabrous, young branches angular, but soon terete. Phyllodia falcate-oblong or lanceolate, narrowed at each end, 6 to 10 in. long, 1 to 1½ in. broad, very oblique at the base, with 3 to 5 prominent nerves, the intermediate ones fine and numerous, but not very closely packed and occasionally anastomosing. Spikes solitary or 2 or 3 together, slender, glabrous, 1 to 2 in. long. Flowers mostly 5-merous, not very close. Calyx sinuate-toothed, not half so long as the corolla. Petals united to the middle. Pod very flexuose, but not spiral, several inches long, 5 to 6 lines broad ; valves flat, thinly coriaceous. Seeds longitudinal in the centre of the pod ; funicle long, dilated

and coloured, the last 2 folds more than half encircling the seed, the next 2 extending along the other side so as nearly to surround it.

N. Australia. Port Essington, *A. Cunningham ;* and probably the same species, in flower only, islands of the Gulf of Carpentaria, *R. Brown.*

Queensland. Port Bowen, *A. Cunningham ;* Endeavour river, *W. Hill.*

Very like *A. leptocarpa,* but phyllodia usually larger and the pod and seed different.

257. **A. holcocarpa,** *Benth.* Glabrous; branchlets slender, terete. Phyllodia broadly falcate, narrowed at both ends and very oblique at the base, 4 to 5 or perhaps 6 in. long, ½ to ¾ in. broad, not glaucous, with 2 or 3 fine rather prominent nerves and very numerous, very fine, closely packed parallel veins between them. Spikes nearly sessile, about 1 in. long, slender but dense. Flowers mostly 5-merous, but sometimes 4-merous, small. Calyx deeply lobed, pubescent. Pod long and slender, straight or slightly curved, nearly terete, longitudinally sulcate-striate, about 2 lines diameter. Seeds oblong, longitudinal, embedded in what appears to be a dried pulp ; funicle short, scarcely folded, dilated into a short more or less oblique aril.

Queensland. Port Bowen and Thirsty Sound, *R. Brown.* Phyllodia precisely like those of *A. crassicarpa,* but the pod very different (*Herb. R. Brown*).

258. **A. plectocarpa,** *A. Cunn. ; Benth. in Hook. Lond. Journ.* i. 375. Glabrous and often somewhat glaucous, with acutely angular branchlets. Phyllodia usually falcate-lanceolate, narrowed at both ends, 4 to 6 in. long, 4 to 8 lines broad, resembling those of *A. julifera,* but varying from 2 or 3 in. long, coriaceous and nearly straight to above 6 in. long, narrow-linear and thin, about 3 nerves fine but more or less prominent, and numerous closely packed very fine parallel veins between them. Spikes slender, not very dense, about 1 in. long, the upper ones often forming a terminal leafy panicle. Flowers small, mostly 5-merous. Calyx short, minutely toothed. Petals smooth. Pod linear, not very long, usually 3 to 4 lines broad, coriaceous, at first flat, with straight slightly thickened margins, but becoming often very much undulate between them, and occasionally varying from under 3 lines broad and quite thin, to almost as broad and thick as in *A. pachycarpa.* Seeds ovate, obliquely transverse, the last 2 or 3 folds of the funicle dilated into an aril under the seed.

N. Australia. Cambridge Gulf and Regent's River, N.W. coast, *A. Cunningham ;* islands of the Gulf of Carpentaria, *R. Brown ;* Sturt's Creek, Roper, Seven-Emu, and Fitzmaurice rivers, *F. Mueller.*

259. **A. pachycarpa,** *F. Muell. in Journ. Linn. Soc.* iii. 139. A tall shrub or tree, glabrous and somewhat resinous, or the foliage glaucous with a minute pubescence ; branchlets slender, scarcely angular. Phyllodia linear-lanceolate or linear, slightly falcate, narrowed at both ends, 3 to 10 in. long or even more, 2 to 4 lines broad, coriaceous, apparently almost nerveless, but with numerous very fine parallel veins seen under a lens, and sometimes 1 to 3 more conspicuous nerves. Spikes solitary or in pairs, pedunculate, rather slender but very dense, about ½ in. long. Flowers small, mostly 5-merous. Calyx short, sinuate-toothed. Petals rather thick, united at the base. Pod straight and undulate as in *A. plectocarpa,* but thinly coriaceous, about 4 lines broad. Seeds orbicular, obliquely transverse ; funicle slightly folded and dilated into a small narrow-oblong obliquely lateral aril.

N. Australia. Sturt's Creek, *F. Mueller.* I had, in the above-quoted paper, referred this plant with doubt to *A. doratoxylon,* of which it has the foliage, and of which the fruit is unknown, but I now think it more probable that the latter has the pod of *A. acuminata,* and that this one is quite distinct, being closely allied to *A. plectocarpa.*

260. **A. tumida,** *F. Muell. in Journ. Linn. Soc.* iii. 144. A glabrous tree ; branchlets terete or nearly so, often glaucous. Phyllodia falcate-oblong or lanceolate, much narrowed and very oblique at the base, usually 4 to 8 in. long, 1 to 1½ in. broad, with very numerous parellel veins or nerves, 5 to 9 of them more prominent, the others very fine, closely packed and very rarely anastomosing. Spikes slender but dense, solitary or in pairs, the upper ones often paniculate, mostly 1 to 1½ in. long when fully out. Flowers mostly 5-merous. Sepals linear-spathulate, ciliate, shortly connected at the base. Petals smooth. Pod falcate, with very coriaceous convex valves so as to be nearly terete, 1½ to 3 in. long, 3 to 4 lines broad, divided inside between the seeds. Seeds ovate, obliquely transverse, funicle short, the last fold dilated into a small turbinate aril under the seed.

N. Australia. Isle Lacrosse, N.W. coast, *A. Cunningham ;* rocky places, Victoria river, Point Pearce, and Sturt's Creek, *F. Mueller ;* Attack Creek, *M'.Douall Stuart.* The foliage is that of *A. crassicarpa,* with the pod of *A. umbellata.*

261. **A. loxocarpa,** *Benth. in Hook. Lond. Journ.* i. 377. Glabrous and somewhat glaucous, the branchlets scarcely angular. Phyllodia long, lanceolate or linear, much narrowed at the base or at both ends, 3 to 6 in. long, 3 to 5 lines broad, much falcate except when short, with several fine but prominent nerves and numerous closely packed smaller ones between them, rarely anastomosing. Spikes slender but dense. Flowers very small, mostly 5-merous. Calyx thin, sinuate-toothed, half as long as the corolla. Petals smooth. Pod linear-cuneate, nearly 3 lines broad above the middle, tapering gradually to the base ; valves flat, hard, obliquely veined, rolling back elastically. Seeds obliquely transverse, very shining ; funicle very short, the last fold expanded into an obliquely cup-shaped aril embracing the base of the seed.

N. Australia. S. Goulburn Island, *A. Cunningham.* The pod and seed are nearly those of *A. drepanocarpa,* but the phyllodia are much broader and falcate, with much finer veins.

262. **A. oncinocarpa,** *Benth. in Hook. Lond. Journ.* i. 378. Glabrous, or slightly hoary with a very minute mealy pubescence ; branchlets terete. Phyllodia lanceolate-falcate, rather obtuse, but narrowed at both ends, 4 to 6 in. long, 6 to 8 lines broad, rather thin, with 5 to 7 fine, but rather prominent nerves, the lower ones often confluent with the lower margin of the base, and numerous fine parallel veins between them. Spikes slender, in pairs or clusters, pedunculate, 1 to 1½ in. long. Flowers often distant, small, mostly 5-merous or 6-merous. Calyx thin, loose, sinuate-toothed, more than half as long as the corolla. Petals smooth, united to the middle. Pod flat, but thick, hard and woody, straight except a hooked obtuse point, about 4 lines broad, much narrowed towards the base, obliquely veined. Seeds oblong, obliquely transverse ; funicle slightly flexuose, thickened nearly from the base into a narrow-turbinate aril under the seed.

N. Australia. Melville Island, *Herb. Fraser ;* Sims' Island, *A..Cunningham.* Near *A. loxocarpa,* with broader phyllodia and remarkable for the hooked pod.

263. **A. retinervis,** *Benth. in Hook. Lond. Journ.* i. 379, partly. Glabrous or slightly hoary, with nearly terete branchlets. Phyllodia falcate-lanceolate, narrowed at both ends, 3 to 6 in. long, 4 to 6 lines broad, with 3 to 7 fine slightly prominent nerves, the intermediate veins longitudinal, very fine, crowded and much anastomosed, forming a dense closely packed network only visible under a lens. Spikes in pairs or clusters, shortly pedunculate, slender but very dense, ½ to ¾ in. long. Flowers small, mostly 5-merous. Calyx with narrow spathulate ciliate lobes, above half the length of the corolla. Petals united to the middle, with prominent midribs. Pod unknown.

N. Australia. Cape Pond, N.W. coast, *A. Cunningham.* The general aspect is that of several falcate *Julifloræ*, but I cannot match the venation with that of any other species. The fruiting specimens I formerly referred here, have the veins much less anastomosed and belong to *A. tumida.*

264. **A. aulacocarpa,** *A. Cunn.; Benth. in Hook. Lond. Journ.* i. 378. Slightly hoary or ashy-glaucous with a minute almost powdery down, which at length disappears; branchlets angular. Phyllodia falcate-lanceolate, narrowed at both ends, 3 to 4 in. long, about ½ in. broad, with a few slightly prominent nerves, the lower ones confluent with the lower margin at the base, and numerous smaller closely packed veins, rarely anastomosing. Spikes slender, loose, 1 to 2 in. long, tomentose-pubescent or glabrous. Flowers mostly 5-merous. Calyx with short broad lobes. Petals united below the middle. Pod falcate-oblong, flat but thick, obtusely recurved at the end, 1 to 2 in. long, ½ to ¾ in. broad, much narrowed at the base, hard, obliquely veined. Seeds obliquely transverse, not seen perfect.

Queensland. Port Bowen, *R. Brown, A. Cunningham;* Rockhampton, *Dallachy;* Cameron's Brush, *Leichhardt?* (specimen not in fruit).
Var. (?) *macrocarpa.* Pods 3 to 5 in. long, ¾ to 1 in. broad, much undulate.—Keppel Bay, Shoalwater Bay, and Broad Sound, *R. Brown.*

265. **A. calyculata,** *A. Cunn.; Benth. in Hook. Lond. Journ.* i. 379. Glabrous or ashy-glaucous; branchlets very flat when young, with acute edges. Phyllodia falcate-obtuse, narrowed at the base, 2 to 3 in. long, about ½ in. broad, rather thick, with a few slightly prominent fine nerves and numerous very fine closely packed parallel veins, rarely anastomosing. Spikes mostly clustered, shortly pedunculate, slender, ½ to ¾ in. long. Flowers very small, mostly 5-merous, probably white (from Solander's ms. name *A. albiflora*). Calyx short, sinuate-toothed, pubescent or villous. Pod falcate-oblong, obtuse, narrowed at the base, flat but thick and hard and obliquely veined, resembling that of *A. aulacocarpa,* but not seen ripe.

Queensland. Endeavour river, *Banks and Solander;* Fitzroy Island, *A. Cunningham.* Pod of *A. aulacocarpa,* with shorter and more obtuse phyllodia, and the branchlets more flattened than in any other *Julifloræ* of the same subseries. The pods in Cunningham's herbarium are not attached, but carefully numbered to prevent their being mismatched.

266. **A. crassicarpa,** *A. Cunn.; Benth. in Hook. Lond. Journ.* i. 379. A handsome tree of 30 to 40 ft. or more, glabrous and somewhat glaucous or hoary with a minute powdery pubescence. Branchlets scarcely angular. Phyllodia falcate-oblong, narrowed at both ends, 5 to 8 in. long, 1 to 2 in. broad, very oblique, some of the principal nerves confluent with the lower margin at the base, and numerous fine parallel veins between them, very

rarely or not at all anastomosing. Spikes solitary or clustered, slender, not very dense, 1 in. long or rather more. Flowers mostly 5-merous. Calyx thin, sinuate-toothed, glabrous, about half as long as the corolla. Petals smooth but with the midrib prominent in the bud, united to the middle. Pod oblong, flat, thick, hard, obliquely veined, 2 to 3 in. long, ¾ to nearly 1 in. broad, obliquely truncate at the base, occasionally slightly twisted. Seeds oblique; funicle not seen.

N. Australia. Arnhem S. Bay, *R. Brown;* Goulburn and Sims Islands, *A. Cunningham;* Point Pearce, *F. Mueller;* Port Essington, *Armstrong;* Goold Island, *M'Gillivray;* Sweers Island, *Henne.*
Queensland. Albany Island, *W. Hill.*

267. **A. auriculiformis,** *A. Cunn.; Benth. in Hook. Lond. Journ.* i. 377. A small tree, glabrous and glaucous, with slightly angular branchlets. Phyllodia falcate-oblong, narrowed at both ends, 5 to 8 in. long, 1 to 2 in. broad, like those of *A. crassicarpa,* but the finer veins less crowded and occasionally anastomosing, the principal nerves, as in that species *A. polystachya* and others, confluent with or near the lower margin at the base. Flowers not seen, unless some of the flowering specimens referred to *A. polystachya* belong to this species. Pod hard, almost woody, as in *A. crassicarpa,* but very much twisted in an irregular spire, with the outer edge often sinuate as in some *Pithecolobiums;* valves obliquely veined, 6 to 8 lines broad.

N. Australia. S. Goulburn Island, *A. Cunningham.*
Queensland. Albany Island, *W. Hill, F. Mueller.*

E. DIMIDIATÆ.—Phyllodia large, broad, very oblique or falcate, with 3 or more distant prominent nerves, more or less confluent at or near the lower margin at the base, pinnately net-veined between them.

The phyllodia are nearly those of the *Plurinerves dimidiatæ,* but usually larger, more oblique and more coriaceous, and the inflorescence always spicate.

268. **A. latifolia,** *Benth. in Hook. Lond. Journ.* i. 382. Glabrous and glaucous; branchlets with 2 or 3 very much raised acute or almost winged angles. Phyllodia obliquely ovate-rhomboid or falcate, 3 to 6 in. long, 1½ to 2 in. broad, with 3 to 5 nerves confluent at the base at or near the lower margin, which is often slightly decurrent, pinnately net-veined between them. Spikes pedunculate, loose, 1 to 2 in. long. Flowers mostly 4-merous. Calyx very short, broad, truncate or obscurely toothed. Petals smooth, above 1 line long, united at the base but readily separating. Pod shortly stipitate, linear, straight or curved, 2 to 4 in. long, nearly 3 lines broad, flat with nerve-like margins, but not seen ripe. Seeds oblong, longitudinal; funicle scarcely folded, thickened into an oblique lateral aril.

N. Australia. Islands of the Gulf of Carpentaria, *R. Brown;* Arnhem's Land, *F. Mueller.*

269. **A. holosericea,** *A. Cunn. in G. Don, Gen. Syst.* ii. 407. Hoary or white with a close silky pubescence; branchlets with 3 much raised angles. Phyllodia obliquely oval-oblong, obtuse or mucronate, 4 to 6 in. long, 1 to 3 in. broad, or the lower ones much larger, with 3 or 4 prominent nerves confluent with the lower margin at the base, and pinnately net-veined

between them. Spikes sessile, often 2 in. long or more. Flowers mostly 5-merous. Calyx small, shortly lobed, pubescent. Petals pubescent, united at the base. Pod long-linear, irregularly or spirally twisted, 2 to 2½ lines broad ; valves convex. Seeds ovate, longitudinal ; funicle folded and dilated into a cup-shaped or turbinate aril at the base.—*A. neurocarpa*, A. Cunn. in Hook. Ic. Pl. t. 168.

N. Australia. Cambridge Gulf, N.W. coast, *A. Cunningham ;* Nichol Bay, *F. Gregory s Expedition ;* Victoria river, *F. Mueller ;* Attack Creek, *M'Douall Stuart ;* islands of the Gulf of Carpentaria, *R. Brown.*

Queensland. Endeavour river, *Banks and Solander ;* Rockhampton and Port Denison, *Thozet, Dallachy,* and others ; Edgecombe Bay, *Dallachy.*

Var. *pubescens,* F. Muell. Everywhere softly pubescent, even the pod.—Victoria river, *F. Mueller.*

270. **A. dimidiata,** *Benth. in Hook. Lond. Journ.* i. 381. Hoary with a minute pubescence or nearly glabrous ; branchlets scarcely angular. Phyllodia broadly and obliquely ovate-rhomboid, obliquely truncate at the base, usually 3 to 4 in. long, 2 to 3 in. broad, but on some barren shoots twice or three times that size, with 4 or 5 prominent nerves more or less confluent with the lower margin at the base, and pinnately net-veined between them. Spikes sessile or shortly pedunculate, usually in pairs, 1 to 2 in. long. Flowers mostly 5-merous. Calyx angular, lobed, readily separating into distinct sepals. Petals united below the middle. Pod linear, nearly straight, 2 to 6 in. long, 2 to 3 lines broad ; valves coriaceous, very convex. Seeds longitudinal, ovoid-oblong ; funicle short, thickened into a turbinate or obliquely cup-shaped aril at the base of the seed.—*A. dolabriformis,* A. Cunn. in Hook. Ic. Pl. t. 169, not of Wendl.

N. Australia. Various parts of the N. coast, *A. Cunningham ;* Victoria river and M'Adam Range, *F. Mueller ;* Port Essington, *Armstrong,* and probably the same species, in leaf only, islands of the Gulf of Carpentaria, *R. Brown.*

271. **A. humifusa,** *A. Cunn. ; Benth. in Hook. Lond. Journ.* i. 382. Diffuse or prostrate, softly pubescent or tomentose ; branchlets nearly terete. Phyllodia broadly and obliquely ovate-rhomboid or almost orbicular, 1½ to 2 in. long and nearly as broad, or in luxuriant shoots nearly twice that size, often undulate, with 3 to 5 nerves more or less confluent with the lower margin at the base, and pinnately reticulate between them. Spikes sessile, oblong, dense, scarcely exceeding ½ in. Flowers mostly 5-merous. Calyx deeply lobed, pubescent. Petals densely pubescent, united at the base. Pod linear, nearly straight, thick, and nearly terete, 1½ to 3 in. long, 2½ to 3 lines broad, coriaceous, pubescent. Seeds oblong, longitudinal ; funicle with the last 1 or 2 folds thickened into an obliquely cup-shaped aril at the base of the seed.

N. Australia. Victoria river, *F. Mueller ;* islands of the Gulf of Carpentaria, *R. Brown.*

Queensland. Endeavour river, *Banks and Solander ;* Cape Cleveland, *A. Cunningham ;* Lizard Island, *M'Gillivray ;* Albany Island, *W. Hill.*

DIVISION II. *BIPINNATÆ.*

Leaves all bipinnate, most frequently with a depressed or shield-shaped gland on the common petiole near the base, and often smaller glands on the partial rhachises at or below the last leaflets, and sometimes numerous glands along the whole general and partial rhachises, but the glands often inconstant in the same species.

SERIES IX. BOTRYOCEPHALÆ.—Unarmed trees or shrubs. Leaves bipinnate. Stipules small or none. Flower-heads globular, in axillary racemes or terminal panicles. Pod (where known) flat. Seeds longitudinal, the last fold of the funicle forming a short lateral or oblique aril, with very small folds below it. The species are all confined to Australia.

272. **A. elata,** *A. Cunn.; Benth. in Hook. Lond. Journ.* i. 383. A handsome tree of 60 ft. or more, the young shoots often tinged with a golden-yellow pubescence. Pinnæ in 2 to 4 distant pairs, 6 to 8 in. long; leaflets 8 to 12 pairs, lanceolate, acutely acuminate, 1 to 2 in. long, minutely silky; gland wart-like on the petiole, and often small ones at the last pairs of leaflets. Flower-heads globular, in racemes often 6 in. long, the upper ones forming a large terminal panicle, often silky with a golden pubescence. Flowers numerous, mostly 5-merous. Calyx fully half as long as the corolla, obtusely toothed. Petals smooth, united to the middle. Pod 4 to 6 in. long, about ½ in. broad. Seeds nearly lenticular.

N. S. Wales. Grose river, *R. Brown ;* shaded ravines of the Blue Mountains, *A. Cunningham, Miss Atkinson,* and others, and southward to Illawarra, *Shepherd.*

273. **A. pruinosa,** *A. Cunn.; Benth. in Hook. Lond. Journ.* i. 383. Glabrous and glaucous, with terete branchlets. Pinnæ 2 to 4 or rarely 5 pairs, 3 to 4 in. long; leaflets 12 to 20 pairs, oblong or linear, oblique or somewhat falcate, obtuse or scarcely acute, attaining ½ in. or rather more; gland prominent on the petiole at a distance from the lowest pair of pinnæ, and smaller ones under several pairs of leaflets. Flower-heads globular, often numerous in axillary racemes or the upper ones paniculate. Flowers numerous, mostly 5-merous. Calyx obtusely toothed. Pod not seen.—*A. schinoides,* Benth. in Hook. Lond. Journ. i. 383.

N. S. Wales. Near Sydney and Liverpool plains, *A. Cunningham ;* New England, *C. Stuart.*

274. **A. spectabilis,** *A. Cunn.; Benth. in Hook. Lond. Journ.* i. 383. A tall shrub, glabrous and glaucous, or the branchlets and petioles shortly hirsute. Pinnæ 2 to 4 pairs; leaflets 4 to 8 pairs, obovate-oblong, very obtuse, 4 to 6 lines long, rather thick and obscurely veined; gland depressed at the lowest pair of pinnæ, often very obscure. Flower-heads in axillary racemes longer than the leaves, the upper ones often paniculate. Flowers mostly 5-merous. Calyx short, obtusely toothed. Petals united at the base only. Pod 3 to 4 in. long, about ½ in. broad, glaucous.—Bot. Reg. 1843, t. 46 ; *A. chrysobotrys,* Meissn. Ind. Sem. Hort. Basil. 1842, from the character in Walp. Rep. ii. 906.

Queensland. Brisbane river, *A. Cunningham ;* between the Severn and Condamine rivers, *Leichhardt ;* forest near Harvey's Range and Maranoa river, *Mitchell.*

N. S. Wales. Between Lachlan and Macquarrie rivers and forest land E. and W. of Wellington valley, *A. Cunningham :* New England, *C. Stuart.*

Var. (?) *Stuartii.* Leaflets 10 to 15 pairs and rather narrower, but glands as in *A. spectabilis.*—New England, *C. Stuart ;* between Byron plains and the M'Intyre, *Leichhardt.*

275. **A. polybotrya,** *Benth. in Hook. Lond. Journ.* i. 384. A tall shrub, the foliage more or less pubescent. Pinnæ usually 2 or 3 pairs, leaflets 6 to 10 pairs, narrow-oblong, obtuse, 3 to 4 lines long, rather thick with a prominent nerve near the lower edge, the rhachis terminating in a recurved deciduous point; a gland at the base of the petiole, those between the leaflets rare and minute. Flower-heads numerous, small, in racemes much exceeding the leaves, the upper ones forming a terminal panicle. Flowers mostly 5-merous. Calyx short, obtusely lobed. Petals united at the base. Pod unknown.

N. S. Wales. Boggy forest land of the N. W. interior, the most beautiful of all the Acacias, *Fraser;* Gwydir river, *A. Cunningham;* near Wellington, *C. Moore*

Var. *foliolosa.* Softly pubescent. Pinnæ 4 to 6 pairs, 2 to 3 in. long; leaflets 15 to 25 pairs, 3 to 6 lines long and less obtuse.

Queensland. Burnett river, *F. Mueller;* S. part of the colony, *Bowman;* limestone hills, *Leichhardt;* Ipswich, *Nernst.*

276. **A. discolor,** *Willd. Spec. Pl.* iv. 1068. A tall shrub or tree, branchlets terete or angular, glabrous or pubescent. Pinnæ 2 to 6 pairs, leaflets 10 to 15 pairs, oblong, obtuse or acute, 3 to 4 lines long, rather firm, 1-nerved, glabrous, pale underneath; gland usually large on the petiole and a few small ones at the upper pairs of leaflets. Flower-heads in axillary racemes, the upper racemes forming a terminal panicle; flowers 6 to 15 in the head, rather large, 5-merous. Calyx short, broadly lobed, ciliate. Petals rather rigid, with prominent midribs, striate in the bud. Pod 1 to 3 in. long, 5 to 6 lines broad. Seeds longitudinal; funicle filiform.—*Mimosa discolor,* Andr. Bot. Rep. t. 235 ; *M. paniculata,* Wendl. Bot. Beob. 57 ; *M. botrycephala,* Vent. Hort. Cels. t. 1 ; *Acacia botrycephala,* Desf. Cat. Hort. Par. ed. 3, 300 ; *A. discolor,* DC. Prod. ii. 468 ; Bot. Mag. t. 1750 ; Lodd. Bot. Cab. t. 601 ; Hook. f. Fl. Tasm. i. 111 ; F. Muell. Pl. Vict. ii. 34 ; *A. maritima,* Benth. in Hook. Lond. Journ. i. 384 (with more glabrous and angular branchlets); *A. Sieberiana,* Scheele in Linnæa, xvii. 337.

N. S. Wales. Port Jackson to the Blue Mountains, *R. Brown, Sieber, n.* 454, and others; Macleay river, *Fraser.*

Victoria. Heath ground as well in the lowlands as in the mountains of Gipps' Land, *F. Mueller.*

Tasmania. Port Dalrymple, *R. Brown.* Common near the seacoast in various localities, *J. D. Hooker.*

Var. (?) *angustifolia.* Branches nearly terete, leaflets linear-oblong, slightly falcate and rather more numerous.—Port Jackson, *Caley, R. Brown.*

277. **A. decurrens,** *Willd. Spec. Pl.* iv. 1072. A handsome tree, glabrous or more or less tomentose-pubescent; branches more or less prominently angled, sometimes almost winged. Pinnæ 8 to 15 pairs or sometimes even more, rarely reduced to 5 or 6, leaflets very numerous (30 to 40 pairs or even more), linear, from under 2 lines to nearly 5 lines long, according to the variety. Flower-heads small, globular in axillary racemes, the upper ones forming a terminal panicle. Flowers 20 to 30 in the head, mostly 5-merous. Calyx short, broadly lobed, ciliate. Petals with slightly prominent midribs. Pod usually 3 to 4 in. long, about ¼ in. broad or rather more, more or less contracted between the seeds. Seeds ovate.—F. Muell. Pl. Vict. ii. 35.

Queensland. Plains of the Condamine, *Leichhardt;* Moreton Bay, *C. Stuart.*

N. S. Wales. Port Jackson to the Blue Mountains, *R. Brown, Sieber, n.* 436, 460, and others; northward to Hastings and Clarence rivers, *Beckler ;* New England, *C. Stuart.*

Victoria. Frequent along river banks, in valleys, etc., ascending to subalpine elevations, forming the main underwood in Eucalyptus forests, *F. Mueller.*

Tasmania. Abundant throughout the island, *J. D. Hooker.*

S. Australia. Mount Gambier, *F. Mueller.*

Of this, the *Black* or *Green Wattle* of the colonists, the following forms appear at first sight very distinct, but pass into each other by many gradations.

a. *normalis.* Glabrous or the young shoots slightly tomentose-pubescent. Leaflets long and narrow, usually 3 to 4 lines ; glands numerous along the primary rhachis.—*Mimosa decurrens,* Wendl. Bot. Beob. 57 ; Vent. Jard. Malm. t. 61 ; *A. decurrens,* DC. Prod. ii. 470 ; *A. angulata,* Desv. Journ. Bot. 1814, ii. 68 ; DC. Prod. ii. 468 ; *A. sulcipes,* Sieb. Pl. Exs. ; *A. adenophora,* Spreng. Syst. iii. 140.—Chiefly about Port Jackson.

b. *mollis,* Lindl. Bot. Reg. t. 371. Foliage softly tomentose-pubescent, the indumentum assuming a golden-yellow tinge on the young shoots. Leaflets 2 to 3 lines long, obtuse ; glands numerous along the primary rhachis.—*A. mollissima,* Willd. Enum. 1053 ; DC. Prod. ii. 470 ; Sweet, Fl. Austral. t. 12 ; Hook. f. Fl. Tasm. i. 117.—The only form in Tasmania and the most common one in Victoria ; less frequent in the northern districts of N. S. Wales. Some of Beckler's specimens from Warwick have the numerous glands of this form with the very small leaflets of the following.

c. *pauciglandulosa,* F. Muell. Pubescent but not so softly so as in the var. *mollis,* and sometimes almost hirsute, with the same golden-yellow tinge on the young shoots. Leaflets small, often under 2 lines ; glands few, often only under the last 1 or 2 pairs of pinnæ.— New England, Clarence and Hastings rivers, Moreton Bay, etc. ; also between Archer's and M'Kenzie's stations in moist places, *Leichhardt.*

Var. (?) *Leichhardtii.* More or less hirsute with spreading hairs. Leaflets small, narrow ; glands few.— Between Archer's station and Birou and towards the Bunya, *Leichhardt.* This seems to connect in some measure *A. decurrens* with *A. pubescens,* but it has the numerous pinnæ of the former.

278. **A. dealbata,** *Link, Enum. Hort. Berol.* 445. A handsome tree, closely resembling the var. *mollis* of *A. decurrens,* and to be added perhaps to the varieties of that species as proposed by F. Mueller, but the branches and foliage are very glaucous or hoary with a minute pubescence not assuming a golden tinge on the young shoots. Pinnæ usually 10 to 20 pairs, leaflets 30 to 40 pairs, linear, crowded, 2 to 3 lines long ; glands usually numerous. Flower-heads small, in axillary racemes paniculate at the ends of branches, as in *A. decurrens.* Pod broader, not contracted between the seeds and more glaucous.—DC. Prod. ii. 470 ; Lodd. Bot. Cab. t. 1928 ; Hook. f. Fl. Tasm. i. 111 ; *A. irrorata,* Sieb. in Spreng. Syst. iii. 141.

N. S. Wales. Port Jackson or Blue Mountains, *Sieber, n.* 446 ; banks of the Macquarrie, and of all the streams falling westerly into the interior, *A. Cunningham, Fraser,* and others.

Victoria. Mostly on river banks or in valleys, flowering usually earlier than *A. decurrens,* var. *mollis, F. Mueller.*

Tasmania. Port Dalrymple, Derwent river, *R. Brown ;* abundant in the island with *A. decurrens,* var. *mollis,* flowering at the same time, but universally distinguished, *J. D. Hooker.*

This, the *Silver Wattle* of the colonists, is unhesitatingly united with *A. decurrens* by F. Mueller; J. D. Hooker considers it as sufficiently distinct, although not easy to characterize from dried specimens. The shape of the pod is different as far as known, but the specimens of the several forms of *A. decurrens,* from many stations, are in flower only.

279. **A. cardiophylla,** *A. Cunn. ; Benth. in Hook. Lond. Journ.* i. 385. A tall shrub ?, pubescent with short rather rigid hairs, branches terete or obscurely angled. Pinnæ 12 to 15 pairs, 3 or 4 lines or the terminal ones

nearly ½ in. long ; the common petiole 1 to 2 in. ; leaflets 6 to 10 pairs, ovate or almost cordate, ½ to ¾ line long ; glands few and minute. Flower-heads small, in axillary racemes, each with above 20 flowers, mostly 5-merous. Calyx turbinate, shortly lobed, ciliate. Petals glabrous, with prominent midribs, but scarcely fully out in the specimens seen. Pod unknown.

N. S. Wales. Eurylean scrub, N. of Macquarrie river, *A. Cunningham*.

280. **A. leptoclada,** *A. Cunn. ; Benth. in Hook. Lond. Journ.* i. 385. A shrub, either glabrous and somewhat glaucous or hispid with scattered short stiff hairs. Pinnæ 3 to 5 pairs, 3 to 4 lines long, on a common petiole of ¼ to ½ in. ending in a recurved point ; leaflets 6 to 10 pairs, oblong, coriaceous, ½ to 1½ lines long ; glands several, usually small. Flower-heads small, hispid with long bristly hairs proceeding from the bracts and sepals, in racemes longer than the leaves. Flowers numerous, closely packed, the corolla not protruding in the bud, apparently ready to open, but not seen fully out. Sepals distinct, linear-spathulate with concave tips. Petals also free in the bud. Pod unknown.

N. S. Wales. Liverpool plains, *A. Cunningham ;* between Wyndham and the M'Intyre, *Leichhardt*.
Var. (?) *polyphylla.* Pinnæ 10 to 12 pairs, leaflets 10 to 20 pairs. Flowers not seen. **Queensland.** E. coast, *R. Brown*.

281. **A. pubescens,** *R. Br. in Ait. Hort. Kew. ed.* 3, v. 467. A shrub, the branches petioles and rhachis hirsute with spreading hairs. Pinnæ 3 to 10 pairs ½ to ¾ in. long, the common petiole about the same length ; leaflets 6 to 20 pairs, crowded, linear, obtuse, 1 to 2 lines long, glabrous. Flowerheads small, in slender racemes longer than the leaves, and paniculate at the ends of the branches. Flowers not numerous, glabrous. Calyx short, sinuate-toothed. Corolla smooth, protruding in the bud, the petals united.— *Mimosa pubescens*, Vent. Jard. Malm. t. 21 ; Bot. Mag. t. 1263 ; *A. pubescens*, DC. Prod. ii. 468 ; Maund, Botanist, t. 48 ; Reichb. Ic. et Descr. Pl. t. 73.

N. S. Wales. Port Jackson to the Blue Mountains, *R. Brown, Sieber, n.* 466 and others.

Series X. Pulchellæ.—Trees or shrubs, unarmed or rarely with axillary spines, without scattered prickles or stipular spines. Leaves bipinnate. Stipules small or none. Flowers in globular heads or rarely in cylindrical spikes, on simple solitary or clustered axillary peduncles. Pods flat, straight or falcate. The species are all Australian.

282. **A. pulchella,** *R. Br. in Ait. Hort. Kew. ed.* 3, v. 464. An elegant shrub, the slender branches quite glabrous or more or less hirsute with spreading hairs, and usually armed with subulate axillary spines (abortive branches or peduncles). Pinnæ 1 pair or very rarely a single one, on a common petiole sometimes exceedingly short, sometimes ¼ in. long, the rhachis of the pinnæ usually under ¼ in. rarely ½ to ¾ in. long ; leaflets 4 to 7 pairs, obovate-oblong or linear-oblong, obtuse, 1 to 2 lines rarely 3 lines long ; gland on a long stipes between the pinnæ or often none. Peduncles axillary, bearing each a globular head of flowers, usually 5-merous. Calyx sinuate-toothed, about half as long as the corolla, often readily separating into distinct sepals. Petals with prominent midribs but not so striate as in *A. strigosa.* Pod flat,

with thickened margins 1 to 2 in. long, 1½ to 2½ lines broad. Seeds longitudinal; funicle thickened into a small club-shaped appressed aril under the
seed, with a short filiform fold below it.—DC. Prod. ii. 455 ; Lodd. Bot.
Cab. t. 212 ; Meissn. in Pl. Preiss. i. 22 ; Paxt. Mag. iv. 198, with a fig.

W. Australia. Very common from the S. coast to Swan and Murchison rivers, *R.
Brown* and others; *Drummond, 1st Coll., n.* 308 to 312, *2nd Coll., n.* 116, 139, 156 ;
Preiss, n. 884, 886, 892 to 900, 904, 907, 908, 909, 911, 912, etc. A polymorphous
species, especially as to the numbers and size of the leaflets, and hairiness, but generally
known by the single pair of pinnæ and the axillary spines, which are rarely entirely wanting,
although often some branches are without them. It then resembles some forms of *A. strigosa*, but the pod is quite different. The following have been distinguished as species, but
are connected by too many intermediate forms to be separable even as varieties :—1. *A. denudata*, Lehm. Del. Sem. Hort. Hamb.; Meissn. in Pl. Preiss. i. 21, quite glabrous, very
spinescent, leaflets usually few ; 2. *A. fagonioides*, Benth. in Hook. Lond. Journ. i. 387,
glabrous or scarcely pubescent, leaflets few, small broad ; 3. *A. grandis*, Henfr. in Gard.
Mag., with a fig., copied into Lemaire, Jard. Fleur. t. 154, glabrous with more numerous
longer leaflets ; 4. *A. hispidissima*, DC. Prod. ii. 455 ; Bot. Mag. t. 4588, copied into
Lemaire, Jard. Fleur. t.·160 (*A. lasiocarpa*, Benth. in Hueg. Enum. 43 ; Meissn. in Pl.
Preiss. i. 22) ; branches very hirsute with long spreading hairs ; leaflets narrow ; pod hirsute ; 5. *A. cycnorum*, Benth. in Hook. Lond. Journ. i. 388 ; *A. cygnorum*, Meissn. in Pl.
Preiss. i. 22 ; pubescent or hirsute ; leaflets narrow, revolute, often very small ; gland generally wanting ; pod flexuose, hirsute.

283. **A. Mitchelli,** *Benth. in Hook. Lond. Journ.* i. 387. A shrub of
a few ft., the branches not much divided, nearly terete, softly pubescent, unarmed. Pinnæ 2 to 3 pairs, the common petiole and partial rhachises pubescent, each usually under 3 lines, rarely above 4 lines long ; leaflets 3 to 6
pairs, oblong, obtuse, 1 to 2 lines long, rather thick. Peduncles slender, as
long as the leaves, bearing each a globular head of numerous small flowers,
mostly 5-merous. Sepals linear-spathulate, ciliate, free or united at the base.
Petals smooth, united to the middle. Pod stipitate, straight or falcate, flat
with nerve-like margins, 1 to 2 in. long, 2½ to 3 lines broad. Seeds ovate,
longitudinal ; funicle dilated into an obliquely-oblong clavate appressed aril,
with a short filiform fold below it.—F. Muell. Pl. Vict. ii. 33, t. suppl. 12.

Victoria. Mount Zero, *Mitchell ;* sterile ridges, Grampians, mouth of the Glenelg,
bushy Iron-bark-tree ridges between Ovens river and May-day hills, *F. Mueller.*

284. **A. pentadenia,** *Lindl. Bot. Reg. t.* 1521. A tall glabrous unarmed shrub ; branchlets usually 4-angular. Pinnæ 2 to 5 pairs, 1½ to 3 in.
long ; leaflets 20 to 30 pairs, very obliquely ovate, broadly oblong or almost
rhomboidal, with a broad oblique base, 1, 2, or nearly 3 lines long, the margins usually recurved ; glands below all the pairs of pinnæ. Peduncles slender,
clustered, ½ in. long or rather more, bearing each a globular head of about 20
flowers, mostly 5-merous. Calyx not half as long as the corolla, shortly
toothed. Petals smooth. Pod flat, with thickened margins, 1 to 1½ in. long,
2 to 2½ lines broad ; valves hard, rolling back elastically. Seeds ovate, longitudinal ; funicle thickened into a small club-shaped aril.—*A. biglandulosa,*
Meissn. in Pl. Preiss. ii. 205.

W. Australia. King George's Sound and adjoining districts, *R. Brown, Drummond,
3rd Coll., n.* 97, *Oldfield.*

285. **A. Gilberti,** *Meissn. in Pl. Preiss.* ii. 204. A glabrous unarmed
shrub, branches scarcely angular. Pinnæ 1 pair on a common petiole of about

½ in., the rhachis of the pinnæ often above 1 in.; leaflets 4 to 6 pairs, obliquely oblong, 4 to 6 lines long. Peduncles 2 or 3 together, slender, about ½ in. long, bearing each a head of 3 to 8 rather large globular 4-merous flowers. Calyx very short, truncate or sinuate-toothed. Petals smooth, nearly 1½ lines long, united at the base. Pod flat, coriaceous, almost woody, acuminate, narrowed at the base, about 3 lines broad, with broad margins. Seeds ovate, longitudinal; funicle small, thickened into a very small oblong club-shaped aril with a short filiform fold below it.

W. Australia, *Drummond, 2nd Coll., n.* 157; Princess Royal Harbour and Mount Bakewell, *Preiss, n.* 887 (*partly*) *and* 891. Foliage nearly of *A. nigricans,* but flowers and pod very different.

286. **A. nigricans,** *R. Br. in Ait. Hort. Kew. ed.* 3, v. 465. A glabrous unarmed shrub, branchlets scarcely angular. Pinnæ 1 or 2 pairs on a short common petiole, the partial rhachis ½ to 1 in. long; leaflets 5 to 10 pairs, or fewer in the lower pinnæ, obovate-lanceolate or linear-oblong, 3 to 5 lines long, flat, or with recurved margins. Peduncles about ½ in. long, bearing each a globular head of 20 to 30 or more 5-merous flowers. Calyx not half as long as the corolla, truncate or sinuate-toothed. Petals rather thick, more or less conspicuously striate. Pod flat, 1 to 1½ in. long, about 3 lines broad, with thickened margins, the valves rather hard, rolling back elastically. Seeds ovate, transverse, the last fold and part of the lower fold of the funicle thickened into a small oblong appressed aril.—*Mimosa nigricans,* Labill. Pl. Nov. Holl. ii. 88, t. 238, not of Vahl; *A. nigricans,* DC. Prod. ii. 466; Bot. Mag. t. 2188; Lodd. Bot. Cab. t. 313; Meissn. in Pl. Preiss. i. 20; *A. rutæfolia,* Link, Enum. Hort. Berol. 444.

W. Australia. King George's Sound and adjoining districts, *Menzies, R. Brown, Drummond, n.* 314 *and 5th Coll. n.* 18, *Preiss, n.* 887 (*partly*), and others.

287. **A. obscura,** *A. DC. Not.* 6. *Pl. Rar. Jard. Gen.* 23. *t.* 3. An unarmed shrub, more or less pubescent or hirsute, branchlets scarcely angular. Pinnæ 1, 2 or very rarely 3 pairs on a short common petiole, leaflets 5 to 10 or rarely more pairs, from ovate to linear-oblong, 2 to 3 or rarely above 4 lines long. Peduncles slender, clustered, bearing each a small globular head of about 12 to 15 5-merous flowers. Calyx short, truncate or sinuate-toothed. Petals somewhat striate. Pod flat, about 1 in. long and 3 lines broad, with thickened margins, the valves rolling back elastically. Seeds ovate, transverse; funicle thickened into a small oblong-clavate aril.—Meissn. in Pl. Preiss. i. 20; *A. Preissiana,* Lehm. Del. Sem. Hort. Hamb. 1842; *A. cycnorum,* Hook. Bot. Mag. t. 4653, copied into Lemaire, Jard. Fleur. t. 322, not of Benth.

W. Australia. Goose Island Bay, *R. Brown;* King George's Sound and adjoining districts, *Baxter, Drummond, 2nd Coll. n.* 153, *4th Coll. n.* 18, *5th Coll. n.* 17, *Preiss, n.* 885, 889 (*partly*); eastward to W. Mount Barren, *Maxwell;* Gordon river, *Oldfield.* Closely allied on the one hand to *A. strigosa,* on the other to *A. nigricans,* differs from the former chiefly in the more numerous leaflets, from the latter in the hispid stems.

288. **A. strigosa,** *Link, Enum. Hort. Berol.* ii. 444. An unarmed shrub of 2 to 4 ft.; branchlets slender, terete, pubescent hirsute or rarely almost glabrous. Pinnæ 2 or rarely 1 pair, the common petiole and partial rhachis each usually 1 to 2 lines long, leaflets 1 to 4 pairs, oblong, obtuse, 1 to 2 lines long, glabrous or ciliate with short rigid hairs. Peduncles slender

longer than the leaves, bearing each a small globular head of 12 to 15 flowers, mostly 5-merous. Calyx about half as long as the corolla, truncate or sinuate-toothed. Petals striate. Pod flat with thickened margins, rarely above 1 in. long and often shorter, about 3 lines broad, the valves rolling back elastically. Seeds ovate, transverse; funicle thickened almost from the base into a small aril of 2 or 3 folds under the seed.—DC. Prod. ii. 466; *A. ciliata*, R. Br. in Ait. Hort. Kew. ed. 3. v. 465, not of Willd.; *A. Browniana*, Wendl. in Flora, 1819, 139.

W. Australia. King George's Sound and adjoining districts, *R. Brown*, *Drummond*, *5th Coll. n.* 16, *Preiss, n.* 902, and others.

Var. *Endlicheri*. Leaflets 5 to 7 pairs, not ciliate, but much smaller than in *A. nigricans.*—*A. Endlicheri*, Meissn. in Pl. Preiss. i. 21. With the common form.

289. **A. Drummondii,** *Lindl. Swan Riv. App.* 15. An unarmed shrub, branchlets furrowed, minutely hoary or pubescent. Pinnæ 2 pairs; common petiole rarely above ¼ in. long; partial rhachises longer, but rarely ½ in., terminating in straight points; leaflets 2 to 6 pairs, oblong-linear, about 3 to 4 lines long in the terminal pair, shorter in the others. Peduncles solitary, often exceeding the leaves, bearing each a cylindrical spike of ½ to 1 in. or rarely longer. Flowers mostly 5-merous. Calyx pubescent or hirsute, more or less lobed, half as long as the corolla. Petals not striate but somewhat angular in the bud, usually sprinkled with a few rigid hairs. Pod not exceeding 1 in. long, about 3 lines broad, flat with thickened margins, glabrous or pubescent. Seeds transverse; funicle short, thickened into a small aril of about 2 folds under the seed.—Meissn. in Pl. Preiss. i. 23; Lemaire, Jard. Fleur. t. 378; Bot. Mag. t. 5191.

W. Australia. Swan River and thence to King George's Sound, *Drummond*, *1st Coll.*, *2nd Coll. n.* 152, *5th Coll. n.* 15; Darling range, *Preiss, n.* 901; Vasse river, *Mrs. Molloy*; Perongerup ranges, *Maxwell*; swamps, King George's Sound, *Oldfield*.

Var. *major*. Leaflets few, obliquely obovate or oblong, 4 to 6 lines long; flowers rather large.—*A. Candolleana*, Meissn. in Pl. Preiss. ii. 206. With the original form, especially at King George's Sound, and often scarcely distinguishable from it.

Var. (?) *parviflora*. Hirsute. Leaflets small with revolute margins. Spikes small, cylindrical, loose. Flowers much smaller than in the ordinary form. Pod unknown.—Arthur river, *Oldfield*; W. Mount Barren, *Maxwell*.

Var. (?) *ovoidea*. Very hirsute. Leaflets small with revolute margins. Spikes ovoid or oblong, ¼ to ½ in. long. Flowers as in the typical form. Pod unknown.—Towards Cape Riche, *Maxwell*.

SERIES IX. GUMMIFERÆ.—Trees or shrubs, without scattered prickles, but more or less armed with persistent spinescent stipules. Flowers in globular heads or (in species not Australian) in cylindrical spikes on simple solitary or clustered peduncles. Pod very various.

The species of this series are numerous in S. America, in Asia, and especially in Africa, and are almost all remarkable for the very great diversity in the size of the stipular spines, even on the same branch.

290. **A. farnesiana,** *Willd. Sp. Pl.* iv. 1083. A much-branched shrub, attaining a considerable size, quite glabrous or slightly pubescent on the petioles and peduncles. Pinnæ 4 to 6 or rarely more pairs; leaflets usually 10 to 20 pairs, linear, about 2 lines long or on luxuriant shoots often much longer. Stipules converted into slender straight thorns, very variable

2 E 2

in length, occasionally ¾ in. long, and sometimes all very minute or almost none, the plant otherwise unarmed. Peduncles usually 2 or 3 together in the older axils, each bearing a globular head of numerous 5-merous flowers. Bracts small, close under the flower-head. Calyx above half as long as the corolla. Pod thick, irregularly cylindrical or spindle-shaped, 2 to 3 in. long, indehiscent, filled with a pithy substance in the midst of which lie the seeds. Seeds obliquely transverse, with short funicles.—DC. Prod. ii. 461 ; *Vachellia farnesiana,* W. and Arn. Prod. 272, with the synonymy adduced ; Wight, Ic. t. 300 ; *A. lenticellata,* F. Muell. in Journ. Linn. Soc. iii. 147.

N. Australia. N.W. coast, *A. Cunningham ;* Nichol Bay, *F. Gregory's Expedition ;* Sturt's Creek and M'Arthur river, *F. Mueller ;* Albert river, *Henne.*
Queensland. Port Denison, *Fitzalan ;* in the interior, *Mitchell.*
N. S. Wales. In the interior, *A. Cunningham ;* Darling river to Cooper's Creek, common, *Victorian and Howitt's Expeditions.*

The species is very common in tropical countries in the New and the Old World, much planted for ornament or for the perfume extracted from its flowers, and readily spreading. Believed by some to be of American origin, by others to be truly indigenous also in Africa and Asia, and has every appearance of being so in Australia.

291. **A. suberosa,** *A. Cunn.; Benth. in Hook. Lond. Journ.* i. 499. Branches terete or nearly so, more or less hirsute, the older ones with a slightly corky bark. Pinnæ 1 or 2 pairs, the common petiole about ¼ in., the partial rhachis nearly ½ in. long ; leaflets 8 to 12 pairs, oblong-linear, thick and rigid, obtuse, 1 to 2 lines long, more or less hirsute or ciliate. Stipules spinescent, short, slender, and straight. Flowers not seen, except some fragmentary remains on a short thick peduncle. Pod flat, but rather thick, with convex valves, about 3 in. long, ¼ to nearly ½ in. broad, pubescent. Seeds longitudinal, broadly ovate, woolly-pubescent. Funicle folded and slightly thickened under the seed.

N. Australia. Vansittart Bay and Careening Bay, *A. Cunningham ;* Glenelg river, *J. Martin.*

292. **A. Bidwilli,** *Benth. in Linnæa,* xxvi. 629. Glabrous ; branches mostly terete. Pinnæ 15 to 20 pairs, the common petiole 2 to 3 in., the partial rhachis ½ to 1 in. long ; leaflets 15 to 25 pairs, oblong, obtuse, rigid, scarcely 1 line long. Stipules spinescent and sometimes 2 to 3 lines long, usually very small or quite obsolete. Peduncles solitary, with an annual deciduous 4-lobed bract about the middle, bearing a globular head of about 20 or rather more 4-merous flowers and sometimes 1 or 2 lower down the peduncle, each flower often 2 lines long. Calyx shortly toothed. Petals smooth, united above the middle. Pod straight, 3 to 6 in. long, about ½ in. broad, narrowed at the base ; valves coriaceous, somewhat convex, reticulate lengthwise. Seeds large, ovate, longitudinal ; funicle slightly thickened from the base upwards, very shortly inflexed or folded under the seed.

N. Australia. Whitsunday and Gloster Islands, *Henne.*
Queensland. Wide Bay, *Bidwill ;* Rockhampton, *Dallachy ;* Port Denison, *Fitzalan.*
Var. (?) *major.* Leaflets sometimes 2 lines long. Seeds broader. Flowers unknown.—
Ridges of the Victoria river, *F. Mueller.* To this belong the fruiting specimens referred to *A. pallida,* F. Muell. in Journ. Linn. Soc. iii. 147.

The flowers of this species have at first sight some resemblance to those of some *Albizzias,* but the stamens are quite free, although inserted on a small prominent disk under the ovary, not united in a tube round the ovary, as in all the *Inga*-flowered genera.

293. **A. pallida,** *F. Muell. in Journ. Linn. Soc.* iii. 147 (*partly*). A tree, quite glabrous, but the foliage of a very pale or glaucous hue. Pinnæ 3 to 10 pairs, the common petiole usually 1 to 2 in. but in the larger leaves 4 to 6 in. long; leaflets 10 to 20 pairs, oblong, coriaceous, rigid, mostly 3 to 4 lines long. Stipules spinescent, with thickened bases or sometimes none. Peduncles solitary, 1 to 1½ in. long, with an annular 4-lobed deciduous bract about the middle as in *A. Bidwilli*, bearing a globular head of about 20 flowers, smaller than in that species, and mostly, if not all, 5-merous. Calyx very short, sinuate-toothed. Corolla scarcely 1½ lines long, shortly lobed. Pod unknown.

N. Australia. M'Adam range, Fitzmaurice river, *F. Mueller.*

91. ALBIZZIA, Durazz.

Calyx campanulate or tubular, 5- or rarely 4-toothed. Corolla 5- or rarely 4-lobed, with a cylindrical tube. Stamens indefinite, usually numerous and long, united at the base in a tube enclosing the ovary. Pod linear or oblong, straight or nearly so, flat, thin, rarely coriaceous, indehiscent or opening without elasticity in 2 valves. Seeds usually orbicular, along the centre of the pod; funicle filiform.—Trees or shrubs, without prickles. Leaves twice pinnate, with a gland on the petiole below the pinnæ and others between or below some or all of the pinnæ and leaflets. Flowers in globular heads or rarely cylindrical spikes, usually hermaphrodite. Stamens white or pink, rarely yellow, much longer than in *Acacia.*

The genus is limited to the Old World, and is chiefly tropical. Of the four Australian species, one is widely dispersed over tropical Asia, the others are endemic. F. Mueller proposes to reunite this and *Pithecolobium* with *Acacia*, but that can scarcely be done without returning to the Linnean genus *Mimosa*, for *Albizzia* as to flower and fruit, and *Albizzia* as to flowers, are undistinguishable from *Inga*, whilst *Lysiloma* as closely connects *Acacia* with *Mimosa*. If the Linnæan *Mimosa* is broken up into distinct genera, there is no character so constant and so easily recognized as that which separates *Acacia* from *Albizzia*. It is, however, very difficult to draw a definite line between *Albizzia* and *Pithecolobium*, which only differ in the fruit.

Flowers in cylindrical spikes on axillary peduncles . . . . 1. *A. lophantha.*
Flowers in globular heads on axillary peduncles.
  Leaflets 5 to 10 pairs, 2 to 3 lines long .   .   .   . 2. *A. basaltica.*
  Leaflets 2 to 4 pairs, ½ to 1 in. long . . . . 3. *A. Thozetiana.*
Flowers in small heads in large terminal panicles.
  Panicle loose. Stamens about ¼ in. long. Pod under 1 in. broad . 4. *A. procera.*
  Flower-heads very numerous and crowded. Stamens about ½ in. long.
    Pod 1½ to above 2 in. broad . . . . 5. *A. canescens.*

A. Lebbeck, Benth., allied to *A. canescens*, but more glabrous, with much larger flowers, not closely sessile, and the pod not so broad, etc., a tree widely dispersed over tropical Asia and Africa, has been introduced into the neighbourhood of Brisbane.

1. **A. lophantha,** *Benth. in Hook. Lond. Journ.* iii. 86. A tall shrub or small tree; branches, petioles, and peduncles usually velvety-pubescent. Pinnæ 8 to 10 pairs; leaflets 20 to 30 or more pairs, linear, 3 to 4 lines long, the nerve near the upper margin, glabrous above, silky-pubescent underneath. Flowers in loose cylindrical axillary pedunculate spikes of from 1¼ to nearly 3 in., each flower 2 to 3 lines long, usually 5-merous. Calyx shortly lobed, not half as long as the corolla. Stamens fully ½ in. long,

united at the base into a tube rather shorter than the corolla. Pod often above 3 in. long, 4 to 6 lines broad, very flat. Seeds transverse, ovate or orbicular.—*Mimosa distachya*, Vent. Jard. Cels. t. 20, not of Cav. ; *M. elegans*, Andr. Bot. Rep. t. 563 ; *Acacia lophantha*, Willd. Sp. Pl. iv. 1070 ; Bot. Reg. t. 361 ; Lodd. Bot. Cab. t. 716 ; Bot. Mag. t. 2108.

W. Australia. Goose Island Bay, *R. Brown* : King George's Sound, *Baxter;* Geographe Bay, *Fraser;* Cape Naturaliste, *Oldfield;* Swan River ?, *Drummond.*

2. **A. basaltica,** *Benth.* A shrub ; branchlets nearly terete, rusty with a minute glandular pubescence. Pinnæ 1 or 2 pairs, the common petiole rarely ½ in. long and often very short ; leaflets 5 to 10 pairs, oblong or almost ovate, very obtuse, mostly 2 to 3 lines long, coriaceous, minutely hoary-pubescent. Peduncles in the upper axils scarcely exceeding the leaves, bearing a dense globular head of about 20 to 30 flowers, mostly 5-merous, about 1½ lines long. Calyx pubescent, shortly lobed, about two-thirds as long as the corolla. Staminal tube nearly as long as the corolla, the filaments much longer. Pod about 3 in. long, 4 to 5 lines broad, coriaceous, very flat, with thickened margins. Seeds flat, orbicular.—*Acacia basaltica*, F. Muell. in Journ. Linn. Soc. iii. 146.

Queensland. Basaltic plains, Peak Downs, *F. Mueller;* Bowen river, *Bowman;* Zamia Creek and Comet river, *Leichhardt.*

3. **A. Thozetiana,** *F. Muell. Herb.* A tree, attaining 50 to 60 ft., with a dense spreading head, glabrous or the young shoots hoary. Pinnæ 1 pair, with a short common petiole; leaflets 2, 3, or very rarely 4 pairs, cuneate-oblong or broadly linear, very obtuse, the end ones ½ to 1 in. long, the lower ones much smaller. Peduncles in the upper axils ½ to 1 in. long, bearing each a globular head of 20 to 30 or more flowers, 5-merous or sometimes 4-merous, minutely hoary-pubescent, about 1½ lines long. Calyx tubular, more than half as long as the corolla, but narrower than in the preceding species. Pod attaining 6 to 8 in. in length and ½ to ¾ in. in breadth, very flat, with thickened margins. Seeds very flat, orbicular, bordered by a narrow wing.—*Acacia Thozetiana*, F. Muell. Fragm. iv. 9.

Queensland. Wide Bay, *Bidwill, Leichhardt;* Fort Cooper, Rockhampton, *Thozet,* Thozet's Creek, *Dallachy.*

4. **A. procera,** *Benth. in Hook. Lond. Journ.* iii. 88. A tall tree, the young shoots slightly silky-pubescent, at length glabrous. Pinnæ usually 3 pairs, distant along a common petiole often ½ ft. long, ; leaflets 6 to 8 pairs, obliquely oval-oblong, usually obtuse, often nearly 1 in. long and ½ in. broad, very unequally narrowed at the base, penniveined but not very prominently so, minutely hoary-pubescent or glabrous above. Peduncles about 1 in. long, in clusters of 2 or 3, in a loose terminal panicle, each bearing a globular head of 15 to 20 sessile flowers, mostly 5-merous, and scarcely more than 3 lines long, including the stamens. Corolla slender, less than 2 lines long, divided to the middle. Pod 5 to 7 in. long, 9 to 10 lines broad, very flat and thin. Seeds very flat, orbicular.—*Mimosa procera*, Roxb. Pl. Corom. ii. 12, t. 121 ; *A. elata*, Roxb. Fl. Ind. iii. 546 ; *Acacia procera*, Willd. Sp. Pl. iv. 1063.

N. Australia. Gloucester Island, *Henne.* Widely distributed over S. E. India and the

Archipelago. The Australian specimens are not complete, but, as far as they go, they do not appear to differ at all from the Indian ones.

5. **A. canescens,** *Benth.* A beautiful spreading tree, the young shoots silky-pubescent, the adult foliage more or less hoary with a very minute appressed pubescence. Pinnæ usually 2 pairs, rarely with a fifth odd pinna or a third pair, the common petiole 2 to 4 in. long, each rhachis 3 to 6 in.; leaflets 5 to 8 pairs on the terminal pinnæ, very obliquely obovate and unequally narrowed at the base, mostly ¾ to 1½ in. long, fewer and smaller on the lower pinnæ, penniveined but the veins much less prominent than in *A. Lebbeck.* Flower-heads small, numerous, on short peduncles in dense terminal panicles much shorter than the last leaves. Flowers sessile, mostly 5-merous. Calyx about 1 line, corolla about 2½ lines long, both silky-pubescent. Stamens about ½ in. long, the united part shorter than the corolla. Pod stipitate, often 8 to 10 in. long and 1½ to 2½ in. broad in our specimens, very thin and flat. Seeds flat, orbicular, along the centre of the pod, not quite ripe in the specimens seen.

Queensland. Burdekin river, *F. Mueller;* Fitzroy and Bowen rivers, *Bowman;* Rockhampton, *Thozet, Dallachy.* Allied on the one hand to *A procera,* on the other to *A. Lebbeck,* differing from the former in the broader flowers, much longer stamens, the panicle more dense, and a much broader pod; from *A. Lebbeck* in the much smaller closely sessile flowers and broader pod, and from both in the general aspect of the foliage.

92. PITHECOLOBIUM, Mart.

(Cathormion, *Hassk.*)

Calyx campanulate or tubular. Corolla 5-lobed, with a cylindrical tube. Stamens indefinite, usually numerous and long, united at the base in a tube enclosing the ovary. Pod flattened, usually rather thick and much curved, annular or spirally twisted, either opening entirely or on the outer edge in 2 valves, or quite indehiscent, very smooth and often coloured inside or with a thin pulp. Seeds ovate or orbicular; funicle filiform.—Trees or rarely shrubs, without prickles. Leaves twice pinnate, usually with a gland on the petiole below the pinnæ, and others between or below some or all of the pinnæ and leaflets; leaflets few and rather large in all the Australian species. Flowers in globular or oblong heads or umbels, or rarely in cylindrical spikes, usually hermaphrodite and white, the stamens rarely red.

A considerable tropical genus, distributed over the New as well as the Old World. Of the three Australian species, one is also in the Indian Archipelago, the two others endemic, one as yet doubtful as to the genus, the pod being unknown. *Pithecolobium,* with the flowers and fruit of some sections of *Inga,* only differs from that genus in the twice-pinnate, not simply pinnate leaves. *Calliandra* and *Albizzia* have the same flowers and only differ in the pod. *Acacia* is at once and constantly distinguished by the stamens never united in a tube round the ovary.

Flowers pedicellate in the head (umbellate).
Leaflets oblong or rhomboidal, acuminate. Pod twisted, 2-valved . 1. *P. pruinosum.*
Leaflets obliquely obovate, obtuse. Pod indehiscent, very hard,
   separating into distinct articles . . . . 2. *P. moniliforme.*
Flowers sessile, the corolla ¾ in. long. Leaflets acuminate . . . . 3. *P. grandiflorum.*

1. **P. pruinosum,** *Benth. in Lond. Journ.* iii. 211. A beautiful tree, the young branches, foliage, and inflorescence rusty with a short pubescence

or glabrous. Pinnæ very irregularly in 1 or 2 pairs, with or without an odd one, the petiole and each rhachis varying from 1 to 6 in. long; leaflets usually 3 or 4 pairs on the terminal pinnæ, but very irregular in number, size, and shape, mostly broadly oblong or rhomboidal and acuminate, rarely very obtuse, the larger ones often 2 to 3 in. long, but mostly smaller. Peduncles 2 or 3 together in the upper axils or shortly racemose. Flowers numerous, in globular umbels, on pedicels of about 2 lines. Calyx small, shortly toothed. Corolla fully 2 lines long. Pod several in. long, 7 to 8 lines broad, flat but much curved and twisted, the upper inner margin thickened and continuous, the outer one much sinuate and undulate, the valves smooth and reddish inside. Seeds ovate, transverse; funicle rather thick, but terete, folded under the seed.

Queensland. Brisbane river, Moreton Bay, *A. Cunningham* and others; Rockhampton, *Thozet, Dallachy.*

N. S. Wales. Hunter's River, Ash Island, *R. Brown* and others; Clarence and Richmond rivers, *C. Moore;* Liverpool plains, *A. Cunningham;* southward to Kiama, *Harvey;* Illawarra, *Ralston.*

The Javanese *P. Junghuhnianum*, Benth., scarcely differs from this species.

2. **P. moniliferum,** *Benth. in Hook. Lond. Journ.* iii. 211. A tree, with the young shoots usually pubescent, at length glabrous. Pinnæ 1 or 2 pairs, the common petiole ½ to 1 in., each rhachis 1 to 2 in. long; leaflets 4 to 7 pairs, obliquely obovate or oval-oblong, obtuse, very oblique at the base, shining and reticulately penniveined above, opaque and less veined underneath, the terminal ones 1, 1½, or nearly 2 in. long, the others smaller. Peduncles clustered in the upper axils, forming a short irregular terminal panicle. Flowers numerous, in globular umbels, the pedicels about 1½ in. long. Calyx nearly 1 line, corolla about 2 lines long, minutely silky-pubescent. Stamens more than twice as long, the united part nearly as long, as the corolla. Pod usually falcate, 3 to 4 in. long, ¾ in. broad, very thick and hard, indehiscent but separating into closed 1-seeded articles. Seeds transverse, oblong, flat but thick; funicle very short.—*Inga monilifera,* DC. Prod. ii. 440; *Cathormion moniliferum,* Hassk. Retzia, 231.

N. Australia. Point Pearce, Victoria river, *F. Mueller;* islands of the Gulf of Carpentaria, *R. Brown;* along all the watercourses round the Gulf of Carpentaria, *Leichhardt.* The species was originally described from Timor, and is in several islands of the Indian Archipelago.

3. **P. (?) grandiflorum,** *Benth.* A beautiful tree of 30 ft., or, according to some, a tall shrub, glabrous or nearly so. Pinnæ 1 or 2 pairs, the common petiole and each rhachis about 2 to 4 in. long; leaflets 2 to 6 pairs, ovate, acutely acuminate, 1½ to 2 in. long, less oblique and firmer than in *P. pruinosum,* penniveined. Flower-heads numerous, on short peduncles, in a terminal panicle scarcely exceeding the leaves. Flowers sessile, much larger than in any other *Pithecolobium.* Calyx campanulate, sinuate-toothed, nearly 3 lines long. Corolla ¾ in. long, funnel-shaped, 5-lobed, silky-pubescent outside. Stamens of a rich crimson, the tubular portion nearly as long as the corolla, the free part exceeding it by about 1 in. Pod unknown.—*Mimosa grandiflora,* Soland. ms.

Queensland. Endeavour river, *Banks and Solander;* Cape York, *W. Hill;* Edgecombe Bay, *Dallachy.*

N. S. Wales. Hastings river, *Tozer*.

Until the fruit shall have been seen, it is impossible to fix the genus of this fine species. It has the foliage of the Indian section *Clypearia* of *Pithecolobium*, the flowers are more like those of several American *Calliandras*.

Order XLI. ROSACEÆ.

Calyx either enclosing the ovary, or adhering to it, or quite free, 5-, rarely 4-lobed, with the addition, in a few genera of as many external accessary lobes. Petals as many as true calyx-lobes, inserted on the calyx at the base of the lobes, or in *Stylobasium* hypogynous. Stamens indefinite, rarely few, inserted with the petals, free. Ovary of 1, 2 or more carpels, usually distinct at the time of flowering, but sometimes combined even then into a single 2- to 5-celled ovary, which is then always inferior or combined with the calyx; ovules 1 or 2, rarely more in each carpel; styles or sessile stigmas distinct. Fruiting carpels either free or variously combined with each other or with the calyx, indehiscent or rarely opening along the inner edge. Seeds without albumen or rarely albuminous; embryo with large cotyledons and a short radicle.—Trees shrubs or herbs. Leaves alternate, simple or compound, almost always with stipules. Flowers in axillary or terminal cymes or solitary, very rarely in simple racemes.

A numerous Order, widely spread over the globe, but more in the temperate and cooler parts of the northern hemisphere than within the tropics or in the southern hemisphere. Of the seven Australian genera; one extends over the tropics of the New and the Old World, four belong to the extratropical flora of the northern hemisphere, with a few tropical species, especially in mountain ranges, one, *Acæna*, is also chiefly extratropical, but in the southern hemisphere, with very few species extending into the tropical mountains or northern temperate regions of America; one only, *Stylobasium*, is endemic in Australia.

Of the several tribes into which *Rosaceæ* are divided, a few only are represented in Australia, and those only by one or two genera each, it is therefore useless entering into any detailed exposition of their characters.

Trees or erect shrubs, with entire leaves. Stipules deciduous or none.
  Carpel solitary, with 2 erect ovules and a basal style (**Chryso-balaneæ**).
    Trees. Petals 5 or 4. Stamens perigynous, with filiform filaments
    and small anthers . . . . 1. Parinarium.
    Shrubs. Petals none. Stamens hypogynous; anthers longer than
    the filaments . . . . 2. Stylobasium.
Herbs or scrambling shrubs. Leaves toothed divided or compound.
  Carpels indefinite, protruding from the open calyx. Petals present.
    Herbs. Calyx with accessary external lobes. Fruit-carpels dry.
      Calyx imbricate. Styles persistent, forming long awns to the
      fruit-carpels . . . . 3. Geum.
      Calyx valvate. Styles deciduous. Carpels seed-like, without
      awns . . . . 4. Potentilla.
    Scrambling shrubs, rarely prostrate or almost herbaceous. Calyx
    slightly imbricate, without accessary lobes. Fruit-carpels succu-
    lent . . . . 5. Rubus.
  Carpels several, enclosed in the calyx-tube. Petals present. Stamens
  numerous. Prickly shrubs with pinnate leaves . . . . . . Rosa (p. 432).
  Carpels 1 to 4, enclosed in the calyx-tube. Petals none. Herbs.
    Ovule erect. Style basal. Leaves palmately lobed or divided.
    Stamens few . . . . 6. Alchemilla.

Ovule pendulous.   Style terminal or nearly so.   Leaves pinnate.
Fruiting-calyx armed with prickles.   Stamens few . . . . .   7. ACÆNA.
Fruiting-calyx without prickles.   Flowers usually monœcious.
Stamens numerous . . . . POTERIUM (p. 434).

1. PARINARIUM, Juss.

(Petrocarya, *Jack ;* Grymania, *Presl.*)

Calyx-lobes 5, imbricate.   Petals 5, rarely 4.   Stamens numerous or rarely few, all perfect or those on one side reduced to small staminodia ; filaments filiform ; anthers small.   Ovary of a single carpel, adnate on one side to the mouth of the calyx-tube and protruding from it, more or less completely 2-celled, with 1 erect ovule in each cell ; style from the base of the ovary. Drupe ovoid or spherical, the endocarp bony.   Seeds 1 or 2, erect.—Trees. Leaves alternate, coriaceous, entire.   Stipules deciduous, usually small. Flowers white or pink, in cymes forming terminal raceme-like or corymbose panicles.

The genus is dispersed over the tropical regions both of the New and the Old World. Of the two Australian species, one is also in the Indian Archipelago, the other is endemic.

Petiole without glands.   Leaves much veined.   Flowers small.   Calyx-
lobes acute.   Perfect stamens about 8 . . . .   1. *P. Nonda.*
Petiole with 2 glands.   Leaves shining, little veined.   Flowers rather
large.   Calyx-lobes obtuse.   Perfect stamens 30 to 50 . . . .   2. *P. Griffithianum.*

1. **P. Nonda,** *F. Muell. Herb.*   Branches rather slender, loosely tomentose when young.   Leaves ovate, obtuse or obtusely acuminate, rounded or almost cordate at the base, 2 to 3 in. long, 1 to 1½ in. broad, rarely narrower and narrowed at each end, glabrous but rather rough above, whitish with a minute tomentum underneath, with many prominent parallel pinnate veins and much reticulate between them.   Flowers small, the terminal panicle or thyrsus loose, the axillary ones smaller and raceme-like.   Bracts shorter than the flowers, deciduous.   Calyx pubescent, nearly regular, about 2 lines long, the lobes acute, rather shorter than the tube and almost as long as the petals. Stamens short, usually about 8 perfect on the same side of the flower as the ovary, the ring completed by 6 to 10 small staminodia.   Drupe ovoid, densely villous inside, 2-celled, 2-seeded.

N. Australia.   From the Upper Lind to Van Diemen's river, Gulf of Carpentaria, *Leichhardt* ; Gilbert river, *F. Mueller.*

Queensland.   Cape York, *McGillivray ;* Albany Island, *F. Mueller.*

The species is nearly allied to the *P. sumatranum* of the Indian Archipelago, and still more to the African *P. curatellæfolium,* Planch., but the flowers appear to be smaller than in either, with some slight differences in the foliage.   It is the one to which Leichhardt gives the name of *Nonda-tree* in his travels.

2. **P. Griffithianum,** *Benth. in Hook. Fl. Nig.* 334.   Branches stout, glabrous or minutely hoary when young.   Leaves elliptical-oblong, acuminate, 3 to 4 in. long or rather more, acute at the base, shining above, paler underneath, but quite glabrous, the veins not very prominent and distant ; 2 small glands at the top of the petiole.   Flowers rather large, in terminal corymbose hoary-pubescent panicles.   Calyx-tube obliquely turbinate, incurved, about 2 lines long ; lobes very obtuse, the largest as long as the tube.   Petals exceeding the calyx.   Stamens very numerous (30 to 50), all perfect.   Ovary

very villous, with a long style. Drupe oblong, very villous inside, 2-celled, 2-seeded.

N. Australia. Port Essington, *Armstrong;* Quail Island, *Flood.*
The species extends over the Indian Archipelago to the Philippine Islands, for *Grymania salicifolia,* Presl, Epimel. Bot. 193 (Cumiug, n. 1057), appears to be the same species, although with rather narrower, more rigid leaves.

2. STYLOBASIUM, Desf.

(Macrostigma, *Hook.*)

Calyx-lobes 5, imbricate. Petals none. Stamens 10, hypogynous; filaments short, persistent; anthers large, linear, erect. Ovary of a single carpel, sessile in the base of the calyx, 1-celled, with 2 erect ovules; style from the base of the ovary; stigma large, peltate. Drupe nearly dry, obovoid or globular, surrounded by the persistent calyx. Seed erect.—Unarmed shrubs. Leaves alternate, entire. Stipules minute or none. Flowers solitary in the axils of the leaves, on short pedicels, the upper ones forming a terminal raceme, usually polygamous, the females with long filiform staminodia, the males with a small abortive ovary.

The genus is exclusively Australian. It differs from the whole Order in its hypogynous stamens, and is connected with the American genus *Leiostemon* alone, by its large anthers.

Leaves cuneate-oblong or obovate. . Drupe twice as long as the calyx . 1. *S. spathulatum.*
Leaves linear. Drupe or nut scarcely exceeding the calyx . . . . 2. *S. lineare.*

1. **S. spathulatum,** *Desf. in Mem. Mus.* v. 37, *t.* 2. An erect bushy shrub of several feet, glabrous or the young shoots hoary-pubescent. Leaves mostly cuneate-oblong or the lower ones obovate, very obtuse or emarginate, ¾ to 1½ in. long, narrowed into a short petiole, rather thick and fleshy; veinless except the midrib. Pedicels short, with 1 or 2 minute bracteoles. Calyx, when in flower, about 3 lines long, rather narrow, the lobes obtuse, shorter than the tube, the anthers shortly protruding, when in fruit very open. Drupe nearly globular, about 5 lines diameter. .

W. Australia. Sharks Bay, *Milne,* and probably from the same locality, *Baudin's Expedition;* Flinders Bay, *Collie;* Port Gregory, *Oldfield.*

2. **S. lineare,** *Nees in Pl. Preiss.* i. 95. An erect bushy shrub, glabrous and somewhat glaucous or the young shoots slightly hoary. Leaves narrow-linear, very obtuse, the lower ones 1 to 1½ in. long, narrowed into a short petiole, thick and fleshy, the upper ones often very small and distant. Calyx in the original form nearly as large as in *S. spathulatum,* but the lobes longer than the tube, the anthers shortly protruding. Drupe exceeding the calyx-lobes, but much smaller and drier than in *S. spathulatum.*

W. Australia. Abandoned fields, near Perth and York, *Preiss, n.* 2383 *and* 2384; Limestone Range, N. of Murchison river, *Oldfield.*
Var. *parviflora.* Branches slender. Calyx scarcely 2 lines long.—*Macrostigma australe,* Hook. Ic. Pl. t. 412.—W. Australia, *Drummond, n.* 16.

3. GEUM, Linn.

(Sieversia, *R. Br.*)

Calyx-tube short, open; lobes 5, imbricate, usually with as many small external accessary lobes alternating with them. Petals 5, broad, spreading. Stamens indefinite. Carpels indefinite, with 1 erect ovule in each; style terminal, filiform, with a hook or twist at or below the end. Fruit a head of small dry achenes, surrounded by the persistent calyx, each one terminating in a long filiform straight hooked or geniculate awn, formed by the persistent style, and either naked hairy or plumose.—Herbs with a perennial rootstock. Leaves chiefly radical, pinnate or pinnatisect, the leaflets or segments toothed, the terminal one much larger than the others; stem-leaves usually small and bract-like. Flowers yellow white or red, solitary and terminal, or few in a loose corymbose terminal panicle.

The genus is dispersed over the temperate regions of the globe. Of the two Australian species, one is common in the northern hemisphere of the Old World, the other is endemic.

Flowers yellow, several on the stem, Radical leaves with 3 or more
    large ovate or lanceolate segments. Styles (in the flower) twisted
    below the end . . . . 1. *G. urbanum.*
Flowers large, white, solitary. Radical leaves with one large reniform
    crenate segment. Styles, even in the flower, twisted at the end only. 2. *G. renifolium.*

1. **G. urbanum,** *Linn.; DC. Prod.* ii. 551. Stems erect, slightly branched, 1 to 2 ft. high, glabrous or softly pubescent. Lower leaves on long petioles, with 3, 5 or more large segments intermixed with small ones, the upper leaves usually with only 3 large segments or a single one divided into 3, and sometimes 2 or 3 small ones along the stalk, all as well as the leafy stipules coarsely toothed or lobed. Flowers yellow, terminating the branches of a very loose panicle. Calyx-lobes entire, acute. Petals often not exceeding the calyx in the northern specimens, considerably larger in the Tasmanian ones. Fruit-carpels covered with silky hairs. Style plumose-hairy, twisted below the middle at the time of flowering, but afterwards the lower part elongates and becomes glabrous under the twist, the extremity at length frequently falls away, leaving the fruit with a terminal recurved style, hooked only at the extremity, as in the section *Sieversia.*—Hook. f. Fl. Tasm. i. 114, with the synonyms adduced.

N. S. Wales. Nepean river, *R. Brown ;* Macleay river, *Beckler,* all with the small petals of the European form.

Victoria. Moist banks of the Mitta-Mitta, Delatite river, etc., *F. Mueller,* with rather large petals.

Tasmania. Port Dalrymple, *R. Brown ;* not uncommon in shady places in various parts of the colony, *J. D. Hooker,* with rather large petals.

The species is common in Europe, temperate Asia and the E. Indian mountains, and naturalized in several other parts of the world.

2. **G. renifolium,** *F. Muell. in Trans. Phil. Inst. Vict.* ii. 66. Root-stock thick and hard. Leaves radical, with a single terminal reniform segment, often 2 to 3 in. broad, coarsely crenate, usually broadly and obscurely lobed, very much wrinkled, the veins very prominent underneath, and sometimes a few very small segments scattered along the petiole. Flowering-stem ¾ to 1 ft. high, with a single large terminal white flower, and usually 3 distant small sessile bract-like deeply toothed leaves, the lowest above the middle of the stems. Calyx-lobes enlarged after flowering and often toothed. Petals longer than the calyx. Styles plumose-villous, not produced beyond the twist even at the time of flowering, elongated and hooked at the end when in fruit. —Hook. f. Fl. Tasm. ii. 361.

Tasmania. Southern Alps, *Oldfield;* Mount Lapeyrouse, *C. Stuart.* This fine species belongs to the section *Sieversia,* by some ranked as a genus, but differing only in the style not produced beyond the twist.

4. POTENTILLA, Linn.

Calyx-tube short, open ; lobes 5 or rarely 4, valvate, with as many external accessary lobes alternating with them. Petals 5, rarely 4, broad, spreading. Stamens indefinite. Carpels indefinite, with 1 erect ovule in each; style terminal or lateral, deciduous. Fruit a head of small dry seed-like achenes, surrounded by the persistent calyx, the receptacle scarcely enlarging.—Herbs with a perennial tufted stock, and occasionally creeping stolons or runners. Flowering stems usually short. Leaves either digitately 3- or 5-foliolate or pinnate; leaflets toothed. Peduncles 1-flowered, solitary or in a loose terminal cyme.

The species are numerous, dispersed over the whole of the northern hemisphere without the tropics, especially in Europe and Asia, extending into the mountains of E. India, and descending along the Andes into S. America, and one or two species only, and those the same as northern ones, are found in the extratropical regions of the southern hemisphere, including New Zealand. The only Australian species extends over the greater portion of the area of the genus.

1. **P. anserina,** *Linn.; DC. Prod.* ii. 582. Stock tufted, with long creeping runners rooting at the nodes. Leaves pinnate, with numerous oblong, deeply notched leaflets, either green and slightly silky on the upper side and of a shining silvery-white underneath from the silky down with which they are covered, or very white on both sides. Peduncles long, solitary at the nodes, bearing a single rather large yellow flower.—Hook. f. Fl. Tasm. i. 113.

Victoria. Searle's Cove, *F. Mueller;* Fitzroy river, *Robertson ;* mouth of the Glenelg, *Allitt.*

Tasmania. Port Dalrymple and Derwent river, *R. Brown;* Circular Head and elsewhere on the N. and W. coasts, rarer to the southward, *J. D. Hooker.*

The species is widely dispersed over the extratropical regions of both the northern and southern hemispheres, including New Zealand.

5. RUBUS, Linn.

Calyx-tube short, open ; lobes 5, imbricate, without external accessary ones. Petals 5, erect or spreading. Stamens indefinite. Carpels indefinite, with 2 pendulous ovules in each, one of them smaller and abortive; styles terminal. Fruit a head of succulent carpels, forming a kind of granulated berry round the conical or shortly oblong dry or spongy receptacle.—Weak scrambling shrubs or rarely prostrate and almost herbaceous, usually prickly. Leaves pinnately or palmately divided into distinct segments or leaflets or lobed only, the lobes or segments toothed. Flowers axillary or in terminal leafy panicles.

A considerable genus, dispersed over most parts of the globe. Of the five Australian species, one extends over Africa and the warmer parts of Asia, one is common in the Indian Archipelago, one is abundant in China, but not in the intervening Achipelago, and two are endemic. The fruits of several are edible but acid.

Unarmed dwarf glabrous creeping plant. Flowers yellow . . . . . 1. *R. Gunnianus.*

Prickly shrubs, scrambling climbing or almost erect. Flowers pink or
white.
  Leaves broad, toothed or lobed, rusty underneath . . . . . . 2. *R. moluccanus.*
  Leaves pinnate, with 3 or 5 leaflets, white-tomentose underneath.
    Fruit with few large carpels . . . . 3. *R. parvifolius.*
  Leaves pinnate, with 5 or 7 leaflets, green on both sides. Fruit with
    numerous small carpels . . . . 4. *R. rosæfolius.*
  Leaves digitate, with 5 leaflets on long petiolules . . . . 5. *R. Moorei.*

1. **R. Gunnianus,** *Hook. Ic. Pl. t.* 291. A dwarf, creeping, tufted,
glabrous and unarmed plant, forming patches of several feet in diameter, the
slender woody stems usually buried in the soil. Leaves mostly ovate, $\frac{1}{2}$ to nearly
1 in. long, deeply crenate-lobed or pinnately divided into 3 segments or leaflets, the lobes obtuse. Peduncles solitary, not exceeding the leaves, 1-
flowered. Calyx-lobes rather obtuse, about 2 lines long. Petals yellow,
narrow, exceeding the calyx. Fruit globular, scarlet, the carpels few, large,
very fleshy, and said to be of an excellent flavour.—Hook. f. Fl. Tasm. i.
112.

Tasmania. Common in the mountains, at an elevation of 3000 to 5000 ft., *J. D.
Hooker.*

2. **R. moluccanus,** *Linn.; DC. Prod.* ii. 566. A tall scrambling
shrub; branches and petioles terete, clothed with a short rusty or white
woolly down, often mixed with longer hairs, and armed with numerous small
recurved prickles. Leaves usually broadly ovate-cordate, toothed, shortly and
broadly 3- or 5-lobed, 2 to 4 in. long, occasionally deeply 3-lobed but not
quite to the midrib, green, somewhat rugose and glabrous or sprinkled with a
few hairs above, rusty or whitish-tomentose underneath, the principal veins
more villous and often armed with prickles. Flowers red, irregularly clustered in short panicles in the upper axils, the upper ones forming a terminal
panicle, usually very silky-villous. Bracts deeply cut, very deciduous. Pedicels usually short when in flower, longer in fruit. Calyx-lobes acuminate, 4
or 5 lines long. Fruit nearly globular, glabrous, scarcely exceeding the
calyx-lobes in our specimens, said to be red, insipid or slightly acid.—*R.
Hillii*, F. Muell. in Trans. Phil. Inst. Vict. iii. 67, and Fragm. iv. 31.

N. Australia. Port Essington, *Armstrong* ; a rather smaller-flowered form, with
the leaves less lobed, almost as in *R. acerifolius,* Wall.
  Queensland. Brisbane river, *W. Hill, Leichhardt* ; Rockhampton, *Dallachy.*
  N. S. Wales. Hunter and Williams rivers, *R. Brown* ; Hastings and Macleay rivers,
Beckler ; Clarence river, *C. Moore* ; Paramatta, *Woolls* ; Illawarra, *A. Cunningham* ;
Kiama, *Harvey.*
  Victoria. Nangatta mountains, Upper Genoa river, *F. Mueller.*
  The species extends over the Indian Archipelago to the Philippines, and the closely allied
R. rugosus, Sm., and *R. reflexus*, Bot. Reg., to E. India and China. The majority of the
Australian specimens belong to a form precisely the same as one common in the Archipelago,
which appears to be that originally described by Rumphius.

3. **R. parvifolius,** *Linn.; DC. Prod.* ii. 564. A scrambling shrub;
branches softly pubescent or woolly, armed with small hooked prickles.
Leaves pinnate, with a common petiole of 1 to 2 in.; leaflets 3 or very
rarely 5, nearly orbicular, about $\frac{3}{4}$ to 1 in. long or in luxuriant shoots nearly
twice as much, deeply and irregularly toothed, glabrous or sprinkled with a
few hairs and deeply wrinkled above, white and tomentose or woolly under-

neath. Flowers few, in short terminal panicles or solitary in the upper
axils. Bracts narrow, entire or rarely lobed. Sepals varying from 2 to 5
lines long, acuminate, softly hairy inside and out. Petals pink, usually erect
and shorter than the calyx, rarely longer and spreading. Fruit globular, red,
said to be of a pleasant flavour, the carpels rather large and not numerous, glabrous or slightly hairy.—Bot. Reg. t. 496 ; *R. ribesifolius,* Sieb. Pl. Exs. ;
R. macropodus, Ser. in DC. Prod. ii. 557 ; Hook. f. Fl. Tasm. i. 112 ; F.
Muell. Fragm. iv. 30, and Pl. Vict. ii. t. 15 ; *R. Zahlbrucknerianus,* Endl.
Atakt. t. 35.

Queensland. Moreton Bay, *Fraser, F. Mueller;* Rockhampton, *Dallachy ;* in the
interior, on the Maranoa, *Mitchell ;* plains of the Condamine, *Leichhardt.*
N. S. Wales. Common about Port Jackson, *R. Brown, Sieber, n.* 192, and others ;
Blue Mountains, *Miss Atkinson ;* Hastings and Macleay rivers, *Beckler ;* New England, *C.
Stuart.*
Victoria. Wooded valleys and banks of streams, common, *F. Mueller.*
Tasmania. Very common in many parts of the colony, *J. D. Hooker.*
S. Australia. Mount Gambier and near Adelaide, *F. Mueller.*
The species ranges from S. China to Loochoo, but I have seen no specimens from the
tropical regions intervening between that and Australia.

4. **R. rosæfolius,** *Sm. Ic. Pl. t.* 60. A shrub, with creeping stolons
and erect and weak but scarcely climbing stems, glandular-pubescent or
rarely glabrous, armed with straight or more frequently recurved prickles.
Leaves pinnate ; leaflets 5, rarely 3 or 7, ovate-lanceolate, acuminate, coarsely
and usually doubly toothed, 1, 2, or even 3 in. long, green and glandular-
pubescent on both sides or rarely glabrous. Flowers white, not numerous,
in a short terminal panicle or in the upper axils. Bracts narrow, mostly entire. Sepals hoary-tomentose, 3, 4, or rarely 5 lines long, with a long subulate point. Petals spreading. Fruit ovoid or rarely globular, with exceedingly numerous small carpels, very little succulent, and said to have a rather
unpleasant resinous flavour.—Ser. in DC. Prod. ii. 556 ; Bot. Mag. t. 1783
(with double flowers); Hook. Ic. Pl. t. 349 ; Lodd. Bot. Cab. t. 158 ; F.
Muell. Fragm. iv. 32 ; *R. eglanteria,* Tratt. ; Ser. in DC. Prod. ii. 556.

Queensland. Moreton Bay, *F. Mueller* and others ; Broad Sound and Mount Elliott,
Dallachy.
N. S. Wales. Hawkesbury, Hunter's, and Paterson rivers, *R. Brown ;* Blue Moun-
tains, *Miss Atkinson ;* northward to New England, *C. Stuart ;* Clarence and Hastings
rivers, *Beckler ;* southward to Illawarra, *Shepherd ;* and Twofold Bay, *F. Mueller.*
Victoria. Wooded valleys, Broadribb and Snowy rivers, *F. Mueller.*
The species is widely spread over the warmer regions of Africa and Asia.

5. **R. Moorei,** *F. Muell. in Trans. Phil. Inst. Vict.* ii. 67, *and Fragm.*
iv. 29. A tall scrambling shrub, the branches and petioles glabrous or
loosely tomentose, with numerous small reflexed prickles. Leaves digitate,
with 5 petiolulate leaflets, ovate-lanceolate, acutely acuminate, 3 to 4 in. long,
and bordered by regular prickly teeth in the glabrous or slightly pubescent
specimens ; in the more tomentose ones shorter, broader in proportion, more
coriaceous, with shorter teeth, glabrous above, softly velvety or villous underneath. Flowers white, unisexual or polygamous, in loose axillary racemes or
raceme-like panicles. Bracts small, entire. Calyx pubescent, very spreading,
the segments ovate or oblong, obtuse, 2 to 3 lines long. Fruit (which I have
not seen) said to be dark-red and insipid.

N. S. Wales. Grose river, *R. Brown;* Blue Mountains, *Miss Atkinson;* Richmond and Clarence rivers, *C. Moore;* from the Creek Brush to Archer's Station, *Leichhardt;* southward to Illawarra, *A. Cunningham, Backhouse, Ralston.*

The nearest affinity of this species is with the New Zealand *R. australis,* Forst., but the leaves of the latter species, although protean in their forms, never quite resemble those of *R. Moorei;* the flowers are much smaller and very much more numerous, in large panicles, etc.

Several species of *Rosa* are cultivated in gardens, and one, *Rosa rubiginosa,* Linn., the *Sweetbriar,* with glandular-pubescent aromatic leaves and pink flowers, is said to have established itself, apparently wild, in South Australia.

6. ALCHEMILLA, Linn.

Calyx-tube ovoid or campanulate; lobes 4, valvate, with 4 small external accessary lobes alternating with them. Petals none. Stamens 4 or fewer, inserted round an annular disk at the mouth of the calyx. Carpels 1 to 4, enclosed in the calyx-tube, with 1 ascending ovule in each; style from the base or inner side of the carpel, protruding from the calyx-tube, with a capitate stigma. Achenes 1 to 4, 1-seeded, enclosed in the calyx-tube.—Herbs, either annual or with a perennial sometimes almost woody stock and annual flowering stems, or, in species not Australian, perennial tufted undershrubs. Leaves palmately lobed or divided. Flowers small, green, in terminal panicles or axillary clusters.

The genus is not numerous in species, widely spread over the northern hemisphere, confined to mountains within the tropics, extending down the Andes to extratropical S. America, one or two species found also in S. Africa and New Zealand. The two Australian species are common in temperate or mountain regions, especially in the Old World.

Perennial. Flowers in terminal panicles. Leaves orbicular, broadly lobed 1. *A. vulgaris.*
Small annual. Flowers minute, in axillary clusters. Leaves small, deeply
lobed . . . . 2. *A. arvensis.*

1. **A. vulgaris,** *Linn.; DC. Prod.* ii. 589. A perennial, either glabrous or more or less hairy. Radical leaves large, on long petioles, broadly orbicular or reniform, divided only to one-fourth or one-third of their depth into 7 or 9 broad regularly toothed lobes, green on both sides. Flowering stems decumbent or ascending, seldom above 6 in. high, bearing a few leaves on short petioles, with large green toothed stipules and a loose panicle of small green flowers, the pedicels usually at least as long as the calyx.—Wight, Ic. t. 229.

Victoria. Haidinger range, sources of the Mitta-Mitta, Murray, and Snowy rivers, in the Australian Alps, *F. Mueller.* The species is common in Europe, N. Asia, and the mountains of E. India.

2. **A. arvensis,** *Scop.; DC. Prod.* ii. 590. A little annual, rarely above 2 or 3 in. high and often smaller, much branched, green, and hairy. Leaves on short petioles, orbicular, usually deeply lobed or divided. Flowers very minute, green, sessile, in little clusters or heads in the axils of the leaves, half enclosed in the leafy stipules.

Victoria. Mount Korong, *F. Mueller;* Wendu Valley, *Robertson.*
S. Australia. Mountain pastures, Rivoli Bay, *F, Mueller.*
Tasmania. Derwent river, *R. Brown;* various parts of the island, probably introduced, *J. D. Hooker.*

The species is common in the northern hemisphere in the Old World, and in various parts of N. and S. America and Africa, and is also found in New Zealand ; but in the latter country, and in some, if not all, the Australian stations, very probably introduced from Europe.

7. ACÆNA, Linn.

(Ancistrum, *Forst.*)

Calyx-tube ovoid or campanulate ; lobes usually 4 or 5, but varying from 3 to 7, valvate. Petals none. Stamens 2 to 10. Carpels 1 or rarely 2, enclosed in the calyx-tube, with 1 pendulous ovule in each ; style terminal or nearly so, protruding from the calyx-tube, usually dilated into an oblique fringed stigma. Achene solitary, dry, enclosed in the hardened tube of the calyx, which is usually closed at the top and more or less awned with subulate or conical spines, often glochidiate at the end.—Herbs, with a perennial tufted stock. Leaves radical or alternate, pinnate, with toothed or cut leaflets. Stipules sheathing at the base. Flowers hermaphrodite or polygamous, small, green or purplish, in a terminal globular head, or in an elongated or interrupted spike, the flowering-stem either leafy or reduced to a leafless scape.

The genus is dispersed over the temperate and colder regions of the southern hemisphere ; it is especially abundant in S. America, and occurs also in California, Mexico, and the Sandwich Islands. Of the three Australian species, two are apparently also natives of S. America and New Zealand, the third is probably endemic.

The genus has been divided into two sections, *Euacæna*, with the fruiting calyx more or less angular, the spines, when present, one only to each angle ; and *Ancistrum*, with the calyx ovoid, irregularly covered with numerous spines or tubercles. In the former the flowers are usually capitate, but spicate in a few species, in the latter they are spicate in most if not all species. There are a few species however, (not Australian,) with the spines not developed, and ambiguous between the two sections.

Spike cylindrical or elongated and interrupted. Stamens 4 to 10.
  Spines of the fruiting calyx numerous, irregularly scattered . . . 1. *A. ovina.*
Heads globular. Stamens 2. Spines of the fruiting calyx 4, 1 to each
  angle.
  Calyx-lobes united at the base, usually persistent. Fruiting head
    above ½ in. diameter, with long glochidiate spines . . . . . 2. *A. sanguisorbæ.*
  Calyx-lobes separately deciduous. Fruiting heads not ¼ in. diameter,
    with very short fine spines. . . . . 3. *A. montana.*

1. **A. ovina,** *A. Cunn. in Field, N. S. Wales,* 358. Stems ascending or erect, leafy, 1 to 2 ft. high, silky-hairy. Leaflets ovate, from orbicular to oblong, ¼ to ¾ in. long, deeply and obtusely crenate or pinnatifid, glabrous above, silky-hairy underneath. Flowers in a long interrupted spike, more dense towards the end, polygamous. Calyx-lobes usually 5, rarely 4, 6, or 7. Stamens in the males either about as many or 8 to 10, in the females reduced to minute staminodia, or 1 or 2 of them filiform, without anthers. Ovary in the females with a single or rarely 2 ovules ; style obliquely dilated at the end, with a broad unilateral fringed stigma. Fruit ovoid, 2 to 3 lines long, glabrous or loosely villous, covered with short prickles, barbed at the end and irregularly arranged, 2 or 3 of them usually longer than the others, with a conical base.—Hook. f. Fl. Tasm. i. 115 ; *A. echinata,* Nees in Pl. Preiss. i. 95 ; *A. Behriana,* Schlecht. Linnæa, xx. 660 (calyx often 6- or 7-lobed, stamens often 10).

Queensland. Near Warwick, *Beckler.*

N. S. Wales. Port Jackson, *R. Brown, Woolls ;* frequent in rather moist pastures in the western country, *A. Cunningham ;* head of the Gwydir, *Leichhardt.*
Victoria. I have seen no specimens from this colony, where, however, it is, no doubt, to be found.
Tasmania. Pastures, especially in the northern parts of the island, *J. D. Hooker.*
S. Australia. Memory Cove, *R. Brown ;* grassy plains, Barossa range, and Spencer's Gulf, *F. Mueller.*
W. Australia. Shady rocks, Mount Brown, *Preiss, n.* 2395 ; King George's Sound, Bald Island, and Cape Naturaliste, *Oldfield.*
The species extends to New Zealand, and probably also to extratropical S. America, for *A. montevidensis,* Hook. f. Fl. Antarct. ii. 265, appears to be quite the same.

2. **A. sanguisorbæ,** *Vahl; DC. Prod.* ii. 492. Stems prostrate or creeping and rooting at tħe nodes, the flowering branches ascending from a few in. to nearly 1 ft., loosely silky-villous, leafy at the base. Leaflets from nearly orbicular and ¼ in. long to oblong and ¾ in., prominently toothed, glabrous or nearly so above, silky-hairy underneath. Flowers numerous, in dense globular heads, on long terminal peduncles, usually under ½ in. diameter at the time of flowering, becoming when in fruit dense globular burrs of ¾ in. diameter or more. Calyx-lobes usually 4. Stamens 2. Style with the fringed stigma of *A. ovina.* Fruiting calyx nearly 2 lines long, turbinate, the lobes 4-angled, with a long prickle barbed at the end, diverging from near the summit of each angle.—Hook. f. Fl. Tasm. i. 114 ; *A. sarmentosa,* Carmich. ; DC. l. c.

N. S. Wales. Port Jackson, *A. Cunningham ;* Blue Mountains, *Miss Atkinson ;* head of Gwydir river, *Leichhardt ;* Hastings river, *Beckler ;* southward to Illawarra, *A. Cunningham ;* Gabo Island, *Maplestone.*
Victoria. Port Phillip, *R. Brown ;* near Skipton, *J. S. Whan.*
Tasmania. Port Dalrymple, *R. Brown ;* common in pastures on roadsides, etc., throughout the colony, *J. D. Hooker.*
S. Australia. Kangaroo Island, *R. Brown ;* Guichen Bay and Torrens river, *F. Mueller.*
The species is also found in New Zealand, Tristan d'Acunha, and Lord Auckland's Isles, and it is perhaps not really distinct from the S. American *A. ovalifolia,* Ruiz and Pav.

3. **A. montana,** *Hook. f. in Hook. Lond. Journ.* vi. 276. A small species, tufted or shortly prostrate, glabrous except a few silky hairs on the petioles, midribs, and margins of the leaflets and sometimes on the peduncles. Leaflets ovate, strongly serrate, mostly about ¼ in. long. Flowering stems ascending, leafless or with 1 or 2 small distant pinnatifid or pinnate leaves and a small terminal globular head, rarely above ¼ in. diameter. Calyx nearly glabrous, the lobes ovate, separating from the base and very deciduous. Stamens 2. Stigma short. Fruiting calyx not 1 line long, obovoid, glabrous, with 4 slender spines not 1 line long, and the top of the calyx-tube conical between them.

Tasmania. Moist summit of the Table Mountain on the Derwent, *R. Brown ;* summit of Mount Wellington, *J. D. Hooker,* and of Mount Lapeyrouse, *C. Stuart.* J. D. Hooker, Fl. Tasm. i. 115, unites this as a variety with *A. sanguisorbæ ;* but, independently of the foliage, the fruiting calyx appears to me to be quite different, and to bring it nearer to *A. adscendens,* Vahl. R. Brown's herbarium contains specimens of a dwarf alpine form of *A. sanguisorbæ,* from the Table Mountain, growing in company with *A. montana,* and quite maintaining its characters, as pointed out in his notes.

Poterium Sanguisorba, Linn., a herb with the habit and globular inflorescence nearly of

A. sanguisorbæ, but with strictly monœcious flowers, more numerous and longer stamens, and the fruiting calyx without prickles, is said to have established itself in some parts of Victoria, introduced from Europe.

Order XLII. SAXIFRAGEÆ.

Calyx free or adnate to the ovary, with 4 or 5 valvate or imbricate lobes or segments. Petals as many as calyx-lobes, valvate or imbricate, sometimes very small or wanting. Stamens as many or twice as many as calyx-lobes, rarely fewer and very rarely indefinite, inserted with the petals on or outside a perigynous or epigynous disk or rarely hypogynous. Ovary more or less adnate to the calyx, or if free usually attached by a broad base, either 2- to 5-celled or with 2 to 5 parietal placentas, very rarely contracted at the base or apocarpous; ovules usually several, very rarely solitary in each cell or to each placenta; styles as many as cells or placentas, distinct or rarely united. Fruit capsular or very rarely succulent and indehiscent. Seeds usually small, with a copious albumen and small or terete straight embryo, very rarely larger and without albumen.—Herbs shrubs or trees. Leaves alternate or opposite, simple or compound, with or without stipules. Flowers usually regular and hermaphrodite.

A large Order, ranging over nearly the whole world, the shrubby or arborescent genera chiefly tropical, the herbaceous ones from the more temperate or colder regions of the northern hemisphere, with a few extratropical southern genera or species. Of the 20 Australian genera, 1 is widely dispersed over the tropical and subtropical regions of the New and the Old World, 1 spreads over E. India and the Archipelago, 1 is represented in extratropical S. America, 2 in New Caledonia, 1 in New Zealand and the antarctic regions generally, 2 in New Zealand alone, and the remaining 12, many of them monotypic, are endemic in Australia. The Order includes a great variety of forms, evidently connected with each other, but difficult to unite by a common character which shall separate them from several other Calyciflorous and some Thalamiflorous Orders into which they appear sometimes to pass. There is especially no one character to distinguish them from *Rosaceæ* which has not some exception, although the greater number of genera differ from that Order in their definite stamens, united carpels with free styles, and copious albumen.

Tribe I. **Escallonieæ.**—*Shrubs or trees. Leaves alternate (except* Polyosma). *Stipules none. Stamens as many as calyx-segments. Styles usually united or cohering, at least under the stigma.*

Ovary 2- to 5-celled.
　Flowers corymbose. Petals valvate.
　　Ovary semi-adnate. Petals fringed inside. Fruit capsular.
　　　Leaves white underneath . . . . 1. Argophyllum.
　　Ovary free. Petals not fringed. Fruit succulent. Leaves
　　　green . . . . 2. Abrophyllum.
　Flowers racemose. Petals imbricate. Ovary inferior . . . . 3. Quintinia.
Ovary with 2 parietal placentas. Flowers racemose.
　Petals valvate. Ovary inferior. Fruit succulent, 1-seeded . . 4. Polyosma.
　Petals imbricate. Ovary free. Capsule many-seeded . . . . 5. Anopterus.

Tribe II. **Cunonieæ.**—*Shrubs or trees. Leaves opposite (except* Tetracarpæa). *Stipules usually present. Stamens twice as many as calyx-segments or indefinite. Styles free, at least at the top.*

Stamens twice as many as calyx-segments.
　Calyx-segments valvate or rarely slightly imbricate. Flowers solitary, cymose, capitate or paniculate.

Petals none or minute, or small and jagged.
Flowers in dense globular heads. Ovary 2- or 3-celled.
  Carpels small, follicular . . . . 6. CALLICOMA.
Flowers solitary. Ovary 2-celled. Fruit oblong, 1-seeded . 7. ANODOPETALUM.
Flowers few in short loose cymes or panicles or solitary.
  Ovary 4-celled. Fruit small, 1-seeded, surrounded by the
  wing-like enlarged calyx-lobes. Stem climbing . . . 8. APHANOPETALUM.
Flowers small, numerous, cymose. Ovary 2-celled. Fruit
  1-seeded.
  Fruit small, surrounded by the wing-like enlarged calyx-
    lobes. Leaflets 1 or 3, articulate on the petiole . . . 9. CERATOPETALUM.
  Drupe 1-seeded, the calyx-lobes small, reflexed. Leaves
    simple, continuous with the petiole . . . . 10. SCHIZOMERIA.
Petals narrow, entire, as long as or longer than the calyx.
Flowers in dense false whorls. Capsule terminating in 2
  diverging awn-like persistent styles . . . . 11. ACROPHYLLUM.
Flowers very small and numerous, paniculate. Capsule small,
  turgid, not awned . . . . 12. ACKAMA.
Calyx-segments more or less imbricate. Flowers racemose.
Ovary and fruit 2- rarely 3-celled . . . . 13. WEINMANNIA.
Carpels 4, quite free from the first . . . . 14. TETRACARPÆA.
Stamens indefinite, more than twice as many as calyx-segments, rarely
  fewer. Sepals 4.
Sepals valvate. Stamens 10 to 15. Ovary 2-celled. Seeds
  winged. Flowers racemose . . . . 15. GEISSOIS.
Sepals much imbricate. Stamens very numerous. Ovary 5- to
  12-celled. Seeds winged. Flowers large, solitary . . . 16. EUCRYPHIA.
Sepals valvate or slightly imbricate. Stamens few or numerous.
  Ovary 2-celled. Seeds not winged. Flowers solitary. Leaves
  3-foliolate (leaflets apparently in whorls of 6) . . . . . . 17. BAUERA.

TRIBE III. **Saxifrageæ.**—*Herbs. Leaves in the Australian genera radical, alternate, or imbricate, without stipules.*

Leaves and pitchers radical. Scape leafless. Flowers paniculate.
  Sepals 6, valvate. Petals 0. Stamens 12. Carpels 6, free . . 18. CEPHALOTUS.
Radical leaves entire, stem-leaves alternate, lobed. Flowers cymose.
  Calyx-lobes petals and stamens 5. Ovary 2-celled . . . . . 19. EREMOSYNE.
Small tufted plant with densely imbricate leaves. Flowers solitary,
  sessile. Calyx-lobes and petals 5. Stamens 2. Ovary inferior,
  2-celled . . . . 20. DONATIA.

TRIBE 1. ESCALLONIEÆ.—Trees or shrubs. Leaves alternate or rarely (in *Polyosma*) more or less opposite. Stipules none. Stamens as many as calyx-segments. Styles usually united or cohering, at least under the stigma.

1. ARGOPHYLLUM, Forst.

Calyx-tube shortly turbinate or hemispherical, adnate to the ovary; lobes 5 or 6, persistent. Petals as many, valvate, persistent, fringed inside below the middle with long hairs, forming a corona. Stamens as many as petals; anthers usually shorter than the filaments. Disk scarcely prominent. Ovary half-adnate, 2- to 5-celled, with numerous ovules in each cell; style shortly conical, with a capitate shortly lobed stigma. Capsule small, coriaceous, 2- to 5-celled, opening loculicidally and sometimes also septicidally into as many or twice as many valves or cells. Seeds minute, globular, reticulate. Embryo minute in a fleshy albumen.—Shrubs. Leaves alternate, usually white underneath. Flowers small, in terminal or axillary corymbose panicles.

Besides the Australian species, which is endemic, the genus comprises 3 or 4 from New Caledonia.

1. **A. Lejourdanii,** *F. Muell. Fragm.* iv. 33. An elegant shrub of 6 to 7 ft., the branches and inflorescence clothed with a close silky-white or reddish tomentum. Leaves ovate, acuminate, serrate, mostly 5 to 6 in. long, on a petiole of about 1 in., not coriaceous, green and glabrous or with scattered hairs above, silvery-white with a close silky tomentum underneath, the pinnate parallel primary veins, the transverse secondary ones, and smaller reticulations prominent underneath. Panicles terminal or in the upper axils, shorter than the leaves, corymbose or more frequently ovate. Flowers, including the small expanded petals, about 3 lines diameter. Capsule depressed-globular, about 2 lines diameter, usually 2-celled and 4-valved, rarely 3-celled and 6-valved.

Queensland. Mount Elliott, *Dallachy.* The large thin serrate leaves, and their elegant venation on the under side, distinguish this species at once from the New Caledonian specimens of *A. nitidum,* Labill., the plate of which it in some respects resembles.

2. ABROPHYLLUM, Hook. f.

(Brachynema, *F. Muell.*)

Calyx-tube exceedingly short, adnate to the broad base of the ovary; segments 5, spreading, deciduous. Petals 5, valvate, spreading, deciduous. Stamens 5; anthers large, on very short filaments. Ovary free, except the broad base, oblong, 5-furrowed, 5-celled, with many ovules in each cell; stigma sessile, 5-lobed. Berry free, ovoid, 5-celled. Seeds numerous, small, nearly globular; embryo (minute?) in a fleshy albumen.—Shrub. Leaves alternate. Flowers in corymbose panicles, terminal or in the upper axils.

The genus is limited to a single species, endemic in Australia.

1. **A. ornans,** *Hook. f. ms.* A tall handsome shrub, the young shoots and inflorescence pubescent with short appressed hairs. Leaves elliptical or ovate-lanceolate, acuminate, 6 to 9 in. long, with a few short broad mucronate teeth in the upper part, narrowed at the base into a petiole of 1 to 2 in., thin and glabrous or sprinkled with a few appressed hairs on the principal veins underneath. Panicles irregularly dichotomous, much shorter than the leaves. Flowers rather small, yellowish. Calyx-segments triangular-lanceolate, about ½ line long. Petals about 2 lines long. Berries 3 to 4 lines.—*Brachynema ornans,* F. Muell. Fragm. iii. 90.

N. S. Wales. Mount Tomah, Blue Mountains, *A. and R. Cunningham;* Richmond river, near Ballina, *C. Moore.* Dr. Hooker had described the genus for the 'Genera Plantarum' when the specimen and character arrived from F. Mueller, whose specific name he consequently adopted, but the generic name was preoccupied by a curious Brazilian genus allied to *Ebenaceæ,* described and figured in the 'Linnean Transactions,' xxii. 126, t. 22.

3. QUINTINIA, Alph. DC.

Calyx-tube obconical, adnate to the ovary, with 5 persistent teeth. Petals 5, imbricate, deciduous. Stamens 5; anthers ovate. Ovary inferior, 3- to 5-

celled, with several ovules in each cell, the free summit broadly conical, tapering into a persistent 3- to 5-furrowed style, with a capitate 3- to 5-lobed stigma. Capsule inferior, opening at the summit in teeth or valves continuous with the styles, which separate up to the stigma. Seeds ascending, long, spindle-shaped, with a loose testa; embryo (very minute?) in a fleshy albumen.—Glabrous trees or shrubs. Leaves alternate, coriaceous, without stipules. Flowers small, white, in racemes, either simple in the axils or several forming a terminal panicle.

Besides the 2 following species, which are endemic in Australia, there are 2 in New Zealand.

Racemes in a terminal leafless panicle . . . . 1. *Q. Sieberi.*
Racemes simple, axillary . . . . 2. *Q. Verdonii.*

1. **Q. Sieberi,** *A. DC. Monogr. Camp.* 90, *and in DC. Prod.* iv. 5. A spreading tree of 30 to 40 ft. Leaves oval-elliptical, shortly acuminate, mostly 3 to 4 in. long, entire, narrowed into a petiole of about ½ in., coriaceous, reticulate. Racemes numerous, in a terminal panicle, scarcely longer than the last leaves. Pedicels very short, rarely 1 line long. Calyx-lobes very short and broad. Petals oval-oblong, spreading, about 2 lines long. Styles separating in the ripe capsule up to the stigmas, which remain united. Seeds obovate or oblong, with a loose reticulate testa, but not winged.—Endl. in Flora, 1832, ii. 389, t. 3, and Atakta, 10, t. 10 (the plate wanting in our copy) ; F. Muell. Fragm. ii. 126.

N. S. Wales. Blue Mountains, *Sieber, n.* 261, *A. Cunningham,* and others ; southward to Illawarra, *A. Cunningham ;* and the dividing range towards the Yowaka, *Leichhardt.*

2. **Q. Verdonii,** *F. Muell. Fragm.* ii. 125. Very near *Q. Sieberi,* the leaves of the same shape and size, but much less reticulate. Racemes in the specimens seen all simple and solitary in the upper axils, 3 to 4 in. long. Flowers rather smaller than in *Q. Sieberi,* on pedicels about 2 lines long. Calyx lobes narrower, about half as long as the petals. Capsule smaller than in *Q. Sieberi.* · Seeds small, ovoid-oblong, obtuse, not winged.

N. S. Wales. Macleay and Hastings rivers, *Beckler.*

4. POLYOSMA, Blume.

Calyx-tube ovoid, adnate to the ovary, the limb small, 4-toothed, persistent. Petals 4, valvate, linear, erect and frequently cohering in a tube, spreading at the end, deciduous. Stamens 4 ; anthers linear, erect. Ovary inferior, 1-celled, with numerous ovules attached to 2 parietal placentas, protruding far into the cavity and almost dividing it into 2 cells ; style filiform, with an entire terminal stigma. Berry ovoid, inferior, with a single large erect seed ; testa rather thick ; embryo small, in the summit of a fleshy albumen.— Trees. Leaves opposite or nearly so, usually turning black in drying. Flowers white or greenish, in terminal simple racemes.

Besides the Australian species, which is endemic, the genus comprises several from E. India, the Archipelago, and S. Pacific Islands.

1. **P. Cunninghamii,** *J. J. Benn. Pl. Jav. Rar.* 196. A tall shrub or small tree, quite glabrous except the inflorescence and flowers. Leaves ovate-

elliptical, acuminate acute or rarely obtuse, 3 to 4 in. long, irregularly notched with callous teeth, much narrowed into a short petiole, somewhat coriaceous, penniveined. Racemes usually shorter than the leaves. Pedicels 1 to 2 lines long, with 2 minute bracteoles under the ovary. Calyx-teeth small. Corolla about 5 lines long, slightly pubescent outside with appressed hairs, the petals remaining long coherent in a narrow tube. Fruit ovoid, above ½ in. long, crowned by the small persistent cup-shaped calyx-limb.

N. S. Wales. Illawarra, *A. Cunningham, M'Arthur;* near Grafton, Clarence river, *Beckler.*

5. ANOPTERUS, Labill.

Calyx-tube very short, adnate to the broad base of the ovary; lobes 6 to 9, short, persistent. Petals as many as calyx-lobes, imbricate, spreading, deciduous. Stamens as many as petals; anthers versatile. Ovary free, except the broad base, 1-celled, with several ovules attached to 2 parietal placentas; style 2-lobed, the lobes stigmatic along the inner side. Capsule oblong-conical, thickly coriaceous, opening in 2 recurved valves with the placentas on their margins. Seeds pendulous, imbricate, flattened; testa membranous, dilated at the hilum end into a broad membranous wing, the nucleus at the opposite end small, oblong; embryo minute, in a fleshy albumen.—Shrubs or small trees, quite glabrous. Leaves alternate, evergreen, coriaceous, without stipules. Flowers white, rather large, in short terminal racemes.

The genus is endemic in Australia.

Flowers mostly 6-merous. Capsule not exceeding ¾ in. . . . . 1. *A. glandulosus.*
Flowers mostly 8- or 9-merous. Capsule 1 in. long or more . . . 2. *A. Macleayanus.*

1. **A. glandulosus,** *Labill. Pl. Nov. Holl.* i. 86, *t.* 112. A handsome evergreen shrub, in some localities growing into a tree of 30 to 40 ft. Leaves chiefly at the ends of the branches, from elliptical-lanceolate to almost obovate, but usually narrow, shortly acuminate, mostly 4 to 8 in. long, obtusely or callously serrate, narrowed into a very short petiole, coriaceous and shining, penniveined. Racemes 3 to 6 in. long, somewhat drooping. Bracts large, membranous and imbricate at the base of the very young raceme, but soon falling away, those in the raceme small and narrow. Pedicels 3 to 5 lines long. Petals usually 6, ovate, 5 to 6 lines long. Capsule ½ to ¾ in. long.— DC Prod. iv. 6; Hook. f. Fl. Tasm. i. 151; Bot. Mag. t. 4377.

Tasmania. In forests, abundant in many, especially the subalpine parts of the colony, *J. D. Hooker.*

2. **A. Macleayanus,** *F. Muell. in Journ. Pharm. Soc. Vict.* 1859. Very closely allied to *A. glandulosus* in aspect as well as in character, and may prove to be a variety only. Leaves usually rather longer and narrower, more acuminate, on petioles of ½ to 1 in. Calyx-lobes petals and stamens often 8 or 9 each. Capsule and seeds much larger than in *A. glandulosus,* the former from 1 to 1⅓ in. long, the seeds ½ to ¾ in., including the wing.

Queensland. Mount Lindsay, at an elevation of 4000 to 5000 ft., *W. Hill.*
N. S. Wales. Hastings river, *Beckler;* Clarence river, *C. Moore.*

TRIBE II. CUNONIEÆ.—Shrubs or trees. Leaves opposite (scattered in *Tetracarpæa*). Stipules usually present, but deciduous. Stamens twice as many as calyx-segments or indefinite. Styles free, at least at the top.

6 CALLICOMA, Andr.

(Calycomis, *R. Br.*)

Sepals 4 or 5, free, valvate or the margins slightly imbricate. Petals none. Stamens twice as many as sepals, hypogynous; anthers ovate, versatile. Ovary 2-celled or rarely 3-celled, with several pendulous ovules in each; styles distinct, filiform, each with a minute terminal stigma. Capsule small, separating into distinct carpels opening along the inner edge. Seeds small, ovoid-oblong, tuberculate; embryo very small, in a somewhat fleshy albumen —Tree or shrub. Leaves opposite, simple. Flowers small, in dense globular heads.

The genus is limited to a single species, endemic in Australia.

1. **C. serratifolia,** *Andr. Bot. Rep. t.* 566. A tall shrub, growing into a tree of 30 to 40 ft., the young shoots often tomentose or villous, the branches soon glabrous. Leaves from elliptical-oblong to ovate-lanceolate, shortly acuminate, coarsely serrate, 2 to 4 in. long, coriaceous, glabrous and shining above, either white underneath with a minute tomentum, or softly tomentose or villous and more rust-coloured, the parallel pinnate veins prominent underneath. Stipules ovate, very deciduous. Flowers numerous, in dense globular heads on peduncles of ½ to 1 in., of which 2 to 4 are usually on a short common peduncle in the upper axils, and several form a terminal cluster or short panicle. Sepals and capsules not above 1½ lines long, tomentose or villous. Stamens more than twice as long.—DC. Prod. iv. 7; Bot. Mag. t. 1811; Lodd. Bot. Cab. t. 1167.

Queensland. Glasshouse Mountains, *F. Mueller* (?) (specimens in leaf only).

N. S. Wales. Port Jackson to the Blue Mountains, *R. Brown, Sieber, n.* 269, and others; New England, *C. Stuart;* Hastings river, *Beckler.*

C. ferruginea, D. Don, Cunon. 11, in Edinb. New Phil. Journ. Apr. to June, 1830, with the leaves softly rusty-tomentose or villous underneath, passes into the common form by every gradation.

7. ANODOPETALUM, A. Cunn.

Calyx divided nearly to the base into 4 or 5 valvate lobes. Petals as many, very small. Stamens twice as many, inserted round a prominent disk; anthers small, the connective produced into a linear or conical appendage. Ovary superior, 2-lobed, 2-celled, with few pendulous ovules in each cell; styles diverging. Fruit oblong, fleshy, 1-seeded, probably indehiscent, but not seen ripe.—Tree or shrub. Leaves opposite, simple. Flowers solitary or 2 together in the upper axils.

The genus is limited to a single species, endemic in Australia.

1. **A. biglandulosum,** *A. Cunn. in Hook. f. Fl. Tasm.* i. 148. A tall bushy shrub, growing also into a tree of 50 to 60 ft., quite glabrous; branches very tough; branchlets angular. Leaves petiolate, narrow-oblong or lanceolate, obtuse, with a few obtuse serratures, 1 to 1½ in. long, coriaceous, shining, with few veins. Stipules lanceolate, acuminate, deciduous. Flowers few and inconspicuous, nearly sessile or very shortly pedicellate in the upper axils. Calyx-segments rather broad, about 2 lines long, usually clothed inside with a minute but dense tomentum. Petals much shorter than the calyx, narrow.

Stamens almost as long as the calyx, the appendage of the connective almost as long as the anther. Young fruit about 5 lines long, oblong, fleshy, with a single pendulous seed.

Tasmania. Subalpine districts, often forming a dense and almost impenetrable scrub, *J. D. Hooker.*

8. APHANOPETALUM, Endl.

(Platyptelea, *J. Drumm.*)

Calyx-tube very short, adnate to the broad base of the ovary; lobes 4, slightly imbricate, 2 opposite ones rather larger than the other 2, persistent and enlarged after flowering. Petals none or very minute. Stamens 8; filaments short, anthers oblong, 2-lobed at the base. Ovary 4-furrowed, 4-celled, with 1 pendulous ovule in each cell, tapering into 4 more or less united styles, shortly divergent at the top; stigmas terminal. Fruit hard, small, indehiscent, surrounded at the base by the horizontally spreading wing-like enlarged calyx-segments. Seed solitary, reniform or horse-shoe-shaped, rugose. Embryo curved, in the axis of the fleshy albumen.—Shrubs or trees, with weak or twining branches, quite glabrous. Leaves opposite, simple. Stipules minute or none. Flowers few in short cymes or leafy panicles, or solitary in the axils of the leaves.

The genus is limited to Australia. It is nearly allied to *Ceratopetalum* in character, and especially in the fruit, but with a very different habit.

Leaves ovate-lanceolate or elliptical (Eastern species) . . . . 1. *A. resinosum.*
Leaves linear (Western species) . . . . 2. *A. occidentale.*

1. **A. resinosum,** *Endl. Nov. Stirp. Dec.* 35, *and Iconogr. t.* 96. Stated by some to be a tree of 30 to 40 ft., by others described as a tall straggling or climbing shrub, quite glabrous, the smaller branches scabrous with raised dots said to be resinous. Leaves ovate lanceolate or elliptical, obtuse or scarcely acuminate, obtusely serrate, 1½ to 3 in. long, acute at the base, on a petiole of 1 to 3 lines, thinly coriaceous, smooth and shining. Peduncles axillary, sometimes 3-flowered, the central pedicel without bracteoles, the 2 lateral ones bracteolate, or all bracteolate and an additional pair lower down, or the inflorescence further developed into a short dense more or less leafy panicle. Calyx-lobes at first small, but soon enlarging, and under the ripe fruit oblong-lanceolate, obtuse, about ½ in. long. Petals, when present, quite microscopic. Fruit without the wings scarcely 1¼ lines diameter.—F. Muell. Fragm. i. 228.

Queensland. Moreton Bay, *F. Mueller, W. Hill, C. Stuart;* Pine river, *Fitzalan.*
N. S. Wales. Hunter's river, *R. Brown;* Hastings and Macleay rivers, *Beckler;* Illawarra, *A. Cunningham, Ralston;* Twofold Bay, *F. Mueller.*

2. **A. occidentale,** *F. Muell. Fragm.* i. 228. A shrub, with slender twining branches. Leaves linear, entire, 1 to 2 in. long, smooth and shining, the midrib prominent underneath, narrowed into a short petiole or almost sessile. Peduncles filiform, usually 1-flowered, with a pair of bracts below the middle, occasionally 3-flowered. Flowers rather smaller than in *A. resinosum,* the calyx-lobes more obtuse. Petals none. Styles separated much lower than in *A. resinosum.* Fruit nearly the same, the spreading calyx-lobes more oblong or almost obovate, about ¼ in. long.—*Platyptelea clematidea,* J. Drumm. and Harv. in Hook. Kew Journ. vii. 55.

W. Australia. Crevices of limestone rocks, Murchison river and Champion Bay, *Drummond, 6th Coll. n. 94, Oldfield.*

9. CERATOPETALUM, Sm.

Calyx-tube short, adnate to the base of the ovary; lobes 5, valvate, persistent and enlarged after flowering. Petals small and laciniate or none. Stamens 10, inserted on a perigynous disk; anthers small, the connective produced into a recurved appendage. Ovary short, half-inferior, 2-celled, with 4 collateral ascending ovules in each cell, tapering into 2 more or less united styles, free and recurved at the top; stigmas terminal. Fruit small hard and indehiscent, surrounded by the 5 wing-like horizontally spreading enlarged calyx-lobes. Seed solitary, slightly curved; embryo green, curved, in the axis of a fleshy albumen.—Trees or shrubs, glabrous and resinous. Leaves opposite, with 1 or 3 digitate leaflets articulate on the petiole. Stipules very small. Flowers small, in terminal trichotomous cymes or corymbose panicles.

The genus is limited to Australia.

Leaflets 3. Petals lobed . . . . 1. *C. gummiferum.*
Leaflets usually solitary. Petals none . . . . 2. *C. apetalum.*

1. **C. gummiferum,** *Sm. Bot. Nov. Holl. t.* 3. A tree attaining 30 to 40 ft. Leaflets 3, lanceolate, in some specimens all under 1½ in. long, in others mostly twice that size, obtuse or obtusely acuminate, obtusely serrulate, narrowed at the base, coriaceous, shining, penniveined and strongly reticulate. Cymes or panicles loosely trichotomous, the common peduncle shorter or longer than the leaves. Calyx-lobes in flower scarcely above 1 line long, in fruit linear-oblong, fully ½ in. long. Petals rather shorter than the calyx, deeply cut into 3 to 5 very narrow lobes. Stamens as long as the calyx. Fruit without the wings above 1½ lines diameter, the adnate calyx-tube strongly ribbed.—DC. Prod. iv. 13.

N. S. Wales. Port Jackson to the Blue Mountains, *R. Brown, Sieber, n.* 260, and others.

2. **C. apetalum,** *D. Don, Cunon.* 11, *in Edinb. New Phil. Journ. Apr. to June,* 1830. A beautiful tree of 50 to 60 ft., with a shining silvery bark. Leaflets usually solitary (occasionally 3 on luxuriant shoots or perhaps young trees), from ovate-lanceolate to narrow-lanceolate, 3 to 5 in. long, or nearly twice that size on luxuriant barren branches, obtusely serrate, coriaceous, shining, narrowed at the base, articulate on a petiole of ½ to 1 in. Flowers numerous in rather dense corymbose cymes, usually shorter than the last leaves, sometimes slightly pubescent. Calyx-lobes acute, above 1 line long in flower, scarcely above ¼ in. in fruit. Petals none. Appendage of the connective of the anthers smaller and straighter than in *C. gummiferum.*

N. S. Wales. Grose river, *R. Brown;* Port Jackson and especially in the Blue Mountains, *A. and R. Cunningham,* and others; Illawarra, *Shepherd.*

C. montanum, D. Don, Cunon. 11, was established on narrow-leaved specimens which do not otherwise differ from the common form.

10. SCHIZOMERIA, D. Don.

Calyx-tube short, adnate to the base of the ovary; lobes 5, valvate, not enlarged after flowering. Petals small, toothed. Stamens 10, inserted outside a lobed disk; anthers ovate, the connective produced into a short conical appendage. , Ovary short, free except the broad base, 2-celled, with 4 ovules in each cell attached to a pendulous placenta; styles distinct, short, recurved, with terminal stigmas. Fruit a drupe, with the small calyx-lobes reflexed from its base; epicarp thick and fleshy; endocarp bony. Seed solitary, somewhat curved; embryo green, rather large, in a fleshy albumen.—Tree. Leaves opposite, simple. Stipules small. Flowers small, in terminal trichotomous cymes.

The genus is limited to a single species, endemic in Australia, and very nearly allied to *Ceratopetalum* in habit and flowers, but the fruit is different, and the leaves truly simple, the lamina continuous with the petiole.

1. **S. ovata,** *D. Don, Cunon.* 12, *in Edinb. New Phil. Journ. Apr. to June,* 1830. A tree attaining 50 ft., with a dense foliage of a light green. Leaves ovate or ovate-lanceolate, obtuse or acuminate, mostly 3 to 4 in. long, nearly entire or with irregular obtuse serratures, shortly narrowed at the base and continuous with the petiole, coriaceous, penniveined and reticulate. Flowers rather smaller than those of *Ceratopetalum apetalum,* and the cymes usually looser, but otherwise much resembling them. Calyx-lobes scarcely above 1 line long. Petals shorter than the calyx, broad and toothed or lobed at the end. Drupe ovoid or globular, under ½ in. diameter.

N. S. Wales. Port Jackson, *R. Brown;* Blue Mountains, *Miss Atkinson;* northward to Macleay and Hastings rivers, *Beckler;* southward to Illawarra, *A. Cunningham.*

11. ACROPHYLLUM, Benth.

(Calycomis, *D. Don, not of R. Brown.*)

Calyx divided almost to the base into 4 to 6, usually 5, valvate segments. Petals as many, entire, exceeding the calyx. Stamens twice as many as petals, inserted round the slightly crenate disk; anthers small. Ovary free, 2-lobed, 2-celled, with several ovules in each cell; styles 2, subulate. Capsule small, septicidally dehiscent, the valves or carpels terminating in the long, persistent, straight but divergent awn-like styles. Seeds few, globular.—Shrub. Leaves opposite or verticillate, simple. Flowers in dense axillary clusters (reduced cymes), the 2 opposite ones forming a false whorl.

The genus is limited to a single species, endemic in Australia.

1. **A. venosum,** *Benth. in Maund, Botanist,* ii. 95. An elegant, erect, glabrous shrub, with slender branches. Leaves opposite or in threes, sessile or very shortly petiolate, ovate or ovate-lanceolate, acute, 2 to 3 in. long, bordered by triangular acute and regular teeth, rounded or truncate at the base, rigid, prominently penniveined and reticulate. Flowers pink, numerous in dense false whorls, each subtended by a pair of floral leaves reduced to bracts, whilst the uppermost leaves of the flowering branch are again large without flowers in their axils. Pedicels filiform, short at the time of flowering, 3 to 4 lines long in fruit. Calyx-segments about 1 line long. Petals rather longer, very narrow. Stamens longer. Capsules slightly exceeding the calyx, the slender rigid divaricate styles at least 2 lines long.—*Weinmannia australis,* A. Cunn. in Field, N. S. Wales, 353; DC. Prod. iv. 9; *Calycomis verticillata,* D. Don, Cunon. 10, in Edinb. New Phil. Journ. April to

June, 1830; *Acrophyllum verticillatum*, Hook. Bot. Mag. t. 4050; *Weinmannia venosa*, Knowl. and Westc. Fl. Cab. t. 65, according to Walp. Rep. ii. 373.

N. S. Wales. Moist shaded rocks, Springwood, Blue Mountains, rare, *A. and R. Cunningham.* Don was in error in supposing this to be the *Calycomis* mentioned by R: Brown in the Appendix to Flinders's Voyage; that was merely the orthography adopted by R. Brown for the genus *Callicoma* of Andrews. The plant described by A. Cunningham has not been hitherto recognized, owing to De Candolle having misunderstood his expression " foliis ternatis," and rendered it " foliis trifoliolatis."

12. ACKAMA, A. Cunn.

Calyx-tube short, campanulate; lobes 5, valvate. Petals 5. Stamens 10, inserted round a crenate disk; anthers small, tipped by a minute gland-like appendage to the connective. Ovary free, 2-celled, with several ovules in each cell; styles filiform, deciduous. Capsule small, turgid, septicidally dehiscent. Seeds few, ovoid, hairy; embryo cylindrical in the axis of a fleshy albumen.—Trees. Leaves opposite, pinnate. Flowers small, very numerous, in compound panicles, in terminal pairs, becoming axillary by the elongation of the central shoot.

Besides the Australian species which is endemic, the genus comprises another from New Zealand. The inflorescence, which is uniformly racemose in *Weinmannia*, being paniculate in both species of *Ackama*, gives them a habit so different from that of *Weinmannia*, that when coupled with the valvate calyx and the shape of the fruit, there seems to be quite sufficient to maintain *Ackama* as a distinct genus rather than as a section of *Weinmannia*, as proposed by A. Gray.

1. **A. Muelleri,** *Benth.* A tree, glabrous or nearly so except the inflorescence. Leaflets usually 5, rarely 7 or 3, ovate-elliptical or ovate-lanceolate, acuminate, obtusely and very shortly serrate, usually 3 to 4 in. long but sometimes much larger, narrowed at the base and more or less petiolulate, somewhat coriaceous, penniveined, with usually a minute tuft of hairs in the axils of the principal primary veins underneath. Flowers very small and numerous, clustered along the short ultimate branches of a very compound panicle, the branchlets all minutely pubescent. Calyx about ½ line long. Petals slightly exceeding the calyx-lobes. Stamens exserted. Capsule ovoid-globular, 1 to 1½ lines long.—*Weinmannia paniculata*, F. Muell. Fragm. ii. 83, altered to *W. paniculosa*, l. c. 175.

N. S. Wales. Hastings and Clarence rivers, *Beckler*.

13. WEINMANNIA, Linn.

Calyx divided almost to the base into 4 or 5 more or less imbricate segments. Petals as many as calyx-segments. Stamens twice as many as petals, inserted round the disk; anthers small. Ovary free, 2- or rarely 3-celled, with several pendulous ovules in each cell; styles distinct, each with a terminal or decurrent stigma. Capsule oblong or ovoid, septicidally dehiscent. Seeds oblong reniform or nearly globular, usually (but not always) hairy; embryo in the axis of a fleshy albumen.—Trees or shrubs. Leaves opposite, simple, or digitately or pinnately compound, with 3 or more leaflets. Flowers in simple racemes, terminal or axillary, solitary or clustered.

A genus widely distributed over the warmer regions of the globe, extending into extratropical S. America, S. Africa, and New Zealand. The only Australian species is endemic, but as yet insufficiently known to be quite certain as to its genus.

1. **W. rubifolia,** *F. Muell.* (*under* Geissois). A small tree, the young branches inflorescence and veins of the leaflets more or less clothed with long fine hairs. Leaflets 3 or 5, digitate, ovate-elliptical, acuminate, sharply serrate, much narrowed into a petiolule, rigid but not thick, the primary parallel veins very prominent underneath, with transverse reticulations, the terminal one usually 2 to 3 in. long, or rarely more, the lateral ones smaller. Stipules large, hairy, deciduous. Racemes axillary, usually several together on a very short common peduncle, 1½ to 3 in. long when in fruit. Flowers not seen. Pedicels very short or scarcely any. Sepals shorter than the fruit. Capsules reflexed, 1½ to nearly 2 lines long, narrow, hairy, with 2, rarely 3, recurved styles, the stigmas shortly decurrent. Seeds 2 or 3 in each carpel, narrow-oblong, the testa more or less extended into a loose wing at one or both ends, or, in some seeds the nucleus appears to extend nearly the whole length.—*Geissois rubifolia*, F. Muell. Fragm. ii. 82.

N. S. Wales. Cloud's Creek, a tributary of Clarence river, *Beckler.* The flowers being unknown, the genus is somewhat doubtful. The inflorescence and fruit are quite those of *Weinmannia*, except that the capsules are more pendulous than is usual in that genus. The foliage and habit scarcely differ from those of a Feejee plant, which we take to be the *Weinmannia spiræoides*, A. Gray, but of which the perfect flowers and fruit are unknown. The *W. rubifolia* is, however, certainly not a *Geissois*.

14. TETRACARPÆA, Hook. f.

Sepals 4, quite free, imbricate. Petals 4, imbricate. Stamens 8, hypogynous, anthers oblong, erect. Carpels 4, distinct, narrowed at the base, with numerous ovules in each; styles short, each with an obtuse stigma. Fruit carpels opening along the inner edge. Seeds numerous, obovoid; testa loose, membranous; embryo minute in a fleshy albumen.—Shrub. Leaves scattered. Flowers in terminal racemes.

The genus is limited to a single species, endemic in Australia. It had been referred by Endlicher to *Dilleniaceæ* on account of the hypogynous stamens and erect anthers, but the foliage and habit, the seeds and several other characters, are entirely those of the *Cunonieæ* or woody *Saxifrageæ*, in other genera of which a gradual passage may be observed from hypogynous stamens and a free apocarpous pistil to epigynous stamens, and an inferior ovary.

1. **T. Tasmanica,** *Hook. f. in Hook. Ic. Pl. t. 264, and Fl. Tasm.* i. 150. A small erect bushy shrub, usually about 1 ft. high, quite glabrous. Leaves obovate-oblong, obtuse, crenate, ¼ to 1 in. long, narrowed into a petiole, coriaceous and shining, the midrib prominent underneath, the veins obscure or inconspicuous. Flowers rather small, white, in erect racemes of 1 to 2 in. Bracts small, narrow. Pedicels at first very short, 2 to 3 lines long when in fruit. Petals nearly orbicular, about 1½ lines diameter, on short slender claws. Carpels narrow, erect, about 2 lines long.

Tasmania. Common in subalpine situations, *J. D. Hooker.*

15. GEISSOIS, Labillardière.

Calyx-tube very short, adnate to the broad base of the ovary; segments 4, valvate, deciduous. Petals none. Stamens indefinite, usually 10 to 15, hypogynous, filaments long, anthers ovate. Ovary oblong-conical, 2-celled, with several ascending ovules in each cell; styles filiform, united at the base. Capsule narrow, coriaceous, 2-celled, opening septicidally. Seeds oblong, flat, imbricate, produced upwards into a short wing; embryo in the axis of a fleshy albumen.—Trees. Leaves opposite, digitately compound; leaflets 3 or 5, petiolulate, coriaceous, entire or with distant serratures. Flowers purple or red, usually larger than in *Weinmannia*, in simple lateral racemes.

The genus is from New Caledonia, where there are 3 or 4 species, but there also appears to be an Australian one, although our specimens are insufficient for defining it. A detached raceme of old capsules with the seeds fallen out, from Cloud's Creek, Hastings river, *Beckler*, much resembles those from New Caledonia. These capsules are cylindrical, about ¾ in. long, on short pedicels, the epicarp minutely tomentose, the endocarp more or less separating from it. To the same species may very likely belong some specimens in leaf only (in Herb. F. Muell.) from Hastings river, *Beckler*, and Duck Creek, *C. Moore*. They are glabrous. Leaves opposite, 3-foliolate; leaflets petiolulate, ovate, 6 to 10 in. long, 3 to 5 in. broad, coriaceous, remotely and not deeply toothed, green on both sides. Stipules orbicular, coriaceous, more persistent than in most *Cunonieæ*.

16. EUCRYPHIA, Cav.

(Carpodontos, *Labill.*)

Sepals 4, free, broad and thin but rigid, much imbricate, cohering into a calyptra and falling off together as the flower opens. Petals 4, broad, oblique, much imbricate. Stamens very numerous, hypogynous; anthers small. Ovary free, 5- to 12-furrowed, 5- to 12-celled, with several ovules in each cell; styles distinct, with small terminal stigmas. Capsule hard, ovoid or oblong, septicidally dehiscent, the carpels remaining long attached by the filiform placentas. Seeds few in each carpel, oblong, compressed, produced upwards into a membranous wing.—Trees or shrubs. Leaves opposite, simple or pinnate. Stipules very deciduous. Flowers large, white, solitary in the upper axils.

Besides the two Australian species, which are endemic, there are two Chilian ones, and there, as in Australia, one has simple, the other pinnate leaves. The genus is placed by most authors in *Hypericineæ*, from which it differs in foliage, in the presence of stipules and in the albuminous seeds. As suggested by Planchon, it appears to be much nearer to the Cunonieæ, notwithstanding the hypogynous stamens and very much imbricate sepals and petals. The fruit and seeds are very nearly those of *Geissois*, except that the carpels are more numerous.

Leaves simple . . . . 1. *E. Billardieri.*
Leaves pinnate . . . . 2. *E. Moorei.*

1. **E. Billardieri,** *Spach; Hook. f. Fl. Tasm.* i. 54. A handsome tree, attaining a very large size, although the smaller forms are often reduced to a bushy shrub, quite glabrous, the buds and young leaves often very gummy. Leaves simple, shortly petiolate, oblong, very obtuse, entire, coriaceous, glaucous or whitish underneath, 1 to 2 in. long in the larger forms. Flowers white, very showy, the broad petals often 1 in. diameter. Peduncles much norter than the leaves. Capsules usually about ½ in. long.—*Carpodontos lucida*, Labill. Voy. t. 18; DC. Prod. i. 556.

Tasmania, *R. Brown;* mountainous districts, especially in the interior and towards the S. and W. coasts, *J. D. Hooker.*

Var. *Milligani.* A shrub or small tree, smaller and more compact in all its parts than the typical form; leaves often all under ½ in. long and more crowded; flowers and fruits smaller, but there appears to be no other difference, and many specimens are quite intermediate.—*E. Milligani*, Hook. f. Fl. Tasm. i. 54. t. 8.

2. **E. Moorei,** *F. Muell. Fragm.* iv. 2. A handsome tree, the young shoots and foliage pubescent, the buds very gummy. Leaves pinnate; leaflets usually 9 to 11, narrow-oblong, entire, coriaceous, the terminal one often 1½ to 2 in. long, the lateral ones shorter, the veins more prominent than in *E Billardieri.* Flowers and fruits quite the same as in the smaller forms of *E. Billardieri.*

N. S. Wales. Wooded hills near the sources of the Clyde and Shoalhaven rivers, *C. Moore.*

17. BAUERA, Banks.

Calyx divided nearly to the base into 6 to 10, rarely 4 or 5, spreading segments, often toothed, valvate or slightly imbricate. Petals as many as calyxsegments. Stamens indefinite, few or numerous, inserted round a narrow disk; anthers short. Ovary wholly or partially free, 2-celled, with 2 or more ovules in each cell; styles distinct, recurved. Capsule superior or half-inferior, broad, truncate, opening loculicidally in 2 valves, or in 4 from the septicidal splitting of the valves. Seeds obovate with a granulate testa; embryo nearly terete, in a fleshy albumen.—Shrubs. Leaves opposite, each with 3 leaflets without any common petiole, so as to have the appearance of a whorl of 6 leaves. Stipules none. Flowers axillary, solitary, but sometimes the pairs crowded in a terminal leafy head.

The genus is limited to Australia. By a mistake of Salisbury's, copied by subsequent authors, the name of the genus has been attributed to Kennedy. In Andrews's 'Botanical Repository,' where it was first published, it is expressly stated that it was named by Banks, without any allusion to Kennedy.

Flowers pedicellate. Ovary superior. Ovules several. Leaves mostly
serrate . . . . 1. *B. rubioides.*
Flowers sessile.
Ovary superior. Ovules several, ascending. Leaves mostly 3-toothed 2. *B. capitata.*
Ovary half-inferior. Ovules 2 in each cell, pendulous.· Leaves entire 3. *B. sessiliflora.*

1. **B. rubioides,** *Andr. Bot. Rep. t.* 198. An elegant shrub, sometimes small slender and prostrate, sometimes erect and bushy, attaining 5 or 6 ft. or even more; branches terete, glabrous or more frequently pubescent or hirsute with long fine hairs. Leaflets oblong or lanceolate, rather acute, rarely exceeding ½ in. and sometimes not ¼ in. long, evergreen and often shining, marked by a few serratures sometimes deep sometimes obscure or almost disappearing. Flowers pink or white, on slender pedicels, sometimes shorter, but more frequently longer than the leaves. Calyx-segments and petals rarely fewer than 6, and often 8 or 9. Petals longer than the calyx, often twice as long, spreading to a diameter of from ½ to ¾ in. Capsule shorter than the persistent calyx, very broad, wholly superior although attached by a broad base, several-seeded.—Vent. Jard. Malm. t. 96; Bot. Mag. t. 715; DC. Prod. iv. 13; Hook. f. Fl. Tasm. i. 149, t. 31; *B. rubiæfolia*, Salisb. in Ann. Bot. i. 514, t. 10; Lodd. Bot. Cab. t. 1313; F. Muell. Fragm. iv. 23; *B. humilis*, Sweet; Lodd. Bot. Cab. t. 1197; DC. Prod. iv. 13; *B. Billardieri*, D. Don, Cunon. 13, in Edinb. New Phil. Journ. Apr. to June, 1830.

N. S. Wales. Port Jackson to the Blue Mountains, *R. Brown, Sieber, n.* 287, and others; very abundant on banks of streams, often in the water, *Lowne;* New England, *C. Stuart.*

Victoria. Wet swampy places and marshy woods, in the southern and eastern parts of the colony, *F. Mueller.*

Tasmania. Derwent river, *R. Brown;* very abundant throughout the colony, generally in poor wet soil, *J. D. Hooker.*

Var. *microphylla,* Ser. Small, slender and prostrate. Leaflets mostly about 2 lines long. Pedicels long and slender. Flowers small. Petals usually 5 or 6. Stamens few.—*B. microphylla,* Sieb. in DC. Prod. iv. 13; D. Don, Cunon. 13. in Edinb. New. Phil. Journ. Apr. to June, 1830; *B. galioides,* Sieb. in Reichb. Icon. Exot. t. 77.—Port Jackson, *R. Brown, Caley, Sieber, n.* 286, and others. The Tasmanian variety figured in the plate above quoted is not quite so small and slender as Sieber's and some other Port Jackson specimens.

2. **B. capitata,** *Ser. in DC. Prod.* iv. 13. A small shrub, either diffuse, with the habit of the shorter specimens of *B. rubioides,* var. *microphylla,* or with a woody stock and numerous slender stems of ½ ft., slightly pubescent. Leaflets narrow, scarcely above ¼ in. long, obtuse, usually with 1 prominent lobe or tooth on each side. Flowers almost sessile, solitary in each axil, but several pairs close together at the ends of the branches, forming little leafy heads. Calyx-segments 4 to 6, usually 5, about 2 lines long, more distinctly 3-fid than the leaves. Stamens not numerous; anthers small. Ovary free but attached by a broad base; ovules several in each cell, ascending from near the base of the partition. Capsule loculicidal but scarcely septicidal. Seeds rather large, pubescent, rugose, with a prominent raphe.—F. Muell. Fragm. iv. 24.

N. S. Wales. Port Jackson, *R. Brown, F. Mueller;* Hastings river, *Beckler;* Newcastle, *Leichhardt.*

3. **B. sessiliflora,** *F. Muell. in Trans. Phil. Soc. Vict.* i. 41, *and Fragm.* iv. 24. A shrub, with the habit of the larger forms of *B. rubioides,* usually hirsute. Leaflets oblong or almost obovate, obtuse, ½ to 1 in. long or rather more, the margins recurved and usually entire, scabrous pubescent or hirsute. Flowers (purple or pink?) sessile in small axillary leafy clusters or terminal leafy heads. Calyx-tube turbinate, adnate to the ovary; lobes 6 to 9, usually 7 or 8, lanceolate, hirsute, 1 line long or rather more. Petals 2 or 3 times as long. Stamens few, rarely twice as many as petals; anthers larger than in *B. rubioides.* Ovary half-inferior with a very thin dissepiment, and 2 ovules in each cell, suspended from the summit. Capsule not seen ripe, but apparently 1-seeded.—F. Muell. Pl. Vict. ii. t. 16.

Victoria. Crevices of rocks on the summit of Mount William and descending along the streams to the base of the Grampians, frequent at Morro-Morro, *F. Mueller.*

Tribe III. SAXIFRAGEÆ.—Herbs. Leaves, in the Australian genera, radical, alternate, or imbricate, without stipules. Stamens various. Styles free or shortly united at the base.

18. CEPHALOTUS, Labill.

Calyx-tube short, free, lobes 6, valvate. Petals none. Stamens 12, inserted round a glandular disk; anthers short, with a thick glandular connective. Carpels 6, free, distinct, with 1 or rarely 2 ascending ovules in each, and tapering into short recurved styles. Fruit-carpels small, erect, verticillate round a globular succulent enlarged torus, opening along the inner edge. Seeds solitary, erect; embryo very small, in the axis of a fleshy albumen.— Herb. Leaves radical, several of them converted into pitchers. Flowers white, in a narrow panicle at the end of a leafless scape.

The genus is limited to a single species, endemic in Australia. The completely apocarpous pistil has induced several botanists to place it in *Rosaceæ*, but the habit, definite stamens, and albuminous seeds, with a small embryo, are much more those of *Saxifrageæ*.

1. **C. follicularis,** *Labill. Pl. Nov. Holl.* ii. 7, *t.* 145. Stock short and perennial. Leaves rosulate, obovate-oblong, obtuse, entire, ½ to 1 in. long, narrowed into a petiole often as long as the lamina, glabrous or ciliate; some converted into ovoid or nearly globular pitchers of about 1 in. diameter or rather more, each with 3 external longitudinal raised nerves, dilated into narrow leaf-like double wings ciliate on the edge; the mouth of the pitcher bordered by a transversely plaited ring; the ovate lid attached to the side next the petiole. Scape 1 to 2 ft. high, silky-pubescent. Flowers scarcely 2 lines diameter, silky-hairy, nearly sessile, in short alternate racemes forming a narrow panicle of 1 to 3 in. Calyx-tube greenish, the lobes white, especially at the edges. Stamens not exceeding the calyx. Ripe carpels very little longer and very hairy, the central torus very small at the time of flowering, at least 1 line diameter when in fruit.—R. Br. App. in Flind. Voy. 600, t. 4; Bot. Mag. t. 3118, 3119; Nees, in Pl. Preiss. i. 278.

W. Australia. Wet marshes, King George's Sound, *Labillardière, R. Brown,* and others.

19. EREMOSYNE, Endl.

Calyx-tube hemispherical, compressed, adnate to the ovary; lobes 5, narrow. Petals 5. Stamens 5. Ovary half-adnate to the calyx-tube, 2-celled, with 1 erect ovule in each cell; styles short, distinct, with capitate stigmas. Capsule half inferior, membranous, compressed, broad, 2-celled, loculicidally dehiscent. Seeds solitary in each cell, erect, albuminous.—Small herb. Leaves alternate, lobed. Flowers minute, in terminal cymes.

The genus is limited to a single species, endemic in Australia.

1. **E. pectinata,** *Endl. in Hueg. Enum.* 53; *Iconogr. t.* 112. A slender diffuse or divaricately branched annual of 3 or 4 in., hirsute with short white spreading hairs. Radical leaves rosulate, petiolate, obovate or spathulate, about ½ in. long; stem-leaves alternate, sessile, deeply divided into 3 to 7 narrow linear lobes, the central one longer and often broader than the others; the upper leaves very small and narrow, entire or 3-cleft. Flowers numerous in terminal corymbose cymes, each flower about ½ line in diameter, the small white petals exceeding the calyx-lobes, and occasionally with 2 or 3 white hairs on the outside. Capsule about ¾ line broad, hirsute with few hairs. Seeds ovoid-conical, with a thin pale mottled testa, the albumen not very copious; embryo apparently oblong, but not seen perfect.—Lehm. in Pl. Preiss. ii. 236.

W. Australia. Swan River, *Huegel, Drummond,* 3rd *Coll. n.* 20; sandy shady places, Plantagenet district, *Preiss, n.* 2046.

20. DONATIA, Forst.

Calyx-tube adnate to the ovary, lobes 4 or 5. Petals 5 to 10, inserted round a broad flat disk. Stamens 2 or 3, inserted on or within the disk; anthers 2-celled, opening outwards. Ovary inferior, 2- or 3-celled, with several ovules in each cell attached to a pendulous placenta; styles short and thick, united at the base, stigmas globular. Fruit . . . —Densely tufted alpine herbs. Leaves small, closely imbricated, persistent. Flowers solitary, sessile amongst the leaves.

Besides the Australian species, which is also in New Zealand, there is one from Fuegia. The genus is very anomalous, but appears to be nearer to *Saxifrageæ* than to any other Order.

1. **D. Novæ-Zelandiæ,** *Hook. f. Fl. Nov. Zel.* i. 81, *t.* 20. A small densely tufted perennial, the short stem completely covered by the crowded imbricated leaves. Leaves linear, about 2 lines long, entire, coriaceous, shining, the fresh ones terminating the tufts of a bright green, the old persisting ones below them brown, with dense tufts of white shining almost scarious hairs in their axils. Flowers solitary and sessile in the tufts, about as long as the leaves. Calyx-tube turbinate, the 5 lobes thick, somewhat obtuse, shining like the leaves. Petals 5. Stamens 2, inserted near the centre of the disk close to the base of the styles and almost cohering with them. Ovules about 12 in each cell, in a dense tuft pendulous from the top.

Tasmania. Summit of Mount Lapeyrouse, *Oldfield;* also in New Zealand.

Order XLIII. CRASSULACEÆ.

Sepals 3 or more, usually 5, but sometimes up to 20, free from the ovary, but occasionally united in a lobed calyx. Petals as many as sepals, free or rarely united in a lobed corolla. Stamens as many or twice as many as petals, inserted with them at the base of the calyx. Ovary superior, the carpels as many as the petals, distinct, usually with a small flat scale at the base of each; with several ovules in each; styles simple, distinct. Ripe carpels capsular. Seeds several, with a thin fleshy albumen and straight embryo.—Herbs or rarely low shrubs or undershrubs. Leaves succulent, without stipules. Flowers in terminal racemes cymes or panicles, or rarely in axillary clusters.

A rather numerous Order, extending over the greater part of the globe, but particularly abounding in S. Africa and in the rocky districts of Europe and Asia. The only Australian genus is generally spread over the area of the Order. The Order is nearly allied to some herbaceous *Saxifrageæ*, but it is more apocarpous, the stamens less perigynous, and is readily known by its succulent leaves and thoroughly isomerous flowers.

1. TILLÆA, Linn.
(Bulliarda, *DC.*)

Sepals, petals, stamens and carpels 3 or 4 each, very rarely, in species not Australian, 5, all distinct. Ripe carpels opening along the inner edge, the seeds often reduced to 1 or 2 in each.—Small, often minute, herbs. Leaves opposite. Flowers minute, axillary or in a terminal leafy panicle.

The genus has very nearly the extensive geographical range of the Order. Of the 4

Australian species wo are also in New Zealand, one of which extends also to extratropical S. America, the two others are endemic.

Flowers under 1 line long, axillary. Carpels short and obtuse.
Flowers in dense leafy clusters. Petals shorter than the sepals . . 1. *T. verticillaris.*
Flowers solitary, mostly pedicellate. Petals as long as or exceeding
 the sepals.
  Leaves not 2 lines long. Pedicels usually longer. No scales un-
   der the carpels . . . . 2. *T. purpurata.*
  Leaves often above ¼ in. long. Pedicels rarely as long as the
   leaves. A scale under each carpel . . . . 4. *T. recurva.*
Flowers above 1 line long, in a broad dichotomous or 3-chotomous
 leafy panicle. Carpels oblong . . . . 3. *T. macrantha.*

I omit *Dasystemon calycinum,* DC. Mem. Crass. 15. t. 3, and Prod. iii. 382, described from a plant from the Jardin des Plantes of Paris, supposed to have been raised from Australian seed, for it is evidently founded on a mistake. The stamens with thick ovoid filaments, forming the chief character of the genus, are not so figured in the plate; probably on a first hasty examination the carpels were taken for stamens. The plant appears to me to be *Crassula expansa,* Ait., a S. African not an Australian species, an error as to origin very common in botanical gardens.

1. **T. verticillaris,** *DC. Prod.* iii. 382. An annual, when first flowering simple and 1 in. high, but when old much branched, forming dense tufts of 3 or 4 in. diameter, or slender and 4 or 5 in. long. Leaves ovate-lanceolate or linear, connate at the base, 1 to 2 lines long. Flowers very small in dense axillary clusters mixed with small leaves, many of them nearly sessile, others on pedicels of 1 or 2 lines. Sepals usually 4, very rarely 5, acute or aristate, about ½ line long. Petals shorter, narrow, acute. Carpels without scales, when ripe very obtuse, not exceeding the calyx, with 1 or 2 seeds in each.—Hook. f. Fl. Tasm. i. 145 ; *T. pedunculata,* Sieb. Pl. Exs., not of Sm. ; *T. adscendens* and *T. colorata,* Nees, in Pl. Preiss. i. 277.

Queensland. On the Maranoa, *Mitchell ;* Brisbane river, Moreton Bay, *F. Mueller.*
N. S. Wales. Port Jackson to the Blue Mountains, *R. Brown, Sieber, n.* 173, and others; northward to Hastings and Clarence rivers, *Beckler ;* southward to Twofold Bay, *F. Mueller ;* in the interior to Barrier range, *Victorian Expedition.*
Victoria. Rocky and gravelly places, common in various parts of the colony, *F. Mueller* and others.
Tasmania. Common on dry rocks and gravel in many parts of the island, *J. D. Hooker.*
S. Australia. From Port Lincoln, *Wilhelmi,* to Bugle and Barossa ranges, *F. Mueller ;* and Lake Gilles, *Burkitt.*
W. Australia. King George's Sound, *R. Brown,* and thence to Swan and Murchison Rivers, *Drummond, 4th Coll. n.* 114, 115, *Preiss, n.* 1931, 1932, *Oldfield,* and others.
The species extends to New Zealand, and also to extratropical S. America, if *T. minima,* Miers, be really the same.

2. **T. purpurata,** *Hook. f. in Hook. Lond. Journ.* vi. 472, *and Fl. Tasm.* i. 145. A very slender decumbent annual of ½ to 1 in., rarely lengthening to 2 in. Leaves linear, connate at the base, 1 to 1½ or rarely 2 lines long. Flowers minute, on slender solitary pedicels mostly longer than the leaves, rarely short. Petals about ½ line long, sepals shorter, acute or obtuse. Carpels obtuse, not longer than the sepals, with several seeds.—F. Muell. Pl. Vict. ii. t. 19.

N. S. Wales. Paramatta, *R. Brown.*
Victoria. Wet pastures, very abundant in many parts of the colony, *F. Mueller.*

Tasmania. Wet places, Formosa, *Gunn;* S. Esk river, *C. Stuart.*
S. Australia. Wet places, foot of Mount Remarkable, and many other parts of the colony, *F. Mueller.*
W. Australia, *Drummond.*
Also in New Zealand.

3. **T. macrantha,** *Hook. f. in Hook. Ic. Pl. t.* 310, *and Fl. Tasm.* i. 145. An erect dichotomous much-branched annual, attaining 2 or even 3 in. Leaves linear, 1 to 3 lines long, connate at the base. Flowers larger than in the other species, numerous, either on long pedicels in the forks, or shortly pedicellate or sessile on the last branches, forming a broad corymbose panicle occupying the greater part of the plant. Sepals lanceolate, acute, about 1½ lines long, sometimes, but not always, ciliate. Petals as long as the sepals or shorter. Carpels acuminate at the time of flowering ; when ripe oblong, obtuse, abruptly pointed by the base of the style, nearly as long as the sepals. Seeds several, but not numerous.

Victoria. Wet sandy places near Brighton, etc., *F. Mueller ;* very common in wet pastures about Melbourne, *Adamson.*
Tasmania. Wet hollows, Georgetown, very common, *Gunn.*
S. Australia. Near Adelaide, *F. Mueller.*

4. **T. recurva,** *Hook. f. Fl. Tasm.* i. 146. A slender plant, densely tufted and 1 or 2 in. high in sandy places, lengthening out to 1 ft. in water. Leaves linear or linear-lanceolate, ¼ in. long or more in the longer specimens, 1 to 2 lines in the smaller ones. Flowers few, small, solitary, on peduncles rarely exceeding the leaves. Sepals about ¾ line long, acuminate. Petals about as long. Carpels acuminate with the recurved styles, with a small cuneate or linear-spathulate scale under each, sometimes half as long as the carpel. Fruit-carpels about as long as the calyx, with 2 or 3 seeds in each. —*T. verticillaris,* Hook. Ic. Pl. t. 295, not of DC. ; *T. intricata,* Nees in Pl. Preiss. i. 278.

Queensland. A specimen not in flower from Moreton Bay, *F. Mueller,* appears to be this species.
N. S. Wales. Inundated forests, Lake George, *A. Cunningham.*
Victoria. Common on wet banks of rivers, etc., *F. Mueller, Robertson.*
Tasmania. Port Dalrymple, *R. Brown ;* common in bogs and inundated places throughout the island, *J. D. Hooker.*
S. Australia. Onkaparinga river and towards Lake Torrens, *F. Mueller.*
W. Australia, *Drummond,* 4*th Coll. n.* 110 ; wet places on the lake, Rottenest Island, *Preiss, n.* 1929.

ORDER XLIV. DROSERACEÆ.

Calyx free or very shortly adnate to the broad base of the ovary, divided to the base or nearly so, into 4 or 5 or rarely 8 segments or sepals. Petals as many as calyx-segments, hypogynous or slightly perigynous. Stamens as many as petals, or rarely, in genera not Australian, twice as many or more, and inserted with them. Ovary either 6-celled, with 2 to 5 parietal placentas or 1 basal placenta, or 2- or 3-celled, with several ovules to each placenta or cell ; styles either as many as placentas, simple or divided to the base so as to appear twice the number, or variously branched ; or rarely the styles united into one ; stigmas various. Capsule opening loculicidally, in as many valves as cells or placentas, the valves rarely splitting septicidally. Seeds several, with a reticulate testa, sometimes produced beyond the nucleus into a loose wing; embryo cylindrical or sometimes minute in a fleshy albumen.—Herbs. Leaves usually ciliate or covered with glandular hairs. Flowers solitary or in. one-sided racemes, either simple or forming a branching cyme.

A small Order, found in nearly all parts of the world, the principal genus closely allied to the herbaceous *Saxifrageæ*, differing chiefly in the insertion of the petals and stamens, being more generally hypogynous; the whole group is easily recognized by the glandular leaves, involute in vernation. Of the two Australian genera, the principal one constitutes nearly the whole Order and ranges over the general area, the other is endemic and very anomalous.

Ovary 1-celled. Styles 2 to 5, distinct or shortly united at the base . . 1. DROSERA.
Ovary 2-celled. Style undivided . . . . 2. BYBLIS.

1. DROSERA, Linn.

(Sondera, *Lehm*.)

Calyx-segments 4, 5, or rarely 8. Petals as many. Stamens as many; anthers opening laterally or outwards in longitudinal slits. Ovary 1-celled, with 2 to 5, usually 3, parietal placentas; styles as many as placentas, simple or variously branched. Capsule opening in as many valves as placentas, with the placentas in their centre.—Herbs. Leaves usually involute in vernation, the lamina more or less covered on the upper side with glandular hairs or cilia and bordered with longer ones, usually irritable and closing over insects or other objects resting on them, the under side and petioles without glandular hairs. Stipules when present scarious and usually lobed or jagged. Flowers solitary or in one-sided racemes or forked cymes, on radical scapes or terminal peduncles.

A large genus, with the extensive geographical range of the Order, and comprising the great majority of its species. Of 41 Australian species, 4 are also E. Indian or in the Archipelago, of which 1 extends to New Zealand, 4 more extend to New Zealand only, the remaining 33 are endemic.

The Australian species may be readily distributed into the two old-established sections *Rorella* and *Ergaleium*, if characterized chiefly by their mode of vegetation. In *Rorella* the stock or stem, very short and completely covered with the leaves, except in *D. indica*, forms at its upper end the winter bud for the following year's vegetation, the lower end dying away either at the close of the season or after having endured several years covered with the old imbricate bases of the leaves, never forming a bulb at the base, but emitting new roots or sometimes stolons immediately under the fresh leaves of the new year. In this section also the styles are usually simple or once branched, very rarely dichotomous, and the stipules are wanting only in 3 species. In the second section, *Ergaleium*, the short stem-like stock forms usually, perhaps always, at its lower end a bulb, and at the upper end either a rosette of leaves with a leafless scape or leafy stems, which appear to be annually renewed, although in what manner this takes place has not been observed. The stock or stem between the bulb and the rosette has frequently loose ragged remains of leaves or petioles, as if it were partially at least perennial. In this section the styles are always short and very much divided, forming usually a dense tuft on the ovary, and the stipules are either entirely wanting, or, in *D. Banksii*, small and very evanescent. In both sections, however, and especially in *Rorella*, there are single exceptional species, which prevent giving any definite character derived from the singular diversities in the styles and other floral characters. Planchon, in his excellent study of the genus, in the 'Annales des Sciences Naturelles,' ser. 3, ix., proposes each of these anomalous species as a distinct section, but that course appears to me not to tend towards clearness of method, but rather to confuse the mind, and I have preferred adopting, with slight modifications, the two old sections, sub dividing them more artificially in the following table :—

SECT. I. **Rorella.**—*Stock not bulbous, the upper end perennial. Scapes leafless (except in* D. indica). *Stipules often present. Styles usually simple or divided into 2 simple branches, or rarely dichotomous.*

No stipules.
    Stems elongated. Leaves alternate, linear. Racemes several-
      flowered. Styles divided to the base into 2 filiform branches .   1. *D. indica.*
    Stems short, covered by the sheathing bases of the linear leaves.
      Peduncles or scapes 1-flowered. Styles short, simple, with capi-
      tate stigmas . . . . 2. *D. Arcturi.*
    Stemless. Leaves obovate, rosulate. Scapes short, several-flowered.
      Styles forked or dichotomous . . . . 3. *D.·glanduligera.*
Stipules scarious. Leaves radical, rosulate (except in *D. binata*), the
    stems or stock dying away below the rosette or rarely persisting
    and densely covered with the dried remains of the old leaves and
    stipules.
    Scape filiform, with 1 minute 4-merous flower. Styles 4, undi-
      vided . . . . 4. *D. pygmæa.*
    Scape filiform, with a short loose almost corymbose raceme of 2 to
      4 flowers. Leaves obovate or orbicular.
      Styles 2, divided to the base into 2 branches (or 3 simple ?), with
        large capitate stigmas . . . . 5. *D. platystigma.*
      Styles 5, filiform, undivided. Stipules short, with fine lobes . 6. *D. pulchella.*
      Styles 3, filiform, undivided. Stipules silvery-white, rather broad,
        densely imbricate in a prominent central bud . . . . 7. *D. leucoblasta.*
    Scape with a one-sided raceme of several flowers, all on short
      pedicels.
      Racemes glabrous or scarcely pubescent. Leaves obovate or
        orbicular.
        Scapes slender, rarely 2 in. high.
          Racemes short, rather loose. Calyx narrow, about 1 line
            long. Styles 3, short, with globular stigmas . . . . 8. *D. nitidula.*
          Racemes slender. Pedicels very short. Calyx not half
            line long. Styles filiform.
            Styles 3. Petiole not twice as long as the lamina . . 9. *D. paleacea.*
            Styles 5, rarely 4. Petiole 3 times as long as the lamina 10. *D. parvula.*
        Scapes attaining several in. Pedicel very short. Calyx above
          1 line long.
          Styles 5, simple, fringed at the stigmatic end. . . . . 11. *D. Burmanni.*
          Styles 3 or 4, divided to the base into 2 branches, entire or
            forked at the end . . . . 12. *D. spathulata.*
      Racemes, especially the calyxes, softly villous.
        Styles 3 or 4, simple. Leaves oblong. Old petioles and sti-
        pules often persistent below the rosette.
        Anthers oblong, on thick filaments. Petioles closely re-
          flexed on the stock . . . . 13. *D. Drummondii.*
        Anthers small, on slender filaments. Petioles not reflexed 14. *D. scorpioides.*
      Styles 3, dichotomous. Leaves orbicular, the petioles long,
        woolly-hairy as well as the stipules . . . . 15. *D. petiolaris.*
    Scape tall, with a loose cyme. Leaves linear, forked or dichoto-
      mous. Styles divided into a dense tuft of numerous lobes . . 16. *D. binata.*

SECT. II. **Ergaleium.**—*Stock short, slender, stem-like, naked or with ragged remains of old petioles, forming (usually if not always) a bulb at the lower end and producing at the upper end a rosette of leaves and leafless scapes, or leafy stems or branches. Stipules none (or in* D. Banksii *small and evanescent). Styles dichotomous or divided into very numerous filiform branches, forming a dense tuft.*

Rootstock bearing a simple rosette of leaves, with leafless scapes or
    peduncles.

Leaves semicircular or almost reniform, cuneate at the base, on a
　long petiole . . . . 17. *D. zonaria.*
Leaves orbicular obovate or oblong, tapering into a very short pe-
　tiole.
　Peduncles, usually several, all 1-flowered.
　　Leaves small, thick, oblong, not half as long as the peduncle . 18. *D. bulbosa.*
　　Leaves obovate, penniveined, nearly as long as or longer than
　　　the peduncle . . . . 19. *D. rosulata.*
　　Leaves obovate, several-nerved, nearly as long as the peduncle 20. *D. Whittakeri.*
　Peduncles several, filiform, 2- or 3-flowered. Leaves obovate,
　　penniveined . . . . 21. *D. macrophylla.*
　Peduncles solitary, bearing a cyme of many flowers.
　　Leaves all reduced to lanceolate membranous scales . . . . 22. *D. squamosa.*
　　Leaves broadly obovate or orbicular . . . . 23. *D. erythrorhiza.*
Rootstock bearing a rosette of leaves, and either leafy flowering-stems,
　or leafless scapes and leafy side-branches. Leaves not peltate.
　Stem-leaves opposite or whorled. Central scape usually leafless,
　　bearing a cyme of several flowers.
　　Leaves of the primary rosette scarcely petiolate, above ⅓ in. long.
　　　Scape (with the cyme) 3 to 6 in. long or more . . . . . 24. *D. stolonifera.*
　　Leaves of the primary rosette distinctly petiolate, orbicular,
　　　under 2 lines diameter. Scape (with the cyme) 2 to 4 in. . 25. *D. humilis.*
　Stem-leaves alternate.
　　Central scape leafless, 1- or 2-flowered, ⅓ to ¾ in. long. Leafy
　　　side-branches short, usually barren . . . . 26. *D. ramellosa.*
　　Stem leafy, with a many-flowered terminal cyme . . . . . 27. *D. flabellata.*
Rootstock terminating in a single or branched leafy flowering-stem.
　Lower leaves reduced to short linear-subulate or linear-lanceolate
　　scales or (in the first 2 species) rosulate and not peltate. Stem-
　　leaves peltate, on filiform petioles, often clustered in the axils.
　Stem-leaves lunar-peltate, *i. e.* broadly crescent-shaped or at least
　　with 2 prominent angles.
　　Lower leaves, when present, rosulate, not peltate. Racemes
　　　simple, the pedicels all short.
　　　Sepals entire, glabrous. Seeds narrow-linear . . . . . 28. *D. auriculata.*
　　　Sepals toothed, villous or nearly glabrous. Seeds ovoid . . 29. *D. peltata.*
　　Lower leaves not rosulate, all reduced to small narrow acute
　　　scales. Racemes branched or the lower pedicels long.
　　　Flowers large, red or yellow, few, in a short loose cyme. Seeds
　　　　linear . . . . 30. *D. Neesii.*
　　　Flowers small, white, in a large divaricate panicle. Seeds
　　　　ovate . . . . 31. *D. gigantea.*
　Stem-leaves orbicular-peltate, without angles, the lower ones not
　　rosulate, often reduced to narrow acute scales.
　　Flowers solitary or very few, mostly 8-merous . . . . . 40. *D. heterophylla.*
　　Flowers few, small, in a simple raceme, lower pedicels short.
　　　Stipules often to the upper leaves. Styles not much divided . 41. *D. Banksii.*
　　Flowers 5-merous, in cymes or loose racemes. Stipules none.
　　　Styles not much divided. Flowers very small, in many
　　　　flowered cymes . . . . 32. *D. myriantha.*
　　　Styles divided nearly to the base into very numerous filiform,
　　　　mostly simple branches.
　　　　Flowers many in the cyme, white. Calyx-segments entire 33. *D. pallida.*
　　　　Flowers in cymes, red. Calyx-segments ciliate-toothed,
　　　　　villous or nearly glabrous . . . . 34. *D. penicillaris.*
　　　　Flowers few, in short, loose, simple or rarely once forked
　　　　　racemes. Calyx-segments ciliate-toothed, glabrous or
　　　　　slightly villous . . . . 35. *D. filicaulis.*

Styles divided into very numerous dichotomous branches.
Flowers in a loose cyme or corymbose panicle.
 Glabrous, usually simple and erect. Leaves rather large 36. *D. Huegelii.*
 More or less glandular-pubescent, usually flexuose or
 twining. Leaves small . . . . 37. *D. macrantha.*
Flowers few, in a simple or very rarely once-forked loose
 raceme . . . . 38. *D. Menziesii.*
Styles divided into very numerous very short branches, form-
ing a dense globose mass. Filaments dilated upwards. Se-
pals rather large, usually glabrous and entire . . . . . 39. *D. calycina.*

Sect. I. RORELLA, *DC.*—Stock not bulbous, the upper end perennial.
Scapes leafless, except in *D. indica.* Stipules often present, scarious. Styles
usually simple or divided into 2 simple branches, rarely dichotomous.

1. **D. indica,** *Linn.; DC. Prod.* i. 319. Leafy stems, from a few in.
to 1 or nearly 2 ft. long. Leaves linear, acuminate, often several in. long,
fringed with the glandular ciliæ of the genus, either quite to the base or
leaving a short glabrous petiole, often half stem-clasping, but not sheathing.
Stipules none. Flowers in loose, lateral, often leaf-opposed racemes, short and
few-flowered, or long with more numerous flowers, glabrous or glandular-
pubescent. Pedicels longer than the calyx. Sepals narrow, about 1½ lines
long in flower, 2 lines in fruit. Anthers oblong-linear. Styles 3, divided to
the base, each into 2 filiform branches, dilated and stigmatic on the inner side
at the end. Seeds obovoid, with a close testa.—Planch. in Ann. Sc. Nat.
ser. 3, ix. 204 ; Wight, Ill. t. 20 C.; F. Muell. Pl. Vict. i. 58 ; *D. serpens,*
Planch. l. c.

N. Australia. Islands of the Gulf of Carpentaria, *R. Brown ;* Upper Victoria river,
F. Mueller ; Port Essington, *Armstrong ;* Attack Creek, *M'Douall Stuart's Expedition.*
 Queensland. Endeavour river, *R. Brown,* *A. Cunningham ;* Shoalwater Bay, Keppel
Bay, *R. Brown ;* Port Curtis, *M'Gillivray ;* Rockhampton, *Thozet ;* Broad Sound, *Bow-
man.*
 Victoria. Moist gravelly places round freshwater lakes, near Eustone, on the Murray,
rare, *F. Mueller.*
 W. Australia. Murchison river, *Oldfield.*
 Common in East India and the Archipelago, extending as far as Amoy, in China, and also
in various parts of tropical Africa. The Australian specimens are usually larger, with longer
racemes and larger flowers than the Indian ones, but not always so, and there is no other
difference.

2. **D. Arcturi,** *Hook. Journ. Bot.* i. 247, *and Ic. Pl. t.* 56. Stock
tufted, sometimes slightly elongated and covered by the sheathing bases of
the old leaves, the plant otherwise stemless. Leaves linear, obtuse, 1 to 2
or rarely nearly 3 in. long and 1 to 2 lines broad in the larger specimens, in
others not half that size, the glandular ciliæ extending from the middle upwards, narrowed below into a glabrous petiole, sheathing at the base. Scape
1-flowered, usually exceeding the leaves, bearing occasionally a small linear
bract above the middle. Calyx 3 to 4 lines long. Petals obovate, rather
large. Styles 3, undivided, short and thick, each with a broad capitate almost
reniform stigma.—Planch. in Ann. Sc. Nat. ser. 3, ix. 189 ; Hook. f. Fl.
Tasm. i. 28 ; F. Muell. Pl. Vict. i. 57.

 N. S. Wales. Mount Kosciusko, *F. Mueller.*

Victoria. Boggy places or mossy banks of rivulets and ponds, at an elevation of 5000 to 7000 ft.; alps of the Bogong and Munyang ranges, *F. Mueller.*
Tasmania. Table Mountain, near the Derwent, *R. Brown ;* bogs, at an elevation of 3000 to 4000 ft., Mount Wellington, and Western Mountains, *J. D. Hooker ;* Mount Lapeyrouse, *Oldfield.* Found also in New Zealand. F. Mueller describes the leaves as sometimes 4 in. long and ½ in. wide. I have never seen them anything near that size.

3. **D. glanduligera,** *Lehm. Pugill.* viii. 37, *and Pl. Preiss.* i. 252. Leaves all rosulate, broadly obovate or orbicular, 2 to 3 or rarely 4 lines diameter, narrowed into a short petiole or the inner ones rounded at the base or almost peltate. Stipules none. Scape leafless, ½ to 1½ in. high, including a loose raceme of 6 to 10 flowers, rarely reduced to 2 or 3. Pedicels recurved, about as long as the calyx, glandular-ciliate as well as the rhachis. Sepals 1 to 1½ lines long. Petals red (or orange ?), not large. Styles 3, very slender, divided to about the middle into 2 branches, sometimes again forked. Seeds ovoid.—Planch. in Ann. Sc. Nat. ser. 3, ix. 206 ; F. Muell. Pl. Vict. i. 55.

N. S. Wales. George's river, *R. Brown ;* Twofold Bay, *F. Mueller.*
Victoria. Heaths and pastures, scattered over the colony, *F. Mueller.*
S. Australia. Near Adelaide, Lynedoch valley, Barossa and Bugle ranges, etc., *F. Mueller.*
W. Australia, *Drummond, 2nd Coll. n.* 85 ; near Perth, *Preiss, n.* 1976 ; Hill river, *Oldfield ;* Hampden, *Clarke.*

4. **D. pygmæa,** *DC. Prod.* i. 317. A minute species, said to be annual, but evidently forming a hybernating bud in the centre of the rosette, like the other species of the section. Leaves rosulate, orbicular, ½ to nearly 1 line diameter, on slender petioles, forming tufts of about ½ in. diameter. Stipules scarious, deeply lobed. Scapes glabrous, filiform, ½ to nearly 1 in. long, bearing a single minute terminal flower. Sepals 4, about ¼ line long in flower, nearly 1 line in fruit. Petals rather larger. Styles 4, slightly club-shaped and stigmatic at the end. Capsule 4-valved. Seeds few, rather large in proportion, ovoid.—Planch. in Ann. Sc. Nat. ser. 3, ix. 289 ; Hook. f. Fl. Tasm. i. 29 ; F. Muell. Pl. Vict. i. 56.

N. S. Wales. Race ground, Paramatta, *R. Brown ;* Jervis's Bay, *Caley.*
Victoria. Sandy heath-ground, occasionally wet, at the base of the Serra Range, Port Albert, etc., *F. Mueller ;* near Melbourne, *Adamson, Robertson.*
Tasmania. Abundant in peaty and sandy soil along the N.W. coast, *J. D. Hooker.*
S. Australia. Encounter Bay, *F. Mueller.*
Also in New Zealand.

5. **D. platystigma,** *Lehm. Pugill.* viii. 37, *and Pl. Preiss.* i. 249. A small delicate species. Leaves rosulate, nearly orbicular, 1½ to 2 lines diameter, the petiole longer, broad and flat. Stipules scarious, finely divided, the central tuft not exceeding the leaves. Scapes filiform, ½ to 1½ in. long, bearing at the end 2 or 3 comparatively large flowers or rarely one only, on short pedicels. Sepals green, about 1 line long. Petals 3 times as long, spreading, of an orange colour. Anthers small. Styles according to Lehmann 3, simple; according to Schlotthauber 2, simple; I find however, in one flower of Preiss's and 2 of Oldfield's, 2 styles, each divided to the base into 2 filiform branches, with rather large capitate stigmas. Placentas 2, with rather numerous ovules.

W. Australia. King George's Sound, *Oldfield, Preiss, n.* 1994.

6. **D. pulchella,** *Lehm. Pugill.* vii. 38, *and Pl. Preiss.* i. 250. A small species, resembling at first sight *D. platystigma.* Leaves rosulate, orbicular, 1 to 2 lines diameter, on a much-dilated oblong petiole, 2 or 3 times as long and often as broad as the lamina. Stipules scarious, divided into filiform segments, the central tuft much shorter than the leaves. Scape filiform, 1 to 1½ in. long, bearing at the end 2 to 4 or rarely more white or pink flowers, on pedicels at length nearly as long as the calyx. Sepals minutely glandular-pubescent, about ¾ line long in flower, 1½ lines in fruit. Petals spreading, scarcely twice as long as the calyx. Anthers small. Styles 5, filiform, undivided, longitudinally stigmatic on the inner side towards the end. Placentas 5 ; ovules not numerous.

W. Australia. King George's Sound, *R. Brown, Preiss, n.* 1992, *Drummond, 1st Coll. and 3rd Coll. n.* 3. It varies much in the breadth of the sepals, rather narrow in Preiss's specimens, very broad in some of Drummond's, intermediate in others. Those specimens of Drummond's which Planchon had referred to *D. micrantha,* Lehm., appear to me to belong to this species.

7. **D. leucoblasta,** *Benth.* A small species, with the aspect and comparatively large flowers of *D. platystigma,* but usually rather stouter and at once known by the stipules and styles. Leaves rosulate, orbicular, 1 to 2 lines diameter, on a petiole usually much longer than the lamina and slightly flattened, but not near so broad as in *D. pulchella.* Stipules scarious, silvery-white, the lobes much broader than in the adjoining species and densely imbricate in an oblong or conical bud in the centre of the rosette, 2 to 4 lines long at the time of flowering. Scape filiform, 1 to 2 in. long, with 2 or 3 flowers at the summit on very short pedicels. Sepals broad, nearly 1 line long at the time of flowering, somewhat enlarged afterwards. Petals spreading, 3 times as long as the calyx. Anthers small, Styles 3, filiform, longitudinally stigmatic on the inner side towards the end. Placentas 3.

W. Australia, *Drummond, 2nd Coll. n.* 14; dry sandy flats, Kalgan river, *Oldfield.*

8. **D. nitidula,** *Planch. in Ann. Sc. Nat. ser.* 3, ix. 285. A very small species, but much more rigid than *D. platystigma.* Leaves densely tufted, the upper ones rosulate, but the old tufts forming stocks of nearly ½ in. under the rosette, densely covered with the old leaves and stipules ; lamina orbicular, rarely above 1 line diameter, on a scarcely dilated petiole often twice as long. Stipules scarious, brown, the lobes filiform, shorter than the petioles. Scapes about 1 in. high, including a short one-sided raceme of 6 to 10 flowers, nearly glabrous. Pedicels as long as the calyx. Sepals very narrow, rather rigid, about 1 line long at the time of flowering. Petals longer. Anthers small. Styles 3, short, each with a globular capitate stigma. Fruiting calyx narrow, erect, nearly 2 lines long. Capsule nearly as long, with 3 valves and placentas.

W. Australia, *Drummond.*

9. **D. paleacea,** *DC. Prod.* i. 318, *according to R. Br. Herb.* A very small species. Leaves densely tufted, the upper ones or all in the young plant rosulate, but the old tufts forming stocks of ½ in. or in one specimen nearly 1 in. long, densely covered with remains of old leaves and stipules ; lamina obovate, oblong or almost orbicular, ½ to 1 line diameter ; petiole longer, scarcely dilated. Stipules scarious, cut into fine filiform lobes, form-

ing a prominent erect bud. Scape filiform, ½ to 1 in. or rarely 2 in. high, the slender, rather dense, one-sided raceme of minute flowers occupying sometimes nearly one-half of it, rarely short and few-flowered. Pedicels erect, nearly as long as the calyx. Sepals nearly glabrous, not ⅓ line long when in flower, and not much above ⅓ line when in fruit. Petals longer. Anthers very small. Styles 3, filiform, rather thickened and stigmatic on the inner side towards the end. Capsule ovoid, with 3 valves and placentas and few seeds.—*D. micrantha,* Lehm. Pugill. viii. 39 ; *D. pygmæa,* Lehm. in Pl. Preiss. i. 250, not of DC.; *D. minutiflora,* Planch. in Ann. Sc. Nat. ser. 3, ix. 286.

W. Australia. King George's Sound, *R. Brown;* sandy places near Perth, *Preiss, n.* 1995 ; also *Drummond.*

10. **D. parvula,** *Planch. in Ann. Sc. Nat. ser.* 3, ix. 287. A small plant, evidently allied to *D. paleacea,* but probably distinct. We have, however, only the single specimen described by Planchon. Leaves and stipules as in *D. paleacea,* and the stock covered with the remains of the old leaves below the rosette, nearly 1 in. long, as in that species. Scape filiform, 1 in. long, with a short loose raceme. Pedicels longer than the calyx. Sepals rather more than ½ line long. Styles and placentas, according to Planchon, 5 or sometimes 4.

W. Australia, *Drummond.* I have been unable to detach any flower for re-examination, without destroying the specimen.

11. **D. Burmanni,** *Vahl; DC. Prod.* i. 318. Leaves all radical, rosulate, obovate-spathulate, about 3 or 4 lines diameter, narrowed into a petiole not so long. Stipules scarious, cut into narrow lobes, not so long as the petiole. Scapes solitary or 2 or 3 from the same tuft, slender, attaining 5 o 6 in. and rarely under 3 in. long, the upper portion occupied by a slender one-sided raceme of several flowers. Pedicels short. Sepals glabrous, 1½ to 2 lines long. Anthers small. Styles 5, undivided, filiform, not branched but slightly dilated and fringed towards the end.—Planch. in Ann. Sc. Nat. ser. 3, ix. 190 ; Wight, Ic. t. 944.

N. Australia. Regent river, N.W. coast, *A. Cunningham;* Upper Victoria river, Providence Hill, and M'Adam range, Nicholson river, Gulf of Carpentaria, *F. Mueller.*

Queensland. Endeavour river, *Banks and Solander;* Brisbane river, *F. Mueller, Henne, C. Stuart.*

The species is widely spread over E. India and the Archipelago, extending to S. China. Without examining the styles, it is very difficult to distinguish it from *D. spathulata.*

12. **D. spathulata,** *Labill. Pl. Nov. Holl.* i. 79, *t.* 106, *f.* 1. A stemless species, not very easy to distinguish from the coarser specimens of *D. Burmanni,* without examining the styles. Leaves rosulate, obovate or spathulate, not usually so broad as in *D. Burmanni* and often ½ in. long, sometimes oblong-spathulate and narrowed into a rather long petiole. Stipules scarious, cut into narrow lobes. Scapes usually 3 to 6 in. high, including the simple or rarely forked, 1-sided raceme. Pedicels short, glabrous as well as the calyx or minutely glandular-pubescent. Sepals about 1½ lines long, often united at the base. Petals pink red or white, as long as or rather exceeding the calyx. Anthers oblong. Styles 3 or rarely 4, but divided to the base

into 2 branches either entire and filiform or slightly dilated, emarginate or shortly forked at the end. Seeds numerous, small.—DC. Prod. i. 318; Planch. in Ann. Sc. Nat. ser. 3, ix. 193 ; Hook. f. Fl. Tasm. i. 29 ; Bot. Mag. t. 5240 ; F. Muell. Pl. Vict. i. 66.

Queensland. Brisbane river, *F. Mueller, C. Stuart.*

N. S. Wales. Common about Port Jackson, *R. Brown* and others ; northward to Hastings river, *Beckler.*

Victoria. Boggy soil towards Brighton, *Howitt ;* Mount Abrupt, *Wilhelmi.*

Tasmania. Wet marshy hollows, Rocky Cape, *J. D. Hooker.*

The species is also in New Zealand. Some specimens of Cuming's, from the Philippine Islands, are also referred to it by Planchon, and do not in fact appear at all different. They are, however, probably the same as the S. Chinese *D. Loureiri,* Hook. and Arn. Bot. Beech. 167, t. 31 ; Benth. Fl. Hongk. 130, which must in that case be united with *D. spathulata.* Among the Australian specimens there appear to be two slightly different forms, one with larger deeper-coloured flowers, and the style-branches usually dilated and emarginate at the end, the other more slender, with paler and smaller flowers, the style-branches divided some way down into two slender forks.

13. **D. Drummondii,** *Lehm. Pl. Preiss.* ii. 235. Variable in aspect according to age, but readily distinguished from the preceding species by the woolly hairs of the raceme. Leaves all in the young plants, the upper ones in the older tufts, rosulate, the stock under the rosette lengthening out to above 1 in., covered with the remains of old leaves and stipules ; lamina from narrow-oblong to almost obovate, rarely above 2 lines long, the petiole much longer and closely reflexed over the old leaves. Stipules scarious, finely cut, but much shorter than in *D. scorpioides.* Scape 1 to 3 or 4 in. long, nearly glabrous or woolly hairy, the raceme simple, short and few-flowered or lengthening to above an inch. Pedicels short, covered as well as the calyx with woolly hairs. Sepals from 1 line long when in flower to 2 lines in fruit, often denticulate and ciliate. Petals rather large. Filaments thick, with oblong anthers. Styles 3 or 4, simple and filiform. Ovules 4.—*D. barbigera,* Planch. in Ann. Sc. Nat. ser. 3, ix. 287.

W. Australia, *Drummond, 3rd Coll. n. 34.*

14. **D. scorpioides,** *Planch. in Ann. Sc. Nat. ser.* 3, ix. 288. Very near *D. Drummondii* and perhaps a variety, the stock lengthening out sometimes to 2 or 3 in., and the scarious stipules much more prominent. Leaves linear-oblong, 2 to 3 lines long, or rarely a few of the inner ones small and almost obovate, the petiole longer than the lamina, but not so reflexed as in *D. Drummondii.* Scape 1 to 2 in. long, woolly or nearly glabrous, the raceme short dense and few-flowered. Sepals often very woolly, but not so much ciliate as in *D. Drummondii,* and rather smaller. Petals apparently twice as long as the calyx. Filaments slender, with small anthers. Styles 3, filiform.

W. Australia. King George's Sound, *R. Brown, Baxter, Wakefield,* and probably from thence to the eastward, *Drummond, 4th Coll. n. 125, and 5th Coll. n. 283* ; Cape Arid and near S.W. Bay, *Maxwell.*

Var. (?) *brevipes.* Leaves smaller, from oblong to almost obovate, the petioles shorter. Scapes short, flowering from below the middle, nearly glabrous. Stamens as in the normal *D. scorpioides.*—*Drummond, 5th Coll. n. 284.*

15. **D. petiolaris,** *R. Br. in DC. Prod.* i. 318. Stock short, densely

tufted, with long silky or rusty hairs covering the persistent bases of the old leaves and stipules. Leaves rosulate, orbicular or broadly obovate, rarely above 2 lines diameter, on a rather broad petiole of ½ to 1 in. in the ordinary form, the under side of the lamina and the petiole clothed with long silky hairs. Stipules scarious, but not prominent. Scapes in the largest specimens 1 ft. high but usually about half that, including the rather loose, often long, 1-sided raceme, the calyx, pedicels, and rhachis more or less villous with soft silky or velvety hairs. Pedicels rather shorter than the calyx, often reflexed. Sepals above 1 line long in flower, 2 lines in fruit. Petals broad, rather large. Anthers small. Styles 3, repeatedly dichotomous, the last branches short and stigmatic.—*D. fulva*, Planch. in Ann. Sc. Nat. ser. 3, ix. 289.

N. Australia. Islands of the Gulf of Carpentaria, *R. Brown, Henne;* Goulburn Island, *A. Cunningham;* Port Essington, *Armstrong;* M'Adam range, *F. Mueller.*

Queensland. Endeavour river, *Banks and Solander, R. Brown.* Banks's specimens are remarkable for their petiole 1 to 2 in. long and less dilated, with a lamina of 1 to 2 lines diameter, which induced Planchon to consider them as belonging to a distinct species; but R. Brown's carefully-selected series of specimens show every gradation from the longest to the shortest petioles.

Planchon describes the styles of this species as twice bifid; I find them 3 or 4 times bifid. It is, however, exceedingly difficult to trace their ramifications from dried specimens. In the bud they form a dense mass which requires great care in unfolding. and after flowering they are so mixed in the withered petals, that it almost impossible to extract them whole. The leaves are, as observed by Planchon (l. c. 289, 290) not peltate, and the association of the species with the very dissimilar *D. Banksii* into one section (*Lasiocephalum*), proposed by Planchon (l. c. 94), and founded partly on this character, can scarcely be admitted.

16. **D. binata,** *Labill. Pl. Nov. Holl.* i. 78, *t.* 105. Stock small, appearing sometimes to emit creeping stolons. Leaves radical, on long petioles, the lamina divided to the base into 2 long linear lobes, sometimes again once or twice forked, and often 2 or 3 in. long, elegantly fringed by the glandular cilia of the genus, glabrous underneath as well as the petioles. Stipules short, broad, brown and scarious, slightly jagged. Scapes exceeding the leaves, often 1 to 1½ ft. high, bearing a loose cyme of large white flowers, consisting usually of 2 or 3 racemose branches, rarely reduced to a short simple raceme. Sepals about ¼ in. long. Petals twice as long. Styles usually 3, divided into numerous dichotomous lobes, some very short, others longer, clavate or forked at the stigmatic end. Capsule globular. Seeds very numerous, small and linear.—DC. Prod. i. 319; Bot. Mag. t. 3082; Planch. in Ann. Sc. Nat. ser. 3, ix. 206; Hook. f. Fl. Tasm. i. 29; F. Muell. Pl. Vict. i. 59; *D. pedata*, Pers. Syn. i. 337; DC. Prod. i. 319; *D. dichotoma*, Sm. in Rees' Cyclop. xii.

N. S. Wales. Port Jackson, *R. Brown, Sieber, n.* 177, and *Fl. Mixt. n.* 625, and others; Blue Mountains, *Miss Atkinson;* Illawarra, *A. Cunningham.*

Victoria. Wet boggy places, often growing in moss, Wilson's Promontory, Buffalo ranges, Grampians, etc., *F. Mueller.*

Tasmania. Formosa, *Lawrence;* abundantly on the coast from Rocky Cape to Woolnorth, *Gunn;* South Port, *C. Stuart.*

S. Australia. Encounter Bay, *Whittaker, F. Mueller;* cataracts near Mount Lofty, *F. Mueller.*

The species is also in New Zealand. The Port Jackson specimens have the leaves usually dichotomous, in the southern ones they are more frequently 2-lobed only; but these differences are by no means constant, and the two forms occur sometimes on the same specimen.

SECT. II. ERGALEIUM, *DC.*—Stock short, slender, stem-like, naked or with ragged remains of old petioles, forming usually, if not always, a bulb at the lower end, and producing at the upper end a rosette of leaves and leafless scapes, or leafy stems or branches. Stipules none, or, in *D. Banksii,* small and evanescent. Styles dichotomous or divided into very numerous filiform branches, forming a dense tuft.

Nearly all the species of this section dye the paper in which they are preserved a rich carmine or purple colour. When growing, they are said to disappear entirely after the fruiting is over, but I find no observation of how much of the underground stock besides the bulb persists till the next season, nor do the specimens show what relation the new shoot has had to the old bulb. It is, indeed, not often that collectors have gathered their specimens with the bulb.

17 ? **D. zonaria,** *Planch. in Ann. Sc. Nat. ser.* 3, ix. 303. Only known from two barren rosettes, about 2 in. diameter. Leaves of a light green colour, broadly orbicular or almost reniform or fan-shaped, above ½ in. broad, shortly cuneate at the base, on a petiole usually longer than the lamina, the margin elegantly fringed by the glandular cilia of the genus, the veins scarcely conspicuous.

W. Australia, *Drummond.* Possibly a barren state of *D. rosulata.*

18. **D. bulbosa,** *Hook. Ic. Pl. t.* 375. Bulbous. Leaves at the end of the slender stock rosulate or apparently verticillate, oblong, slightly spathulate, narrowed at the base but not distinctly petiolate, 3 to 4 lines or rarely ¼ in. long, rather thick, with 1 broad nerve. Peduncles or scapes 1-flowered, few or numerous, twice as long as the leaves, glabrous or nearly so. Sepals about 2 lines long. Petals twice as long, apparently white. Styles deeply divided into numerous filiform branches, slightly dilated and stigmatic at the end.

W. Australia. Swan River, *Drummond,* 1*st Coll.*; Murchison river, *Oldfield.*

19. **D. rosulata,** *Lehm. Pugill.* viii. 36, *and Pl. Preiss.* i. 251. Bulbous. Leaves at the end of the slender rootstock rosulate, obovate, tapering at the base, rarely above 1 in. long including the short broad petiole, with a broad central nerve and a few lateral veins diverging from it above the middle. Peduncles or scapes 1-flowered, slender, often filiform, rarely exceeding the leaves. Sepals scarcely above 2 lines long at the time of flowering, longer in fruit. Petals white. Anthers ovate. Styles 3, divided to the base into numerous filiform branches, slightly dilated and stigmatic at the end.—*Planch.* in Ann. Sc. Nat. ser. 3, ix. 301.

W. Australia. Sandy boggy places near Perth, *Preiss, n.* 1983 ; Champion Bay and Cape Leschenault, *Oldfield;* Vasse river, *Mrs. Molloy;* King George's Sound, *Harvey.*

20. **D. Whittakerii,** *Planch. in Ann. Sc. Nat. ser.* 3, ix. 302. Bulbous, with rosulate leaves at the end of the stock, as in *D. rosulata,* which this species closely resembles, the leaves of the same size, but showing, besides the midrib, 2 or 3 lateral nerves on each side, distinct in the petiole and diverging in the lamina. Scapes 1-flowered, not much longer than the leaves. Sepals at least 3 lines long at the time of flowering and nearly 5 lines in fruit, more acute than in *D. rosulata.* Petals white, half as long

again as the calyx. Anthers ovate. Styles 3, divided to the base into numerous filiform branches, slightly dilated and stigmatic at the end. Capsule shorter than the calyx. Seeds ovoid.—F. Muell. Pl. Vict. i. 57, t. suppl. 6 ; *D. rosulata*, Behr, in Linnæa, xx. 628, not of Lehm.

Victoria. Rather frequent in the southern parts of the colony, *F. Mueller ;* Wimmera, *Dallachy.*

S. Australia. Rich boggy flats, entirely disappearing after the cessation of the winter rains, *Behr.*

The species is scarcely to be distinguished from *D. rosulata,* except by the venation of the leaves.

21. **D. macrophylla,** *Lindl. Swan Riv. App.* 20. Bulbous with a slender more or less scaly rootstock. Leaves at the end rosulate, obovate, tapering at the base, thin, glandular, the veins few, diverging from the central nerve much above the base, as in *D. rosulata,* and slightly reticulate, varying from ½ to 2 in. long, including the very short petiole. Scapes or peduncles several, each with 2 or 3 flowers, apparently white and rather large, on slender pedicels. Calyx glabrous, 2 to 3 lines long in flower, longer in fruit. Styles 3, deeply divided into numerous filiform branches. Seeds nearly globular.—Hook. Ic. Pl. t. 376 ; Lehm. in Pl. Preiss. i. 251.

W. Australia. Swan River, *Drummond, 1st Coll. and 3rd Coll. n.* 40 ; Princess Royal Harbour, *Preiss, n.* 1986.

22. **D. squamosa,** *Benth.* Bulbous, the old scales on the short slender rootstock often numerous. Leaves forming a tuft at the end, as in the allied species, but all reduced to erect or scarcely spreading, lanceolate, membranous scales of 2 to 4 lines, acute or obtuse and scarcely ciliate, without any true lamina. Scape solitary, 1 to 2 in. long, leafless, bearing a compact cyme of numerous flowers, rather smaller than those of *D. erythrorhyza,* but otherwise resembling them.

W. Australia. Between Perth and King George's Sound, *Preiss (Herb. Sonder), Harvey ;* towards the Great Bight, *Maxwell.* It is possible that this may prove to be a variety of *D. erythrorhiza,* bearing the same relation to it as *D. bulbosa* to *D. rosulata,* with undeveloped leaves, but the difference appears constant in all the specimens seen from different collectors.

23. **D. erythrorhiza,** *Lindl. Swan Riv. App.* 20. Bulbous, the stemlike rootstock slightly scaly. Leaves rosulate at the end, very broadly obovate or almost orbicular, tapering at the base, thin, penniveined, but one vein or nerve at each side starting from near the base, mostly ¾ to 1 in. long, including the very short petiole. Scape solitary, 1 to 3 in. long below the inflorescence, bearing a rather loose cyme of numerous flowers apparently white. Sepals glabrous, not 2 lines long when in flower, above 3 lines in fruit. Petals longer. Anthers ovate. Styles divided to·the base into numerous filiform branches.—Lehm. in Pl. Preiss. i. 251 ; *D. primulacea,* Schlotthaub. in Bonplandia, iv. 110.

W. Australia. Swan River, *Drummond, 1st Coll., Preiss, n.* 1987, *Collie ;* King George's Sound, *Maclean ;* between Perth and King George's Sound, *Harvey.*

24. **D. stolonifera,** *Endl. in Hueg. Enum.* 5. Bulbous, with few scales on the stem-like rootstock. Radical leaves rosulate at the end, obovate, tapering at the base, rarely above ½ in. long, including the very short petiole.

From this rosette, in the ordinary form, proceeds a leafless scape of 3 to 6 in., bearing a loose cyme of rather numerous flowers of the size and form of those of *D. erythrorhiza*, and 3 or 4 barren branches shorter than the scape each with several whorls of 2 to 4 leaves with a broadly obovate or orbicular lamina, on a petiole sometimes longer than the lamina and very narrow, sometimes shorter and dilated. In other specimens the stem is continued beyond the first rosette, producing a second or even a third rosette or tuft of leaves with a more orbicular lamina and longer petiole than the lowest, the scape and barren branches proceeding from the uppermost rosette. In others again the lateral leafy branches terminate in a small cyme, or 2 or 3 scapes each with a cyme proceed from the primary tuft, or rarely the side-branches are again branched, but in all the forms assumed the leaves are all opposite or in whorls or rosettes. Calyx mostly under 2 lines long in flower, nearly 3 lines in fruit. Styles 3, with very numerous filiform branches and seeds ovoid as in the preceding species.—Lehm. in Pl. Preiss. i. 253; Hook. Ic. t. 389; *D. porrecta*, Lehm. Pugill. viii. 41, and Pl. Preiss. i. 252; *D. purpurascens*, Schlotthaub. in Bonplandia, iv. 111.

W. Australia. Wet sandy places, Swan River, *Preiss, n.* 1984, 1985, *Drummond,* 1*st Coll., and* 3*rd Coll. n.* 45, *Oldfield,* and others ; near Mount Wuljenup, *Preiss, n.* 1977; Stirling Terrace, *Maxwell.*

25. **D. humilis,** *Planch. in Ann. Sc. Nat. ser.* 3, ix. 300. This may prove to be a small variety of *D. stolonifera*. It is much more slender, usually 2 to 4 in. high including the cyme. All the leaves including those of the primary rosette have a rather long petiole, with a small orbicular lamina about 1 to 1½ lines diameter, and the scapes are generally several from the tuft. Flowers as in *D. stolonifera*, but rather smaller, and the variations in the development of the lateral leafy branches are the same.

W. Australia, *Drummond;* Murchison river, *Oldfield.*

26. **D. ramellosa,** *Lehm. Pugill.* viii. 40, *and Pl. Preiss.* i. 252. Bulbous, with a slender rootstock. Leaves at the end rosulate, broadly obovate, tapering at the base, 3 to 6 lines long including the broad petiole. Scapes solitary or several, rarely above ½ in. long, 1- or 2-flowered, with 2 or 3 lateral leafy shoots, very short at the time of flowering, but lengthening out to 2 or 3 in. Leaves on these shoots all alternate, broadly orbicular, not peltate, 2 to 3 lines diameter, narrowed into a petiole about as long as the lamina. Sepals 1½ lines long when in flower, 2 lines in fruit. Petals longer, white. Styles 3, divided into very numerous filiform branches.—*D. penduliflora*, Planch. in Ann. Sc. Nat. ser. 3, ix. 301.

W. Australia. Sandy wet places near the lake in Rottenest Island, *Preiss, n.* 1990 ; Swan River, *Oldfield, Drummond.*

27. **D. flabellata,** *Benth.* Bulb not seen, but probably as in the allied species. Stems in our specimens simple or slightly branched, leafy, about ½ ft. high or rather more. Lower leaves rosulate, stem-leaves alternate, all broadly fan-shaped, not peltate, 2 to 4 lines diameter, narrowed into a short broad petiole. Flowers rather small, numerous, in a terminal branching cyme. Sepals attaining 2 lines or rather more after flowering, slightly toothed, glabrous. Styles divided nearly to the base into very numerous

filiform branches. Seeds not very small, globose or slightly angular, tuberculate.

W. Australia. Towards Cape Riche, *Drummond, 5th Coll. n.* 281.

28. **D. auriculata,** *Backh. ; Planch. in Ann. Sc. Nat. ser.* 3, ix. 295. Bulbous, with a slender stock. Leafy stem erect, simple or slightly branched, ½ to 1½ ft. high, glabrous. Lower leaves at the summit of the stock either all reduced to short linear scales, or forming a small rosette, with orbicular almost reniform or peltate laminæ and short petioles. Stem-leaves scattered, peltate, broadly crescent-shaped or at least truncate on one side, the 2 angles more or less produced into glandular-ciliate appendages, the petiole filiform. Flowers several, white, in a terminal simple raceme. Pedicels at length exceeding the calyx, the lower ones not much longer than the others. Sepals attaining 2 lines or rather more in fruit, glabrous, entire or scarcely glandular-toothed. Styles divided from a little below the middle into a dense tuft of short dichotomous linear lobes. Seeds very numerous, narrow-linear, the loose testa extending beyond the nucleus at one or both ends.—Hook. f. Fl. Tasm. i. 30 ; F. Muell. Pl. Vict. i. 61.

N. S. Wales. Port Jackson, *R. Brown, Sieber, n.* 176 (with *D. peltata*), and others ; northward to Clarence river, *Beckler ;* southward to Twofold Bay, *F. Mueller.*
Victoria. Sandy poor pasture land and sterile ridges, not rare, *F. Mueller.*
Tasmania. Abundant in rocky grassy heathy places throughout the island, *J. D. Hooker.*
S. Australia. Bugle Range, *F. Mueller ;* Encounter Bay, *Whittaker.*
Also in New Zealand. This species scarcely differs, except in the seed, from those forms of *D. peltata* which have nearly glabrous sepals.

29. **D. peltata,** *Sm. in Willd. Spec. Pl.* i. 1546. Bulbous, with a slender rootstock. Leafy stem erect or flexuose, ½ to 1½ ft. high. Lower leaves at the summit of the rootstock usually rosulate, orbicular or reniform, not peltate, 2 to 3 lines diameter, on a broad petiole often longer than the lamina ; stem-leaves peltate, semiorbicular or broadly crescent-shaped, on slender or filiform petioles. Flowers white, in loose simple racemes. Pedicels usually exceeding the calyx. Sepals attaining about 2 lines, or more in the large-flowered specimens, ciliate-toothed, and more or less clothed with rather long soft hairs. Styles short, densely dichotomous from below the middle, the ultimate branches shortly linear-clavate. Seeds very numerous, small, ovoid or globular, the testa not produced beyond the nucleus.—Sm. Exot. Bot. i. 79, t. 41; DC. Prod. i. 319 ; Hook. f. Fl. Tasm. i. 30 ; F. Muell. Pl. Vict. i. 60 ; *D. petiolaris,* Sieb. Pl. Exs. (which includes also *D. auriculata*) ; *D. lunata,* Hook. Ic. Pl. t. 54, and probably also Hamilt. (Buchan.) in DC. Prod. i. 319.

N. S. Wales. Port Jackson, *R. Brown, Sieber, n.* 176, and *Fl. Mixt. n.* 523 (partly), and others.
Victoria. Fertile pastures and meadows, not rare, *F. Mueller.*
Tasmania. Moist places and grassy lands throughout the island, but not so common as *D. auriculata, J. D. Hooker.*
Var. *gracilis.* Stems slender. Flowers much smaller.—*D. gracilis,* Hook. f. in Planch. Ann. Sc. Nat. ser. 3, ix. 297, and Fl. Tasm. i. 30, t. 5.—Paramatta, *Woolls ;* mountain districts, Tasmania, *J. D. Hooker.* This form is represented by Labillardière, Pl. Nov. Holl. t. 106, f. 2.
Var. *foliosa.* Short and stout, with larger leaves and fewer flowers.—*D. foliosa,* Hook. f.

in Planch. Ann. Sc. Nat. ser. 3, ix. 298, and Fl. Tasm. i. 30, t. 6.—Grassy plains near Ballarat, *F. Mueller ;* marshy places, Tasmania, *J. D. Hooker.*

The species appears to extend over E. India and the Archipelago to S. China, for I can find no character whatever to distinguish the common *D. lunata*, Ham., of that country. The rosulate leaves are indeed less frequently present at the time of flowering, but are to be found in some specimens, and are not always constant in the Australian ones. The sepals and styles are the same in both.

30. **D. Neesii,** *Lehm. Pugill.* viii. 42, *and Pl. Preiss.* i. 254. Bulbous, with a slender rootstock. Leafy stem erect flexuose or perhaps sometimes twining, 1 to 1½ ft. high, glabrous or slightly glandular-pubescent under the inflorescence. No rosulate leaves ; lower leaves reduced to small scattered linear acute scales ; stem-leaves on slender petioles, peltate, broadly crescent-shaped, with 2 acuminate ciliate angles, sometimes small, but often 3 lines broad. Flowers rather large, red or purple in the original form, not numerous, in a loose cyme or in a once-branched or simple raceme, with the lower pedicels long. Sepals above 2 lines long in flower, often 3 lines in fruit, slightly toothed, glandular-ciliate, villous outside. Styles deeply divided into very numerous filiform slightly dichotomous branches. Seeds very numerous, narrow-linear.

W. Australia. Princess Royal Harbour, *Preiss, n.* 1978 (not seen); Hill river, *Oldfield ;* between Moore and Murchison rivers, *Drummond, 6th Coll. n.* 113 ; Champion Bay, *Oldfield.*

Var. *sulphurea.* Flowers yellow or straw-colour, but without any other difference.—*D. sulphurea*, Lehm. Pugill. viii. 43, and Pl. Preiss. i. 254.—King George's Sound and adjoining districts, *R. Brown, King, Preiss, n.* 1981, *Oldfield ;* Vasse river, *Oldfield.*

31. **D. gigantea,** *Lindl. Swan Riv. App.* 20. Probably bulbous ; the stock slender at the base, usually thickened at the crown and often covered with a dense mass of old remains of leaves. Leafy stem tall, erect, branching upwards, glabrous. Lower leaves reduced to lanceolate-subulate scattered scales ; stem-leaves on slender petioles, peltate, broadly crescent-shaped, with 2 prominent subulate-acuminate angles, 2 to 3 lines diameter. Flowers small, white, in a large loose divaricately branched terminal panicle. Pedicels longer than the calyx. Sepals 1 to 1½ lines long, entire, glabrous. Styles short and thick, shortly divided into numerous branches forming a dense almost globular mass. Seeds obovoid or almost globular, with a close testa. —Lehm. Pl. Preiss. i. 255 ; Planch. in Ann. Sc. Nat. ser. 3, ix. 298.

W. Australia. Wet bogs and swamps, Swan River, *Drummond, 1st Coll., Preiss, n.* 1991, *Oldfield ;* Blackwood river, *Oldfield.*

32. **D. myriantha,** *Planch. in Ann. Sc. Nat. ser.* 3, ix. 291. Bulbous, with a slender rootstock. Leafy stem slender, simple or slightly branched, under 1 ft. high, glabrous. Lower leaves few, reduced to small narrow acute scales ; stem-leaves on filiform petioles, peltate, orbicular, not 2 lines diameter. Flowers apparently white, smaller than in any other leafy species, rather numerous in a branched cyme. Pedicels as long as the calyx. Sepals scarcely 1 line long, acute, minutely glandular-notched. Petals twice as long. Styles 3, divided to the base into 2 forked branches or into 3 or 4 simple ones. Seeds numerous, oblong-linear, the testa produced beyond the nucleus.

W. Australia, *Drummond.*

33. D. pallida, *Lindl. Swan Riv. App.* 20.   Bulbous, with a slender rootstock.   Leafy stem flexuose or twining, often 1 to 2 ft. long or even more, glabrous or glandular-pubescent.   Lower leaves few, reduced to linear acute scales; stem-leaves on slender petioles, peltate, orbicular, often above 2 lines diameter.   Flowers apparently white, several in a loose cyme, larger than in *D. myriantha,* but usually smaller than in the following species.   Sepals from under 2 lines to at least 3 lines long, glabrous and entire or very shortly glandular-ciliate.   Styles divided to the base into extremely numerous very slender and acute branches, stigmatic a considerable way down.—Lehm. Pl. Preiss. i. 253.

W. Australia. Swan River, *Drummond, 1st Coll., and 3rd Coll. n.* 46, *Preiss, n.* 1996, *Oldfield;* Murchison river, *Oldfield;* King George's Sound, *Harvey, Oldfield.*

34. D. penicillaris, *Benth.*   Bulbous, with a slender rootstock.   Leafy stem slender, flexuose or twining, usually glabrous.   Lower leaves reduced to linear acute scales; stem-leaves on slender petioles, peltate, orbicular, rather small.   Flowers apparently red, rather large, several in a loose cyme or short branched raceme.   Sepals pubescent or villous, ciliate-toothed, 2 to 3 lines long.   Anthers oblong.   Styles divided, as in *D. pallida,* into exceedingly numerous very slender branches, stigmatic a considerable way down.   Seeds numerous, linear.—*D. Drummondii,* Planch. in Ann. Sc. Nat. ser. 3, ix. 293, not of Lehm.

W. Australia. Swan River, *Drummond, 1st Coll., and 3rd Coll. n.* 44; between Moore and Murchison rivers, *Drummond, 6th Coll. n.* 112; Oldfield river, *Maxwell.* This may possibly prove to be a variety of *D. filicaulis,* with the inflorescence rather of *D. pallida.*

35. D. filicaulis, *Endl. in Hueg. Enum.* 6.   Bulbous, with a slender rootstock.   Leafy stem slender, flexuose, glabrous, usually simple and under 1 ft. long.   Lower leaves few, reduced to narrow acute scales; stem-leaves on slender petioles, peltate, orbicular, rarely 2 lines diameter.   Flowers rather large, apparently red, few in short loose simple racemes, very rarely once-branched.   Sepals glabrous or more frequently slightly villous, ciliate-toothed, from about 2 to nearly 4 lines long when in fruit.   Anthers oblong.   Styles divided, as in *D. pallida,* nearly to the base into extremely numerous slender simple branches, stigmatic a considerable way down.—Lehm. Pl. Preiss. i. 255.

W. Australia, *Drummond, 3rd Coll. n.* 47; marshy places, Swan River, *Preiss, n.* 1988; King George's Sound, *Huegel;* Gales Brook, *Maxwell.* This species has quite the aspect of *D. Menziesii,* except that the leaves are usually smaller and the calyx less villous. The style-branches are quite different, being the same as in *D. pallida,* but it remains to be ascertained how far these differences in the style are really good specific characters.

D. microphylla, Endl. in Hueg. Enum. 6, is known only from a short diagnosis, in which there is nothing to distinguish it from *D. filicaulis,* nor from several allied species.

36. D. Huegelii, *Endl. in Hueg. Enum.* 6.   Bulbous, with a slender rootstock.   Leafy stem erect or flexuose, slender but rigid, usually simple, under 1 ft. high, glabrous.   Lower leaves few, reduced to very small fine scales; stem-leaves few, on slender petioles, peltate, orbicular, very concave, almost campanulate and reflexed, rather large.   Flowers large, apparently red, in a loose cyme.   Sepals fully 3 lines long, fringed at the end with long

2 H 2

cilia. Styles 3, repeatedly dichotomous, with short almost clavate branches, shortly stigmatic at the end.—Lehm. in Pl. Preiss. i. 253.

W. Australia. King George's Sound and neighbourhood, *Huegel, Preiss, n.* 1980, *Collie, Drummond, 5th Coll. n.* 280 *;* between Moore and Murchison rivers, *Drummond, 6th Coll. n.* 111. I have not seen either Huegel's or Preiss's specimens.

37. **D. macrantha,** *Endl. in Hueg. Enum.* 6. Bulbous, with a slender rootstock. Leafy stem erect flexuose or climbing to the length of 2 ft. or more, glandular-pubescent or hirsute. Lower leaves few, reduced to fine scales ; stem-leaves on slender petioles, peltate, orbicular, mostly 2 to 3 lines diameter, flat or slightly concave. Flowers often large, but variable in size, in a loose cyme, but not usually numerous, white or pink. Sepals about 3 lines long, fringed at the end with long cilia. Styles repeatedly branched, the ultimate branches extremely numerous and slender, but short. Seeds linear.—Lehm. in Pl. Preiss. i. 254.

W. Australia. Swan River, *Drummond, 1st Coll., Oldfield ;* Vasse river, *Mrs. Molloy ;* near Guildford, *Preiss, n.* 1982 ; between Moore and Murchison rivers, *Drummond, 6th Coll. n.* 108 ; Kojonup and Cape Arid, *Maxwell.*

Var. *minor.* Leaves and flowers much smaller, but the other characters the same.—*D. subhirtella,* Planch. in Ann. Sc. Nat. ser. 3, ix. 292.—In *Drummond's* and in *Burges's* collections.

38. **D. Menziesii,** *R. Br. in DC. Prod.* i. 319. Bulbous, with a slender rootstock. Leafy stem slender, erect flexuose or almost twining, glabrous or nearly so. Lower leaves few, reduced to small slender scales ; stem-leaves on slender petioles, peltate, orbicular, usually small. Flowers rather large, pink or red in the original form, few in a short simple raceme very rarely once-branched. Sepals 2 to 3 lines long, pubescent or villous or very rarely nearly glabrous, more or less ciliate. Filaments not dilated. Styles repeatedly divided into very numerous slender dichotomous branches. Seeds linear.

W. Australia. King George's Sound and adjoining districts, *Menzies, A. Cunningham, Drummond, 2nd Coll. n.* 5, and others.

Var. *flavescens.* Flowers pale-yellow.—*D. intricata,* Planch. in Ann. Sc. Nat. ser. 3, ix. 293.

W. Australia, *Drummond, 2nd Coll. n.* 7 ; Hill and Vasse rivers, *Oldfield.*

Var. *albiflora.* Flowers white.—*D. Planchoni,* Hook. f. Fl. Tasm. i. 29 ; Planch. in Ann. Sc. Nat. ser. 3, ix. 294 ; F. Muell. Pl. Vict. i. 62.

Victoria. In many localities towards the Murray, *F. Mueller.*

Tasmania. N. coast, Rocky Cape, George Town, etc., *J. D. Hooker.*

S. Australia. Throughout the greater part of the colony as far north as Flinders Ranges, *F. Mueller.*

I can see no difference whatever between the three varieties, except the colour of the flowers. The styles of the white variety are precisely the same as in the true *D. Menziesii,* although different from those of *D. filicaulis,* which had been confounded with it.

39. **D. calycina,** *Planch. in Ann. Sc. Nat. ser.* 3, ix. 299. Bulbous, with a slender rootstock. Leafy stem usually erect and slender, glabrous. Lower leaves few, reduced to small slender scales ; stem-leaves on slender petioles, peltate orbicular or rarely slightly truncate and 2-angled on one side. Flowers apparently red or white, few in loose simple or rarely once-forked racemes. Sepals herbaceous, thin, very obtuse, 3 to 4 lines long after flowering, entire or scarcely denticulate, glabrous. Petals usually scarcely ex-

ceeding the calyx or not half as long again. Filaments more or less dilated
under the anthers. Styles short, divided from the middle into exceedingly
numerous short dichotomous branches forming dense globular tufts, often not
longer than the entire portion.

W. Australia. King George's Sound, *Collie ;* between Moore and Murchison rivers,
Drummond, 6th Coll. n. 109. The specimens from both localities very similar.

Var. *minor.* Leaves and flowers much smaller, but the same style ; filaments rather less
dilated.—Between Moore and Murchison rivers, *Drummond, 6th Coll. n.* 110.

40. **D. heterophylla,** *Lindl. Swan Riv. App.* 20. Bulbous, with a
slender stock, usually enclosed in numerous old scaly remains of leaves.
Leafy stem simple, slender, glabrous, the short narrow-linear or subulate
scales or leaves at the base more numerous than in other species ; stem-leaves
on slender petioles, usually small, peltate orbicular or slightly truncate and
2-angled on one side. Flowers solitary or rarely 2 or 3 in a simple raceme,
large. Calyx-lobes usually 8, 2 to 3 lines long, nearly glabrous, fringed with
prominent glands, but not ciliate. Petals 8, twice as long as the calyx-lobes,
narrow, not so readily twisting up together after flowering as in other species.
Styles repeatedly forked into very numerous short slender branches, stigmatic
at the end, forming a short very dense tuft.—*Sondera Preissii,* Lehm. Pugill.
viii. 45, and Pl. Preiss. i. 256 (in flower) ; *S. macrantha,* Lehm. l. c. (in
fruit).

W. Australia. From King George's Sound to Swan River and Champion Bay,
Drummond, 1st Coll. and 2nd Coll. n. 18, *Preiss, n.* 1989, *Oldfield,* and others.

41. **D. Banksii,** *R. Br. in DC. Prod.* i. 319. Stem filiform, leafy,
glabrous, 2 to 4 in. long, very slender at the base, but possibly forming a bulb
as in the preceding species. Leaves all scattered, peltate, orbicular, on slender petioles, the lower ones with a lamina of ½ line diameter, on a petiole of
1 to 2 lines, the upper ones twice as large or rather more. Stipules to some
of the upper leaves very thin, narrow, scarious and deciduous. Flowers few,
small, in a simple raceme like the smaller specimens of *D. peltata.* Pedicels
nearly as long as the calyx. Sepals villous, 1 to 1½ lines long. Petals
spreading, longer than the calyx. Styles (not seen by myself and imperfectly
observed by Planchon) 3, divided to the base into 2 deeply 3-fid branches.—
Planch. in Ann. Sc. Nat. ser. 3, ix. 291.

Queensland. Endeavour river, *Banks and Solander.*

2. BYBLIS, Salisb.

Calyx-segments or sepals 5. Petals 5, broad, oblique, united in a ring at
the base, contorted-imbricate. Stamens 5, hypogynous, often declinate ;
anthers attached by the base, opening at the end in oblong pores or short
slits. Ovary 2-celled, with several ovules in each cell attached to the dissepiment ; style undivided, with a terminal oblong or capitate stigma. Capsule somewhat compressed, 2-celled, opening in 2 valves, bearing the dissepiment in their centre. Seeds oblong, albuminous. Embryo . . .—Herbs,
more or less glandular-pubescent. Leaves linear-subulate, involute in vernation, without stipules. Peduncles axillary, bearing a single blue flower.

The genus is limited to Australia. It is very anomalous in the Order, with which it is chiefly connected by the glandular pubescence and the leaves involute in vernation. The flowers, especially those of *B. gigantea*, have a remarkable resemblance in structure to those of *Cheiranthera* in *Pittosporeæ*.

Stems slender. Leaves filiform, not above 2 in. long. Petals under ¼ in.
Anthers oblong or almost ovate, nearly equal . . . . 1. *B. liniflora.*
Stems stout. Leaves often above 6 in. long. Petals ¼ to 1 in. Anthers
oblong-linear, often unequal . . . . 2. *B. gigantea.*

1. **B. liniflora,** *Salisb. Parad. Lond. t.* 95. Glabrous or viscid with a glandular pubescence, sometimes copiously so. Stems slender, rarely 6 in. high and often only 2 or 3 in. Leaves filiform, 1 to 2 in. long. Peduncles slender, usually exceeding the leaves. Sepals lanceolate, acute, 2 to 3 lines long. Anthers varying from ¾ line to 1½ lines in length, the filaments longest where the anthers are shortest.—DC. Prod. i. 319 ; Endl. Iconogr. t. 113 (incorrect as to the anthers) ; *B. filifolia*, Planch. in Ann. Sc. Nat. ser. 3, ix. 305.

N. Australia. N.W. coast, *Bynoe ;* Hooker and Sturt's Creeks, Upper Victoria river, *F. Mueller ;* islands of the Gulf of Carpentaria, *R. Brown ;* adjoining mainland, *F. Mueller.*

Queensland. Shoalwater Bay, *R. Brown ;* Port Denison, *Herb. F. Mueller.*

B. cærulea, Planch. in Ann. Sc. Nat. ser. 3, ix. 306, is founded on Bauer's drawing published by Endlicher of R. Brown's specimens, in which the short anthers are represented as attached by the middle of the back, and Planchon thought he recognized these anthers in the specimen glued down in the Banksian herbarium. The excellent specimens in Brown's own herbarium show however that this is a mistake. The anthers are often as short as figured by Bauer, sometimes as long as figured by Salisbury, but always attached by the base, and varying much in intermediate lengths in different specimens.

2. **B. gigantea,** *Lindl. in Swan Riv. App.* 21. More or less glandular-pubescent and viscid. Rootstock hard. Stems erect, stout, ½ to 1½ or even 2 ft. high. Leaves linear-subulate, terete or channelled above, often 6 in. to 1 ft. long. Peduncles mostly shorter than the leaves. Flowers much larger than in *B. liniflora*, sometimes twice as large, but otherwise like them and variable in size. Sepals lanceolate, acute, 3- to 7-nerved, either much shorter than the petals or produced into a glandular point sometimes exceeding the petals. Anthers usually linear, declinate, unequal, the shorter ones 2 lines, the longer 3 lines long, but variable in size and proportion, sometimes nearly equal and very little longer than the longest forms of *B. liniflora*.—Lehm. in Pl. Preiss. i. 257 ; Planch. in Ann. Sc. Nat. ser. 3, ix. 306 ; *B. Lindleyana,* Planch. l. c. 307.

W. Australia, *Drummond, 1st Coll. ;* sandy places, Canning river, *Preiss, n.* 1993 ; Port Gregory, *Oldfield ;* Hampden, *Clarke.* I can see no difference in the numerous specimens in different herbaria, except in the size as well of the plant as of the parts of the flower.

ORDER XLV. **HALORAGEÆ.**

Calyx-tube adnate to the ovary ; lobes 2, 4 or none, or rarely 3. Petals 2, 4 or none, valvate induplicate or slightly imbricate. Stamens 2 to 8, rarely 1 or 3 ; filaments short ; anthers erect, 2-celled, opening longitudinally. Ovary inferior, flattened or angular, either 2- or 3- or rarely 4-celled, with 1 pendulous ovule in each cell, or 1-celled with 1 to 4 pendulous ovules ; styles as many as ovules, quite distinct, with papillose or plumose stigmas. Fruit inferior, small, indehiscent, with 1 to 4 cells and seeds, or divisible into 2 to 4 1-seeded indehiscent carpels. Seeds pendulous, with a membranous testa; embryo cylindrical, in the axis of a fleshy albumen; radicle long, superior; cotyledons small.—Herbs, often aquatic, or undershrubs. Leaves opposite whorled or alternate, without stipules. Flowers small, often unisexual or incomplete, axillary or rarely in terminal corymbs racemes or panicles.

The Order is dispersed over nearly the whole globe. Of the Australian genera, 2 small ones are endemic; the principal one is also chiefly Australian, but two or three of the species extend also into Eastern Asia, or are, with a fourth genus, widely spread over the extra-tropical regions of the southern hemisphere. The three others are aquatic plants, represented nearly all over the globe.

A. **True Halorageæ.**—*Flowers with petals, at least in the males, answering to the ordinal characters given above.*

Petals, at least in the males, induplicate, keeled. Fruit a nut-like or rarely spongy, undivided drupe.
  Flowers in dense terminal corymbose panicles, 2- 3- or 4-merous.
    Dissepiments of the ovary evanescent . . . . 1. LOUDONIA.
  Flowers solitary or clustered within each bract, along the rhachis
    of simple or paniculate terminal racemes.
    Flowers 3- or 4-merous . . . . 2. HALORAGIS.
    Flowers 2-merous . . . . 3. MEIONECTES.
Petals in the males imbricate. Fruit separable into 2 or 4 nut-like
  carpels. Aquatic or mud plants . . . . 4. MYRIOPHYLLUM.

B. *Anomalous genera of a very reduced type allied to* Halorageæ, *but often referred to* Monochlamydeæ. *Flowers unisexual.*

Calyx-teeth minute. Petals (in Australian species) none. Stamens
2. - Ovary 1-celled, with 1 ovule. Styles 2 or rarely 4. Terrestrial stemless plants, with broad radical leaves and radical scapes 5. GUNNERA.
Perianth none. Flowers surrounded by bracts. Stamens several.
Ovary 1-celled, with 1 ovule. Styles 2. Floating plant, with
verticillate dichotomous leaves. . . . . 6. CERATOPHYLLUM.
Perianth none. Flowers with or without 2 bracteoles. Stamen 1.
Ovary 4-celled, with 1 ovule in each cell. Styles 2. Aquatic or
mud plant, with opposite entire leaves . . . . 7. CALLITRICHE.

A. TRUE HALORAGEÆ.

1. LOUDONIA, Lindl.

(Glischrocaryon, *Endl.*)

Calyx-tube or ovary with 2 to 4 longitudinal wings or angles; lobes 2 to 4, short, alternating with the wings. Petals as many as calyx-lobes, induplicate, deciduous. Stamens twice as many as petals; filaments filiform, persistent; anthers oblong or linear, deciduous. Ovary 1-celled, with 2 to 4 pendulous ovules, or imperfectly divided into as many cells; styles short, rather thick, with terminal obtuse stigmas. Fruit a small 1-seeded nut, the adnate calyx winged or inflated.—Glabrous herbs, with a perennial rootstock and erect stems. Leaves alternate, linear. Flowers yellow, in dense terminal corymbose panicles.

The genus is limited to Australia. It differs from *Haloragis* more in habit and inflorescence than in floral characters. The characters by which the species are distinguished from each other may possibly not be found to be really constant.

Flowers 2-merous, rarely 3-merous . . . . 2. *L. Behrii.*

Flowers 4-merous.
Calyx and epicarp closely adnate to the endocarp, with 4 prominent
  wings. . . . . 1. *L. aurea.*
Calyx and epicarp inflated and connected with the endocarp by a loose
  spongy substance, with 4 scarcely prominent wings or angles . . . 3. *L. Roei.*

1. **L. aurea,** *Lindl. Swan Riv. App.* 42, *with a woodcut.* Quite glabrous. Rootstock woody. Stems erect, simple or slightly branched, 1 to 2
or even 3 ft. high, glaucous or yellowish, often turning black in drying.
Leaves linear, quite entire, distant or more crowded towards the base of the
stem, rarely 2 in. long and often much smaller, sometimes 1 to 2 lines broad
and flat, sometimes very narrow and thick, almost terete. Flowers golden-
yellow, in terminal corymbose panicles. Calyx-tube 4-winged, about 2 lines
long in the ordinary form, the lobes short and broad. Petals about as long
as the calyx-tube. Stamens in the perfect flowers 8. Styles 4, short, thick,
club-shaped, with ovoid stigmas. Ovules 4, but only 1 enlarges after flowering. Fruit varying from 2 to 3 lines in length, the wings usually broad.
In some specimens the flowers are smaller and mostly females, without any
or with very few stamens.—Nees in Pl. Preiss. i. 159 ; *L. flavescens,* J.
Drumm. in Hook. Lond. Journ. i. 396 (with smaller flowers) ; *L. citrina,*
F. Muell. in Linnæa, xxv. 385.

S. Australia. Rocks and gravelly banks of streams, Flinders range and near Cudnaka, *F. Mueller.* I am unable to discover any difference between these specimens and
some of Drummond's.

W. Australia. Swan River, Darling range, etc., *Drummond, 1st Coll., Collie, Preiss,*
n. 2067, 2068, 2079 ; Champion Bay, *Oldfield ;* Eyre, Phillips, and Fitzgerald ranges,
Maxwell.

2. **L. Behrii,** *Schlecht. Linnæa,* xx. 648. Very near the poorer specimens of *L. aurea,* with the same habit. Stems generally shorter. Leaves
narrow and small, often few and distant, rarely crowded. Panicles small and
dense. Flowers of the size of the smaller varieties of *L. aurea,* usually with
2 wings to the calyx-tube, 2 petals, 4 stamens, 2 styles and ovules, and a
broadly 2-winged fruit, but sometimes a third part is added to each.

Victoria. Mount Corong and N.W. desert to the Murray, *F. Mueller ;* Wimmera,
Dallachy.

S. Australia. Gregarious in barren sandy soils, *Behr ;* Mount Barker Creek, Gawler-
town, etc., *F. Mueller ;* Kangaroo Island, *Waterhouse.*

3. **L. Roei,** *Schlecht. Linnæa,* xx. 648. Stems erect, simple and glabrous, as in the other two species. Leaves few, small, linear, distant. Panicle small and dense. Flowers of the size of the smaller varieties of *L.
aurea,* or still smaller. Calyx-tube with 4 narrow shortly decurrent wings.
Petals as in *L. aurea,* but smaller. Stamens 8 (or sometimes 12 ?) ; anthers
not much longer than the filaments. Fruit yellow, almost globular, with 4
prominent angles or narrow wings, about 3 lines diameter, the inflated calyx-
tube and epicarp connected with the endocarp by a very loose, spongy,
almost fibrous substance. Seeds as in *L. aurea.*

W. Australia. Fitzgerald range, *Maxwell.* Endlicher's description of *Glischrocaryon
Roei,* Endl. in Ann. Wien. Mus. ii. 210, agrees precisely with Maxwell's specimens after the
petals and anthers have fallen away, except that the number of " stamina sterilia " (persistent filaments after the anthers have fallen) is 12 instead of 8. The remarkable spongy
inflated fruit, if really normal, is very characteristic of the species.

2. HALORAGIS, Forst.

(Cercodia, *Murr.*; Goniocarpus, *Kœn.*)

Calyx-tube or ovary with as many or twice as many nerves as lobes, those alternating with the lobes occasionally expanded into angles or wings; lobes 4, rarely 3 or abnormally 5, short. Petals as many as calyx-lobes, induplicate and boat-shaped or hood-shaped, deciduous, often wanting in female flowers. Stamens twice as many as petals or fewer, those opposite the petals and enclosed in them always present in complete or male flowers, one or more of the alternate ones occasionally wanting, and female flowers usually without any; anthers oblong or linear, deciduous; filaments short. Ovary 2- to 4- or rarely 5-celled, with 1 pendulous ovule in each cell; styles short and thick, stigmatic at the top, often plumose in the female flowers. Fruit a small, 2- to 4- or rarely 5-celled drupe or nut, the adnate calyx either smooth or variously ribbed, angled, winged, or muricate.—Herbs or undershrubs, glabrous scabrous or hispid. Leaves alternate or opposite, entire toothed or lobed. Flowers small, solitary or several together in the axils of the floral leaves or bracts, forming leafy or leafless racemes, either simple or in a branching terminal panicle. Pedicels usually very short, with two small opposite often deciduous bracteoles under the flower.

The genus is chiefly Australian, but a few species are also found in New Zealand, in Eastern Asia, in S. Africa, and extratropical S. America. Of the 36 Australian species, 1 extends to New Zealand and the island of Juan Fernandez, 2 to New Zealand and Eastern Asia, 1 to New Zealand only, the remaining 32 are endemic. The characters derived from the ribs and wings of the fruit, upon which the genus had been divided into three, are either too little in accordance with other distinctions, or too variable in certain species, to be available even as sectional. Most of the species are monœcious, the female flowers variously mixed in with the males, and although I have frequently had specimens with the flowers all of one or the other kind, I have not been able to ascertain that any species is constantly diœcious. The males have never plumose stigmas, but I always find small obtuse styles and their corresponding ovules, which appear often to come to perfection. The females have usually smaller petals or none at all, fewer stamens or none or filiform filaments only. As the differences between the two are probably the same in nearly all the species, I have not alluded to them in the specific characters.

SERIES 1. **Alternifóliæ.**— *Leaves all alternate or rarely here and there irregularly opposite, or (in some specimens of* H. hexandra *and* H. ceratophylla) *a few of the lower ones, or those of barren shoots only, opposite.*

Leaves narrow-linear, entire.
  Glabrous, small and slender. Leaves very small. Flowers solitary, minute. Styles and ovules 4 . . . . 22. *H. pusilla.*
  Glabrous. Flowers mostly clustered. Fruit not ribbed.
    Styles and ovules 2, rarely 1. Fruit globular . . . . . ; 1. *H. digyna.*
    Styles and ovules 4 or sometimes 3. Fruit ovoid . . . . 2. *H. mucronata.*
  Sprinkled with a few hairs. Flowers solitary or 2 together.
    Fruit 8-ribbed.
      Calyx-lobes ovate. Petals deciduous . . . . 3. *H. pithyoides.*
      Calyx-lobes cordate. Petals reflexed, persistent . . . . . 4. *H. cordigera.*
  Densely hirsute. Fruit ovoid, muricate . . . . 5. *H. elata.*
Leaves linear or lanceolate, the larger ones with a few coarse teeth or pinnatifid.
  Western species.
    Styles and ovules 1, 2, or 3.
      Glabrous. Calyx-lobes, petals, styles, and ovules usually 3, stamens 6. Fruit smooth . . . . 6. *H. tenuifolia.*

Glabrous. Calyx-lobes and petals 4; stamens 8; styles and
ovules 1 or 2. Fruit 8-ribbed . . . . 7. *H. scoparia.*
Leaves bordered with minute tooth-like points. Calyx-lobes
and petals 4; styles and ovules 2 or 3. Fruit smooth . 8. *H. aculeolata.*
Styles and ovules 4.
Slightly hairy. Calyx-lobes cordate. Ovary acutely 4-
angled . . . . 9. *H. foliosa.*
Glabrous. Calyx-lobes almost cordate. Fruit depressed-
globular, almost spongy, not angled. Flowers minute . 10. *H. platycarpa.*
Glabrous. Leaves linear-lanceolate, the teeth short . . . 2. *H. mucronata.*
Eastern species. Glabrous or scabrous. Styles and ovules 4.
Leaves nearly sessile. Flowers mostly solitary.
Fruit ovoid-globular, often muricate, not angled . . . . 11. *H. ceratophylla.*
Fruit acutely angled . . . . 12. *H. acutangula.*
Leaves lanceolate, distinctly petiolate. Flowers clustered.
Fruit muricate and angled . . . . 14. *H. odontocarpa.*
Leaves linear to oblong, entire or minutely toothed, those of barren ·
branches often opposite.
Calyx-lobes, petals, styles, and ovules mostly 3; stamens 6 . . 13. *H. hexandra.*
Calyx-lobes, petals, styles, and ovules 4; stamens 8 . . . . 11. *H. ceratophylla.*

SERIES 2. **Oppositifoliæ.**—*Stem-leaves all opposite or rarely the uppermost alternate*
(*or nearly all in* H. pusilla). *Floral-leaves or bracts alternate or rarely the lowest opposite.*

Styles and ovules 2. Leaves lanceolate. Flowers mostly clustered.
Leaves serrate . . . . 15. *H. serra.*
Leaves entire. . . . . 16. *H. glauca.*
Styles and ovules 4.
Leaves distinctly petiolate, lanceolate or oblong, serrate. Flowers
mostly clustered.
Leaves broadly lanceolate or oblong. Fruit ovoid, not inflated,
terete or 4-winged . . . . 17. *H. alata.*
Leaves narrow-lanceolate. Fruit 4-angled, the epicarp inflated
and connected with the endocarp by a loose network . . . 18. *H. racemosa.*
Leaves nearly sessile (except in *H. nodulosa*). Flowers soli-
tary or rarely 2 together (clustered in *H. stricta* and ·*H.
lanceolata*).
Western species. Flowers minute. Racemes paniculate (ex-
cept in *H. lanceolata*).
Hirsute with spreading hairs.
Leaves broadly ovate-cordate, regularly toothed . . . . 19. *H. rotundifolia.*
Leaves obovate-oblong or lanceolate, almost entire . . . 20. *H. rudis.*
Glabrous or with a few scattered hairs.
Leaves few, narrow-linear, scarcely toothed. Bracts mi-
nute or none.
Tall and erect. Leaves ½ to 1 in. long. Racemes pa-
niculate . . . . 21. *H. paniculata.*
Small and slender. Leaves small and few, mostly alter-
nate. Racemes simple. . . . . 22. *H. pusilla.*
Leaves ovate-lanceolate or oblong, deeply toothed. Ra-
cemes flexuose. Bracts often as long as the flowers.
Fruit globular . . . . 23. *H. intricata.*
Leaves petiolate, oblong, obscurely toothed. Racemes
flexuose. Bracts petiolate. Fruit urceolate, muricate,
with a smooth neck . . . . 24. *H. nodulosa.*
Leaves oblong or cuneate, nearly entire. Racemes fili-
form, paniculate. Bracts minute or none . . . . . 25. *H. trichostachya.*
Leaves lanceolate or oblong, entire. Racemes slender,
simple. Bracts longer than the minute flowers . . 26. *H. lanceolata.*

Eastern species.
  Glabrous or nearly so. Leaves ovate or orbicular. Flowers
    minute, iu filiform leafless panicles . . . . 27. *H. micrantha.*
  Scabrous or hirsute.
    Leaves linear or linear-lanceolate, entire or with small
      distant teeth . . . . : . . . . 28. *H. stricta.*
    Lower leaves divided into narrow linear lobes.
      Lobes above the middle of the leaf almost digitate . . 29. *H. heterophylla.*
      Lobes pinnately disposed along the rhachis . . . . 30. *H. pinnatifida.*
    Leaves broadly toothed or crenate.
      Leaves oblong, often 1 iu. long. Fruit small, narrow.
        Bracts minute . . . . 31. *H. acanthocarpa.*
      Leaves ovate or oblong, under ½ in. long, narrowed at
        the base. Fruit small, nearly globular. Upper bracts
        minute . . . . 32. *H. tetragyna.*
      Leaves broadly ovate, rounded or cordate at the base.
        Fruit globular. Bracts exceeding the flower . . . 32. *H. teucrioides.*

SERIES 3. **Oppositifloræ.**—*Floral leaves and flowers all or nearly all opposite, as well as the stem-leaves. Flowers solitary in each axil.*
Stems hirsute. Leaves deeply serrate, ovate or oblong . . . . 34. *H. scordioides.*
Minutely scabrous. Leaves ovate or ovate-lanceolate, small, entire
  or slightly toothed . . . . 35. *H. depressa.*
Nearly glabrous. Leaves narrow-linear or terete, mostly entire . . 36. *H. salsoloides.*

H. cyathiflora, Fenzl, in Hueg. Enum. 44, described from a Swan River specimen of Huegel's with male flowers only, which I have not seen, can scarcely belong to this genus. The habit and foliage must be nearly that of *H. digyna,* but with a disk-shaped, 5- to 8-toothed calyx, twice as many stamens as calyx-teeth, and neither petals nor rudiment of the ovary, both of which exist in the male flowers of all the other species.

H. trifida, Walp. Rep. v. 672, or *Goniocarpus trifidus,* Nees in Pl. Preiss. i. 159, is described as having the lower leaves in whorls of three, the upper ones opposite, all filiform and trifid. Iu the fragments of Preiss, n. 2401, in Souder's herbarium, the leaves are alternate and there are no flowers. The species must therefore remain doubtful until further investigated from better specimens.

1. **H. digyna,** *Labill. Pl. Nov. Holl.* i. 101, t. 129. Tall and glabrous, with terete branches. Leaves alternate, linear, usually terminating in a minute white callous point, rarely exceeding 1 in. in length, rather thick, entire or very obscurely toothed; the floral ones smaller but exceeding the flowers. Flowers shortly pedicellate, pendulous, clustered in the upper axils forming terminal leafy racemes. Calyx-lobes ovate-lanceolate, not cordate. Petals 4, at least 1 line long, hood-shaped, glabrous or slightly ciliate on the keel. Stamens 8 or occasionally 6. Styles and ovules 2, or rarely 1 only. Fruit nearly globular, smooth, not ribbed, crowned by the small erect calyx-lobes.

W. Australia, *Labillardière, Maxwell,* in both cases probably from King George's Sound or to the eastward.

2. **H. mucronata,** *Benth.* Usually glabrous, with the aspect of *H. digyna,* but smaller and more slender. Leaves alternate, usually narrow-linear, rather thick or semiterete, quite entire, ½ to above 1 in. long, usually but not always terminating in a minute callous point; the floral ones similar but shorter; in the barren shoots the leaves are sometimes broader, with a few teeth. Flowers glabrous, much smaller than in *H. digyna,* nearly sessile, clustered in the upper axils, forming leafy racemes. Calyx-lobes short,

triangular. Petals 4, not ¾ line long. Stamens usually 8. Styles and ovules 4 or rarely 3. Fruit small, ovoid, glabrous, smooth or slightly 8-ribbed, crowned by the small erect calyx-lobes.—*Goniocarpus mucronatus,* Nees in Pl. Preiss. ii. 225.

Victoria. Heaths near Fitzroy river and marshes near. Portland, *Robertson.*

S. Australia. Kangaroo Island, *R. Brown;* Murray scrub and Onkaparinga river, *F. Mueller.*

W. Australia. Bald Head, King George's Sound, *R. Brown;* muddy soil near Vasse river, *Preiss, n.* 1221.

I can find no difference between the S. Australian and western specimens. It is possible they may both prove to be varieties of *H. digyna;* but in the latter, although the flowers are considerably larger, they appear never to have more than two styles and ovules.

3. **H. pithyoides,** *Benth.* Slender but rigid, erect, with numerous virgate branches, ½ to 1 ft. high, glabrous or sprinkled with a few hairs, usually of a glaucous or blackish tint when dry. Leaves alternate, narrow-linear, thick or semiterete, rarely above ½ in. long, entire, the floral ones small and mostly reduced to bracts shorter than the flowers. Flowers small, usually solitary, in slender racemes, forming a terminal panicle scarcely leafy at the base. Calyx-tube turbinate, 8-ribbed, usually hispid; lobes ovate, not cordate, quite glabrous. Petals 4, hood-shaped, about ¾ line long, ciliate on the keel, not reflexed. Stamens 8. Styles and ovules 4. Fruit small, ovoid-globular, with 8 prominent ciliate almost aculeate ribs, and crowned by the glabrous calyx-limb.—*Goniocarpus pithyoides,* Nees in Pl. Preiss. ii. 225.

W. Australia. ⸱ Swan River, *Preiss, n.* 1224, *Drummond, Clarke.*

4. **H. cordigera,** *Fenzl, in Hueg. Enum.* 45. Erect, with slender but rigid terete branches, attaining 1 ft. or more, sprinkled with a few hairs. Leaves alternate, narrow-linear, thick or semiterete, under 1 in. and often under ½ in. long, entire or rarely obscurely toothed, the floral ones reduced to minute bracts. Flowers mostly solitary, pendulous, on short pedicels, in slender racemes, forming a terminal leafless panicle. Calyx-lobes rather large, remarkably cordate and almost peltate, erect. Petals 4, above 1 line long, hood-shaped, ciliate on the keel, reflexed immediately on expanding and remaining long persistent. Stamens usually 8. Styles 4, longer than in most species. Young fruit nearly globular, 8-ribbed, but not seen ripe.—Hook. Ic. Pl. t. 598; *Goniocarpus cordiger,* Nees in Pl. Preiss. ii. 226.

W. Australia. Swan River, *Huegel, Drummond,* 1*st Coll. and* 4*th Coll. n.* 83, *Preiss, n.* 1223. I have not seen Huegel's own specimens, but the description is very accurate.

5. **H. elata,** *A. Cunn.; Fenzl, in Hueg Enum.* 45. Rather coarse, with erect or ascending branches, ½ to 1⅓ ft. high, hirsute with spreading hairs. Leaves alternate, or a few very rarely irregularly opposite, linear, acutely acuminate, ½ to 1 in. long, with revolute margins, entire or rarely with a few short teeth, the floral ones smaller but mostly exceeding the flowers. Flowers solitary, not very small, in terminal racemes, forming a narrow leafy panicle. Calyx hirsute, the lobes not cordate. Petals 4, ciliate on the keel. Stamens usually 8. Styles and ovules 4. Fruit small, ovoid, prominently muricate.—Schlecht. Linnæa, xx. 648.

N. S. Wales. Barren rocky ridges, W. from Wellington valley, *A. Cunningham.*

S. Australia. Barossa range, *Behr ;* Mount Lofty ranges and near Lake Torrens, *F. Mueller.*

6. **H. tenuifolia,** *Benth.* Tall, glabrous, and erect, from a shortly creeping base; branches terete and smooth. Leaves alternate, narrow-linear, the larger ones 1 to 2 in. long with a few narrow-linear lobes, the floral ones gradually reduced to small bracts. Flowers on very short pedicels, mostly in clusters of 2 or 3, forming terminal racemes leafy at the base. Calyx-lobes 3, broad but scarcely cordate. Petals 3, above 1 line long. Stamens 6. Styles and ovules 3. Young fruits ovoid, above 1 line long, not ribbed, crowned by the calyx-lobes, 3-celled or very rarely 2-celled.

W. Australia, *Drummond, 4th Coll. n.* 86. With the habit of *H. scoparia,* this species has a more slender foliage, almost approaching that of *Meionectes,* with the ternary flowers of *H. hexandra,* and the fruit apparently of *H. aculeolata.*

7. **H. scoparia,** *Fenzl, in Hueg. Enum.* 45. Tall, glabrous, and erect; branches terete and smooth. Leaves alternate, linear or linear-lanceolate, acute, the larger ones 1 to 2 in. long with a few remote very prominent teeth or lobes, the upper ones entire, the floral ones gradually reduced to small bracts. Flowers on very short pedicels, in clusters of 2 or 3, forming long loose terminal racemes leafy at the base. Calyx-lobes short, broad, almost cordate. Petals 4, about 1 line long. Stamens 8. Styles and ovules 1 or 2, the stigmas apparently not plumose even in the females. Ovary after flowering prominently 8-ribbed. Fruit not seen.

W. Australia, *Huegel, Drummond, 4th Coll. n.* 82. I have not seen Huegel's specimen, but the description leaves no doubt as to its identity.

8. **H. aculeolata,** *Benth.* Erect, virgate, about 1 ft. high in our specimens, the stem terete and glabrous or with slightly prominent ciliolate or aculeolate angles. Leaves alternate, narrow-linear, mucronate entire or more frequently with a few distant prominent teeth or lobes, as in *H. scoparia,* but also bordered with very short, rigid, cartilaginous points or minute teeth, the floral leaves smaller, but all much exceeding the flowers. Flowers apparently solitary or 2 together, in loose terminal leafy racemes, but not seen perfect. Fruit ovoid, almost corky, above 1 line long, quite glabrous, without prominent ribs, crowned by the 4 connivent, triangular, not cordate calyx-lobes; cells 2 or very rarely 3.

W. Australia, *Clarke.* The specimens are not good, but I cannot match them with any of the allied species. The foliage is nearly that of *H. scoparia* and *H. foliosa,* the shape of the fruit and calyx-lobes that of *H. tenuifolia,* but with a different number of parts.

9. **H. foliosa,** *Benth.* Tall and erect, sprinkled with a few short spreading hairs. Leaves alternate, linear-lanceolate, acute, mostly ¾ to 1½ in. long, with a few remote, acute, very prominent teeth, or the upper ones entire, narrowed at the base but scarcely petiolate, the floral ones shorter and more lanceolate, but nearly all exceeding the flowers. Racemes terminal, leafy, rather dense. Flowers usually clustered in the axils, nearly sessile. Calyx-lobes cordate, acutely acuminate. Petals 4, above 1 line long, often ciliate on the keel. Stamens 8 or sometimes 6 only. Styles and ovules 4. Ovary after flowering acutely 4-angled. Ripe fruit not seen.

W. Australia. Between Moore and Murchison rivers, *Drummond, 6th Coll. n.* 82.

10. **H. platycarpa,** *Benth.* Erect, glabrous, under 1 ft. high in our specimens, the branches rather slender, terete. Leaves alternate, linear or lanceolate, acute, the larger ones ¾ to 2 in. long, with a few remote prominent teeth or lobes, narrowed at the base into a short petiole, the floral ones gradually reduced to small bracts. Flowers very small, pedicellate, solitary or 2 or 3 together under each bract, in slender terminal racemes leafy at the base. Calyx-lobes ovate, acuminate, almost cordate. Petals 4, scarcely above ¼ line long in the females, ½ line or rather more in the males. Stamens usually 6. Styles and ovules 4. Fruit depressed-globular, nearly 2 lines diameter, with a loose almost spongy epicarp and a crustaceous endocarp, 4-nerved, the top almost disk-like, with the small connivent calyx-lobes in the centre. Rarely the parts of the flowers and fruit are in fives instead of fours.

W. Australia. Swan River, *Drummond, 1st Coll. and n.* 705.

11. **H. ceratophylla,** *Endl. Atakta,* 16, *t.* 15. Glabrous and glaucous or scabrous, with minute rigid hairs, rather coarse but not usually tall, the decumbent or ascending angular stems rarely exceeding 1 ft. Leaves alternate or rarely a few of the lower ones or those of barren side-shoots opposite, either linear or linear-lanceolate with coarse distant teeth or lobes, or shortly pinnatifid, or sometimes nearly all entire, linear-oblong, and obtuse, usually rather thick with very scabrous margins, from ½ to 1½ in. long, the floral ones gradually reduced to small bracts. Flowers nearly sessile, solitary or 2 together, usually much larger than in *H. tetragyna* and *H. heterophylla,* in long terminal racemes leafy at the base. Calyx-lobes lanceolate-triangular. Petals 4, about 1 line long, the keel scabrous-hirsute. Stamens usually 8. Styles and ovules 4. Fruit ovoid or globular, much larger than in *H. tetragyna,* scarcely ribbed but sometimes very rugose or muricate.—*H. aspera,* Lindl. in Mitch. Trop. Austr. 306 ; *H. pinnatifida,* Hook. f. Fl. Tasm. i. 119, but not of A. Gray.

Queensland. Bargoo (Victoria) river, *Mitchell ;* Rockhampton, *Dallachy ;* Warwick, *Beckler.*

N. S. Wales. From the Darling to Cooper's Creek, *Victorian Expedition ;* very common in all the swamps of the interior, *Fraser.*

Victoria. Scrub of the N.W. parts of the colony towards the Murray, *F. Mueller ;* Wimmera, *Dallachy.*

Tasmania. Herdsman's Cove, Derwent river, *R. Brown ;* N. coast and S. Esk river, *J. D. Hooker.*

S. Australia. Frequent on the Murray, Gawler river, Cudnaka, Mount Remarkable, etc., *F. Mueller ;* Spencer's Gulf, *Warburton ;* Port Lincoln, *Wilhelmi.*

I have not seen authentic specimens of Endlicher's plant, but the figure appears to me to represent this species rather than *H. heterophylla,* which is also in R. Brown's collection, and of which some coarse specimens, not well in flower, resemble the more slender ones of *H. ceratophylla.*

12. **H. acutangula,** *F. Muell. in Trans. Vict. Inst.* 1855, 125. A glabrous, glaucous, rather coarse species, with the habit, narrow toothed leaves, inflorescence, and flowers of *H. ceratophylla,* of which it may prove to be a variety, differing in the fruit prominently and acutely 4-angled, and smooth between the angles.

S. Australia. Port Lincoln, *Wilhelmi.*

13. **H. hexandra,** *F. Muell. Fragm.* iii. 31. Glabrous, diffuse, much-

branched, some specimens under ½ ft., others above 1 ft. long, the branches slightly angular. Leaves alternate, or a few of the lowest or on some barren shoots opposite, from oblong-lanceolate to linear, acute, ½ to 1½ in. long, entire or with a few minute remote teeth, narrowed at the base but scarcely petiolate, rather thick, the floral ones smaller and narrower, but all much exceeding the flowers. Flowers very small, pendulous, usually 2 together in each axil, forming slender terminal leafy racemes. Pedicels usually short, but sometimes longer than the flowers. Calyx-lobes 3 or rarely 4, ovate-triangular. Petals usually 3, even where the calyx is 4-merous, little more than ¼ line long. Stamens 6. Styles and ovules 3. Fruit very small, ovoid, with 6 ribs, occasionally prominent and tubercular-rugose.

W. Australia, *Drummond, 4th Coll. n.* 84; bogs near Wilson's Inlet, *Oldfield.*

14. **H. odontocarpa,** *F. Muell. Fragm.* i. 108. Apparently tall, glaucous and nearly glabrous or loosely hairy. Leaves alternate, distinctly petiolate, lanceolate, mostly 1 to 1½ in. long, coarsely serrate and rather thick, the floral ones very small, mostly reduced to bracts shorter than the flowers. Flowers clustered, in terminal racemes, smaller than in *H. ceratophylla.* Calyx-lobes short, acute, ciliate. Petals 4, scarcely 1 line long. Stamens 8. Styles and ovules 4. Fruit ovoid, above 1 line long, prominently 4-angled or almost winged, more or less muricate, and often with 1 or 2 thick prominent conical or tooth-like protuberances on each of the 4 sides.

N. S. Wales. Kulkyne on the Darling, *Goodwyn and Dallachy.* The habit and petiolate leaves are those of *H. alata,* but the leaves appear to be all alternate, and the echinate points, if constant, are quite characteristic.

15. **H. serra,** *Brongn. in Duperr. Voy. t.* 69. Erect, branching, quite glabrous or the angles of the stem and edges of the leaves minutely scabrous. Stem-leaves opposite, lanceolate, acute, regularly and sharply serrate, narrowed at the base but scarcely petiolate, the floral ones alternate, mostly reduced to small bracts. Flowers glabrous, rather small, distinctly pedicellate, mostly clustered, in slender racemes, leafy at the base. Calyx-lobes short. Petals 4, nearly 1 line long. Stamens 8 or fewer. Styles and ovules 2. Fruit small, 2-celled, smooth or obscurely rugose.

N. S. Wales. Common about Clifton, New England, *C. Stuart;* near Castlereagh, *C. Moore;* Liverpool range, *Leichhardt.*

16. **H. glauca,** *Lindl. in Mitch. Trop. Austr.* 91. Apparently annual, tall and erect, slightly branched, quite glabrous and glaucous with terete stems. Stem-leaves opposite, lanceolate, acute, entire or slightly serrate, narrower than in *H. serra,* the floral ones alternate and gradually reduced to small bracts. Flowers glabrous, pendulous but scarcely pedicellate, mostly clustered, in terminal racemes leafy at the base. Calyx-lobes lanceolate or oblong. Petals 4. Stamens 6 to 8. Styles and ovules 2. Fruit globular, rugose, 2-celled.

N. S. Wales. Swamps of the Narran, *Mitchell (Herb. Lindley).* Very near *H. serra,* and perhaps a variety, but the aspect too dissimilar to justify the uniting it without seeing intermediate specimens.

17. **H. alata,** *Jacq. Ic. Pl. Rar.* i. 7, *t.* 69. A tall erect species, apparently glabrous, but scabrous with minute asperities only visible under a lens,

the branches acutely angular. Stem-leaves opposite, distinctly petiolate, from ovate-lanceolate to oblong, ¾ to 1½ in. long, or the lower ones sometimes twice that size, regularly and sharply serrate, the floral ones mostly alternate and small. Flowers shortly pedicellate, clustered and drooping, forming terminal racemes leafy at the base. Calyx-lobes broad. Petals 4, about 1 line long, glabrous. Stamens 8. Styles and ovules 4. Fruit rather small, globular or ovoid, with 4 ribs scarcely prominent in most of the Australian specimens, more or less dilated into wings in most of the New Zealand ones, but variable in both countries, smooth or rugose between the ribs.—Hook. f. Fl. N. Z. i. 62 ; *Cercodia erecta*, Murr. ; DC. Prod. iii. 67.

N. S. Wales. Grose river, *R. Brown;* Nepean river, *Woolls;* Castlereagh river, *C. Moore ;* Clarence and Richmond rivers, *Beckler ;* New England, *Leichhardt* (the latter doubtful, leaves narrower and almost entire).

Victoria. Port Phillip, *Gunn.*

Also in New Zealand and in the island of Juan Fernandez.

18. **H. racemosa,** *Labill. Pl. Nov. Holl.* i. 100, *t.* 128. A glabrous erect herb or undershrub attaining 5 or 6 ft., with the acutely angular branches and general aspect of *H. alata.* Stem-leaves opposite, distinctly petiolate, narrow-lanceolate, regularly and acutely serrate, often above 2 in. long, the floral ones alternate, gradually reduced to bracts. Flowers pedicellate, clustered in the upper axils, forming terminal leafy thyrsoid panicles. Calyx-tube acutely 4-angled, the lobes broad and short. Petals 4, about 1 line long, mucronate. Stamens 8. Styles 4, rather long; ovules 4. Fruit 3 or even 4 lines long, acutely 4-angled or winged, crowned by the short connivent calyx-lobes, quite smooth, the thin somewhat inflated epicarp connected with the endocarp by a loose network or spongy substance as in *Loudonia Roei.*— *Cercodia racemosa*, DC. Prod. iii. 67.

W. Australia. S. coast, *Labillardière, R. Brown;* dry sandy places, Nornalup inlet, *Maxwell.*

19. **H. rotundifolia,** *Benth.* Apparently annual, erect, branching from the base, attaining 1 ft. or rather more, hirsute with spreading hairs. Stem-leaves opposite, nearly sessile, ovate-orbicular, mostly ½ to ¾ in. long, regularly crenate-serrate, cordate at the base, the floral ones and minute bracts alternate. Racemes filiform, in a terminal panicle, leafy only at the base. Flowers minute, solitary and distant, glabrous or minutely pubescent. Calyx-lobes broadly ovate, almost cordate. Petals 4, about ¼ line long. Styles and ovules 4. Fruit ovoid-oblong, not ½ line long, 8-ribbed, smooth or obscurely rugose.

W. Australia. Swan River, *Drummond, 1st Coll. ;* Flinders Bay, *Collie.*

20. **H. rudis,** *Benth.* Low and diffuse but coarse, densely hirsute with spreading hairs, the stems hard and almost woody at the base, but perhaps annual. Stem-leaves opposite, from obovate-oblong to almost lanceolate, entire or with a few small teeth, under ½ in. long, narrowed at the base, rather thick, hispid on both sides, the floral ones small narrow and alternate as well as the minute bracts. Racemes short and slender, in small terminal panicles, leafy at the base only. Flowers very small, the males not seen. Calyx-lobes 4, ovate-triangular. Styles and ovules 4. Fruit very small, nearly globular, slightly constricted under the persistent calyx-limb.

W. Australia, *Drummond, 4th Coll.* . 81.

21. H. paniculata, *R. Brown, Herb.* Erect, very slender, and but little branched below the inflorescence, glabrous or slightly hairy. Stem-leaves opposite, few in distant pairs, linear, obtuse, ½ in. long or more, entire or obscurely crenate, narrowed at the base but not petiolate, the floral ones reduced to minute alternate bracts. Racemes slender, divaricate, in a loose terminal panicle. Flowers very small, solitary or 2 together. Calyx-tube 8-ribbed, shortly ciliate, the lobes ovate, not cordate, glabrous. Petals 4, about ¼ line long. Stamens 8. Styles and ovules 4. Fruit only seen when young.

W. Australia. King George's Sound, *R. Brown, Harvey.*

22. H. pusilla, *R. Br. Herb.* A small slender annual, branching at the base only, most of the specimens not above 2 or 3 in. high and quite glabrous. Lower leaves opposite, linear, entire, 2 to 3 lines long, upper ones smaller and alternate, the floral ones reduced to minute bracts. Flowers solitary under each bract, very small and distant, sometimes scarcely forming a terminal raceme. Petals 4, not above ½ line long. Stamens 8. Styles and ovules according to R. Brown's notes 4. Fruit as small as in *H. micrantha,* nearly globular, 8-ribbed.

W. Australia. To the E. of King George's Sound, *R. Brown* (*Herb: R. Br.*)
Var. (?) *subaphylla.* Stems slender and wiry, 3 to 6 in. long, almost leafless. Flowers few and very distant.—S. coast, *R. Brown.*

23. H. intricata, *Benth.* Diffuse or ascending and very much branched, glabrous or sprinkled with a few rather rigid hairs, branchlets filiform, very flexuose. Stem-leaves opposite, ovate-lanceolate or oblong, acutely and coarsely serrate, under ½ in. long, the floral ones alternate, mostly reduced to small narrow bracts often as long as the flower. Racemes filiform, flexuose, forming very much branched terminal panicles leafy at the base. Flowers minute, solitary, distant, hispid with a few stiff hairs. Calyx-lobes small, cordate. Petals 4, scarcely ½ line long. Stamens usually 8. Styles and ovules 4. Fruit minute, globular, obscurely angled, but not seen ripe.

W. Australia, *Drummond, 5th Coll. n.* 39.

24. H. nodulosa, *Walp. Rep.* v. 672. Apparently annual, much branched, rarely exceeding ½ ft., sprinkled with a few short rigid hairs, branches slender, flexuose. Stem-leaves opposite, oblong, entire or obscurely toothed, under ½ in. long, narrowed into a distinct petiole; the floral ones alternate, small, yet all petiolate and exceeding the flowers. Racemes slender, flexuose, paniculate, leafy at the base. Flowers minute, solitary, distant, and nearly sessile. Calyx-tube urceolate, about ½ line long, lobes small, not cordate. Petals 4, scarcely ½ line long. Stamens in the flower examined 4 only. Styles and ovules 4. Fruit about ¾ line long, globular and muricate at the base, with a narrow smooth neck crowned by the calyx-lobes.—*Goniocarpus nodulosus,* Nees in Pl. Preiss. i. 158.

W. Australia. Swan River, *Drummond, 1st Coll., Preiss, n.* 2378. The urceolate calyx distinguishes this from all the other species; the reduction in the number of stamens may not be constant.

25. H. trichostachya, *Benth.* A small, rather rigid, erect species, probably annual, branched at the base, not 6 in. high including the panicle, sprinkled with a few short appressed hairs. Stem-leaves rather crowded at the base of the branches, opposite, linear-cuneate or oblong, entire or nearly so, mostly about ¼ in. long, narrowed into a short petiole, the floral ones all reduced to minute alternate scale-like bracts. Racemes filiform, forming much branched leafless terminal panicles more slender than in any other species. Flowers very small, distant, nearly sessile, pendulous. Calyx-lobes broad but scarcely cordate. Petals 4, about ½ line long, glabrous, hood-shaped. Stamens 6 in the flowers examined, but perhaps sometimes 8. Styles and ovules 4. Fruit small, 8-ribbed, sometimes 4-angled by the prominence of 4 of the ribs, and quite glabrous and smooth, sometimes hirsute and nearly globular.

N. Australia, *Drummond, n.* 205.

26. **H. lanceolata,** *R. Brown, Herb.* A diffuse glabrous annual, with the habit nearly of the smaller forms of *H. hexandra,* the slender ascending branches rarely above 3 or 4 in. high. Stem-leaves opposite, oblong or lanceolate, from under ½ in. to nearly ¾ in. long, quite entire, narrowed at the base, rather thick, the floral ones alternate and much reduced, but all longer than the flowers, which, however, I have not seen perfect. Fruiting racemes slender, terminal, leafy, the fruits as small as in *H. micrantha,* shortly pedicellate, 2 or 3 together in each axil, reflexed, very small, ovoid, 4-angled but otherwise smooth, crowned by the 4 calyx-lobes.

W. Australia. Marshes, King George's Sound, *R. Brown* (*Herb. R. Br.*).

27. **H. micrantha,** *R. Brown in Flind. Voy. App.* 550. Glabrous or slightly scabrous, much branched and diffuse or slender and erect, usually under 6 in. high, but when very luxuriant twice that height, the greater part occupied by the panicle. Stem-leaves opposite, orbicular-cordate or very broadly ovate, serrate-crenate, 3 to 4 lines or rarely ½ in. diameter, the floral ones reduced to minute alternate bracts. Racemes filiform, in a loose terminal panicle. Flowers minute, solitary. Calyx-lobes short, not cordate. Petals 4, about ½ line long. Styles and ovules 4. Fruit small, nearly globular prominently 8-nerved, otherwise smooth and shining.—Hook. f. Fl. Tasm. i. 121 ; *H. tenella,* Brongn. in Duperr. Voy. t. 68 B ; *Goniocarpus micranthus,* Thunb.; DC. Prod. iii. 66 ; *G. microcarpus,* Thib.; DC. Prod. iii. 66 (from the diagnosis).

N. S. Wales, Port Jackson, *R. Brown* and others ; New England, *C. Stuart* ; Clarence and Hastings rivers, *Beckler.*
Victoria. Dandenong and Haidinger ranges, Ovens river, Mount Buller, Mount Useful, *F. Mueller;* Portland, *Allitt.*
Tasmania. Port Dalrymple, *R. Brown;* abundant in moist sandy soil in several parts of the colony, *J. D. Hooker.*
S. Australia. Mount Lofty range, *F. Mueller.*
Also in New Zealand, Khasia, and Japan.

28. **H. stricta,** *R. Br. Herb.* Erect, rigid but slender, rather tall, nearly glabrous in appearance but very scabrous. Stem-leaves opposite, linear or linear-lanceolate, acute, entire or with small distant teeth, the larger ones 1 to 2 in. long, the floral ones alternate, mostly reduced to small bracts. Flowers clustered within each bract, shortly pedicellate, forming slender terminal racemes leafy at the base. Calyx-lobes small, acute. Petals 4, about

1 line long.　Stamens 8.　Styles and ovules (according to R. Brown's notes) 4.　Fruit small, but not seen ripe.

Queensland.　Broad Sound, *R. Brown* (*Herb. R. Br.*)　The inflorescence is that of *H. serra*, but the foliage is different and the pistil 4-merous.

29. **H. heterophylla,** *Brongn. in Duperr. Voy. t.* 68 A.　A rather slender species, usually small, but sometimes 1 ft. high, glabrous or minutely scabrous, with erect or ascending stems.　Stem-leaves all or mostly opposite, deeply divided above the middle into 3, 5 or 7 linear or rarely lanceolate acute lobes almost digitate; a few of the upper ones often alternate linear entire or nearly so, the floral ones smaller, the uppermost reduced to small bracts.　Flowers like those of *H. tetragyna*, small, solitary, or 2 together within each bract, in slender terminal leafy racemes.　Calyx scabrous, with short lobes.　Petals 4, in the males oblong, boat-shaped, about 1 line long, present also in some of the females, but shorter and hood shaped.　Styles and ovules 4.　Fruit small, globular or nearly so, tubercular-rugose.

Queensland.　Keppel Bay, *R. Brown;* Burdekin river, *F. Mueller;* Moreton Bay, *C. Stuart;* Warwick, *Beckler.*

N. S. Wales.　Port Jackson, *R. Brown* and others; New England, *C. Stuart;* Arne river, *Beckler,* also in *Leichhardt's* collection.

Victoria.　Portland, *Allitt;* Snowy River, *F. Mueller* (rather doubtful).

S. Australia.　Barossa range, *Behr.*

Var. (?) *filiformis.*　More-slender, with narrower leaf-lobes.—*H. filiformis,* A. Gray, Bot. Amer. Expl. Exped. i. 628.—Hunter's River, *American Exploring Expedition.* I have not seen these specimens, but none of our N. S. Wales ones agree well with the description. Generally speaking, this species is readily distinguished from *H. ceratophylla* by its slender habit, opposite stem-leaves only divided above the middle, and by the small flowers of *H. tetragyna;* but some specimens, mostly in an imperfect state, appear almost to connect the two.

30. **H. pinnatifida,** *A. Gray, Bot. Amer. Expl. Exped.* i. 627.　Tall and glabrous or minutely scabrous on the angles of the stem and margins of the leaves.　Stem-leaves opposite, deeply divided from nearly the base into linear lobes not broader than the rhachis, the larger leaves 1½ to 2 in. long, the lobes ¼ to ½ in. or even nearly 1 in. long; the floral leaves alternate, linear, the upper ones entire and small.　Flowers of *H. ceratophylla,* but rather larger, solitary within each bract or floral leaf, forming terminal leafy racemes.　Fruit only seen young.

N. S. Wales.　Hunter's River, *American Exploring Expedition;* near Cassilis, *C. Moore,* a single specimen in *Herb. F. Mueller,* agreeing well with the description given by A. Gray, whose specimens I have not seen.

31. **H. acanthocarpa,** *Brongn. in Duperr. Voy. t.* 70.　Stems decumbent or erect, 1 to 2 ft. long, scabrous-hirsute as well as the leaves.　Stem-leaves opposite, sessile or shortly petiolate, oblong or broadly lanceolate, mostly ¾ to 1½ in. long, regularly and acutely serrate as in *H. alata* and *H. serra,* rounded at the base.　Flowers very small, alternate along the filiform branches of a long loose terminal panicle, with small leaves at the base of the primary branches, the others reduced to small bracts.　Calyx-lobes short. Petals 4, glabrous, rather above ½ line long.　Stamens 8.　Styles and ovules 4.　Fruit nearly 1 line long, narrow-oblong, muricate with 2 or 3 transverse

rows of tubercles, crowned by the small smooth calyx-limb.—*H. leptotheca*, F. Muell. Fragm. iii. 32.

N. Australia. Victoria river, *F. Mueller*; Port Essington, *Armstrong*; islands of the Gulf of Carpentaria, *R. Brown*; Sims Island, *A. Cunningham*; Gould Island, *M'Gillivray*.

32. **H. tetragyna,** *Hook. f. Fl. Nov. Zel.* i. 63, *and Fl. Tasm.* i. 120. Rootstock apparently perennial, more or less scabrous with appressed hairs; stems branching, diffuse decumbent or ascending, sometimes all under 6 in., rarely above 1 ft. long. Stem-leaves all or mostly opposite, linear-lanceolate, elliptical or the lower ones ovate, rarely above ½ in. long, except in tall luxuriant forms, not cordate, and usually narrowed at the base, the floral ones all or almost all alternate and mostly reduced to small bracts shorter than the flowers. Flowers small, nearly sessile, solitary within each bract, in slender usually one-sided terminal racemes, often branching into narrow panicles. Calyx-tube not ½ line long. Petals in the males rather above 1 line long, smaller or none in the females. Stamens 8. Styles and ovules 4. Fruits nearly globular, 4-angled, transversely rugose, attaining about ¾ line diameter. —F. Muell. Fragm. iv. 26; *Goniocarpus tetragynus*, Labill. Pl. Nov. Holl. i. 39, t. 53; DC. Prod. iii. 66; *Haloragis gonocarpus*, Spreng. Syst. ii. 261; *Goniocarpus tenellus*, DC. Prod. iii. 66.

Queensland. Moreton Bay, *C. Stuart.*

N. S. Wales. Port Jackson to the Blue Mountains, *R. Brown* and others; New England, *C. Stuart*; Head of the Gwydir, *Leichhardt*; Hastings and Macleay rivers, *Beckler.*

Victoria. Australia Felix, *F. Mueller*; Creswick, *J. S. Whan*; probably over the whole colony.

Tasmania. Common in dry stony places, fields, etc., *J. D. Hooker.*

S. Australia. Barossa and Mount Lofty ranges, *F. Mueller.*

Var. *micrantha*. Leaves longer than in the southern specimens, and mostly lanceolate ; racemes more slender and more branching, flowers nearly as small as in *H. micrantha*. To this belong most of the northern specimens, and *Goniocarpus scaber*, Kœn. (*Haloragis scabra*, Benth. Fl. Hongk. 139), from Khasia, the Indian Archipelago, and China, appears not to be specifically distinct.

Var. *hispida*. More hirsute. Flowers small. Leaves rather broad, but all narrowed at the base.—Mount Mitchell, Clarence river, *Beckler*; head of the Gwydir, *Leichhardt*. The southern form of the species is also in New Zealand.

33 **H. teucrioides,** *A. Gray, Bot. Amer. Expl. Exped.* i. 625. A perennial, usually much coarser than *H. tetragyna*, scabrous-pubescent or hispid with decumbent or erect stems often 1 to 2 ft. long. Stem-leaves opposite, ovate or orbicular, deeply and acutely serrate, rounded or cordate at the base, the larger ones ¾ in., but mostly not above ½ in. long and broad. Flowers rather larger than in *H. tetragyna*, solitary under each bract, the lower ones often opposite, the upper ones alternate, forming much shorter racemes and a much more leafy panicle than *H. tetragyna*, all the bracts usually exceeding the flowers. Fruit of *H. tetragyna*, but the angles usually tuberculate and smoother between them.—*Goniocarpus teucrioides*, DC. Prod. iii. 66; *H. elata*, Hook. f. in Hook. Lond. Journ. vi. 475, not of A. Cunn.; *H. Gunnii*, Hook. f. Fl. Tasm. i. 120.

N. S. Wales. Port Jackson to the Blue Mountains, *Sieber, n.* 544, *Leichhardt*, and others.

Victoria. Australia Felix, without the precise station, *F. Mueller.*
Tasmania. Port Dalrymple, *R. Brown ;* abundant in wet shady places, *J. D. Hooker.*
S. Australia. S. coast, *R. Brown,* with smaller leaves.
W. Australia. Princess Royal Harbour, *R. Brown, Preiss,* n. 2087, the specimens in both cases imperfect and scarcely in flower, and therefore doubtful. Preiss's are referred by Nees in Pl. Preiss. i. 158, to *Goniocarpus tetragynus,* together with Preiss's specimens n. 2390, which are also bad, but do not appear to be of the same species. The whole species is united by F. Mueller with *H. tetragynus.*

34. **H. scordioides,** *Benth.* Rather a coarse species, probably tall ; branches loose, hirsute with spreading hairs. Leaves opposite, the lower ones petiolate, oblong, deeply and sharply serrate, mostly ¾ to above 1 in. long, glabrous or nearly so ; the floral ones much smaller and ovate, but all opposite in the specimens seen. Flowers nearly sessile and solitary in the axils of the floral leaves. Calyx-tube globular, slightly 4-angled ; lobes ovate-lanceolate. Petals 4, attaining nearly 1½ lines, oblong, ciliate on the keel. Stamens 8. Styles and ovules 4. Fruit not seen.

W. Australia. Thomas river, *Maxwell.*

35. **H. depressa,** *Walp. Rep.* ii. 99. A small species, diffuse or prostrate, very much branched, glabrous in appearance, but scabrous with minute asperities. Leaves all opposite, ovate, often cordate, usually broad, under ½ in. and often not ¼ in. long, the upper floral ones gradually smaller, but all opposite or very rarely the upper ones of side-branches alternate. Flowers almost sessile, forming short interrupted terminal racemes, and similar to those of the smaller forms of *H. tetragyna,* except that the calyx-tube and fruit are smooth and shining, with 4 or 8 prominent nerves, not tuberculate.—Hook. f. Fl. Tasm. i. 120.

Victoria. Mount Useful and Mount Cobberas, at an elevation of 4500 to 5000 ft., *F. Mueller.*
Tasmania. Port Dalrymple, Table mountain, and Cataract river, *R. Brown ;* abundant in alpine and subalpine situations, *J. D. Hooker.*
Also in New Zealand.
There are two forms of this species : 1. *serpyllifolia.* Leaves mostly under 3 lines long and rather narrow.—*Goniocarpus serpyllifolius* and *G. vernicosus,* Hook. f. in Hook. Ic. Pl. t. 290 and 311, *H. serpyllifolia* and *H. vernicosa,* Walp. Rep. ii. 90 ;—and 2. *montana.* Leaves broader, often cordate, 3 to 5 lines long.—*H. montana,* Hook. f. in Hook. Lond. Journ. vi. 475, united with *H. depressa* in Fl. Tasm. i. 120.

36. **H. salsoloides,** *Benth.* Erect, often much branched, ½ to 1 ft. high, nearly glabrous or scabrous with minute hairs, branches terete or nearly so, the short flowering summits almost always nodding. Leaves all opposite, narrow-linear, almost terete, rarely above ½ in. long, quite entire, or obscurely and minutely toothed. Flowers opposite and solitary in the upper axils, rather larger than in *H. tetragyna.* Calyx-lobes short. acute. Petals 4, about 1 line long. Stamens 8, usually persistent long after the fall of the petals, the filaments exceeding the calyx-lobes. Styles and ovules 4. Fruit not seen.—*Goniocarpus salsoloides,* Reichb. in Sieb. Pl. Exs., and in Steud. Nom. Bot. ed. 2.

N. S. Wales. Port Jackson to the Blue Mountains, *R. Brown, Sieber, n.* 249, and others.

3. MEIONECTES, R. Br.

Calyx-tube or ovary somewhat compressed, lobes 2. Petals 2, induplicate and boat-shaped. Stamens 4; filaments short. Ovary 2-celled with 1 pendulous ovule in each cell; styles 2. Fruit small, 2-celled.—Diffuse herb with alternate once or twice pinnatifid leaves, and the inflorescence of *Haloragis*.

The genus is limited to a single species endemic in Australia, only differing from *Haloragis* in the binary not ternary or quaternary numbers of the parts of the flower.

1. **M. Brownii**, *Hook. f. in Hook. Ic. Pl. t. 306, and Fl. Tasm. i. 123.* A glabrous slender herb, creeping and rooting at the base, the branches ascending sometimes to the height of 6 in., or when luxuriant to 1 ft. Leaves alternate, deeply pinnatifid with linear lobes not broader than the common rhachis, the larger ones often again lobed, the upper floral leaves smaller and more entire. Flowers in clusters of 2 or more in the upper axils, the upper ones forming a more or less leafy raceme. Calyx-lobes short. Petals glabrous, about 1 line long. Fruit ovate, slightly compressed, 1½ to 2 lines long, including the erect connivent persistent calyx-lobes, more or less rugose and usually with the 2 ribs prominent on the edge.—*M. Preissii*, Nees in Pl. Preiss. i. 224.

Victoria. Wet pastures and swamps in various parts of the colony, *F. Mueller* and others.

Tasmania. Pools of fresh water at Circular Head and other places in the northern parts of the island, *J. D. Hooker.*

S. Australia. Valleys of Mount Lofty ranges where inundated in time of rain, *F. Mueller.*

W. Australia. Swamps, King George's Sound and to the eastward, *R. Brown;* Swan River, *Preiss, n.* 2385; Flinders Bay, *Collie.* These western specimens are generally coarser and rather larger in all their parts than most of the Tasmanian and Victorian ones, but some of F. Mueller's are almost if not quite as large as Preiss's, and some of Brown's are quite as slender as those from Tasmania.

4. MYRIOPHYLLUM, Linn.

Flowers mostly unisexual. Male fl. : Calyx-tube very short or scarcely any, lobes short, petal-like or scarcely any. Petals 4, concave, imbricate or half induplicate. Stamens 4, 6 or 8. Styles minute and rudimentary, without any ovules. Female fl. : Calyx-tube ovoid, lobes minute or none. Petals usually none. Ovary 2- or 4-celled, with one pendulous ovule in each cell; styles as many as ovules, usually short and stigmatic from the base, often plumose. Fruit small, usually furrowed between the 2 or 4 carpels, which at length separate into as many small 1-seeded nuts. Aquatic herbs, the lower leaves when submerged often pinnately divided into capillary lobes; those of the flowering extremities usually less divided or entire. Flowers very small, in the axils of the exserted flowering leaves or rarely also or entirely in the submerged axils, the upper ones usually males, the lower ones females, sometimes dioecious, but perhaps not constantly so in any species.

The genus is found in fresh waters nearly in every part of the globe. Of the 12 Australian species, 3 are also in New Zealand, and one of these extends to extratropical S. America, the remaining 9 are endemic.

Leaves all in whorls of 3 to 8, the submerged ones pinnatisect with
capillary segments, the emerged floral ones entire toothed or
shortly lobed.

Leaves usually more than 4 in the whorl, the emerged ones narrow-linear. Calyx-lobes conspicuous . . . . 1. *M. variæfolium.*

Leaves usually 4, the emerged ones oblong or broadly lanceolate,
sessile. Calyx-lobes minute.

Emerged leaves entire or slightly toothed. Plant rather large 2. *M. elatinoides.*

Emerged leaves pinnatifid. Plant small or slender . . . . 3. *M. verrucosum.*

Leaves usually 3, the emerged ones linear-lanceolate, above ½ in.
long, serrulate, narrowed into a petiole . . . . 4. *M. latifolium.*

Leaves all opposite or rarely in whorls of 3, pinnatisect with capillary lobes. Male flowers 1 or 2 together on a distinct peduncle,
each enclosed in a hood-shaped bract . . . . 5. *M. Muelleri.*

Leaves all opposite, entire. Small plants creeping in mud.

Leaves oblong. Carpels smooth . . . . 6. *M. amphibium.*

Leaves linear. Carpels tuberculate or muricate . . . . . . 7. *M. pedunculatum.*

Leaves all alternate.

Submerged leaves pinnatisect with capillary segments. Stamens 8.

Emerged floral leaves entire, linear.

Carpels 2, smooth . . . . 8. *M. dicoccum.*

Carpels 4, tuberculate . . . . 9. *M. trachycarpum.*

Leaves all pinnatisect with fine segments. Carpels 4 . . . 10. *M. gracile.*

Leaves all linear and entire, or rarely with a few lobes in *M.
Drummondii.* Minute filiform plants.

Stamens 8, with oblong anthers longer than the filaments . . 11. *M. filiforme.*

Stamens 4 or fewer, with ovate anthers much shorter than the
filaments.

Leaves all entire. Carpels smooth . . . . 12. *M. integrifolium.*

Leaves occasionally with a few lobes. Carpels muricate . 13. *M. Drummondii.*

1. **M. variæfolium,** *Hook. f. in Hook. Ic. Pl. t.* 289, *and Fl. Tasm.*
i. 122. Usually a rather large species, the erect flowering summits assuming
almost the aspect of *Hippuris.* Leaves in whorls of from 4 to 8, usually 5
or 6, the lower submerged ones divided into capillary lobes, the emerged floral
ones narrow-linear, all entire or the lower ones toothed, ¼ to above ½ in. long,
Male fl. : Calyx-lobes conspicuous and sometimes above ½ line long. Petals
1¼ to 1½ lines. Stamens 8. Female fl. small without apparent calyx-teeth
or petals. Carpels 4, small, tuberculate or almost echinate, or rarely quite
smooth.

N. S. Wales. Port Jackson, *R. Brown, Woolls;* Lachlan river, *A. Cunningham;*
Glendon and Newcastle, *Leichhardt;* Hastings river, *Beckler.*
Victoria. Swamps, Portland, *F. Mueller;* Glenelg and Wendu rivers, *Robertson.*
Tasmania. Abundant in fresh waters throughout the colony, *J. D. Hooker.*
S. Australia. Torrens river, *F. Mueller.*
W. Australia. Swan River to King George's Sound, *Drummond, 1st Coll.; 4th Coll.
n.* 75, *Harvey, Oldfield.*
Also in New Zealand.

2. **M. elatinoides,** *Gaud.; DC. Prod.* iii. 68. A rather large species.
Leaves in whorls of 4 or rarely 5 or 6, the lower submerged ones divided
into capillary lobes, the emerged floral ones broadly lanceolate, obtuse entire
or scarcely toothed, quite sessile, ¼ to nearly ½ in. long. Male fl. : Calyx-
teeth scarcely perceptible. Petals about 1 line long. Stamens 8. Female

fl. very small, without apparent calyx-teeth or petals. Carpels 4, small, smooth or slightly tuberculate.—Hook. f. Fl. Tasm. i. 121.

Victoria. Australia Felix, *F. Mueller ;* Wendu river, *Robertson.*
Tasmania. Fresh and brackish waters, Georgetown and in the Derwent, *J. D. Hooker.*
S. Australia. Murray river, Lake Victoria, pools on Mount Barker, *F. Mueller.*
Also in New Zealand and in extratropical S. America.

3. **M. verrucosum,** *Lindl. in Mitch. Trop. Austr.* 384. Usually much smaller and more slender than the last two species. Leaves mostly in whorls of 4, the lower submerged ones divided into capillary lobes, the emerged floral ones sessile, oblong or lanceolate, all pinnatifid with short obtuse lobes, more or less glaucous, mostly about 2 lines long. Flowers rather smaller than in *M. elatinoides.* Calyx-lobes very small, but perceptible in both sexes, very deciduous in the females. Petals in the males under 1 line long. Stamens 8. Females without petals. Styles 4, very short. Carpels 4, rarely above ½ line long, obtuse on the back, more or less tuberculate.

N. Australia. Victoria river, *Bynoe ;* Albert river, Gulf of Carpentaria, *F. Mueller.*
Queensland. Balonne river at St. George's Bridge, *Mitchell ;* Moreton Bay, *C. Stuart.*
N. S. Wales. Port Jackson to the Blue Mountains, *R. Brown* and others ; water-holes in the Severn, *Leichhardt ;* in the interior towards the Barrier Range, *Victorian Expedition.*
Victoria. Waters on the Grampians, muddy places by the Barwan, etc., *F. Mueller.*
S. Australia. S. coast, *R. Brown.*
W. Australia, *Drummond,* 3rd *Coll. ?, n.* 186, 4th *Coll. n.* 80; pools, Murchison river, *Oldfield.*
The species has some affinity to the northern *M. verticillatum* as well as to the Asiatic *M. indicum,* but besides the differences in the floral leaves, the fruit is much smaller than in the former, much less furrowed between the carpels than in the latter. As in the allied species, flowers are occasionally found also in the axils of the submerged leaves.

4. **M. latifolium,** *F. Muell. Fragm.* ii. 87. A large species. Leaves in whorls of 3, the lower submerged ones divided into capillary lobes, but in the specimens seen always few, those of the tall erect emerged summits lanceolate, serrulate, ¾ to 1 in. long, narrowed into a petiole. Flowers rather large, all sessile. Male fl. : Calyx-lobes small. Petals 1½ lines long. Stamens 8. Female fl. : Calyx-teeth inconspicuous. Carpels 4, styles short, very plumose. Fruit not seen.

N. S. Wales. Port Jackson, *R. Brown ;* Clarence river, *Beckler.*

5. **M. Muelleri,** *Sond. Herb.* The plant appears to be entirely submerged. Leaves opposite or rarely in whorls of three, all pinnately divided into capillary lobes, those of the upper floral leaves however not quite so fine. Flowers monœcious in the upper axils, the males rather large, solitary or 2 together, on a peduncle of about 2 or 3 lines, each one enclosed before expanding in an almost petal-like hood-shaped bract, the bracteoles remaining small. Calyx-lobes very small. Petals above 1 line long. Stamens 8. Rudimentary styles conspicuous. Female fl. sessile. Calyx-teeth almost imperceptible. Petals none. Styles 4, erect, usually as long as the ovary, papillose from the base, but not plumose. Fruit-carpels about ½ line long, smooth.

Victoria. Watery marshes, Melbourne, *Adamson.*
S. Australia. Near Holdfast Bay, *F. Mueller.*
W. Australia. King George's Sound or to the eastward, *Baxter.*

The glands observable in the axils of the submerged leaves of many *Myriophyllums* are particularly conspicuous in this species.

6. **M. amphibium,** *Labill. Pl. Nov. Holl.* ii. 70. *t.* 220. A small plant, creeping in the mud, without capillary-lobed leaves. Leaves opposite, oblong or sometimes almost obovate, entire, mostly 3 to 4 lines long. Flowers solitary in the upper axils. Males: Calyx-lobes ⅓ to ½ line long. Petals about 1 line long, narrower than in the other species. Stamens 8. Female flowers very small, the minute calyx-teeth scarcely perceptible. Styles often rather long, plumose at the extremity. Fruit about ⅓ line long; carpels ovoid, smooth or scarcely punctate.—Hook. f. Fl. Tasm. i. 122:

Tasmania. Wet places about Recherche Bay, . *D. Hooker.*

7. **M. pedunculatum,** *Hook. f. in Hook. Lond: Journ.* vi: 474, *and Fl. Tasm.* i. 122, *t.* 23 B. Very closely allied to *M. amphibium;* and perhaps a variety, differing in the leaves from narrow-linear to linear-oblong, rarely 3 lines long, the flowers rather smaller, the males sometimes, but not always, shortly pedunculate, the styles in the females very short, the carpels rather smaller, covered with prominent tubercles.

Victoria. Boggy pastures, Australian Alps; *F. Mueller.*
Tasmania. King s Island, *R. Brown ;* shallow parts and inundated water-banks, ascending to 4000 ft., *J. D. Hooker.*
W. Australia, *Drummond, n.* 204.
Also in New Zealand.

8. **M. dicoccum,** *F. Muell. in Trans. Phil. Inst. Vict.* iii. 41. A rather slender species. Leaves all alternate, the submerged ones pinnatisect with capillary segments, the emerged floral ones linear, entire, ¼ to ½ in. long, narrowed at the base. Upper flowers male or perhaps hermaphrodite, but not seen perfect. Calyx-teeth inconspicuous. Petals about ¾ line long. (Stamens 4 ?) Female flowers very small, with a 2-celled ovary and 2 short plumose styles. Carpels 2, rather above ½ line long, very smooth:

N. Australia: Robinson river, *F. Mueller.*

9. **M. trachycarpum,** *F. Muell. Fragm.* ii. 87. A slender species with the habit of *M. dicoccum:* Leaves all alternate, the submerged ones pinnatisect with capillary segments, the emerged floral ones linear, entire or scarcely toothed, 2 to 3 lines long, narrowed at the base. Flowers small, solitary, the upper ones male, but not seen very perfect. Calyx-teeth very small. Petals under 1 line long. Stamens apparently 8. Female flowers very small. Calyx-teeth quite inconspicuous. Carpels 4 or rarely 3, about ½ line long, conspicuously verrucose:

N. Australia. Gulf of Carpentaria, opposite Groote Island, *R. Brown ;* ponds near Macadam Range, *F. Mueller* (*Herb. R. Br. and F. Muell.*).

10. **M. gracile,** *Benth.* A small slender almost filiform species, nearly allied to *M. trachycarpum,* and perhaps a variety. Leaves alternate, all, even the uppermost floral ones, deeply pinnatifid or pinnatisect, with few (3, 5 or 7), narrow, rather short lobes. Flowers and fruit of *M. trachycarpum.* Stamens in the males 8. Carpels 4, small, tuberculate.

Queensland, *Bowman ;* Moreton Bay, *F. Mueller.*

11. **M. filiforme,** *Benth.* Stems filiform, simple or scarcely branched,

about 2 in. or rarely 3 in. long. Leaves all alternate, linear-subulate, entire, 1 to 2 lines long. Flowers minute, mostly 2 or 3 together, the upper ones male; Petals about ½ line long. Stamens 8, with oblong anthers and short filaments as in all the preceding species. Females very minute. Styles scarcely any. Carpels 4, about ¼ line long, muricate.

N. Australia. Gulf of Carpentaria, mainland opposite Groote Island, *R. Brown* (*Herb. R. Br.*). With some resemblance to *M. Drummondii*, this species is much more slender, and has the male flowers quite different.

12. **M. integrifolium,** *Hook. f. Fl. Tasm.* i. 123, *t.* 23 A. A minute, slender, simple or branched plant, rarely exceeding 1 in. in height. Leaves all alternate or here and there irregularly opposite, linear-subulate, entire, 1 to 2 lines long. Flowers minute, sessile in the upper axils, the uppermost 2 usually males. Calyx-teeth inconspicuous. Petals 4, scarcely ¼ line long. Stamens 4, often persistent after the petals have fallen, and differing from all others of the genus, except *M. Drummondii*, in their filiform filaments, much longer than the small ovoid anthers. Females without calyx-teeth or petals. Styles scarcely prominent, with papillose stigmas. Carpels 4, when ripe ¼ to nearly ½ line long, ovoid, separated by deep furrows, smooth or minutely papillose.—*Pelonastes integrifolia,* Hook. f. in Hook. Lond. Journ. vi. 475.

Victoria. Wet places on the Murray, Darebin Creek, Emu flats, etc., *F. Mueller.*
Tasmania. Wet places, lagoons, etc., frequent, *J. D. Hooker.*
S. Australia. Murray river, Plenty Creek, etc., *F. Mueller.*
W. Australia, *Drummond, n.* 686.

13. **M. Drummondii,** *Benth.* A little plant of about 1 in. in height, closely resembling *M. integrifolium.* Leaves alternate, narrow-linear, entire or here and there with a few linear lobes. Flowers entirely of *M. integrifolium,* the uppermost 2 males, with 4 stamens, long filaments, and small anthers. Carpels 4, of the same size as in that species, but prominently tuberculate or muricate.—*Pelonastes tuberculata,* Hook. f. in Hook. Lond. Journ. vi. 474.

W. Australia, *Drummond, n.* 18 ; Geographe Bay, *Oldfield.* Perhaps a variety only of *M. integrifolium.*

B. Anomalous Genera.

5. GUNNERA, Linn.

(Milligania, *Hook. f.*)

Flowers mostly unisexual. Male fl. : Calyx of 2 or 3 minute teeth, often scarcely perceptible, without any tube. Petals none in the Australian species. Stamens 2, with very short filaments. Female fl. : Calyx-tube terete or angular; lobes 2 or 3, very small. Petals none. Ovary 1-celled, with 1 pendulous ovule; styles 2, linear, stigmatic from the base or rarely 4 styles, more or less connate in pairs. Fruit a small nut-like drupe, the seed often adhering to the pericarp. Embryo very minute.—Stemless herbs, with a tufted creeping rhizome. Leaves radical, usually broad, in some S. American species attaining an immense size. Flowers small, in clusters or in exotic species in spikes or racemes, crowded along the rhachis of a radical scape.

The genus is spread over the cooler regions of the southern hemisphere, extending northward along the Andes to the Gulf of Mexico. The only Australian species is endemic, but is nearly allied to some others from New Zealand and Antarctic America.

1. **G. cordifolia,** *Hook. f. Fl. Tasm.* i. 125. A small succulent fleshy herb, with a tufted rootstock, emitting creeping stolons, loosely hairy on the scapes and the edges and ribs of the leaves. Leaves radical, broadly ovate or orbicular, ½ to 1 in. diameter, crenate and sometimes obscurely lobed, on petioles nearly as long as the lamina. Scapes usually unisexual, the males attaining 2 or 3 in., bearing in their upper half clusters of flowers, which appear to consist each of 2 thick ovoid anthers, about 1 line long, surrounded by 4 or 5 linear or spathulate, slightly jagged bracts, the calyx-teeth so minute as to be difficult to find. Occasionally there are a few female flowers below the rhachis, but usually the females are upon separate scapes, nearly sessile, in a short dense almost globular head. Calyx-tube narrow, about ½ line long, the lobes short. No petals or stamens. Styles almost filiform. Drupes ovoid, slightly compressed, about 1 line long, with 4 ribs of which 2 more prominent. —*Milligania cordifolia,* Hook. f. in Hook. Ic. Pl. t. 299.

Tasmania. Abundant about springs and in marshes in alpine situations, *J. D. Hooker.*

6. CERATOPHYLLUM, Linn.

Flowers unisexual, without any perianth, the males consisting of several (12 to 20) almost sessile anthers, the females of a 1-celled ovary, with 1 pendulous ovule and a simple filiform style. Fruit an oval nut, tipped by the persistent style and often bearing 2 or 4 reflexed prickles or surrounded by a toothed or crest-like wing.—Aquatic floating herbs. Leaves whorled, divided into linear dichotomous segments. Flowers small, axillary, each one surrounded by a whorl of minute bracts.

The genus is found in fresh waters in most parts of the world, and like *Callitriche,* is considered by some as containing but one species, by others divided into several, characterized by the excrescences on the ripe fruit. The Australian form is one frequent both in Europe and E. India.

1. **C. demersum,** *Linn. ; DC. Prod.* iii. 73. A glabrous perennial, floating like the submerged species of *Myriophyllum,* and the leaves whorled in the same manner, but dichotomously, not pinnately, divided into linear segments either fine and subulate or rather broader and denticulate. Flowers small, sessile in the axils. Anthers of the males oblong, mucronate. Fruit in the Australian specimens ovoid, slightly compressed, 2 to 3 lines long, more or less covered with minute tubercles, the margin not winged, but bearing below the middle 2 to 4 reflexed prickles, very variable in length.—*C. submersum,* Linn. ; DC. l. c.

Queensland. Suttor and Burdekin rivers, *F. Mueller.*
S. Australia. Murray river, *F. Mueller.*
The very few fruits I have seen correspond to a form not uncommon in the northern hemisphere, and figured in Wight, Ic. t. 1948, f. 3, as *C. tuberculatum,* Cham.

7. CALLITRICHE, Linn.

Flowers unisexual, without any perianth, the males consisting of a single stamen, with a conspicuous filament and small 4-celled anther, the females of a sessile or stalked 4-celled ovary, with 2 filiform erect or recurved styles, stigmatic from the base ; ovules 1 in each cell, laterally suspended from near the summit of the cell. Fruit small, more or less flattened; notched at the top, 4-celled and 4-lobed, that is surrounded by a double edge, the edges obtuse acute or winged, and consisting of 2 2-celled disk-shaped carpels, united by their inner faces. Embryo in the axis of an oily albumen.—Slender aquatic herbs. Leaves opposite; entire. Flowers axillary, solitary or a male and female from the same axis, each one between 2 small bracteoles, which are sometimes wanting.

The genus is found in almost every part of the globe, and, according to some botanists, consists but of a single species, others divide it into two, the Australian one belonging in that case to the one which has the most universal geographical range. Those who variously extend the genus to from 13 to 20 species, describe the commonest Australian form as endemic.

1. **C. verna,** *Linn.; DC. Prod.* iii. 70. A glabrous slender perennial, either floating in water or creeping and rooting in mud, flowering young so as to appear annual, varying in length according to the depth of the water. Leaves either all obovate-orbicular or oblong, 1 to 6 lines long or the lower submerged ones narrow-linear and obtuse or notched at the end, the upper ones obovate and spreading in little tufts on the surface of the water, or all submerged and linear: Flowers minute. Fruit from ½ to 1 line diameter, with obtuse acute or winged edges.—*C. autumnalis,* Linn.; DC. l. c., partly at least.

Queensland. Brisbane river, *F. Mueller.*
N. S. Wales. Port Jackson to the Blue Mountains, *R. Brown,* and others.
Victoria. Abundant in pools, *Adamson, F. Mueller,* and others.
Tasmania. Port Dalrymple; *R. Brown;* common in still freshwater, margins of rivers, etc., *J. D. Hooker.*
S. Australia. Plenty river, towards Mount Disappointment, *F. Mueller*
W. Australia, *Drummond, n.* 185, 2nd *Coll. n.* 99 (*or* 66).
The Eastern specimens have generally the wings of the fruit broader than in almost any Northern or tropical specimens, and constitute the *C. macropteryx,* Hegelm. Monogr. Callitr. 59 ; the Western and some of the Tasmanian specimens cannot be distinguished from some of our European forms ; in Brown's Tasmanian ones the fruit lobes are quite obtuse on the edge.

Order XLVI. RHIZOPHOREÆ.

Calyx-tube usually adnate to the ovary, sometimes prolonged above it or rarely quite free ; the limb of 4 to about 12 lobes, valvate in the bud. Petals as many as the calyx-lobes, alternate with them, notched cut or jagged or rarely entire, the margins usually induplicate and embracing the anthers. Stamens as many or twice as many as petals or more, inserted with them at the base of the free part or lobes of the calyx ; anthers erect or versatile, 2-celled, opening longitudinally. Ovary more or less inferior or rarely quite superior, 2- or more-celled, with 2 or few pendulous ovules in each cell, or rarely 1-celled by the obliteration of the partition ; style undivided; with an entire or lobed stigma. Fruit inferior or enclosed in the calyx. Seeds solitary or few, with or without albumen—Trees or shrubs. Leaves opposite, simple, entire or slightly toothed, coriaceous. Stipules often large, very deciduous. Flowers axillary, solitary, clustered or in cymes.

A small Order, almost entirely tropical, and chiefly Asiatic or African, with a few American species. The four Australian genera are all Asiatic, one only extending also to Africa and America. The Order is divided into two distinct tribes, by some considered as independent families :—1. *Rhizophoreæ proper*, including the following genera : *Rhizophora*, *Ceriops*, and *Bruguiera*, consists of the *Mangroves*, all maritime evergreen trees, the seeds without albumen, and almost always germinating before falling off, the thick radicle enlarging rapidly, and protruding to a great length from the summit of the capsule. 2. *Legnotideæ*, trees or shrubs, not strictly maritime, with usually smaller flowers, and the seeds albuminous, not germinating before they fall. To this tribe belongs the genus *Carallia*.

Calyx-segments longer than the tube. Seeds without albumen, germinating before falling.

    Calyx-segments and petals 4. Stamens 8 to 12. Fruit more than half superior . . . . 1. RHIZOPHORA.

    Calyx-segments and petals 5 or 6. Stamens twice as many. Fruit more than half superior . . . . 2. CERIOPS.

    Calyx-segments and petals 8 to 15. Stamens twice as many. Fruit inferior . . . . 8. BRUGUIERA.

Calyx campanulate, with short teeth. Petals 5 to 8. Stamens twice as many. Fruit inferior. Seeds albuminous, not germinating before falling . . . . 4. CARALLIA.

1. RHIZOPHORA, Linn.

Calyx-tube adnate, segments 4. Petals 4, entire. Stamens 8 to 12 ; filaments short ; anthers long, acuminate, connivent. Ovary half-inferior, 2-celled, with 2 pendulous ovules in each cell ; style filiform ; stigma 2-toothed. Fruit ovoid or conical, the persistent calyx-segments reflexed from near the base. Seed solitary, without albumen, the rapidly enlarged radicle penetrating through the summit of the fruit.—Trees. Leaves entire. Cymes axillary.

The genus consists of three species only, ranging over tropical seacoasts, two, including the Australian one, in the Old World, the third in America.

1. **R. mucronata,** *Lam. ; DC. Prod.* iii. 32. A glabrous evergreen tree, with thick branches. Leaves from broadly ovate to oblong-elliptical, obtuse, with a projecting point (often worn off from the old leaves), 3 to 4 in. long in the Australian specimens, but sometimes longer and narrower, coriaceous, entire. Stipules rather large, oblong, obtuse, very deciduous. Flowers in axillary dichotomous cymes shorter than the leaves, with a pair of short thick concave bracts, connate at the base under each fork and under each flower. Calyx sessile within the bracts, about ½ in. long, the segments separating down to the adnate part. Petals shorter than the calyx, the induplicate margins fringed with long hairs. Anthers 8, nearly sessile, 4 embraced by the petals, 4 between them. Style rather thick, nearly as long as the petals. Fruit ovoid, 1 to 1½ in. long.—Arn. in Ann. Nat. Hist. i. 362 ; Wight, Ic. t. 238.

N. Australia. Port Essington, *Leichhardt ;* shores and islands of the Gulf of Carpentaria, *F. Mueller.*

Queensland. Along the coast and islands within the tropics, *R. Brown, F. Mueller,* and others.

The species extends over the tropical shores of Africa and Asia.

2. CERIOPS, Arn.

Calyx-tube adnate; segments 5, rarely 6. Petals as many, emarginate and usually with 1 or more clavate setæ at the top. Stamens twice as many as petals; filaments filiform, longer than the oblong or linear anthers. Ovary half-inferior, 3-celled, with 2 pendulous ovules in each cell; style filiform; stigma undivided. Fruit ovoid or conical, the persistent calyx-segments surrounding it below the middle. Seed solitary, without albumen, the rapidly enlarged funicle penetrating through the summit of the fruit.—Trees or shrubs, with the habit and inflorescence of *Rhizophora*, but usually with smaller leaves and smaller more numerous flowers.

A small genus, limited to the tropical seacoasts of the Old World, the Australian species being the commonest one in Asia. Although the genus appears to be universally adopted, it can scarcely be considered as more than a section of *Rhizophora*.

1. **C. Candolleana,** *Arn. in Ann. Nat. Hist.* i. 364. A tall evergreen glabrous shrub or small tree. Leaves obovate or broadly oblong, 1½ to 2 in. or in luxuriant specimens 3 in. long, coriaceous, entire. Flowers in small dense almost capitate cymes, on short recurved axillary peduncles, with a pair of small short thick concave bracts under each ramification, and rather larger ones under each flower. Calyx sessile within the bracts, about 3 lines long, divided down to the adnate part. Petals shorter than the calyx, emarginate, with 2 to 5 clavate setæ irregularly placed in the notch or on the lobes. Stamens alternately inserted opposite the petals and between them, but the anthers embraced in pairs by the induplicate margins of the petals. Fruit conical, about ½ in. long.—Wight, Ic. t. 240; *Rhizophora Timoriensis*, DC. Prod. iii. 32.

N. Australia. Careening Bay, N.W. coast, *A. Cunningham;* mangrove beach of the Victoria river, *F. Mueller;* Arnhem N. Bay, *R. Brown;* Port Essington, *Armstrong.*
Queensland. Islands of the N.E. coast, *F. Mueller,* and others.
The species extends over the seacoasts of E. India and the Archipelago.

3. BRUGUIERA, Lam.

Calyx-tube turbinate or campanulate, adnate at the base to the ovary, the upper portion free, lined by the disk; lobes 10 to 15 or rarely 8 or 9, narrow and thick. Petals as many, 2-lobed, with or without setæ at the top. Stamens twice as many as petals; anthers linear, but usually shorter than the filaments. Ovary inferior, 2- to 4-celled, with 2 pendulous ovules in each cell; style filiform, with 2 to 4 minute stigmatic lobes. Fruit turbinate, crowned by the persistent calyx-lobes. Seeds solitary, without albumen, the rapidly enlarged radicle penetrating through the summit of the fruit.—Trees, with the habit of *Rhizophora*. Flowers solitary or few together, on short axillary recurved peduncles.

The genus is widely spread along the tropical seacoasts of the Old World. The two Australian species are both common Asiatic ones.

Flowers above 1 in. long. Petals with several setæ at the end . . . 1. *B. Rheedii.*
Flowers under 1 in. long. Petals without setæ at the end . . . . 2. *B. gymnorrhiza.*

1. **B. Rheedii,** *Blume, Enum. Pl. Jav.* 92. A glabrous evergreen tree. Leaves ovate or oblong-elliptical, very shortly acuminate, 3 to 5 in. long, narrowed into a rather long petiole, coriaceous. Stipules oblong, 1 to 2 in. long, very deciduous. Flowers solitary, on short recurved axillary peduncles, without bracts. Calyx very thick and rigid, from a little more than 1 in. to 1½ in. long, the narrow turbinate tube about one-third the whole length, the

angles scarcely prominent or sometimes quite obscure, the lobes usually about 12, but variable in number. Petals shorter than the calyx, densely hairy at the base, and the induplicate margins more or less hairy to the end; setæ usually 1 in the notch and 3 or 4 at the end of each lobe. Anthers embraced in pairs by the induplicate margins of the petals. Ovary very short and wholly inferior. Fruit at first crowned by the calyx-limb, which often falls off as the radicle protrudes, the latter assuming a narrow spindle-shaped form, obscurely notched, with about 6 prominent angles.—Arn. in Ann. Nat. Hist. i. 367 ; *B. australis,* A. Cunn. in Arn. l. c.; *B. Rheedii* and *B. Rumphii,* Blume, Mus. Bot. i. 138.

N. Australia. Port Essington, *A. Cunningham* (rather doubtful); islands of the Gulf of Carpentaria,·*Henne.*

Queensland. Shoal Bay passage, *R. Brown ;* along the coast, from Moreton Bay to Torres Straits, *A. Cunningham, F. Mueller,* and others,

Wight's figure of *B. Rheedii,* Ic. t. 239 A, as well as his specimens, differ in some slight respects, and are considered¹ by Blume as constituting a distinct species. The setæ of the petals appear to be pretty constant in the Australian specimens, but it remains to be proved how far their presence and number are really good specific characters.

2. **B. gymnorrhiza,** *Lam. ; Blume, Mus. Bot.* i. 136. An evergreen tree, closely resembling *B. Rheedii.* Leaves usually smaller. Flowers as in·that species, solitary on short recurved axillary peduncles, but smaller, varying from ¾ to nearly 1 in. in length. Calyx-tube marked with very prominent acute angles ; lobes usually 8 to 10. Petals shorter than the calyx, hairy on the margins, the lobes obtuse, without setæ, but a very short seta, often, although not always, in the notch between them.

N Australia. N.W. coast, *A. Cunningham ;* N. coast, *R. Brown ;* Roper river, *F. Mueller ;* Port Essington and Limmen Bight river, *Leichhardt.*

Queensland. Broad Sound, *R. Brown.*

4. **CARALLIA,** Roxb.

Calyx-tube adnate at the base, campanulate above the ovary, lined by the thin disk, with 5 to 8 very short lobes or teeth. Petals as many as calyx-lobes, clawed, orbicular, jagged or slightly toothed. Stamens twice as many as petals, inserted with them at the base of the calyx-lobes round the undulated margin of the disk. Ovary inferior or adnate as high as the insertion of the ovules, 4-celled or rarely 3- or 5-celled, with 2 pendulous ovules in each cell. Fruit succulent, globular. Seed solitary, with a copious albumen ; embryo curved, not growing before the seed falls.—Trees or shrubs. Flowers small, in axillary, pedunculate, usually trichotomous cymes.

A small genus, extending over tropical Asia, the Australian species the commonest over the whole range of the genus.

1. **C. integerrima,** *DC. Prod.* iii. 33. Usually a tree, glabrous in all its parts. Leaves sessile, obovate, elliptical or oblong, in the Australian specimens obtuse or obtusely acuminate, thinly coriaceous, 3 to 5 in. long, in Asiatic ones very variable in breadth and consistence, and often very obtuse or much acuminate. Cymes axillary or from old leafless nodes, on short peduncles, each short branch bearing 3 to 5 sessile flowers. Calyx shortly and broadly campanulate,· not 3 lines diameter. Fruit globular, about 3 lines

diameter, crowned by the short connivent teeth of the calyx.—Benth. in Journ: Linn. Soc. iii. 74, with the synonyms there adduced; *C. zeylanica*, Arn. ; Wight, Illustr. t. 90.

N. Australia. Brunswick Bay and York Sound, N.W. coast, *A. Cunningham;* N. coast, *R. Brown;* Upper Roper river, M'Adam Range and Nicholson river, *F. Mueller.* **Queensland.** Endeavour river, *R. Brown.*
Widely spread over E. India and the Archipelago, extending to S. China.

Order XLVII. COMBRETACEÆ.

Calyx-tube adnate to the ovary at the base, narrowed above it and some-times elongated; limb usually campanulate, with 4 or 5, rarely more, teeth lobes or segments, valvate or very rarely induplicate or imbricate. Petals none or as many as calyx-lobes, usually small, imbricate or valvate. Stamens as many or twice as many as calyx-lobes, rarely indefinite, inserted on the calyx; anthers opening in longitudinal slits or (in *Gyrocarpeæ*) in 2 valves. Ovary inferior, 1-celled, with 2 or more pendulous ovules, or (in *Gyrocarpeæ*) with 1 only; style filiform or scarcely any, with an entire terminal stigma. Fruit coriaceous, chartaceous or drupaceous, indehiscent (except in a few species not Australian). Seed solitary, pendulous, without albumen; coty-ledons convolute or folded, very rarely flat inside and furrowed outside; ra-dicle short, superior.—Trees shrubs or woody climbers. Leaves alternate or opposite, entire, without stipules. Flowers in axillary or terminal racemes spikes or heads, or (in *Gyrocarpeæ*) in cymes. Bracts usually small; brac-teoles sometimes larger, often wanting.

The Order is distributed over the tropical regions of the New and the Old World, a very few species extending beyond the tropics in S. Africa or in N. India. Of the four Austra-lian genera, three are common to America, Africa, and Asia, one of them restricted to sea-coasts, the fourth is endemic.

Anthers opening in slits. Ovules 2 or more. Flowers in ra-
cemes spikes or heads (**Combretaceæ** proper).
Calyx-tube not produced above the ovary. Petals none. Sta-
mens 10 . . . . 1. TERMINALIA.
Calyx-tube produced above the ovary. Petals 5. Stamens 10
or fewer.
Bracteoles small. Ovules 2 to 5. Maritime shrubs . . . 2. LUMNITZERA.
Bracteoles enlarged and forming wings to the fruiting-calyx.
Ovules 10 to 12. Silky or tomentose shrubs . . . . 3. MACROPTERANTHES.
Anthers opening in 2 valves. Ovules solitary. Flowers small, in
cymes. Petals none (**Gyrocarpeæ**) . . . . 4. GYROCARPUS.

1. TERMINALIA, Linn.
(Chuncoa, *Ruiz and Pav.*)

Calyx-tube not produced above the ovary; limb campanulate or urceolate, 5-cleft. Petals none. Stamens 10, longer than the calyx. Style filiform. Ovules 2, rarely 3. Fruit ovoid, terete, angular, compressed or with 2 or (in species not Australian) 3 to 5 longitudinal wings. Cotyledons convolute.— Trees or erect shrubs. Leaves alternate or rarely opposite, usually marked with minute pellucid dots, often only visible under a strong lens. Flowers hermaphrodite or polygamous, small, green, white or rarely coloured, sessile in loose spikes, rarely contracted into dense heads, either axillary oʼr clustered on the old nodes. Calyx-tube usually small and narrow, the limb much broader.

The genus extends over nearly the whole range of the Order, but is most abundant in Africa and Asia. The Australian species appear to be all endemic, with the exception of *T. microcarpa*, which is also in Timor. Several of them however are as yet insufficiently known. They are often not to be distinguished without the fruit, which, when succulent and not winged, is rarely perfect in herbarium specimens, and we do not as yet know how far the fruit may vary in the same species. Some with broadly winged fruits have precisely the foliage and flowers of others which have wingless fruits. The circumscription of species here given may therefore require much revision when more perfect materials are obtained.

The subdivision of the genus into sections, or with some botanists into distinct genera, has been founded on the fruit alone, and although the line of demarcation is often very indefinite, no better character has as yet been found. The Australian species are included in *Chuncoa*, with 2 or 3 distinct wings to the fruit, *Catappa* with 2 wings, confluent above and below so as completely to encircle the drupe, and *Myrobalanus* without wings; but in *T. volucris* the wings are often slightly confluent so as to do away with all real distinction between *Chuncoa* and *Catappa*, and even between that and *Myrobalanus*, the acute angles of the fruit of *T. melanocarpa* almost pass into the wings of *Catappa*. The section *Pentaptera* with 4 or 5 wings to the fruit, is as yet unknown in Australia. Among the following species several are only known from very imperfect specimens, and may henceforth require much correction in their circumscription, although I do not think they will be much reduced in number.

Sect. I. **Catappa.**—*Fruit with 2 longitudinal membranous or coriaceous wings, or rarely, in the first 3 species, with a third narrow wing or prominent nerve.*

Fruit, including the wings, much broader than long.
  Fruit, including the wings, three times as broad as long ; wings
    quite distinct. Leaves velvety-pubescent underneath . . . . 1. *T. platyptera.*
  Fruit, including the wings, not twice as broad as long ; wings often
    confluent above and below. Leaves nearly glabrous. Spike
    slender, interrupted.
    Leaves obovate, much reticulate. Fruit-wings scarcely confluent 2. *T. volucris.*
    Leaves cuneate-oblong, much reticulate. Fruit-wings shortly con-
    fluent . . . . 3. *T. oblongata.*
Fruit, including the wings, rather longer than broad and quite sur-
  rounded by the confluent wings.
  Leaves lanceolate or oblong, silky-pubescent. Spikes elongated,
    dense. Fruits under ½ in. long . . . . 4. *T. bursarina.*
  Leaves lanceolate or oblong, mostly silky-pubescent. Spikes short,
    dense. Fruits ¾ to above 1 in. long . . . . 5. *T. circumalata.*
  Leaves obovate, glabrous. Spikes short, dense. Fruits ¾ to above
    1 in. long . . . . 6. *T. pterocarpa.*
Fruit orbicular, quite surrounded by a narrow wing. Leaves obovate,
  much reticulate, glabrous. Spikes slender . . . . 7. *T. Thozetii.*

Sect. II. **Myrobalanus.**—*Fruit globular or more frequently ovoid, terete or slightly compressed, or surrounded by a prominent acute angle, but not distinctly winged.*

Leaves very obtuse, usually broad. Flowers rather small; stamens
  not above 3 lines long.
  Leaves quite glabrous.
    Leaves large, narrowed into a short petiole.
      Calyx-tube white, with appressed hairs. Drupe glabrous, sur-
      rounded by a very prominent angle . . . . 8. *T. melanocarpa.*
      Calyx-tube quite glabrous. Drupe ovoid, without any angle . 9. *T. Muelleri.*
    Leaves large, with a short broad flat petiole. Calyx tomentose.
      Drupe ovoid, without any angle . . . . 10. *T. latipes.*

Leaves large with a petiole of 2 to 3 in. Drupe acuminate,
with 2 slightly prominent angles . . . . 11. *T. edulis.*
Leaves minutely hoary underneath. Drupe ovoid-globular, without
angles . . . . ˙ . . . . 12. *T. discolor.*
Leaves loosely tomentose-pubescent, at least underneath.
Drupe ovoid glabrous . . . . 13. *T. porphyrocarpa.*
Drupe ovoid or oblong, often acuminate, tomentose . . . . 14. *T. platyphylla.*
Leaves mostly shortly acuminate. Flowers rather small. Stamens
not above 3 lines long.
Leaves ovate.
Leaves three or four times as long as the petiole, the pellucid
dots very conspicuous under a lens . . . . 15 *T. microcarpa.*
Leaves not twice as long as the petiole, the pellucid dots quite
microscopic . . . . 16. *T. petiolaris.*
Leaves lanceolate or narrow oblong-elliptical. Drupe acuminate . 17. *T. erythrocarpa.*
Leaves narrow, obtuse. Flowers large. Stamens 5 to 6 lines long . 18. *T. grandiflora.*

Section I. Catappa, *DC.*—Fruit with 2 longitudinal membranous or
coriaceous wings, or rarely with a third narrow wing or prominent nerve.

1. **T. platyptera,** *F. Muell. Fragm.* ii. 151. A tree, the young
branches and petioles hoary-pubescent or almost velvety. Leaves crowded at
the ends of the flowering branches, obovate or obovate-oblong, very obtuse,
1½ to 2½ in. long in our specimens, on a rather long petiole, velvety-pubescent on both sides when young, at length nearly glabrous above, the reticulate
veins prominent. Spikes tomentose, slender, interrupted, exceeding the
leaves. Calyx softly tomentose inside and out, the adnate tube about as long
as the broad campanulate limb ; lobes short and broad. Filaments glabrous.
Style villous. Fruit 2-winged, tomentose-pubescent, about 1 in. long and 3
in. broad, including the horizontally divaricate wings, which are quite distinct,
broadly obovate, plicately veined.

N. Australia. Arnhem's Land, *F. Mueller* (in flower) ; Lynd river, *Leichhardt* (in-
fruit).

Var. (?) *glabrata.* Minutely hoary or nearly glabrous ; leaves more coriaceous and rather
larger.— Gilbert river, *F. Mueller.*

2. **T. volucris,** *Herb. R. Br.* Branches divaricate, the young shoots
very minutely hoary or silky-pubescent. Leaves from broadly obovate to
oval-elliptical, 1½ to 3 in. long, narrowed at the base and often decurrent on
the rather long petiole, thin, pale underneath, the primary veins more numerous and less oblique than in *T. pterocarpa,* which this species resembles
without the fruit, and much and finely reticulate between them. Spikes slender, interrupted, usually longer than the leaves, especially when the flowers
are chiefly males, the more female spikes shorter and denser. Calyx minutely
pubescent, the broad limb as long as the ovary. Disk villous. Filaments
glabrous. Style glabrous or hairy at the base. Fruit 2-winged, about ¾ in.
long, and twice that breadth including the broad wings, which are either distinct or slightly continuous above or below the drupe or both ; there are also
frequently on one face of the drupe 1 or 2 prominent longitudinal angles.

N. Australia. Port Keats and Cambridge Gulf, N.W. coast, *A. Cunningham ;* Victoria river, *F. Mueller ;* islands of the Gulf of Carpentaria, *R. Brown ;* Sweers Island,
Henne ; in the interior, lat. 18° 35', *M'Douall Stuart's Expedition.* R. Brown's specimens are the only ones in good fruit, and are those alluded to by him in the Appendix to

Flinders's Voyage under the name of *Chuncoa.* I have little doubt of A. Cunningham's and F. Mueller's specimens belonging to the same species; the others are very imperfect.

Queensland? Some specimens from Broad Sound and Endeavour river, *R. Brown*, without fruit, appear to belong to the same species.

Var. (?) *coriacea.* Leaves larger, broader, more coriaceous; spikes long; lowest bracts sometimes leafy.—Upper Victoria river, *F. Mueller.* Specimens not in fruit and therefore doubtful.

3. **T. oblongata**, *F. Muell. Fragm.* ii. 152. A small tree with spreading branches, glabrous or the young shoots minutely hoary-pubescent. Leaves often clustered at the old nodes or on the short branchlets, cuneate-oblong, very obtuse or emarginate, 1 to 2 or rarely 3 in. long, narrowed into a short petiole, thin and much reticulate. Spikes slender, interrupted, shortly exceeding the leaves. Calyx minutely hoary-pubescent outside, very hairy inside. Stamens and style glabrous, not 3 lines long. Fruit 2-winged, about 8 to 9 lines long and twice as broad, including the wings, which are very shortly continuous both above and below the drupe; drupe in the centre flattened on one face, the other with a projecting longitudinal angle sometimes dilated into a third narrow wing.

N. Australia? A specimen, in leaf only, from the scrub, lat. 17° 30', in *M'Douall Stuart's* collection, appears to belong to this species.

Queensland. Fitzroy, Suttor, Dawson, and Burdekin rivers, *F. Mueller;* Rockhampton, *Thozet.*

T. grandiflora has much the foliage of this species, but the flowers are much larger and the fruit is not winged.

4. **T. bursarina,** *F. Muell. Fragm.* ii. 149. A shrub or small tree, the young branches and foliage softly silky-pubescent. Leaves usually crowded, mostly narrow-oblong or lanceolate, obtuse, 1 to 1½ in. long, but occasionally passing into obovate or ovate, narrowed into a short petiole, the primary veins very oblique and reticulate between them. Spikes pedunculate, dense, exceeding the leaves and sometimes 3 to 4 in. long, the rhachis and flowers softly silky. Calyx-tube about 1 line long, the limb about as long, not so broad and more deeply divided into narrower lobes than in the allied species. Drupe, according to F. Mueller, 2- or rarely 3-winged, rather longer than broad, 2½ to 4 lines long.

N. Australia. Dry gravelly banks of Victoria river and frequent in low places round the Gulf of Carpentaria, *F. Mueller.*

5. **T. circumalata,** *F. Muell. Fragm.* iii. 91. Closely allied to *T. pterocarpa,* with the same flowers and fruit, and perhaps a narrow-leaved variety with the foliage and inflorescence more or less clothed with a soft silky pubescence. Leaves oblong-cuneate or elliptical, 1 to 2 or sometimes nearly 3 in. long, with very oblique primary veins, the reticulate veinlets few and scarcely prominent. Flowers very silky, crowded in short pedunculate spikes. Fruits including the wings, obovate, ¾ to above 1 in. long, the drupe entirely surrounded by a continuous wing.

N. Australia. Cape Pond, N.W. coast, *A. Cunningham;* Depuech Island, *Bynoe;* maritime rocks, Nichol Bay, *F. Gregory's Expedition;* in the interior, lat. 18° 35', *M'Douall Stuart's Expedition.* The latter specimens and some of A. Cunningham's are less pubescent with broader leaves, and seem to connect the species with *T. pterocarpa.* Other specimens from the islands of the Gulf of Carpentaria, *R. Brown,* may belong to *T. circumalata,* but are not in fruit.

6. **T. pterocarpa,** *F. Muell. Fragm.* ii. 152. A small tree, perfectly glabrous or the young shoots very minutely pubescent under a lens. Leaves obovate, of a pale somewhat glaucous hue, very obtuse, about 1½ to 2 in. long, narrowed at the base and shortly decurrent on the petiole, the very oblique primary veins and reticulate veinlets few and not very prominent. Spikes short, pedunculate, very minutely pubescent or glabrous. Calyx-limb much shorter than the adnate tube. Disk hairy; styles and stamens glabrous. Fruit 2-winged, about ¾ to 1¼ in. long and ½ to ¾ in. broad including the wings, which are confluent above and below, completely surrounding the hard almost dry drupe.

N. Australia. Copeland Island, *A. Cunningham;* sandstone table-land between the Upper Victoria river, Alligator river, and Macadam Range, *F. Mueller.*

7. **T. Thozetii,** *Benth.* Of this I have only seen a single specimen resembling *T. volucris* in foliage, except that it is perfectly glabrous, and the fruit is very different. Leaves narrow-obovate, crowded at the ends of the branches, 2 to 3 in. long, reticulate as in *T. volucris.* Flowers not seen. Fruiting-spikes slender, glabrous. Fruits quite smooth and glabrous, nearly orbicular, about ½ in. diameter including the two narrow. confluent wings, which completely encircle the drupe. They may not, however, be perfectly ripe in the specimen.

Queensland. Rockhampton, *Thozet.*

SECTION II. MYROBALANUS, *DC.*—Fruit globular or more frequently ovoid, terete or slightly compressed, or surrounded by a prominent acute angle, but not distinctly winged.

There is no difference whatever in inflorescence or flowers in the two sections, and no constant one in foliage, although in general there is a greater tendency to dry black in *Myrobalanus* than in *Catappa,* and the primary veins diverging from the midrib are more prominent parallel and distant.

8. **T. melanocarpa,** *F. Muell. Fragm.* iii. 92. A tree, usually glabrous, except the silky-white young buds and the flowers. Leaves obovate, very obtuse or rarely obscurely and very obtusely acuminate, 3 to 6 or even 8 in. long and sometimes above 6 in. broad, narrowed into a short petiole, coriaceous, the primary veins prominent underneath and rather distant, transversely reticulate between them. Spikes loose, about as long as the leaves, the rhachis nearly glabrous. Flowers numerous but not crowded. Calyx-tube or ovary white with appressed hairs; limb nearly glabrous outside, above 2 lines broad, densely woolly inside. Stamens and style glabrous. Drupes ovoid, somewhat compressed, obtuse or acuminate, about 1 in. long, surrounded usually by a prominent acute angle, which sometimes in the dried state almost assumes the appearance of a narrow thick wing, but in other specimens is scarcely prominent.

N. Australia. Shaded valleys, islands of the N. Coast, *R. Brown.*
Queensland. Snapper Island, *A. Cunningham;* Port Denison and Edgecombe Bay, *Fitzalan, Dallachy.*

9. **T. Muelleri,** *Benth.* A small tree in the scrub, growing to a considerable height in the ranges, glabrous or the young buds minutely silky. Leaves undistinguishable from those of *T. melanocarpa,* broadly obovate, ob-

tuse, usually 3 to 4 in. long, narrowed into a short petiole, rather coriaceous, the primary veins prominent underneath and rather distant. Spikes loose as in *T. mélanocarpa*, flowers rather larger, and the calyx-tube as well as the limb glabrous outside. Drupe ovoid, said to be blue when fresh and rather acid, about ¾ in. long, without wings or angles.—*T. microcarpa*, F. Muell. Fragm. iii. 92, not of Decaisne.

Queensland. Islands of Howick's Group and off Cape Bedford and Cape Flattery, *F. Mueller :* Cape York, *M'Gillivray ;* Edgecombe Bay, *Dallachy.*

Var. *minor.* Leaves narrower. Fruit smaller.—*T. glabra,* R. Br. Herb., but scarcely of Roxb.—Endeavour river, *Banks and Solander ;* islands of Carpentaria (no fruit), *R. Brown.*

10. **T. latipes,** *Benth.* Branchlets glabrous with a loose bark. Leaves broadly obovate, 3 to 5 in. long, very obtuse, coriaceous, glabrous and glaucous, abruptly narrowed into a very short petiole, which as well as the midrib is very broad and flat, with the primary veins prominent and very divaricate. Spikes rusty-tomentose, about as long as the leaves. Flowers small, rather numerous, tomentose. Drupe ovoid, straight, without wings or angles.

N. Australia. Victoria river, *Bynoe.*

A. Cunningham's herbarium contains specimens of a species apparently allied to the above, but with longer and more slender petioles and slender glabrous spikes. They cannot, however, be determined for want of the fruit.

11. **T. edulis,** *F. Muell. Fragm.* ii. 151. A tree, the fruiting specimens quite glabrous. Leaves very broadly ovate, very obtuse at both ends, 4 to 8 in. long, coriaceous with prominent distant primary veins, on petioles of 2 or 3 in. Flowers unknown. Drupes ovoid-oblong, acuminate, sometimes surrounded by a slightly prominent angle and said to be yellowish when fresh.

N. Australia. Victoria, Fitzmaurice, and Alligator rivers, *F. Mueller ;* South Goulburn Island, *A. Cunningham.* The specimens are insufficient for distinguishing them satisfactorily from *T. melanocarpa* and several others ; the petioles are, however, longer than in any other Australian species except *T. petiolaris,* which has very differently shaped leaves.

12. **T. discolor,** *F. Muell. Fragm.* iii. 92. A tall shrub, the branches and young leaves hoary with a very minute pubescence. Leaves ovate or obovate, obtuse or shortly and obtusely acuminate, mostly 2 to 3 in. long, more narrowed at the base than in *T. Muelleri* and the primary veins less prominent, coriaceous and at length shining above, pale or whitish with a minute tomentum underneath. Flowers not seen, but from the scars on the old rhachis the spikes are probably loose. Fruit only seen imperfect, ovoid-globular, without wings or angles.

N. Australia. Hearson Island, Nichol Bay, *F. Gregory's Expedition.* The specimens are much too imperfect for a satisfactory diagnosis.

13. **T. porphyrocarpa,** *F. Muell. Herb.* A handsome tree, the young branches and petioles densely tomentose. Leaves crowded on the short branchlets, obovate, 2 to 3 in. long, on petioles rarely exceeding ½ in., loosely and softly tomentose-pubescent on both sides or becoming glabrous above when old, the primary veins prominent underneath. Spikes usually shorter than the leaves, rather dense. Calyx glabrous outside, the adnate tube about 1½ lines long, the limb fully 2 lines diameter, densely woolly inside. Fruit ovoid, glabrous, without wings or angles, said to be blue or purple.

Queensland. Mount Archer, Rockhampton, *Dallachy ;* Fitzroy river, *Bowman.*
Var. (?) *eriantha.* Ovary and calyx densely silky-tomentose.—Mount Archer, *Dallachy.*
These specimens are in flower only, and resemble in foliage *T. platyptera* as much as *T. por-phyrocarpa,* but have the larger flowers of the latter species.

14. **T. platyphylla,** *F. Muell. Fragm.* ii. 150. A moderate-sized tree,
the young branches and petioles more or less hoary or rusty with a short soft
tomentum or sometimes densely tomentose and almost woolly. Leaves broadly
obovate or ovate, very obtuse, 4 to 6 in. long, 2 to 4 in. broad, shortly narrowed into a petiole never exceeding 1 in. in some specimens, rather longer in
others, coriaceous, softly pubescent on both sides or nearly glabrous above.
Spikes usually shorter than the leaves, with numerous rather small flowers,
loose or crowded. Calyx silky-pubescent or villous outside, densely villous
inside. Drupes tomentose, ovoid or oblong, obtuse or acuminate, not winged.

N. Australia. Islands of the Gulf of Carpentaria, *R. Brown, Henne ;* Victoria,
Fitzmaurice, and Roper rivers, *F. Mueller ;* Port Essington, *Armstrong.* The species
appears to be chiefly distinguished amongst other large obtuse-leaved ones by its soft pubescence and by the tomentose drupes. From the few specimens seen, the latter appear to be
variable in shape. In R. Brown's specimens they are ovoid-oblong, obtuse, often surrounded
by a slightly prominent or obscure angle ; in one of F. Mueller's from Roper river, they are
obliquely acuminate, with a prominent angle, and shortly contracted at the base ; in another
of F. Mueller's, they are straight, quite terete, oblong, rounded at both ends, but terminating
abruptly in a narrow straight beak of about 2 lines.
A specimen from the N.W. coast, *Bynoe,* has the foliage of *T. platyphylla,* but the
flowers in long loose glabrous spikes. It cannot, however, be determined for want of the
fruit.

15. **T. microcarpa,** *Dcne. Herb. Tim. Descr.* 129. Young shoots minutely pubescent. Leaves broadly ovate-elliptical, rarely slightly obovate,
shortly and obtusely acuminate, 3 to 5 in. long, narrowed into a petiole of
about 1 in., glabrous or slightly hoary underneath with a minute pubescence,
thinly coriaceous, with distant primary veins and copious reticulations, the
pellucid dots although small, yet more conspicuous than in most species.
Spikes attaining the length of the leaves. Flowers numerous but not densely
crowded. Calyx rusty outside with a minute tomentum, densely villous inside, but not seen fully expanded. Drupe, according to Decaisne, olive-
shaped, acuminate, glabrous.

N. Australia ?, *Baudin's Expedition.* Also in Timor. I have not seen the Australian specimens mentioned by Decaisne as having been gathered on the S. coast, probably
from one of those mistakes in the labels which occur in so many instances in the Australian
collections in the Paris Herbarium, owing in a great measure to the illegible handwriting and
absurd orthography of the original labels of the gardener who accompanied Baudin's Expedition. The above description is taken from a Timor specimen communicated by Decaisne.
The species may possibly prove to be a variety of *T. Belerica,* Roxb., which extends over
E. India and the Archipelago. The leaves are ovate, as stated in Decaisne's description,
rather than obovate, as they are said to be by some mistake in the diagnosis.

16. **T. petiolaris,** *A. Cunn. Herb.* A tree, closely resembling *T. microcarpa* in foliage and inflorescence, but the petioles are much longer in proportion to the lamina, the pellucid dots are quite microscopic, and the smaller
reticulations appear pellucid when seen against the light. Young shoots minutely pubescent. Adult leaves quite glabrous, broadly ovate, shortly and
obtusely acuminate, 2 to 3 in. long, narrowed into a petiole of from 1½ to

2 in. Spikes slender, with numerous flowers, only seen in bud. Fruit unknown.

N. Australia. Point Cunningham, Cygnet Bay and York Sound, N.W. coast, *A. Cunningham.*

17. **T. erythrocarpa,** *F. Muell. Fragm.* ii. 150. A tree, the fruiting specimens quite glabrous. Leaves oblong-lanceolate or narrow-elliptical, shortly acuminate, 4 to 6 in. long. narrowed towards the base into a petiole of ½ to 1 in., penniveined and finely reticulate, the midrib very prominent underneath. Spikes shorter than the leaves, but not seen in flower. Drupes red, ovoid, glabrous, ending in a long beak, but not yet ripe in our specimens.

N. Australia. Upper Victoria river, *F. Mueller.*

18. **T. grandiflora,** *Benth.* Branches and foliage silky or the leaves at length glabrous. Leaves linear-oblong or cuneate, obtuse or retuse, 1½ to 3 in. long, coriaceous, very obliquely veined and reticulate, narrowed into a short petiole. Spikes usually exceeding the leaves, with flowers much larger than in any other *Terminalia* known to me. Calyx-tube or ovary above 2 lines long, and the limb of the calyx as much in diameter, the lobes acuminate. Stamens 5 to 6 lines long. Drupe nearly globular, about 1 in. long, tapering into a conical beak of about ¼ in., smooth and glabrous, without wings or angles.

N. Australia. Islands of the Gulf of Carpentaria and Arnhem S. Bay, *R. Brown;* Port Essington, *Armstrong :* between Fitzmaurice and Victoria rivers, *F. Mueller.* There are two forms, one with long narrow leaves, quite glabrous except when very young, the spikes glabrous or slightly silky, and the stamens fully ½ in. long; the other much more silky, the leaves broader shorter and more cuneate, and the silky flowers rather, but not much, smaller.

2. LUMNITZERA, Willd.

Calyx-tube produced above the ovary but scarcely contracted, the limb campanulate, shortly 5-lobed or 5-toothed. Petals 5, spreading. Stamens 10 or fewer. Ovules 2 to 5 ; style filiform, with a minute stigma. Fruit ovoid-oblong, crowned by the persistent calyx, narrowed and flattened at the base, hard and almost woody. Seed linear, with convolute cotyledons.— Maritime trees or shrubs. Leaves crowded at the ends of the branches, obovate or cuneate, thick, entire or slightly crenate. Flowers in short racemes. Bracteoles 2, adnate to the base of the calyx-tube, persistent but not enlarged after flowering.

The genus is limited to the two following species, both of them widely dispersed along the seacoasts of tropical Asia, extending from E. Africa to the Pacific Islands.

Flowers scarlet, in terminal racemes. Calyx fully ½ in. long. Stamens twice as long as the petals . . . . 1. *L. coccinea.*
Flowers white, in axillary racemes. Calyx about 4 lines long. Stamens scarcely exceeding the petals . . . . 2. *L. racemosa.*

1. **L. coccinea,** *W. and Arn. Prod.* 316. A glabrous bushy shrub or small tree. Leaves obovate or oblong-cuneate, very obtuse, often 2 in. long, thick and fleshy. Flowers scarlet, in dense terminal racemes, of which occasionally 2 or 3 form a small corymb. Calyx in the Australian specimens fully

½ in. long at the time of flowering, continuous with and narrowed into a some-what flattened pedicel of 2 or 3 lines; lobes of the limb short, broad and obtuse.   Petals exceeding the calyx-lobes by about 2 lines; stamens twice as long.   Fruiting-calyx above 1 in. long.

Queensland.   Endeavour river, *Banks and Solander;* edges of mangrove swamps, Cape York, *M'Gillivray.*   We have precisely the same form from the Feejee Islands; the common Malayan specimens have usually rather smaller flowers.

2. **L. racemosa,** *Willd.; DC. Prod.* iii. 22.   A glabrous tree or tall shrub, with the foliage of *L. coccinea,* but the racemes are all axillary, usually about as long as the leaves, and the flowers are smaller and white.   Calyx at the time of flowering about 4 lines long, and not above ½ in. when in fruit, the lobes or teeth very short.   Petals about 1½ lines long, and the stamens very little longer.

N. Australia.   Islands of the Gulf of Carpentaria, *R. Brown, Henne.*
Queensland.   Cairncross Island, Torres' Straits, *M'Gillivray, Henne;* Fitzroy river and near Keppel Bay, *Thozet.*
This appears to be the commonest of the two species on the coasts of tropical Asia.

3. MACROPTERANTHES, F. Muell.

Calyx-tube produced above the ovary and scarcely contracted, the limb rather broader, shortly 5-lobed or 5-toothed.   Petals 5.   Stamens 10 or fewer.   Ovules 10 to 12, pendulous; style filiform, with a minute stigma. Fruit (oblong?) crowned by the persistent calyx.   Seed . . .—Silky-white or tomentose shrubs or small trees.   Leaves opposite or clustered at the nodes, small obovate or oblong, entire.   Flowers in pairs on axillary (or terminal?) peduncles.   Bracteoles adnate in the centre to the base of the calyx; the margins free, much enlarged after flowering, forming wings to the fruiting calyx.

The genus is endemic in Australia.   It is very closely allied to *Lumnitzera,* differing in the wings of the fruiting-calyx and in the number of ovules, and the species are not strictly maritime.

Leaves almost sessile, clustered at the nodes, oblong, silvery-white.
   Bracteoles or calyx-wings much shorter than the fruiting-calyx   1. *M. montana.*
   Bracteoles or calyx-wings as long as or longer than the fruiting-
    calyx . . . . 2. *M. Kekwickii.*
Leaves distinctly petiolate, opposite, obovate, tomentose . . . . 3. *M. Leichhardtii.*

1. **M. montana,** *F. Muell. Fragm.* iii. 91.   A small tree, with rigid divaricate branchlets, occasionally spinescent.   Leaves clustered at the nodes, narrow-oblong, obtuse, narrowed into a very short petiole, silvery-tomentose on both sides.   Flowers only seen loose and not perfect.   Calyx after flower-ing attaining nearly 1 in. but not yet ripe, densely silky-pubescent outside, with short lobes.   Petals apparently oblong or obovate, about ¼ in. long. Stamens longer.   Bracteoles about two-thirds as long as the calyx, nearly orbicular, the broad almost scarious free margins folded back.—*Lumnitzera montana,* F. Muell. Fragm. ii. 149.

Queensland.   Arid hills, Newcastle Range, *F. Mueller.*

2. **M. Kekwickii,** *F. Muell. Fragm.* iii. 151.   Branches rigid, probably spinescent, and leaves small, oblong, silvery-white, clustered at the nodes, as

iñ *M. montana.* Flowers in pairs, sessile at the end of axillary peduncles, rather shorter than the leaves. Calyx silky-white, scarcely 4 lines long at the time of flowering, with the orbicular bracteoles about half as long, enlarged after flowering to 6 or 7 lines, with the bracteoles reticulate, almost scarious, and quite as long as the calyx, or even exceeding it. Petals ovate, exceeding the calyx-lobes by about 1 line. Stamens rather longer.

N. Australia. Newcastle Water, lat. 17° 30', *M'Douall Stuart's Expedition.*

3. **M. Leichhardtii,** *F. Muell. Fragm.* iii. 91. Apparently more branched and not so rigid as the other two species. Leaves less crowded, all opposite, obovate, very obtuse, mostly ½ to ¾ in. long, narrowed into a petiole of 1 to 2 lines, softly silky-tomentose on both sides, but not so white as in the other species, and becoming nearly glabrous above with age. Peduncles shorter than the leaves, bearing at the end 2 pedicellate flowers, of which I have only seen the calyx, enlarged after flowering to from 4 to 6 lines, with the adnate bracteoles nearly as long.

Queensland. Ruined Castle Creek, *Leichhardt.*

4. GYROCARPUS, Jacq.

Calyx-tube adnate to the ovary, or none in male flowers ; limb 4- to 7-cleft. Petals none. Stamens 4 to 6, alternating with as many club-shaped staminodia, or fewer or none in the female flowers. Ovary inferior, with 1 pendulous ovule and a sessile stigma, abortive in the male flowers. Drupe dry, crowned by 2 much elongated, erect, spathulate, wing-like calyx-lobes. Seed oblong, pendulous, without albumen ; cotyledons petiolate, convolute round the radicle.—Tall tree. Leaves alternate, broad, entire or lobed. Flowers polygamous, very small, crowded in dense corymbose cymes.

The genus consists of a single species common to the tropical regions of Central America and tropical Asia. It forms one of the small group of *Gyrocarpeæ,* Dumort., or *Illigereæ,* Blume, associated by many botanists with *Laurineæ,* chiefly on account of the dehiscence of the anthers, but which Lindley is no doubt more correct in adding as a suborder to *Combretaceæ.* The same dehiscence of the anthers is exemplified in *Berberideæ* and *Hamamelideæ,* without being constant in either Order. The fruit and seeds are quite those of *Combretaceæ,* and there is considerable affinity in many other respects between *Illigera* and *Combretaceæ* on the one hand and *Hamamelideæ* on the other.

1. **G. Jacquini,** *Roxb. Pl. Corom.* i. 2, *t.* 1, *copied into Lam. Illustr. t.* 850. A tall tree. Leaves deciduous, crowded at the ends of the thick branchlets, broadly ovate or orbicular, on young trees often 8 to 10 in. long and broad and deeply 3-lobed, on older trees usually smaller and entire or broadly and shortly lobed, usually more or less acuminate, truncate or cordate at the base, glabrous or tomentose underneath or on both sides, the petioles varying from 1 to 4 in. Peduncles in the upper axils or close above the last leaves, rarely exceeding the petioles, bearing each a repeatedly branched cyme with densely crowded exceedingly small flowers, forming little globular heads before expanding, sometimes entirely males, sometimes with a few hermaphrodite or female flowers scattered in the cyme or chiefly in the forks. Drupes ovoid, usually about ¾ in. long, the wings erect, oblanceolate, rounded at the end, much narrowed below the middle, varying in the Australian specimens from under 2 in. long and about ½ in. broad to 2½ in. long and about

5 lines broad.—Pers. Syn. i. 143 ; *G. americanus*, Jacq. ; Meissn. in DC. Prod. xv. 247 ; *G. asiaticus*, Willd. ; Meissn. l. c. 248 ; *G. acuminatus*, Meissn. l. c. ; *G. sphenopterus*, R. Br. ; Endl. Iconogr. t. 43 ; Meissn. l. c. ; *G. rugosus*, R. Br. ; Meissn. l. c.

N. Australia. Victoria river, *Bynoe, F. Mueller.*
Queensland. Gilbert river, *F. Mueller ;* Port Denison, *Fitzalan ?*

Also in Columbia and Central America, in tropical Asia, the eastern Archipelago, and islands of the Pacific. All the writers who, unwilling to believe that the same species should have so wide a geographical range, have distinguished several species of *Gyrocarpus*, have expressed some hesitation in doing so, for the characters assigned all break down when applied to other specimens than those actually described. The differences in the indumentum and shape of the leaf are often much greater in different specimens from the same locality than between those gathered at the greatest distances. None are more striking than in two specimens from the Feejee Islands which, according to Seemann's notes, represent the young and the old trees. In the former the leaves are large, broadly cordate and deeply 3-lobed as figured in Jacq. Ic. Amer. t. 178, f. 80, and loosely tomentose on both sides ; in the latter they are quite entire, glabrous, more acuminate and more acute at the base than in the form characterized as *G. acuminatus*, Meissn. The fruit-wings are usually longest in the American, shortest in the Australian specimens, but not uniformly so even in the comparatively few specimens preserved in herbaria. Glabrous and more or less hairy filaments occur in India as well as in Australia. The tomentum of the leaves is even more inconstant than any other character. R. Brown's specimens have been unfortunately mislaid, but from Endlicher's figure engraved from Bauer's drawing, and from the variety of Australian specimens I have seen, I have no doubt that he was right in the suspicion he expressed that his species might not be different from the common one.

INDEX OF GENERA AND SPECIES.

———◆———

The synonyms and species incidentally mentioned are printed in italics.

2 L

END OF VOL. II.

JOHN EDWARD TAYLOR, PRINTER,
LITTLE QUEEN STREET, LINCOLN'S INN FIELDS.

For EU product safety concerns, contact us at Calle de José Abascal, 56–1°,
28003 Madrid, Spain or eugpsr@cambridge.org.